AF862922

Lecture Notes in Civil Engineering

Volume 193

Series Editors

Marco di Prisco, Politecnico di Milano, Milano, Italy

Sheng-Hong Chen, School of Water Resources and Hydropower Engineering, Wuhan University, Wuhan, China

Ioannis Vayas, Institute of Steel Structures, National Technical University of Athens, Athens, Greece

Sanjay Kumar Shukla, School of Engineering, Edith Cowan University, Joondalup, WA, Australia

Anuj Sharma, Iowa State University, Ames, IA, USA

Nagesh Kumar, Department of Civil Engineering, Indian Institute of Science Bangalore, Bengaluru, Karnataka, India

Chien Ming Wang, School of Civil Engineering, The University of Queensland, Brisbane, QLD, Australia

Lecture Notes in Civil Engineering (LNCE) publishes the latest developments in Civil Engineering - quickly, informally and in top quality. Though original research reported in proceedings and post-proceedings represents the core of LNCE, edited volumes of exceptionally high quality and interest may also be considered for publication. Volumes published in LNCE embrace all aspects and subfields of, as well as new challenges in, Civil Engineering. Topics in the series include:

- Construction and Structural Mechanics
- Building Materials
- Concrete, Steel and Timber Structures
- Geotechnical Engineering
- Earthquake Engineering
- Coastal Engineering
- Ocean and Offshore Engineering; Ships and Floating Structures
- Hydraulics, Hydrology and Water Resources Engineering
- Environmental Engineering and Sustainability
- Structural Health and Monitoring
- Surveying and Geographical Information Systems
- Indoor Environments
- Transportation and Traffic
- Risk Analysis
- Safety and Security

To submit a proposal or request further information, please contact the appropriate Springer Editor:

- Pierpaolo Riva at pierpaolo.riva@springer.com (Europe and Americas);
- Swati Meherishi at swati.meherishi@springer.com (Asia - except China, and Australia, New Zealand);
- Wayne Hu at wayne.hu@springer.com (China).

All books in the series now indexed by Scopus and EI Compendex database!

More information about this series at https://link.springer.com/bookseries/15087

H. R. Pasindu · Saman Bandara ·
W. K. Mampearachchi · T. F. Fwa
Editors

Road and Airfield Pavement Technology

Proceedings of 12th International Conference on Road and Airfield Pavement Technology, 2021

Springer

Editors
H. R. Pasindu
Department of Civil Engineering
University of Moratuwa
Moratuwa, Sri Lanka

Saman Bandara
Department of Civil Engineering
University of Moratuwa
Moratuwa, Sri Lanka

W. K. Mampearachchi
Department of Civil Engineering
University of Moratuwa
Moratuwa, Sri Lanka

T. F. Fwa
Department of Civil and Environmental Engineering
National University of Singapore
Singapore, Singapore

ISSN 2366-2557 ISSN 2366-2565 (electronic)
Lecture Notes in Civil Engineering
ISBN 978-3-030-87378-3 ISBN 978-3-030-87379-0 (eBook)
https://doi.org/10.1007/978-3-030-87379-0

© The Editor(s) (if applicable) and The Author(s), under exclusive license to Springer Nature Switzerland AG 2022
This work is subject to copyright. All rights are solely and exclusively licensed by the Publisher, whether the whole or part of the material is concerned, specifically the rights of translation, reprinting, reuse of illustrations, recitation, broadcasting, reproduction on microfilms or in any other physical way, and transmission or information storage and retrieval, electronic adaptation, computer software, or by similar or dissimilar methodology now known or hereafter developed.
The use of general descriptive names, registered names, trademarks, service marks, etc. in this publication does not imply, even in the absence of a specific statement, that such names are exempt from the relevant protective laws and regulations and therefore free for general use.
The publisher, the authors and the editors are safe to assume that the advice and information in this book are believed to be true and accurate at the date of publication. Neither the publisher nor the authors or the editors give a warranty, expressed or implied, with respect to the material contained herein or for any errors or omissions that may have been made. The publisher remains neutral with regard to jurisdictional claims in published maps and institutional affiliations.

This Springer imprint is published by the registered company Springer Nature Switzerland AG
The registered company address is: Gewerbestrasse 11, 6330 Cham, Switzerland

Reviewers for the 12th ICPT

1. Dr. Niranga Amarasingha, Sri Lanka Institute of Information Technology, Sri Lanka
2. Mr. Moe Aung Lwin, Singapore
3. Prof. Saman Bandara, University of Moratuwa, Sri Lanka
4. Dr. Gary Chai, Griffith University, Australia
5. Prof. Rajan Choudhary, Indian Institute of Technology Guwahati, India
6. Mr. Udaya Dampage, Kotelawala Defense University, Sri Lanka
7. Prof. Animesh Das, IIT Kanpur, India
8. Dr. Ishani Dias, Kotelawala, Defense University, Sri Lanka
9. Prof. Ludfi Djakfar, Brawijaya University, Indonesia
10. Dr. Ju Fenghua, Consultant, Singapore
11. Dr. Prageeth Gunarathna, Victoria State Government, Australia
12. Mrs. Shankani Gunarathna, University of Moratuwa, Sri Lanka
13. Dr. Chamara Jayalath, Queensland University of Technology, Australia
14. Dr. Dhanushika Gunatilake, University of South Florida, USA
15. Dr. Vincent Guwe, Shell Specialties, Malaysia
16. Prof. Sen Han, Chang'an University, China
17. Dr. Chamod Hettiarachchi, Tongji University, China
18. Dr. Sampath Hewage, University of Moratuwa, Sri Lanka
19. Prof. Dinh Hiep, Vietnam National University of Civil Engineering, Vietnam
20. Dr. Ajayshankar Jagadeesh, National University of Singapore, Singapore
21. Dr. Kunal Jain, Punjabi University, India
22. Dr. Shiran Jayakody, Senior Lecturer, South Eastern University of Sri Lanka, Sri Lanka
23. Dr. Bhagya Jayasinghe, Nanyang Technological University, Singapore
24. Mrs. Chamindi Jayasuriya, University of Wollongon, Australia
25. Dr. Lu Jia, Kraton Polymer Singapore Pte. Ltd., Singapore
26. Prof. Chen Jingyun, Dalian University of Technology, China
27. Mr. Sothinathan Kapilan, Singapore Land Transport Authority, Singapore
28. Dr. Kamal Karunananda, SAITM, Sri Lanka
29. Dr. Aruna Karunarathne, Swinburne University of Technology, Australia
30. Dr. I Dewa Made Alit Karyawan, University of Mataram, Indonesia

31. Dr. Qasim Khan, National University of Singapore, Singapore
32. Dr. Hooi Ling Khoo, Universiti Tunku Abdul Rahman, Malaysia
33. Dr. Thanuja Kulathunga, General Sir John Kotelawala Defence University, Sri Lanka
34. Dr. Anupam Kumar, TU Delft, Netherlands
35. Dr. Pradeep Kumar, CSIR—Central Road Research Institute, New Delhi, India
36. Dr. Rakesh Kumar, Central Road Research Institute, India
37. Dr. Vasantha Kumar S, School of Civil Engineering, VIT, India
38. Mr. Sakitha Kumarage, University of Queensland, Australia
39. Dr. Kang-Won Lee, The University of Rhode Island, Sri Lanka
40. Dr. Kelvin Lee, Samwoh Innovation Centre Pte Ltd, Singapore
41. Prof. Ming-Gin Lee, Chaoyang University of Technology, Taiwan
42. Dr. Zhang Lei, Changi Airport Group, Singapore
43. Prof. Zhen Leng, Polytechnic University, Sri Lanka
44. Dr. Chu Longjia, Chang'an University, China
45. Mrs. Rosmawati Mamat, Johor Bahru, Malaysia
46. Prof. W. K. Mampearachchi, University of Moratuwa, Sri Lanka
47. Dr. Fauzan Mohd. Jakarni, Malaysia
48. Dr. Ghim Ping Ong, National University of Singapore, Singapore
49. Dr. HR Pasindu, University of Moratuwa, Sri Lanka
50. Dr. Darshana Perera, Aurecon, Australia
51. Dr. Loshaka Perera, University of Melbourne, Australia
52. Dr. Piyal Wasantha, Monash University, Australia
53. Mr. Gautam Prakash, National Institute of Technology Patna, India
54. Mrs. Rajapakshe Premarathne, University of Moratuwa, Sri Lanka
55. Mrs. Piyumee Premarathne, University of Moratuwa, Sri Lanka
56. Dr. Thushara Priyadarshana, OUSL, Sri Lanka
57. Prof. A U Ravi Shankar, NITK, SURATHKAL, India
58. Prof. Boonchai Sangpetngam, Chulalongkorn University, Thailand
59. Dr. Jaritngam Saravut, Prince of Songkla University, Thailand
60. Dr. Bagus Setiadji, Universitas Diponegoro, Indonesia
61. Ms. Sambriddhi Shrestha, Civil Aviation Authority of Nepal, Nepal
62. Dr. Hansinee Sitinamaluwa, University of Moratuwa, Sri Lanka
63. Dr. Arooran Sounthararajah, Monash University, Australia
64. Dr. Santosh Srirangam, HSL Constructor Pte Ltd, Singapore
65. Dr. Yu-Min Su, National Kaohsiung University of Science and Technology, Taiwan
66. Dr. Sanjeev Suman, National Institute of Technology, Patna, India
67. Dr. Jun Yew Tan, Samwoh Innovation Centre Pte. Ltd., Singapore
68. Mr. Teoh Tean, ECO2 BUILDER PTE LTD, Singapore
69. Dr. Kazuya Tomiyama, Kitami Institute of Technology, Japan
70. Mr. Sajith Udayanga, Wayamba University, Sri Lanka
71. Prof. Sri Wardani, Universitas Diponegoro, Indonesia
72. Dr. Greg White, University of the Sunshine Coast, Australia
73. Dr. W Wickramasinghe, University of Peradeniya, Sri Lanka
74. Dr. Zhanping You, Michigan Technological University, USA

12th ICPT Conference Organizing Committee

Conference co-chairs:

- Prof. Saman Bandara
- Prof. W. K. Mampearachchi

Program chair:

- Dr. H.R.Pasindu

Track chair(s):

- Dr. Ishani Dias
- Dr. Shiran Jayakody
- Dr. Chamod Hettiarachchi

Finance chair:

- Dr. Loshaka Perera

Publicity chair:

- Dr. Dimantha de Silva

Contents

Pavement Design and Geotechnics

Pavement Construction, Performance Evaluation and Management

A Framework for Selecting Safety Treatments for Rural Roads

R. K. T. K. Ranawaka, H. R. Pasindu, and T. W. K. I. M. Dias

Abstract Road safety is a vital element of the road's overall function, which is often neglected in decision-making for road maintenance management. As a result, the safety issues, especially in rural roads, remain without funding to implement the necessary countermeasures. One constraint faced by local authorities is the lack of analysis tools to select appropriate safety treatments within the available budget. This study presents a methodology to logically determine the safety treatment criteria for a selected road to increase the safety performance at project level. The decisions regarding the safety treatments are taken based on a linear programming model which optimizes the safety performance of the selected road. Cumulative Safety Index (CSI) represents the safety performance of the road, which is determined based on the prevailing issues on that road. The model comprises the objective function by which maximizes the safety performance of the selected road concerning the number of prevailing safety issue types. This model is used to identify the optimal safety treatment scheme for the road chosen, ensuring prevailing safety issues of the road are effectively addressed. The objective function consists of the Initial CSI of the selected road and the safety improvement after treating relevant issue type coupled with a binary decision variable. This model can also be considered an input to road asset management systems where multi-objective optimization (MOO) models maximize the network pavement condition and maximize overall network safety performance.

Keywords Safety treatments · Linear optimization · Rural roads · Project level

R. K. T. K. Ranawaka (✉) · H. R. Pasindu
University of Moratuwa, Moratuwa, Sri Lanka
e-mail: 188102d@uom.lk

T. W. K. I. M. Dias
General Sir John Kotelawala Defence University, Colombo, Sri Lanka

© The Author(s), under exclusive license to Springer Nature Switzerland AG 2022

H. R. Pasindu et al. (eds.), *Road and Airfield Pavement Technology*, Lecture Notes in Civil Engineering 193, https://doi.org/10.1007/978-3-030-87379-0_1

1 Introduction

Roads with poor safety performance are the key factors that indicate how unsafe the road users are in the infrastructure system of a country. Especially in developing countries, this situation has worsened due to improper road maintenance strategies. An unsafe infrastructure system also contributes to the increase of operational costs of the system. The above scenario occurs mainly due to the negligence of relevant authorities, caused by not having a proper framework to guide them towards an adequate road safety management consideration.

Although various public awareness programs, advocacy programs, driver training programs, etc., are carried out to make society aware of road safety, it can still be seen that the accidents trends have not been decreased worldwide. In a situation where the United Nations have announced the second decade of action for road safety, it clearly shows that the required reduction of the world's number of accidents has not been achieved.

A road accident can result from various causal factors related to roadway characteristics such as geometric issues, pavement issues, lack of road signage and markings, improper road management, etc. Based on the mentality, reaction speeds, fatigue, negligence, and other psychological factors, the human body is constantly exposed to commit an error. But, if the prevailing safety conditions of the road can reduce the risk caused by human error, it can decrease the probability of a crash occurring. That is identified as the concept behind forgiving roads [1]. Forgiving road is a road that will forgive the driver if he has committed a driving error, rather than punishing the driver for his or her mistake. Forgiving roads are achieved by proper road maintenance and proper safety management [2]. On the other hand, using an adequate safety management system is required to achieve cost-optimized decision-making criteria [3]. Especially in developing countries, due to the problems related to budget allocation, obtaining safety treatment decisions have become a serious consideration [4].

The literature review shows that various approaches have been developed to incorporate road safety into pavement management systems and pavement conditions as multi-objective optimization models. As per the suggestion of the Federal Highway Administration, using Pavement Management Systems (PMS) is required to achieve cost-optimized decision-making for road maintenance [3]. Various methodologies have been developed considering pavement management optimization techniques. Linear programming, integer linear programming, dynamic programming, and heuristics can be identified as the powerful mathematical programming techniques used in pavement management optimization models [5, 6]. Previous studies have shown that linear programming is an ideal mathematical programming technique to deal with day-to-day problems rather than other programming techniques.

Road safety analysis has now become a primary concern than ever. Therefore, an essential requirement has been raised to incorporate safety modelling techniques in pavement management techniques [7]. Recent studies show that researchers have focused on incorporating risk assessment in M&R budget allocations rather than traditional PMSs [8]. Highway Safety Manual (HSM) always encourages researchers and road agencies to develop state-specific safety performance evaluation methodologies and incorporate them in funding schemes [9]. The primary reason for that is that the ground conditions where the existing methods have been developed may not always correspond to another country's ground conditions. Therefore, this study proposes a novel approach to improve the safety performance of rural roads based on a linear programming approach. Improvement of the safety performance is measured by the reduction of the Cumulative Safety Index (CSI) [10] after treating each type of safety issue. The improvement of safety performance coupled with a binary decision variable is used in the objective function, which minimizes the CSI of the respective project after applying the relevant safety treatment strategies.

2 Literature Review

Safety issues prevailing in rural low-volume roads are different from the safety issues prevailing in the urban context. Especially in developing countries like Sri Lanka, the allocation of funding for safety improvements in rural areas are significantly lower when compared to the urban context. Table 1 analyzes safety issues related to roadway characteristics in rural and low-volume roads as they have been considered in different studies.

The three major road safety countermeasures applied in road safety management can be categorized as follows [15]:

- Engineering countermeasures: These are the measures that involve roadway treatments. They vary from installing warning signs, implementing traffic calming measures to road re-alignments.
- Enforcement countermeasures.

Table 1 Safety issues in rural low-volume roads

Safety issue	Addressed in			
	[11]	[12]	[13]	[14]
Lack of roadside space	√	√	√	√
Issues at curve alignments			√	√
Signalized intersections	√	√		√
Entrance from by roads	√		√	
Roadside edges				√
Pavement marking	√		√	√
Roadway issues	√	√	√	
Roadside slopes and edges				√
Pedestrian facilities			√	
Improper signage			√	

Table 2 Accident reduction percentages by different countermeasures

Countermeasure	Accident reduction (%)
Installation of general warning signs	25
Upgrading of roadway delineation	15
Installation of the no-passing zone through delineation	40
Lighting of at grade intersections (overall)	30
Lighting of at grade intersections (night)	50
Installation of right-turning lanes	35
Installation of rumble strips	25
Improvement of road surface	25
Installation of crash barrier (fatal accidents)	65
Sight distance improvements	30
Improvement of horizontal alignment	40
Construct grade-separated interchange	55

- Education countermeasures.

Table 2 illustrates typical accident reduction percentages from engineering countermeasures [15].

A pavement management system is an effective tool when cost-effective decisions must be taken [3]. To guide the respective user towards a logical conclusion, different approaches can be used. Specialized algorithms, logistics regressions, manual rules, and linear programming are frequently used to obtain proper analytical solutions [6].

One of the significant drawbacks in current safety treatment selection processes is the lack of knowledge on the effectiveness of available treatment types or treating each safety issue. It is essential to identify the point of the crashes and use that information on the treatment selection process [16]. Although there are studies that consider the effect of treatment types, they do not provide a comparative analysis of the methodological problem's effectiveness.

Highway Safety Manual has developed Safety Performance Functions to predict crashes in rural road segments as follows [17]:

$$N_{spf} = AADT \times L \times 365 \times 10^{6} \times e^{-0.312} \tag{1}$$

where,

N_{spf}—Predicted number of annual crashes.

L—Lane width.

AADT—Annual Average Daily Traffic.

The following model developed by Sharma and Landge [18] predicts vehicle crashes in Indian rural highways.

$$\text{Accident/year/km} = 2.913 + (0.326 * \text{AD}) - (0.478 * \text{SW}) - (0.173\text{LW}) \quad (2)$$

where,

AD—Access density.

SW—Shoulder width.

LW—Lane width.

These models always indicate proper engineering judgments are essential to get the maximum benefits of the statistical tools [18].

Sandamal et al. [4] have developed a multi-objective optimization model to incorporate pavement condition and safety performance in pavement management systems. The objective function of the model is to maximize the pavement condition and maximize safety performance. Pavement condition and safety performance are evaluated by the International roughness index and Cumulative Safety Index, respectively [4]. The objective functions are as follows.

Condition (Q);

$$\text{Minimize Q } F_1) = \left[\sum\nolimits_{k=1}{}^{n} \sum\nolimits_{r=1,m} (Q_{rk} \,.\, L_k \,.\, X_{rk})\right] / L \quad (3)$$

Safety Performance (S);

$$\text{Minimize S } F_2) = \left[\sum\nolimits_{k=1,n} \sum\nolimits_{s=1,p} (CSI_{sk} \,.\, L_k \,.\, Y_{sk})\right] / L \quad (4)$$

Constraints;

Budget (B);

$$B \geq \sum\nolimits_{k=1,n} \sum\nolimits_{r=1,m} (C_{rk} \,.\, L_k \,.\, X_{rk}) + \sum\nolimits_{k=1,n} \sum\nolimits_{s=1,p} (C_{sk} \,.\, L_k \,.\, Y_{sk})$$

where,

Q—Average network IRI value.

S—Average cumulative safety index of the network.

CSI_{sk}—Cumulative safety index after applying the safety improvement strategy ‘s’ in road section ‘k.‘

CSI_i—Safety index for the issue type ‘i’ for a given road section ‘k.‘

Q_{rk}—IRI of road section ‘k’ after applying the operation ‘r’ in m/km.

L_k—The length of road section ‘k’ in km.

X_{rk}—The decision variable (if operation ‘r’ is applied to road section ‘k’ then $X_{rk} = 1$ and operation ‘r’ is not used to road section ‘k’ then $X_{rk} = 0$).

Y_{sk}—The element in sth row and kth column of matrix $Y = [Y_{sk}]p \times n \in M$ where Y_{sk} is the binary decision variable subjected to constrains.

L—The total length of the road network.

B—The total budget available for the financial year.

C_{rk}—The cost per km for applying operation 'r' to road section 'k'.

C_{sk}—The cost per km for using safety improvement strategy 's' to road section 'k'.

n—The number of road sections in the network.

m—The number of maintenance operations used for road section 'k'.

p—The number of safety improvement strategies used for road section 'k'.

Pasindu et al. [10] developed a multi-objective optimization model to incorporate safety performance and pavement condition in a pavement management system. They have used the stochastic random-based evolutionary computation technique, Genetic Algorithm (GA) is used to find the optimization problem. The chromosome element consists of three mutually exclusive pavement maintenance and safety improvement strategies in the proposed method. The optimum solution has been generated considering the Pareto Front developed from the available solution schemes [10]. Augeri et al. [19] have used average safety improvement and six other objective functions in their pavement management system model. The safety improvement for a given section has been calculated considering the traffic impact factor (TI), expected number of accidents (AC), and crash modification factor (CMF), as shown below.

$$F_3(x_{i.u}) = \{\sum_{1,n} \sum_{u=1,s} \left[(\sum_{y=1,t} ACi\,.\,CMF_{i.u.y}).x_{i.u} \right] / L.t\} \quad (5)$$

F3(xi.u)—Average safety improvement during the planning period.

ACi—The expected number of accidents on-site 'i'.

CMFi.u.y—Crash modification factor associated with the maintenance strategy 'u' on the section 'i' in the year 'y'.

xi.u—The decision variable (if operation 'u' is applied to road section 'i' then $xi.u = 1$ and operation 'u' is not applied to road section 'i' then $xi.u = 0$).

L—Total length of the road network.

t—Analysis period in years.

When considering the research related to pavement management systems in the recent past, many studies have been carried out on pavement management optimization techniques [20]. Linear programming, integer linear programming, dynamic programming, and heuristics can be identified as the powerful mathematical programming techniques used in pavement management optimization models.

Researchers can apply the mixed integer programming (MIP) technique when a solution space consists of discrete and continuous variables to find complex problems. Decisions variables are constrained to be integer values in a mixed integer programming problem [21–23]. With the issues involved in isolated or invisible variables, MIP performs well [24]. But integer linear programming has not been identified as an ideal tool to deal with problems exposed to combinatorial explosions [25]. Although many engineers prefer MIP over linear programming to make decisions on available solutions, MIP is not considered to be not very useful to provide solutions for the challenges in day-to-day problems [26].

A linear programming model provides solutions by maximizing a linear objective along with one or more constraints. Linear programming models have been developed to produce optimal maintenance and rehabilitation strategies [6, 27].

Correira [6] has developed a linear programming model: The multi-year pavement maintenance problem. The objective function of the model is to maximize the overall network condition over the planning horizon.

$$\sum_{t=1}^{T}\sum_{s=1}^{S}\sum_{l=1}^{L} Z_{tsl}xl \tag{6}$$

$$X_{tsr} = \begin{cases} 1, & \begin{array}{l} if\, maintenance\, alternative\, r\, is\, selected \\ for\, segment\, s\, in\, year\, t; \end{array} \\ 0, & otherwise, \end{cases}$$

$$Z_{tsl} = \begin{cases} 1, & if\, at\, year\, t\, the\, segment\, s\, is\, in\, condition\, l; \\ 0, & otherwise, \end{cases}$$

$$Y_{tslr} = \begin{cases} 1, & \begin{array}{l} if\, maintenance\, alternative\, r\, is\, selected\, for \\ segment\, s\, which\, is\, in\, condition\, l\, in\, year\, t; \end{array} \\ 0, & otherwise, \end{cases}$$

whereas in this study, the safety performance of a given road is maximized according to a given budget. The majority of the optimization models have been developed, considering the budget as the major constraint.

The budget allocation can be provided in two ways. One way is to allocate a single budget period or allocate a multi-period improvement budget [16]. Mahalel [16] has proposed a problem formulation that gives a budget for a single period.

$$Minimize\left[\sum_{i=1}^{N}\sum_{j=1}^{N_i} b_{ij}X_{ij} + \sum_{k=1}^{K}\sum_{j=1}^{n_k} B_{kj}Y_{kj}\right] \tag{7}$$

where,

b_{ij}—number of accidents at location i after execution of project *j*.

B_{ki}—number of accidents along road stretch k after execution of project j.

N, k—number of black spots and road stretches, respectively.

X_{ij}, Y_{ki}—integer O-1 decision variables indicating whether project j related to "black spot" or road stretch is executed.

Following are the constraints used in the model.

- The total sum spent on improving "black spots" or road stretch is smaller than the entire budget available.
- Authorities can select only one improvement level at each location.
- The exact location will not be improved twice, once a "black spot" and once a part of a road stretch.

One of the significant differences between the model proposed by Mahalel [16] and the model presented in this study is that if a particular location consists of more than one safety issue, they will be improved if the Linear Programming model selects them. Whereas in the model proposed by [16], there is a constraint not to adopt more than one improvement level for a single location.

3 Methodology

One of the prime objectives of the road governing agencies/ authorities is to ensure that the road provides a safe, comfortable, and efficient travel experience for the road users. When it comes to ensuring a safe trip, it is their responsibility to implement comprehensive road safety strategies as they help to reduce the crash occurrence of the considered road. When developing such techniques, there should be a proper guideline or a management system to guide the agencies/authorities towards optimizing the safety performance of the road they govern, according to the budgetary constraints. Therefore, the approach should indicate the estimated safety performance improvement in a crash reduction after adopting the selected safety treatment measures. Suppose a separate budget can be allocated to improve the safety performance of the road. In that case, a linear programming model will favor optimizing the safety performance based on the available resources.

This study proposes a linear programming model to determine the types of safety issues prioritized for a given project. Let us consider a road R, with its safety treatment scheme planned for a given year. In this considered road, there may be several safety issues requiring treatment to improve the safety performance of that road. There may be an annual monetary budget that can restrict the repairs to manage these treatments. When there is such a budgetary limitation, it is not possible to treat all the safety issues that require treatment. So budgetary limitations can be a crucial factor to restrict possible treatment schemes, and there should be a proper framework to guide the decision-makers on this.

The methodology is developed considering several working principles as follows.

- There should be a quantitative approach to evaluate the safety performance of the selected road
- In the selected road, there should be prevailing safety issues that require a treatment
- Each treatment has a defined cost to implement
- The treatment will contribute to improving the safety performance of that selected road
- The available budget should be less than the cost to implement all the required treatments.

Therefore, the objective function should be to improve the safety performance of the road considering the budget constraint. Prevailing safety issues on the road and the safety issues that need to be treated can be identified by conducting a road safety audit with a multi-disciplinary team. The number of treatment schemes that can be implemented is not considered as a constraint here.

The methodology can be summarized in Table 3.

3.1 Evaluation of Safety Performance

In the process of scheduling the treatment schemes, it is necessary to define the performance indicators. The indicator should have the capability of measuring the safety performance of the road at a given time.

"Existing data on accidents and exposure can be used in terms of microscopic control of safety. There does not seem to exist a deterministic relationship between conditions at a location and its annual number of crashes." [16].

When considering the safety evaluation models and approaches discussed in the literature review, there are two significant limitations for not using those models in the Sri Lankan context for rural roads. One limitation is the lack of crash data for rural low-volume roads, and the other is the lack of traffic data. These can be identified in most developing countries. Therefore, an alternative safety evaluation model is required, which can overcome the limitations mentioned above. Also, the social and economic environment where those models have been developed cannot be adopted with the social and economic environment status in developing countries. The methodology used in this study does not directly rely on crash data and traffic data. This approach is carried out based on on-time field inspections, overcoming the limitations raised with the models discussed under the literature review.

Safety performance is evaluated by the cumulative safety index developed by [10] in this study. The cumulative safety index (CSI) has been validated as a reasonable approach to evaluate the safety performance of a given road considering the

Table 3 Summary of the methodology

Objective	Improving the safety performance of the selected road
Required data and indicators	Safety Performance indicator Safety Performance model Prevailing safety issues Available budget
Constraints	Minimum safety performance level Allocated budget for the road
Results	Safety treatment plan Estimated cost Safety performance improvement

equivalent property damage only factor (EPDO). The CSI has been developed considering the crash factors: exposure, probability, and consequence of ten different safety issues.

1. Improper signage and lane marking at intersections
2. Improper signage and lane marking at access roads
3. Limited roadside space availability
4. Issues at horizontal and vertical curves
5. Inadequate pedestrian facilities near schools, town areas
6. Issues at rail crossings
7. Excessive pavement edge drop-offs
8. Roadside vertical drops—steep roadside slope
9. Unprotected bridges and culverts
10. Open drains.

These safety issues are following the safety issues identified by [4, 28]. CSI was then calculated as proposed by [10] as in (8) and (9)

$$CSI_k = \sum_{1}^{I} \sum_{0}^{N} (SI_{in})_k \tag{8}$$

$$CSI_i = \sum_{1}^{K} \sum_{0}^{N} (SI_{nk})_i \tag{9}$$

where,

CSI_k = Cumulative safety index for road *k*.
CSI_i = Cumulative safety index for road safety issue *i*.
SI_{ink} = safety index for the nth occurrence of the ith safety issue in road *k*.
I = Number of road safety issue types.
N = Number of occurrences for each type of safety issue.
K = Number of roads in the network.

3.2 *Formulation of the Linear Programming Model*

When there is a list of locations that poses a safety risk for the travel of a particular road, it becomes necessary to identify and determine the safety issues that need safety treatments. Ideally, the decision-makers would like to know the effectiveness of each safety issue treatment on the safety performance of the road and use that information in the budget allocation process. The essential thing in this process is obtaining clear information on the safety improvement capability of treating each safety issue type.

The objective function in this study is defined to minimize the CSI after the execution of safety improvements. The linear model is developed to suggest the safety issues that need to be prioritized to improve the project's safety performance, considering the available budget. The total amount spent on improving the safety performance is always smaller than the total available amount. The budget allocation can be carried out considering two period types; single period improvement and multi-period improvement budgets. In this study, a single period budget allocation is adapted. The variables needed to adapt a multi-period improvement budget may associate with a high level of uncertainty. Variables such as inflation, discount rate, traffic growth rate, accident growth rate, etc., include them.

This solution can be used to allocate several different safety treatments for a given project. As per the budget availability, any road is not limited to select only one issue to get treated. The proposed mathematical model maximizes the objective function that corresponds to the overall project condition.

3.2.1 Scope of the Model

Improvement of the roads' safety performance is measured by the reduction of the CSI after the treatments are applied. That is governed by the available treatment types and their CSI reduction capability based on the existing ground conditions. Based on the condition of each safety treatment strategy's prevailing safety issue and adaptability at the desired location, the quantified amount of safety improvement depends. This model guides us to select the safety issues that require the treatment, not to choose the treatment type for each location. The advantage of using this approach to determine the safety improvement is, the methodology depends on on-field inspections. Limitations prevailing in developing countries such as lack of crash data, lack of traffic data are overcome by this approach. An auditor can adopt the guidelines provided by Pasindu et al. [10] and determine the existing safety performance and safety improvement in terms of CSI.

The basis for the operation of the model is the estimation of the effectiveness of selected road safety countermeasures. Estimation of CSI's efficacy provides a clear indication of crash reduction since CSI is validated as a viable approach for safety evaluation in EPDO [10].

Indexes

$$X_i \in \{1, 2 \ldots\ldots I\} = index\ for\ the\ safety\ issue\ types$$

Model Parameters

$$Imp(X_i) : Improvement\ accounted\ when\ the\ issue\ type\ X_i\ is\ treated\ in\ a\ given\ road$$

$$CSI_i = Initial\ CSI\ of\ the\ road$$

$$CSI_z = Minimum\ CSI\ allowed\ for\ a\ given\ road$$

$$B = Available\, budget$$

$$C_i = Cost\, to\, treat\, the\, selected\, safety\, issue\, X_i$$

Decision Variable

$$U_i = \begin{cases} 1, & if\, the\, safety\, issue\, type\, X_i \\ & is\, selected\, to\, be\, treated\, in\, the\, selected\, road \\ 0, & Otherwise \end{cases}$$

Objective Function.
Minimize

$$CSI_i - \sum_{i=1}^{I} Imp(X_i)xU_i \tag{10}$$

Subjected to

$$\sum_{i=1}^{10} C_i U_i \le B \tag{11}$$

$$CSI_z > 50 \tag{12}$$

The objective function expresses the overall safety condition of the selected road at the project level. U_i is in the binary form, which tells the decision-maker to select or not to select the particular safety issue type for treatments to improve the safety performance of the road. Meanwhile, X_i represents the safety performance improvement adjacent to each safety issue after treating is undergone.

The set of constraints guides the decision-maker whether that road requires treatment at any place and guides the decision-maker to select issue types to treat based on the available budget. Constraints do not guide the decision-maker to choose only one issue type to get treated.

4 Illustrative Example

Below presented is an illustrative example to provide a better understanding of the developed linear programming model.

Two individual projects are considered in this illustrative example. They have initial safety performance of 302 and 601 based on the Cumulative Safety Index considered in this study. It is assumed that treatments are carried out at the same time when the budget is allocated.

The model selects the most feasible overall treatment plan based on the safety performance improvement and cost. This example uses a smaller budget than the total cost to treat all the safety issue types prevailing on each road. Cost estimations for the proposed treatment schemes were selected based on the findings of [10, 29] and based on the knowledge of local construction practices.

Table 4 summarizes the issue types along with possible implementation strategies.

Table 4 Road safety issues and possible treatments

Issue type	Safety issue	Possible safety treatments
1	Lack of signage, road markings at intersections	Sight distance improvement Lane marking Installing warning signs Providing pedestrian facilities Removing roadside objects
2	Poor visibility at access roads	Sight distance improvement Lane marking Installing warning signs Removing roadside objects
3	Limited roadside space availability	Clearing roadside vegetation Removing roadside objects
4	Road alignment, Sight distance issues, and Lack of warning signs at curves	Sight distance improvement Installing warning signs Removing roadside objects Geometrical improvements Improving side friction Superelevation improvement
5	Inadequate pedestrian facilities near Schools, Town areas	Sight distance improvement Installing warning signs Removing roadside objects Providing pedestrian facilities
6	Issues at railroad crossings	Sight distance improvement Providing rail crossing hardware and warning system
7	Roadside vertical drops—Unprotected steep roadside slope	Installing guardrails Geometrical improvements Installing warning signs Measures to increase the distance between the road edge and slope Increasing lane/shoulder width (if applicable)
8	Unprotected bridges and culverts	Installing warning signs Increasing lane/shoulder width (if applicable) Installing guardrails
9	Open drains	Providing drain covers
10	Excessive pavement edge drop offs	Pavement edge correction

The summary of the treatment schemes proposed by the model is presented below in Tables 5, 6, 7 and 8. It contains details about the initial CSI and final CSI of the issue types, percentage improved if the treatment is applied for each issue type, cost of treating each issue type, and the decision provided by the model along with the safety performance improvement.

Figure 1 demonstrates the reduction of CSI according to different budget levels on both roads.

Future Avenues Related to the Model

Further, one of the benefits of this approach is that it is possible to carry out sensitivity analyses in future endeavors. Carrying out sensitivity analysis is very important since budget allocation problems may be exposed to uncertainties. The costs used in the model are sometimes estimated values. The safety performance improvements may be approximated sometimes, and the available budget is subjected to changes.

This model was initially developed considering a single–period budget allocation. But the model can be further calibrated as it adopts the uncertainties raised when considering a multi-period budget allocation such as inflation, discount rate, traffic growth rate, accident growth rate.

Also, this model addresses the roadway conditions, social and environmental parameters related to rural low-volume roads in developing countries. Therefore, the model can be further calibrated as it addresses the parameters mentioned above of developed countries and/or urban environments.

Table 5 Initial Conditions of Road 1

Road 1										
Initial CSI 601	Total Cost $34,780									
Issue Type	1	2	3	4	5	6	7	8	9	10
Initial CSI of issue	60	133	34	30	47	0	256	8	16	17
Imp (X_i)	18	53	10	12	14	0	77	3	5	7
Cost ($ '1000)	5.2	1.5	0.5	0.6	4	0	18.2	1.5	1.08	2.2

Table 6 Treatment schemes for Road 1

Budget ($)	Issues to be treated	Final CSI
0	None	601
5000	2,3,4,10	519
10,000	2,3,4,5,9,10	500
15,000	1,2,3,4,5,8,9	486
20,000	1,2,3,4,5,8,9,10	479
25,000	2,3,4,5,7	435
30,000	1,2,3,4,5,7	417

Table 7 Initial Conditions of Road 2

Road 2										
Initial CSI 302	Total Cost $18,680									
Issue Type	1	2	3	4	5	6	7	8	9	10
Initial CSI of issue	54	81	12	54	32	8	8	27	8	18
Imp (X_i)	5.2	0.9	0.55	1.2	2	0.6	4.55	1.5	1.08	1.1
Cost ($'1000)	5.2	1.5	0.5	0.6	4	0	18.2	1.5	1.08	2.2

Table 8 Treatment schemes for road 2

Budget	Issues to be treated	Final CSI
0	None	302
$5000	2,3,4,6,8	217
10,000	2,3,4,5,6,8,9,10	196
15,000	1,2,3,4,5,6,8,9,10	180

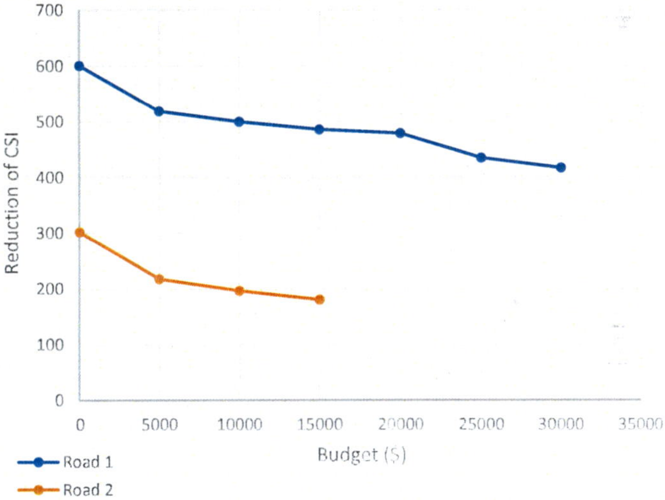

Fig. 1 Reduction of CSI according to different budget levels

5 Conclusion

The literature indicated that there had been only a small amount of effort to develop safety performance optimization methodologies. Yet, multi-objective optimization models have been developed considering both safety performance and pavement condition, where both the strategies have been given equal priority. It might provide beneficial results in terms of monetary conditions. But, being in a situation where the whole world is looking to halve the traffic deaths and injuries by the end of this decade, those multi-objective models will not end up being favorable in terms of safety performance. Therefore, a safety performance optimization technique is required, which will operate under a given budget.

This paper presents a novel approach to guide the decision makers on selecting required safety treatment criteria for a selected road. The decision-making criteria are based on solving a linear programming model. The objective function improves the safety performance of a selected project limited to the allocated budget. In this study, the safety performance of the road is evaluated by the Cumulative Safety Index (CSI), which is calculated based on the risk parameters of prevailing safety issues of the selected road. CSI is a mathematical rating that varies within a numerical scale of 1–5. Another novelty of this methodology is, this methodology focuses on safety issues of a road more precisely, rather than in network-level models.

The model guides the decision-maker on selecting/ not selecting an issue type to be treated based on the available budget and reducing safety performance. That is achieved by the binary decision variable included in the objective function and existing CSI of the road and improvement of CSI. The model's major constraint is that the cost to treat all the prevailing safety issues should be greater than the total available budget. Validation of the model was included in the paper using an illustrative example. The illustrative example consisted of two different roads that have two different safety performance levels. A comparison between the models presented in the literature and our model is not shown since the required data and performance conditions are not equal.

This methodology solves the administrative problems prevailing in developing countries on safety management of roads, such as lack of accident data, traffic data, black spots, etc. Since this methodology is based on field inspections, this approach can be easily adapted in developing countries with the contribution of a skilled multi-disciplinary team. The findings of this model can be incorporated into pavement management systems to develop strategies to maintain the overall performance of pavements.

References

1. Exiting the forgiving highway for the self explaining road. https://www.pps.org/article/what-can-we-learn-from-the-dutch-self-explaining-roads. Last accessed 01 Jun 2021
2. Forgiving roads: designed for nothing more serious than a small fright. https://www.ontheroadtrends.com/forgiving-roads-small-fright/?lang=en. Last accessed 02 Jun 2021
3. Federal Highway Administration (1989) Federal-aid highway program manual—Volume 6: engineering and traffic operations. U.S. Department of Transportation, The United States
4. Sandamal RMK, Ranawaka RKTK, Pasindu HR (2020) A framework to incorporate safety performance for low volume roads in pavement management systems. In: 2020 Moratuwa engineering research conference (MERCon). IEEE, Moratuwa, Sri Lanka, pp 407–412. https://doi.org/10.1109/MERCon50084.2020.9185235
5. Yepes V et al (2016) Optimal pavement maintenance programs based on a hybrid greedy randomized adaptive search procedure algorithm. J Civ Eng Manag 22(4):540–550

6. Correira MG, Bonates TDOE, Prata BDA, Junior EFN (2021) An integer linear programming approach for pavement maintenance and rehabilitation optimization. Int J Pavement Eng. https://doi.org/10.1080/10298436.2020.1869736
7. Silva RJ, Ferreira A, Flintsch G (2016) Incorporating road safety into pavement management. Ph.D. Thesis, Universidade de Coimbra, Coimbra
8. Alberti S, Fiori F (2019) Integrating risk assessment into pavement management systems. J Infrastruct Syst 25(1)
9. Wali B, Khattak AJ, Waters J, Chimba D, Li X (2018) Development of safety performance functions: incorporating unobserved heterogeneity and functional form analysis. Transp Res Rec. https://doi.org/10.1177/0361198118767
10. Pasindu HR, Ranawaka RKTK, Sandamal RMK, Thenuwara HV, Dias TWKIM (2021) Incorporating road safety in rural road network level pavement management. In: Transportation research board 100th annual meeting. Transportation Research Board, Washington DC
11. Cafiso S, Cava GL, Montella A (2007) Safety index for evaluation of two-lane rural highways. Transp Res Rec 2019(1):136–145
12. Mehta G, Lou Y (2013) Safety performance function calibration and development for the state of Alabama: two-lane two-way rural roads and four-lane divided highways. In: Transportation research board 92nd annual meeting. Transportation Research Board, Washington DC
13. Montella A (2005) Safety reviews of existing roads: quantitative safety assessment methodology. Transp Res Rec 1922(1):62–72
14. Atkinson JE, Chandler BE, Betkey V, Wejss K, Dixon K, Giragosian A, Donoughe K, O'Donnell C (2014) Manual for selecting safety improvements on high risk rural roads. U.S Department of Transportation
15. O'Cinneide D, Harrington L (2007) The development of a road safety management model. In: Urban transport XIII: urban transport and the environment in the 21st century
16. Mahalel D, Hakkert AS, Prashker JN (1982) A system for the allocation of safety resources on a road network. Accid Anal Prev 14:45–56
17. Brimley B, Saito M, Schultz G (2012) Calibration of highway safety manual safety performance functions. Transp Res Rec: J Transp Res Board 2279(1):82–89
18. Sharma A, Landge V (2013) Zero inflated negative binomial for modelling heavy vehicle crash rate on Indian rural highways. Int J Adv Eng Technol 5:292–301
19. Augeri MG, Greco S, Nicolosi V (2019) Planning urban pavement maintenance by a new interactive multi-objective optimization approach. Eur Transp Res Rev 11–17
20. Yoo J (2004) Multi-period optimization of pavement management systems. Ph.D. Thesis, Texas A&M University, Texas
21. Sinha KC, Labi S (2011) Transportation decision making: principles of project evaluation and programming. Wiley, New Jersey
22. Bixby RE (2012) A brief history of linear and mixed-integer programming computation. In: Documenta Mathematica, extra volume ISMP. Allen Institute for Artificial Intelligence, Seattle
23. Meyer P, Olteanu A (2017) Integrating large positive and negative performance differences into multicriteria majority-rule sorting models. Comput Oper Res 81:216–230. https://doi.org/10.1016/j.cor.2016.11.007
24. Mitra G, Lucas C, Moody S, Hadjiconstantinou E (1994) Tools for reformulating logical forms into zero-one mixed-integer programs. Eur J Oper Res 72(2):262–276
25. Bradley SP, Hax AC, Magnanti TL (1977) Applied mathematical programming, 1st edn. Addison-Wesley, Boston
26. Soylu B, Yıldız GB (2016) An exact algorithm for bi-objective mixed-integer linear programming problems. Comput Oper Res 72:204–213
27. Wang KCP et al (1993) Revisions to Arizona department of transportation pavement management system. Transp Res Rec: J Transp Res Board 1397:68–76

28. Ranawaka RKTK, Pasindu HR, Dias TWKIM (2019) Evaluation of causal factors related to accidents in low volume roads. In: International conference on road and airfield pavement technology. Kuala Lumpur
29. Pasindu HR, Sandamal RMK, Perera MYI (2020) A framework for network level pavement maintenance planning for low volume roads. In: 9th International conference on maintenance and rehabilitation of pavements—Mairepav9. Zurich, Switzerland

Analysis of Skidding Potential and Safe Vehicle Speeds on Wet Horizontal Pavement Curves

Jia Peng, L. Chu, and T. F. Fwa

Abstract Skidding on wet horizontal pavement curves is a major traffic safety concern. The maximum safe driving speed against skidding is an important threshold for safe driving. However, because of the complex tire-pavement-fluid interaction mechanism and the large number of variables involved (including curve geometric parameters, pavement surface properties, properties of tire in motion, and water film thickness), currently there is no practical working procedure that allows pavement engineers to determine the maximum safe driving speed on a horizontal curve under a given wet weather condition. This paper presents a finite element model to predict the maximum safe driving speed on a wet curved roadway section based on solid mechanics and hydrodynamics. The numerical simulation model was developed and validated against experimental skid resistance values on slip angles from 0° to 90°. Based on skidding analysis, the maximum safe driving speed on a horizontal curve is derived by comparing the available tire-pavement frictional resistance and the required friction to prevent skidding caused by the centrifugal force of the vehicle concerned. An illustrative case study is presented to compare the calculated maximum safe vehicle speed with AASHTO design speed. The analysis presented suggested that the proposed approach offers a useful tool to calculate maximum safe speeds on in-service pavement curves for safe driving.

Keywords Vehicle skidding potential · Horizontal pavement curve · Safe vehicle speed · Finite element model · Skid resistance

1 Introduction

Horizontal curves are accident-prone sections in a road network. According to the Fatality Analysis Reporting System (FARS) database, approximately 25% of all fatal

J. Peng · L. Chu (✉) · T. F. Fwa
School of Highway, Chang'an University, Xi'an, China
e-mail: longjiachu@chd.edu.cn

T. F. Fwa
National University of Singapore, Singapore, Republic of Singapore

© The Author(s), under exclusive license to Springer Nature Switzerland AG 2022

H. R. Pasindu et al. (eds.), *Road and Airfield Pavement Technology*, Lecture Notes in Civil Engineering 193, https://doi.org/10.1007/978-3-030-87379-0_2

highway crashes occurred along horizontal curve sections, and the average crash rate for horizontal curves is about three times that of other types of highway segments [1, 2]. In addition, the vast majority of these crashes on horizontal curves are roadway departures [3]. The main reason contributing to this type of crash is vehicle skidding, due to decreased available friction and increased friction demand when vehicles negotiate a horizontal curve [4]. Speed limits posted on highways would serve as a safety instruction against vehicle skidding on horizontal curves. However, these speed limits can ensure safe driving on dry pavements, they may not be adequate for wet and polished pavements. This is because the tire-pavement skid resistance on a pavement decreases rapidly as water film thickness increases and pavement macrotexture deteriorates under traffic polishing actions [5–7].

In this paper, the maximum safe driving speed is defined as the critical speed at which a vehicle is about to skid on a horizontal pavement curve, and skidding potential is defined as the difference between vehicle speed and the maximum safe speed against skidding. The determination of skidding potential and safe vehicle speed can be made using skid resistance values obtained from experimental studies, empirical relationship or numerical model. However, because of the complex tire-pavement-fluid interaction mechanism and the large number of variables involved (including curve geometric parameters, pavement surface properties, tire characteristics in motion, and water film thickness), currently there is no practical working procedure that allows pavement engineers to determine the maximum safe driving speed on a horizontal curve under a given wet weather condition.

This study proposes a computer-based procedure for skidding potential analysis and calculation of maximum safe driving speeds on wet horizontal pavement curves. The procedure comprises (i) finite element simulation analysis to calculate the available tire-pavement friction and (ii) tire skidding potential analysis to determine the maximum safe driving speed during vehicle cornering maneuver. The finite element skid resistance simulation model is developed based on solid mechanics and fluid dynamics, and then validated against skid resistance values over the full range of slip angle from 0° to 90°. Tire skidding potential analysis is performed by comparing the available tire-pavement frictional resistance with friction demand caused by the centrifugal force of the vehicle concerned. A numerical example demonstrates the application of the model and skidding risk analysis to calculate the maximum safe driving speeds for horizontal pavement curves under different wet weather conditions. The calculated maximum safe vehicle speeds are compared with AASHTO design speeds in the corresponding curve geometry characteristics.

2 Past Studies on Skidding Potential and Safe Vehicle Speeds on Wet Pavement

The complexity of skidding potential and safe vehicle speeds analysis arises from the dynamic tire-pavement-fluid interaction and the large number of variables involved. The interaction mechanisms are governed by solid mechanics and fluid dynamics theories. The variables involved include pavement, vehicle and environmental factors, such as curve geometric parameters, pavement surface properties, properties of tire in motion, and water film thickness. Currently, there is no practical working procedure that allows pavement engineers to determine skidding risk and safe vehicle speeds on a horizontal curve under a given wet weather condition. So far, past research studying skidding potential and safe vehicle speed mainly relied on analysis of pavement skid resistance characteristics. These efforts can be classified into three major categories: (i) Experimental studies, (ii) empirical relationship based on experimental or observed data, and (iii) computer-based numerical models.

Experimental studies provide direct measurements of skid resistance on wet pavement, and determinate skidding potential and maximum safe vehicle speeds based on these data. Full-scale experimental studies were conducted by the National Aeronautics and Space Administration (NASA) and Federal Aviation Administration (FAA) in the 1960s and 1970s [8–11]. Experimental measurements can provide relatively accurate results matching those of the actual pavement environment. However, due to the high cost of the experiments, they are rarely conducted nowadays.

Today, most highway agencies adopt a conservative constant friction in the geometric design of horizontal curves and the determination of the corresponding design speeds. These conservative friction values were derived mainly from either (i) past experience, field observations, experimental data, and accidents records; or (ii) statistical analysis based on experimental or observed field data. The horizontal curve geometric dimensions and design speed recommendations made by AASHTO are examples of this approach [12]. The AASHTO side friction recommendations were obtained from field studies of past research [13–15].

In recent years, with the rapid development of computer technology, numerical simulation has emerged as a powerful tool for predicting pavement skid resistance values under different operating conditions. Based on the theories of solid mechanics and hydrodynamics, computer-based numerical techniques can simulate dynamic tire–pavement–fluid interaction and solve for tire–pavement skid resistance as a function of vehicle operating variables (i.e. pavement surface properties, vehicle speed and water film thickness). Representative finite element solutions include those by NUS and Delft University of Technology researchers [16–20]. Similar finite element solutions have also been obtained by other researchers [21–23]. All these researchers only dealt with maximum safe vehicle speeds on either straight tangent road sections or on curved sections involving small slip angles up to 10°.

The skid resistance simulation model for curved pavement sections is more complex than that for straight pavement sections. This is because the existence of slip angle results in non-symmetrical tire deformation about its central plane, and non-symmetrical fluid motion around the tire-pavement contact area. Previous research by the author has validated the friction coefficients by the simulation model over the full range of slip angles from 0° to 90° [24]. In this paper, the simulation model was developed and validated on both straight and curved pavement sections. Tire skidding analysis was performed for the cornering maneuver of a vehicle to determine the available tire-pavement skid resistance values and the maximum safe vehicle speeds.

3 Development of Finite Element Simulation Model

3.1 Simulation Approach and Technique

The finite element skid resistance model has been developed earlier by the authors for tires moving on both straight and curved pavement section [16, 24–27]. Based on these previous research, the simulation model is validated against braking coefficients and lateral force coefficients on straight and curved pavement sections. The simulation approach consists of the following two main steps: Building the tire model and simulating the dynamic tire-fluid-pavement interaction process.

In the tire modeling process, tire rubber was represented by Lagrangian element types in ABAQUS. The tire rubber was considered as hyperelastic material, and simulated by neo-Hookean strain energy potential [21, 28, 29]. The viscoelastic properties of rubber materials were characterized by structural damping in the simulation. The radial plies and belts were modeled using surface membrane elements with embedded rebar layers. These reinforcement materials were modeled as elastic materials by Young's modulus and Poisson's ratio. The rim was assumed to be rigid. Figure 1 shows the bald-tread tire modeling process. First, 2D finite-element mesh of tire cross section was developed. Secondly, a partial 3D combination of the tire inner structure and tread was formed using rotation option of small angle. Then, a full 3D tire model was generated by making use of the revolving and reflecting features in ABAQUS. In the development of finite element tire model, mesh convergence analysis and calibration were performed to ensure the accuracy of tire characteristics based on measured tire deflections and footprint areas under different static wheel loads.

In simulating tire-fluid-pavement interaction process, water was modeled as Eulerian element and represented by the Us-Up equation of state material model, and pavement was represented as a rigid surface in the analysis [30]. The Eulerian mesh is typically a simple rectangular grid of elements constructed to extend well beyond

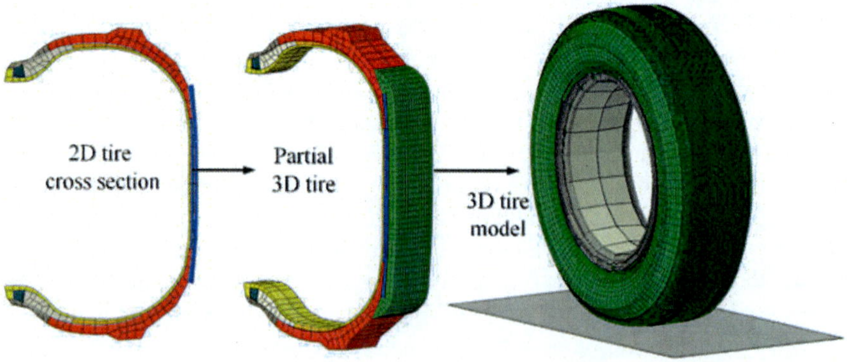

Fig. 1 The diagram of tire modeling process

the Eulerian material boundaries, giving the material space in which to move and deform. Therefore, it is necessary to set a certain height proportion between the inlet and void part in building the fluid model process. Generally, the void height proportion should be at least ten times than that of the water inlet mesh, and the height of the fluid model at different water thicknesses should be chosen by this scale. The Coupled Eulerian–Lagrangian (CEL) method in ABAQUS was applied to allow the Eulerian fluid materials to be combined with the traditional nonlinear Lagrangian tire analyses. The tire-pavement contact algorithm used in the simulation process were the Coulomb friction model and exponential decay friction model. To reduce simulation running time, a rolling tire was fixed at a given location, water as well as rigid pavement was modeled to flow toward the tire in the simulation.

Compared with vehicle movements on straight pavements, cornering movements cause asymmetric tire deformation (see Fig. 2), and this will need a more accurate tire model for skid resistance analysis on horizontal curves. The simulation model on wet straight and curved pavements adopted in this paper are shown in Fig. 3, and

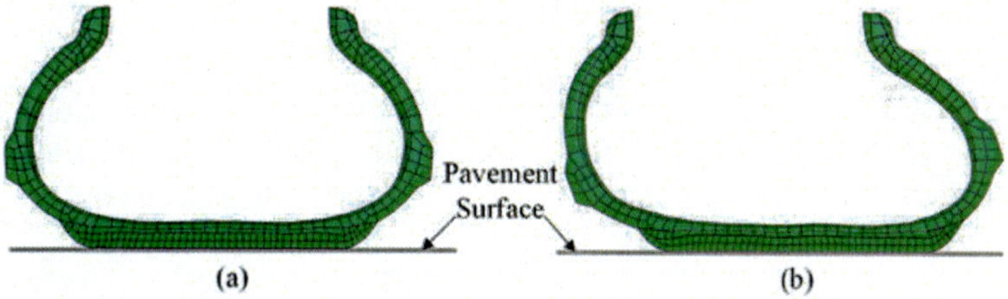

Fig. 2 Diagram of Cross-sectional deformation of the finite element tire model; **a** symmetrical deformation of tire moving on straight pavement; **b** asymmetric deformation as tire cornering movement

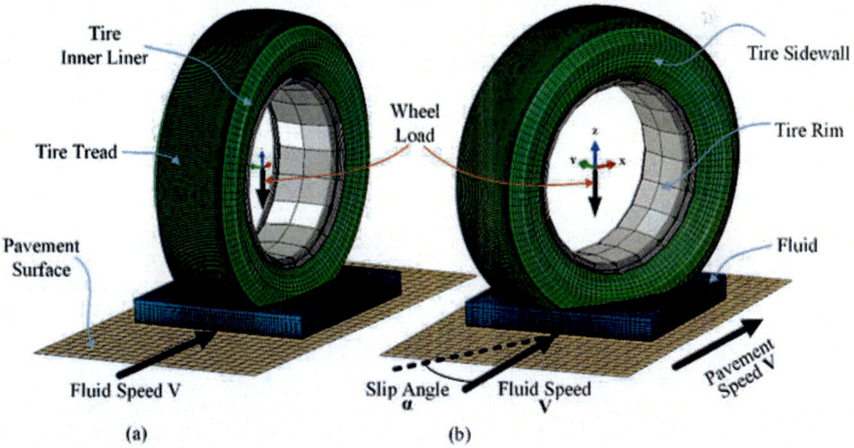

Fig. 3 The finite element skid resistance simulation model; **a** tire moving on straight pavement with zero slip angle; **b** tire moving on curved pavement with a slip angle

the main difference between them is the presence of slip angle. The slip angle causes transverse deformation of the tire which increases the difficulty for vehicle steering, and this deformation leads to the variation of the distribution of skid resistance values. The finite element skid resistance simulation model involves a large number of mesh elements with small size, and enough physical running time for stable results should be ensured during the calculation process. All of these lead to significant computational costs and time. Therefore, High Performance Computer Cluster is necessary to analyze the skid resistance problem to improve numeration efficiency.

Tire-pavement frictional characteristics is usually expressed in terms of skid number (SN). For a tire moving on a pavement at a slip angle, longitudinal skid resistance SN(x) and lateral skid resistance SN(y) in the x and y directions are defined as follows:

$$SN(\mathrm{x}) = 100 \times \frac{F_x}{F_z} \tag{1}$$

$$SN(\mathrm{y}) = 100 \times \frac{F_y}{F_z} \tag{2}$$

where F_x and F_y are tire-pavement frictional resistance in x and y directions, F_z the applied wheel load in the z direction, x the horizontal forward direction coincides with the plane of the tire, y the direction perpendicular to the plane of the tire, and z the vertical direction, as shown in Fig. 3.

3.2 *Validation of Simulation Model*

The validation of the finite element simulation model developed in this study was made by checking with skid resistance measured in two perpendicular directions, namely the longitudinal frictional resistance SN(x) and lateral frictional resistance SN(y). The measured skid resistance values of straight and curved pavements by Horne and Bergman [10, 31] were adopted for validating the model. The three test surfaces, test conditions and tires considered in this section were as follows:

- **Test Surface 1:** Straight section ungrooved asphalt surface;

  Tire 1: ASTM bald-tread tire;
  Condition 1: (a) skid resistance of surface 1 rated by NASA diagonal braking car under locked-wheel braking conditions, wet and puddled surface; (b) by FAA Swedish Skiddometer under steady-state peaking braking conditions, wet surface with 0.02-inch water layer.

- **Test Surface 2:** 500-ft (152.4 m) radius curve section ungrooved concrete surface;

  Tire 2: ASTM bald-tread tire;

Condition 2: Locked wheel braking and peaking braking by General Motors braking trailer, flooded pavement condition.

- **Test Surface 3:** "Safety Walk" surface manufactured by the 3 M Company at Calspan Laboratories, SN40R = 30 (i.e. SN is 30 at a test speed of 40 mph (65 km/h) with a Specification E501 Standard Rib Tire [32]);

  Tire 3: J78-15 tire, vertical load was 6.222 KN and the tire inflation pressure was 165.5 kPa;
  Condition 3: Skid resistance tested in the cornering and braking-in-a-turn modes; Test speed was 48.3 km/h on a wet surface with a water film thickness of 0.5 mm.

To validate the model, the SN_0 and skid resistance-related surface texture properties of the test pavement are required. According to measured skid resistance values, the simulation model can be calibrated using the back-analysis procedure developed by Fwa and Chu [33]. After the skid resistance calibration, the predicted values of SN(x) and SN(y) could be calculated for different slip angles and vehicle speeds.

The validation of longitudinal frictional resistance SN(x) was performed against measured skid resistance values on test straight surface 1 and curved pavement 2. The results of test surface 1 and 2 were all derived from Horne's study [10], and the tire was ASTM smooth standard tire E524 [34]. In order to facilitate the numerical model verification, test conditions were quantified as input parameters. For test straight surface 1, test conditions were (a) tire slip ratio of 100% at a water depth of 0.1 in. (2.54 mm), and (ii) tire slip ratio of 13% at a water layer of 0.02 in. (0.5 mm). For test curved surface 2, the experimental data were validated by tire moving with a small slip angle of 4°, tire slip ratio of 100% and 15% at a water depth of 0.2 in (5 mm). The validation results of test surfaces 1 and 2 are shown in Table 1. It can be seen from Fig. 4 that only the maximum predicted skid number had a difference larger than 5 SN units and percentage error larger than 5% in these 16 test cases. Therefore, the simulation model was able to predict the longitudinal skid resistance within practically acceptable limits, for either straight or curved pavement sections.

Measured skid resistance values on surface 3 by Bergman was adopted for the validation of lateral frictional resistance SN(y) [31]. Lateral force coefficients of J78-15 tire were determined by applying nine slip angles from 1° to 90°. Additionally, tire slip ratios were 100% and 20% at a wet surface with a water depth of 0.5 mm. The comparison of experimental and calculated SN(y) values at speed of 48.3 km/h are presented in Table 2 and Fig. 5. As shown in Table 2, errors for SN(y) at slip angles of 1°, 4° and 8° for locked wheel, and a slip angle of 1° for 20% slip ratio are more than 10%. For these four cases, the skid resistance values were all small and differences were within measurement error margins. Hence, the validation confirms that the simulation model is able to predict skid resistance of a tire moving on wet straight and curved pavements.

Table 1 Comparison of experimental and calculated SN(x) values on test surface 1 and 2

Test surface	Water film thickness (mm)	Slip ratio (%)	Test speed (km/h)	Skid Number SN(x)			
				Measured $SN_m(x)$	Predicted $SN_p(x)$	Difference $\Delta SN(x)$	Error (%) $\delta SN(x)$
Test surface 1 (straight section asphalt surface)	2.54	100	32	64	63.6	−0.40	−0.62
			48	45	45.6	0.60	1.33
			64	32	32.4	0.40	1.25
			80	22	21.94	−0.06	−0.27
			96	13	12.92	−0.08	−0.62
Test surface 1 (straight section asphalt surface)	0.5	13	32	76	69.14	−6.86	−9.03
			48	62	60.93	−1.07	−1.73
			64	51	50.7	−0.30	−0.59
			80	42	41.63	−0.37	−0.88
			96	33	32.49	−0.51	−1.55
Test surface 2 (curved section concrete surface)	5	100	16	25	24.92	−0.08	−0.32
			32	14	14.26	0.26	1.86
			48	10	10.47	0.47	4.70
		15	16	48	47.80	−0.20	−0.42
			32	40	39.78	−0.22	−0.55
			48	24	24.58	0.58	2.42

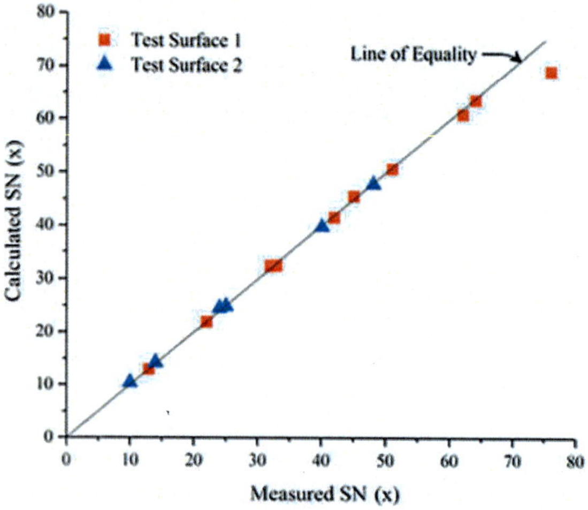

Fig. 4 Comparison of measured data and model predicted values of SN(x) on test surfaces 1 and 2

Table 2 Comparison of experimental and calculated SN(y) values on test surface 3

Slip angle (°)	Measure data by Bergman (1977)		Predicted value by simulation model		Difference		Error (%)	
	$SN(y)^L$	$SN(y)^S$	$SN(y)^L$	$SN(y)^S$	$\Delta SN(y)^L$	$\Delta SN(y)^S$	$\delta SN(y)^L$	$\delta SN(y)^S$
1	1.45	6.93	1.06	5.62	−0.39	−1.31	−26.57	−18.90
4	2.89	22.95	3.58	22.7	0.69	−0.25	24.01	−1.09
8	6.17	43.31	6.90	42.28	0.73	−1.03	11.84	−2.38
16	14.02	55.03	14.09	54.66	0.07	−0.37	0.49	−0.67
32	26.63	57.27	27.18	56.14	0.55	−1.13	2.07	−1.97
48	38.76	54.32	39.15	55.66	0.39	1.34	1.01	2.47
60	45.24	55.24	47.52	55.01	2.28	−0.23	5.05	−0.42
72	49.42	53.57	51.16	53.55	1.74	−0.02	3.52	−0.04
90	52.34	51.93	52.71	51.1	0.37	−0.83	0.71	−1.60

Note $SN(y)^L$ represents lateral skid number of locked wheel, $SN(y)^S$ is lateral skid number of rolling wheel with 20% slip ratio, test tire is J78-15, speed is 48.3 km/h, water film thickness is 0.5 mm

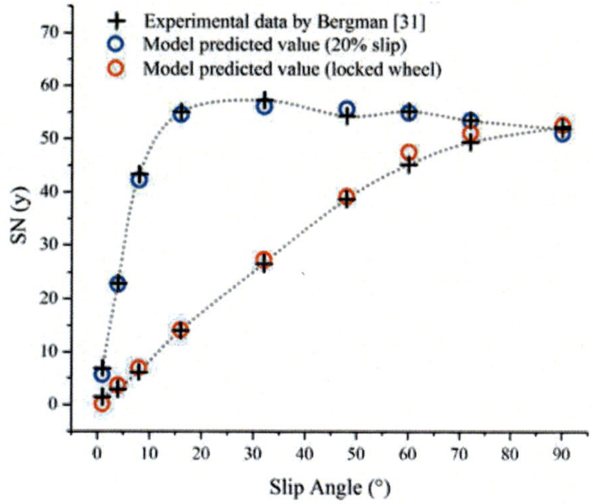

Fig. 5 Comparison of experimental data and model predicted values of SN(y) on test surface 3

4 Skidding Potential Analysis and Determination of Maximum Safe Vehicle Speed

According to Lamm's third criterion [35], the analysis of skidding potential was performed by comparing the available friction that the road could provide to a vehicle under certain conditions, with the friction demand that a vehicle required from the road when it went around a curve at a designated operating speed. The risk of skidding potential depends on the comparison between friction demand f_d and available

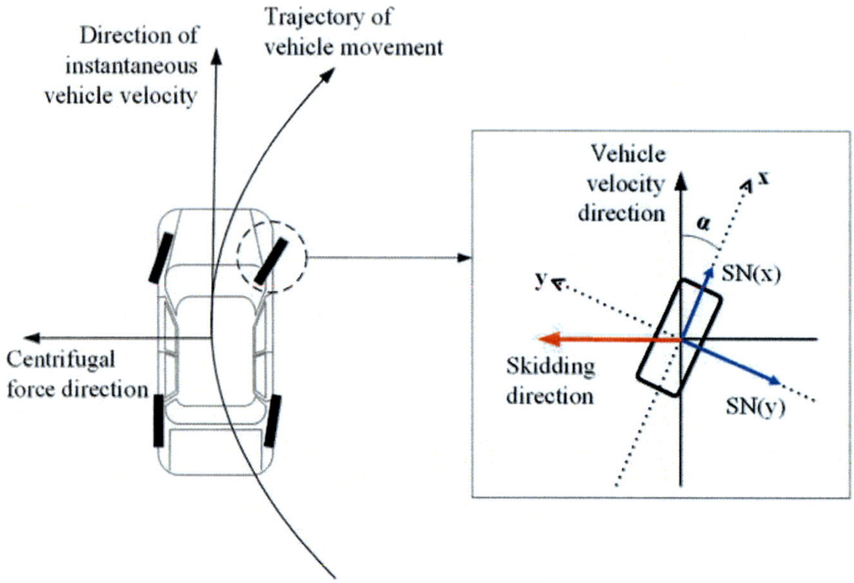

Fig. 6 The force schematic diagram of vehicle traveling on a horizontal curve

friction f_a. When $f_d > f_a$, skidding would occur, when $f_d < f_a$, no skidding would occur. The velocity at the intersection point is the critical speed of impending skidding, i.e. the maximum safe speed against skidding. The skidding potential of a vehicle is defined as the difference between its speed and the maximum safe speed against skidding.

Figure 6 is a schematic diagram of vehicle traveling on a horizontal curve. Slip angle α is the angle between the direction of vehicle velocity and the central plane of the tire. As a result of the angular movement of the vehicle, there is a transverse force produced by the centrifugal force acting on the vehicle. This transverse force has to be resisted by the frictional resistance at the tire-pavement contact in order to keep the vehicle driving safely within the lane. Skidding will occur if the frictional resistance is insufficient to resist the transverse force. This transverse force governs the friction demand that a vehicle requires from the road when it goes around a curve at a designated operating speed. The formula is as follows:

$$f_d = (\frac{V_{85}^2}{gR} - e) \times 100 \tag{3}$$

where V_{85} is the 85th percentile speed of the traffic stream, R is the horizontal curve radius, g is acceleration of gravity, and e is the superelevation rate in fraction.

Compared to the coefficient of friction demand f_d, the available tire-pavement friction f_a is determined by the finite element skid resistance simulation model developed in this study. It is defined by Eq. (4).

$$f_a = \sqrt{SN(x)^2 + SN(y)^2} \quad (4)$$

where α is the slip angle, SN(x) is the longitudinal skid number in the direction parallel to the tire plane, and SN(y) is the lateral skid number in the direction perpendicular to the tire plane.

5 Example: Calculating Maximum Safe Vehicle Speed on Horizontal Curves

This section demonstrates the application of the skid resistance simulation model for calculating the maximum safe vehicle speed on a horizontal pavement curve, and compares with the horizontal curve design speeds recommended by AASHTO [12]. The AASHTO design recommendation determines the design speeds by assuming a constant coefficient of tire-pavement friction f_a based on experimental data. In practice, as described in the literature, f_a varies with vehicle speed, tire slip angle, water film thickness and pavement surface properties. With the assumption of a low constant f_a, the AASHTO design speeds are conservative for most cases, but could be inadequate for some specific cases, such as cases where a pavement surface has been badly polished and worn by traffic, or where a deep water film thickness exists on the pavement surface. This part mainly analyzes the effect of water on vehicle skidding and determines the critical wet pavement condition for safe driving through performing skidding potential analysis based on f_a calculated from the simulation model.

In this section, the skid resistance state of test pavement surface 1 (ungrooved asphalt surface) mentioned in the validation process is adopted as a case of skidding analysis. According to back-calculation results of the skid resistance model, the test surface 1 tested by Horne [10] has a SN_0 value of 80 and SN_{64} value of 40. The tire model adopted is ASTM standard smooth tire (E524) [34] with a wheel load of 4800 N and an inflation pressure of 165.5 kPa. The following ranges of variables are considered:

Values of horizontal curve radius	200, 300, 400 m
Values of superelevation rate	2, 4, 6, 8, 10%
Values of vehicle speed	40, 60, 80,100 km/h
Values of water film thickness	1, 3, 5, 10 mm
Values of slip angle	13°

Values of curve radii and superelevation rates are chosen from the usual range of AASHTO recommendation [12]. A superelevation rate of 2% is used as a representative value of normal cross slope, and the highest superelevation rate for highways in common use is 10%. Four water thicknesses on pavement surface are considered to represent different rainfall intensities. For tire slip angle, past studies have found that its magnitudes would vary within a rather narrow range, and usually do not exceed

13° under normal vehicle driving [36–39]. Hence, the maximum safe vehicle speed analysis is performed for the slip angle of 13°.

Figure 7 illustrates the skidding analysis for the determination of the maximum safe vehicle speeds. The solid curves plotted in colors show the variation of available tire-pavement skid resistance f_a with vehicle speed for different water film thicknesses. The series of dashed-line curves represent the skid resistance demand f_d as a function of vehicle speeds, pavement curve radii and superelevation rates. The intersection point of a solid-line curve and a dashed-line curve defines the maximum safe vehicle speed for the specified wet pavement condition and curve geometry.

Table 3 presents the contrast between AASHTO design speeds and calculated maximum safe vehicle speeds for the test pavement surface, different pavement curve radii, superelevation rates and water film thicknesses. The shaded cells in the table identify those instances where the AASHTO design speed exceeds the maximum safe vehicle speed. The result indicates that the AASHTO design speeds are safe under light to medium rain conditions (i.e. water film thickness is 1 to 3 mm). It also reveals that design speeds exceed the corresponding safe threshold velocities under heavy rainfall intensity (i.e. water film thickness of 5 to 10 mm). That is to say, lateral skidding will take place when vehicles driving on badly wetted curved pavement at AASHTO design speed. These analysis outcomes are consistent with actual crash investigation. The maximum safe vehicle speeds were calculated by the numerical simulation of a bald-tread tire, which represents the worst possible condition for tire completely wearing out. Tire tread patterns can increase the effect of drainage and maintain good contact between tire and pavement. Therefore, compared with actual tread tires, this analysis gives the most conservative results and permits more

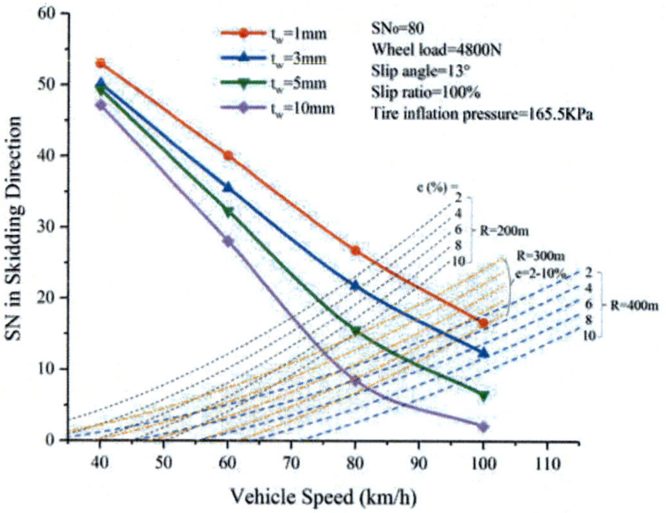

Fig. 7 Determination of maximum safe vehicle speeds for example problem based on skidding analysis

Table 3 Comparison of AASHTO design speeds and calculated maximum safe vehicle speeds

Curve radius (m)	e (%)	AASHTO design speed (km/h)	Calculated maximum safe vehicle speed (km/h)			
			$t_w = 1$ mm	$t_w = 3$ mm	$t_w = 5$ mm	$t_w = 10$ mm
200	2	66	82.94	78.91	74.42	70.18
	4	69	84.55	80.52	75.87	71.47
	6	73	86.16	82.14	77.33	**72.77**
	8	75	87.78	83.77	78.80	**74.08**
	10	78	89.40	85.41	80.29	**75.41**
300	2	78	91.93	87.00	80.71	**75.00**
	4	83	94.00	89.07	**82.52**	**76.56**
	6	85	96.10	91.17	**84.39**	**78.16**
	8	89	98.26	93.33	**86.32**	**79.81**
	10	94	100.46	95.54	**88.32**	**81.53**
400	2	87	98.68	92.97	**85.11**	**78.16**
	4	93	101.28	95.52	**87.31**	**79.97**
	6	96	104.00	98.20	**89.65**	**81.88**
	8	101	106.88	101.03	**92.16**	**83.90**
	10	106	109.96	**104.06**	**94.91**	**86.09**

Notes (1) e = superelevation rate, t_w = water film thickness on pavement surface. (2) Shaded cells (Bold values) refer to those cases where the AASHTO design speed exceeds the corresponding calculated maximum safe vehicle speed

safety margin for vehicle driving on horizontal curves. This example demonstrates that the skidding potential analysis and maximum safe vehicle speed determination procedure can be applied to identify high-skidding risk roads and weather conditions.

6 Conclusion

This paper has presented a procedure for the determination of the maximum safe vehicle speeds on a wet horizontal pavement curve, based on skidding analysis. A finite element simulation model has been developed to predict the available tire-pavement friction of a wet horizontal pavement curve. The simulation model was validated against experimental data on straight and curved pavement sections. Compared with the assumption of constant friction coefficient commonly adopted in practice, this skid resistance simulation model provides a useful method to predict the available tire-pavement friction under given tire slip angle, vehicle speed and water film thickness. Based on skidding potential analysis, the maximum safe vehicle speed for a given wet–pavement operating condition is determined by comparing the available tire-pavement friction and friction demand. It offers a potentially useful tool for identifying high-skidding risk pavement curves for safety improvement, instead of the traditional practice of relying on crash data records. This application of skidding analysis has practical significance in traffic safety management of wet horizontal pavement curves, especially those pavement sections that have suffered surface deterioration under traffic polishing and wearing.

Acknowledgements The authors gratefully acknowledge financial support by Shannxi Science and Technology, China. Project 2021JQ-261.

References

1. National Highway Traffic Safety Administration (2012) 2010 Motor vehicle crashes—overview. U.S. Department of Transportation, Washington, DC
2. Fatality Analysis Reporting System (FARS) National Highway Traffic Safety Administration. http://www.nhtsa.gov/FARS. Last accessed 18 Feb 2021
3. Federal Highway Administration Horizontal curve safety—FHWA safety program. https://safety.fhwa.dot.gov/roadway_dept/horicurves/cmhoricurves/horiz_curve.pdf. Last accessed 18 Feb 2021
4. National Academies of Sciences, Engineering, and Medicine (2004) A guide for reducing collisions on horizontal curves. The National Academies Press, Washington, DC. https://doi.org/10.17226/13545
5. Rose JG, Gallaway BM (1977) Water depth influence on pavement friction. J Transp Eng 103(4):491–506
6. Henry JJ (2000) Evaluation of pavement friction characteristics. NCHRP Synthesis 291, Transportation Research Board, Washington, DC
7. Fwa TF, Ong GP (2008) Wet-pavement hydroplaning risk and skid resistance: analysis. J Transp Eng 134(5):182–190
8. Horne WB, Dreher RC (1963) Phenomena of pneumatic tire hydroplaning. Report NASA T.N.D-2056. National Aeronautics and Space Administration, Langley Research Center, Hampton, VA
9. Leland TJW, Taylor GR (1965) An investigation of the influence of aircraft tire-tread wear on wet-runway braking. In: NASA technical note TN D-2770. NASA, Washington, DC
10. Horne WB (1969) Results from studies of highway grooving and texturing at NASA Wallops station. Pavement grooving and traction studies, NASA SP-5073. National Aeronautics and Space Administration, Washington, DC, pp 425–464
11. Yager TA, Byrdsong TJ (1973) Some effects of grooved runway configurations on aircraft tire braking traction under flooding runway conditions. National Aeronautic and Space Administration, Hampton, VA, TN D-7215
12. AASHTO (2018) A policy on geometric design of highways and streets, 7th ed. American Association of State Highway and Transportation, Washington DC
13. Barnett J (1936) Safe side friction factors and superelevation design. In: Proceedings HRB, vol. 16. Highway Research Board (currently Transportation Research Board), National Research Council, Washington, DC, pp 69–80
14. Moyer RA, Berry DS (1940) Marking highway curves with safe speed indications. In: Proceedings of HRB, vol. 20. Highway Research Board, Washington, DC, pp 399–428
15. Stonex KA, Noble CM (1940) Curve design and tests on the Pennsylvania Turnpike. In: Proceedings of HRB, vol. 20. Highway Research Board, Washington, DC, pp 429–451
16. Ong GP, Fwa TF (2007) Wet-pavement hydroplaning risk and skid resistance: modeling. J Transp Eng 133(10):590–598

17. Ju F, Fwa TF, Ong GP (2013) Evaluating wet weather driving benefits of grooved pavements. Int J Pavement Res Technol 6(4):287–293
18. Zhang L, Ong GP, Fwa TF (2013) Developing an analysis framework to quantify and compare skid resistance performances on porous and non-porous pavements. Transp Res Rec: J Transp Res Board 2369:77–86
19. Anupam K, Srirangam SK, Scarpas A, Kasbergen C (2013) Influence of temperature on tire-pavement friction. Transp Res Rec: J Transp Res Board 2369(1):114–124. https://doi.org/10.3141/2369-13
20. Srirangam SK, Anupam K, Kasbergen C, Scarpas A, Cerezo V (2015) Study of influence of operating parameters on braking friction and rolling resistance. Transp Res Rec: J Transp Res Board 2525:79–90. https://doi.org/10.3141/2525-09
21. Ding Y, Wang H (2018) Evaluation of hydroplaning risk on permeable friction course using tire-water-pavement interaction model. Transp Res Rec: J Transp Res Board 2672(40):408–417. https://doi.org/10.1177/0361198118781392
22. Liu X, Cao Q, Wang H, Chen J, Huang X (2019) Evaluation of vehicle braking performance on wet pavement surface using an integrated tire-vehicle modeling approach. Transp Res Rec: J Transp Res Board 2673(3):295–307. https://doi.org/10.1177/0361198119832886
23. Anupam K, Srirangam SK, Scarpas A, Kasbergen C, Kane M (2014) Study of cornering maneuvers of a pneumatic tire on asphalt pavement surfaces using the finite element method. Transp Res Rec: J Transp Res Board 2457(1):129–139. https://doi.org/10.3141/2457-14
24. Peng J, Chu L, Fwa TF (2020) Determination of safe vehicle speeds on wet horizontal pavement curves. Road Mater Pavement Des 1–13
25. Chu LJ, Fwa TF (2018) Pavement skid resistance consideration in rain-related wet-weather speed limits determination. Road Mater Pavement Des 19(2):334–352. https://doi.org/10.1080/14680629.2016.1261723
26. Fwa TF, Pasindu HR, Ong GP (2011) Critical Rut depth for pavement maintenance based on vehicle skidding and hydroplaning consideration. J Transp Eng 138(4):423–429
27. Peng J, Chu L, Wang T, Fwa TF (2021) Analysis of vehicle skidding potential on horizontal curves. Accid Anal Prev 152:105960
28. Rivlin RS (1948) Large elastic deformations of isotropic materials. I. Fundamental concepts. Philos Trans R Soc Lond Ser A Math Phys Sci 240(822):459–490
29. Rivlin RS (1948) Large elastic deformations of isotropic materials. II. Some uniqueness theorems for pure, homogeneous deformation. Philos Trans R Soc Lond Ser A Math Phys Sci 240(822):491–508
30. ABAQUS (2010) ABAQUS user's manual. Pawtucket, RI, USA
31. Bergman W (1976) Skid resistance properties of tires and their influence on vehicle control. Transp Res Record: J Transp Res Board 621:8–18
32. ASTM E274/E274M–11 (2011) Standard test method for skid resistance of paved surfaces using a full-scale tire. ASTM International, West Conshohocken, PA
33. Fwa TF, Chu L (2019) The concept of pavement skid resistance state. Road Mater Pavement Des 1–20. https://doi.org/10.1080/14680629.2019.1618366
34. ASTM E524-08 (2020) Standard specification for standard smooth tire for pavement skid resistance tests. ASTM International, West Conshohocken, PA
35. Lamm R, Psarianos B, Choueiri EM, Soilemezoglou G (1998) A practical safety approach to highway geometric design international studies: Germany, Greece, Lebanon and the United States. In: Transportation research circular E-C003: International symposium on highway geometric design principles. TRB, National Research Council, Washington, DC, pp. 9:1–14
36. Melzi S, Sabbioni E (2011) On the vehicle sideslip angle estimation through neural networks numerical and experimental results. Mech Syst Signal Process 25(6):2005–2019
37. Ryu J, Nardi F, Moshchuk N (2013) Vehicle sideslip angle estimation and experimental validation. In: ASME 2013 International mechanical engineering congress and exposition. American Society of Mechanical Engineers, V04AT04A052-V04AT04A052

38. Cheli F, Sabbioni E, Pesce M, Melzi S (2007) A methodology for vehicle sideslip angle identification: comparison with experimental data. Veh Syst Dyn 45(6):549–563
39. Todoruţ A, Cordoş N (2018) Evaluation of the vehicle sideslip angle according to different road conditions. In: International congress of automotive and transport engineering. Springer, Cham, pp 814–819

Asphalt Pavement Texture Level and Distribution Uniformity Evaluation Using Three-Dimensional Method

Shihao Dong and Sen Han

Abstract To supplement the research on the evaluation method of asphalt pavement texture, novel three-dimensional (3D) methods are proposed. First, pavement textures were measured in laboratory from asphalt mixture specimens using laser texture scanner (LTS), and the macro-texture and micro-texture were extracted based on spectrum analysis techniques. Then, macro-texture level evaluation indices f_{8mac} and f_{9mac} together with micro-texture level evaluation indices f_{8mic} and f_{9mic} were proposed based on the gray level co-occurrence matrix (GLCM) method, and the hyperparameters existing in GLCM were discussed. Through the correlation analysis with mean texture depth (MTD) measured by sand patch method (SPM) and friction coefficient μ measured by walking friction tester (WFT), the optimum pavement texture level evaluation indices were determined. Additionally, the evaluation index σ of distribution uniformity of pavement texture (DUPT) was proposed based on the uniformity of deviations between sub-surfaces and the average surface of pavement texture. Finally, the correlations of σ with texture profiles were studied. The results show that f_{8mac} and f_{8mic} are the optimum indices for pavement texture level. MTD has significant correlation with f_{8mac}, and the correlation coefficient R is 0.9348; friction coefficient μ has significant correlation with f_{8mic}, and the R is 0.8030. The hyperparameters of GLCM selected in this study were proved effective. Moreover, the effectiveness of σ is also validated by calibrating with standard grooved surface. It can be concluded that the proposed indices in this study are suitable to the evaluation of pavement texture level and pavement texture distribution.

Keywords Highway engineering · Pavement texture · Gray level co-occurrence matrix · Three-dimensional evaluation index · Distribution uniformity

S. Dong · S. Han (✉)
Highway School, Chang'an University, Xi'an 710064, Shaanxi Province, China
e-mail: hyram_hs@chd.edu.cn

S. Dong
e-mail: shihaodong@chd.edu.cn

© The Author(s), under exclusive license to Springer Nature Switzerland AG 2022
H. R. Pasindu et al. (eds.), *Road and Airfield Pavement Technology*, Lecture Notes in Civil Engineering 193, https://doi.org/10.1007/978-3-030-87379-0_3

1 Introduction

Pavement texture determines factors such as noise emission from the tire-pavement interface, acoustic comfort inside vehicles, friction between the tire and road, rolling resistance and tire wear [1–7]. Thus, it is necessary to measure pavement texture and test its effect on pavement performance and quality [8]. According to the different wavelengths and amplitudes, pavement texture can be divided into micro-texture, macro-texture, mega-texture and unevenness [9–11]. Appropriate macro-texture and micro-texture can improve the pavement surface characteristics and provide good skid resistance and noise reduction performance for the pavement surface [12], while the mega-texture and unevenness will cause irregular vibration of the rubber tires and bumps of vehicle, which should be avoided as much as possible [13].

Comprehensive evaluation of pavement texture is the key to studying pavement surface characteristics. So far in the literature, ample researches on pavement texture evaluation have been conducted on three aspects: pavement texture (profile) level, pavement texture (profile) distribution and the effect of surface profile asymmetry [4, 14, 15]. Generally, the indices proposed based on the two-dimensional (2D) texture profile are called the 2D evaluation index, and the indices proposed based on the three-dimensional (3D) pavement texture are called the 3D evaluation index. As for the evaluation of pavement texture (profile) level, there are many typical 2D evaluation indices like macro-texture profile level $L_{TX,0.5\rightarrow 32}$, micro-texture profile level $L_{TX,0.03\rightarrow 0.5}$ and mean profile depth (MPD); except for the mean texture depth (MTD), few indices focus on the evaluation of pavement texture from 3D aspect. The evaluation of pavement texture distribution is mainly based on texture profile, and commonly used indices are texture profile level amplitude A_{max}, characteristic wavelength W_{max}, and characteristic wavelength texture profile level $L_{TX,ch}$ [16, 17]. Pavement density sometimes was measured to evaluate the distribution uniformity of pavement texture (DUPT) indirectly [18]. Some researchers also used the uniformity of gray-scale image of pavement texture to evaluate the DUPT by converting gray level of gray-scale image into corresponding texture depth [19, 20], but there is no theory to prove its rationality and accuracy. Thereby, the research about 3D evaluation of DUPT still needs to be further conducted. As for the evaluation of the effect of texture profile asymmetry, conventional indices like skewness R_{sk} and shape factor g were calculated based on texture profile [17, 21], and few researches study the 3D evaluation method for the effect of texture profile asymmetry.

In summary, the evaluation of pavement texture mainly focuses on 2D evaluation indices, and the 3D evaluation indices for pavement texture are rarely studied. At present, the 3D pavement texture is often evaluated by the mean of 2D evaluation indices calculated based on texture profiles. Utilization of mean to estimate the overall is vulnerable to extreme value. To supplement 3D evaluation methods and indices for pavement texture, some investigations proposed indices based on gray level co-occurrence matrix (GLCM) method [22, 23]. GLCM is a method proposed by Haralick to analyze the texture characteristics of gray-scale images [24]. There are 14 feature indicators $f_1 \sim f_{14}$ of GLCM which can be used to describe image texture characteristics [22, 24, 25]. Among them, researchers found that f_8 and f_9 have significant correlation with MTD, and the correlation coefficients of MTD with f_8 and f_9 were both above 0.9 [22, 23]. The Entropy theory was also used as an 3D method to evaluate the pavement texture level [23]. The calculation of GLCM is determined by several hyperparameters. Even for the same pavement texture, if different hyperparameters are selected, the calculation results of GLCM will be different, which is also the reason why the 3D evaluation method of pavement texture level based on GLCM is difficult to be popularized.

The main objective of this study is to provide a novel and practical 3D evaluation method and indices for pavement texture level and distribution. The rest of the paper is organized as follows. Section 2 at first describes the acquisition method of pavement texture data. Then, the pretreatment for pavement texture is introduced to acquire accurate pavement macro-texture and micro-texture. Thereafter, the establishment procedure of evaluation indices for pavement texture level and the distribution uniformity of pavement texture are illustrated, and the method for the validation of proposed indices effectiveness is also introduced. Section 3 presents the effectiveness analysis of proposed evaluation indices compared with conventional indices. Finally, in Sect. 4, the conclusion is summarized.

2 Materials and Methods

The 3D pavement texture is often evaluated by the mean of 2D evaluation indices calculated based on texture profiles today. In this study, the proposed 3D evaluation indices are compared with the existing 2D evaluation indices, and the effectiveness of 3D indices is validated by correlation analysis. The analysis program of evaluation indices is shown in Fig. 1.

The 2D evaluation indices $L_{TX,0.5\to32}^{mean}$ and $L_{TX,0.03\to0.5}^{mean}$ are calculated relying on texture profiles, which represent the mean of the texture profile level. Meanwhile, the 3D evaluation indices proposed in this study which are f_{8mac}, f_{9mac}, f_{8mic} and f_{9mic} are calculated based on pavement texture, which can evaluate the pavement texture level. MTD and friction coefficient μ are measured as reference for the correlation analysis with the evaluation indices mentioned above. For the purpose of evaluating the DUPT, index σ is proposed and calibrated by standard grooved surface.

2.1 *Acquisition of Pavement Texture Data*

Four types of asphalt mixture which are asphalt concrete (AC) mixture, stone mastic asphalt (SMA), porous asphalt (PA) and open graded friction course (OGFC) were prepared to obtain the pavement texture data. Four nominal maximum sizes of aggregate (NMAS) which are 16, 12.5, 9.5 and 4.75 mm were selected for AC mixture, while three NMAS except for 4.75 mm were used for SMA, OGFC and PA mixture.

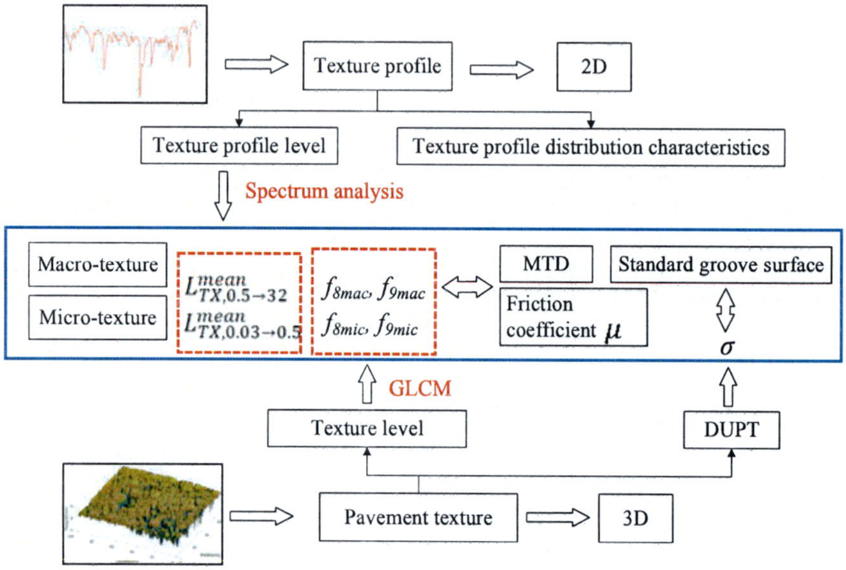

Fig. 1 Indices analysis program

Therefore, total 13 asphalt mixture specimens with different pavement texture were formed in this study. The size of formed specimen is 500 × 500 × 50 mm (Length × Width × Height). One specimen is divided into 12 rectangular regions by perforated steel plate, as is shown in Fig. 2. The length and width of the divided rectangular region is 115 mm and 86 mm respectively. Therefore, total 156 different pavement textures were obtained in this study, and the pavement texture was measured using Laser texture scanner (LTS) Model 9400 produced by AMES Engineering, as is shown in Fig. 2.

(a) Measuring texture data (b) Scanned pavement texture

Fig. 2 Asphalt mixture specimen preparation and texture measurement

A pavement texture acquired by LTS consists of 146 texture profiles in this study. The scanned area is 104 × 72.01 mm (Length × Width). In addition, the distance between two texture profiles is 0.4966 mm, and the distance between two scanning points in each texture profile is 0.0063 mm. Most notably, the vertical resolution of LTS used in this study is 0.003 mm.

2.2 *Pretreatment of Pavement Texture*

Drop-out Removal. Drop-outs are inevitable for laser scanning device, and they are often detected by LTS automatically. As described in the instruction of LTS Model 9400, the scanning points whose amplitude is equal to 15 mm are regard as drop-outs [17]. In this study, linear interpolation technique was employed to eliminate the detected drop-outs.

Tilt Correction. After the drop-out removal of pavement texture, the tilt correction is needed to acquire the horizontal pavement texture. Generally, it is hard to place the LTS in real horizontal plane. Moreover, if the texture in sloping pavement needs to be measured, it is difficult to keep the LTS parallel to the slope, and the texture tilt will appear. The PC data obtained based on tilted pavement texture is detrimental to the calculation of indices like MTD and MPD. To obtain the pavement texture without distortion, a binary linear function representing a 3D plane was fitted, as is shown in Fig. 3a, and the pavement texture could be corrected by subtracting the fitted plane, as is shown in Fig. 3b.

Spike Removal. Sometimes drop-outs are not eliminated as much [4], and this can lead to spikes in the texture profiles, which can seriously bias the calculation of evaluation indices. Figure 4 shows the procedure for the pavement texture spike removal in this study. First, based on the tilt corrected pavement texture, pavement macro-texture and micro-texture were separated using filter processing technique. Thereafter, the spikes existing in pavement macro-texture and micro-texture were

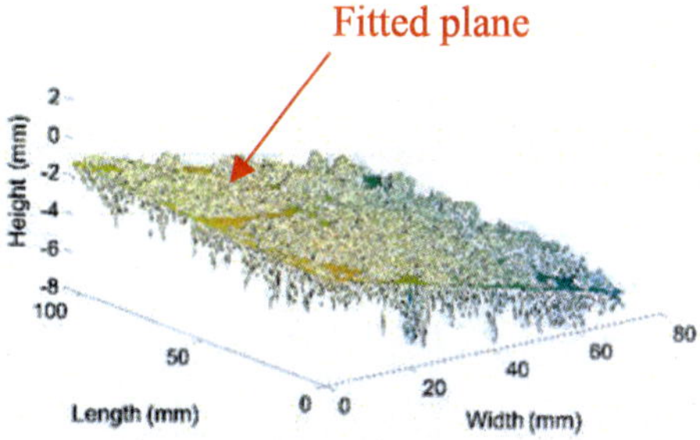

(a) tilted pavement texture

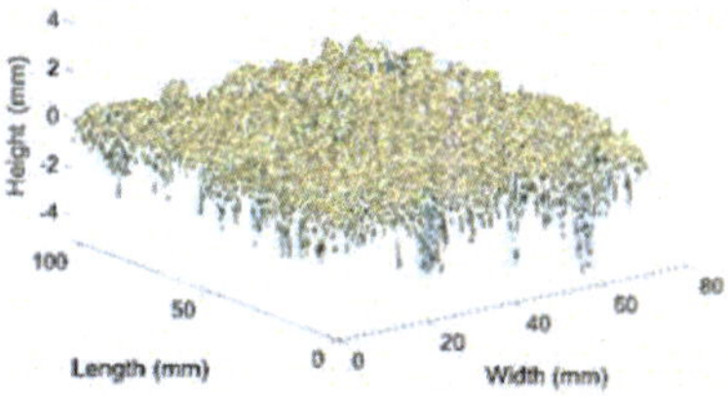

(b) tilt corrected pavement texture

Fig. 3 The proposed tilt correction method

Fig. 4 The procedure of extracting macro-texture and micro-texture

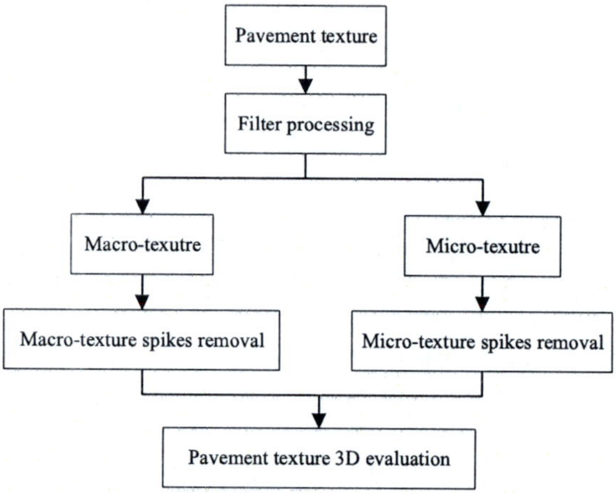

removed respectively. Based on the pavement texture free of spikes, the pavement texture 3D evaluation indices were developed.

As for the macro-texture spikes removal, asymmetric generalized Gaussian distribution (AGGD) and False Discovery Rate (FDR) method were used to detect the spikes, and the linear interpolation was used to remove the detected spikes. The probability density function of the AGGD is shown in Eq. (1).

$$p(x) = \begin{cases} \frac{\beta_P}{2\alpha_P \Gamma(\beta_P^{-1})} \exp(-(\frac{|x_P - \mu_P|}{\alpha_P})^{\beta_P}) \\ \frac{\beta_N}{2\alpha_N \Gamma(\beta_N^{-1})} \exp(-(\frac{|x_N - \mu_N|}{\alpha_N})^{\beta_N}) \end{cases} \tag{1}$$

where β is shape parameter, $\alpha = \sigma\sqrt{\frac{\Gamma(1/\beta)}{\Gamma(3/\beta)}}$; σ is the standard deviation; μ is the average; the subscript P represents positive texture, and N represents negative texture. Similar to the three-sigma threshold with normal distribution, bilateral test based on FDR method was used to detect spikes. Three-sigma threshold is a fixed choice of threshold which is not appropriate when the PC data is big, while FDR can provide an adaptive threshold for spikes detection according to the specific PC data [26]. Additionally, specification [4] recommends a method for removing spikes existing in pavement texture profile. The method claims that if the point z_i in texture profile satisfies Eq. (2), the point z_i will be considered as spike.

$$|z_i - z_{i-1}| \geq \alpha \cdot \Delta x \tag{2}$$

where α is a constant factor, and $\alpha = 3$ is recommanded by the specificaiton; Δx is the distance between neighboring two points in one pavement texture profile.

Figure 5 presents one of 156 patches of pavement texture PC data distribution and corresponding FDR thresholds. Combined with the method recommended in specification [4], the procedures of the proposed spike removal method are elaborated as follows.

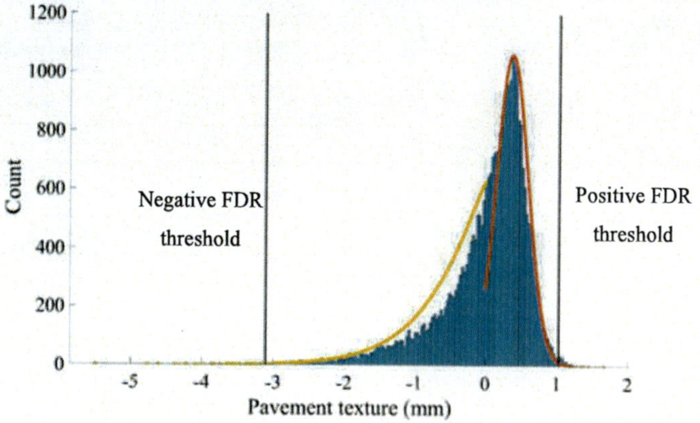

Fig. 5 The AGGD of pavement texture data

(1) The tilt corrected pavement texture PC data were divided into positive texture and negative texture.
(2) Positive texture and negative texture were used to fit β_P and β_N of AGGD respectively.
(3) The *p*-values of each scanned point was calculated using the determined AGGD.
(4) The *p*-values of positive texture points and negative texture points were reordered in increasing order respectively. For example, a patch of pavement texture PC data contains *n* points totally. Positive texture consists of n_1 points, and negative texture consists of n_2 points ($n_1 + n_2 = n$). The *p*-values of positive points should be ordered as $p_{\mathrm{P1}} \leq \ldots \leq p_{\mathrm{Pi}} \leq \ldots \leq p_{\mathrm{Pn1}}$, and that of negative points should be ordered as $p_{\mathrm{N1}} \leq \ldots \leq p_{\mathrm{Ni}} \leq \ldots \leq p_{\mathrm{Nn2}}$.
(5) q_P and q_N were selected to control the positive and negative FDR threshold, and the *p*-values of spikes should satisfy Eq. (3). The corresponding pavement texture of maximum *p*-values of both positive and negative spikes were regarded as the FDR thresholds. In this study, q_P was recommended to select from 0.1 ~ 0.3, and q_N was recommended to select from 10^{-6} ~ 10^{-8}. After spikes were detected, linear interpolation was used to remove spikes.

$$p_i \leq \frac{i}{n} q \tag{3}$$

(6) Equation (2) was used to further detect the spikes, and the detected spikes were removed by linear interpolation. $\alpha = 3$ was selected in this study.

As for the micro-texture spikes removal, for the reason that the peak-to-peak amplitudes range of micro-texture is 0.001 ~ 0.5 mm [4], the scanned points whose amplitude are out of range −0.25 to 0.25 mm in pavement micro-texture are regarded as spikes, and the spikes are removed by linear interpolation method, as is shown in Fig. 6.

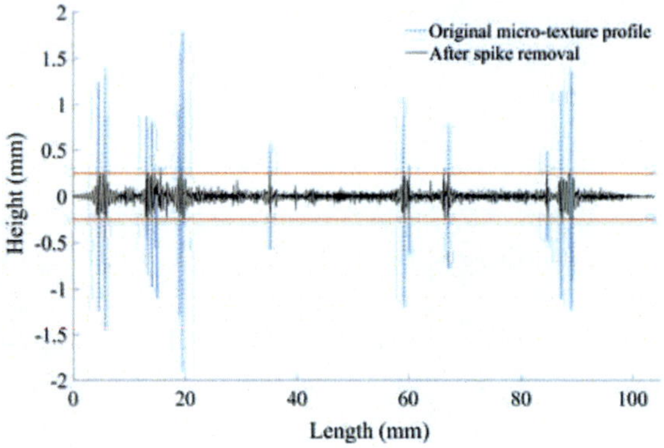

Fig. 6 Micro-texture spike removal

2.3 *Three-Dimensional Evaluation Method for Pavement Texture Level*

Gray level co-occurrence matrix (GLCM) of pavement texture. The pavement texture height data first need to be converted into gray-scale image by Eq. (4) to calculate the corresponding GLCM.

$$G(x, y) = floor(\frac{Z(x, y) - \min(Z)}{v}) + 1 \tag{4}$$

where *floor (x)* represents the downward rounding of *x*; $Z(x, y)$ is the texture depth at (*x*, *y*); and *v* is the vertical resolution.

GLCM describes the frequencies or probabilities of the gray level combinations among the pixel pairs with given spatial relationship in a gray level image [22]. GLCM in frequency can be calculated by Eq. (5), and it in probability can be calculated by Eq. (6).

$$N_d^\theta(i, j) = \#\left\{ \begin{array}{l} [G(x, y), G(x + d\cos\theta, y + d\sin\theta)] \\ : G(x, y) = i, G(x + d\cos\theta, y + d\sin\theta) = j \end{array} \right\} \tag{5}$$

$$P[(i, j); \theta, d] = \frac{N_d^\theta(i, j)}{\sum_{m=1}^{L} \sum_{n=1}^{L} N_d^\theta(m, n)} \tag{6}$$

where $N_d^\theta(i, j)$ denotes the frequencies of pixel pairs (*i*, *j*) occurrence; # is the number of pixel pairs $[G(x, y), G(x + d\cos\theta, y + d\sin\theta)]$; $G(x, y)$ is the value (gray level) at (*x*, *y*) position in the gray-scale image; $P[(i, j); \theta, d]$ is the probability of pixel pairs (*i*, *j*) occurrence; θ is the angle between the line connecting the centers of the 2 pixels and the horizontal direction, which is usually taken as 0°, 45°, 90° and 135°; *d* is the distance between pixels; *L* is the maximum gray level of gray-scale image,

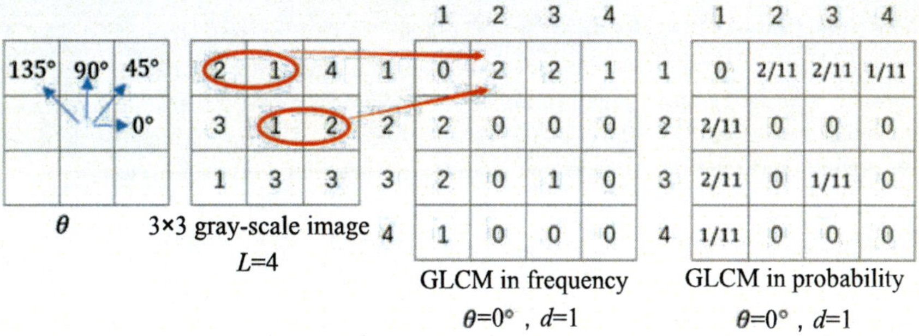

Fig. 7 Calculation method of GLCM

$1 \leq i \leq L$; $1 \leq j \leq L$. The calculation process of GLCM of gray-scale image is shown in Fig. 7.

The calculation of GLCM is related to θ, d and L. θ and d are the hyperparameters that need to be determined artificially, while L depends on the peak-to-peak amplitudes range of pavement texture and LTS's vertical resolution v. Therefore, it can be concluded that GLCM is determined by hyperparameters θ, d and v. Even for the same pavement texture, if different hyperparameters are selected, the calculation results of GLCM will be different, which is also the reason why the 3D evaluation method of pavement texture level based on GLCM is difficult to be popularized. Only with the same hyperparameters, the 3D evaluation indices of pavement texture level based on GLCM are comparable.

According to the wavelength range, the peak-to-peak amplitudes range and the vertical resolution of LTS scanner, the selection of hyperparameters θ, d and v is discussed in this study.

When the precision of the LTS is the same both in transverse and longitudinal directions, the scanning results of pavement texture have rotation invariance in statistics, and the influence of θ on GLCM can be ignored. Due to the higher precision in longitudinal direction (along 90°) than that in transverse direction in this study, θ is set to 90°.

There is no unified standard for the determination of hyperparameter d, and d needs to be determined by specific research objective [22, 27, 28]. For the reason that the wavelength ranges of pavement macro-texture and micro-texture are 0.5 ~ 50 mm and 0 ~ 0.5 mm respectively and the horizontal precision of LTS is 0.0063 mm, in this study, d was set to 79 pixels (0.4977 mm) for macro-texture evaluation and 1 pixel (0.0063 mm) for micro-texture evaluation.

The peak-to-peak amplitudes range of macro-texture is 0.1 ~ 20 mm, and that of micro-texture is 0.001 ~ 0.5 mm [4]. Moreover, considering that the maximum vertical resolution of LTS used in this study is 0.003 mm, so the hyperparameter v in Eq. (4) is set to 0.01 mm for macro-texture evaluation and 0.003 mm for micro-texture evaluation.

The Selection of GLCM Indicators. There are 14 feature indicators of GLCM: $f_1 \sim f_{14}$ [24]. The correlations between these 14 feature quantities and MTD have been studied, and the f_8 and f_9 feature indicators have been proved that both have good correlation with MTD [22]. Index f_8 is also called Sum Entropy, which is calculated by Eq. (7), and index f_9, also known as Entropy, is calculated by Eq. (8) [24].

$$f_8 = -\sum_{i=2}^{2L} P_{x+y}(i) \log_2^{[P_{x+y}(i)]} \tag{7}$$

$$f_9 = -\sum_{i=1}^{L}\sum_{j=1}^{L} P(i,j) \log_2^{[P(i,j)]} \tag{8}$$

where $P_{x+y}(k) = \sum_{i=1}^{L}\sum_{j=1}^{L} P[(i,j);\theta,d]$, and $i + j = k$. Based on the above research results, indices f_8 and f_9 are both applied into macro-texture and micro-texture level evaluation in this study, which are named as f_{8mac}, f_{9mac} and f_{8mic}, f_{9mic} respectively.

2.4 *Three-Dimensional Evaluation Index for DUPT*

Two asphalt pavement textures with the same texture profile level may have completely different pavement texture distribution characteristics [2]. In this study, index σ is proposed to evaluate the DUPT from 3D aspect. DUPT can be regarded as the uniformity of deviation between pavement texture and its mean plane M. Therefore, after the pretreatment of pavement texture introduced in Sect. 2.2, the pavement texture was first divided evenly into four sub-textures A_1, A_2, A_3 and A_4 as is shown in Fig. 8. Then the mean plane M_i of every sub-texture was calculated. The standard deviations σ_M of sub-textures were calculated based on Eq. (9). To avoid the effect of extreme value on calculation, the sub-texture which deviates from M mostly was removed, and then the standard deviation $\sigma_{M'}$ denoting standard deviation of the rest of sub-textures was recalculated according to Eq. (10). Finally, the harmonic mean σ was used as DUPT evaluation index and calculated by Eq. (11). The smaller the σ is, the more uniform the texture distribution is. For completely uniform surface texture, the σ is equal to 0.

$$\sigma_M = \frac{1}{2}\sqrt{\sum\nolimits_{i=1}^{4} (M_i - M)^2} \tag{9}$$

$$\sigma_{M'} = \sqrt{\frac{1}{3}\sum\nolimits_{i=1}^{3} (M_i - M')^2} \tag{10}$$

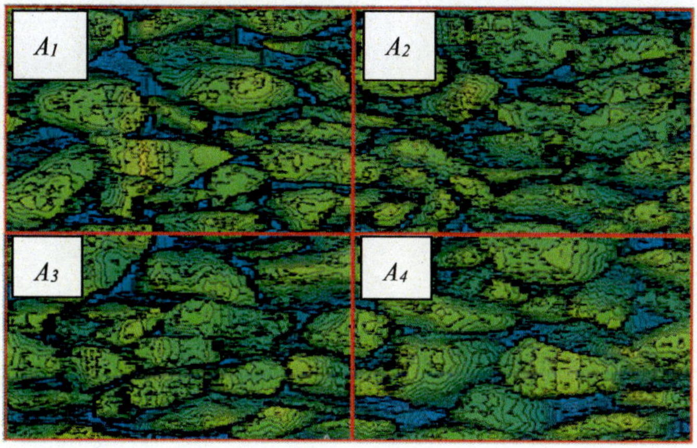

Fig. 8 Top view of the pavement texture to be measured

$$\sigma = \frac{2}{\sigma_M^{-1} + \sigma_{M'}^{-1}} \tag{11}$$

where M is the mean plane of whole pavement texture; M_i is the mean plane of sub-texture A_i; M' is the mean plane of pavement texture which removes the sub-texture deviating from M mostly.

2.5 *Effectiveness Analysis of 3D Evaluation Indices for Pavement Texture*

Analysis on the Validity of 3D Evaluation Indices for Macro-texture Level. The correlations of MTD with f_{8mac}, f_{9mac} and 2D macro-texture profile level $L_{TX,0.5\rightarrow 32}$ were studied to verify the effectiveness of f_{8mac} and f_{9mac}. To compare the MTD with other indices more directly, the circular patch of traditional SPM was replaced with the rectangle patch, as is shown in Fig. 9. The length and width of rectangular patch was fixed to 115 mm and 86 mm respectively. First, a measuring cylinder was filled with 40cm^3 standard sand. The standard sand was poured to the target region gradually. Meanwhile, the disk tool was used to spread the standard sand carefully. As for the corner of rectangle, steel ruler was used to smooth the sand in the corner. The finial reading V' of the rest of standard sand in the measuring cylinder was recorded, and the MTD of target region can be calculated by Eq. (12).

$$\text{MTD} = \frac{(40 - V') \times 1000}{115 \times 86} \tag{12}$$

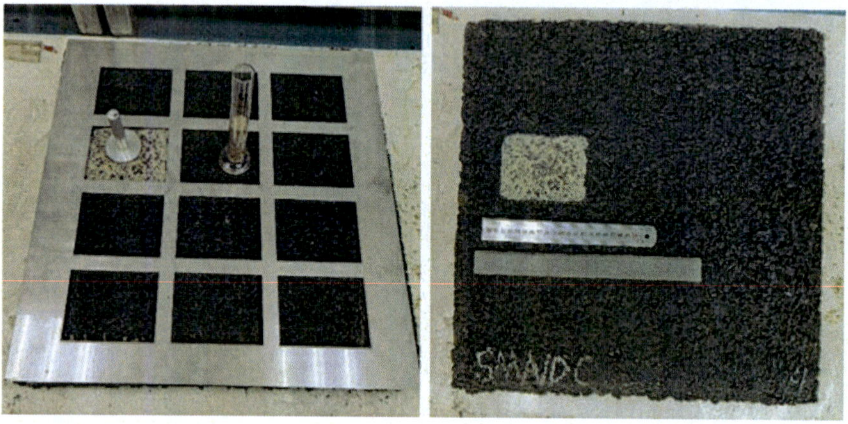

Fig. 9 The revised SPM

The frequency spectrum distribution of pavement texture can represent the relative distribution of different wavelength textures, and the texture profile shown in octave band is presented in Fig. 10.

As shown in Fig. 10, the index $L_{TX,0.5\to32}$ represents the texture profile level whose center wavelength in octave band is within the range of 0.5 ~ 32 mm, and the included wavelength range is 0.35 ~ 44.50 mm, covering most of the macro-texture specified by specification [29]. Similarly, 2D micro-texture profile level $L_{TX,0.03\to0.5}$ represents the texture profile level whose center wavelength in octave band is within the range of 0.03 ~ 0.5 mm, and the included wavelength range is 0.02 ~ 0.71 mm, covering most of the micro-texture specified by specification [29]. $L_{TX,0.5\to32}$ and

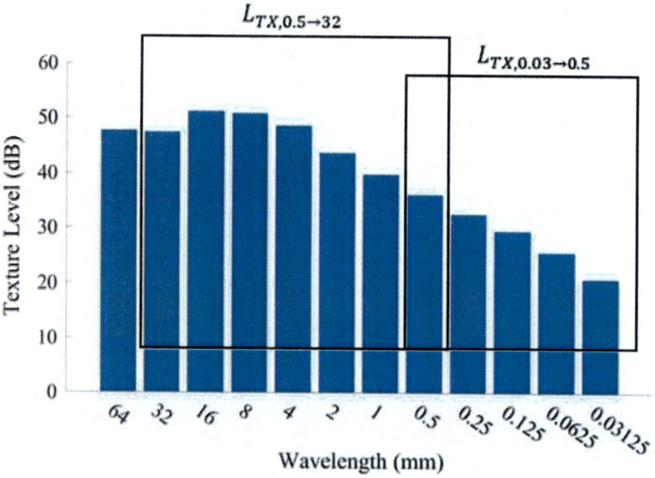

Fig. 10 The texture profile shown in octave band

$L_{TX,0.03\to0.5}$ can be calculated through Eq. (13).

$$L_{TX,i\to j} = 10\lg(\sum_{m=i}^{j} 10^{\frac{L_{TX,m}}{10}}) \tag{13}$$

The complete procedure for the calculation of $L_{TX,0.5\to32}$ could be referred from specification [29] and literature [14, 17], and the mean of $L_{TX,0.5\to32}$ of 146 texture profiles named $L_{TX,0.5\to32}^{mean}$ were taken as the 2D macro-texture profile level evaluation indices and used in the correlation analysis.

Analysis on the Validity of 3D Evaluation Indices for Micro-texture Level. The micro-texture of the mixture is usually represented by the friction coefficient measured at low test speed, as the direct testing of mixture's micro-texture is difficult due to the extremely fine texture waves [16]. The walking friction tester (WFT) (Fig. 11) is used to validate the effectiveness of proposed 3D micro-texture evaluation indices, and the slip ratio is fixed to 10% during the measurement.

WFT is a device to test the tire-pavement friction coefficient μ at low speed, which is used to represent the micro-texture of the mixture. The WFT has two rear wheels and a front wheel which is the test wheel. The WFT is equipped with a speed sensor to measure the test speed and a torque sensor to record the torque of the test wheel. The details of WFT can be found in study [30].

The correlations of u with f_{8mic}, f_{9mic} and $L_{TX,0.03\to0.5}$ were studied to verify the effectiveness of f_{8mic} and f_{9mic}. This paper referred to specification [29] and literature [14, 17] to calculate $L_{TX,0.03\to0.5}$, and the mean of $L_{TX,0.03\to0.5}$ of 146 texture profiles named $L_{TX,0.03\to0.5}^{mean}$ were taken as the 2D micro-texture profile level evaluation indices and used in the correlation analysis.

Analysis on the Validity of the 3D Evaluation Index for DUPT. In order to verify the effectiveness of the 3D evaluation index of DUPT, three kinds of standard grooved

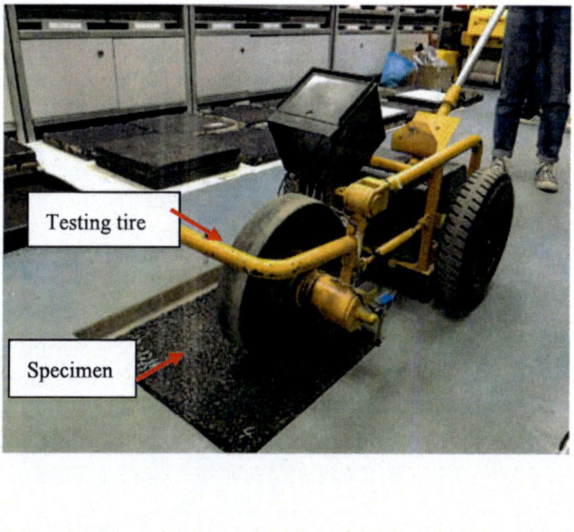

Fig. 11 Walking friction tester

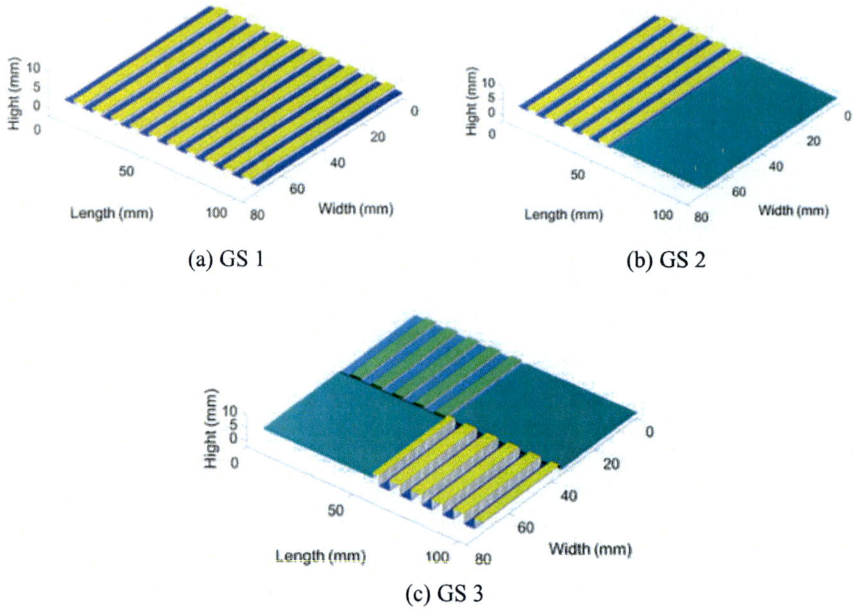

Fig. 12 Standard GS

surfaces (GS) with the same size as the scanning size of LTS were constructed, as shown in Fig. 12.

The width and spacing of the standard grooved texture in Fig. 12 are both 5 mm. The groove depths of GS 1 and GS 2 are both 1 mm and that of GS 3 are 1 mm and 4 mm. According to the uniformity of the grooves on the plane, the order of the GSs uniformity is (a) > (b) > (c). If $\sigma_1 < \sigma_2 < \sigma_3$ is obtained by the calculation method in Sect. 2.4, it means that the proposed index σ for DUPT is effective.

3 Results and Discussion

3.1 *Effectiveness Analysis Results of 3D Evaluation Method for Pavement Texture Level*

The effectiveness of proposed 3D evaluation method was evaluated based on 156 different pavement textures. The correlation analysis of MTD with f_{8mac}, f_{9mac}, and $L^{mean}_{TX,0.5\to32}$ is shown in Fig. 13a, and the correlation analysis of μ with f_{8mic}, f_{9mic}, and $L^{mean}_{TX,0.03\to0.5}$ is shown in Fig. 13b. Due to the limitation of space, only the calculation results of pavement texture evaluation indices for 13 pavement textures with different asphalt types and NMAS are shown in Table 1.

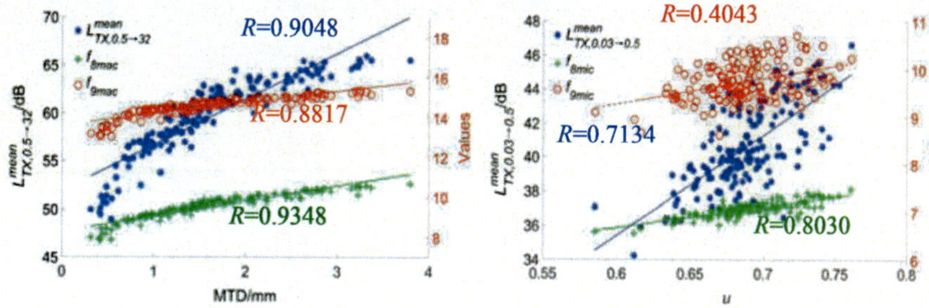

(a) Correlation between f_{8mac}, f_{9mac}, $L_{TX,0.5\to32}$ and MTD

(b) Correlation between f_{8mic}, f_{9mic}, $L^{mean}_{TX,0.06\to0.5}$ and μ

Fig. 13 Correlation analysis for 3D evaluation indices of pavement texture

Figure 13a shows that $L^{mean}_{TX,0.5\to32}$, f_{8mac} and f_{9mac} all have significant positive correlations with MTD, moreover, the correlation coefficients of f_{8mac} and f_{9mac} are larger than that of $L^{mean}_{TX,0.5\to32}$. The correlation between f_{8mac} and MTD is the most significant, and the correlation coefficient R is 0.9348; the correlation coefficient R between f_{9mac} and MTD is 0.8817. These results are consistent with the conclusion in the literature [22], and also indicate that the proposed indices f_{8mac} and f_{9mac} could characterize the pavement texture better than $L^{mean}_{TX,0.5\to32}$. Thus, the indices f_{8mac} and f_{9mac} can be proved effective. Additionally, since $L^{mean}_{TX,0.5\to32}$ is the average value of $L_{TX,0.5\to32}$, it is easily affected by abnormal values, and when the dataset of texture profiles obtained is small and its distribution is skewed, $L^{mean}_{TX,0.5\to32}$ is not capable of characterizing the 3D pavement texture accurately.

Figure 13b shows that $L^{mean}_{TX,0.03\to0.5}$, f_{8mic} and f_{9mic} all have significant positive correlations with μ: the correlation coefficient R between $L^{mean}_{TX,0.03\to0.5}$ and μ is 0.7138; the correlation coefficient R between f_{8mic} and μ is 0.8030; the correlation coefficient R between f_{9mic} and μ is 0.4043.

From the results of the correlation analysis, it can be concluded that it is practical to evaluate pavement texture level using GLCM method, and the hyperparameters θ, d and v selected for GLCM in this study is reasonable. Moreover, the index f_8 obviously shows more significant correlation with both MTD and μ. Therefore, f_{8mac} and f_{8mic} are recommended to be the 3D evaluation indices for pavement texture level.

3.2 *Effectiveness Analysis of 3D Evaluation Indices for DUPT*

Figure 14 shows the calculation results of σ for standard GS1 to GS3 and 13 pavement textures with different asphalt types and NMAS. It can be seen from Fig. 14 that for a completely uniform GS1, σ is equal to 0, and the uniformity of the three GSs is consistent with expectations: $\sigma_1 < \sigma_2 < \sigma_3$. Therefore, it indicates that the 3D

Table 1 Calculation results of pavement texture evaluation indices

Type	MTD (mm)	$L_{TX,0.5\to32}^{mean}$(dB)	f_{8mac}	f_{9mac}	μ	$L_{TX,0.03\to0.5}^{mean}$(dB)	f_{8mic}	f_{9mic}
AC-5	0.41	48.9154	7.8775	13.0071	0.6953	36.8354	6.8844	9.2549
AC-10	1.12	57.5308	9.2019	14.4506	0.7163	39.7544	7.0844	9.9843
AC-13	0.94	57.6903	9.0960	14.3575	0.7101	40.6315	7.2798	10.4951
AC-16	1.84	61.1174	9.8439	14.8985	0.6439	41.9476	7.2740	9.8488
SMA-10	1.07	65.3745	9.7256	14.8219	0.6919	42.7475	7.2902	10.5781
SMA-13	1.59	60.6370	9.4140	14.3943	0.6518	38.3331	6.9475	10.0444
SMA-16	2.55	59.8099	9.8132	14.7534	0.7237	40.3616	6.8845	9.3377
OGFC-10	3.16	57.7571	9.1747	14.4122	0.7518	40.2632	7.0245	9.8227
OGFC-13	2.16	61.1504	9.7585	14.6566	0.7076	40.5009	6.7237	8.6128
OGFC-16	2.86	63.9477	10.1940	14.9893	0.7108	44.3725	7.2325	9.8931
PA-10	1.74	64.6529	10.4176	15.2314	0.7069	45.5839	7.3771	10.1533
PA-13	1.31	63.7677	10.0936	14.9133	0.6511	44.0530	7.1262	9.5327
PA-16	1.82	65.3745	10.4327	15.1573	0.6642	44.4089	7.1830	9.6739

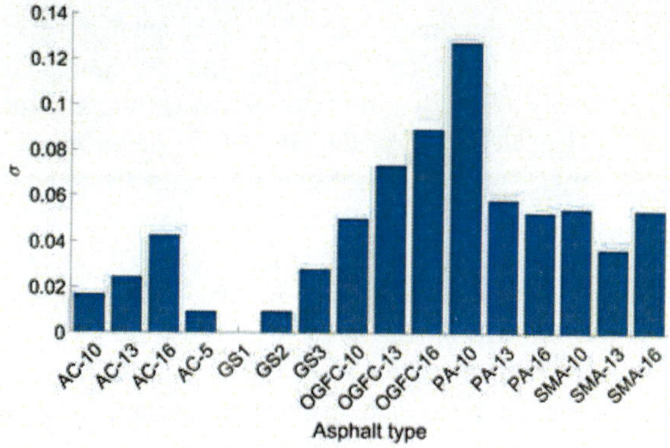

Fig. 14 Calculation results of DUPT

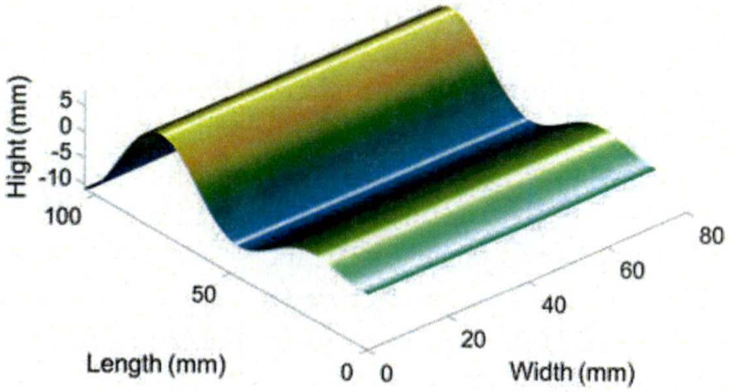

Fig. 15 Special surface texture

evaluation index of surface structure distribution uniformity proposed in this paper is effective.

2D evaluation indices evaluate the pavement texture distribution based on the 2D texture profiles, and they cannot accurately describe the DUPT. For example, as for the special surface texture shown in Fig. 15, if the distribution is evaluated through the profiles extracted along the width direction, which is straight line, it can be concluded that the surface texture is uniformly distributed, but the actual distribution of the special surface texture is not uniform. Therefore, it is meaningful to propose 3D evaluation indices for DUPT.

4 Conclusions

This paper evaluated the pavement texture level from 3D aspect using GLCM method, and a new method to evaluate the DUPT was also determined relying on the mean plane of pavement texture. Additionally, the effectiveness of proposed methods was verified by correlation analysis with existing evaluation indices and standard GS. The following conclusions could be drawn from this study.

(1) The proper pretreatment for pavement texture is necessary to calculate accurate pavement texture indices, and the pretreatment of pavement texture used in this study is effective for both macro-texture and micro-texture.
(2) The calculation of the GLCM was determined by three hyperparameters: the angle θ, the distance between pixels d and the vertical resolution v. Only with the same hyperparameters, the 3D evaluation indices of pavement texture level based on GLCM are comparable.
(3) θ was recommended to be 90°; d was set to 79 pixels (0.4977 mm) and v was set to 0.01 mm for macro-texture level evaluation; d was set to 1 pixel (0.0063 mm) and v is set to 0.003 mm for micro-texture level evaluation; d and v should be as small as possible for micro-texture level evaluation.
(4) The macro-texture level evaluation indices f_{8mac} and f_{9mac} had more significant correlation with MTD than the existing 2D macro-texture profile level evaluation index $L_{TX,0.5\to32}$, and the correlation coefficients R were 0.9348 and 0.9048 respectively. The micro-texture level evaluation indices f_{8mic} and f_{9mic} had significant correlation with pavement friction coefficient μ. Compared with the existing 2D micro-texture level evaluation index $L_{TX,0.03\to0.5}$, f_{8mic} had more significant correlation with μ, whose correlation coefficients R was 0.8030. Due to the significant correlation with MTD and μ, f_{8mac} and f_{8mic} were recommended to be the 3D evaluation indices for pavement texture level.
(5) This paper only conducted a quantitative study on the DUPT and has not yet conducted a qualitative study. Further research should be conducted to classify the DUPT qualitatively based on the 3D evaluation indices proposed in this paper. Additionally, due to the limitation of experimental condition and the article space, the correlation between the proposed 3D indices and pavement surface characteristics such as noise and skid resistance will be studied in the future works.

References

1. Fwa TF (2017) Skid resistance determination for pavement management and wet-weather road safety. Int J Transp Sci Technol 6(3):217–227
2. Chen D (2015) Study on image-based texture analysis method and prediction of skid-resistance & tire/pavement noise reduction of HMA. Chang'an University (In Chinese)
3. Del Pizzo A, Teti L, Moro A, Bianco F, Fredianelli L, Licitra G (2020) Influence of texture on tyre road noise spectra in rubberized pavements. Appl Acoust 159
4. ISO13473-1 (2019) Characterization of pavement texture by use of surface profiles—Part 1: determination of mean profile depth
5. Alhasan A, White DJ, De Brabanter K (2016) Wavelet filter design for pavement roughness analysis. Comput-Aided Civ Infrastruct Eng 31(12):907–920
6. Sohaney RC, Rasmussen RO (2013) Pavement texture evaluation and relationships to rolling resistance at MnROAD. Department of Transportation, Research Services Section
7. Šernas O, Zofka A, Vaitkus A, Gražulytė J (2020) The effect of exposed aggregate concrete gradation on the texture characteristics and durability. Constr Build Mater 261:119921

8. Liu C, Li J, Gao J, Yuan D, Gao Z, Chen Z (2021) Three-dimensional texture measurement using deep learning and multi-view pavement images. Measurement 172:108828
9. Puzzo L, Loprencipe G, Tozzo C, D'Andrea A (2017) Three-dimensional survey method of pavement texture using photographic equipment. Measurement 111:146–157
10. Dong N, Prozzi JA, Ni F (2019) Reconstruction of 3D pavement texture on handling dropouts and spikes using multiple data processing methods. Sensors 19(2):278
11. Li W, Huyan J, Tighe SL, Shao N-N, Sun Z-Y (2019) An innovative primary surface profile-based three-dimensional pavement distress data filtering approach for optical instruments and tilted pavement model-related noise reduction. Road Mater Pavement Des 20(1):132–150
12. Peng J, Chu L, Fwa T (2020) Determination of safe vehicle speeds on wet horizontal pavement curves. Road Mater Pavement Des 1–13
13. Zhang Z, Luan B, Liu X, Zhang M (2020) Effects of surface texture on tire-pavement noise and skid resistance in long freeway tunnels: from field investigation to technical practice. Appl Acoust 160:107120
14. Chen D, Ling C, Wang T, Su Q, Ye A (2018) Prediction of tire-pavement noise of porous asphalt mixture based on mixture surface texture level and distributions. Constr Build Mater 173:801–810
15. ISO13473-4 (2008) Characterization of pavement texture by use of surface profiles—Part 4: spectral analysis of texture profiles
16. Chen D (2020) Evaluating asphalt pavement surface texture using 3D digital imaging. Int J Pavement Eng 21(4):416–427
17. Ren WY (2019) Study on the abrasion characteristic of surface texture and its effect on noise for asphalt pavements (PhD). Chang'an University (In Chinese)
18. Zhang K, Zhang Z, Luo Y, Huang S (2017) Accurate detection and evaluation method for aggregate distribution uniformity of asphalt pavement. Constr Build Mater 152:715–730
19. Lan ZZ (2017) Study on surface texture structure and anti-skid performance of asphalt pavement based on digital image technology. Chongqing Jiaotong University, Chongqing
20. Song ZY, Yan GX, Sui YQ, Li FC (2014) Texture structure distribution of asphalt pavement surface based on digital image processing technology. J Central South Univ (Sci Technol) 45(11):4075–4080
21. Liu Q, Shalaby A (2015) Relating concrete pavement noise and friction to three-dimensional texture parameters. Int J Pavement Eng 18(5–6):450–458
22. Miao Y, Wang L, Wang X, Gong X (2015) Characterizing asphalt pavement 3-D macrotexture using features of co-occurrence matrix. Int J Pavement Res Technol
23. Miao Y, Wu J, Hou Y, Wang L, Yu W, Wang S (2019) Study on asphalt pavement surface texture degradation using 3-D image processing techniques and entropy theory. Entropy 21(2)
24. Haralick RM, Shanmugam K, Dinstein IH (1973) Textural features for image classification. IEEE Trans Syst Man Cybernet SMC-3(6):610–621
25. Jain S, Gruteser M (2019) Recognizing Textures with Mobile Cameras for Pedestrian Safety Applications. IEEE Trans Mob Comput 18(8):1911–1923
26. Katicha SW, Mogrovejo DE, Flintsch GW, de León Izeppi ED (2015) Adaptive spike removal method for high-speed pavement macrotexture measurements by controlling the false discovery rate. Transp Res Rec 2525(1):100–110
27. Clausi DA (2002) An analysis of co-occurrence texture statistics as a function of grey level quantization. Can J Remote Sens 28(1):45–62
28. Eleyan A, Demirel H (2011) Co-occurrence matrix and its statistical features as a new approach for face recognition. Turkish J Electr Eng Comput Sci 19(1):97–107
29. ISO13473–2 (2002) Characterization of pavement texture by use of surface profiles—Part 2: terminology and basic requirements related to pavement texture profile analysis
30. Han S, Liu M, Fwa TF (2018) Testing for low-speed skid resistance of road pavements. Road Mater Pavement Des 1–14

Assessment of Operating Speeds of Rehabilitated Rural Roads with Asphalt Surfacing

D. D. S. Wijeratne, M. Dishan, and W. K. Mampearachchi

Abstract In Sri Lanka, rural roads are not explicitly designed on technical requirements. So far restoration of rural roads went through with mere Construction enhancements and appropriate geometric improvements have not been adopted. It was discovered that the actual operating speeds can be higher after restoration which is affecting the safety of road users. Aim of this research is to survey operating speeds, posted speed limits and to find design operating speeds that expected to be after the rehabilitation. The majority of the rural roads in Sri Lanka does not have posted speed limits indicated by the geometric, road environment and functionality level of the road. Consequently, the speed decision of the rural road drivers relies upon variety of different factors. The 85th percentile speed was taken as the operating speed. This speed was utilized as a basic for recommending rational speed limit. Selected roads were divided into a few segments; straight and curved. Operating speed on straight segments were given priority in suggesting rational speed limits. As operating speed in curves are considerably lower and enforcing lower posted speeds only based on operating speeds on curves for entire road and would not be practical. The recommended speed limit for all the roads under purview is 50 km/h and this speed limit will be superseded to a lesser speed limit at a curved segment, dependent on the operating and design speed. This speed limit should be notified using sign boards and shall be enforced only for the curve itself.

Keywords Traffic speed management · Rural roads · Horizontal curve · Posted speed limit

1 Introduction

Rural roads are restored to improve the living standard of people in rural towns and villages just as significant as urban cities in the country. However, it has been a typical

D. D. S. Wijeratne · M. Dishan (✉) · W. K. Mampearachchi
Department of Civil Engineering, University of Moratuwa, Moratuwa, Sri Lanka

W. K. Mampearachchi
e-mail: wasanthak@uom.lk

© The Author(s), under exclusive license to Springer Nature Switzerland AG 2022

H. R. Pasindu et al. (eds.), *Road and Airfield Pavement Technology*, Lecture Notes in Civil Engineering 193, https://doi.org/10.1007/978-3-030-87379-0_4

practice to restore the roads with simple construction enhancements. A restored road usually follows the existing road having same horizontal and vertical alignments. Geometrical features such as Intersections, Horizontal and vertical curves, approaches, etc. are more often remain unchanged while road surface and road structures are improved or reconstructed in rehabilitation. Even though the road surface and different infrastructure has improved, roads are not geometrically improvised which leads lower design speeds.

It was observed that road accidents in rural roads have gone up after restoration occurred. The actual speeds were observed to be altogether more prominent after rehabilitation than the actual design speeds of those roads. The objective of the research follows assessment of actual operating speeds, posted speed limits and design operating speeds that supposed to be after the rehabilitation. Additionally, it recommends reasonable speed limits dependent on operating speeds (85th percentile speed) and the design speeds that conform to the existing geometric features in restored country roads with asphalt surfacing. The Scope of the data collection was limited to North Western provincial and rural roads in Sri Lanka.

2 Literature Review

Good geometric design means providing the appropriate level of mobility and land use access for road users while maintaining a high degree of safety. The roadway design must also be cost effective in today's financially compelled environment. While balancing these design decisions, the designer needs to provide consistency along a roadway alignment to prevent abrupt changes in the alignment that do not match motorists' expectations. Speed is used both as a design criterion to promote this consistency and as a performance measure to evaluate highway and road designs [1].

Operating Speed is the speed at which drivers are observed operating their vehicles during free flow conditions. The 85th percentile of the distribution of observed speeds is the most frequently used measure of the operating speed associated with a particular location or geometric feature [2]. There were three main speed measures that were recorded. This allows for a more detailed interpretation and analysis of issues relating to the speed in the Speed assessments. The measures include the speed limit, mean speed with 85th percentile speed as well.

- Speed limit: The sign-posted legal speed limit or, where no speed limit signs are posted, the speed limit that applies as a matter of law.
- Mean speed: The average speed of vehicles past a nominated point.
- 85th percentile speed: The speed at or below which 85% of all vehicles are observed to travel under free flowing conditions past a nominated point.

This data can be used to calculate mean speeds and 85th percentile speed and provide a valuable source of information to help guide any discussion on speed management needs [3].

The risk of crash and the probability of serious injury increases with the increased speed of the traffic. Hence the speed is regarded one of basic factors in traffic crash injuries. Vulnerable road users such as pedestrians, cyclist and motorcyclists have a high risk of severe or fatal injury when motor vehicles collide with them. This is because they are often completely unprotected or, in the case of a motorcyclist, have very limited protection and the probability that a pedestrian will be killed if hit by a motor vehicle increases dramatically with speed. The studies from Organization for Economic Co-operation and Development (OECD) Annual Report, 2006 indicates that while most vulnerable (unprotected) road users survive if hit by a car travelling 30 km/h, the majority are killed if hit by a car travelling at 50 km/h or more [4].

2.1 Speed and Horizontal Curves

This research is associated with the vehicle speeds moving along horizontal curves. Hence it is important to discuss about how speed influences a moving vehicle along a curve, oppose to a vehicle moving along a straight section.

The relationship of travelling speed of a vehicle moving along a bend can be derived by analysis of the forces acting on the vehicle as it moves around a curve. Assuming the radius in constant, it can be expressed in Eq. (1).

$$(0.01\text{e} + \text{f})/(1 - 0.01\text{ef}) = \text{v}^2/\text{gR} \tag{1}$$

where;

e = rate of roadway superelevation
f = side-friction factor
v = vehicle speed ft/s (m/s)
g = gravitational constant, 32.2 ft/s^2 (9.81 m/s^2)
R = curve radius, ft, (m).

The value of 'ef' (in the equation denominator) is always small, so the 1 − 0.01 ef terms is approximately equal to 1.0. As such, a simplified curve formula can be used for highway design and this equation can be further developed to accommodate common speed units as Eq. (2) [5].

$$0.01\text{e} + \text{f} = \text{V}^2/127\text{R} \tag{2}$$

where;
V = velocity, km/h.

Research and experience have established limiting values for e and f. Use of the maximum 'e' and 'f' value in the formula permits determination of minimum curve radii for various design speeds. The limiting value of the side friction factor, f, at which the tires begin to skid, may be as high as 0.6 or higher. In design, engineers use only a portion of the side friction a driver begins to feel uncomfortable and react instinctively to avoid higher speed. Empirical studies have determined that such f-values range from about 0.17 at 20 mph (30 km/h) to 0.08 and 80 mph (130 km/h) [5].

3 Selection of Rural Roads for the Study

Following are some of the basic characteristics of North Western Province rural roads in general,

- Paved or unpaved carriage way
- Low average daily Traffic
- Carriageway width ranges from 3.2 m to 4.2 m
- Shoulder width 0-1 m
- Intensive pedestrian activity in some sections
- Higher number of Horizontal curves per unit length
- Poor traffic control and the lesser enforcement over traffic speed
- Hardly having walkways
- Higher accessibility and land use activities
- Higher roughness
- Majority of road users are residential or regional to the area.

North Western provincial road network mainly Consist of a mix of Asphalt, Macadam, DBST, Concrete and Gravel roads. 66% of entire road length has become unsuitable to cater for expanding transportation needs of road users. Most of road bases had been designed decades ago and are weak hence not suitable for current traffic demand.

The top speed limits of 70 km/h and 50 km/h are usual speed limits for ordinary traffic in outside city limit and city limit respectively in arterial roads which are belong to class A and B.

North western provincial road network includes of 2752 km of road length. There are two districts, Kurunegala and Puttalam, with road lengths of 1738 km and 1014 km respectively. For this research, roads with comparatively higher ADT were concerned. Roads with higher crash rates, specially recently rehabilitated were considered for the study as well. Table 1 shows the Provincial roads selected for the speed assessment that covers nearly 50 km of road length. The selected roads are single lane and with relatively similar lane widths.

Table 1 Selected NW Provincial roads for the speed assessment

No	District secretarial division	Name of the Road	Total Length (km)
1	Kuranagala	Rangama—Gonagama	10.25
2	Kuranagala	Kawdawaththa—Alakoladeniya	6.3
3	Kuranagala	Lakeround—Kudagalgamuwa	8.8
4	Mallawapitiya	Thorayaya—Kudakowana	3.46
5	Mallawapitiya	Mallawapitiya—Katupitiya	8.4
6	Maspotha	Hanhamuna—Dampitiya	6.1
7	Maspotha	Pallandeniya—Dikwehera	7.6
Total Length			50.91

4 Methodology

4.1 Segmentation of Road Sections

A road could be subdivided into number of segments in order to collect speed data. Initially each road was divided into segments using two simple basic types. Those were straight segments and curved segments. Following road features were also considered in categorizing segments of the road under purview.

- Roads having wide verges, good overtaking opportunities, large radii bends, good visibility, etc.
- Roads which may include some bends and undulations, average quality verges
- Roads which have sharp bends, narrow verges, undulating roads with low forward visibility
- Higher local activity and ingresses.

Figure 1 shows a road (Thorayaya—Kudakowana in Mallawapitiya divisional secretariat) that was sub divided into number of straight and curved sections. Segments 25, 28, 29, 30 were the significant horizontal curved sections while segments 26 and 27 were recognized as straight segments.

4.2 Data Collection and Analysis

Each selected road was inspected throughout its length and crucial sections that needed to be assessed were identified. GPS coordinates were taken, and the entire road was mapped using digitized maps so that all identified sections could be marked and clearly visualized.

Road section were identified as straight and curves and they were numbered for identification purpose. Following data were collected for each numbered section.

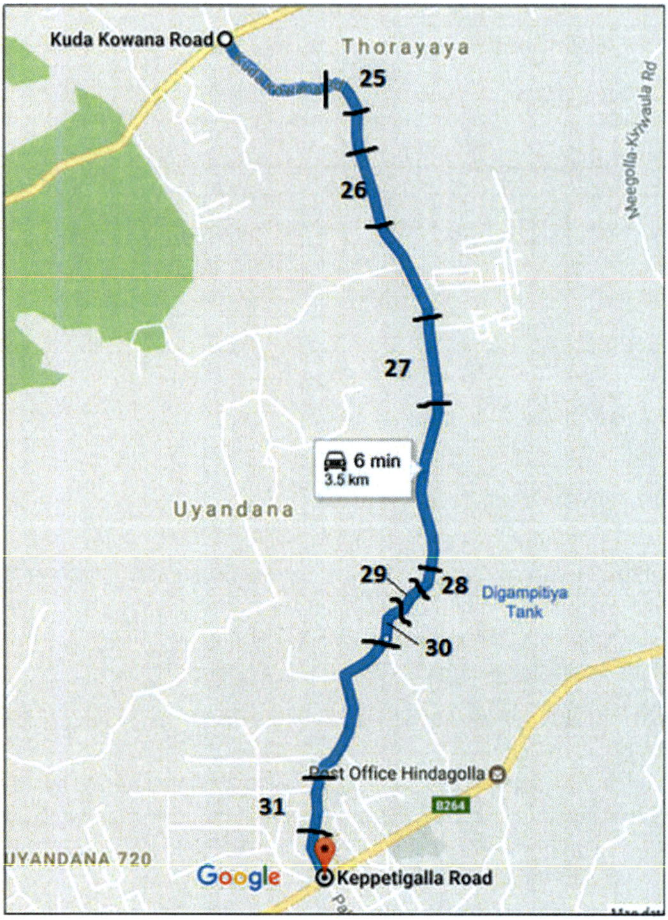

Fig. 1 Example segments as in Thorayaya—Kudakowana road

- Geometrical Data
  - Section Length or Curved length (Arc Length) depending on the section type
  - Angle of change of direction in curved sections (The angle between the tangents of the curve. This angle is equal to the central angle of the curvature by geometry)
  - Superelevation.
- Speed Data.

Sample size of 50 speed data at specific locations were considered in the study. Speed data was recorded for both ways of the sections. Usually, a speed radar meter is used to find instantaneous speed of traffic by law agencies (Sri Lanka Police). For this research video cameras mounted on tripods were used to find the time periods of each vehicle that travelled a constant section length. Data was recorded for each vehicle type. Speed data was collected during daytime over a week. The data was collected with resemblance to traffic during the entire year. Speeds were then calculated for each vehicle and tabulated. The type of vehicle is not considered in this research as average speed data does not vary considerably in rural roads. However, in some curved sections, where the sight distances were lesser, maximum applicable segment that could be covered plus adjacent road section were considered in measuring speed data. Maximum super elevations of curved sections were measured using an Engineer's Level Instrument. Technically, the maximum super elevation is attained at the middle of the curve. Average value was calculated by taking several readings in the curved location.

- First, all the speed data were tabulated from smallest to largest for all vehicles.
- If number of speed data points is n, the actual frequency of each data point is 1/n.
- Speed distribution curve can be constructed by plotting actual frequency against speed data.

Figure 2 illustrate the cumulative speed distribution curve for station 26 at Thorayaya–Kudakowana road. Using the speed distribution curve, 85th percentile value was found. For the road section 26, the 85th percentile speed of the section is 56.5 km/h, which means 85% drivers of the sample travelled at a speed of 56.5 km/h or below.

The mean and the standard deviation speed of the sample section are 45.24 km/h and 10.7 km/h respectively if the speed data sample follows a normal distribution within one standard deviation of the mean, or in other words in the range of 10.7 km/h to the left (i.e. 34.54 km/h) and 10.7 km/h to the right (i.e. 55.94 km/h), almost 70% of travelling speeds are within these limits. Almost 70% of drivers travelling this section were in between 34.54 and 55.94 km/h speeds.

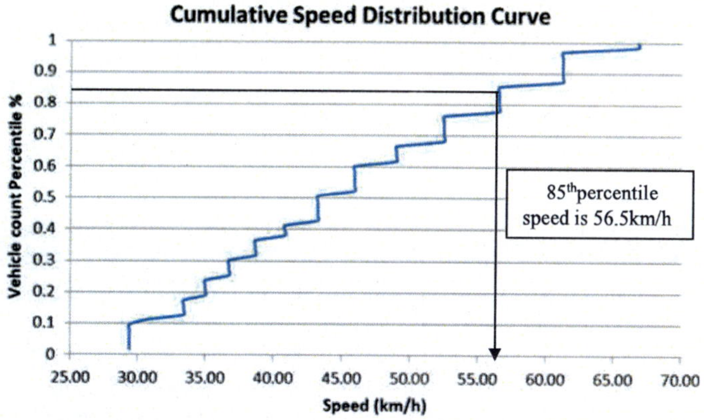

Fig. 2 Cumulative speed distribution curve for station 26

4.3 *Calculation of Available Design Speeds of Curved Sections*

As opposed to the actual speeds collected for a curved section, operating speed that supposed to be travelled along the curve based of its geometrical features (or the available design speed) can be calculated based on the curve radius, available super elevation and the design friction factor. Relationship among radius of a curve (R), Length of curve (L) and projected angle at center (θ) can be described as in Eq. (3).

$$R = L/\theta \tag{3}$$

Safe operating travelling speed along the curve could be calculated using Eq. (2).

$$0.01e + f = V^2/127R$$

Which is

$$v = \surd(127R(0.01e + f))$$

where,

v = velocity in km/h.

For station 25 in Thorayaya – Kudakowana road. (See Fig. 1).

$L = 42.02$ m
$\theta = 84° = 1.466$ radians
$R = 42.02/1.466 = 28.66$ m
$e = 2\%$
$f = 0.15$
$v = \surd(127 \times 28.66(0.01 \times 2.5 + 0.15)) = 24.23$ km/h.

The friction factor was limited to the design friction factor, 0.15 which was below the upper limit of the actual friction factor. The design f value is corresponding to the point at which a driver begins to feel uncomfortable and react instinctively to avoid higher speed [2, 5].

Hence the safe design speed of the above curved section under measured geometric properties was nearly 25 km/h. However, by analysing observed speed data, it was calculated that the Mean speed was 30 km/h and 85th percentile speed was 36.2 km/h (represents 85% of drivers were travelling equal or below this speed) which were more than the safe operating speed (i.e. 25 km/h) of the curve.

5 Results and Discussion

As discussed in Sect. 4, speed data was analysed for each road sections identified. For each road, there were several sections or speed data collection stations that could be either a straight section or a curved section.

5.1 *Speed Data Analysis at Curved Section*

In this study not only speed data analysis was done but safe design speed of a vehicle that travels along the curve was calculated using curve's actual geometrical properties. Hence this gives vital comparison opportunity on each segment of how traffic actually behaves as opposed to the design speeds of curves.

Comparison of Speeds in Rangama—Gonagama

Speed comparison of curved sections of Rangama—Gonagama road is shown in Fig. 3. There were five curved sections studied in this road (stations 5, 2, 4, 7, 10. See Fig. 1). The mean speed of stations 2, 4 and 10 are below the design speed of those corresponding curves.

In station 2, almost 70% of travelling speeds are lying in between the range 32.09 and 44.89 km/h which the design speed 42 km/h lies within that range. However, the 85th percentile value on this section is 46.0 km/h. This means that 85% of speeds in the sample are equal or below 46.0 km/h or 15% of the drivers speed more than 46.0 km/h which is not favourable with safe design speed. Same scenario can be explained with station 4 and 10, (See Table 2).

Comparison of Speeds in Kawdawaththa—Alakoladeniya

Speed comparison of curved sections of Kawdawaththa—Alakoladeniya road is shown in Fig. 4. In case of station 14 curve, both 85th percentile speed and the

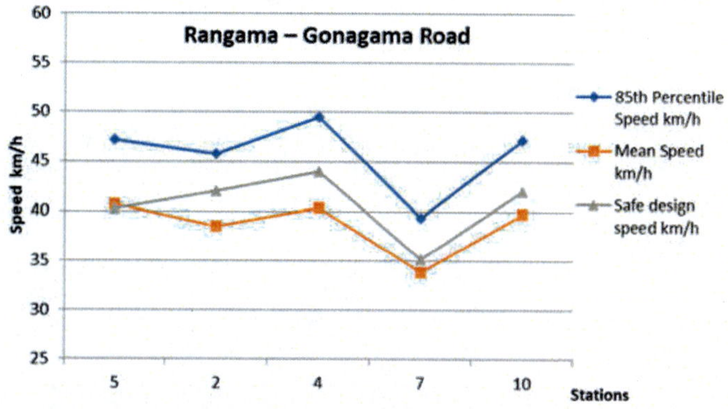

Fig. 3 Comparison of speeds in Rangama—Gonagama

Table 2 Comparison of design speed with statistical results

Station	Mean speed (km/h)	Std. deviation (km/h)	Mean-STD	Mean + STD	Safe design speed (km/h)	85th percentile speed (km/h)
2	38.49	6.4	32.09	44.89	42	46.0
4	40.34	7.3	33.04	47.64	44	49.5
10	39.67	7.35	39.15	42	42	47.2

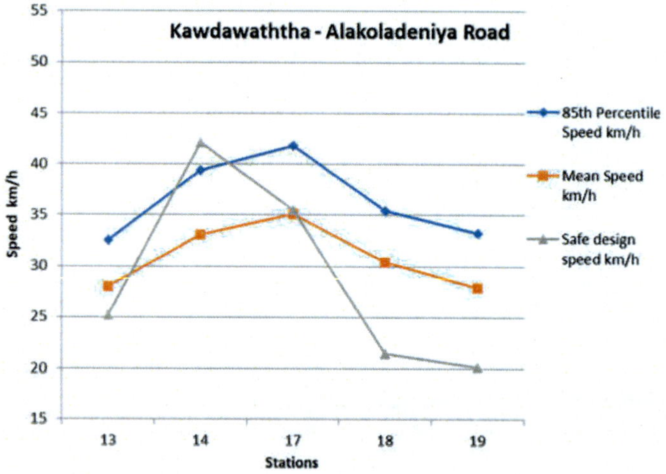

Fig. 4 Comparison of speeds Kawdawaththa—Alakoladeniya,

mean speed are below the design speed. This was because this section had poor line of sight due to vegetation and the roadside constructions. Also, the right of way was limited causing a narrow carriageway. Drivers had chosen lower operating speed due to these reasons when passing this curve. However, it is crucial to investigate this kind of sections together with crash statistics, since drivers who do not have experience about the area can encounter traffic conflicts unknowingly.

Comparison of Speeds in Lakeround—Kudagalgamuwa

Speed comparison of curved sections of Lakeround—Kudagalgamuwa road is shown in Fig. 5.

All the curved sections have similar pattern as discussed in Speed comparison of curved sections of Rangama—Gonagama road.

Comparison of Speeds in Thorayaya—Kudakowana

Speed comparison of curved sections of Thorayaya—Kudakowana road is shown in Fig. 6. In sections 28, 29 and 30 the design speed has increased, and this was clearly because of the increasing curve radius of those bends. However, it is interesting that mean speed and the 85th percentile speed do not follow the trend of the design speed,

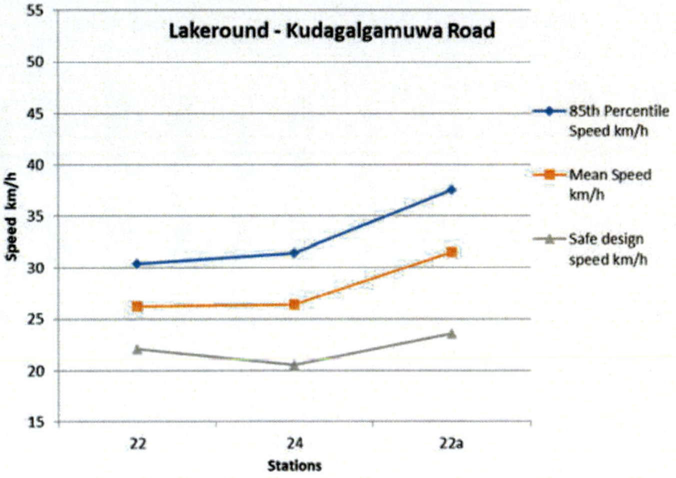

Fig. 5 Comparison of speeds in Lakeround—Kudagalgamuwa

but they have their own gradually increasing trend. The reason is that sections 28, 29 and 30 were curves with one after another (reverse curves). Even though individual design speeds are higher, small tangent between the curves do not facilitate for acceleration of the speed.

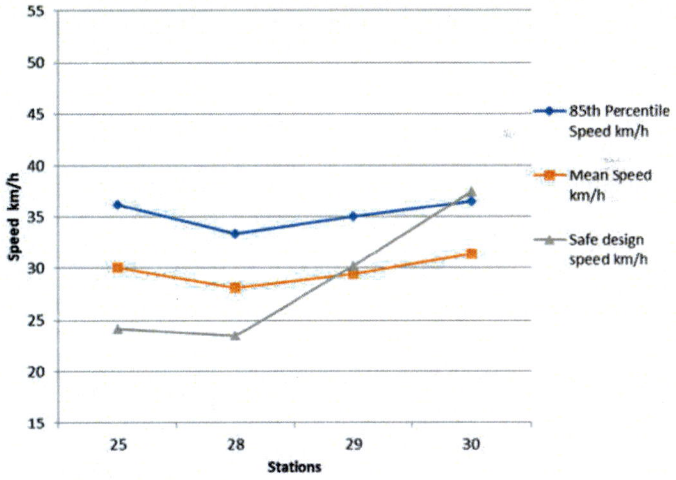

Fig. 6 Comparison of speeds in Thorayaya—Kudakowana

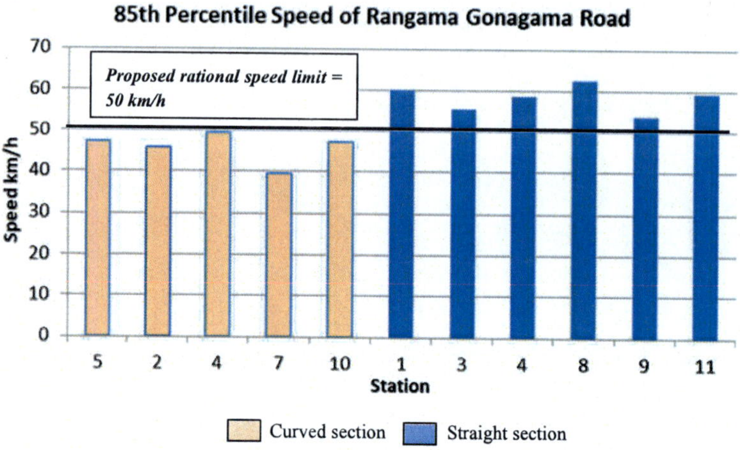

Fig. 7 85th percentile speed of Rangama Gonagama Road

5.2 *Speeds in Straight Sections and Curved Sections*

Figure 7 shows the 85th percentile speeds in 11 number of road stations in Rangama—Gonagama road. First five stations (5, 2, 4, 7 and 10) are curved sections while the rest are in straight sections. The maximum 85th percentile speed is 62.45 km/h in station 8 which is a straight section. All measured speeds are above 50 km/h. 85th percentile speeds of all the curved sections are slightly lower than 50 km/h except in station 7. Hence 50 km/h speed can be assigned as the rational speed limit provided the posted speed for bends. For an example the maximum speed limit for the bend at station 7 can be assigned as 40 km/h by considering the operating speed (i.e. 85th percentile speed) and the available design speed of that curve. This speed limit is imposed only for the curve itself so that the posted speed for the entire road is overridden at that section.

Similarly, 85th percentile speed calculated for other roads were graphed and rational speed limits could be estimated graphically as above. Table 3 shows summary of results and suggested speed limits for the selected roads in the study.

6 Limitations and Conclusion

6.1 *Limitations*

It is important to discuss about the limitations of this study. The main activity of this process is speed data collection. An accurate study of operating speeds depends on unbiased sample of data set. However, vehicle speeds can vary throughout the day

Table 3 Rational Speed limits proposed

Name of the road	Maximum 85th percentile speed in straight sections (km/h)	Minimum 85th percentile speed in straight sections (km/h)	Proposed rational speed limit (km/h)
Rangama Gonagama	62.45	53.8	50
Kawdawatta—Alakoladeniya	53.8	49.2	50
Lakeround—Kudagalgamuwa	61.2	48.3	50
Thorayaya—Kudakowana	55.7	48.6	50
Mallawapitiya—Katupitiya	60.1	52.3	50
Pallandeniya—Dikwehera	58	55.5	50
Hanhamuna—Dampitiya	55.21	45.4	50

as a function of traffic conditions, volume, traffic flow, weather conditions and other factors. Vehicle speeds may be significantly higher than the speed limit at nighttime than of congested peak hour conditions. At nighttime, there is less traffic and there is no enforcement. However, only daytime speed data was collected in this study. Hence the study is limited to daytime traffic behavior.

This assessment is based on the speed data of provincial rural roads of North western province of Sri Lanka. Only several roads from North Western Province were selected for the assessment due to lack of resources and time restrictions.

6.2 Conclusion

Rational Speed Limits based on operating speeds, promote public safety by helping drivers choose a reasonable and prudent speed that is appropriate for normal traffic, and roadway conditions. The recommended rational speed limit for all the roads under purview is 50 km/h and this speed limit will be overridden to a lower speed limit (or limits) at curved sections based on the design speed and the 85th percentile speed of that curves where necessary.

However, 85th percentile speed as a basis of rational speed limit has some limitations since it does not consider formal review of traffic flow, roadway design, crash information and local development activities.

Due to various reasons discussed above, it has become a mandatory requirement that provincial or rural roads also need speed limits associated with law enforcement for the betterment of the public safety. Rational speed limits based on 85% percentile speed or operating speed will be a good basis as identified in the study.

References

1. Fitzpatrick K, Carlson P, Brewer MA, Wooldridge MD, Miaou SP (2003) Design speed, operating speed, and posted speed practice. NCHRP Report 504. Transportation Research Board, National Research Council
2. AASHTO (2001) A policy geometric design of highways and streets. American Association of State Highway and Transportation Officials, Washington DC
3. Howard E, McInerney R (2010) Vehicle speeds and the iRAP protocols. International Road Assessment Programme (iRAP), London
4. OECD Annual Report (2006) OECD Publishing, Paris
5. Wright PH, Dixon K (2004) Highway engineering. Wiley

Assessment of Present Pavement Condition Using Machine Learning Techniques

Madhavendra Sharma and Pradeep Kumar

Abstract Quantification of present pavement condition in terms of an index term i.e., Pavement Condition Index (PCI) is one of the most important and primary steps while taking decision related to Maintenance and Rehabilitation of Pavements. PCI as proposed by ASTM D6433 rates pavement in seven conditions viz. Good, Satisfactory, Fair, Poor, Very Poor, Serious and Failed. Determination of rating condition of pavement using distress severity and extent turns out to be tedious process. Hence, present study investigates application machine learning techniques for assessment of present pavement condition. Three different algorithms i.e., Logistic Regression, Naïve Bayes and K-Nearest Neighbor have been tested in the present study using Long Term Pavement Performance database consisting of over 10,000 datapoints. The dataset was divided into 7:3 ratio for training and testing phase. Employed algorithms were tested based on accuracy, precision, recall and f-measure. Logistic Regression Classifier was found to have highest accuracy of 0.92 among three classifiers used in the study.

Keywords Pavement condition index · Naïve Bayes · Logistic regression · K-nearest neighbor · Pavement condition assessment

1 Introduction

Pavement Management System (PMS) was conceptualized in 1960s to develop proper coordination in the activities involved in decision making for selecting best alternative in developing and maintaining pavement infrastructure [1]. Pavement Condition Assessment (PCA) is a key component of PMS [2, 3]. Pavement Condition Assessment (PCA) is the process of collecting and processing indicators of pavement condition. PCA helps the pavement authorities to assess the maintenance measures needed to prevent accelerated and future distress along with the maintenance and rehabilitation measures to be taken for improving pavement quality [4]. To assess the present pavement condition, different indicators have been proposed

M. Sharma (✉) · P. Kumar
Pavement Evaluation Division, CSIR Central Road Research Institute, New Delhi 110025, India

© The Author(s), under exclusive license to Springer Nature Switzerland AG 2022
H. R. Pasindu et al. (eds.), *Road and Airfield Pavement Technology*, Lecture Notes in Civil Engineering 193, https://doi.org/10.1007/978-3-030-87379-0_5

and used widely. International Roughness Index (IRI), Pavement Condition Rating (PCR), Present Serviceability Index (PSI) and Pavement Condition Index (PCI) are some of the indicators accepted worldwide. Considering different aspects of pavement functional condition and structural integrity, these indicators tend to provide a quantitative measure of present pavement condition [5–7].

1.1 Pavement Condition Index

Pavement Condition Index (PCI) was developed by Shahin et al. [8]. PCI is based on a 0–100 scale, where 100 corresponds to excellent condition and failed condition of pavement is denoted by 0. PCI rates the pavement condition in seven different categories based on distress data. Schematic representation of PCI scale along with its interpretation is presented in Fig. 1.

The determination of PCI is done after visual inspection of pavement sections under consideration. The process of determination of PCI using ASTM:D6433 [9] is itself a tedious task while dealing with a large road network. The tedious task pertains to obtaining the deduct values from deduct value curves (as shown in Fig. 2) as per the field evaluation from the deduct value curves. Even using software like Micro Paver still is time consuming, as the distress data for each section has to be entered manually [10].

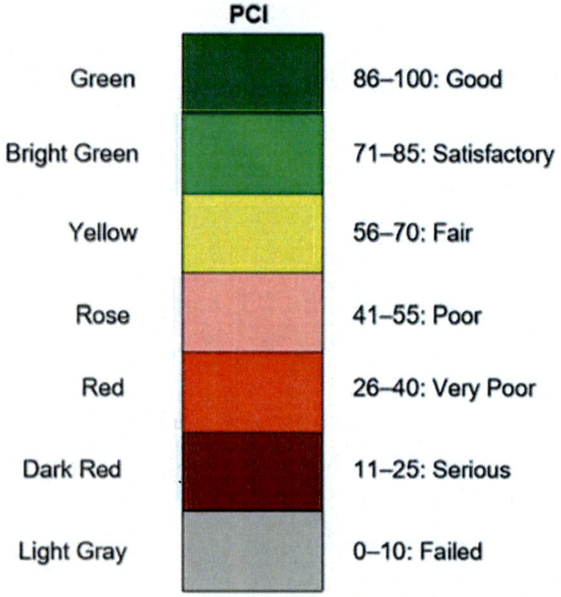

Fig. 1 PCI scale and its interpretation [9]

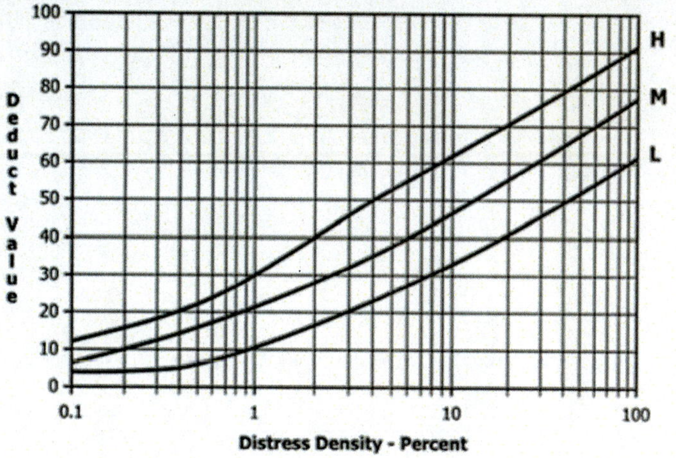

Fig. 2 Typical deduct value curve [9]

With development in computational algorithms in recent times, several alternative soft computing techniques have been applied for determination of PCI. Such techniques include Artificial Neural Networks (ANN), Fuzzy Inference Systems (FIS), Genetic Programming (GP), Support Vector Machines (SVM) etc. The underlying advantage of these techniques are based on the training and prediction capabilities delivering higher degree of precision. Lately such techniques have also been used in various civil engineering applications [11–20].

2 Scope

Research conducted in the domain of predicting PCI from distress data using machine learning techniques have been usually confined to use of regression techniques and artificial neural networks (ANN). Consideration of a performance indicator as a real number (continuous variable) might have resulted in such an approach whereas the maintenance and rehabilitation (M&R) provided over a certain range of indicator are same. Hence, despite of PCI being a continuous variable, prediction of PCI category can be done using classification. Present study is an attempt to test the applicability of Logistic Regression, Naïve Bayes and K-Nearest Neighbor techniques in predicting the pavement condition. The models for the study were trained based on dataset obtained from Long Term Pavement Performance (LTPP). The input for model comprises of distress type, severity and quantity while output consists of PCI category.

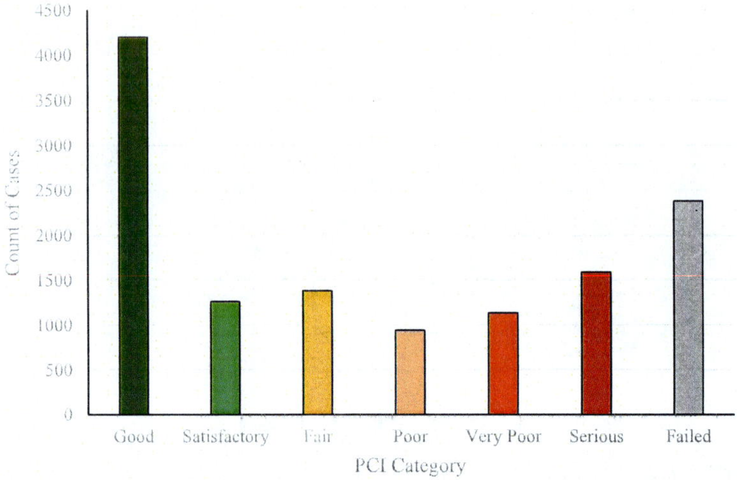

Fig. 3 Count of cases of different PCI categories in dataset

3 Database Collection and Processing

In the present study, the LTPP dataset was used for development and testing of PCI models. The selected dataset corresponds to pavement sections having "Asphalt Concrete" surface and distress data was collected later than 2000. A total of 12,881 data points were obtained corresponding to different sites maintained within LTPP database.

The dataset thus obtained have distress data for Alligator Cracking, Bleeding, Block Cracking, Edge Cracking, Longitudinal Cracking, Transverse Cracking, Patching, Pothole, Shoving and Ravelling for all levels of severity. Pavement Condition Index (PCI) for this dataset was obtained using PAVER software which was observed to be laborious and time consuming. Around 30% of the datapoints were observed to be in Good category. PCI Category wise distribution of cases in the dataset is presented in Fig. 3.

4 Material and Methods

Classification is a supervised machine learning process that maps input data into predefined groups or classes [21]. The analysis was done using scikit-learn library in Python [22]. This section presents a brief description on the three algorithms and performance measures used in present study. The methodology used in the study is presented in Fig. 4.

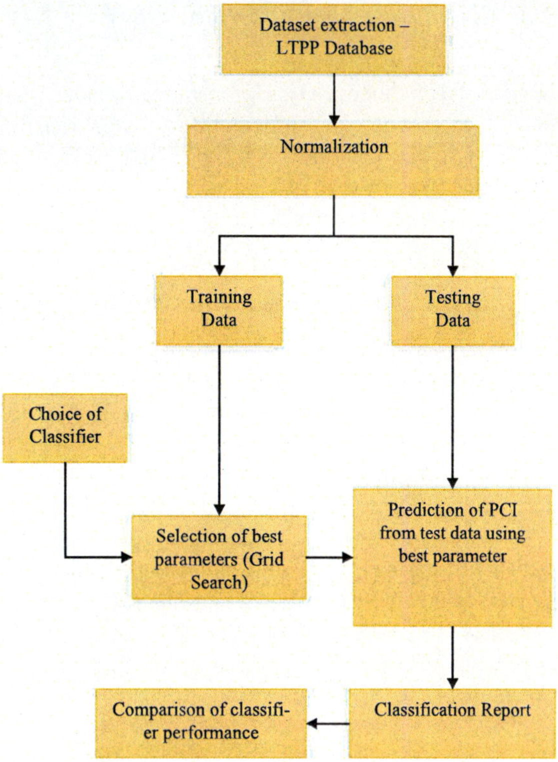

Fig. 4 Methodology adopted in the study

4.1 Logistic Regression

Logistic regression is a linear classifier. The Logistic regression is a special case of the generalized linear model (GLM), which generalizes the ordinary linear regression by allowing the linear model to be related with a response variable that follows the exponential family via an appropriate link function [23, 24].

4.2 Naïve Bayes Classifier

Bayes theorem with independence assumption forms basis of Naïve Bayes classifier to classify objects. A typical assumption is that the continuous values associated with each class are distributed according to one dimensional normal distribution [25].

4.3 K-Nearest Neighbor Classifier

K-Nearest Neighbor is a type of instance-based classification method. KNN is beneficial in cases where the sufficient information regarding the underlying distribution is not available. It works on the principle of minimum distance of instance from neighboring instances.

4.4 Performance Evaluation of Classifier

The classification performance of the proposed models will be judged based on confusion matrix using the four fundamental statistical measures i.e., True Positive (TP), False Positive (FP), True Negative (TN) and False Negative (FN). TP is defined as the case classified as positive which is observed as positive. FP corresponds to case where the predicted value is positive whereas it has been observed as negative. TN and FN are defined as TP and FP, respectively. On the basis of above mentioned statistical measures, precision and recall is defined; precision is the measure of accuracy associated with developed classification rules while recall is the percentage of examples of a designated class that are correctly predicted [26, 27]. For a multiclass classification model above mentioned parameters can be defined as per Eqs. (1), (2), (3) and (4) [28].

$$Precision_M = \frac{\sum_{i=1}^{l} \frac{TP_i}{TP_i + FP_i}}{l} \tag{1}$$

$$Recall_M = \frac{\sum_{i=1}^{l} \frac{TP_i}{TP_i + FN_i}}{l} \tag{2}$$

$$Average\ Accuracy = \frac{\sum_{i=1}^{l} \frac{TP_i + TN_i}{TP_i + TN_i + FP_i + FN_i}}{l} \tag{3}$$

$$F - measure = \frac{\sum_{i=1}^{l} \frac{2 \times Recall_{M_i} \times Precision_{M_i}}{Recall_{M_i} + Precision_{M_i}}}{l} \tag{4}$$

where i represents the considered class and l is the total number of classes.

5 Results and Discussion

This section reports and discusses the results achieved in predicting the category of PCI using three different algorithms i.e., K-Nearest Neighbor, Naïve Bayes and Logistic Regression. Determination of optimal architecture for an algorithm is governed by a set of hyperparameters affecting behavior of algorithm in a particular problem. Determination of optimal set of hyperparameters is of significant importance as it improves performance and reduces overfitting error. Grid Search was used in the present study for K-Nearest Neighbor and Logistic Regression while Naïve Bayes does not have any hypermeter to be tuned, hence was trained with default architecture. A three fold cross validation was also employed to generalize the training process of classifiers. The dataset was divided in the ratio of 7:3 for training and testing. Random seed was fixed at zero to have repeatability in the training and testing set for all three classifiers.

Hyperparameters used in the study to fine tune K-Nearest Neighbor classifier included number of nearest neighbor and leaf size. Number of neighbor equals the number of samples created in the algorithm to define metrics of a point from its centroid. While leaf size governs the speed of construction and query, as well as the memory required [22]. A total of around 146 iterations was done in the grid search to determine best hyper parameters. Mean test score, mean train score and mean fit time for the K-Nearest Neighbor classifier is shown in Fig. 5. Best parameters observed in grid search were {'leaf_size': 20, 'n_neighbors': 1} with best score of 0.7992 for training. Table 1 shows the precision, recall and F1 score for different categories of PCI from the trained KNN Classifier. High accuracy of 0.826649 suggests that trained model can be used for predicting the category of PCI from distress data. Figure 6 presents the confusion matrix showing the actual and predicted class on the test data set using the trained KNN model.

Second algorithm used in the study was Naïve Bayes classifier. Naïve Bayes classifier as trained using scikit learn library doesn't have any hyperparameters to fine tune, hence it was trained and tested in the default form in present study. Table 2 presents classification report for the trained Naïve Bayes Classifier model. The trained model was found to have a good accuracy of 0.710479. The confusion matrix for the Naïve Bayes classifier model on test dataset is presented in Fig. 7.

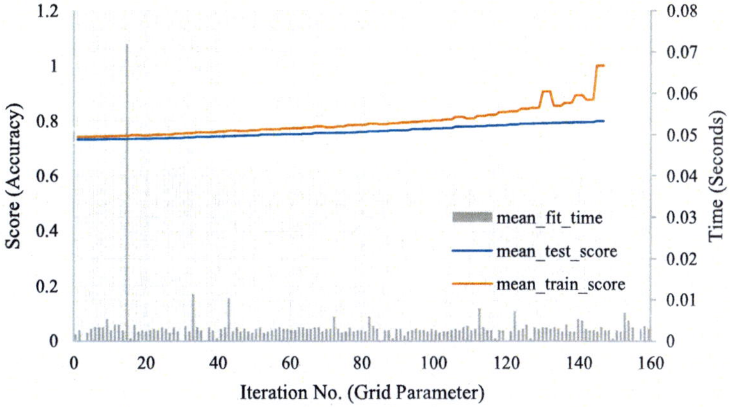

Fig. 5 Train and test accuracy for K-Nearest Neighbor classifier

Table 1 Classification report for K-Nearest Neighbor classifier

	Precision	Recall	F1-Score
Good	0.904628	0.893352	0.898955
Satisfactory	0.751082	0.797701	0.77369
Serious	0.968127	0.981422	0.974729
Very Poor	0.606557	0.510345	0.554307
Fair	0.815657	0.861333	0.837873
Poor	0.699552	0.67679	0.687982
Failed	0.587393	0.59593	0.591631
Accuracy	0.826649		

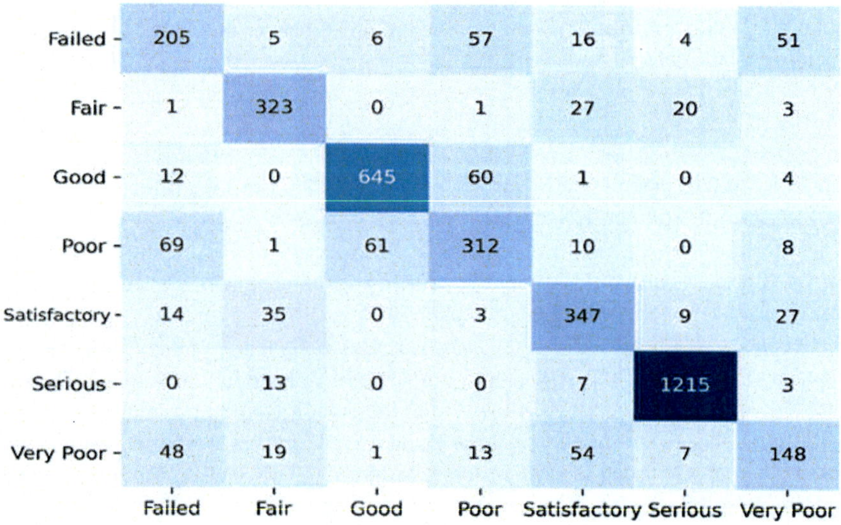

Fig. 6 Confusion matrix for K-Nearest Neighbor classifier

Table 2 Classification report for Naïve Bayes classifier

	Precision	Recall	F1-Score
Good	0.896853	0.710526	0.79289
Satisfactory	0.809278	0.36092	0.499205
Serious	0.951638	0.98546	0.968254
Very Poor	0.397887	0.77931	0.526807
Fair	0.67366	0.770667	0.718905
Poor	0.460674	0.444685	0.452539
Failed	0.362667	0.395349	0.378303
Accuracy	0.710479		

Fig. 7 Confusion matrix for Naïve Bayes classifier

Further, Logistic Regression Classifier was used for training. Grid Search was used for determining optimal hyperparameters of the classifier. Considered parameters in the grid search include regularization constant (C), penalty function and solver. C denotes inverse of regularization where smaller values of C means higher regularization. Penalty function governs the norm for penalization while solver deals with the optimization algorithm used in classifier [22]. A total of 119 iterations were performed by grid search algorithm. The training and testing accuracy with different iterations in grid search is presented in Fig. 7. Best parameters determined by grid search for the logistic regression classifier were {'C': 1e-05, 'penalty': 'none', 'solver': 'newton-cg'} with best training score of 0.9157. Classification report with respect to test data for Logistic Regression classifier is presented in Table 3. Figure 8 presents the confusion matrix for the test dataset. As shown in Table 3, highest accuracy of 0.92 among the three classifiers was obtained for the Logistic Regression classifier (Fig. 9).

6 Conclusion

The objective of this study was to develop a classification model to predict the present pavement condition based on Pavement Condition Index Category. Three different classification algorithms were employed in the study for prediction of PCI category based on dataset comprising of various types of distresses including, Alligator

Table 3 Classification report for Logistic Regression classifier

	Precision	Recall	F1-Score
Good	0.945792	0.918283	0.931834
Satisfactory	0.903002	0.898851	0.900922
Serious	0.998384	0.998384	0.998384
Very Poor	0.835616	0.841379	0.838488
Fair	0.951482	0.941333	0.946381
Poor	0.789691	0.830803	0.809725
Failed	0.814493	0.81686	0.815675
Accuracy	0.918758		

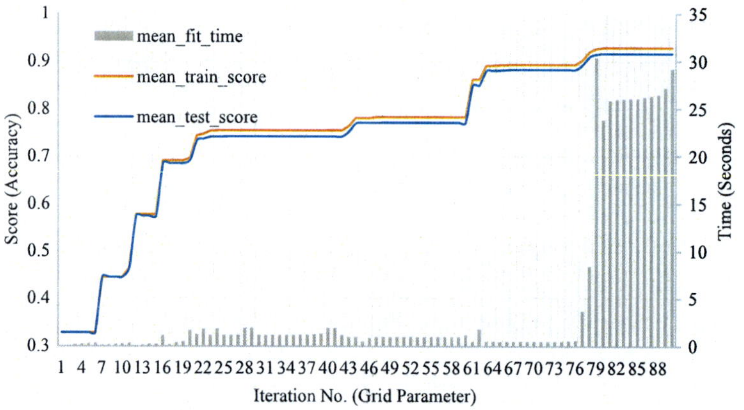

Fig. 8 Train and test accuracy for Logistic Regression classifier

Cracking, Bleeding, Block Cracking, Edge Cracking, Longitudinal Cracking, Transverse Cracking, Patching, Pothole, Shoving and Ravelling. The three robust classification algorithms tried in the study includes K-Nearest Neighbor, Naïve Bayes and Logistic regression. The major findings of the study are enlisted below:

- All the three-classifier employed in the study for prediction of PCI category showed reliable prediction with accuracy in excess of 0.71.
- Logistic Regression Classifier with parameters {'C': 1e-05, 'penalty': 'none', 'solver': 'newton-cg'} was found to provide highest accuracy of 0.915 on training and 0.918 on testing dataset.
- Present pavement condition assessment, being a key component of decision making in PMS usually is a time-consuming process when dealing with network level. The proposed classifier model helps in significantly reducing the time consumed in the process.

Further research is underway to improve the classifier model by considering more datapoints to increase the incorporated variance. As the proposed model provides the

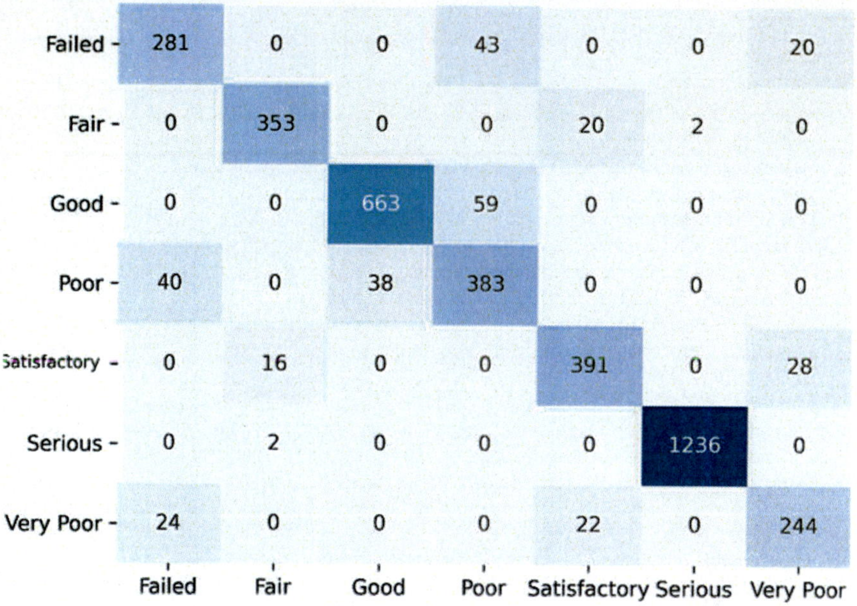

Fig. 9 Confusion matrix for Logistic Regression classifier

interpreted category of PCI rather than providing just index value it can be useful for the field engineers in easy interpretation of pavement condition from quantified data while saving the tedious effort of determining condition using deduct value curves manually.

References

1. Ziari H, Maghrebi M, Ayoubinejad J, Waller ST (2016) Prediction of pavement performance: application of support vector regression with different kernels. Transp Res Rec 2589:135–145. https://doi.org/10.3141/2589-15
2. Sun L, Gu W (2010) Pavement condition assessment using fuzzy logic theory and analytic hierarchy process. J Transp Eng 137:648–655. https://doi.org/10.1061/(asce)te.1943-5436.0000239
3. AASHTO (2012) Pavement management guide. Second. American Association of State Highway and Transportation Officials
4. Mubaraki M (2014) Third-order polynomial equations of municipal urban low-volume pavement for the most common distress types. Int J Pavement Eng 15:303–308
5. Shahin MY (2005) Pavement management for airports, roads and parking lots. Springer US
6. Zimmerman KA (1995) Pavement management methodologies to select projects and recommend preservation treatments. Washington DC
7. Sharma M, Kumar P, Suman SK (2019) A review of pavement condition rating models for flexible pavements. Int Res J Eng Technol 6:2218–2223
8. Shahin MY, Darter MI, Kohn SD (1978) Development of a pavement condition index for roads and streets

9. ASTM:D6433-15 (2015) Standard practice for roads and parking lots pavement condition index surveys
10. Elhadidy AA, El-Badawy SM, Elbeltagi EE (2019) A simplified pavement condition index regression model for pavement evaluation. Int J Pavement Eng 1–10. https://doi.org/10.1080/10298436.2019.1633579
11. Sadrossadat E, Heidaripanah A, Osouli S (2016) Prediction of the resilient modulus of flexible pavement subgrade soils using adaptive neuro-fuzzy inference systems. Constr Build Mater 123:235–247. https://doi.org/10.1016/j.conbuildmat.2016.07.008
12. Koduru HK, Xiao F, Amirkhanian SN, Juang CH (2010) Using fuzzy logic and expert system approaches in evaluating flexible pavement distress: case study. J Transp Eng 136:149–157. https://doi.org/10.1061/(asce)0733-947x(2010)136:2(149)
13. Tien Bui D, Moayedi H, Abdullahi MM et al (2019) Prediction of Pullout behavior of belled piles through various machine learning modelling techniques. Sensors 19:3678. https://doi.org/10.3390/s19173678
14. Yadollahi MM, Benli A, Demirboga R (2017) Application of adaptive neuro-fuzzy technique and regression models to predict the compressive strength of geopolymer composites. Neural Comput Appl 28:1453–1461. https://doi.org/10.1007/s00521-015-2159-6
15. Gandomi AH, Alavi AH, Mirzahosseini MR (2011) Nonlinear genetic-based models for prediction of flow number of asphalt mixtures. J Mater Civ Eng 248–263. https://doi.org/10.1061/(ASCE)MT.1943-5533.0000154
16. Sollazzo G, Fwa TF, Bosurgi G (2017) An ANN model to correlate roughness and structural performance in asphalt pavements. Constr Build Mater 134:684–693. https://doi.org/10.1016/j.conbuildmat.2016.12.186
17. Shahnazari H, Tutunchian MA, Mashayekhi M, Amini AA (2012) Application of soft computing for prediction of pavement condition index. J Transp Eng 138:1495–1506. https://doi.org/10.1061/(ASCE)TE.1943-5436.0000454
18. Issa A, Samaneh H, Ghanim M (2021) Predicting pavement condition index using artificial neural networks approach. Ain Shams Eng J. https://doi.org/10.1016/j.asej.2021.04.033
19. Bashar MZ, Torres-Machi C (2021) Performance of machine learning algorithms in predicting the pavement international roughness index. Transp Res Rec J Transp Res Board 036119812098617. https://doi.org/10.1177/0361198120986171
20. Kumar R, Suman SK, Prakash G (2021) Evaluation of pavement condition index using artificial neural network approach. Transp Dev Econ 7:20. https://doi.org/10.1007/s40890-021-00130-7
21. Han J, Kamber M, Pei J (2012) Data mining: concepts and techniques
22. Pedregosa F, Varoquaux G, Gramfort A et al (2011) Scikit-learn: machine learning in Python. J Mach Learn Res 12:2825–2830
23. Jeong H, Jang Y, Bowman PJ, Masoud N (2018) Classification of motor vehicle crash injury severity: a hybrid approach for imbalanced data. Accid Anal Prev 120:250–261. https://doi.org/10.1016/j.aap.2018.08.025
24. Nour MK, Naseer A, Alkazemi B, Jamil MA (2020) Road traffic accidents injury data analytics. Int J Adv Comput Sci Appl 11. https://doi.org/10.14569/IJACSA.2020.0111287
25. Górecki T, Krzyśko M, Wołyński W (2019) Variable selection in multivariate functional data classification. Stat Transit 20. https://doi.org/10.21307/stattrans-2019-018
26. Davis J, Goadrich M (2006) The relationship between precision-recall and ROC curves. ACM Int Conf Proc Ser 148:233–240. https://doi.org/10.1145/1143844.1143874
27. He H, Ma Y (2013) Imbalanced learning—foundations, algorithms, and applications. IEEE Press, Wiley
28. Sokolova M, Lapalme G (2009) A systematic analysis of performance measures for classification tasks. Inf Process Manage 45:427–437. https://doi.org/10.1016/j.ipm.2009.03.002

Automatic Pavement Crack Rating for Network-Level Pavement Management System

Jun Yew Tan, Teron Nguyen, S. Kapilan, Than Than Nyunt, Yeow Hoo Lim, and Yin Fong Leong

Abstract Cracking is one of the important distresses that can be used to trigger pavement maintenance treatments. Traditional crack rating is commonly based on the Pavement Condition Index (PCI) approach involving time-consuming visual inspection and manual classification processes. The emergence of automatic and high-speed laser imaging devices significantly improves the efficiency and productivity of pavement crack data collection but it requires suitable methods and concepts for automatic crack ratings. This paper discusses the development of an automatic pavement crack rating using crack data collected from a high speed 3D sensor. Two levels of crack ratings are proposed: Level 1 provides detailed crack information including cracking extent, crack types and severity, and Level 2 is a macro-indicator of general/overall cracking extent on a pavement section of 10 m length. The method and concept were developed and tested initially for Singapore expressway network under the effort of the Land Transport Authority (LTA) of Singapore to integrate crack data into the pavement management system (PMS).

Keywords Pavement crack · LCMS-2 · Automatic classification · Crack rating · Pavement management system

1 Introduction

1.1 Background

Pavement is an expensive road infrastructure asset that require suitable maintenance and rehabilitation to maintain its service life. To manage the road infrastructure, transportation agencies often use Pavement Management Systems (PMS) to store,

J. Y. Tan (✉) · T. Nguyen
Samwoh Innovation Centre Pte. Ltd. (SWIC), 25E Sungei Kadut Street 1, Singapore 729333, Republic of Singapore
e-mail: junyew.tan@samwoh.com.sg

S. Kapilan · T. T. Nyunt · Y. H. Lim · Y. F. Leong
Singapore Land Transport Authority, Singapore, Republic of Singapore

© The Author(s), under exclusive license to Springer Nature Switzerland AG 2022
H. R. Pasindu et al. (eds.), *Road and Airfield Pavement Technology*, Lecture Notes in Civil Engineering 193, https://doi.org/10.1007/978-3-030-87379-0_6

analyse and visualise large datasets in the network scale. PMSs require accurate and reliable pavement condition data as critical input in order to evaluate the health of the pavement network [1, 2]. Surface distresses such as pothole, crack, rutting and roughness are important parameters that define the health of the pavement network in a PMS [3, 4].

In Singapore, pavement surface defects or distresses are typically gathered by teams of inspectors as they drive through the road network at low speed. Surface defects that require urgent attention such as potholes are marked and work orders are issued subsequently to carry out rectification works within a stipulated time. As most of the surface defects are repaired almost immediately, these are not captured efficiently into the pavement management system.

The emergence of automated equipment (semi or full) offers the opportunity to significantly improve the productivity and efficiency of pavement surface condition data collection in the current dynamic road environment [5]. Roads can now be surveyed at a much faster rate and more thoroughly, and the data is available digitally for all types of analysis. This has opened up the opportunity to use surface crack data as additional input to Singapore's PMS.

This paper discusses the development of an automatic crack rating for asphalt pavement that is suitable for network-level pavement management systems.

1.2 Crack Rating Systems

Crack rating is one of the key parameters of a wider pavement condition rating system. There are many pavement condition rating systems used around the world such as the Pavement Condition Index (PCI), Pavement Serviceability Rating (PSR), Pavement Condition Rating (PCR), etc.

The Pavement Condition Index (PCI) system is the most widely known pavement rating system developed by the US Army Corps of Engineers [6]. The PCI system rates the pavement condition as a range of numerical value between 0 and 100, with 100 representing the best possible condition and 0 representing the worst possible condition. The PCI system includes 19 types of pavement distresses including non-crack parameters such as ride quality, rutting, depressions, polished aggregates and weathering and raveling (see Fig. 1).

1.3 Motivation

A rating system for crack-only data is required for Singapore's pavement management system. The PCI system is the most widely used but it often involves time-consuming, laborious visual inspection and manual classification processes as some of the distresses in the PCI system is not suited for fully automatic processing. Studies

Distress types for asphalt pavements	
1. Alligator cracking	11. Low ride quality
2. Bleeding	12. Patching and utility cut patching
3. Block cracking	13. Polished aggregate
4. Bumps and sags	14. Potholes
5. Corrugations	15. Rutting
6. Depressions	16. Shoving
7. Edge cracking	17. Slippage cracking
8. Joint reflection cracking	18. Swelling
9. Lane/shoulder drop-off	19. Weathering and raveling
10. Longitudinal and transverse cracking	

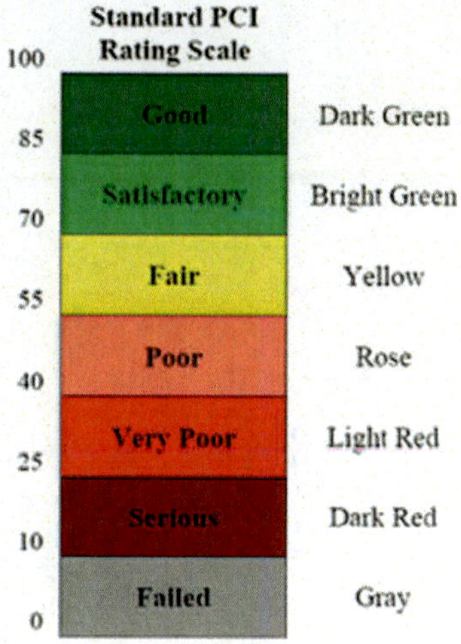

Fig. 1 Asphalt pavement distresses considered by the PCI system and the Standard PCI rating scale [6]

have also shown that manual survey results do not correlate well with automatic classification and rating [7, 8].

In researching and developing a suitable crack rating system for Singapore condition, consideration was given to the needs of the operational as well as the management level. The operational level requires more detailed information on the crack in order to carry out the appropriate repair action. On the other hand, the management level is interested in a simple linear ranking, for example 0–5, that can be quickly and easily compared from one pavement section to the other. Therefore, the resulting crack index system must be reliable and able to meet the conflicting needs of both the management and the operational level.

An automated classification procedure needs to be developed and suited to the capabilities of the automated survey equipment within the Singapore dynamic road environment and traffic loading based on the types of crack that commonly occur here.

2 Approach and Methodology

2.1 Automated Pavement Condition Survey

In recent years, the emergence of advanced image sensor technology allows pavement condition survey to be carried out automatically and quickly. One advanced system being used is the Laser Crack Measurement System-2 (LCMS-2) shown in Fig. 2.

Fig. 2 An LCMS-2 vehicle used in Singapore

The LCMS-2 sensor uses laser line projectors, high speed camera and advanced optic to acquire high resolution 3D profiles of the lane of a road [9]. This unique 3D vision technology allows for automatic pavement condition assessment of all types of asphalt and concrete surfaces. The LCMS-2 acquires both 3D and 2D image data of the road surface with 1 mm interval over a 4 m lane width in transverse direction and also 1 mm interval in the longitudinal direction at survey speeds up to 100 km/h [10].

The LCMS-2 system comes with a processing software, called RoadInspect, that automatically detects surface distresses. Figure 3 shows a grey-scale image on the left known as the range image which shows the distance from the sensors to the road surface corrected to the same reference level. The image on the right is a 3D image which shows pavement cracks visualized in color-coded overlays. The color coding is based on a user-defined range of crack width.

The colour coding overlay for the cracks used in this image and all subsequent 3D images is: Green (crack width < 3 mm), Orange (crack width 3-20 mm) and Red (crack width > 20 mm). The color-coded overlay produced by RoadInspect software shown in Fig. 3 is useful at micro-level but not suitable for network-level pavement management system. In order to categorize cracks in a meaningful manner for pavement management purpose, the crack data from the LCMS-2 needs to be organized and rated in such a way that is consistent with the pavement management functions.

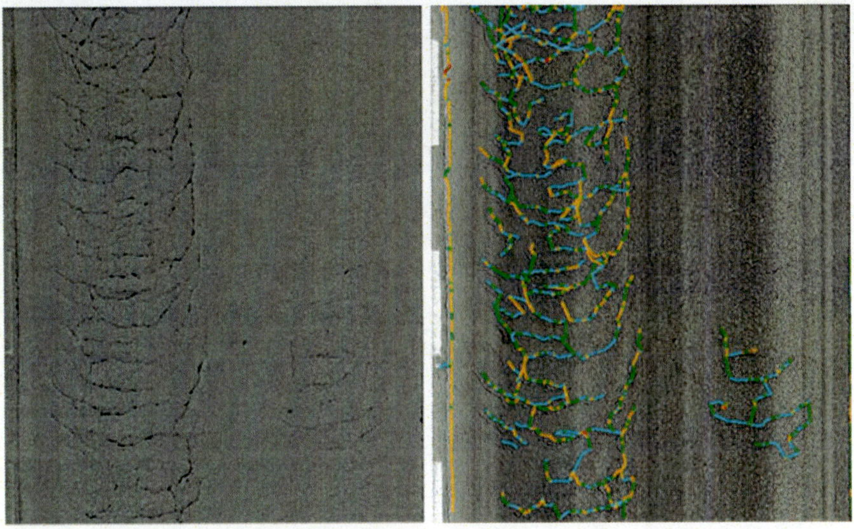

Fig. 3 LCMS-2 rating of cracking width (left) range image and (right) using color overlay

2.2 Classification of Crack Types

The first step in developing a crack rating system is to consider the types of cracks to be classified and rated. The PCI system classifies six types of cracks for asphalt pavement, namely alligator cracking, block cracking, edge cracking, joint reflection cracking, longitudinal and transverse cracking, and slippage cracking [6]. It was decided that the number of crack types shall be simplified considering factors such as the frequency of occurrence in Singapore roads and the main causal factor.

Proper classification of crack types is vital in the assessment works. To determine the types of cracks that most commonly occur in Singapore pavement condition, distress data of over 300 lane-km of asphalt pavements was collected and studied. The following observation can be summarized:

(a) Three main classification of cracks can be observed: transverse cracks, longitudinal cracks, and alligator-type cracks.
(b) The transverse cracks occur mainly as construction joints and the several different types of transition joints at bridges/flyovers.
(c) Longitudinal cracks are mostly constructions joints/resurfacing of a lane occurring outside the wheelpath.
(d) Alligator cracks and slippage cracks can be found mostly within the wheelpath.

Pavement surface distresses can be broadly regarded as either functional or structural related. Transverse and longitudinal cracks are in-principle functional distress types that are climate or environmental related, whereas alligator, block and slippage cracking are structural distress types caused by or associated with traffic loading.

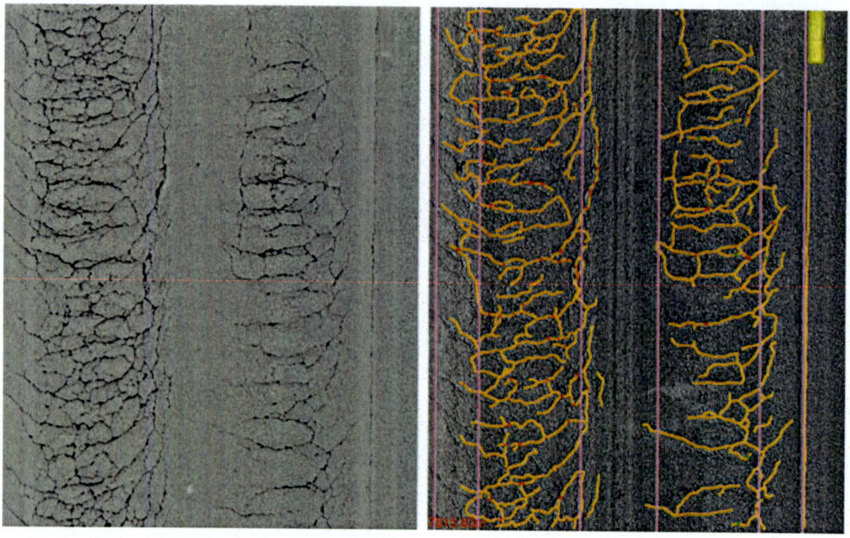

Fig. 4 Example of high severity structural cracking (slippage crack characterized by crescent shape) (left) range image and (right) using color overlay

Figure 4 is an example of high severity structural cracking consisting of slippage cracks characterized by its crescent shape.

2.3 *Definition of Wheelpath*

Transverse and longitudinal cracks are generally defined based on their angle and orientation. However, some longitudinal cracks can be load-induced or structurally-related cracks.

To make a distinction between structural cracks and longitudinal cracks (defined as functional distress), reference can be made to whether the cracks occur within the wheelpath or outside the wheelpath. AASHTO R 85-18 [11] defines five zones across the lane width based on the crack location (see Fig. 5). Zones 1 and 5 are located outside the wheelpaths, zones 2 and 4 are the wheelpaths and zone 3 is in between the wheelpaths. The width of zones 1 and 5 can be varied according to the width of a lane and local requirements. The centre zone width is 750 mm and both wheelpaths (zones 2 and 4) are 1000 mm.

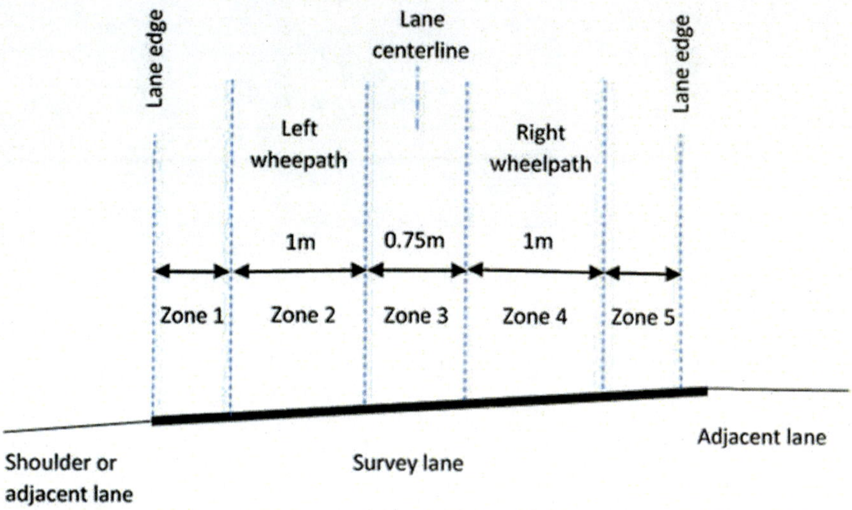

Fig. 5 Definition of wheelpath on Singapore roads based on AASHTO R 85-18 [11]

3 Rating Procedure

3.1 Sample Unit

A pavement is rated based on a unit area referred to as sample unit. Sample unit can be defined based on a length or an area that is logical and consistent with the PMS functions. In Singapore's context, a sample unit for crack rating is defined as every 10 m length of a road in order to be consistent with the data interval used in the PMS.

3.2 Rating Process

Two levels of crack rating are proposed, from general to detailed information, to serve different purposes of the local transport authority (see Fig. 6).

Level 2 crack rating is a macro-indicator of general/overall cracking extent on a pavement section. This is a management-level index which gives a quick overview of the general crack health of the road network.

Level 1 crack rating contains detailed crack information including cracking extent, crack types and severity. This is an operational-level index which can be used for decision-making when it comes to repair action, scheduling and allocation of budget.

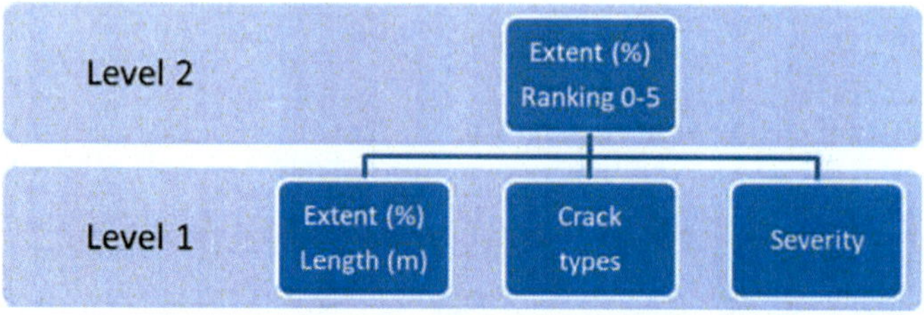

Fig. 6 Two levels of crack rating indicator and the components of crack data within each level

Table 1 Level 2 crack rating: thresholds and ranking

Overall extent, E (%)	Ranking (0–5)	Meaning
$E > 25$	5	Very Bad
$15 < E \leq 25$	4	Bad
$10 < E \leq 15$	3	Fair
$5 < E \leq 10$	2	Good
$0 < E \leq 5$	1	Very Good
$E = 0$	0	Excellent

3.3 *Level 2 Crack Rating*

Level 2 crack rating is calculated as the total crack areas over 40 m^2 (4 m width × 10 m length) of the pavement unit section, and regardless of crack types, extent and severity. Based on the extent (%) of all cracks on a pavement section, the equivalent rating in a five-scale (0–5) is proposed as shown in Table 1. At crack ranking = 0, the pavement section is in excellent condition, whereas ranking = 5 means the pavement section is in very bad condition.

3.4 *Level 1 Crack Rating*

The rating process is shown in a flowchart in Fig. 7. In summary, the rating process can be organized into the following steps:

Step 1: Determine whether it is functional or structural crack based on the location of the crack.
Step 2: For functional crack, determine whether it is transverse or longitudinal crack based on orientation.
Step 3: Determine severity level according to the type of crack.
Step 4: Compute extent of the crack for each severity level.

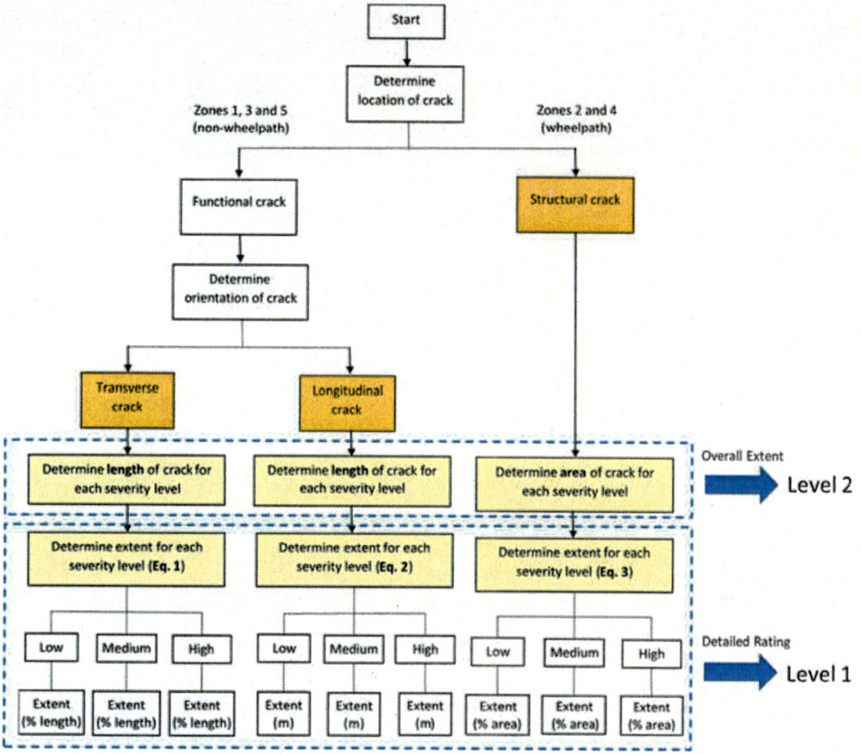

Fig. 7 Flowchart of crack rating process

3.4.1 Severity and Extent

The outcome of the rating process should provide the following vital information for the road agency:

(a) Type of crack (transverse, longitudinal, or structural)
(b) Severity level
(c) Extent.

Severity level refers to how far the cracking has progressed and is often directly related to the crack width, whereas extent is the amount of road surface area which is affected. Severity is categorized into three levels: low, medium and high.

The severity for linear cracks such as transverse and longitudinal cracks is defined by the crack width and categorized into low, medium, or high severity level. The range of crack width for the categories differs from country to country depending on local specifications and requirements. The following severity levels for transverse and longitudinal cracks is proposed for Singapore and this can be further refined as more data is available:

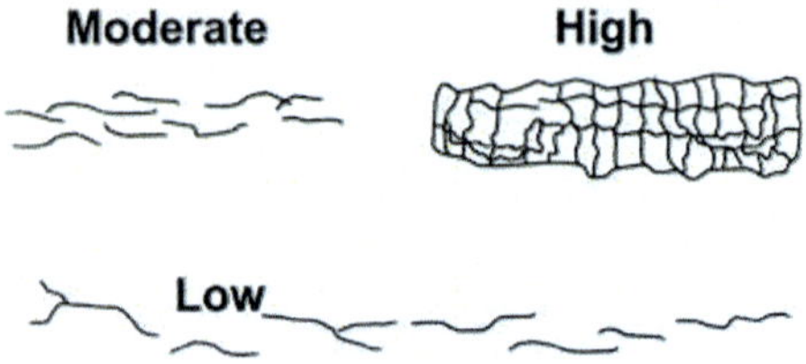

Fig. 8 Illustration of severity level for structural cracks

Severity Level for Transverse land Longitudinal Cracks

Low: average crack width $\leq$ 3 mm
Medium: 3mm < average crack width $\leq$ 20 mm
High: average crack width > 20 mm.

Severity level for structural cracking is more complicated as it consists of multiple interconnecting cracks. The following description is used to define structural cracks and this is shown graphically in Fig. 8.

Severity Level for Structural Cracks

Low: area of cracks sealed or unsealed with no or a few interconnections
Medium: area of interconnected lightly spalled cracks
High: area of interconnected medium to highly spalled cracks with a well-defined pattern. Some of the pieces may rock under traffic resulting in potholes.

The extent of a crack, except for transverse crack, is the percentage of surface area or length that is affected by the crack. The extent for linear cracks (transverse and longitudinal) is computed using length while the extent for area cracks (structural) is computed using area.

For transverse crack, the extent is not computed into percentage because there is no definable limit to the number of transverse cracks that may occur within a sample unit. The equation for transverse crack is shown in Eq. (1).

$$E_{u,s}^{T} = \sum_{i}^{n} T(length)_{s,i} \tag{1}$$

where

E^T is the extent for transverse crack (expressed in meter) for sample unit u and severity level s.
$T(length)_{s,i}$ is the length of the i-th transverse crack with severity s.
n is the total number of transverse cracks with severity s.

For longitudinal crack, the extent is computed as a percentage of longitudinal crack length to the total length of zones 1, 3 and 5 within a sample unit, as shown by Eq. (2).

$$E^L_{u,s} = \frac{\sum_i^n L(length)_{s,i}}{l_1 + l_3 + l_5} \times 100\% \tag{2}$$

where

E^L is the extent for longitudinal crack (expressed in percentage) for sample unit u and severity level s.
$L(length)_{s,i}$ is the length of the i-th longitudinal crack with severity s.
n is the total number of longitudinal cracks with severity s.
$l_1 + l_3 + l_5$ is the sum of the length of the three functional non-wheelpath zones (zones 1, 3, and 5) within a sample unit. This shall be equal to 10 m + 10 m + 10 m = 30 m.

For structural crack, the extent is computed as a percentage of total structural crack area to the total area of zones 2 and 4 within a sample unit, as shown by Eq. (3).

$$E^S_{u,s} = \frac{\sum_i^n S(area)_{s,i}}{a_2 + a_4} \times 100\% \tag{3}$$

where

E^S is the extent for structural crack (expressed in percentage) for sample unit u and severity level s.
$S(area)_{s,i}$ is the area of the i-th structural crack with severity s.
n is the total number of structural cracks with severity s.
$a_2 + a_4$ is the sum of the area of the two wheelpath zones 2 and 4 within a sample unit. This shall be equal to 10 m^2 + 10 m^2 = 20 m^2.

4 Result and Discussion

For validation purpose, the proposed automated cracking rating was applied to every 10 m of LCMS-2 data collected in Singapore expressway network. After data collection, crack information was analysed, classified and quantified according to the flowchart in Fig. 7. Table 2 shows samples of validation carried out for the overall extent and level 2 crack ranking, ranging from 0 to 5, on the Kranji Expressway. Medium and high severity structural cracks were seen clearly on pavement surfaces with ranking 3, 4 and 5. High severity cracks were observed on both wheelpaths of crack rating 5 along 10 m segments. The calculated crack extent (E in %) and the appearance of cracking along 10 m length of pavement unit section from RoadInspect output are quite similar. The simplified 5-scale level 2 crack rating can express pavement crack condition in a relatively linear relationship. Maintenance and rehabilitation activities should be planned immediately for those segments with level 2 crack rating from 3 to 5.

Table 2 Validation samples of overall extent (E%) and Level 2 crack ranking on Kranji Expressway

Ranking: 5	Ranking: 4	Ranking: 3	Ranking: 2	Ranking: 1	Ranking: 0
E = 47.51 %	E = 20.81 %	E =12.5 %	E = 8.41 %	E = 2.55 %	E = 0 %
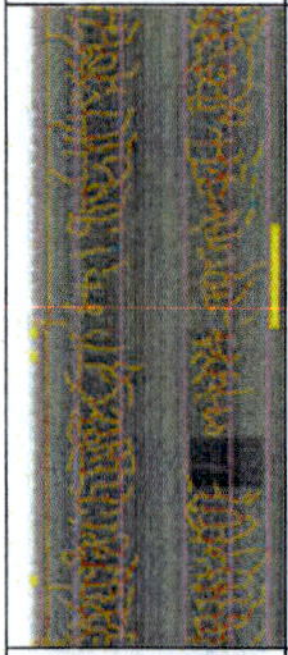	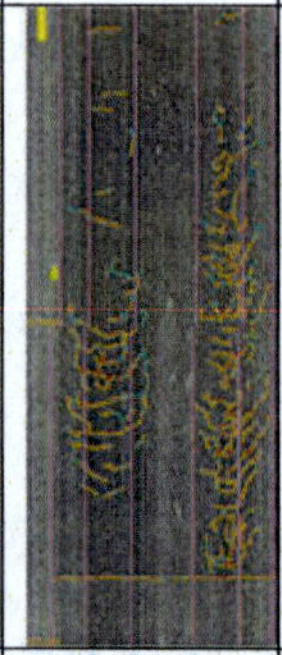	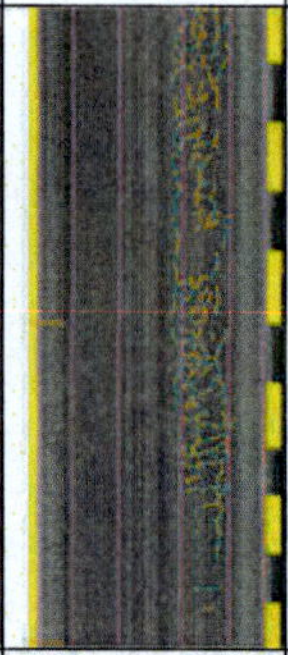	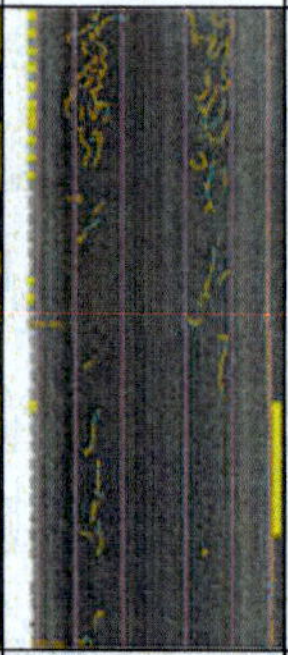	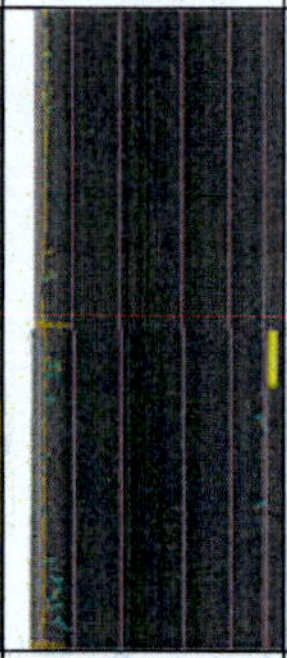	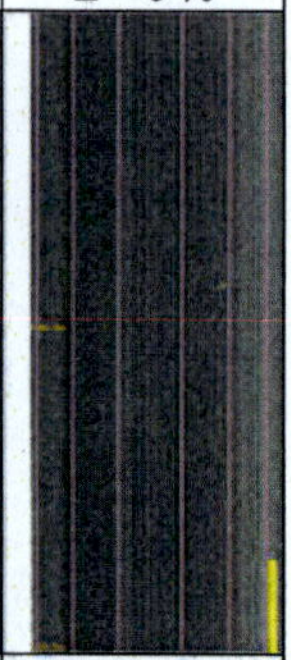
E = 44.26 %	E = 18.88 %	E = 14.35 %	E = 9.21 %	E = 2.10 %	E = 0 %
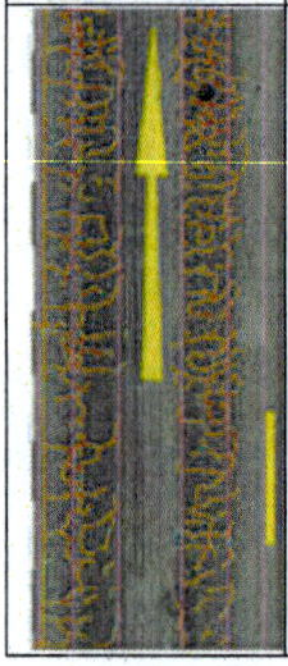	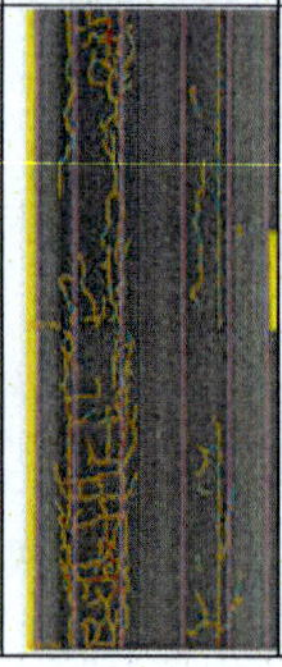	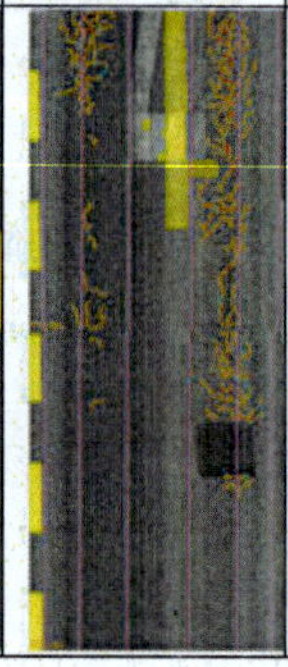	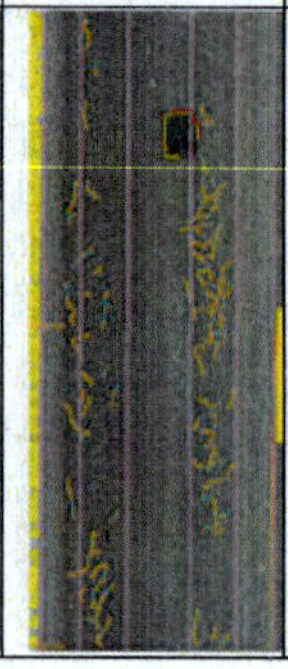		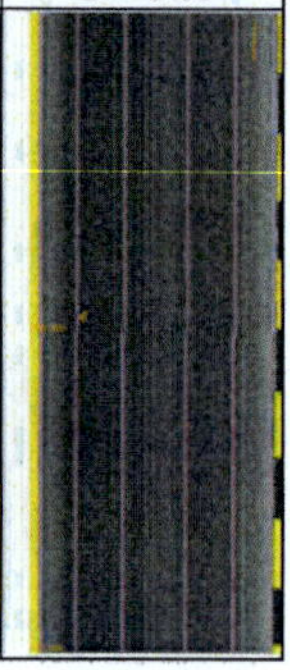

From level 2, cracks are broken down into structural, longitudinal and transverse cracks with three severity levels for each crack type. Table 3 lists the examples of both crack rating levels which are proposed for LTA PMS. Level 2 crack index can be used as macro-level indicator to quickly tell the cracking extent of a pavement segment. It should be noted that the severity or "seriousness" of a crack may not be equal to the "extent". The crack severity can be different even within the same cracking index (e.g. index 1, 2, 3, 4 or 5) due to direction and various connectivity levels. For example, a section with ranking = 1, where cracking extent ranges from 0 to 5%, may contain transverse or longitudinal cracks at high severity that requires a specific crack sealing/filling preventive maintenance. Therefore, the simplified crack index at level 2 can provide an overview of the cracking condition. Further investigation into level 1 is always recommended for detailed consideration of suitable maintenance treatments.

Table 3 Data format incorporating both levels of crack rating

Road Code	Section	Str-Low	Str-Mid	Str-High	Long-Low	Long-Mid	Long-High	Tran-Low	Tran-Mid	Tran-High	Overall Extent	Ranking
BUE01Q	020	0.00	0.00	90.57	0.00	18.30	0.00	0.00	0.00	0.00	46.66	5
BUE01Q	020	9.50	0.00	75.00	0.00	12.90	0.00	0.00	0.00	0.00	43.22	5
BUE01Q	020	0.00	24.70	50.00	0.00	0.00	0.00	0.00	0.00	0.00	37.35	5
BUE01Q	020	14.01	0.00	49.59	0.00	0.00	0.00	0.00	0.00	0.00	31.80	5
BUE01Q	020	0.00	10.58	49.63	0.00	0.00	0.00	0.00	0.00	0.00	30.11	5
BUE01Q	030	28.49	0.00	0.00	0.00	30.90	0.00	0.00	5.35	0.00	17.90	4
BUE01Q	020	5.41	0.00	14.90	0.00	0.00	0.00	0.87	1.05	0.00	10.63	3
BUE01Q	020	16.77	0.00	0.00	0.00	0.00	0.00	0.00	3.11	0.00	9.16	2

Note Str-Low, Str-Mid and Str-High: Structural crack at low, medium, and high severity. Long-Low, Long-Mid and Long-High: Longitudinal crack at low, medium, and high severity. Tran-Low, Tran-Mid and Tran-High: Transverse crack at low, medium, and high severity

5 Conclusion

The emergence of automatic data collection method for cracking requires reconsideration of pavement condition evaluation which was built conventionally on visual inspection and manual survey [12]. The usage of 3D imaging sensor in LCMS-2 in Singapore has resulted in higher data collection efficiency and the opportunity to utilise crack data as part of LTA PMS.

This study has initially developed an automatic pavement crack rating for network-level pavement management system by incorporating two levels of crack rating for different usage purposes. Crack rating should start from general ranking—Level 2 to detailed rating—Level 1 as recommended in recent practices [7]. The proposed crack rating scales are simple enough to be analyzed in combination with pavement structural, roughness, and rut depth data for final decision making on pavement maintenance at network level. Validation of the data sampled from two expressways have shown promising results. Further study is on-going to validate and improve proposed methodology based on larger datasets.

Acknowledgements The authors would like to acknowledge the contributions of Anggraini Zulkati & Siew Kim Foo from the Singapore Land Transport Authority, Kars Drenth of Samwoh Innovation Centre BV, and Ju Kit Ong, a former employee of Samwoh Innovation Centre Pte. Ltd. for their valuable assistance in making this paper possible.

References

1. AASHTO (2012) Pavement management guide, 2nd edn. American Association of State Highway and Transportation Officials
2. Nguyen T, Lechner B, Wong YD (2019) Response-based methods to measure road surface irregularity: a state-of-the-art review. Eur Transp Res Rev 11(1):1–18
3. Haas R, Hudson WR (2015) Pavement asset management. Wiley
4. Nguyen T, Lechner B, Wong YD, Tan JY (2021) Bus ride index–a refined approach to evaluating road surface irregularities. Road Mater Pavement Des 22(2):423–443
5. NCHRP Synthesis 531 (2019) Automated pavement condition surveys—a synthesis of highway practice
6. ASTM D6433-18 (2018) Standard practice for road and parking lots pavement condition index surveys (PCI)
7. National Academies of Sciences, Engineering, and Medicine (2020) Standard definitions for common types of pavement cracking. The National Academies Press, Washington, DC. https://doi.org/10.17226/25928
8. Henning T, Mia M Did we get what we wanted?—Getting rid of manual condition surveys, Pavemetrics Systems Inc. website http://www.pavemetrics.com
9. Laurent J, Hébert J, Talbot M Automated detection of sealed cracks using 2D and 3D road surface data, Pavemetrics Systems Inc. Website http://www.pavemetrics.com
10. Drenth K, Tan JY, Drenth M, Kit OJ (2020) LCMS-2 measurements of the quality of road markings. In: Proceedings of the 9th International conference on maintenance and rehabilitation of pavements—Mairepav9. Springer, Cham, pp 943–952

11. AASHTO R 85-18 (2018) Standard practice for quantifying cracks in asphalt pavement surfaces from collected pavement images utilizing automated methods
12. FHWA Distress identification manual for the long-term pavement performance program (FHWA-HRT-13-092). Revised May 2014

Challenges for the Compaction and Proving of Granular Fills and Layers in Airport Pavements

Greg White and Hudson Anstee

Abstract The compaction and proving of granular materials during construction is fundamentally import to airport pavement design practice. This is particularly so in countries that make significant use of granular base courses with relatively thin asphalt surfaces, such as Australia. The compaction of dredged sand fills, up to 1500 mm deep is also important where airport pavements are constructed over low bearing capacity marine clays, primarily in areas of reclaimed land. The large 180 tonne Supercompactor for compacting sand fills is no longer available in Australia and the 50 tonne Macro rollers for proving granular pavement layers were downgraded from 1400 to 1000 kPa tyre pressure. Meanwhile, large commercial aircraft have ever-increasing tyre pressures and wheel loads. This has created a challenge in the form of a gap between roller capability and aircraft demand. The gap is relatively minor for the compaction of deep sand fills, but the inability to theoretically compare the effect of static rollers with that of vibrating and impact rollers makes this difficult to quantify. The alternate is an expensive field trial to demonstrate acceptable sand density at depths to 1500 mm. In contrast, a significant gap exists for the proving of fine crushed rock layers used in upper base courses under thin (60–100 mm) asphalt surfaces. This challenge requires rollers with higher tyre pressure or significantly thicker asphalt surface courses to be adopted, both of which are expensive.

Keywords Compaction · Proof rolling · Granular · Fill · Airport · Pavement · Construction

G. White (✉) · H. Anstee
University of the Sunshine Coast, Sippy Downs, QLD, Australia
e-mail: gwhite2@usc.edu.au

H. Anstee
Symal, Melbourne, VIC, Australia

© The Author(s), under exclusive license to Springer Nature Switzerland AG 2022

H. R. Pasindu et al. (eds.), *Road and Airfield Pavement Technology*, Lecture Notes in Civil Engineering 193, https://doi.org/10.1007/978-3-030-87379-0_7

1 Introduction

Airport pavements are thick, stiff and expensive structures designed to support the safe operation of large aircraft traffic loadings [1]. Furthermore, airport pavements are located in high-risk environments where the cost of disrupting operations to affect maintenance and rehabilitation is significant. Therefore, airport pavements are generally designed in a conservative manner, with high specification requirements, significant quality assurance testing and active verification during construction [2]. One of the important elements of airport pavement design, specification and construction is the proving of the granular material layers, including deep fills and granular pavement layers.

Deep fills can include many materials, with hydraulically or mechanically placed dredged sand over naturally occurring low strength marine clays common where airports are located in areas of reclaimed land. In Australia this includes Sydney, Brisbane and Cairns airports. Singapore, Hong Kong and Japanese reclaimed island airports provide other examples. Because the dredged sand fill is economically placed in thick layers, and is usually placed years prior to the construction of the pavement, allowing the underlying marine clay to consolidate, proving the fill prior to pavement construction is particularly important [3].

Granular pavement layers generally include unbound base and sub-base courses, commonly comprised of fine crushed rock or uncrushed natural gravel [1]. In some European countries, full depth asphalt or concrete pavements are preferred, and the use of granular materials is generally restricted to working platforms [4]. In contrast, Australia commonly uses crushed rock and natural gravel as the base and sub-base layers of flexible airport pavements, with only thin (50–60 mm) asphalt [5] or sprayed seal [6] surfacing. Although these pavement structures are economical to construct, their performance relies on high quality granular materials and high levels of compaction during construction. As a result, proving granular pavement layers is more important in Australia than in jurisdictions that prefer deep or full depth asphalt pavement options [4].

Proving granular fill and pavement layers during airport pavement construction is the process of applying a roller load to the layer, at the time of construction, which will induce stresses throughout the layer that will exceed the stresses expected to be induced by the design aircraft loadings, when applied to the surface of the pavement during service [7]. For granular pavement layers, the stresses are higher near the pavement surface, and therefore larger and heavier proof rollers, with higher tyre pressures, are required for the uppermost granular pavement materials. In contrast, granular fills are located underneath the pavement, so the required level of proving stress is lower, but the thickness of fills is usually much greater than the thickness of granular pavement layers, so the proving stress must penetrate further, requiring higher roller wheel loads.

As aircraft increase in weight, wheel load and tyre pressure, heavier and higher tyre pressure proof rollers are required [8]. Also, some proof rollers have either been lost, become unserviceable or have had their capacity reduced over time [9].

Therefore, the ability to prove granular fills and pavement materials has become more challenging in recent years.

This research defines the challenge associated with proving both granular fills and granular pavement layers during the construction of typical airport pavements. The context is focused on Australian airport pavements, but the findings are applicable to other jurisdictions that also prefer proven granular materials over deep or full depth asphalt pavements, or concrete pavements, for the construction of runways, taxiways and parking aprons. Once the challenge is defined, potential solutions are explored and further work to address this issue is outlined. Demonstrative stress with depth plots, calculated using the pavement thickness design software known as Airport Pavement Structural Design System (APSDS), using the method described by White [7], for various aircraft and rollers are presented to illustrate the issues.

2 Granular Fills and Pavement Layers

Granular fills take many forms. The most common is select fill, which typically has a minimum CBR of 15, placed between the natural subgrade material and the formal pavement structure [3]. Select fill is usually used to correct the shape of the natural subgrade so that a generally uniform pavement structure can be built with a finished pavement surface level that meets the regulated longitudinal and transverse grades, and the rate of change between grades. Select fill is usually constructed and compacted in layers up to 300 mm thick and each layer is generally tested for density and then proof rolled before the next layer is compacted. Consequently, proof rolling select fill is no more challenging than proof rolling other granular pavement layers deeper in the pavement.

In contrast, hydraulically or mechanically placed dredged sand fill, which is commonly used in coastal areas with marine clay subgrades, is usually placed in a thick layer, up to 1.5 m in depth, and then allowed to drain, while consolidating the underlying marine clay over many months or years. The sand is usually single sized, with the majority of particles typically 300–600 μm in size [3]. The main use of sand fills is to protect the soft marine clays from subsequent aircraft loading, to minimise post-construction differential settlement of the fill, and to provide a working platform for the construction of the formal pavement structure. Because the sand fill is usually placed in a single, thick, layer, compaction and subsequent density testing at depth is disruptive and expensive [10]. Therefore, greater reliance is placed on compacting and proving the whole fill thickness, rather than on proving thin layers. Consequently, a much larger roller is required, with the capacity to create significant stresses up to 1.5 m below the surface of the fill.

Regardless of the granular material type, it is important to prove the layer when it is in its most compactable moisture condition. For select fill and granular materials, the optimum moisture content-maximum dry density (OMC-MDD) relationship is well established and compaction and proof rolling is performed when the material is within 1% of OMC [9].

In contrast, single sized sands used in thick fills over marine clay do not follow the OMC-MDD trend that is common in soils, select fill, natural gravel and crushed rock materials. Rather, single sized sands are most compactable when they are either completely dry, or when they are completely saturated [3]. In practice, it is impossible to completely dry a thick sand fill. Therefore, sand fills are compacted and proof rolled in a saturated condition [1]. This requires a bund to be constructed around the area to be compacted and proven, and water to be pumped into the bunded area until the water level is above the top of the sand. The compaction and proving processes are then performed while the water level is continually maintained above the top of the sand surface. Although the sand must be saturated at the time of compaction and proving, it is also important that the water can draw-down under gravity, to prevent a quick-sand effect preventing the sand from densifying.

3 Proof Roller Capacity

It is well established that proving airport pavement granular fills and pavement layers requires significantly larger rollers than those commonly used in road construction, and even those used to compact the granular pavement layers in airport pavements. Consequently, the Australian airport industry has used large, purpose-built rollers for proof rolling during airport pavement construction. The original rollers were imported from the USA [7] but similar rollers were later constructed locally. In recent times, some airport projects have developed bespoke rollers, to address safety issues and operational inefficiencies, as well as aiming to increase the capacity of the rollers to prove granular pavement layers. However, a gap has developed and subsequently increased, between the proof roller capacity and the capacity required [8].

There are two types of proof roller germane to this issue. The first is known as the Macro Roller, while the other is known as the Supercompactor.

The Macro Roller (See Fig. 1) has a loaded mass of up to 50 tonnes, which is distributed on four earthmoving vehicle tyres that were originally inflated to 1400 kPa. The Macro roller was developed for the proving of 150–300 mm thick granular base and sub-base layers, primarily in flexible airport pavements [7]. The 1400 kPa tyre inflation pressure exceeded the official pressure rating of the tyres but was conditionally allowed by the tyre manufacturers subject to partial filling with water, speed limits and loaded standing-time limits [7]. However, around the year 2000, the conditional over-inflation was revoked by the tyre manufacturers, reducing the tyre inflation pressure to a maximum of 1000 kPa [9]. This created a gap between the proof roller capacity and the required capacity, for adequately proving the upper base course layers when the asphalt surface was limited to a thickness of 50–60 mm, even for the aircraft operating at that time [4].

The Supercompactor (See Fig. 2) is a larger roller with up to 180 tonne loaded mass, evenly distributed across four wheels, with a tyre inflation pressure up to 1000 kPa [3]. It was intended specifically for the compaction of up to 1.5 m thick sand

Fig. 1 Example 50 tonne Macro roller

Fig. 2 Example 180 tonne Supercompactor

fill layers, most commonly associated with new airport and runway developments, such as the new runways at Sydney airport, the new Brisbane airport, and the new Hong Kong airport. There were only two Supercompactors used in Australia, and one was provided to Hong Kong, but was never returned. The other was used for development at Perth airport and was then mothballed in Perth.

The Supercompactor is large and in practice, it can only be used on large projects where the area of fill to be compacted and proven is significant. Because smaller developments are common at Australian airports, alternates to the Supercompactor have been considered. The most significant of these alternates was the 15 tonnes vibrating steel drum roller (See Fig. 3), which was trialled at Sydney airport in the 1960s [3]. The trials concluded that the vibrating roller, which was far more maneuverable and convenient in confined areas, was 'probably' adequate for compacting 1.5 m of sand fill, but this conclusion was flawed. During all the trial sections, not one was compacted without the Supercompactor, and therefore the conclusion that the vibrating roller was 'probably' adequate was never conclusive. Despite this, the

Fig. 3 Example 15 tonne vibrating steel drum roller

15 tonne, vibrating, steel drum roller has generally been used since, commonly in combination with the Macro Roller, as a replacement to the Supercompactor.

Figure 4 compares the stress with depth of the 180 tonne Supercompactor with the 50 tonne Macro Roller at 1400 kPa tyre pressure and 1000 kPa tyre pressure. It is clear that reducing the tyre pressure of the roller reduces the stress induced near the top of the material being rolled, while the rate of stress dissipation with depth is similar. Furthermore, the unavailability of the Supercompactor has a significant impact on the capacity to prove thicknesses exceeding 200 mm, where the Macro roller stress is much lower than the Supercompactor stress. Although the reduction

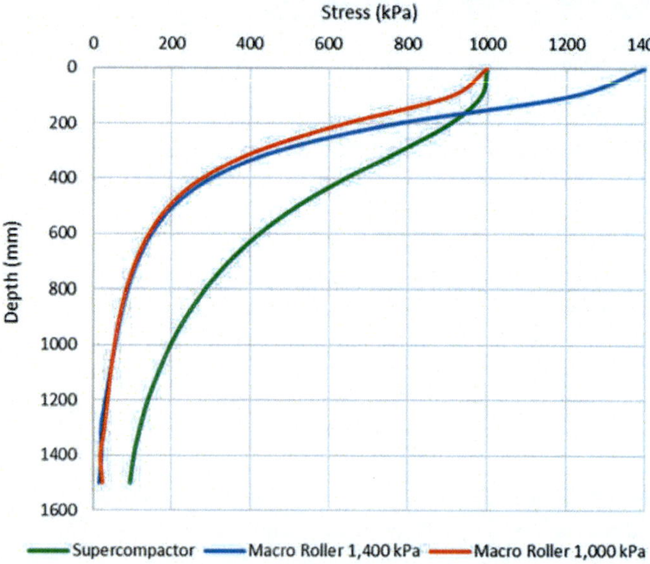

Fig. 4 Comparison of stress with depth for Supercompactor and Macro rollers

in Macro Roller tyre pressure from 1400 to 1000 kPa does not significantly impact the induced stress at depth, the 50 tonne Macro roller was developed to prove thin layers of fine crushed rock, typically 150–300 mm, and in the upper 200 mm, the reduction in the tyre pressure has a significant impact on the capacity of the rollers.

4 Aircraft Developments

While the airport pavement proof rollers have become less available and their capacity has been reduced by the tyre manufacturers, the aircraft using airport pavements have continued to evolve and become more damaging to airport pavements [11]. Since the rollers were developed in the 1950–1960s, aircraft tyre pressures (See Fig. 5) and individual wheel loads (See Fig. 6) have increased significantly. The rollers were designed and introduced in response to the aircraft that were contemporary to airport pavement design in that era, such as the DC8-50 (1959), B727-200 (1960) and B747-100 (1958). However, since that time, the B747-400 (1988), A340-600 (2002), B777-300ER (2004), A380-800 (2006), B787-8 (2011), A350-900 (2015) and A350-1000 (2018) have become the common design aircraft for significant airport pavement infrastructure. These new aircraft have higher tyre pressures and wheel loads and in

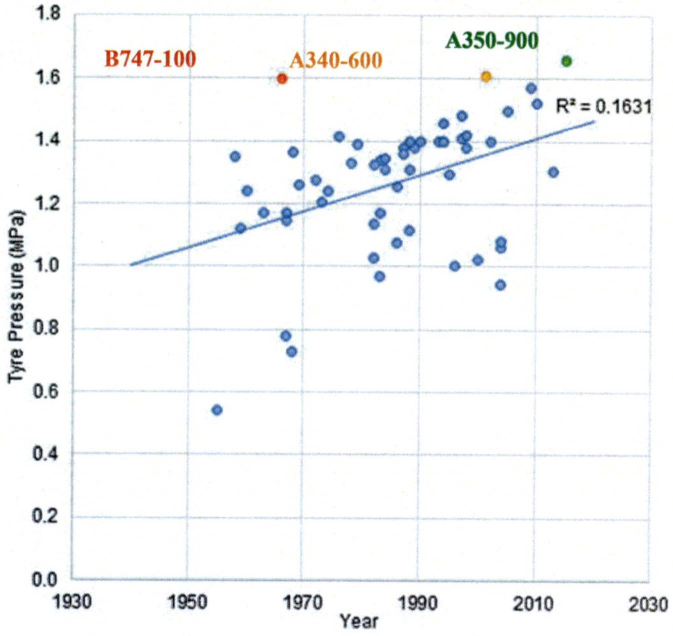

Fig. 5 Comparison of stress with depth for Supercompactor and Macro rollers

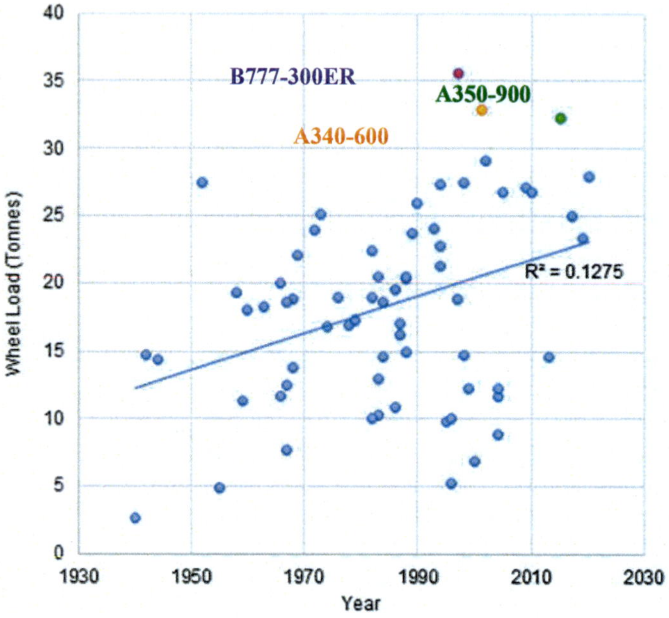

Fig. 6 Comparison of stress with depth for Supercompactor and Macro rollers

some cases, have many more wheels to support the substantially larger aircraft mass (Table 1).

The estimated stress with depth induced by various aircraft through 1000 mm of crushed rock is shown in Fig. 7. Although this is not representative of a realistic pavement, it is clear that the high tyre pressure of the A350-900 is the most demanding near the top of the pavement, while the high wheel load, which is applied to 12 closely spaced wheels, associated with the B777-300ER, is critical at depth.

Table 1 Germane aircraft weights, tyre pressures and number of main wheels

Aircraft	Weight (tonnes)	Tyre pressure (kPa)	Number of wheels
DC8-50	162	1350	8
B727-200	84	1020	4
B747-400	398	1379	16
A340-600	366	1610	12
A380-800	575	1500	20
B777-300ER	352	1525	12
B787-8	228	1572	8
A350-900	269	1660	8
A350-1000	317	1520	12

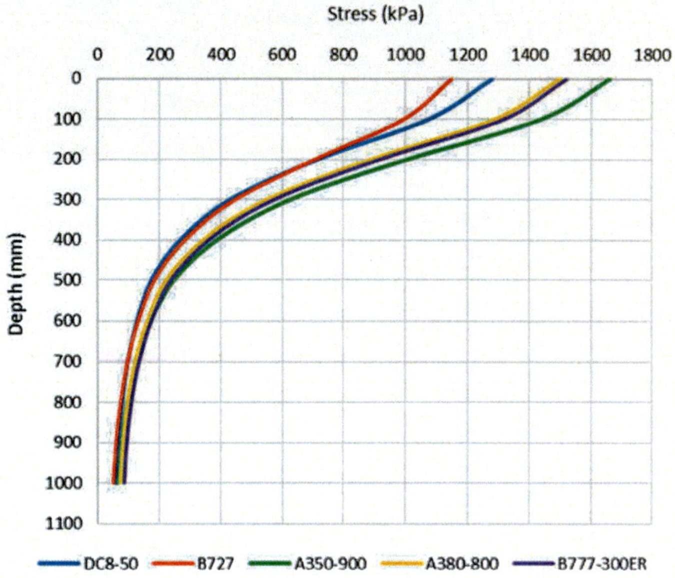

Fig. 7 Comparison of stress with depth for large aircraft

The increase in aircraft wheel load and tyre pressure has a similar effect to the reduction in the roller capacity. As the roller capacity has reduced over time, the demand that aircraft place on granular pavement materials has increased. These two trends have combined to create a significant but unquantified challenge for airport pavements constructed with granular materials.

5 Quantifying the Challenge

To quantify the emerging challenge associated with compacting deep sand fills and granular crushed rock layers, a typical pavement was designed. The stresses induced by the B777-300ER and the A350-900 operating on the surface of the pavement were calculated in APSDS and the levels of stress were compared to those associated with the currently available rollers. The pavement comprised 100 mm of asphalt on 600 mm of fine crushed rock on 1500 mm of sand fill on CBR 1% subgrade. This is typical of a new flexible pavement constructed on sand fill over marine clay, such as the new runway recently constructed at Brisbane airport. The A350-900 induced higher stress near the pavement surface, while the B777-300ER stress exceeded that of the A350-900 for depths greater than 800 mm below the pavement surface (Fig. 8).

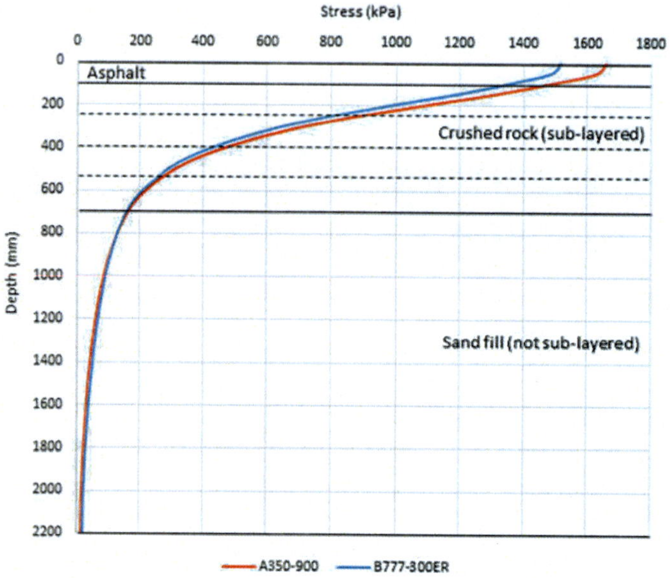

Fig. 8 Comparison of stress with depth for large aircraft

5.1 Sand Fill Compaction

The 50 tonne Macro roller at 1000 kPa induced stress in the sand fill layer that exceeds the B777-300ER and A350-900 stresses down to 2100 mm from the pavement surface (Fig. 9). That is, the use of the Macro roller, even at its reduced tyre pressure, still proves a 1500 mm thick sand fill, indicating there is no significant gap. However, the Supercompactor was used to both compact and prove sand fills, so the reduced ability of the Macro roller, compared to that of the Supercompactor, may still impact on the achieved density of the sand fill at depth. This is because the Supercompactor induced stresses approximately seven times higher than the Macro roller at depth (Fig. 10).

5.2 Base Course Proving

Unlike sand fills, fine crushed rock base course layers do not rely on the Macro roller for compaction. Rather, the Macro roller is primarily used to prove these granular pavement layers after they have been compacted with conventional pneumatic tyred and steel drum vibrating rollers.

When applied to the bottom two sub-layers, the 50 tonne Macro roller at 1000 kPa tyre pressure adequately proves the underlying 300 mm (two layers) of base course (Fig. 11). Furthermore, when applied to the third later (from the bottom) it is also

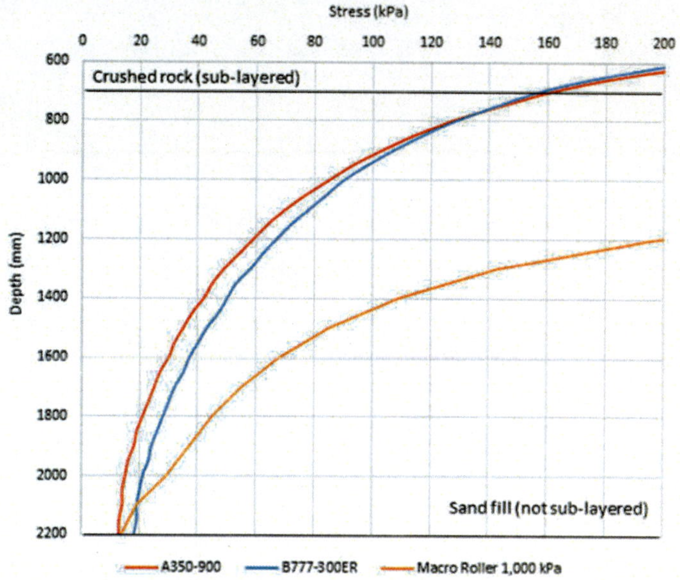

Fig. 9 Comparison of 50 tonne Macro roller and aircraft stress in sand fill

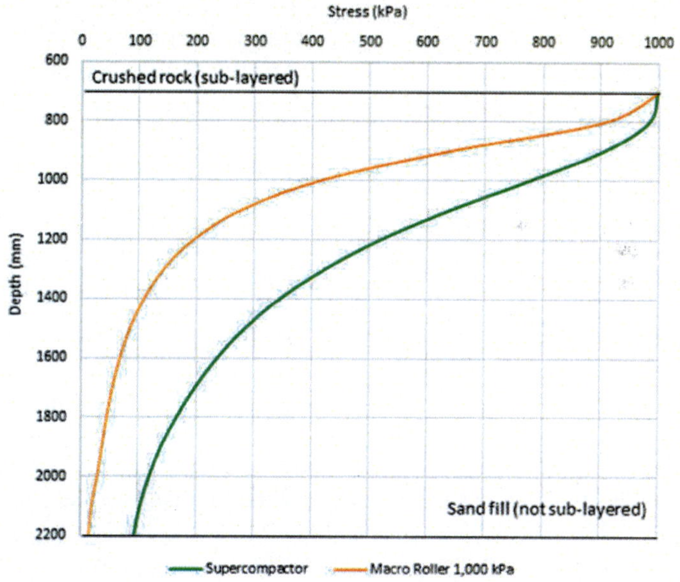

Fig. 10 Comparison of 50 tonne Macro and Supercompactor stress in sand fill

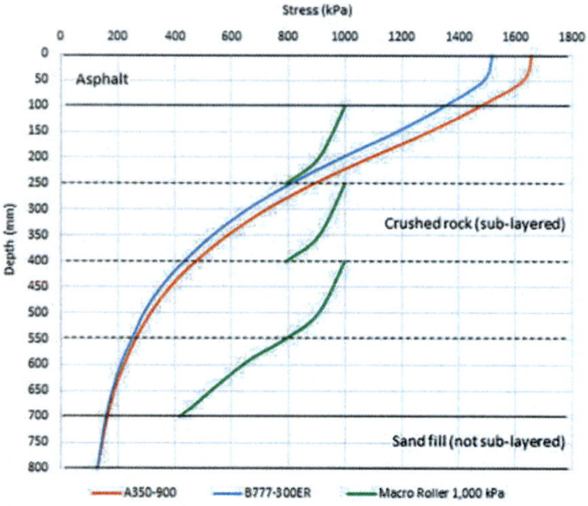

Fig. 11 Comparison of 50 tonne Macro roller and aircraft stress in base course

adequate. However, the uppermost layer of fine crushed rock, placed under 100 mm of asphalt, can not be adequately proven by the 50 tonne Macro roller at 1000 kPa tyre pressure, because the A350-900 induced stress at the bottom of the asphalt is 1477 kPa (Fig. 11). Even the B777-300ER induced stress is 1351 kPa, which also significantly exceeds the 1000 kPa capacity of the Macro roller.

6 Potential Solutions

6.1 Sand Fill Compaction

The reliance on the Supercompactor to compact the sand, as well as prove it, is one of the reasons the 15 tonne vibrating steel drum roller was used in combination with the Macro roller, in areas where the Supercompactor was not practical. However, the analysis performed in APSDS can not take into account the additional densification of the sand that results from the vibrating nature of the steel drum roller, and therefore the effect of any roller that uses impact or vibration energy, requires verification by field trials or the development of new theoretical analysis methods that allow reliable comparison or the relative effect of static (non-vibrating) and dynamic (vibrating) roller forces.

Field trials would largely replicate the sand compaction trials from the 1960s [3] but would focus on rollers other than the Supercompactor. Because measuring sand density at depths exceeding 600 mm is challenging, a gauge-based approach to measuring relative sand density should be considered. However, given the relatively small size of the gap and the challenge associated with its quantification, a theoretical approach is preferred, as long as an appropriate method of analysis can be identified. This could be energy based, or discrete element based, and might include shakedown theory.

6.2 *Base Course Proving*

The fine crushed rock challenge is simpler to quantify than the sand fill compaction challenge, because all the proof rollers are static in nature. However, the gap associated with the upper base course under relatively thin asphalt surfacing is significant. This can be overcome by increasing the stress induced at the surface of the layer being proven, by increasing the roller tyre pressure, because increasing the roller wheel load will not increase the stress immediately under the tyre. Fitting second-grade or second-hand aircraft tyres to the Macro roller is the most likely solution. Aircraft tyres are routinely inflated to 1700 kPa tyre pressure, which would be adequate to address the current challenge.

The only other option is to reduce the stress at the top of the granular layer by either reducing the aircraft tyre pressure of increasing the thickness of the asphalt surface course. Changing large commercial aircraft characteristics is not viable. However, increasing the thickness of the asphalt surface, to approximately 250 mm, would reduce the stress at the top of the fine crushed rock to less than 1000 kPa. This is the approach taken in the USA, where 250 mm is a typical total asphalt thickness for airport pavements designed for large aircraft [4]. However, the additional cost is significant and must be compared to the cost of increasing the capacity of the roller.

7 Conclusion

It is concluded that the historical gap between the capability of the heavy rollers for Australian airport pavement construction, and the aircraft that use the pavement, has increased over time. This is the result of a combination of aircraft with higher wheel loads and tyre pressures, and the reduction in the capability of the current rollers. The gap is relatively minor for the compaction of deep sand fills, but the inability to theoretically compare the effect of static rollers, to that of vibrating and impact rollers, makes this difficult to quantify. The alternate is an expensive field trial to demonstrate acceptable sand density at depths to 1500 mm. In contrast, a significant gap exists for the proving of fine crushed rock layers used in the upper base course under thin (60–100 mm) asphalt surfaces. This challenge requires rollers with a higher tyre pressure, or the use of significantly thicker asphalt surface courses, both of which are expensive.

References

1. Australian Airports Association (2017) Airport practice note 12—airfield pavement essentials. Australia
2. White G (2018) Regional airport pavement challenges and innovations. In: Institute of Public Works Australia—Queensland National Conference. Gold Coast, Queensland, Australia
3. Rodway B (1976) Compaction of sand in thickness layers. In: ARRB conference proceedings, session 13, vol 8
4. White G, Kelly G, Fairweather H, Jamshidi A (2020) Theoretical socio-enviro-financial cost analysis of equivalent flexible aircraft pavement structures. In: 99th Annual meeting of the transportation research board. Washington, District of Columbia, USA, 12–16 January 2020
5. White G (2018) State of the art: asphalt for airport pavement surfacing. Int J Pavement Res Technol 11(1):77–98
6. White G (2019) A standardised sprayed sealing specification for Australian airports. In: 12th Conference on asphalt pavements for Southern Africa. Sun City, South Africa, 13–16 October 2019, pp 806–821
7. White G (2005) Design of proof rolling regimes for heavy duty aircraft pavements. In: 7th International conference on the bearing capacity of roads, railways and airfields. Trondheim, Norway, 27–29 June 2005
8. White G (2016) Challenges for Australian flexible airport pavements. Aust Geomech 51(3):39–46
9. White G (2008) Developments in proof rolling of granular layers in aircraft pavements in Australia. In: 6th International conference on road and airfield pavement technology. Sapporo, Japan, 20–23 July 2008
10. Jamieson S, White G (2021) Defining Australian rigid aircraft pavement design and detailing practice. In: International airfield and highway pavement conference, 8–10 June 2021, a virtual event
11. White G (2017) Managing the impact of more demanding aircraft. In: 10th International conference on the bearing capacity of roads, railways and airfields. Athens, Greece, 28–30 June 2017

Development of Statistical Deterioration Model for Low Volume Roads in Indian Scenario

B. G. Vijay and Raghavendra S. Sanganaikar

Abstract Pavement serviceability is the concept representing the level of service that pavement structure offers to the road users. International Roughness Index (IRI) and Present Serviceability index (PSI) are the parameters which are used to represent the performance of pavement. Road roughness is considered to be the most important parameter affecting vehicle operating costs which account for a large proportion of total transport costs. Roughness is concerned with vehicle vibration, operating speed and wear & tear of the wheels. It affects the road user cost to a significant extent. Therefore, roughness of the pavement plays the decisive role in exercising the option of implementing the optimum maintenance and rehabilitation strategies of the road network at appropriate time. Four study stretches, which include conventional and treated, each of 400 m long were selected. The pavement distresses are collected, processed and analysed for developing a deterioration model. Pavement distress viz cracking, rutting, ravelling, potholes, patching were measured. Rating (both visual and ride rating) studies were also carried. An attempt is made to develop International Roughness Index equation for the selected study stretches. A model is developed to correlate International Roughness Index (IRI) with different types of pavement distresses.

Keywords Pavement distress · Multiple linear regression model · IRI · PSI

1 Introduction

Efficient road transportation plays a vital role in the economy of any nation. Road transport in India, occupies a dominant position in the overall transportation system of the country due to its advantages like easy availability, flexibility of operation, door to door service and reliability. India owns the second largest network of roads in the world, next to USA. Out of the total road network of India, village and other

B. G. Vijay (✉)
KLE Dr. M. S. Sheshgiri College of Engineering and Technology, Belagavi, India

B. G. Vijay · R. S. Sanganaikar
Vidyavardhaka College of Engineering, Mysore, India

© The Author(s), under exclusive license to Springer Nature Switzerland AG 2022

H. R. Pasindu et al. (eds.), *Road and Airfield Pavement Technology*, Lecture Notes in Civil Engineering 193, https://doi.org/10.1007/978-3-030-87379-0_8

roads (Low Traffic Volume Roads) consist of 80% share. Low traffic volume roads are mainly rural roads in India carrying daily traffic less than 450 Commercial Vehicles per Day (CVPD) The thickness of sub-base layer is around 200 mm and the total thickness of each layer of base course varies from 100 to 120 mm. Surface course is usually 20 mm thick premix carpet layer. Low volume roads serve as one of the key infrastructure work, placed for integrated rural development, which has become a matter of growing urgency for considerations of social justice, national integration and economic uplift of the rural areas. The importance of preserving road network in good condition is widely recognized and therefore, performance evaluation of the existing roads is an absolute necessity. Performance of flexible pavements has long been recognized as an important parameter in their design and maintenance. In order to measure and prepare model for pavement performance, it is necessary to clearly define the pavement performance. Deterioration of pavement can be attributed to various factors like age, traffic, environment, material properties, pavement thickness, strength of pavement as well as subgrade properties which affect the mechanical characteristics of a pavement. These factors affect the performance of the pavement in a complex manner. To understand the mechanism and to forecast the future condition of pavement, these deterioration models are necessary. Pavement deterioration model is a mathematical relationship between the pavement condition and the factors listed above. The pavement deterioration model predicts the future condition of the pavement, which is helpful in development of Maintenance Management Model or Maintenance Priority Index (MPI). This index is a rating used to prioritize the maintenance schedule of pavement based on the severity of distresses and its condition.

2 Literature Review

Hernán de Solminihac et al. [1] studied on relating serviceability results obtained by a 9-member evaluation panel, representing the general public as closely as possible, to parameters (particularly of roughness) measured with instruments on road sections of asphalt concrete, Portland cement concrete, and asphalt overlay, respectively. Results show that prediction of serviceability is quite accurate based on roughness evaluation, while also revealing that, by comparison to studies in more developed countries, Chileans are seemingly more tolerant, in that they assign a somewhat higher rating to ride quality. Jorge et al. [2] have developed an incremental Nonlinear Model for predicting Serviceability. The objective of this research was the development of sound flexible pavement performance models to be used primarily for the management of the road infrastructure because accurate prediction of pavement performance is important for effective management of the infrastructure. The model highlights some of the advantages of relaxing the linear restriction that is usually placed on the specification form of pavement performance models. Bektas et al. [3] developed a pavement condition rating system that provides a consistent, unified approach in rating pavements in Iowa is being proposed. The proposed 100-scale system is based on five individual indices such as cracking, ride, rutting, faulting, and friction. The researchers compared PCI-2 results to PCI results and found that, in general, PCI-2 which is established by combining individual indices with weight factors offers fairly good correlation to PCI condition results, particularly, for the pavement types where PCI utilizes distress and roughness data. The poorly related ones are due to the fact that some of the current PCI is heavily characterized by pavement age with various other data, such as material property and traffic and is characterized less than PCI-2 by the pavement distress and roughness data. Bin Ab. Latif et al. [4] worked on developing relationship between International Roughness Index (IRI) and Present Serviceability Index (PSI). IRI was measured by using the walking profilometer. PSI data was collected manually. Both IRI and PSI were measured along the 100 m section of road. This study only focused on the asphaltic pavement. The objectives of this study were to determine the IRI, PSI and relationship between IRI and PSI. The statistical analysis which is R^2 value was used to evaluate the relationship between IRI and PSI. From this study, it was found that IRI value increased when PSI value decreased. While the PSI rating shows the tested road for this study is still in good condition.

3 Objectives of the Study

- Collection and processing of pavement deterioration data for analysing.
- Analysis and development of suitable pavement deterioration model.
- Evaluation of effect of various pavement distresses on IRI.

4 Data Collection

Data collection is done for analysis from all the considered road stretches. The structural evaluation and functional evaluation data is collected and is processed for further studies. The data includes values for 3 cycles.

4.1 Test Stretches for Field Study

For the study, totally 4 road stretches each of length 400 m with different treatment given. The following criteria were used for the selection of the test stretches.

- The test stretches were straight without horizontal curves and steep gradient.
- The test stretches have fairly uniform riding quality and surface condition throughout the length of stretch.

Table 1 Selected test stretches for the present study

Stretch	Name of test stretches	Type of construction
Section 1	Ganjalaghatta road	Conventional
Section 2	Ganjalaghatta road	Conventional
Section 3	Rayashettyhalli road	Coir treated
Section 4	Rayashettyhalli road	Coir treated

- Cross drainage works, over bridge are to be avoided within the selected test stretches.

The test stretches selected in the present study are shown in Table 1.

4.2 Structural Condition Evaluation

Tables 2, 3, 4, 5, 6 and 7 give the details of Structural condition evaluation for the stretches mentioned above.

Table 2 Structural condition evaluation of Ganjalaghatta road for cycle-I

Sl. No.	Stretch	Rutting measured in the middle of the section (mm)		Cracking of bituminous layer severity and area (%)	Pot holes (number)	Ravelling area (%)	Patching (% of area of subsection)
		LWP	RWP				
1	Section 1	18	25	1	0	0	0
2	Section 2	20	13	1	1	0	0
3	Section 3	31	16	1	0	1	0
4	Section 4	23	13	3	0	1	0

Table 3 Structural condition evaluation of Ganjalaghatta road for cycle-II

Sl. No.	Stretch	Rutting measured in the middle of the section (mm)		Cracking of bituminous layer severity and area (%)	Pot holes (number)	Ravelling area (%)	Patching (% of area of subsection)
		LWP	RWP				
1	Section 1	24	34	10	0	10	0
2	Section 2	39	30	13	5	27	9
3	Section 3	38	28	18	0	13	0
4	Section 4	40	25	16	0	22	0

Table 4 Structural condition evaluation of Ganjalaghatta road for cycle-III

Sl. No.	Stretch	Rutting measured in the middle of the section (mm)		Cracking of bituminous layer severity and area (%)	Pot holes (number)	Ravelling area (%)	Patching (% of area of subsection)
		LWP	RWP				
1	Section 1	64	62	30	2	43	20
2	Section 2	77	39	30	11	46	19
3	Section 3	86	42	46	10	39	0
4	Section 4	55	38	35	0	39	0

Table 5 Structural condition evaluation of Rayashettyhalli road for cycle-I

Sl. No.	Stretch	Rutting measured in the middle of the section (mm)		Cracking of bituminous layer severity and area (%)	Pot holes (number)	Ravelling area (%)	Patching (% of area of subsection)
		LWP	RWP				
1	Section 1	14	15	1	0	1	0
2	Section 2	13	5	0	0	3.5	0
3	Section 3	18	0	2	0	0	0

Table 6 Structural condition evaluation of Rayashettyhalli road for cycle-II

Sl. No.	Stretch	Rutting measured in the middle of the section (mm)		Cracking of bituminous layer severity and area (%)	Pot holes (number)	Ravelling area (%)	Patching (% of area of subsection)
		LWP	RWP				
1	Section 1	0	42	0	0	0	0
2	Section 2	0	47	0	3	0	0
3	Section 3	20	49	58	0	0	0

Table 7 Structural condition evaluation of Rayashettyhalli road for cycle-III

Sl. No.	Stretch	Rutting measured in the middle of the section (mm)		Cracking of Bituminous layer Severity and area (%)	Pot Holes (number)	Ravelling Area (%)	Patching (% of area of subsection)
		LWP	RWP				
1	Section 1	9.9	17.0	0.0	1.4	73.1	0.0
2	Section 2	23.0	32.0	0.0	3.0	65.0	0.0
3	Section 3	63.2	82.1	6.0	2.2	99.1	30.0

4.3 *Benkelman Beam Rebound Deflection Studies*

Performance of flexible pavements is closely related to the elastic deflection of pavement under the wheel loads. The deformation or elastic deflection under a given load depends upon subgrade soil type, its moisture content and compaction, the thickness and quality of the pavement courses, drainage conditions, pavement surface temperature etc.

The deflection survey essentially consists of two operations:

(i) Condition survey for collecting the basic information of the road structure and based on this, the demarcation of the road into sections of more or less equal performance; and
(ii) Actual deflection measurements.

Deflection measurements are conducted as per IRC 081-1997 [5].

4.4 *Functional Condition Evaluation*

4.4.1 Pavement Condition Rating

A rating scale with 0 to 5 points, as adopted on CGRA studies was selected in this study. The raters were given adequate training regarding assessment of the riding quality and deciding the rating value in a 5 point rating scale.

Two types of ratings were conducted namely, Visual Rating and Riding Rating.

Visual Rating Technique

The members of the rating panel were trained to walk along the left and right wheel paths on the selected stretches and condition of the pavements was assessed based on the visual judgment of surface characteristics (Table 8).

Ride Rating Technique

For the rating by riding technique the raters were taken in test vehicle driven along the stretches at a speed of 30 kmph and are trained to assess the PSR value according to comfort condition (Table 9).

The Functional Condition Evaluation details are as given in Tables 10 and 11.

Table 8 Description of visual rating scale

Sl. No.	Description based on visual rating	Numerical scale
1	Perfectly even surface Without undulation, Cracking Patching	4–5
2	Slightly uneven Surface with some undulation, no pot holes and slight cracking	3–4
3	Moderately uneven surface, Visible patching and medium cracking	2–3
4	Uneven surface with improper patched, potholes medium to heavy cracking	1–2
5	Uneven surface with different type undulation, badly patched potholes, heavy cracking	0–1

Table 9 Description of ride rating scale

Sl. No.	Description based on ride rating	Numerical scale
1	Without discomfort perfect smoothness	4–5
2	Little distortion, fairly smooth riding	3–4
3	Medium distortion fair to uneven riding	2–3
4	Heavy distortion uncomfortable riding	1–2
5	Intolerable, very discomfort in riding	0–1

Table 10 Functional condition evaluation for Ganjalaghatta road

Ganjalaghatta						
Stretch	PSR values					
	Cycle I		Cycle II		Cycle III	
	VR	RR	VR	RR	VR	RR
Section 1	4.6	4.6	4.0	4.2	3.5	3.9
Section 2	4.3	4.7	3.9	4.1	3.5	3.8
Section 3	4.6	4.8	4.0	4.0	3.6	3.6
Section 4	4.7	4.8	4.1	4.2	3.4	3.6

Table 11 Functional condition evaluation for Rayashettyhalli road

Rayashettyhalli						
Stretch	PSR Values					
	Cycle I		Cycle II		Cycle III	
	VR	RR	VR	RR	VR	RR
Section 1	4.5	4.3	4.0	4.0	3.5	3.7
Section 2	4.6	4.6	3.8	4.0	3.4	3.6
Section 3	4.6	4.5	3.8	3.6	3.5	3.4

5 Analysis of Data

5.1 Variation of Individual Distresses for Different Cycles

The variation of individual distresses for all the four sections of Ganjalaghatta Road stretch is tabulated as under (Table 12; Fig. 1, Table 13; Fig. 2, Table 14; Fig. 3, Table 15; Fig. 4, Table 16; Fig. 5).

Table 12 Variation of rutting for different cycles

Cycle	Sub-section No.	Section number	Rutting measured in the middle of the section (mm)	
1	1	Section 1	18	25
2	1		24	34
3	1		64	62
1	2	Section 2	20	13
2	2		39	30
3	2		77	39
1	3	Section 3	31	16
2	3		38	28
3	3		86	42
1	4	Section 4	23	13
2	4		40	25
3	4		55	38

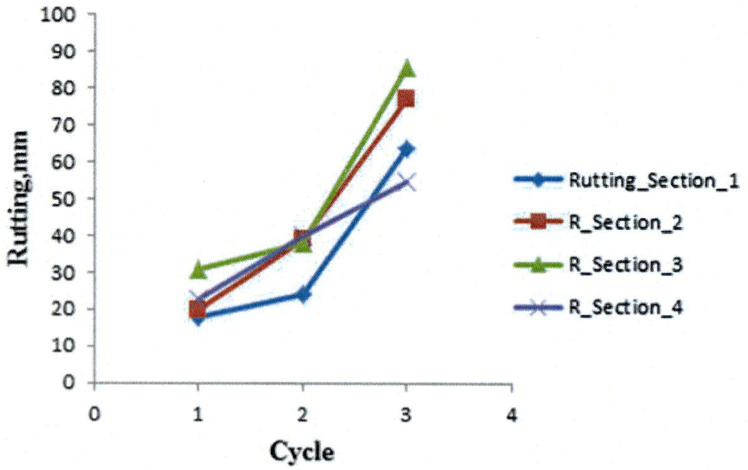

Fig. 1 Variation of rutting for different cycles

Table 13 Variation of cracking for different cycles

Cycle	Sub-section No.	Chainage of sub-section (50 m length)	Cracking of bituminous layer severity and area (%)
1	1	Section 1	18
2	1		24
3	1		64
1	2	Section 2	0
2	2		13
3	2		30
1	3	Section 3	1
2	3		18
3	3		46
1	4	Section 4	3
2	4		16
3	4		35

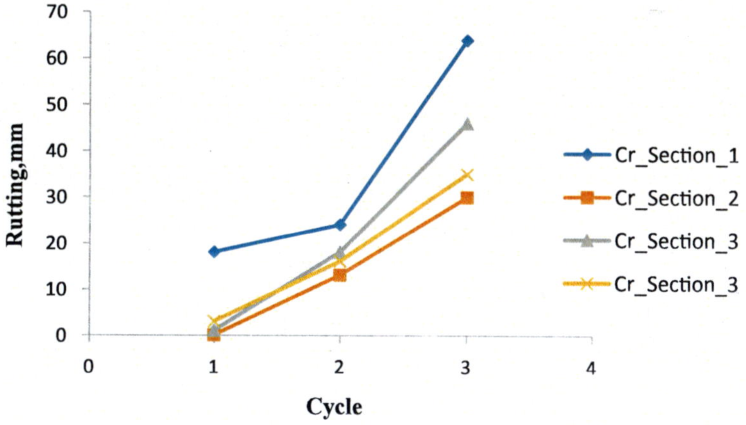

Fig. 2 Variation of Cracking for different cycles

5.2 *Effect of Various Distresses on IRI for Ganjalaghatta Road Section*

See Table 17; Figs. 6, 7, 8 and 9.

Table 14 Effect of number of pot holes on different cycles

Cycle	Sub-section No.	Chainage of sub-section (50 m length)	Pot holes (number)
1	1	Section 1	0
2	1		0
3	1		2
1	2	Section 2	1
2	2		5
3	2		11
1	3	Section 3	0
2	3		0
3	3		10
1	4	Section 4	0
2	4		0
3	4		0

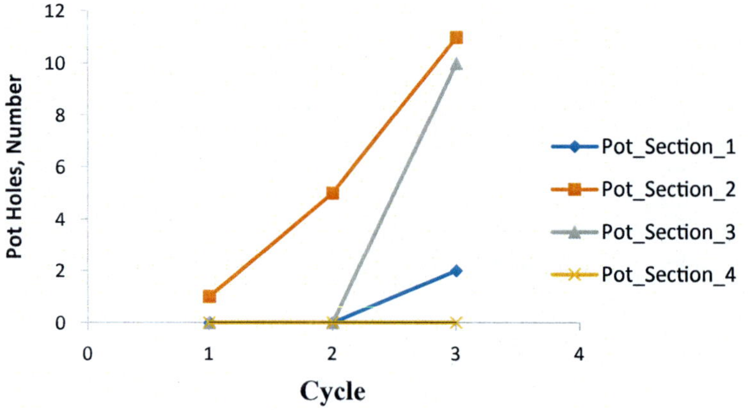

Fig. 3 Effect of number of pot holes on different cycles

5.3 *Multiple Linear Regression Model for IRI*

The model developed for IRI is Eq. (1) and their respective model summary and Anova given in Tables 18 and 19.

$$IRI = 4.095\sqrt{CRK} - 0.465RAV + 0.322PAT \tag{1}$$

CRK = Cracking
RAV = Ravelling
PAT = Patching
IRI = International Roughness Index.

Table 15 Effect of ravelling on different cycles

Cycle	Sub-section No.	Chainage of sub-section (50 m length)	Ravelling area (%)
1	1	Section 1	0
2	1		10
3	1		95
1	2	Section 2	0
2	2		27
3	2		90
1	3	Section 3	1
2	3		13
3	3		70
1	4	Section 4	1
2	4		22
3	4		39

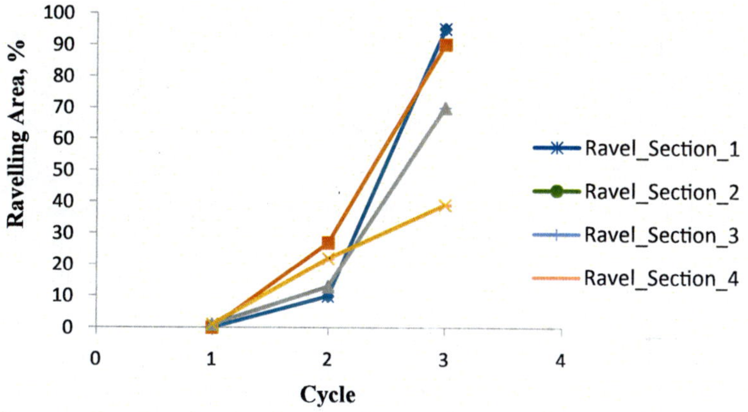

Fig. 4 Effect of ravelling on different cycles

6 Discussions and Conclusions

6.1 Discussions

(i) Quantification of Distresses

a. Rutting has varied linearly between cycles 1, 2, 3 and in case of Section 4, whereas Rutting, between cycles 1 and 2 has varied a little and variation is higher between cycle 2 and 3 for Sections 1,2 and 3.

b. Cracking has a linear variation for Sections 2, 3 and 4; variation of cracking is little between cycle 1 and 2 and higher between cycles 2 and 3.

Table 16 Effect of patching on different cycles

Cycle	Sub-section No.	Chainage of sub-section (50 m length)	Patching (% of area of subsection)
1	1	Section 1	0
2	1		0
3	1		20
1	2	Section 2	0
2	2		9
3	2		19
1	3	Section 3	0
2	3		0
3	3		0
1	4	Section 4	0
2	4		0
3	4		0

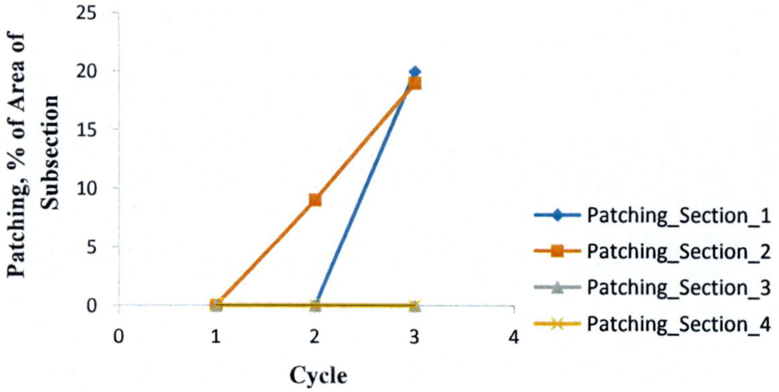

Fig. 5 Effect of patching on different cycles

c. Number of potholes has increased linearly for Section 2 and no distresses are found for all the cycles in case of Section 4. For Sections 1 and 3, potholes have appeared only after cycle 2.

d. In case of Section 4, variation in ravelling almost linear over 3 cycles, and for Sections 1, 2 and 3 variation is a little till cycle 2 and higher variation is observed between cycle 2 and 3.

e. For Section 1, no patching is observed till cycle 2 and patching is observed after cycle 2. There is a linear variation observed for Section 2. No patching is observed for Sections 3 and 4 for all cycles.

Table 17 Data compiled from three cycle reports of TDP

Cycle	Sub-section No.	Cracking of bituminous layer severity and area (%)	Pot holes (number)	Ravelling area (%)	Patching (% of area of subsection)	IRI
1	1	1	0	0	0	7.124
1	2	1	1	0	0	7.265
1	3	1	0	1	0	7.304
1	4	3	0	1	0	7.47
2	1	10	0	10	0	7.23
2	2	13	5	27	9	7.24
2	3	18	0	13	0	7.74
2	4	16	0	22	0	7.71
3	1	30	2	43	20	7.54
3	2	30	11	46	19	7.61
3	3	46	10	39	0	7.95
3	4	35	0	39	0	7.83

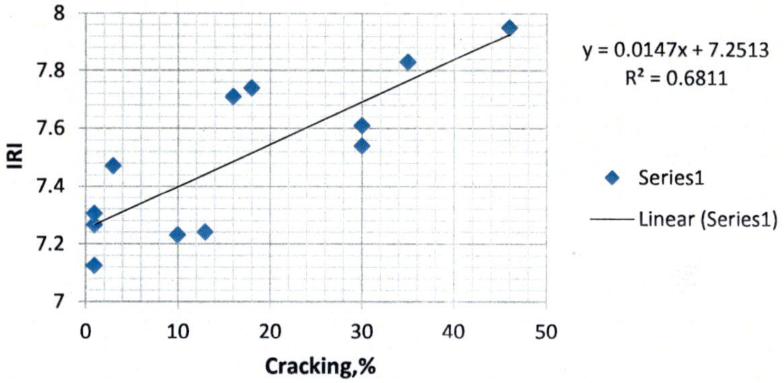

Fig. 6 Variation of IRI with respect to cracking

(ii) Effect of Distress on IRI

Model has been developed to assess the effect of Distresses on IRI. The model developed is as given below:

$$IRI = 4.095\sqrt{CRK} - 0.465RAV + 0.322PAT$$

With R^2 value of 0.907.

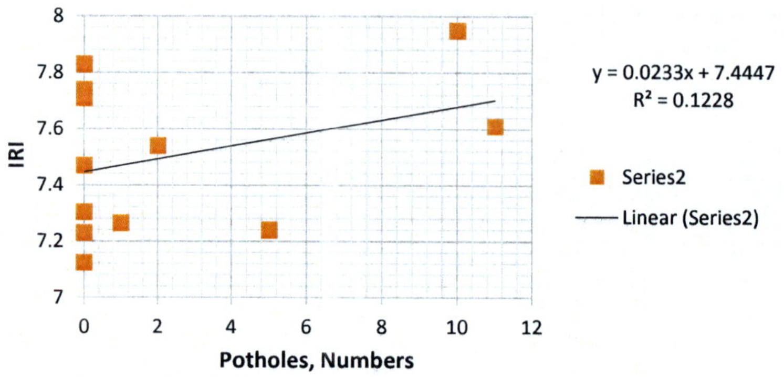

Fig. 7 Variation of IRI with respect to pot holes

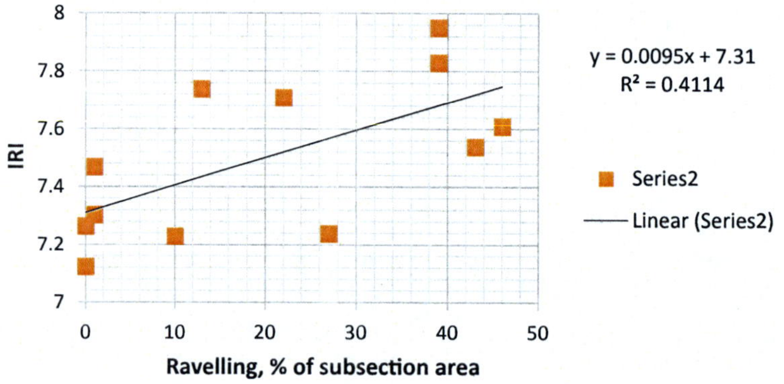

Fig. 8 Variation of IRI with respect to ravelling

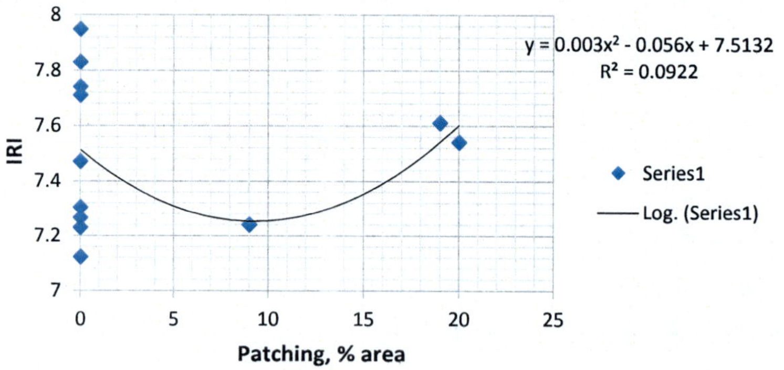

Fig. 9 Variation of IRI with respect to patching

Table 18 Model summary

Regression Statistics					
Model	Multiple R	R Square	Adjusted R Square	Standard error	Observations
10	0.952	0.907	0.776	2.63	12

Table 19 ANOVA

	Df	Sum of Square	Mean Square	F	Significance F
Regression	3	613.71	204.57	29.56	0.000112
Residual	9	62.29	6.92		
Total	12	676.00			

(iii) The observations were made, such as repairs and patching shall have high effect on IRI. The model is to be validated.

6.2 *Conclusions*

(i) Compared to the previous cycles of testing, the distress values have increased. This is due to lack of proper maintenance.
(ii) The data collected in previous 3 cycles show abnormality. The significance of the data variation is to be established. Hence further data collection shall be made.
(iii) The model shows high dependency on the cracking parameter whereas lowest dependency on patching.
(iv) The statistical approach used in the present study does not give satisfactory results in terms of statistical modulus.

References

1. Hernán de Solminihac T et al (2003) Analysis of pavement serviceability for the aashto design method: the chilean case
2. Prozzi JA et al A non linear model for predicting serviceability
3. Bektas F et al (2014) Pavement management performance modelling: evaluating the existing PCI equations
4. Bin Ab. Latif A Relationship between International Roughness Index (IRI) and Present Serviceability Index (PSI)
5. IRC: 81-1997 Guidelines for strengthening of flexible road pavements using Benkelman beam deflection technique
6. Rejani VU et al (2021) Upgradation of pavement deterioration models for urban roads by non-hierarchical clustering. Int J Pavement Res Technol 243–251

7. Svenson K et al (2017) Detecting road pavement deterioration with finite mixture models. Int J Pavement Eng 20(4)
8. Jain SS et al (2005) HDM4 pavement deterioration models for Indian National Highway Network. J Transp Eng 131(8)
9. Keith Kay R et al (1992) Pavement surface condition rating manual
10. North Carolina Department of Transportation Pavement Management Unit (2007) NCDOT digital imagery distress evaluation handbook
11. IRC: SP: 16-2004 Guidelines for the surface evenness of highway pavements
12. IRC: 115-2014 Guidelines for structural evaluation and strengthening of flexible road pavements using falling weight deflectometer (fwd) technique
13. Nakamura VF (1962) Serviceability ratings of highway pavements
14. KaanOzbay et al (2001) Models for pavement deterioration using LTPP

Evaluation of the Structural Condition of Composite Pavement

Endang Budiman, Bagus Hario Setiadji, and Sri Prabandiyani Retno Wardani

Abstract One of the road tolls at the borderline of Jakarta, the capital of Indonesia, i.e., Tangerang-Merak toll road, was found in a damaged condition in several locations. The toll's pavement structure comprises an asphalt layer (blacktopping) over a concrete slab (or a composite structure). An evaluation is required to find out the last condition of the pavement to propose the remedy for the problem. An eight-step evaluation was conducted systematically in this study, and the 1993 AASHTO method was the method used to analyze the results. The first step was to conduct traffic analysis by calculating ADT from toll gate transaction data validated with CCTV data, while vehicle axle loads were measured by static load surveys to calculate the vehicle equivalent factor. Another analysis of the pavement's structural condition was conducted using backcalculation analysis using FWD deflection data to produce the moduli of subgrade reaction and the elastic modulus of the concrete pavement. The analysis showed that several locations of the toll road suffered a medium to severe deterioration as indicated by high deflection values. This fact was also supported by the findings that the pavement system's remaining life is 22% on average. Two recommendations were made by implementing 7.12 in. or 18 cm overlay thickness on the in-service pavement for road segments with good prediction of moduli, and 2 in. or 5 cm repairing treatment to strengthen the existing pavement may be considered as planned stage of construction for the rest. Before the overlay and treatments were conducted, various repairs were taken place on the damaged spots along the road segment.

Keywords AASHTO 1993 · Evaluation of structural conditions · Composite pavement structure · Deflection

E. Budiman · B. H. Setiadji (✉) · S. P. R. Wardani
Universitas Diponegoro, Jl. Prof. Soedarto, SH, Semarang 50275, Indonesia
e-mail: bhsetiadji@ft.undip.ac.id

© The Author(s), under exclusive license to Springer Nature Switzerland AG 2022

H. R. Pasindu et al. (eds.), *Road and Airfield Pavement Technology*, Lecture Notes in Civil Engineering 193, https://doi.org/10.1007/978-3-030-87379-0_9

1 Introduction

A toll road is a national-level road that requires strict technical requirements and has to meet the aspects of safety, security, comfort, and smoothness of traffic. The toll road is planned to be able to serve long-distance traffic flows with high mobility. According to the Toll Road Regulatory Agency (BPJT), the aim of toll road construction is to smoothen traffic in developed areas and improve goods and services distribution services to support economic growth and alleviate the burden on Government funds through the participation of road users. Therefore, a toll road must have a solid pavement structure and minimal damage during its service life. A toll road maintenance program is one program that should be carried out periodically to ensure the pavement is always at its high performance.

One of the toll roads with the highest traffic load at the borderline of Jakarta, the capital of Indonesia, is Tangerang-Merak toll road. At present, the surface of pavement of this toll road, a composite pavement structure or asphalt concrete (AC) over jointed plain concrete pavement (JPCP), is in damaged condition due to high traffic loads, so it can affect vehicles' travel time and increase the number of accidents. Evaluating the existing pavement condition is the first and critical process to develop a road pavement rehabilitation plan. A comprehensive evaluation of pavement conditions will provide critical input into pavement rehabilitation design. Therefore, research is needed to evaluate the pavement's structural conditions, and it was expected to end up with proposed possible solutions to the problems encountered to enable the road serving traffic loads up to the planned service life. Especially for toll roads with the rigid pavement that have been overlaid with asphalt pavement (blacktopping) or AC/JPCP, there were not many studies that explain how to evaluate this pavement's structural conditions [1].

2 Structural Condition of Road Pavement

According to AASHTO [2], the evaluation of the structural and functional condition of road pavement is the information about the performance of pavement structures against traffic loads and environmental conditions. The pavement's structural condition will experience a decrease in performance due to the repetition of traffic loads, inadequate drainage systems, high rain intensity, changes in temperature, unstable subgrade conditions, and poor performance of maintenance works.

An overlay is designed to be applied over the entire surface of the existing pavement to increase the structural or functional capacity significantly. An overlay is used when restoration techniques are no longer adequate or cost-effective but necessary before reconstruction is required.

AASHTO [2] illustrates reduced service levels and structural capacity and the contribution of overlay to return the service level to approaching the previous level (Fig. 1). When newly built, the pavement structural capacity is denoted as SC_0 (or

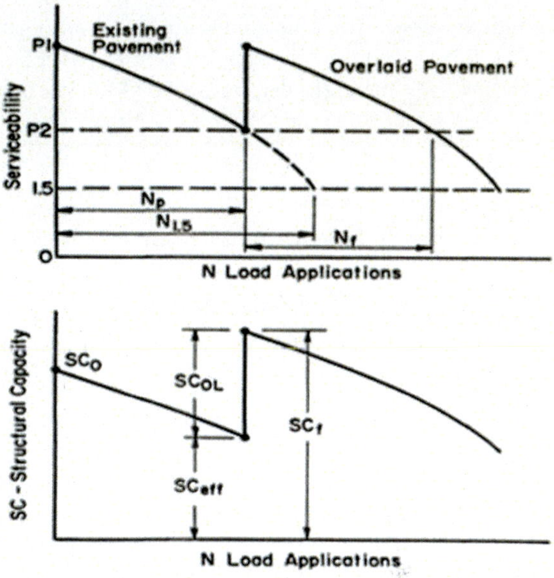

Fig. 1 Illustration of the structural capacity of road pavement and the contribution of overlay to improve the reduced service level of existing pavement [2]

initial service capacity for flexible pavement) or D (or initial slab thickness for rigid pavement) with service level P_1. The structural capacity can be expressed as the equivalent slab thickness (D) as for composite pavements. Over time and loading (N_p), the service level will decrease to P_2, and structural capacity is reduced to SC_{eff} (or D_{eff}). The overlay is carried out to increase the structural capacity (SC_{ol}) to withstand the predicted traffic load during the design period with the required structural capacity (SC_f).

Several measurements or surveys are needed to evaluate in-service pavement's structural capacity, such as deflection measurement [3, 4] and road condition survey. A backcalculation procedure is carried out to determine the layer moduli [5], which, together with the service load prediction (N_f), is used to calculate the required slab thickness (D_f).

Besides, a road condition survey is performed to check whether there are deteriorations in the existing pavement. The presence of damages could reduce the existing structural capacity (D_{eff}). The difference between D_f and D_{eff} is the additional capacity in terms of overlay to maintain or increase the structural capacity of the composite structure.

3 Results and Discussions

The structural evaluation procedure in this study was conducted on a composite pavement structure, i.e., AC overlay over the AC/JPCP layers, which can be described in 8 steps: (a) existing pavement structure overview; (b) traffic analysis; (c) road condition survey; (d) deflection testing; (e) coring and material testing; (f) determination of required pavement thickness (D_f); (g) determination of effective pavement thickness (D_{eff}); and (h) determination of thickness overlay (D_{ol}). Some steps were described in details in the following sections.

3.1 Existing Pavement Structure Overview

The toll road evaluated in this study is Tangerang-Merak toll road at 32.100–36.300 km heading to Tangerang with a total length of 4.2 km (see Fig. 2). The structure of the road consists of five layers as follows.

1. Asphalt mixture wearing course (AC-WC) layer as blacktopping with a thickness of 10 cm. The thickness came from two times overlay works of 5 cm each on the top of concrete pavement. AC-WC is a dense-graded asphalt mixture with 12.5 mm-maximum size aggregate. This mixture was designed using Marshall Method and produced an optimum asphalt content of 5.85% which determined as the asphalt content required to meet the specification of Asphalt Concrete Wearing Course (AC-WC) in General specification of Ministry of Public Works, Republic of Indonesia [6].

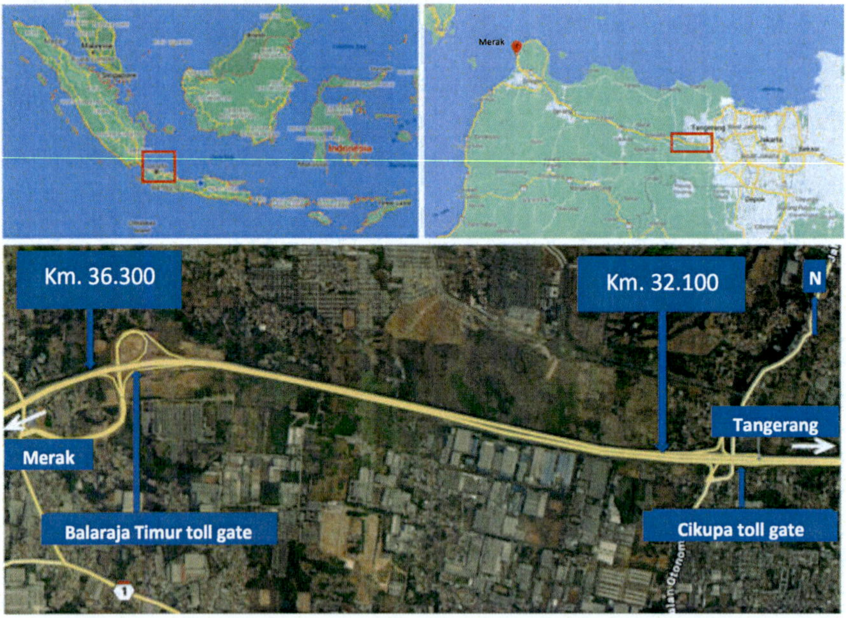

Fig. 2 Location of road segment evaluated

2. Concrete pavement surface without reinforcement or Jointed Plain Concrete Pavement (JPCP), with a thickness of 29 cm and a quality of the concrete *fc* 37.35 MPa.
3. A 10-cm lean concrete and the quality of concrete *fc* 10.4 MPa.
4. Class A granular base layer with a thickness of 25 cm has a CBR design value of 90%.
5. The subgrade on the existing pavement layer has a CBR design value of 6%, the maximum dry density 2.315 g/cm^3 and the optimum water content is 6%.

The rigid pavement of the road was firstly constructed in 1990, and two asphalt-mixture overlays were conducted in 2005 and 2007. During this period, no detail historical data regarding the condition of the pavement was found.

3.2 *Traffic Load Analysis*

Using traffic data in terms of the number of vehicles per day in 2019 in the direction of Tangerang, the cumulative prediction of 18-kip ESAL (Equivalent Standard Axle Load) in the design lane during the predicted service life (N_d) is shown in Table 1.

This calculation used the value of the DL (Design Lane) value equals 0.65 (because it consists of 3 lanes per direction), the service life of 5 years, and a growth factor of 6%. The Vehicle Damage Factor (VDF) formula was used to calculate the equivalency factor based on the static scale survey conducted by the Directorate General of Highway, Ministry of Public Works and Housing.

Table 1 Traffic load analysis (heading to Tangerang)

Group	Vehicle/day	VDF	ESAL in 2019	Cum. ESAL up to 2024
I (Non Bus)	33,063	0.0024	18,826	105,069
I (Small bus)	265	1.8252	114,753	640,445
I (Large bus)	911	3.7796	816,903	4,559,209
II	8713	2.5030	5,236,114	29,223,231
III	3210	16.0490	16,587,421	92,575,905
IV	1173	6.4680	2,318,760	12,941,213
V	1005	16.2430	3,940,346	21,991,432
Total			29,033,124	162,036,505

3.3 Calculation of Asphalt Concrete and Slab Modulus Using Backcalculation Procedure

The method of backcalculation procedure used in this research was the closed-form method. This approach could provide a unique solution by solving the equations. The accurateness of the results produced by the closed-form method can outperform of that produced by iterative or database backcalculation methods [7]. This procedure is used by AASHTO Guide for rehabilitation design of existing AC/PCC structure [2], by adopting AREA method [8] in determining the layer moduli, as shown in Eqs. 1–10. All the variables in the equations were expressed in English units. The conversion to Metrics unit may be possible to be applied once the modulus values were obtained.

Deflection measurements using FWD and temperature measurements were carried out at each point in the evaluated locations. Temperature, together with some properties of the asphalt mixture, was used to calculate the elastic modulus of the mix (E_{AC}) (in psi) using Eq. (1).

$$\log(EAC) = 5.553833 + 0.028829\left(\frac{P_{200}}{F^{0.17033}}\right) - 0.33476Vv + 0.070377_{70F,10^6} + 0.000005t_p^{(1.3+0.49825\log(F))}P_{AC}^{0.5} + \frac{0.00189}{F^{1.1}}t_p^{(1.3+0.49825\log(F))}P_{AC}^{0.5} + 0.931757\left(\frac{1}{F^{0.02774}}\right) \quad (1)$$

in which: P_{200} is percentage of aggregate that passes through sieve # 200; F is load frequency (Hz); V_v is air voids (%); $\eta_{70°F,10^{\wedge}6}$ is absolute viscosity at 70° F, 10^6 poise; P_{AC} is asphalt content (%); and t_p is AC mix temperature (°F).

Using typical values for the several parameters: F = 18 Hz at duration 25–30 ms, $P_{200} = 4\%$, $V_v = 5\%$, $\eta_{70F,10^{\wedge}6} = 2$ poise, $P_{AC} = 5\%$, Eq. (1) can be simplified is as follows.

$$\log(E_{AC}) = 6.451235 - 0.000164671t_p^{1.9254} \quad (2)$$

Deflections from four FWD sensors were selected for backcalculation analysis using the 1993 AAHTO method to obtain the concrete pavement layer's elastic modulus, as follows.

$$d_{0compress} = -0.0000328 + 121.5006\left(\frac{D_{AC}}{E_{AC}}\right)^{1.0798} \quad (3)$$

$$d_{0JPCP} = d_0 - d_{0compress} \quad (4)$$

$$AREA_{JPCP} = 6\left[1 + 2\left(\frac{d_{12}}{d_{0JPCP}}\right) + 2\left(\frac{d_{24}}{d_{0JPCP}}\right) + \left(\frac{d_{36}}{d_{0JPCP}}\right)\right] \quad (5)$$

$$\lambda_k = \left[\frac{\ln\left(\frac{36 - AREA_{JPCP}}{1812.279133}\right)}{-2.559340}\right]^{4.387009} \quad (6)$$

$$k_d = \left(\frac{P}{8d_{0JPCP}\lambda_k^2}\right)\left\{1 + \left(\frac{1}{2\pi}\right) \times \left[\ln\left(\frac{a}{2\lambda_k}\right) + \gamma - 1.25\right]\left(\frac{a}{\lambda_k}\right)^2\right\} \quad (7)$$

$$k_s = 0.5k_d \quad (8)$$

$$E_{JPCP} = \frac{\lambda_k^4[12(1 - \mu_{JPCP}^2)k]}{D_{JPCP}^3} \quad (9)$$

$$S_c = 43,5\left(\frac{E_{JPCP}}{10^6}\right) + 488,5 \quad (10)$$

where: $d_{compress}$ is a deflection that occurs due to pressure in the AC layer in the middle of the slab (mils); d_{AC} is layer thickness of asphalt layer (in.); d_0, d_{12}, d_{24}, d_{36} are deflection at a distance 0, 12, 24, 36 in., respectively from the load (mils); λ_k is the radius of the relative stiffness of slab to subgrade (in.); γ is Euler constant; k_d and k_s is effective dynamic- and static-k value (pci), respectively; E_{JPCP} is slab elastic modulus (psi); S_c is the modulus of rupture (psi).

The calculation of the asphalt, slab, and subgrade moduli at each point evaluated are tabulated in Table 2. The grey-shaded parts of the table indicated that it not possible to calculate the moduli due to the *AREA* values greater than 36 (see Eq. 6).

The lack of providing satisfactory results for all segments showed that the AASHTO procedure has a limitation in practice. This could be contributed using four or more deflections in AREA method to predict the moduli, i.e., the cumulative deviations or errors between the corresponding measured and calculated deflections will increase when unusual pattern of the deflections is found. Therefore, Fwa and Setiadji [9] proposed the use of any two-deflection combination to overcome this problem and to produce better prediction of the moduli. The use of only two deflections in the backcalculation procedure will minimize the cummulative errors between the measured and calculated deflections.

In Table 2, the static *k-value*, which is medium to high, means that the subgrade bearing capacity is quite good. There is a low *E* value due to the low pavement-structure carrying capacity indicated by large deflection measured.

Table 2 Calculation of asphalt, slab, and subgrade moduli

Km	D_{AC} (in.)	D_{JPCP} (in.)	t_p (° F)	E_{AC} (psi)	$d_{compress}$ (mils)	D_0 (in.)	D_{0JPCP} (in.)	D_{12} (in.)	D_{24} (in.)	D_{36} (in.)	$AREA$	λ_k (in.)	k_d (pci)	k_s (pci)	E_{JPCP} (psi)	S_c (psi)
32.175	3.937	9.055	108.20	123,501.15	1.66	1.73	0.10	1.30	1.10	1.00	529.90					
32.425	3.937	9.055	101.20	180.280.54	1.09	3.06	2.00	1.40	1.00	0.90	22.92	17.77	1729.73	864.86	2724.09	607.00
32.600	3.937	9.055	124.20	47,662.67	4.71	10.69	6.00	6.50	4.40	3.20	31.09	39.36	120.04	60.02	4552.57	686.54
32.825	3.937	9.055	109.27	116,355.62	1.78	3.14	1.40	1.00	0.70	0.70	24.78	20.32	1931.49	965.75	5204.27	714.89
33.150	3.937	9.055	109.27	116,355.62	1.78	1.93	0.20	1.10	1.00	0.90	205.31					
33.250	3.937	9.055	105.22	145,475.50	1.39	2.30	0.90	1.60	1.60	1.50	57.81					
33.375	3.937	9.055	105.37	144,353.74	1.40	6.05	4.60	2.20	2.00	1.60	18.92	13.92	1170.26	585.13	693.548	518.67
33.600	3.937	9.055	103.09	163,196.78	1.22	1.94	0.70	1.20	1.20	1.10	55.57					
33.775	3.937	9.055	106.81	133,419.65	1.53	2.35	0.80	1.60	1.40	1.30	58.35					
34.025	3.937	9.055	106.81	133,419.65	1.53	4.23	2.70	2.10	1.60	1.40	25.54	21.57	867.21	433.60	2968.05	617.62
34.200	3.937	9.055	106.81	133,419.65	1.53	2.23	0.70	1.40	1.30	1.10	61.91					
34.425	3.937	9.055	103.81	157,036.91	1.28	2.97	1.70	1.50	1.30	1.10	30.10	34.27	557.37	278.69	6057.19	751.99
34.650	3.937	9.055	112.07	99,195.76	2.12	1.40	-0.70	0.90	0.90	0.80	-31.50					
34.800	3.937	9.055	99.81	193,852.31	1.01	2.44	1.40	1.30	1.10	0.90	29.93	33.52	690.03	345.02	6868.67	787.29
35.025	3.937	9.055	99.81	193,852.31	1.01	2.19	1.20	0.90	0.90	0.80	28.07	27.14	1261.44	630.72	5391.04	723.01
35.200	3.937	9.055	99.81	193,852.31	1.01	2.70	1.70	1.50	1.30	1.10	29.62	32.24	629.05	314.53	5358.83	721.61
35.425	3.937	9.055	108.71	120,034.28	1.72	1.73	0.00	1.10	0.90	0.70	2865.5					
35.575	3.937	9.055	108.71	120,034.28	1.72	1.78	0.10	1.00	0.90	0.80	433.45					
35.850	3.937	9.055	105.37	144,353.74	1.40	1.61	0.20	1.40	1.20	1.00	186.13					
36.250	3.937	9.055	99.81	193,852.31	1.01	2.22	1.20	1.00	0.90	0.80	28.75	29.16	1068.72	534.36	6089.90	753.41

3.4 Calculation of Overlay Thickness (D_{OL})

The overlay thickness was calculated as a deviation between effective- and required-slab thickness, D_{eff} and D_f, respectively. The required slab thicknesses (D_f) were determined using nomograph in the AASTHO method [2]. Several parameters were determined based on field justification: overall standard deviation (S_o) = 0.39 1; reliability (R) = 0.96; drainage coefficient (C_d) = 0.8; load transfer (J) = 2.5; initial serviceability (p_o) = 4.5; terminal serviceability = 2.5; while effective static *k-value*, slab elastic modulus and modulus of rupture were obtained from Table 2, and the proposed traffic volume was as calculated in Table 1.

On the other hand, the effective-slab thickness (D_{eff}) was determined using Eq. (11) as follows.

$$D_{eff} = \left(D_{JPCP} \times F_{jc} \times F_{dur}\right) + \left[\left(\frac{D_{AC}}{2}\right)F_{AC}\right] \quad (11)$$

in which: F_{jc} is joint and cracks adjustment factor; F_{dur} is durability adjustment factor; F_{AC} is AC quality adjustment factor.

Some adjustments were made for the parameters to calculate D_{eff}, i.e., F_{jc} was in between 0.90 and 0.95 (depends on the number of deterioration), while F_{dur} and F_{AC} were selected to equal 0.96 to indicate minor PCC and AC distresses, respectively.

Finally, the overlay thickness (D_{ol}) was calculated using the following equations.

$$D_{ol} = \mathrm{A}\left(D_f - D_{eff}\right) \quad (12)$$

$$\mathrm{A} = 2.2233 + 0.0099\left(D_f - D_{eff}\right)^2 - 0.1534\left(D_f - D_{eff}\right) \quad (13)$$

Another interesting parameter that could be calculated besides the overlay thickness is the remaining life (RL) of the pavement structure. According to AASHTO [2], the remaining life can be determined as a ratio of total traffic to date (N_p) to total traffic to pavement failure ($N_{1.5}$) (in ESAL) as follows.

$$\mathrm{RL} = 100\left[1 - \left(N_p/N_{1.5}\right)\right] \quad (14)$$

N_p can be obtained from Table 1, while $N_{1.5}$ was determined using nomograph in the AASHTO method [2] with $p_t = 1.5$.

As mentioned previously, not all thicknesses at points evaluated can be determined due to there were no moduli resulted from the calculation as the effect of large AREA values, as shown in Table 2. After carefully examining the calculation, only ten thicknesses can be produced, as depicted in Table 3.

Table 3 shows that the average remaining life of the pavement of Tangerang-Merak toll road (heading to Tangerang) is 26%, although the pavement surface visually may not show severe damage. An overlay of 7.12 in. or 18 cm (maximum thickness in Table 3) will be applied to the toll road segment STA 32 + 425 to STA 36 + 250.

Table 3 Calculation of overlay thickness and remaining life

Km	Effective pavement thickness				D_f *(in.)*	Overlay thick		Remaining Life		
	F_{jc}	F_{dur}	F_{AC}	D_{eff} (in.)		*A*	D_{OL} (in.)	$N_{1,5}$ (ESAL)	N_p (ESAL)	*RL (%)*
32.425	0.90	0.96	0.96	11.75	15.53	1.79	6.75	40.563.280	29.033.124	28
32.600	0.95	0.96	0.96	12.30	16.33	1.77	7.12	35.656.937	29.033.124	19
32.825	0.90	0.96	0.96	11.75	14.66	1.86	5.41	41.770.119	29.033.124	30
33.375	0.95	0.96	0.96	12.30	15.71	1.82	6.19	40.174.614	29.033.124	28
34.025	0.95	0.96	0.96	12.30	16.04	1.79	6.68	39.010.255	29.033.124	26
34.425	0.94	0.96	0.96	12.19	15.13	1.86	5.45	39.328.606	29.033.124	26
34.800	0.90	0.96	0.96	11.75	14.73	1.85	5.52	39.251.934	29.033.124	26
35.025	0.90	0.96	0.96	11.75	14.90	1.84	5.79	38.222.538	29.033.124	24
35.200	0.95	0.96	0.96	12.30	15.32	1.85	5.79	40.014.391	29.033.124	27
36.250	0.9	0.96	0.96	11.75	14.77	1.85	5.58	39.675.227	29.033.124	27

Before the overlay is applied, the damage that occurred has to be repaired first by proposing different treatments, as shown in Table 4. The treatments selected were those recommended based on the solutions in AASHTO Guide [2] for rehabilitation methods other than overlay.

In the case of the overlay thickness and remaining life cannot be evaluated (as showed by grey-shaded parts in Table 2), a treatment for repairing the damage should be applied and a minimum thickness of 2 in. or 5 cm for strengthening of existing pavement structure may be considered as a planned stage of the construction. The development of a closed-form backcalculation algorithm for composite pavement requires a lot of work due to different fundamental concept of two systems, i.e.,

Table 4 Proposed treatment on the damage before overlying the road

Km	Type of distress	Proposed treatment
32.425	Raveling	Patching
	Potholes	Shallow patching
32.600	Upheaval at patching	Re-patching
32.825	Depression at patching	Re-patching
33.375	Depression at patching	Re-patching
34.025	Upheaval at patching	Re-patching
34.425	Upheaval at patching	Re-patching
34.800	Transversal cracking	Grouting
35.025	Raveling	Patching
	Block cracking	Shallow patching
35.200	Upheaval	Patching
36.250	Block cracking	Shallow patching

flexible and rigid pavements. The use of iterative best-fit trial and error backcalculation method by simulating the pavement structures into three layers (AC, PCC and subgrade) could be one proposed solution in practice [10], although the use of the method in evaluating the composite structure need a careful consideration.

4 Conclusions

This paper evaluated composite pavements' functional and structural conditions on the Tangerang-Merak toll road, especially at 32.100–36.300 km. Based on the road condition survey results, several damages were found on the pavement surface along the section evaluated. The study indicated high deflection values produced by FWD, which led to the summary that the pavement was in medium to severely damage at several locations. It was also supported by the remaining life analysis, which showed a relatively low bearing capacity on those points. A calculation was conducting using AASHTO 1993 method and proposed two recommendations, that is, an overlay thickness of 7.12 in. or 18 cm to be implemented for the road sections with good prediction of moduli using backcalculation analysis, and a minimum thickness of 2 in. or 5 cm of repairing treatment to strengthen the existing pavement may be considered as planned stage of construction for the rest. All treatments and overlay were carried out after the damages in the sections were repaired first.

References

1. Khadarusno H (2017) Evaluation of the structural performance of continuous cement concrete pavements with reinforcement and concrete asphalt (AC/CRCP) using the AASHTO 1993 method. Master Thesis, Institut Teknologi Bandung, Indonesia (in Bahasa Indonesia)
2. AASHTO (1993) Guide for design of pavement structure
3. Reddy PV, Kumar RS (2018) Structural evaluation of rigid pavement using falling weight deflectometer. Int J Tech Innov Modern Eng Sci 4(12):345–349
4. Barman M, Pandey BB (2008) Backcalculation of layer moduli of concrete pavement by falling weight deflectometer. Paper No 545
5. Ioannides M (1990) Dimensional analysis in non-destructive test rigid pavement evaluation. J Transp Eng 116(1):23–26
6. Directorate General of Highway (2014) 2010 general specification for road and bridge constructions. 3rd revision. Ministry of Public Works and Housing Republic of Indonesia, Jakarta
7. Hall KT, Darter MI (1994) Improved methods for asphalt-overlaid concrete pavement backcalculation and evaluation. In: Quintus HL, Bush III AJ, Baladi GY (1994) Non-destructive testing of pavements and backcalculation of moduli (second volume) STP 1198. ASTM, Philadelphia
8. Ellis T (2008) A comparison of nondestructive testing backcalculation techniques for rigid and flexible pavements. Thesis, University of Arkansas
9. Fwa TF, Setiadji BH (2006) Evaluation of backcalculation methods for nondestructive determination of concrete pavement properties. Transp Res Rec 1949:83–97
10. Hu L, Chou E (2017) Development of an overlay design procedure for composite pavements. Federal Highway Administration Technical Report No. FHWA/OH-2017-23. Ohio

Experimental Study on MOC Cement Based Micro Concrete for Repairing of Wide Cracks in Concrete Pavement Slabs

Avanish Singh, Rakesh Kumar, and Pankaj Goel

Abstract Cracks of variable width and types can develop due to various factors like overloading, thermal expansion and contraction, moisture and temperature stresses, poor construction quality, loss of support underneath the slab, and other similar causes. Often a combination of these factors contributes to the propagation and widening of cracks over time resulting in surface roughness if not repaired properly. Wide crack even requires reconstruction or a full-depth repair. Cracks should be repaired with a durable material lasting for longer life. To repair these cracks, several materials and procedures have been developed. The efficiency of these materials and procedures depends on weather conditions and application efficiency. This study aims to develop a cost-effective magnesium oxychloride cement-based micro concrete for a faster repair of wide cracks; 25 mm or more. To simulate the conditions in the laboratory beam specimens cast and tested for flexure strength failure were used. Then broken beams were repaired by laboratory-developed MOC cement-based micro concrete and again tested for flexure strength. It was found that up to about 80% flexure strength can be restored in one day.

Keywords Micro concrete · Flexure strength · Repair material · Compressive strength

1 Introduction

Concrete pavements often suffer cracking during their lifetime due to various reasons. Concrete cracks, when the tensile stress generated due to the physical and/or chemical processes exceeds its tensile strength. Some of the more typical causes in concrete pavement cracking include drying shrinkage, chemical reaction, thermal expansion, poor quality materials, lack of appropriate control joints and overload conditions, loss of underneath support, and the combinations of the causes that produce flexural,

A. Singh (✉)
Maulana Azad National Institute of Technology, Bhopal, India

R. Kumar · P. Goel
Rigid Pavement Division, CSIR-CRRI, New Delhi, India

© The Author(s), under exclusive license to Springer Nature Switzerland AG 2022

H. R. Pasindu et al. (eds.), *Road and Airfield Pavement Technology*, Lecture Notes in Civil Engineering 193, https://doi.org/10.1007/978-3-030-87379-0_10

Fig. 1 Typical view of cracked concrete slab

tensile, or shear cracks in concrete. Cracks may not affect the appearance only but also causes the loss of structural integrity and functional characteristics of the pavement by making the surface rough. Repair of deteriorated concrete pavement is essential to restore structural capacity and assure serviceability and safety of road users (Fig. 1).

The concrete repair material industry shows a rapid development, the cases where the concrete repair materials that cannot meet their needs are quite common. It is estimated that almost half of the materials used for concrete repair fails in the application site conditions [1]. Repair materials are divided into 4 types: ordinary Portland cement (OPC)-based materials (cement paste, mortar, and self-compacting concrete), modified cement-based materials (cement mortars with high aluminum content, with phosphate, and concrete with fibers), polymer modifiers for OPC based products (styrene butadiene, acrylic, vinyl acrylic, or other copolymers modified cement-based materials), and pure polymers (epoxy resins, polyesters and some polyurethane-based systems) [2–4]. Park and Yang [5] reported that OPC mortars developed satisfactory performance to repair of reinforced concrete structures; however, polymer-modified mortars provided better mechanical properties of the repaired structural element [6]. The choice of repair materials with poor performance generates a significant spend of money and it is a big challenge in the civil construction industry [7]. Repair materials currently in use have serious problems due to generally early-age poor performance or incompatibility with the substrate or low endurance properties for a long time.

2 Scope of Study

The main objective of this study is to experiment with the use of magnesium oxychloride cement-based micro concrete to restore the integrity of a cracked slab (crack width >= 25 mm) under near gravity filling of the crack, emphasizing the physical characteristics of MOC cement. As a result, the objective of this laboratory work is to restore structural integrity and increase the flexure strength of cracked concrete beams by manual filling of MOC-based micro concrete.

3 Materials and Method

3.1 MOC Based Micro Concrete

To prepare the MOC-based micro concrete magnesium oxide of 85% purity, magnesium chloride solution, and locally available filler material was used. Two samples of MOC-based micro concrete were prepared. The detailed characteristics of MOC-based micro concrete are given in Table 1

Table 1 Properties of MOC based micro concrete

Characteristics	Sample 1	Sample 2
MgO:Sand	1:2	1:3
Concentration of $MgCl_2$ solution	26°Be	26°Be
Pot life		
20 °C	140 min	150 min
30 °C	100 min	110 min
40 °C	60 min	65 min
Flow (IS 5512)		
5 min	185 mm	190 mm
15 min	175 mm	185 mm
30 min	160 mm	160 mm
Compressive Strength (ASTM C 109)		
1 day	48 MPa	30 MPa
3 days	56 MPa	48 MPa
7 days	66 MPa	56 MPa
Flexure Strength (ASTM C 348)		
1 day	11 MPa	8 MPa
3 days	15 MPa	10 MPa
7 days	17 MPa	13 MPa
Bond Strength (MDOT)		
1 day	3.2 MPa	3.0 MPa
Drying shrinkage (ASTM 490)		
7 days	0.038%	0.036%
14 days	0.040%	0.041%
21 days	0.056%	0.054%
Abrasion Resistance (IS 9284)	0.26%	0.32%

Table 2 Mix details of substrate concrete

Description	M30	M40
Cement (kg)	375	400
20 mm Aggregate (kg)	613	575
10 mm Aggregate (kg)	695	705
Sand (kg)	629	658
Water (kg)	168	160
Super Plasticizer (%)	0.5	0.35

Table 3 Physical properties of substrate concrete

Test	M30 (MPa)	M40 (MPa)
Compressive strength (28 days)	34	45
Flexure strength (28 days)	3.8	5.2

3.2 Substrate Concrete

For the substrate concrete two grades of concrete M30 and M40 were prepared and cured for 28 days.

- Coarse aggregate: Crushed quartzite coarse aggregate of nominal maximum size of 20 mm ($\gamma = 1580$ kg/m^3 for the size range 10–20 mm, $\gamma = 1600$ kg/m^3 for the size range 4.75–10 mm) was used.
- Fine aggregates: fine aggregate used in the mix was river sand (4.75 mm) size, ($\gamma = 1640$ kg/ m^3).
- Cement: OPC 43 grade was used for both M30 and M40 grades.

The mix details and physical properties of the substrate concrete are given in Table 2 and Table 3, respectively.

3.3 Crack Repair by Micro Concrete Filling

If repair is required to restore structural integrity for cracks with surface widths of 25 mm and extending downward from nearly horizontal surfaces, they should be repaired by micro concrete filling incorporating a long pot-life material. First of all, the cracked surfaces should be cleaned. Contaminants such as oil, grease, dirt, or fine particles of concrete prevent bonding and diminish the effectiveness of repairs. Preferably, contamination should be removed by vacuuming or flushing with water (water blasting) or other especially effective cleaning solutions (air blasting). Then MOC based micro concrete should be filled in the crack and compacted by tamping rod and left in the atmosphere for curing for 24 h (Fig. 2).

Fig. 2 Repaired broken beams of substrate concrete

3.4 *Compressive Strength of Concrete Cubes*

Compressive strength cube specimens with dimensions 150 × 150 × 150 mm were cast from the concrete mix with and without fibers for the determination of compressive strength at various ages. The cubes were demolded after 24 h of casting and moist curing. Thereafter the molded specimens were marked for identifications and kept submerged in curing tanks at room temperature for 28 days.

3.5 *Flexure Strength of Concrete Beams*

The design of concrete pavement is based on the flexural strength of concrete. The flexural strength of concrete is determined by the use of beam specimen under 4-point loading standard test procedure. The beam specimen with the dimension 100 × 100 × 500 mm were cast from concrete mix. The beams were demolded after 24 h of casting and cured for 28 days.

4 Interpretation of Results

The compressive strength and flexure strength of substrate concrete after 28 days of curing are presented in Table 3. The broken beams tested for flexure strength were cut 1 in. from the longer portion to ensure a required gap between the two pieces of the tested beam. Repaired beams were air-cured for 1 day, 3 days, and 7 days, then tested for the flexure strength, and the results are presented in Table 4 (Figs. 3 and 4).

Table 4 Flexure strength of beams after repair

	Initial strength (MPa)	After repair (Sample1) (MPa)			After repair (Sample2) (MPa)		
		1 day	3 day	7 day	1 day	3 day	7 day
M30	3.8	3.0	3.1	3.1	2.5	2.5	2.6
M40	5.2	3.2	3.5	3.5	2.9	3.0	3.0

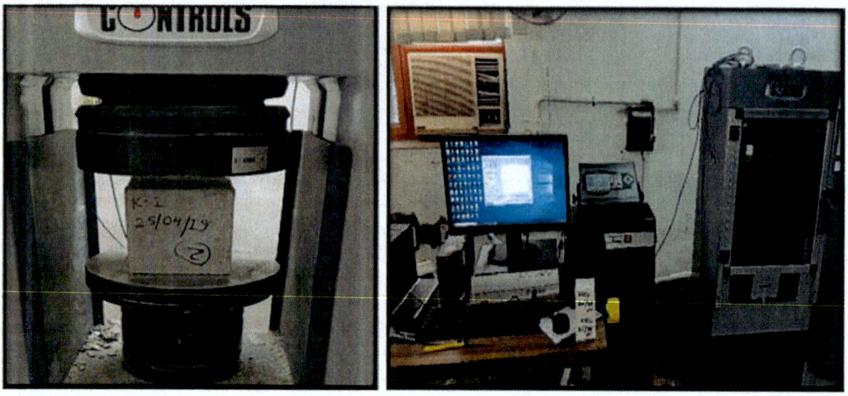

Fig. 3 Testing of compressive strength of substrate concrete

Fig. 4 Testing of flexure strength of substrate concrete under 4-point loading

5 Observations and Conclusions

Cracking of concrete is a casual process, highly variable, and influenced by numerous factors. However, the longer the crack, the higher the stress concentrations induced

Fig. 5 Percentage regain in flexure strength of M30 grade repaired concrete beams

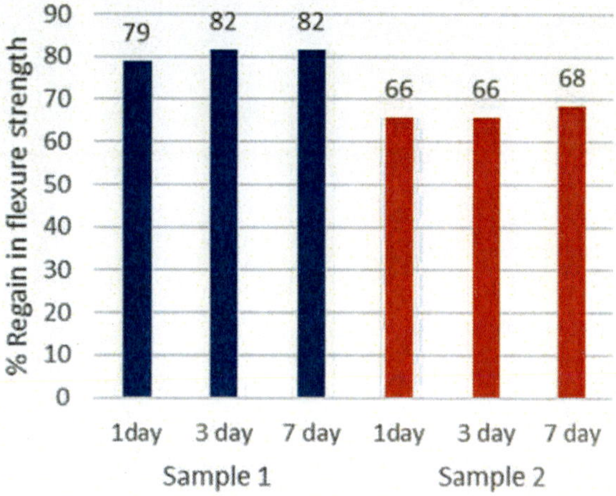

by it. Due to the presence of a crack in a structural element, the strength of the structure will decrease progressively with the increase of the size of the crack. As a result, the structure will be subject to failure when its strength becomes too low that fracture occurs under normal loading. This experimental study proved that completely fractured beams and concrete paving slabs can be repaired successfully to restore up to 80% of their original strength by using MgO-based repair material.

Beams repair by sample 1 (MOC based micro concrete having MgO: Sand is 1:2) shown an appreciable amount of regain in flexure strength. It is also evident from the results that the flexure strength of repaired beams is almost constant with the age. For M30 grade 82% and for M40 grade 67% flexure strength was regained by broken beams. Similarly, beams repair by sample 2 (MOC based micro concrete having MgO: Sand is 1:3) also shown a considerable amount of regain in flexure strength. For M30 grade 68% and for M40 grade 58% flexure strength was regained. Figures 5 and 6 shows the % regain in the flexure strength of M30 and M40 grade beams repaired by MOC-based micro concrete. Figure 7 shows the failure of repaired beams. It is clear that all the beams are failed in bond. That is why flexure strength is not increasing with the age.

Thus, the use of MOC-based micro concrete can be a cost-effective solution for the repair of wide cracks, as it can regain up to 80% flexure strength. It is an economical alternative for easily available repair materials in the market such as epoxy-based and polymer-based. It costs about 10 times cheaper than epoxy-based repair materials.

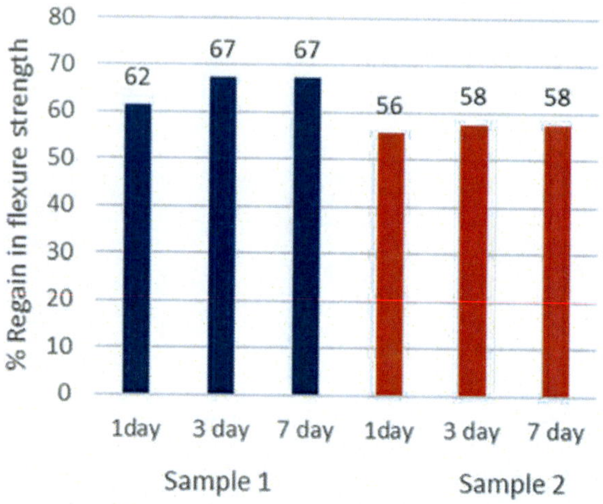

Fig. 6 Percentage regain in flexure strength of M40 grade repaired concrete beams

Fig. 7 Typical view of failure in repaired beams of substrate concrete

References

1. Issa CA, Debs P (2007) Experimental study of epoxy repairing of cracks in concrete. Constr Build Mater 21(1):157–163
2. Delatte N (ed) (2009) Failure, distress and repair of concrete structures. Elsevier
3. Van Balen K, Papayianni I, Van Hees R, Binda L, Waldum A (2005) Introduction to requirements for and functions and properties of repair mortars. Mater Struct 38(282):781–785
4. Hassan KE, Brooks JJ, Al-Alawi L (2001) Compatibility of repair mortars with concrete in a hot-dry environment. Cement Concr Compos 23(1):93–101
5. Park SK, Yang DS (2005) Flexural behavior of reinforced concrete beams with cementitious repair materials. Mater Struct 38(3):329–334

6. ACI Committee 224—Cracking (2007) Causes, evaluation, and repair of cracks in concrete structures. American Concrete Institute
7. Smoak WG (1997) Guide to concrete repair. US Department of the Interior, Bureau of Reclamation, Technical Service Center

Feasibility Study on Pavement Rutting Evaluation Method Based on Smartphone

Jin-Xi Zhang, Pei-Rong Wang, Dan-Dan Cao, and Jing-Xiang Zeng

Abstract With the continuous increase of constructions in highway, road maintenance has become more and more important. Thus, it is of great significance to develop the rapid, intelligent and real-time detection technologies for road surface conditions. This paper used the self-developed driving data acquisition APP to collect the vibration acceleration data during driving, and carried out the feasibility study on the evaluation method of pavement rutting using smartphones. Firstly, the collected vibration acceleration data are de-noised, and the vibration characteristics under different working conditions are analyzed. Secondly, seven time-domain vibration acceleration indexes with high correlation with pavement rutting are extracted, and the dimensions of seven primary indexes are reduced to two independent principal components by principal component analysis. Finally, the rutting evaluation model based on convolutional neural network is established and compared with the results of back propagation neural network and multilayer perceptron neural network. The results show that the average relative error of the rutting evaluation model based on the convolutional neural network is 16.6%, which is lower than the other two models. It indicates that the pavement rutting can be evaluated satisfactorily by smartphones. In addition, this paper divided the evaluation results of rutting into four grades (Excellent, Good, Medium and Poor) and displayed them in different colors on the map. This study is of great significance to improve the level of intelligent detection of road rutting and road maintenance management.

Keywords Rutting · Smartphone · Vibration acceleration · Principal component analysis · Convolutional neural network

---

J.-X. Zhang (✉) · P.-R. Wang · D.-D. Cao · J.-X. Zeng
Department of Road and Urban Railway Engineering, Beijing University of Technology, No. 100, Pingleyuan, Chaoyang District, Beijing 100124, China
e-mail: zhangjinxi@bjut.edu.cn

P.-R. Wang
e-mail: wangpeirong2018@163.com

D.-D. Cao
e-mail: dandan_cao@bjut.edu.cn

© The Author(s), under exclusive license to Springer Nature Switzerland AG 2022

H. R. Pasindu et al. (eds.), *Road and Airfield Pavement Technology*, Lecture Notes in Civil Engineering 193, https://doi.org/10.1007/978-3-030-87379-0_11

1 Introduction

Rutting is a kind of permanent deformation of asphalt pavement under the combined action of load and environment factors [1, 2]. In the maintenance and management of asphalt pavement, pavement rutting is an important technical condition index [3]. Relevant studies [4, 5] show that rutting has become one of the main diseases of highways, and its harm mainly includes the following aspects:

(1) Rutting may lead to a decrease in road roughness performance and affect driving comfort;
(2) When driving on the road section with serious rutting, the vehicle steering is difficult to control, and the operation stability of driving is reduced;
(3) In rainy or snowy days, the road with rutting is easy to accumulate water or freeze, which lead to vehicle skidding and threaten the driving safety;
(4) Rutting is also the inducement of other pavement diseases. If rutting is not repaired in time, it may lead to problems such as pavement wrapping, looseness, block cracking and water damage.

Therefore, rutting not only affects the ride comfort and safety, but also is closely related to the vehicle operating cost, tire wear and service life of the road. The overall development trend of pavement rutting detection technology has experienced three stages: from traditional manual detection to semi-automatic detection, and then to fully automatic detection [6]. At present, the widely used automatic detection technology of pavement rutting mainly includes ultrasonic [7, 8], laser [9] and digital image [10] detection technology, among which laser detection is more widely used. Although automated detection can greatly improve detection efficiency and accuracy, it usually relies on high-cost hardware equipment and skilled professional operators. In addition, periodic detection is usually used to acquire road conditions, demanding a lot of manpower and resources.

In recent years, the digital measurement methods have been developed more and more in road engineering. Rapid measurement of large scale pavement condition has become a hot topic in pavement condition evaluation and maintenance. With the development of smartphones, they are equipped with a variety of sensors, and have the capabilities of efficient information processing and data transmission. For vehicle drivers and passengers, devices such as triaxial acceleration sensors, gyroscopes, GPS and gravity sensors in smartphones can effectively reflect the movement status of the driver or vehicle during driving [11]. A large number of vehicle drivers and passengers can use their smartphones as important "tools" to obtain big data on road conditions. This crowdsourced dataset is convenient, extensive coverage, inexpensive, and has high sampling frequency, making it a suitable source for large-scale road condition evaluation [12]. Therefore, considering the cost and efficiency, many scholars try to use smartphones to evaluate and monitor the road condition.

In this paper, a low-cost method is proposed to evaluate the road rutting in real-time using smartphone sensors and machine learning algorithms. The paper is organized as follows. Section 2 introduces the related work. Section 3 gives the details about our experiment. Section 4 analyzes the vibration characteristics under different working conditions. Section 5 determines the driving vibration index through correlation analysis and principal component analysis. Section 6 establishes the rutting evaluation model. Lastly, Sect. 7 presents the conclusions and future works.

2 Related Work

In the last years, there have been significant progress in using smartphones to assess and monitor the pavement quality. These devices represent a substantial reduction in the costs involved to accomplish this task compared to usual equipment such as laser profilers.

Mohan et al. [13] monitored road and traffic conditions through the Nericell system. They used the accelerometer, microphone, GSM radio, and/or GPS sensors in these phones to detect potholes, bumps, braking, and honking. Nericell addressed several challenges including virtual reorienting the accelerometer on a phone that is at an arbitrary orientation, and performing honk detection and localization in an energy efficient manner. Kyriakou et al. [14] presented a study on the detection of roadway pavement anomalies using smartphone sensors and on-board diagnostic (OBD-II) devices. The proposed approach also utilized artificial neural network (ANN) techniques in the analysis, captured a vehicle's interaction with a roadway pavement while the vehicle is moving, and utilized the observed interaction patterns for the detection of potholes in the pavement. Huijun et al. [15] proposed a road condition detection method based on the Mahalanobis-Taguchi System, which can be used to distinguish well covers, potholes and speed bumps. Yi et al. [16] proposed a smartphone-based system to objectively assess bumping caused by road anomalies. The proposed system included the development of a signal processing heuristic that relied on the vertical acceleration component to identify road surface conditions such as potholes and bumps.

Islam et al. [17] collected vehicle acceleration data with an Android-based application. In the study, acceleration data were double integrated numerically to obtain a pavement profile, which was input into the software program ProVAL. The pavement roughness was then calculated. Douangphachanh et al. [18] found that in the frequency range of 40–50 Hz, the road roughness has a strong relationship with the vibration amount of each axis of accelerometer and gyroscope. Aleadelat et al. [19] used smartphones' 3D accelerometers to collect a vehicle's vertical acceleration data. Through the use of various signal processing and pattern recognition techniques, such as cross correlations, Welch periodograms, and variance analyses, the measured signals (time series acceleration data) were identified and correlated with the actual international roughness index (IRI) values. It was found that the variance among the vertical acceleration measurements was the key feature for classifying the measured signals.

In previous studies, there exist lots of studies in detecting road bumps and anomalies using mobile sensors, and most of them focused on identifying and locating road bump and anomalies. At the same time, there are many researches on the measurement and estimation of pavement roughness. However, there are very few studies that have directly explored the use of smartphone to estimate rutting of road pavement, where this study used machine learning to improve the accuracy of the evaluation results.

3 Driving Experiment

3.1 Experimental Equipment

The data collection equipment of this experiment mainly include an Android smartphone (Fig. 1a) installed with the driving data collection APP [20] and a multifunctional road condition detection vehicle (Fig. 1b). The driving data collection APP was designed and developed by the author and others. The model of the experimental phone was Huawei Honor 10. The experimental vehicles included Dongfeng Honda CRV (Fig. 1c) and Kia Freddy (Fig. 1d).

3.2 Experimental Roads

In order to carry out safe driving experiments and obtain effective experimental data, the driving experiment roads need to have different rutting depth, low traffic volume, smooth route and dry conditions. This experiment selected three asphalt concrete roads located in the eastern suburb of Jinan, Shandong Province, China. Table 1 shows the test road conditions.

3.3 Experimental Scheme

The smartphone installed with data acquisition APP was fixed horizontally on the central armrest box in the front row of the car with transparent tape. The acquisition frequency was set to 10 Hz. Data such as three-axis vibration acceleration, time, latitude and longitude, speed during driving process were collected. For the collected three-axis vibration acceleration, the left and right direction was recorded as the X axis, the front and rear direction was recorded as the Y axis, and the vertical direction was recorded as the Z axis. The two vehicles were driven three times on each experimental road at the speed of 30 km/h, 50 km/h and 70 km/h respectively. During the experiment, the vehicles were kept as steady as possible. In addition, the multifunctional road detection vehicle was used to detect the experimental roads to obtain the corresponding rutting depth data.

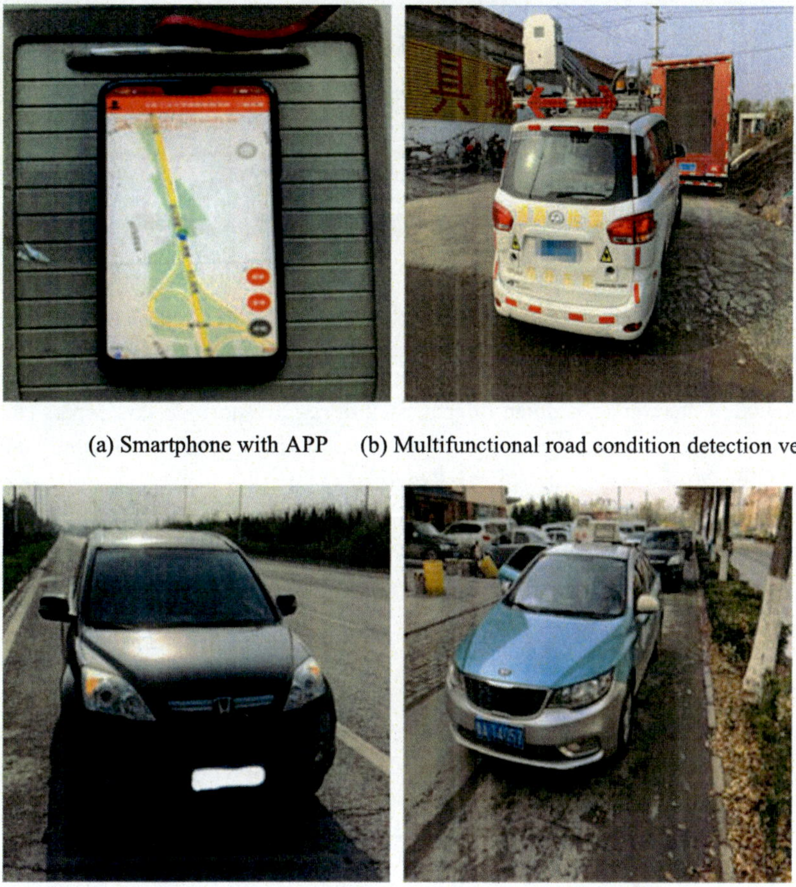

(a) Smartphone with APP (b) Multifunctional road condition detection vehicle

(c) Dongfeng Honda CRV (d) Kia Freddy

Fig. 1 Experimental equipment

Table 1 Test roads

Road name	Length (m)	Average rutting depth (mm)
Feiyue avenue	1100	5.83
Chunbo road	2000	7.66
Daliu line	300	10.24

The experimental data includes two parts: the data collected by smartphone and the data of rutting depth. The detection unit length of rut depth was 20 m. To match the two types of data, the data collected by the smartphone was divided into 20 m units.

4 Analysis of Vibration Characteristics Under Different Working Conditions

4.1 Data Preprocessing

In engineering applications, due to the influences of various internal and external factors of the system, the data collected by sensors are often mixed with many unnecessary components [21]. In addition to the vehicle vibration caused by the road itself, the three-axis vibration acceleration data collected in this paper also contains the interference signals such as vehicle self-vibration, acceleration sensor itself and external noise. Therefore, it is necessary to filter the data to eliminate the noise and interference in the signal. Kalman filter is used to remove noise in this paper. The filtered data can not only retain the fluctuation trend of the original data, but also eliminate the burr fluctuation of the original data.

4.2 Analysis of Vibration Characteristics Under Different Working Conditions

The vibration caused by the same road at different driving speeds and different vehicles also has different characteristics. Note that the positive and negative vibration acceleration only represents the direction, and has nothing to do with the magnitude. In order to compare the characteristics of vibration variation under different working conditions, the mean value of the absolute value of vibration acceleration is selected as the characteristic index reflecting the driving vibration.

In this paper, according to the scheme in Sect. 3.3, the driving experiments were carried out under different speeds and vehicles. The mean value of absolute value of vibration acceleration of some roads under different working conditions are calculated. Figure 2 shows that the greater the velocity, the greater the vibration acceleration, but the trends of vibration variation at different velocities are consistent. Figure 3 shows the vibration acceleration of Kia Freddy is larger than that of CRV. This may be due to differences in vehicle suspension systems. However, they have similar trends. The standard deviation of the absolute value of the vibration acceleration is calculated for comparison, and the results are the same as above. Considering that it is impossible for vehicles to travel along the exact same path in the experiment, the experimental speeds also have a certain deviation. Therefore, the fluctuation of vibration acceleration is inevitable. To sum up, although the magnitude of driving vibration under different working conditions are different, their variation trends are highly consistent.

Due to the different suspension systems of different vehicles, this paper mainly studies the driving vibration data of Dongfeng Honda CRV.

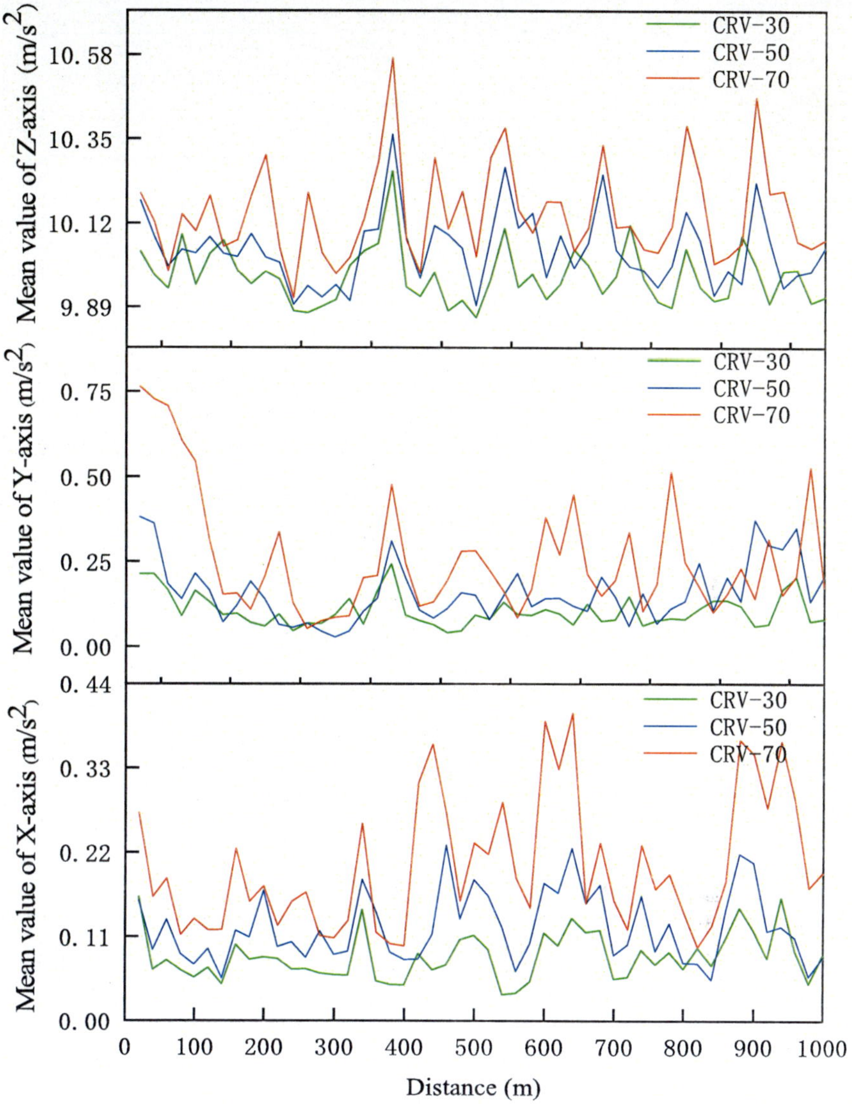

Fig. 2 Comparison diagram of vibration characteristics of different driving speed

5 Analysis and Extraction of Vibration Characteristic Index

The vibration generated by vehicles on the road is related to many factors, not only related to the performance of the vehicle itself, but also more related to the pavement performance. The pavement performance that affects driving vibration includes

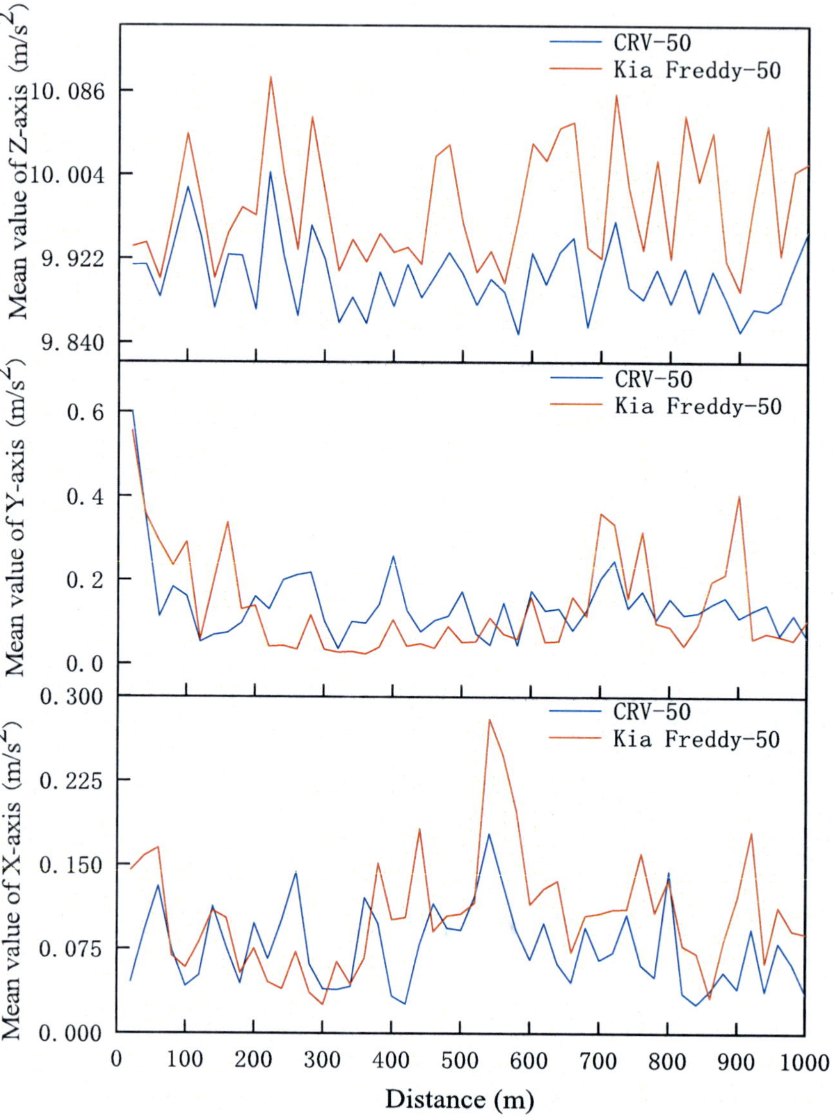

Fig. 3 Comparison diagram of vibration characteristics of different vehicles

roughness, rutting, breakage, etc. Different pavement characteristics have different effect on driving vibration. Relevant research shows that [20] the road roughness has a greater impact on the vertical vibration acceleration of driving, while the impact on the vibration acceleration of other directions is relatively small. Rutting is different from pavement roughness. Vehicles swing from side to side and bump back and forth

Table 2 Primary index set

Serial number	Index	Symbol	Serial number	Index	Symbol
1	X-axis standard deviation	Xstd	5	X-axis mean	Xmean
2	X-axis maximum	Xmax	6	Y-axis rms value	Yrms
3	X-axis peak-to-peak value	Xp-p	7	Y-axis mean	Ymean
4	X-axis rms value	Xrms			

on rutted road. In order to establish the evaluation model of pavement rutting, firstly, the correlation analysis and extraction of driving vibration indexes are carried out.

In order to select the optimal characteristic parameters of the driving vibration acceleration of the rutting evaluation model. The paper first extracts 18 time-domain features in three axes, including peak-to-peak value, rms value, maximum, minimum, mean and standard deviation of the absolute value of acceleration. Then Pearson correlation test was carried out between 18 vibration indexes and rutting depth. The first 7 of all significantly related indexes were selected as the primary index set (Table 2). Table 2 shows the top 5 indexes are vibration acceleration indexes of X-axis, and the indexes ranked 6–7 are vibration acceleration indexes of Y-axis. Therefore, pavement rutting mainly affects the vibration acceleration of X-axis, and also has a certain influence on Y-axis.

The above seven indexes all reflect the vibration characteristics of vehicle. Through the correlation test of the seven indicators, we found that not only the indexes in the same axis are highly correlated, but also there is a certain correlation between the indexes in different axes. When there is a certain correlation between variables, it will cause the impact of information overlap on the evaluation model. If it is directly included in the analysis, multi-collinearity may occur, which will affect the final evaluation results. If only a few indicators are selected for analysis, important data information may be lost and incomplete, leading to the reduction of the generalization performance of the model. In order to determine the final evaluation index and improve the accuracy and generalization ability of the model, this paper uses principal component analysis (PCA) to reduce the dimension of the primary index.

Principal component analysis is a statistical analysis method which converts multiple variables into a few principal components (comprehensive variables) by dimensionality reduction technology. It is one of the most important feature extraction methods. The converted principal components retain the main features of the original variables as much as possible with as few dimensions as possible. The first two principal components were extracted by principal component analysis. The two principal components are computed according to Eqs. (1) and (2).

$$F_1 = 0.442\overline{X}_{std} + 0.463\overline{X}_{\max} + 0.380\overline{X}_{p-p} + 0.449\overline{X}_{rms} + 0.429\overline{X}_{mean} + 0.179\overline{Y}_{rms} + 0.167\overline{Y}_{mean} \quad (1)$$

$$F_2 = -0.126\overline{X}_{std} - 0.188\overline{X}_{\max} - 0.173\overline{X}_{p-p} - 0.076\overline{X}_{rms} - 0.061\overline{X}_{mean} + 0.678\overline{Y}_{rms} + 0.686\overline{Y}_{mean} \quad (2)$$

where F_1 is the first principal component, F_2 is the second principal component; $\overline{X}_{\text{std}}$ is the normal normalized value of X_{std}, otherwise the same.

6 Rutting Evaluation Model

Artificial neural network is a kind of machine learning technology that realizes artificial intelligence by simulating the neural network of the human brain. In recent years, artificial neural network has played a very significant role in pattern recognition, automatic control, signal processing, assistance decision-making and other aspects. Artificial neural network has the ability of autonomous learning and finding optimal solutions in data processing. Therefore, it is much simpler and more accurate to use artificial neural network for regression than previous regression processing methods. In this paper, convolutional neural network is used to establish the rutting evaluation model and compared with other neural network models.

6.1 Construction of Convolutional Neural Network Model

Convolutional neural network is one of the most classic deep neural network structures, belonging to feedforward neural network [22, 23]. It can efficiently obtain useful feature descriptions from data. At present, convolutional neural network has been widely used in speech recognition, image recognition and classification, natural language processing and other fields. The construction process of the rutting evaluation model based on convolutional neural network in this paper is as follows.

(1) Input Characteristic Data

The collected vibration acceleration is processed and calculated by the above method to obtain F_1 and F_2. Section 3.2 proved that the greater the speed, the greater the vibration on the same road. Therefore, F_1, F_2 and speed are taken as input, and rutting depth is taken as output. The input vibration characteristics F_1 and F_2 have different unit dimensions from the speed. In order to eliminate the differences between the indicators, the Z-score standardization method is used to process the data, which is helpful to improve the accuracy of the model. The calculation formula of the Z-score standardization method is shown in Eq. (3). In this study, a total of 1020 groups of sample data are obtained, 70% of which are randomly selected as the training set and 30% as the test set.

$$X = \frac{x - \mu}{\sigma} \quad (3)$$

where x is the original data, X is the standardized data; μ and σ are the mean and standard deviation of the original data respectively.

(2) Determine Model Structure

The model structure selected in this study is mainly composed of four CNN layers, one flatten layer and two fully connected layers. The number of convolution kernel of the four CNN layers is 16, 32, 64 and 128 in turn, and the size of their convolution kernel is all 3. Their activation functions are all set to ReLU for better gradient and faster training. The flatten layer converts the two-dimensional vector in the convolution layer into one-dimensional vector. The number of neurons in the first fully connected layer is 32, and the activation function is ReLU. The number of neurons in the second fully connected layer is 1, and the activation function is Linear activation function. The selected optimizer is Adam Optimizer. The optimal learning rate is determined to be 0.0004. The loss function adopts the mean square error. The number of training is set to 1500.

6.2 *Results and Analysis*

Figure 4 shows the comparison of the rutting depth estimated by the convolutional neural network model and the measured rutting depth. In Fig. 4, the estimated values are close to the measured values on the whole. Figure 5 shows the absolute errors of the rutting depth estimated by the convolutional neural network model, most of which are less than 1.5 mm. To measure the deviation between the rutting depth estimated by the convolutional neural network model and the measured rutting depth. The mean absolute error (MAE) and the mean relative error (MRE) are selected as indexes to

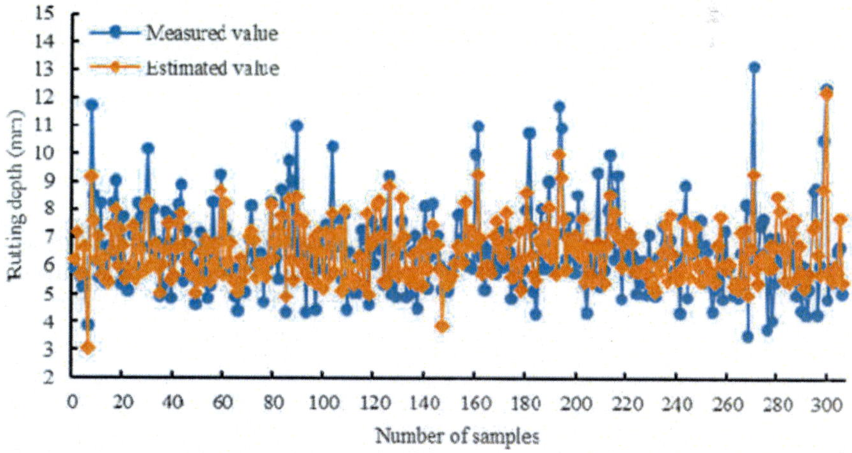

Fig. 4 Comparison of rutting depth estimated by the convolutional neural network model and measured rutting depth

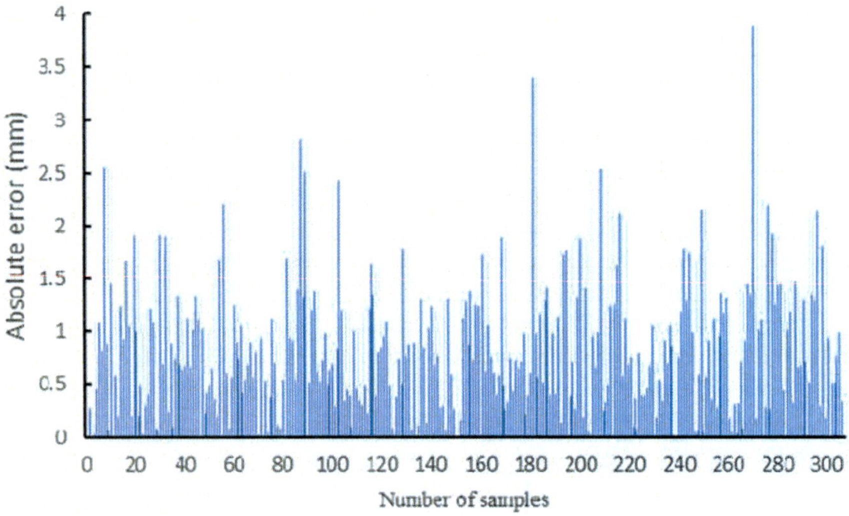

Fig. 5 Absolute errors of rutting depth estimated by the convolutional neural network model

evaluate the accuracy of the model. MAE is 1.14 mm and MRE is 16.6% according to Eqs. (4) and (5).

$$MAE = \frac{1}{n}\sum_{i=1}^{n}\left|X_i - X_i^{'}\right| \tag{4}$$

$$MRE = \frac{1}{n}\sum_{i=1}^{n}\left|\frac{X_i - X_i^{'}}{X_i}\right| \times 100\% \tag{5}$$

where X_i is the measured value of rutting depth, X_i' is the estimated value of rutting depth.

In this study, the back propagation neural network (BP-NN) and multi-layer perceptron Neural Network (MLP-NN) were used as control models to verify the performance of the convolutional neural network model. Table 3 shows the mean absolute error and mean relative error of the three models. In Table 3, the mean absolute error and the mean relative error of the convolutional neural network model are lower than the other two models, and the mean absolute errors of the three models are less than 1.5 mm, the mean relative errors are less than 20%. Figure 6 shows the cumulative distribution of the relative errors of the three models. In Fig. 6, the

Table 3 MAE and MRE of the three models

	CNN	BP-NN	MLP-NN
MAE (mm)	1.14	1.35	1.26
MRE (%)	16.6	17.8	18.4

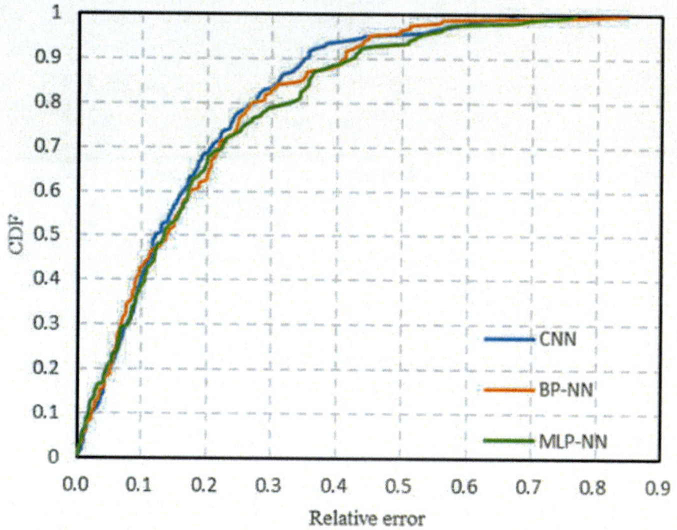

Fig. 6 CDF of relative errors of the three models

40% relative error of the three models is less than 10%; while the 70% relative error of CNN and MLP-NN is less than 20%, and the 60% relative error of BP-NN is less than 20%. To sum up, it is reliable and feasible to evaluate road rutting through smartphones.

6.3 *Application*

In order to show the evaluation results of the convolution neural network rutting evaluation model more intuitively, this paper used the GPS position information collected by the driving data acquisition APP to display the evaluation results on the map. The rutting depth can be divided into four grades according to the Highway Performance Assessment Standards (JTG 5210–2018). In this paper, the four grades were represented by different colors on the map. Table 4 shows the grading range of rutting depth and corresponding colors. We selected some roads in Beijing as the test objects. Figure 7 displays the rutting evaluation results of these roads on the map.

Table 4 Rutting depth grade

Evaluation grade	Rutting depth(mm)	Color
Excellent	≤10	Green
Good	>10, ≤13.33	Yellow
Medium	>13.33, ≤16.67	Orange
Poor	>16.67	Red

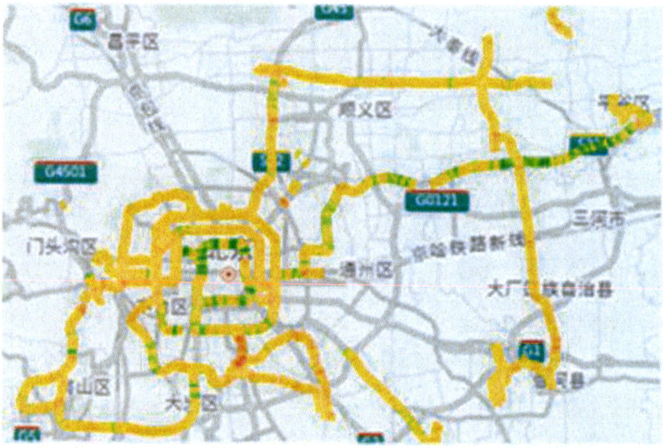

Fig. 7 Display of rutting evaluation results in the map of Beijing city

Due to the channelization of traffic, the increase of traffic load and traffic volume, rutting has become a major disease of asphalt pavement and one of the main decisive factors of pavement maintenance. The results of this study can help the pavement engineerings conduct a better road maintenance.

7 Conclusions and Future Work

The large variety of sensors in most modern smartphones makes these devices useful tools to collect a large amount of data in many situations, allowing the emergence of new systems and applications. This paper studied the feasibility of using smartphones to evaluate pavement rutting. First, it was found that the variation trends of vibration acceleration under different vehicles and speeds are highly similar. Then, by analyzing the correlation between rutting and vibration acceleration, the first 7 indexes were selected from the 18 time domain indexes of vibration acceleration as the primary index set. The result showed that rutting mainly affects the vibration acceleration of x-axis. Next, the principal component analysis was used to transform the seven indexes into two independent principal components to reduce the interference of redundant indexes on the evaluation model. Finally, the rutting evaluation model based on convolutional neural network was established and compared with the models based on BP neural network and MLP neural network. The result showed that the error of convolution neural network model is the smallest, the MAE is 1.14 mm, and the MRE is 16.6%. This study proved that the smartphone can be used to evaluate the road rutting conditions. In order to show the evaluation results of the model more intuitively, the evaluation results were displayed in different colors on the map.

The accuracy of the method proposed in this paper is lower than that of the current automatic detection equipment. In future works, we intend to introduce more vehicle and road types for extending the research application and to improve the accuracy of the evaluation method.

Acknowledgements This research was funded by the National Key Research and Development Program of China (2018YFB1600302).

References

1. Jones R, Horner D, Sullivan P et al (2005) A methodology for quantitatively assessing vehicular rutting on terrains. J Terramech 42(3–4):245–257
2. Cui W-S, Tao J, Zhang Z-Q et al (2009) Test on rut causes of expressway asphalt pavement. J Chang'an Univ (Nat Sci Ed) 29(04):8–12
3. Shi-xuan LEI (2019) Research on performance evaluation and maintenance countermeasures of highway asphalt pavement in Shaanxi Province. Chang'an University, China
4. Xu T, Huang X (2012) Investigation into causes of in-place rutting in asphalt pavement. Constr Build Mater 28(1):525–530
5. Start M, Kim J, Berg W (1998) Potential safety cost-effectiveness of treating rutted pavements. Transp Res Rec 1629:208–213
6. Ma J, Zhao X-M, He S-H et al (2017) Review of pavement detection technology. J Traffic Transp Eng 17(5):121–137
7. Pan X-J (2014) A dissertation submitted for the Degree of Master. Chang'an University, China
8. En-jun WANG, Xian-qiao CHEN, Xiu-min CHU et al (2008) Data acquisition method by ultrasonic wave distance measurement for rut detection. J Wuhan Univ Technol 01:138–141
9. Si Y-W (2014) Research on integrated multi-point laser pavement rutting detection technology. Chang'an University, China
10. Zhang Y, Gao TT (2012) VC-based rutting digital imaging automatic detection technology research and design for road construction. Adv Mater Res 461:370–372
11. Wang D-Y, Chen K (2017) Research review of smartphone applications in pavement condition survey. Highway Eng 42(05):97–101
12. Liu C, Wu D, Li Y, et al (2021) Large-scale pavement roughness measurements with vehicle crowdsourced data using semi-supervised learning. Transp Res Part C Emerg Technol 125:103048
13. Mohan P, Padmanabhan VN, Ramjee, Nericell R (2021) Rich monitoring of road and traffic conditions using mobile smartphone. ACM SenSys 2008, Raleigh, North Carolina, USA
14. Kyriakou C, Christodoulou SE, Dimitriou L (2019) Smartphone-based pothole detection utilizing artificial neural networks. J Infrastruct Syst 25(3)
15. Huaijun W, Na H, Junhuai L et al (2018) A road quality detection method based on the Mahalanobis-Taguchi system. IEEE Access 6:29078–29087
16. Yi C-W, Chuang Y-T, Nian C-S (2015) Toward crowdsourcing-based road pavement monitoring by mobile sensing technologies. IEEE Trans Intell Transp Syst 16(4):1905–1917
17. Islam S, Buttlar WG, Aldunate RG et al (2014) Measurement of pavement roughness using android-based smartphone application. Transp Res Rec J Transp Res Board 2457(3):30–38
18. Douangphachanh V, Oneyama H (2014) Exploring the use of smartphone accelerometer and gyroscope to study on the estimation of road surface roughness condition. In: Proceedings of the 11th international conference on informatics in control, automation and robotics. Science and Technology Publications, Vienna, Austria, pp 783–787
19. Aleadelat W, Ksaibati K, Wright CHG et al (2018) Evaluation of pavement roughness using an android-based smartphone. J Transp Eng Part B: Pavements 144(3)

20. Wang L (2019) Study on pavement performance intelligent recognition method based on driving data. Beijing University of Technology, China
21. Meng L (2018) Analysis and application of smartphone acceleration sensor signal for pavement leveling performance. Southeast University, China
22. Hu H-Y, Liu J-R, Gao F et al (2020) Driver identification based on 1-D convolutional neural networks. China J Highway Transp 33(08):195–203
23. Kiranyaz S, Zabihi M, Rad AB et al (2020) Real-time phonocardiogram anomaly detection by adaptive 1D convolutional neural networks. Neurocomputing 411:291–301

Feasible Treatment Selection for Routine Maintenance of Flexible Pavement Sing Fuzzy Logic Expert System

Rajnish Kumar, Sanjeev Kumar Suman, and Ankita Singh

Abstract Pavement maintenance management system motivates to provide a scientific tool for maintenance and rehabilitation of roads pavement at desired serviceability levels. In view of the fund's constraints and the need for judicious spending of available resources, the maintenance planning and budgeting are required to be done based on scientific methods. Unfortunately, the current maintenance practices are ad-hoc and subjective in nature. Pavement condition responsive maintenance is very useful for judicious disbursement of maintenance funds. The objective of this paper is to select a feasible treatment for routine maintenance based on pavement condition parameters of flexible pavement using Fuzzy Logic Expert System (FLES). Six different national highways have been selected to provide the maintenance based on the PCI, traffic volume, pavement age, precipitation, temperature and budget. FLES offers a convenient tool to better represent the uncertainty involved in pavement condition rating and assessment. The pavement maintenance treatment needs are generally determined based on the results of visual inspection, which in most of the cases does not give an adequate representation of pavement condition. Treatment selection FLES model has considered anticipated distresses-based condition index, anticipated traffic, and prevailing climate, age of the pavement and budget for treatments. Model predicts treatment types based upon their expected life. The triangular membership function for all the parameter is considered and analyzed with sufficient number of fuzzy rules as suggested by the maintenance engineers. The predicted result was compared with the twenty-five maintenance engineer's responses, which shows homological results. Hence, this approach may provide an appropriate and economically viable maintenance treatment.

R. Kumar (✉) · S. K. Suman · A. Singh
National Institute of Technology Patna, Patna 800005, India
e-mail: rajnish.ce17@nitp.ac.in

S. K. Suman
e-mail: sksuman@nitp.ac.in

A. Singh
e-mail: ankitas.pg19.ce@nitp.ac.in

© The Author(s), under exclusive license to Springer Nature Switzerland AG 2022

H. R. Pasindu et al. (eds.), *Road and Airfield Pavement Technology*, Lecture Notes in Civil Engineering 193, https://doi.org/10.1007/978-3-030-87379-0_12

Keywords Pavement treatments · FLES model · Pavement condition index · Budget · Climate · Expert response

1 Introduction

The systematic planning of maintenance strategies is an integral part of the pavement management system's decision-making process. The premature failure of pavement section before its design life due to unavoidable uncertainties (occurrence of traffic, variation in drainage conditions, rainfall, weather conditions, and non-uniform construction) is an issue that needs to be addressed strategically. The decisive process can be data-centric, an expert system, or both. A timely updated PMS database is not readily available in developing countries like India. The maintenance of India's extensive road network largely depends on expert opinion. Deciding the most appropriate maintenance strategy for a large number of pavement sections is complex. The only efficient way for effective pavement management is to employ a computerized system [1].

The pavement maintenance management system is carried out considering the pavement rating system or PCI which reflects the structural and functional properties. The maintenance issues are always a major concern for the maintenance engineer over a period of time [2]. There are several researchers who have done a considerable amount of work in the field of pavement maintenance. The overall condition index is developed considering the expert opinion, traffic information and pavement distress to prioritized the pavement section. The prioritization model is developed based on the budget constraints for maintenance [3]. Saaty's scale is used to provide the weightage or relative importance among the assigned variables or alternatives subjectively. The Analytical Hierarchy Process (AHP) is one of the prioritization model which is adapted to provide the pavement maintenance based on the relative importance [4]. The pavement maintenance strategy is also prioritised at the network level using the AHP [5]. The AHP has shown the strength in the selection of alternative but it is very complex and time cosuming technique when the number of alternatives are more than eight [6]. Multi-attribute utility theory (MAUT) is also used in the prioritisation of pavement in terms of utility function. MAUT is advantageous in the selection of most suitable preferences among the availabe alternatives [7]. However, MAUT is unable to produce individual utility function [7].

The fund allocation for maintenance strategy is determined on the basis of increment in the vehicle operation cost (VOC). The VOC increases with the time due to increment in the traffic loading [3]. A deterioration model is developed considering cracking, raveling, potholes and roughness. A comparison is made between developed deterioration model and HDM-4 deterioration [8]. In HDM-4 software, the higher value of NPV/CAP ratio represents the requirement of a higher maintenance strategy [9]. The remaining service life is also considered as an indicator for the maintenance work of a pavement [10]. However, HDM-4 sotware is an expensive software which is not affordable for everyone. Fuzzy logic is a powerful inference system between linguistic and numerical data. It establishes the relationship between symbolic and numerical spaces [11]. Now, a day fuzzy logic expert system (FLES) receives the attention of the researcher due to incorporation of human expert knowledge with its nuances. Fuzzy logic reveals the result in an interpretable way for a researcher. Fuzzy logic is a powerful inference system between linguistic and numerical data. It establishes the relationship between symbolic and numerical spaces. Now, a days FIS got the attention of the researcher due to incorporation of human expert knowledge with its nuances [12, 13].

The FLES has been adopted by the researchers to reveal the pavement condition. For the development of pavement rating system, FLES is used [14–16]. However, pavement roughness is determined using the fuzzy set theory [15]. Furthermore, it is used to determine the future condition of pavement distresses and finally determine the pavement performance [16, 17]. The maintenance and rehabilitation is evaluated using fuzzy logic [18]. Fuzzy logic systems (FLS) have the advantage of being able to integrate experts' basic information into decision-making processes through the use of implication laws capable of representing imprecise knowledge and qualitative data. Fuzzy structures are useful for modelling expert logic because they effectively manage linguistic rules and are fault-tolerant to subtle adjustments in input or device parameters. Based on the previous study of FLES, it is concluded that the pavement treatment process is determined using the the various pavement deteriorating parameters.

The objective of this paper is to develop a multiobjective optimisation technique for the pavement treatment selection process using the various deteriorating parameters.

2 Methodology

FLES has been proposed for the evaluation of optimal pavement treatment using the various pavement affecting parameters. There is always an issue regarding the selection of treatment process for a pavement managers from various set of solution available. The pavement deteriorating parameters are selected based on the literatures and from the expert advise. In the formulation of FLES, each solution reflected an optimal pavement treatment based on various objectives. In this study, there are six different parameters selected as the input of the FLES model whereas the pavement treatment process is used as the output. The input parameters of the model are PCI, traffic condition, precipitation, temperature, pavement age and budget allotted. If–then rules are used in the FLES model to establish a relationship between input parameters and output parametrs. The fuzzy rules are developed using the expert knowledge but it may changes from expert to expert. The expert opinion is the best suited option in the formulation of fuzzy rules. In this study, a group of twenty five pavement pavement maintenance engineers having working experience more than ten years are selected to define the fuzzy rules.

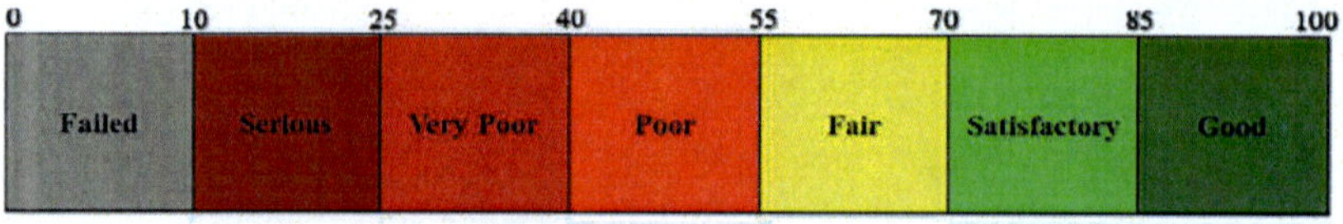

Fig. 1 Pavement condition index (PCI) rating [10]

2.1 *Pavement Condition Index (PCI)*

PCI is widely accepted for highways, airfields and parking lots to determine the pavement surface condition. The PCI is a numerical index, ranging from 0 (for a failed pavement) to 100 (for a pavement in perfect condition) which provides pavement's structural integrity and surface operation condition [9]. The PCI value is calculated based on the distress type, severity and quantity of pavement. The distresses in pavement are grouped into four categories such as surface defects, cracks, deformation and disintegration [10]. The degree of pavement deterioration is a function of distress types, distress severity and amount or density of distress. Deduct value was introduced as a type of weighing factor to indicate the effect of these factors [19] (Fig. 1).

2.2 *Traffic Volume*

The pavement distresses are mainly caused by pavement condition and traffic moving through it, so the maintenance mainly depends upon these two factors. As per IRC:37 (2018), the traffic volume is the number of commercial vehicles that passes through particular road segment, this traffic volume is measured in the form of cumulative vehicle per day (CVPD). The average annual growth rate of commercial vehicle is considered as 5% while designing the highways [20]. Different CVPD is obtained for different pavement section, hence, it is required to characterised the traffic volume in terms of CVPD (as shown in Table 1).

Table 1 Traffic volume factors [20]

Traffic	Commercial vehicle per day (CVPD)
Low	0–2000
Medium	2000–5000
High	5000–7000

Table 2 Categories of rainfall departure from normal [22]

Category	Rainfall amount (in cm)
Low rainfall	0–100
Medium rainfall	100–250
High rainfall	250–350

2.3 Precipitation

Precipitation in terms of rain, fog, snow, results in the change of moisture level of pavement and consequently due to this rise moisture level, distresses are formed in the pavement. Dawson [21] explore the various ways in which the moisture affects the pavement. The precipitation in the form of rainfall is generally drain out from the pavement through the gullies, drains and pipes that are constructed along with the road drainage whereas the surface runoff might enter inside the pavement through the distresses (cracks and potholes). Local water table also plays important role in formation of pavement distresses, due to near subgrade water table, the bonding between the subgrade and pavement layers degrades the bonding between aggregates, this results in the formation of potholes. During the flood and heavy rain, pavement subgrades might be completely submerged in water which also result in the negative impact in pavement.

India Meteorological Department (IMD) define rainy day with precipitation of 2.5 mm or more whereas extremely heavy rainy day is explained with rainfall greater than 650 mm [22]. During the entire monsoon period, from June to September, India received approximately 86% (102.08 cm) of its annual rainfall. The rainfall in India is categorised as High, Medium and Low (shown in Table 2).

2.4 Temperature

In the hot climate regions, due to elevated environmental temperature, the pavement surface gets heated up (up to 50 °C) which results in disintegration of bond between bitumen and fine aggregate, this tendency is known as ravelling whereas in the cold region, freezing low temperature of pavement renders the brittle property of surface which is prone to generate cracks. The stress–strain response of pavement is hugely depending on pavement temperature. The stress–strain response helps in spreading the wheel load over the surface and limiting this stress–strain response may results in load related deterioration [23]. Due the elevated pavement temperature, the resistance to permanent deformation is reduced for the pavement material. During this situation, thermal stress in asphalt layers is developed which results in thermal cracking [24]. According to IMD, most of the region of Indian states having 0–5 °C temperature in the winter season whereas 39–42 °C temperature is recorded in the summer season. The utility scores are developed using IMD report of temperature variation.

Table 3 Different categories of pavement age [20]

Pavement age	Age in years
Category 1	0–4
Category 2	4–8
Category 3	8–12
Category 4	12–15

Table 4 Categorisation of budget allotted for pavement treatment

Budget	Fund allotted (in lakh)
Low	0–15
Medium	15–30
High	30–75
Very High	>75

2.5 *Pavement Age*

The target National Highways in this study are having design period of 20 years as well as corridors, having design period of 30 years are also considered. More detailed design period of different categories of road are prescribed in IRC:37 (2018). A long-life pavement is designed for more than 50 years which is also termed as perpetual pavements. The pavement age is categorised in to four different categories and they are tabulated in Table 3.

2.6 *Budget*

The treatment is provided in the flexible pavement to maintain the pavement condition using the fund allotted by the Government. The budget is categorized in to four different group as low, medium, high and very high (as shown in Table 4). The fund allotted for maintenance for each one kilometer below 15 lakhs is termed as low budget whereas the fund allotted greater than 75 lakhs is termed as very high budget. The budget is utilized to improve road condition and extend its life.

2.7 *Pavement Treatment Process*

The treatment process is categorized in to four different group such as preventive treatment, maintenance/renewal, strengthening/rehabilitation and reconstruction (as discussed in IRC:82- (2015). The preventive treatment is mainly provided in the flexible pavement to fill the crack. The treatment process is adopted based on the types of distresses and their severity level. The periodic renewals treatment process

Table 5 Different types of flexible pavement treatment process [25]

Treatment process	Membership range (in terms of expected life in years)
Crack fill (CF)	0–2
Fog seal (FS)	1–3
Slurry seal (SS)	2–5
Chip seal (CS)	3–7
Micro-surfacing (MS)	5–8
Ultra-thin friction course (UTFC)	7–10
Thin hot mix bituminous overlay (THO)	8–12
Hot in place recycling (HIR)/cold in place recycling (CIR) + thin HMB overlay (THO)	10–13
Ultra-thin white topping (UTWO)	12–15
Reconstruction (R)	13–15

is a provision to provide a surfacing layer over the pavement after certain interval of time. The renewal treatment rectifies the camber, super elevation which get flattened due to repeated traffic. The suggested periodic renewal treatment is provided to eliminate the pavement roughness based on the experience and pavement condition. The pavement is strengthened using the hot mix bitumen mixture overlay and ultra-thin white topping. Once the pavement is completely deteriorated, the pavement is reconstructed or replaced [25].

The different types of treatment process are selected from the various literatures and various maintenance engineers advise. The range of MF for treatment process are suggested by the experts. The selection of treatment process for an older pavement are varied with expert to expert. Total twenty-five experts have been participated and their opinion have been considered. The treatment process with respect to the age of pavement is changes with expert to expert. Hence, the average range of already constructed pavement age is mentioned in Table 5 corresponding to various treatment process.

3 Data Collection

Six different national highways (NH) of Bihar, India are selected for this study. Pavement distress data, traffic volume, pavement age, precipitation, temperature, and the budget available for maintenance are all gathered in order to offer the optimal treatment to the pavement. The pavement distresses and severity level have been acquired from the field during the visual examination. All data required for this study is gathered from six distinct NHs situated in Bihar, India. NH-30 connects Bakhtiyarpur to Patna, NH-19 connects Patna to Chhapra, NH-28 connects Patna to Dhobi, NH-102 connects Chhapra to Muzaffarpur, and NH-57 is in Dabhanga

(a) NH-102 (b) NH-19 (c) NH-83 (d) NH-30

Fig. 2 Pavement condition of four national highways of India

(Bihar), India. The pavement condition of four NH is shown in Fig. 2. The pavement condition index (PCI) is computed utilising distress data in accordance with ASTM D6433-11 guidelines [26]. The traffic volume, pavement age and budget available for all six NH are collected from the national highway authority of India (NHAI) whereas temperature and precipitation has been tabulated from the Indian Meteorological Department (IMD) rainfall statistical report 2019 [22].

4 Model Formulation

FLES model is developed using the Fuzzy Inference System Program (FISPro). The FISPro software is developed using C + + language having a graphical Java interface for automatic learning. This software helps in developing FIS from the numerical data. The FISPro software has different learning methods not like machine learning or deep learning which have a black box system. FISPro is easy to operate and produce the entire fuzzy rules [27]. The basic working structure of FIS is explained in Fig. 3. The FIS is widely accepted for the classification problems. The learning approach consists of fuzzy partitioning, fuzzy rule induction and fuzzy defuzzification. The semantics is maintained while implementing the FISPro software. For a multi-objectives optimization model, an index is developed to maintain the semantic interpretability [28].

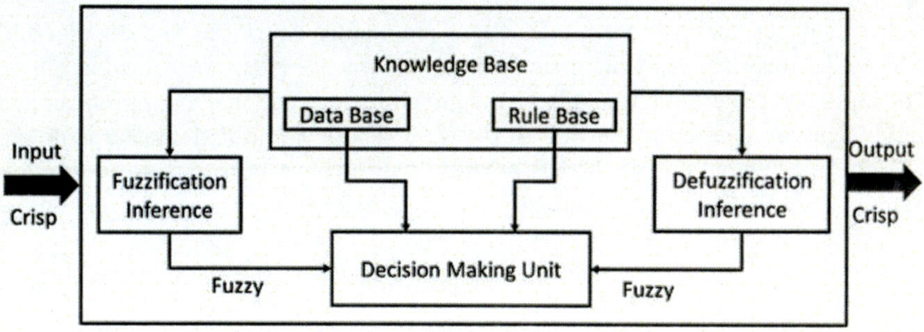

Fig. 3 The basic structure of FIS

4.1 Generation of Membership Functions

A fuzzy set is deliberated as a smooth boundary which generalizes the membership functional notation from black to white binary category. This binary category provides partial membership on the basis of classical set theory. In the fuzzy set, the membership is categorized between 0 to 1 [29], where 0 represents the non-membership function and 1 is represented for complete membership function. In mathematical terms, a fuzzy set is described by a mapping of the interval [0,1] from its discourse universe. MF is graphically represented and also define the membership degree through the mapping from variable space. The generated membership function in this model is shown in Fig. 4.

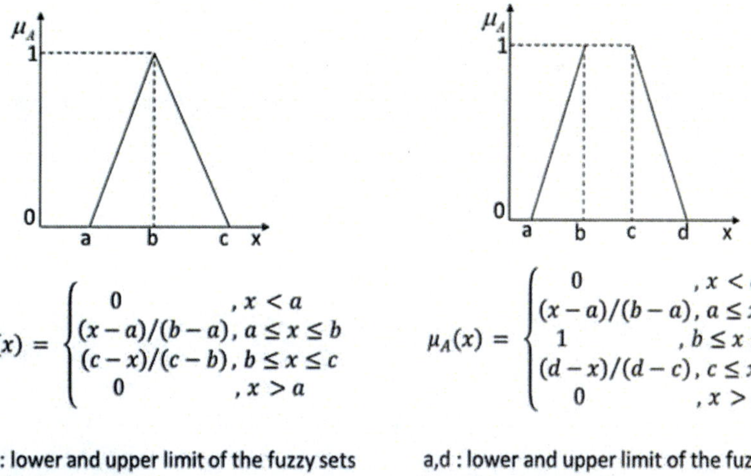

$$\mu_A(x) = \begin{cases} 0 & , x < a \\ (x-a)/(b-a) & , a \le x \le b \\ (c-x)/(c-b) & , b \le x \le c \\ 0 & , x > a \end{cases}$$

a,c : lower and upper limit of the fuzzy sets
b : the average limit
(a)

$$\mu_A(x) = \begin{cases} 0 & , x < a \\ (x-a)/(b-a) & , a \le x \le b \\ 1 & , b \le x \le c \\ (d-x)/(d-c) & , c \le x \le d \\ 0 & , x > d \end{cases}$$

a,d : lower and upper limit of the fuzzy sets
b,c : the average limit
(b)

Fig. 4 Shape of membership function **a** triangular member function **b** trapezoidal member function

In this paper, the membership functions are mainly developed in triangular and trapezoidal shapes. The mathematical expression for MF is also explained in Fig. 4. The input parameters such as PCI, traffic, precipitation, temperature, pavement age and budget are defined in the FISPro software. The low and high cluster is drawn as semi trapezoidal shape whereas medium cluster have been expressed through triangular shape. Each input parameter is categorized and their ranges are explained in the Sect. 2.

4.2 Membership Function

The defined ranges of input variables are entered in the FISPro software to generate the membership function (MF). Each parameter has their different range of variation and according to them MF is developed by the FISPro. The lower and higher range of input parameter are generated in semi-trapezoidal shape whereas mid-range value is represented by triangular shape in developing the MF for all input parameter. The remaining input parameter such as traffic, precipitation, temperature, age and budget has been represented by triangular membership function, since, the input parameter value has been defined in a range bound by the maintenance engineers. The graphical representation of MF generated using FISPro is shown in Fig. 5. The MF for the pavement treatment process is also generated and shown in Fig. 6. All ten-treatment process has been defined with respect to the age of pavement and their corresponding age has been represented by Table 5. The MF for treatment process has been generated by FISPro and their graphical representation is shown in Fig. 6. The MF for each pavement treatment process is represented as triangular shape by FISPro software, since they lie in a range bound. The variation in their ranges of each input and output parameter is expressed by x-axis whereas the y-axis represents the MF in the range of [0,1]. The zero is explained as the no relation whereas the maximum relation is shown through 1.

4.3 Formulation of Fuzzy Rule

The fuzzy rule is generated using simple and logical IF–THEN rules [30]. Fuzzy rule formulation is a major challenge for a high dimensional problem. The selection of every possible rule is not as easy as it looks. The fuzzy rules are designed based on expert knowledge or from the numerical data that restricts a number of possible situations [14]. For a more precise rule, one may have a higher number of fuzzy rules with more information. The successive refinement and sequential set of fuzzy rules will provide a precise rule. The complete fuzzy rule for any model is equal to $\prod_{i=1}^{n} m_i$ [14] where m represents the number of MF for i input and numbers of input are indicated by n. The *i*th pair rule is as follow:

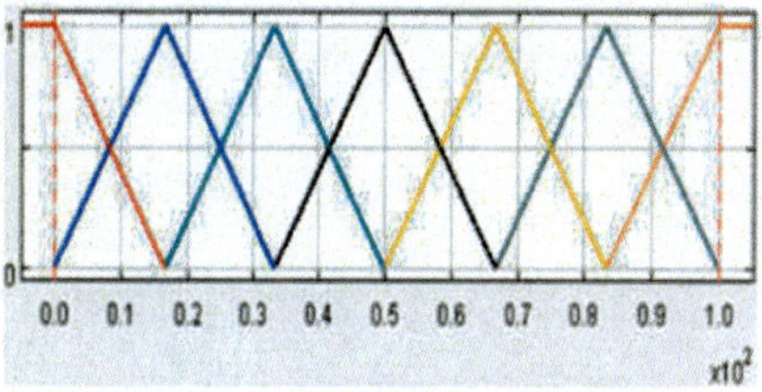

(a) Membership Function for PCI

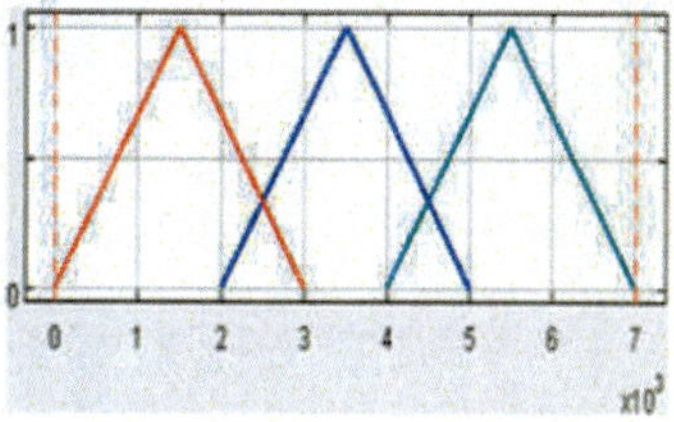

(b) Membership Function for traffic

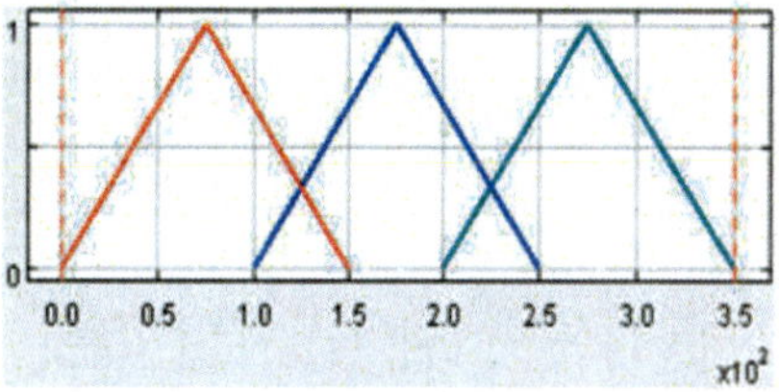

(c) Membership Function for precipitation

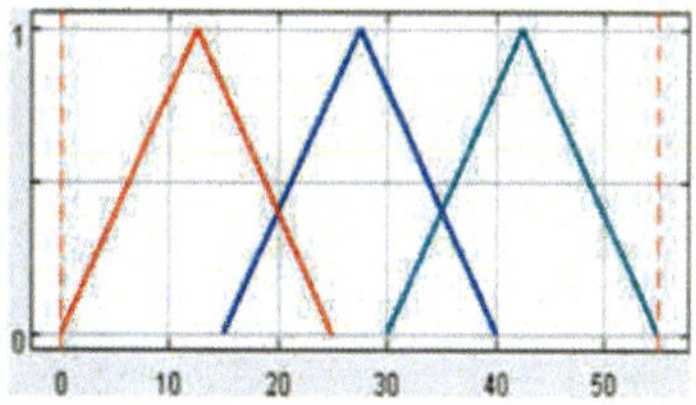

(d) Membership Function for temperature

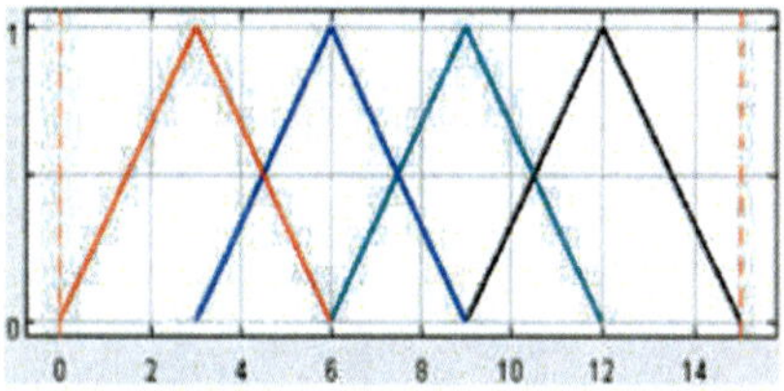

(e) Membership Function for Age

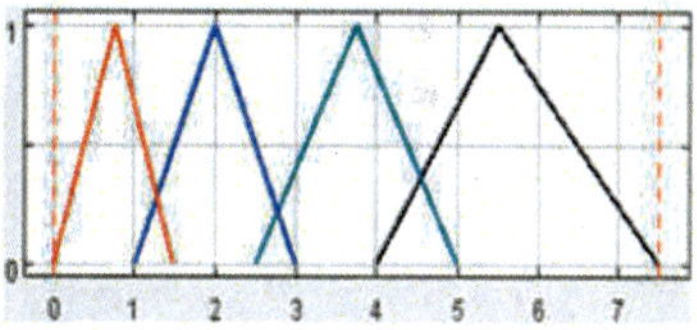

(f) Membership Function for budget

Fig. 5 Membership function for input parameter

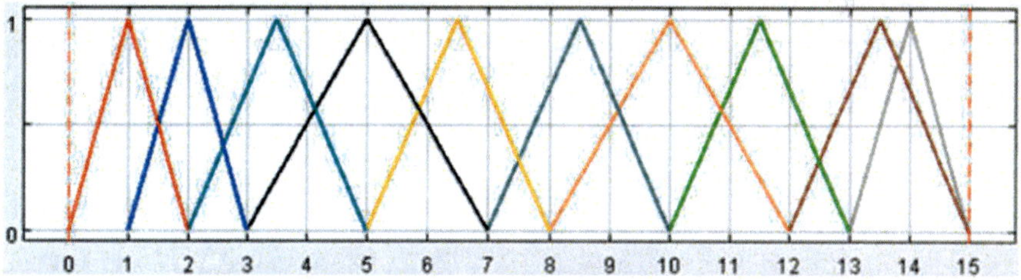

Fig. 6 Membership function for the output variable (treatments)

$$If x_1 is A_1^i AND x_2 is A_2^i \ldots\ldots AND x_p is A_p^i THEN y_1 is C_1^i$$

The rule formulation in this paper is complex due to the consideration of six input variables with different ranges of variation and one output variables. FISPro software is used to overcome this issue. Fuzzy system is generated with the numerical

Rule	Active	IF PCI	AND TRAFFIC	AND PRECI...	AND TEMP...	AND BUDGET	AND AGE	THEN TREA...
1688	✔	EXCELLENT	L	H	M	M	A-6-12	CRACK FILL
1689	✔	EXCELLENT	L	H	M	H	A-0-6	FOG SEAL
1690	✔	EXCELLENT	L	H	M	H	A-3-9	FOG SEAL
1691	✔	EXCELLENT	L	H	M	H	A-6-12	FOG SEAL
1692	✔	EXCELLENT	L	H	M	H	A-9-15	FOG SEAL
1693	✔	EXCELLENT	L	H	H	L	A-0-6	CRACK FILL
1694	✔	EXCELLENT	L	H	H	L	A-3-9	CRACK FILL
1695	✔	EXCELLENT	L	H	H	L	A-6-12	CRACK FILL
1696	✔	EXCELLENT	L	H	H	M	A-0-6	CRACK FILL
1697	✔	EXCELLENT	L	H	H	M	A-3-9	CRACK FILL
1698	✔	EXCELLENT	L	H	H	M	A-6-12	CRACK FILL
1699	✔	EXCELLENT	L	H	H	H	A-0-6	FOG SEAL
1700	✔	EXCELLENT	L	H	H	H	A-3-9	FOG SEAL
1701	✔	EXCELLENT	L	H	H	H	A-6-12	FOG SEAL
1702	✔	EXCELLENT	L	H	H	H	A-9-15	FOG SEAL
1703	✔	EXCELLENT	M	L	L	L	A-0-6	CRACK FILL
1704	✔	EXCELLENT	M	L	L	L	A-3-9	CRACK FILL
1705	✔	EXCELLENT	M	L	L	L	A-6-12	CRACK FILL
1706	✔	EXCELLENT	M	L	L	M	A-0-6	CRACK FILL
1707	✔	EXCELLENT	M	L	L	M	A-3-9	CRACK FILL
1708	✔	EXCELLENT	M	L	L	M	A-6-12	CRACK FILL
1709	✔	EXCELLENT	M	L	L	H	A-0-6	FOG SEAL
1710	✔	EXCELLENT	M	L	L	H	A-3-9	FOG SEAL
1711	✔	EXCELLENT	M	L	L	H	A-6-12	FOG SEAL
1712	✔	EXCELLENT	M	L	L	H	A-9-15	FOG SEAL
1713	✔	EXCELLENT	M	L	M	L	A-0-6	CRACK FILL
1714	✔	EXCELLENT	M	L	M	L	A-3-9	CRACK FILL
1715	✔	EXCELLENT	M	L	M	L	A-6-12	CRACK FILL
1716	✔	EXCELLENT	M	L	M	M	A-0-6	CRACK FILL
1717	✔	EXCELLENT	M	L	M	M	A-3-9	CRACK FILL
1718	✔	EXCELLENT	M	L	M	M	A-6-12	CRACK FILL
1719	✔	EXCELLENT	M	L	M	H	A-0-6	FOG SEAL

Fig. 7 A part of designed rule base for the fuzzy-based treatment selection system

data using FISPro [27]. Wang and Mendel's method is used to generate the fuzzy rule considering the predefined MF for used input and output variables [31]. It starts generating the rule for each training dataset. The defined fuzzy rule with expert knowledge is shown in Fig. 7. The FISPro software have generated more than three thousand rules on considering the input parameter and output parameter with their variation in the ranges. The expert advises have been proven very helpful in identifying the most appropriate rule in the selection of the pavement treatment process using the flexible pavement deteriorating parameters. Total 1882 set of legitimate rules are identified for the selection of various treatment process.

4.4 *Defuzzification*

The defuzzification is always necessary for the gradual conjunctive rules. The findings are obtained using maxima and centroid defuzzification. The continuous function is revealed only by the centroid defuzzification [32]. This function is not linear and generally monotonic in nature. But, when two rules are triggered at the same level then defuzzified values shift towards the wider one. The fuzzy linguistic output is converted into a crisp output using the defuzzification process considering the centroid defuzzication. The fuzzy rule is analyzed through rule viewer as shown in Fig. 8. The effects of individual MF of each input for the output are observed through the rule viewer.

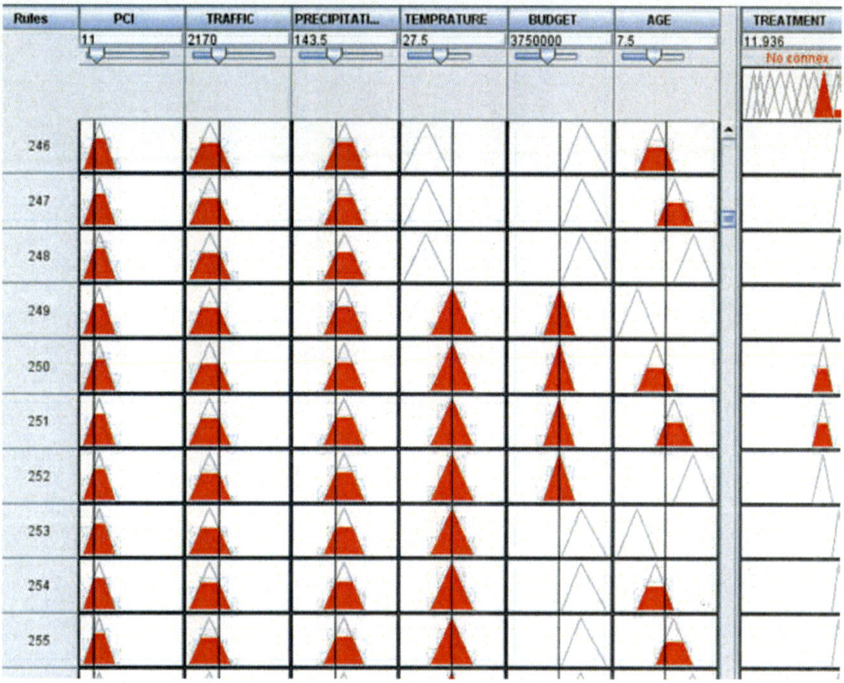

Fig. 8 Fuzzy rule viewer window for the inference system using FISPro software

5 Results and Discussion

The developed FLES model is able to optimise a flexible pavement treatment process using the various input parameter such as PCI, traffic, age, precipitation, temperature and budget. Each input parameter has been categorised as mentioned in section II and their range of variation is also defined. The MF for each parameter is developed as their range of variation mentioned in section II. The fuzzy rules are developed by FISPro based on the input parameter and output parameter.

5.1 Validation of FLES Model

The developed fuzzy model has been used to perform a case study over six national highways. The data has been collected from NH-30, NH-19, NH-28, NH-83, NH-57 and NH-102 at the locations such as Fatuha, Hazipur, Gopalganj, Gaya, Darbhanga and chhapra situated in Bihar. Using generated rules and membership functions, the FLES is analyzed of its efficiency. For each chosen section at six different national highways, data corresponding to various parameters is recorded as mentioned in Table 6. The inference results are obtained from the FLES model and compared with

Table 6 Selection of treatment process using the expert advise

Highway	PCI	Traffic (CVPD)	Age (Years)	Precipitation (cms)	Temperature (^{0}C)	Budget (in Lakh)	Inference result	Treatment FLES	Treatment expert advise
NH-30	90	2321	10	271	41	12.00	3.58	SS/CS	CF
NH-19	63	3080	9	227	16.5	25.50	9.32	UTFC/THO	UTFC
NH-28	23	756	12	245	47.8	58.50	14	UTW/R	R
NH-83	45	6649	5	266	12.65	65.25	13.7	UTW/R	UTW
NH-57	36	1060	12	230	43	47.56	10.9	THO/HIR/CIR	HIR/CIR
NH-102	53	3549	7	98	36.85	30.75	9.54	UTFC/THO	UTFC

the expert's choices. The FLES model has suggested us two treatment choices for a given set of conditions due to overlapping discourse for treatments membership defined by expected life. The maintenance engineer or the expert has suggested the treatment process depending upon the budget available for undertaking maintenance works. The decision can also be based on pavement conditions accompanied by the pavements' age. Due to limited fund allocation, the budget followed by pavement conditions is given more priority than other parameters.

For the data sets in Table 6, NH 30, the inference results have provided two choices—slurry Seal and chip seal. However, used FLES for the first dataset has shown some inconsistency with the expert's decision that suggests the use of crack fill for the given set of varying conditions. For the second section (NH 19), the inference results have been represented the use of an Ultra-thin friction course or Thin HMB overlay which describe the agreement with expert's treatment choice of the ultra-thin friction course. In case there is availability of higher budget and if the condition and age demands, Thin HMB overlay can also be feasible. The expert's opinion for the section chosen corresponding to NH 28 was the reconstruction, and the inference also suggests the use of Ultra-thin white topping or reconstructing the pavement section. NH 83, NH 57, and NH 102 have represented the compatibility with the expert's choices. Out of 6 sections, the five-pavement section have shown the consistency which defends the developed FLES approach's applicability.

6 Conclusions

The inferred results have shown the homogeneity with the expert's judgment while making choices for the treatment, considering the parameters used while developing FLES. Maintenance engineers have mostly stressed over the existing condition, budget, anticipated traffic, and age for the rule designing followed by other factors' influence. However, the developed FLES has provided two suggestive choices which lead to obscure choices. The available budget and resources are followed by pavement condition and then pavement age is used as the supportive parameter by the concerned experts or engineers in the selection of appropriate choices. The study has mainly focused to make treatment selection more efficient and effortless. Simultaneously, appropriate and feasible choices can be made as introduced FLES helped make the selection process closer to human reasoning.

Further study can look into each parameter's sensitivity concerning the choices made. Other structural and functional factors can also be incorporated. A hierarchical system can be developed that considers both pavement condition assessment and treatment selection.

Acknowledgements This research did not receive any specific grant from funding agencies in the public, commercial, or not-for-profit sectors. The authors declare that the contents of this article have not been published previously. All the authors have contributed to the work described, read and approved the contents for publication in this journal.

Conflict of Interest All the authors have no conflict of interest with the funding entity and any organization mentioned in this article in the past three years that may have influenced the conduct of this research and the findings.

References

1. Chassiakos AP (2006) A fuzzy-based system for maintenance planning of road pavements 2006:535–540
2. Ramachandran S, Rajendran C, Amirthalingam V (2019) Decision support system for the maintenance management of road network considering multi-criteria. Int J Pavement Res Technol 12:325–335. https://doi.org/10.1007/s42947-019-0039-7
3. Reddy BB and AV (2001) Priority ranking model for managing flexible pavements at network level. J Indian Road Congress, Paper. pp 379–394
4. Ahmed S, Vedagiri P, Krishna Rao KV (2017) Prioritization of pavement maintenance sections using objective based Analytic Hierarchy Process. Int J Pavement Res Technol 10:158–170. https://doi.org/10.1016/j.ijprt.2017.01.001
5. Farhan J, Fwa TF (2011) Use of analytic hierarchy process to prioritize network-level maintenance of pavement segments with multiple distresses. Transp Res Rec 11–20. https://doi.org/10.3141/2225-02
6. Torres-Machi C, Pellicer E, Yepes V, Chamorro A (2017) Towards a sustainable optimization of pavement maintenance programs under budgetary restrictions. J Clean Prod 148:90–102. https://doi.org/10.1016/j.jclepro.2017.01.100
7. Wu Z, Flintsch G, Ferreira A, de Picado-Santos L (2012) Framework for multiobjective optimization of physical highway assets investments. J Transp Eng 138:1411–1421. https://doi.org/10.1061/(ASCE)TE.1943-5436.0000458
8. Jain SS, Aggarwal S, Parida M (2005) HDM-4 pavement deterioration models for Indian national highway network. J Transp Eng 131:623–631. https://doi.org/10.1061/(ASCE)0733-947X(2005)131:8(623)
9. Jain K, Jain SS, Chauhan MS (2013) Selection of optimum maintenance and rehabilitation strategy for multilane highways. Int J Traffic Transp Eng 3:269–278. https://doi.org/10.7708/ijtte.2013.3(3).04
10. Suman SK, Kumar R (2018) Pavement performance modelling using markov chain with and without application of maintenance. In: 2nd international conference on recent multidisciplinary research. AIT Conference Center. Thailand, pp 1–9
11. Zadeh LA (2008) Is there a need for fuzzy logic? Inf Sci (Ny) 178:2751–2779. https://doi.org/10.1016/j.ins.2008.02.012
12. Alonso JM, Magdalena L, Guillaume S (2004) KBCT: a knowledge extraction and representation tool for fuzzy logic based systems. IEEE Int Conf Fuzzy Syst 2:989–994. https://doi.org/10.1109/FUZZY.2004.1375543
13. Alonso JM, Magdalena L (2010) Guaje—a java environment for generating understandable and accurate models. XV Congr Español sobre Tecnol y Lógica Fuzzy 399–404
14. Mahmood M, Rahman M, Nolle L, Mathavan S (2013) A fuzzy logic approach for pavement section classification. Int J Pavement Res Technol 6:620–626. https://doi.org/10.6135/ijprt.org.tw/2013.6(5).620
15. Loizos A (2001) A simplified application of fuzzy set theory for the evaluation of pavement roughness. Road Transp Res 10:21–32
16. Karaşahin M, Terzi S (2014) Performance model for asphalt concrete pavement based on the fuzzy logic approach. Transport 29:18–27. https://doi.org/10.3846/16484142.2014.893926
17. Jeong H, Kim H, Kim K, Kim H (2017) Prediction of flexible pavement deterioration in relation to climate change using fuzzy logic. J Infrastruct Syst 23. https://doi.org/10.1061/(ASCE)IS.1943-555X.0000363

18. Moazami D, Behbahani H, Muniandy R (2011) Pavement rehabilitation and maintenance prioritization of urban roads using fuzzy logic. Expert Syst Appl 38:12869–12879. https://doi.org/10.1016/j.eswa.2011.04.079
19. Shahin MY (2005) Pavement management for airports, roads, and parking lots, 2nd edn. Springer, New York
20. IRC:37 (2018) Guidelines for the design of flexible pavements, fourth Indian road congress, New Delhi, India
21. Dawson A (2009) Water in road structures: movement, drainage and effects. In: Geotechnical, geological and earthquake engineering, vol 5. Springer. https://doi.org/10.1007/978-1-4020-8562-8
22. IMD (2019) Rainfall Statistics of India—2019. India Meteorol Dep 1:1–110
23. Dawson A, Mallick R, García HA, Dehdezi PK (2014) Energy harvesting from pavements. Clim Chang Energy Sustain Pavements. https://doi.org/10.1007/978-3-662-44719-2
24. Dave EV, Hoplin C (2015) Flexible pavement thermal cracking performance sensitivity to fracture energy variation of asphalt mixtures. Road Mater Pavement Des 16:423–441. https://doi.org/10.1080/14680629.2015.1029697
25. IRC:82 (2015) Code of practice for maintenance of bituminous surfaces of highways, Indian Road Congress, New Delhi, India
26. ASTM D6433–07 (2008) Standard practice for roads and parking lots pavement condition index surveys. ASTM International, West Conshohocken, PA 19428–2959, United States, pp 1–48
27. Guillaume S, Charnomordic B (2011) Learning interpretable fuzzy inference systems with FisPro. Inf Sci (Ny) 181:4409–4427. https://doi.org/10.1016/j.ins.2011.03.025
28. Gacto MJ, Alcalá R, Herrera F (2010) Integration of an index to preserve the semantic interpretability in the multiobjective evolutionary rule selection and tuning of linguistic fuzzy systems. IEEE Trans Fuzzy Syst 18:515–531. https://doi.org/10.1109/TFUZZ.2010.2041008
29. Negnevitsky M (2002) Artificial intelligence a guide to intelligent systems. Harlow, Engl Addison-Wesley
30. Nelles O, Fischer M, Mueller B (1996) Fuzzy rule extraction by a genetic algorithm and constrained nonlinear optimization of membership functions. IEEE Int Conf Fuzzy Syst 1:213–219. https://doi.org/10.1109/fuzzy.1996.551744
31. Wang L, Mendel JM (1992) Rules by learning from examples. October 22:1414–1427
32. Jones H, Charnomordic B, Dubois D, Guillaume S (2009) Practical inference with systems of gradual implicative rules. IEEE Trans Fuzzy Syst 17:61–78. https://doi.org/10.1109/TFUZZ.2008.2007851

Functional Condition Evaluation of Airfield Pavements Using Automated Road Survey System—A Case Study of a Small Sized Airport

Pradeep Kumar and Madhavendra Sharma

Abstract Airports are vital national resources. Airfield Pavements within an airport represents a large capital investment in infrastructural development made by a country. Timely and appropriate maintenance and rehabilitation of such in-service facilities are essential to provide an all-weather surface for safe and regular operations of the aircraft. Pavement maintenance is done based on functional and structural pavement condition evaluation. This paper presents a case study dealing with functional evaluation of airfield pavements of a small sized airport in India. The considered airport consists of two runway, seven link taxiways, three apron areas and an isolation bay with different surface types. Present paper reports the methodology adopted for the functional condition evaluation of the airfield pavements with help of Automated Road Survey System. Further, the evaluated pavement condition was quantified in terms of Pavement Condition Index (PCI) as per the ASTM D5340 with help of PAVER and GIS based software. GIS was used for preparing inventory database and base maps for the concerned pavements which were then used in PAVER software for determining the PCI. The airfield pavement network within the airport was divided into a four-level hierarchy consisting of the network, branch, section and sample. The obtained PCI rating shows that the overall condition of the airfield pavements within the considered airport is satisfactory to good, however some of the areas have distresses that needs to be repaired by localized maintenance.

Keywords Airfield pavement management system · Pavement condition index · GIS · PAVER

1 Introduction

Air transport is growing manifolds in terms of number of operations worldwide. As a result of growing air transport, new airfield infrastructure is being developed worldwide. Airports serves as a source of fast and reliable mode of transportation in present times. Airfield pavements are a major investment with a new runway typically

P. Kumar · M. Sharma (✉)
Pavement Evaluation Division, CSIR—Central Road Research Institute, New Delhi 110076, India

© The Author(s), under exclusive license to Springer Nature Switzerland AG 2022

H. R. Pasindu et al. (eds.), *Road and Airfield Pavement Technology*, Lecture Notes in Civil Engineering 193, https://doi.org/10.1007/978-3-030-87379-0_13

costing between $100 M and $500 M to plan, design and construct, including all the associated infrastructure [1, 2]. These airfield pavements act as the backbone for economic development in the region, not only in developed countries but also in developing countries [3].

Immediately after the construction, airfield pavements begin a gradual deterioration attributable to weather and loading. With higher deterioration when pavements are in poor condition, the maintenance cost would increase manifolds as compared to pavements in good condition [4]. Airfield Pavement Management System (APMS) is employed for assisting the airport authorities in effective decision making in Maintenance and Rehabilitation (M&R) of pavements. Collection, assessment and reporting of present pavement condition are some of the major tasks to be performed while developing an APMS. Condition assessment of airfield pavements is done based on condition indicators obtained from visual distress survey. Visual distress surveys are usually performed by walking (Foot on Ground) or windshield method consuming time and labor.

This paper presents a methodology adopted for condition evaluation of the airfield pavements in India using Automated Road Survey System (ARSS). Use of ARSS for condition surveys of highway pavements is accepted worldwide but use of such techniques for airfield pavements is not widespread. One of the best practices in the state of the art in airport infrastructure management is maintenance management with a combination of Global Positioning System (GPS) and Geographic Information System (GIS) Technologies [5]. The work reported within the present paper incorporate both GPS and GIS for the functional condition evaluation of the airfield pavements.

APMS provides a consistent, objective and systematic procedure for establishing facility policies, setting priorities and schedules, allocating resources, and budgeting for pavement maintenance and rehabilitation [6]. APMS is widely used techniques for planning and assessment of airfield pavements around the globe [3, 7–10]. APMS uses the concept of Pavement Condition Index (PCI), Foreign Object Damage Potential (FOD Index), Structural Index and Friction properties of Airfield Pavements for condition assessment [11]. Determination of PCI is done based on pavement surface distresses as specified in ASTM D5340 [12]. FOD Index is determined based on subset of distresses mentioned in ASTM D5340 and can be calculated using the PAVER software. APMS as reported in different articles [8, 9, 11, 13–18], used FOG visual survey technique for distress survey of airfield pavements, however FAA guidelines [6] doesn't lay specific guidelines for the visual condition survey. Hence, to reduce the time and labor consumption, the present study used ARSS for condition survey.

2 Materials and Methods

The methodology adopted for the study make use of technologies like Geographic Information System (GIS), Global Positioning System (GPS), Remote Sensing systems in form of Automated Road Survey System (ARSS) and PAVER.

2.1 *Airfield Pavement Network Definition*

For the purpose of the inspection and evaluation, airfield pavements within the airport are divided in a hierarchical structure of network, branches and sections which is considered to be the first step in establishing a APMS [12, 19]. The inventory data was obtained from the airport authority. The obtained data was used as secondary data for preparation of GIS map and classification of considered pavements into network, branch and section. The pavements within the considered airport were classified into three different network, one each of runway (RW), taxiway (TW) and Apron (AP). Pavement within network were divided further into branch and section based on crust and surface characteristics. Pavement area distribution within airport as per use and surface type is presented in Figs. 1 and 2 respectively. The inventory GIS map prepared is presented in Fig. 3.

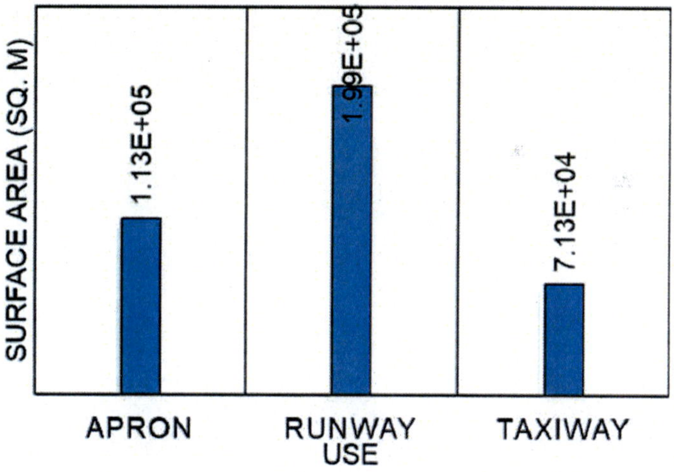

Fig. 1 Pavement area distribution as per use

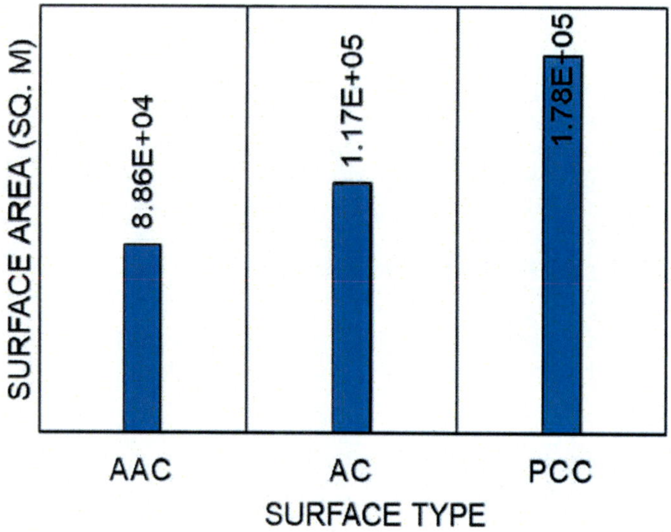

Fig. 2 Pavement area distribution as per surface type

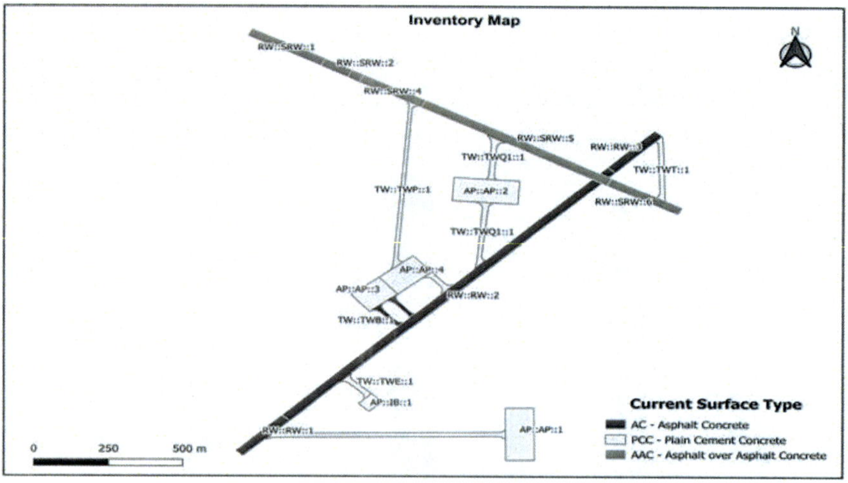

Fig. 3 Prepared inventory map in GIS of airfield pavements

2.2 *Airfield Pavement Distress Condition Evaluation*

Airfield Pavement Distresses and condition were surveyed in accordance with the methods outlined in FAA Advisory Circular 150/5380-6C and ASTM D 5340—18 using Automated Road Survey System (ARSS). The procedure is based on the identification and measurement of distress at the pavement surface.

Fig. 4 Automated Road Survey System (ARSS) used for survey of Airfield Pavements

Traditionally, Pavement condition surveys have been conducted using foot-on-ground (FOG) approach on the airfields where inspectors walk on the pavement sections and collect the detailed distress data [9, 11, 21]. In this study, the pavement condition surveys have been conducted using Pavement Surface Imaging System (PSIS) installed in Automated Road Survey System (ARSS).

ARSS used in present study as shown in Fig. 4 is based on the latest survey techniques utilizing Global Positioning System and pavement surface imaging systems. Using PSIS of ARSS condition survey is conducted while travelling at speeds up to 80 kmph.

PSIS of ARSS involve the use of two pavement cameras as shown in Fig. 4. These cameras were used to generate the calibrated images (with known length and width) of the pavement surface. The measurement of various surface defects viz. cracking, raveling, and potholes was done by processing the captured calibrated pavement surface images, using the Image Processing software. To fulfill the objective of this study, emphasize were made on GPS data and pavement surface distress. Some of the prominent distresses observed in the study are presented in Table 1.

2.3 *Preparation of Distress Dataset*

As per the ASTM D 5340—18, for the calculation of Pavement Condition Indices; the pavement area needs to be divided into small inspection samples. The considered sample area for the flexible pavements was 450 ± 200 m^2 and for rigid pavements was 25 ± 10 slabs. Hence, the base maps in the GIS were divided into samples using small polygons. The ARSS data collected was further processed frame by frame using image processing technique to quantify the distresses observed on the airfields in terms of distress type, distress severity and extent as shown in Table 1.

Table 1 Prominent distresses observed on airfield pavements

Distress image	Pavement surface type	Distress type	Distress severity
	AC	Weathering	Low
	PCC	Longitudinal cracking	Medium
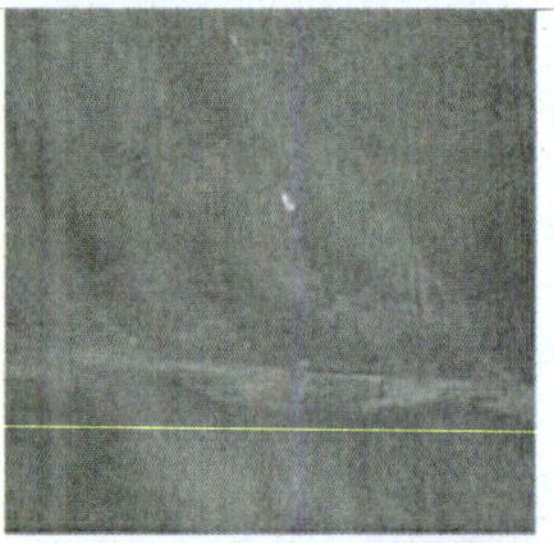	PCC	Joint spalling	Medium
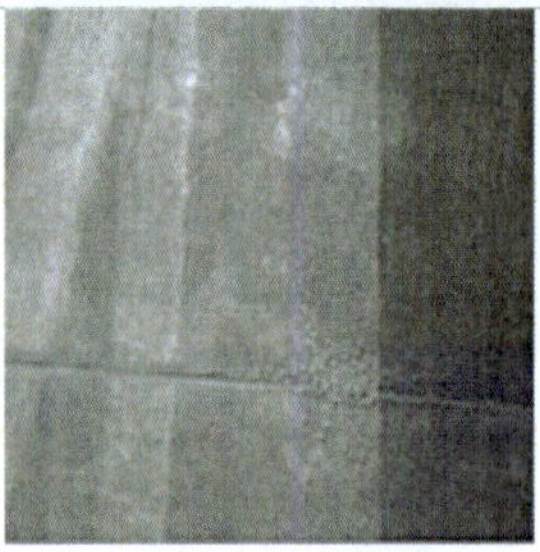	PCC	Joint Seal damage	High

(continued)

Table 1 (continued)

Distress image	Pavement surface type	Distress type	Distress severity
	PCC	Corner break	Low

The distress data prepared in such a way is a GPS enabled dataset having locations of distresses in terms of latitude and longitude which was further included in the GIS database at sample level in accordance with the ASTM D 5340—18. Figure 5 shows the schematics of the process followed in GIS for preparation of base map.

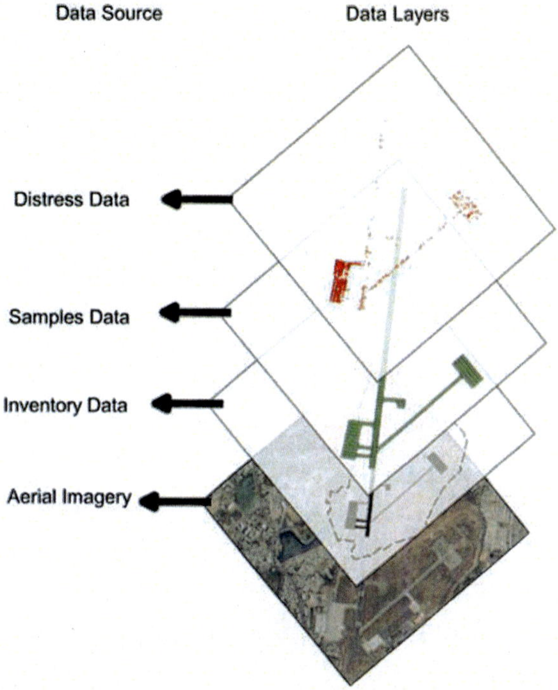

Fig. 5 Schematic diagram for distress database preparation in GIS

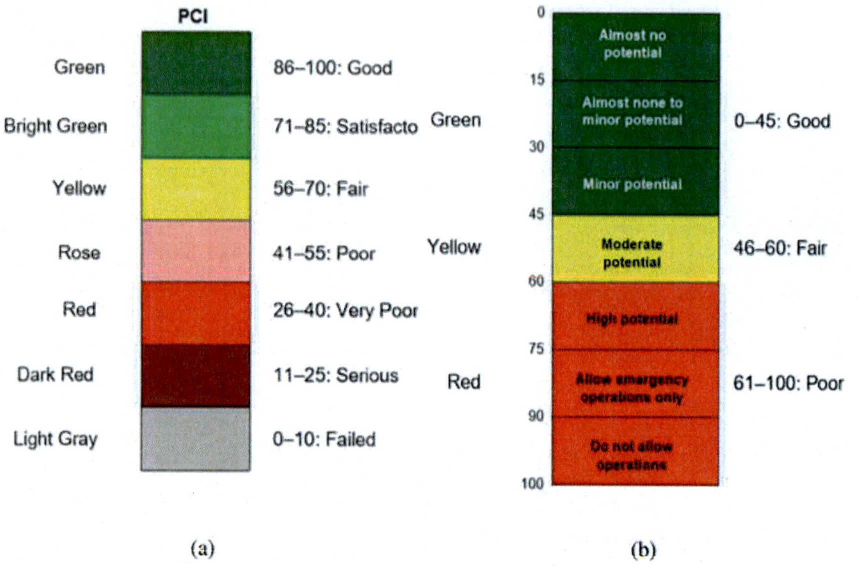

Fig. 6 Representation of condition indices; **a** Pavement condition index, **b** FOD potential index [23]

3 Result

As a part of functional condition evaluation two types of functional indices a) Pavement Condition Index (PCI) and b) Distress based Foreign Object Damage (FOD) Potential Index, were determined in the present study to quantify the present functional condition of the airfield pavements.

3.1 Pavement Condition Index (PCI)

Pavement Condition Index (PCI) as shown in Fig. 6a, is a score from 0 (failed) to 100 (good) that rates the ability of a pavement to perform its function effectively and safely [22].

3.2 Foreign Object Damage (FOD) Potential Index

The loose material resulting from surface distresses can cause damage to aircraft propellers and is called Foreign Object Damage (FOD). A FOD potential rating scale as shown in Fig. 4b ranging from 0 to 100 is used to indicate the potential for FOD problems [23].

In the present study PAVER software was used for determination of PCI and FOD potential index. Subsequent sections summarize the process of PCI and FOD potential determination using PAVER.

The distress data at sample level was imported in the PAVER database after preparation of airfield database. Distress data was imported using inspection wizards within the PAVER for rigid and flexible pavement sections as shown in Figs. 7 and 8 respectively.

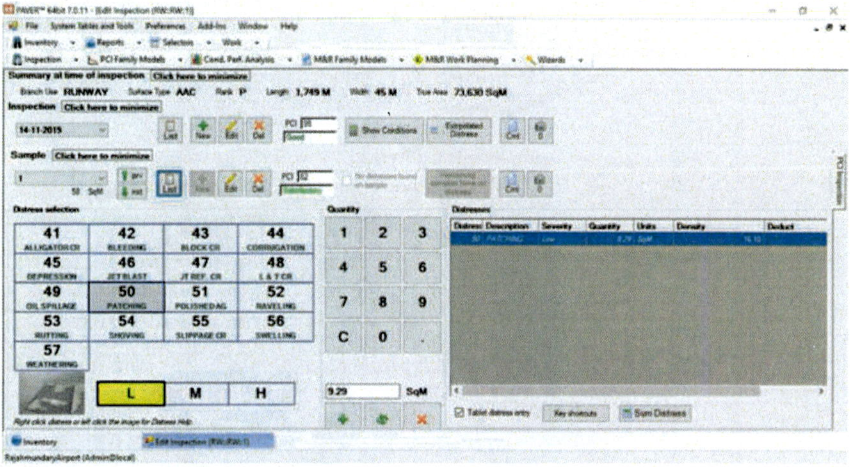

Fig. 7 Rigid Pavement Surface Inspection Import wizard

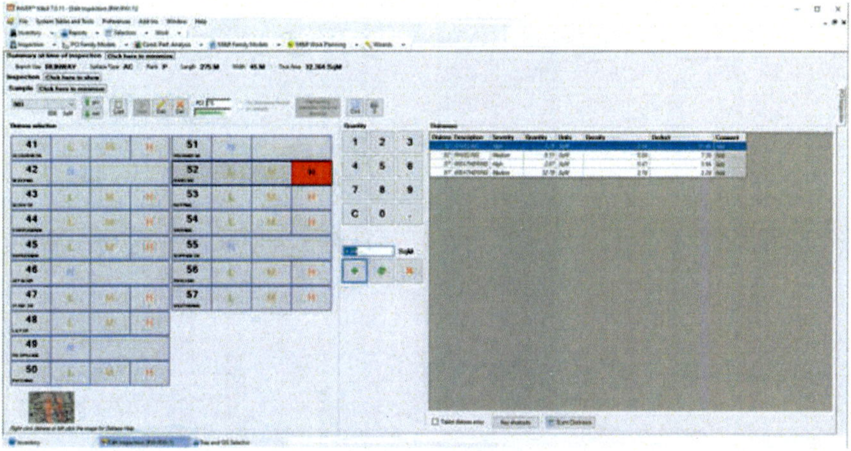

Fig. 8 Flexible pavement surface inspection import wizard

4 Conclusion and Discussions

Pavement Condition Data collection is considered to be the most expensive and time consuming task in establishing a Pavement Management System (PMS) [24]. To save time and labor in collecting condition data for airfield pavements, presented study worked on using ARSS for collecting condition data. Based on inspected sample data, the FOD Potential Rating and PCI were calculated using the PAVER condition analysis wizard. Section wise results are presented in Figs. 9 and 10.

While considering Pavement Condition Index (PCI), Runway pavements were found to be in satisfactory to good condition. While the condition of apron was

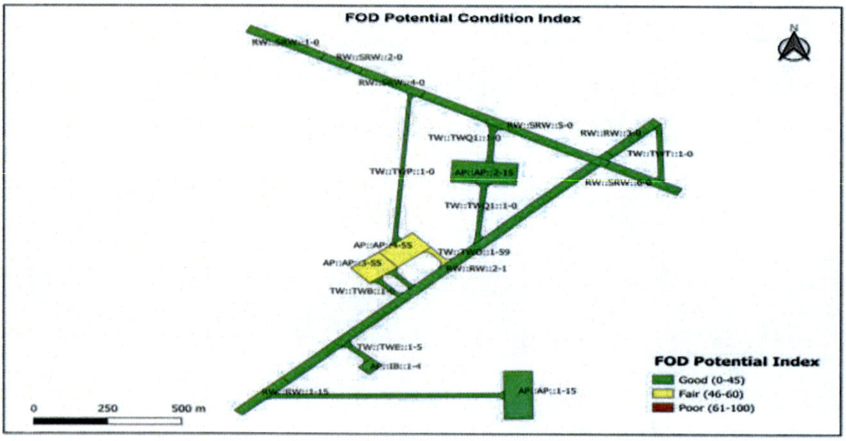

Fig. 9 Observed FOD Potential Index for different airfield pavement sections

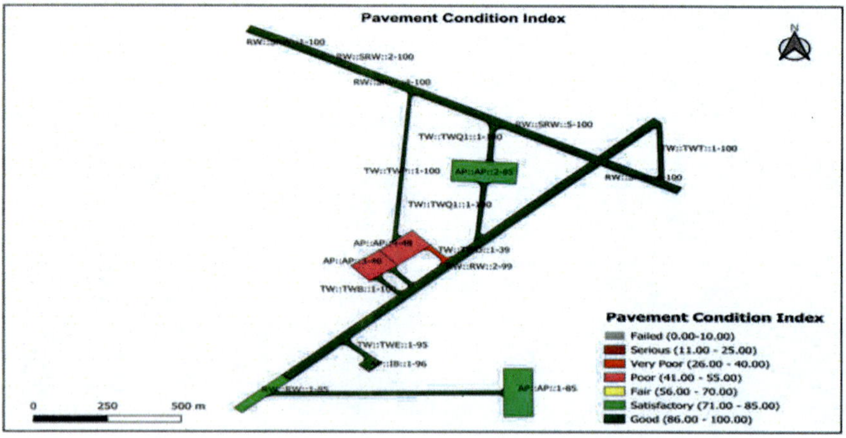

Fig. 10 Observed Pavement Condition Index (PCI) for different airfield pavement sections

found to be in very poor and poor condition. In case of FOD potential, all pavements are in fair to good condition as seen in Fig. 9.

The study performed was aimed at the use of ARSS as a fast measure of evaluating airfield pavements for functional aspects. The salient conclusions that can be drawn from the present study are as follows:

- The use of ARSS and GIS subsequently has greatly reduced the effort for visual inspection and processing of distress data.
- There are several advantages observed with use of PSIS (ARSS) over the use of FOG approach in the present study including minimum interruption to aircraft operations.
- Earlier the location of a distress needs to be measured from a benchmark point, using tape and compass method. While in the present study the geotagged images of pavement were obtained with use of GPS technology in ARSS, significantly reducing the manual effort.
- Use of GIS to prepare inventory and GPS to subsequently mark distressed locations will further help in preparing spatial temporal pavement condition data.
- The use of ARSS, helped in reducing the subjective factor linked with traditional methods, hence improving productivity and repeatability.

Referenes

1. White G, Kitchen R (2019) Parametric comparison of the whole of life cycle cost of rigid parametric comparison of the whole of life cycle. In: Eighteenth annual international conference on pavement engineering, asphalt technology and infrastructure. Liverpool, England
2. Grothaus JH, Helms TJ, Germolus S et al (2009) ACRP report 16: guidebook for managing small airport
3. Irfan M, Khurshid MB, Iqbal S, Khan A (2015) Framework for airfield pavements management—an approach based on cost-effectiveness analysis. Eur Transp Res Rev 7. https://doi.org/10.1007/s12544-015-0165-5
4. FAA (2006) AC 150/5380-7A, airport pavement management program (PMP)
5. McNerney MT (2000) Airport infrastructure management with geographic information systems: state of the art. Transp Res Rec 58–64. https://doi.org/10.3141/1703-08
6. FAA (2014) AC 150/5380-7B, airport pavement management program (PMP)
7. Weso M, Iwanowski P (2020) APCI evaluation method for cement concrete airport pavements in the scope of air operation safety and air transport participants life. 1–14. https://doi.org/10.3390/ijerph17051663
8. Fuselier GK, Orandello BJ, Arze JM (2001) Customizing work planning with micropaver and paver-GIS Washington Dulles international airport case study. Proc - Int Air Transp Conf 181–191. https://doi.org/10.1061/40579(271)15
9. Butt AAD, Shahin MYD, Grovetti MR (1993) Airfield pavement management system. Proc Infrastruct Plan Manag 207–211
10. McNerney MT, Keegan KA (2011) The benefits of using a geospatial airport pavement management system at Denver and Tampa International Airports. TDI Congr 2011 Integr Transp Dev a Better Tomorrow - Proc 1st Congr Transp Dev Inst ASCE 255–270. https://doi.org/10.1061/41167(398)26

11. Greene J, Shahin MY, Alexander DR (2004) Airfield pavement condition assessment. Transp Res Rec J Transp Res Board 1889:63–70. https://doi.org/10.3141/1889-08
12. ASTM D 5340 (2018) Standard test method for airport pavement condition index surveys
13. Lam JC, Bryce JM, Priddy LP, Flintsch GW (2013) Development of infrastructure management strategies for small and mid-size airfields. Airf Highw Pavement 2013 Sustain Effic Pavements - Proc 2013 Airf Highw Pavement Conf 597–608. https://doi.org/10.1061/9780784413005.048
14. Noruzoliaee M, Zou B (2019) Airfield infrastructure management using network-level optimization and stochastic duration modeling. Infrastructures 4:2. https://doi.org/10.3390/infrastructures4010002
15. Hein DK, Aho B (2011) Preserving our airfield pavements. TDI Congr 2011 Integr Transp Dev a Better Tomorrow - Proc 1st Congr Transp Dev Inst ASCE 244–254. https://doi.org/10.1061/41167(398)25
16. Rada GR, Schwartz CW, Witczak MW, Rabinow SD (1992) Integrated pavement management system for Kennedy International Airport. J Transp Eng 118:666–685. https://doi.org/10.1061/(ASCE)0733-947X(1992)118:5(666)
17. Chen W, Yuan J, Li M (2012) Application of GIS/GPS in Shanghai airport pavement management system. Procedia Eng 29:2322–2326. https://doi.org/10.1016/j.proeng.2012.01.308
18. Yanti, Sunarjono S, Riyanto A, et al (2019) Visual assessment deterioration analysis of runways at Sultan Aji Muhammad Sulaiman Sepinggan Airport Balikpapan. AIP Conf Proc 2114. https://doi.org/10.1063/1.5112445
19. Shahin MY (2005) Pavement management for airports, roads, and parking lots, second. Springer US, Boston, MA
20. FAA (2014) AC 150/5380-6C: guidelines and procedures for maintenance of airport pavements
21. Shahin MY, Kohn SD, Lytton RL, Japel E (1981) Development of a pavement maintenance management system. Maint Manag
22. Freeman TJ, Borowiec JD, Wilson B, et al (2016) ACRP report 159 - pavement maintenance guidelines for general aviation airport management
23. AFI32-1041 (2017) Pavement evaluation program
24. Cafiso S, Di Graziano A, Goulias DG, D'Agostino C (2019) Distress and profile data analysis for condition assessment in pavement management systems. Int J Pavement Res Technol 12:527–536. https://doi.org/10.1007/s42947-019-0063-7

Ice-Breaking and Skid Resistance Performance Evaluation of Asphalt Pavement Groove-Filled with Elastomer

Tengfei Yao, Sen Han, Changpeng Men, Jianchao Zhang, Jianrong Luo, and Yang Li

Abstract Icing on roads in cold regions seriously affects vehicle driving safety. Therefore, this study aims to propose a pavement surface treatment technology with groove-filled elastomer. Through theoretical analysis, the ice-breaking mechanism of groove-filled elastomer asphalt pavement was clarified. Then, through the indoor ice-breaking test, the ice-breaking performance of the asphalt pavement groove-filled with elastomeric materials under three filling forms of above, level and lower than the asphalt pavement surface and different ice thickness were studied. The results show that the filling form where the elastomer material is flush with the asphalt pavement has the best ice-breaking effect, and the groove-filled elastomer asphalt pavement still has an excellent ice-breaking effect when the thickness of the ice layer is less than 2 mm. Finally, through BPN and friction coefficient test, the skid resistance performance of groove-filled elastomer asphalt pavement under dry, wet and icing conditions is verified. Compared with AC-13 asphalt mixture, the skid resistance of groove-filled asphalt pavement is improved by 4.5, 10.8 and 24.6%.

Keywords Groove-filled asphalt pavement · Elastomer · Ice-breaking performance · Skid resistance performance

1 Introduction

In the cold winter, many highways and urban roads are often affected by rain, snow and freezing weather. When the snowfall is heavy or the ambient temperature continues to decrease, the road surface may be covered by ice for a long time,

T. Yao · S. Han (✉) · C. Men · Y. Li
Key Laboratory for Special Area Highway Engineering of Ministry of Education, Chang'an University, Xi'an City 710064, PR China
e-mail: hyram_hs@chd.edu.cn

T. Yao
e-mail: yaotengfei@chd.edu.cn

J. Zhang · J. Luo
Shaanxi Expressway Engineering Consulting Co., Ltd, Xi'an City 710054, PR China

© The Author(s), under exclusive license to Springer Nature Switzerland AG 2022

H. R. Pasindu et al. (eds.), *Road and Airfield Pavement Technology*, Lecture Notes in Civil Engineering 193, https://doi.org/10.1007/978-3-030-87379-0_14

resulting in a serious lack of skid resistance of the pavement. The research data show that the traffic accident rate of the frozen section is twice as high as that under the normal climate conditions, and the casualty rate is as high as 70% [1, 2]. Especially in special sections such as tunnel entrances and exits, long steep slopes and bends, the accident rate is higher. The anti-skid problem of the road surface in icy and snowy environments has become a key problem restricting safe road construction [3, 4].

First of all, traditional passive deicing methods on pavement include manual and mechanical deicing or spraying snow melting agents, but these methods are time-consuming and laborious, and deicing agents such as chloride pollute the environment and corrode road ancillary facilities [5]. Thermal snow melting method has less pollution to the environment and high snow melting efficiency. However, there are problems such as high construction difficulty, high energy consumption, and high maintenance costs in the later period, which are difficult to popularize and use on a large scale [6]. The super-hydrophobic anti-icing pavement is mainly made by preparing a layer of super-hydrophobic material on the surface of the pavement. This super-hydrophobic structure could reduce the surface energy of the pavement and weaken the adhesion between the ice and the pavement so that the ice on the road surface is easy to remove [7, 8]. However, the super-hydrophobic structure of this layer may lead to a decrease in the initial skid resistance performance of the pavement and the wear resistance of the hydrophobic material needs further study. Self-stressing elastic paving technology could be divided into elastic-filled pavement and inlaid paving pavement. It mainly adds a certain amount of elastic material to the pavement material to cause uneven deformation of the pavement under the load of the wheel, thereby causing the ice layer to generate self-stress to break the ice [9]. Kiyoshi Takeichi paved a variety of inlaid pavements on urban roads respectively. Through long-term observation, it was found that the grooved filled elastomer pavement had good snow removal ability and skid resistance performance, but the interfacial bond durability between elastomer and pavement still needs further study [10].

Polyurethane (PU) is a kind of polymer material with repeated carbamate segments (-NHCOO-) in its molecular structure, which is famous for its high strength, excellent elasticity, oil resistance and low-temperature resistance [11]. Polyurethane elastomer (PUE) is composed of a rigid hard segment and a flexible soft segment block among them, the hard segment is formed by diisocyanate and small molecular diol or diamine (chain extender), and the soft segment is oligomer polyol [12]. The thermodynamic incompatibility of the soft and hard phases in the chain segment causes the polyurethane to appear microphase separation, that is, the hard segment phase is evenly distributed in the soft segment phase to play a role of elastic crosslinking. This structural feature makes PUE possess extremely excellent performance [13]. Many scholars have introduced polyurethane into the research of pavement engineering materials. Sun et al. [14] prepared a polyurethane mixture with a skeleton chain structure. The properties and functional groups of PU were evaluated by Brookfield rotational viscosity test, Fourier transform infrared spectroscopy (FTIR) and Dynamic mechanical analysis (DMA), and the pavement performance of the polyurethane mixture was evaluated. Gao et al. [15] used different types of rubber particles and polyurethane to prepare rubber particle mixtures, and evaluated the high-temperature resistance, aging resistance, and skid resistance of the rubber particle mixtures. Cong et al. [16] studied the effect of polyurethane hard segment content on polyurethane properties. They tested the hardness, elongation, elastic modulus and other properties of polyurethane and studied OGFC13 through freeze–thaw splitting test, rutting test, friction coefficient test, and three-point bending test. In order to improve pavement performance, polyurethane with 41% to 51% hard chain segment content is recommended. Wang et al. [17] used polyurethane as cement to prepare porous elastic pavements, and evaluated their road performance and acoustic properties. The applicability of this kind of pavement in urban roads in cold regions has also been verified.

This study aims to propose a pavement surface treatment technology with groove-filled the elastomer. Through theoretical analysis, the ice-breaking mechanism of groove-filled elastomer asphalt pavement was clarified. Then, through the indoor ice-breaking test, the ice-breaking performance of the asphalt pavement groove-filled with elastomeric materials under three filling forms of above, level and lower than the asphalt pavement surface and different ice thickness were studied. Finally, through BPN and friction coefficient test, the skid resistance performance of groove-filled elastomer asphalt pavement under dry, wet and icing conditions is verified.

2 Mechanism and Experiment Procedure

2.1 Ice Breaking Mechanism of the Asphalt Pavement Groove-Filled with Elastomer

As shown in Fig. 1, the basic ice breaking principle of the asphalt pavement groove-filled with elastomer is that under the action of vehicle load, the difference of modulus

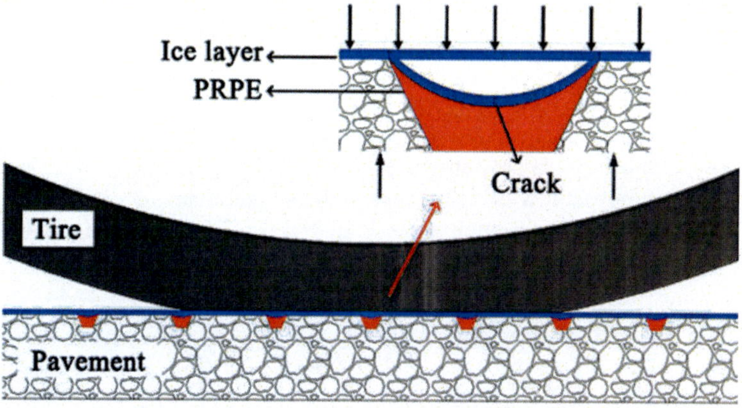

Fig. 1 Ice breaking mechanism of the asphalt pavement groove-filled with elastomer

between road material and elastic material causes the stress concentration in the ice layer at the junction of the two. The self-stress generated would be much greater than the internal stress of the ice layer under the normal pavement structure. When the stress value is greater than the strength of the ice layer, the ice layer would automatically break. The edges and corners of the groove would be exposed at this moment, which could immediately play a certain anti-skid effect. With the vertical compressive stress and transverse shear stress of the tire, the ice layer located on the upper part of the asphalt concrete between the grooves would gradually be worn by the tire, thereby achieving the effect of the overall road surface ice layer being broken and the original road surface exposed.

2.2 *Preparation Process of Groove-Filled Asphalt Pavement Specimens*

In this study, referring to related papers [18], rectangular grooves with a width of 10 mm and a depth of 10 mm were selected, and the groove spacing is 50 mm. The elastomer material is made of polyurethane and rubber particles in a certain ratio. The filling container is a 400 ml hose, and the filling equipment is an electric glue gun. The specific preparation process is as follows:

a. Form the specimen plate. The 30 cm × 30 cm × 5 cm AC-13 asphalt mixture specimens were formed in the laboratory.
b. Groove. The sample plate was grooved with a grooving machine in the laboratory according to the size of 10 mm × 10 mm × 50 mm, and then the dust in the groove was washed. After washing, it was placed in a ventilated place to dry.
c. Stick antifouling tape. In order to prevent the pavement from being polluted by the elastomer material, the plastic tape was pasted on the groove edge of the filling front edge.
d. Preparation of the elastomer. The elastomer was prepared according to the optimum ratio of elastomer material determined by the experiment.
e Filled with the elastomer. Put the prepared the elastomer into the hose, install the nozzle and the back cover into the electric gun, set the filling speed of the gun at 0.5–1 m/min, and slowly and evenly fill along the groove edge.
f. Compaction. First, lay a non-woven fabric along the filled groove, and then manually compact it with a steel pipe with an outer diameter of 30 mm.
g. Curing. After the compaction is completed, tear off the non-woven fabric strip and tape, and place the test piece plate at room temperature for 24 h.

2.3 *Evaluation of the Ice-Breaking Performance of Groove-Filled the Elastomer Asphalt Pavement*

This test uses a self-developed pavement ice-breaking simulation tester. The equipment is shown in Fig. 2. The whole system is mainly composed of a wheel mill and a refrigeration system. The refrigeration system could automatically and accurately control the temperature in the equipment, and the wheel rolling device could simulate the stress situation of the road under the repeated action of the wheel dynamic load. The wheel rolling device is a smooth tire, and the interaction mode with the road surface is pure rolling. The main parameters of the experimental instrument are as follows: rubber wheel width: 50 mm; rubber wheel grounding pressure: 0.7 MPa; walking times: 42 times/min; temperature control range: 0 ~ −5 °C; temperature control accuracy: ± 0.5 °C; specimen size: 300 mm × 300 mm × 50 mm. The test operation process is as follows:

a .Firstly, the corresponding ice thickness was frozen in the refrigerator.
b .Turn on the power and set the test temperature to—5°C. After the temperature is constant, the frozen specimen is placed on the test platform for 30 min, so that the temperature of the specimen is consistent with the temperature in the equipment.
c .Start the wheel rolling system to roll the test piece, observe the ice surface breaking condition of the test piece at all times, take out the test piece according to different rolling times, take photos of it, and record the corresponding rolling times.

This section mainly explores the influence of three kinds of the elastomer filled height on the ice-breaking performance of groove-filled structure through the ice-breaking test. The filled height is 1 mm lower than the groove height, flush with the groove and 2 mm higher than the groove(Abbreviation: the elastomer and groove level, the elastomer lower than groove and the elastomer higher than groove) as

Fig. 2 Pavement ice-breaking simulation tester

Fig. 3 Grooving filled elastomer specimen plates

shown in Fig. 3. Under the test temperature of—5 °C, the ice layer thickness (1 mm, 2 mm, 3 mm, 4 mm) was changed to explore the change of ice-breaking performance of different structures.

2.4 *Evaluation of Skid Resistance of Groove-Filled the Elastomer Asphalt Pavement*

In order to verify the skid resistance performance of the grooved asphalt pavement filled with the elastomer, the BPN and friction coefficient of the groove-filled the elastomer asphalt pavement were tested under dry, wet and icing conditions respectively. The pendulum friction tester is used for the BPN test, and the walking friction coefficient tester (WFT) developed by ourselves is used for the friction coefficient test. At the same time, the common asphalt pavement, grooved pavement (unfilled) and groove-filled asphalt pavement were tested and compared.

As shown in Fig. 4 is a picture of WFT. It has two rear wheels and one front wheel. The front wheels are test tires. The diameter of the three wheels is the same, which is 400 mm. The center distance between the front and rear wheels is 540 mm,

Fig. 4 The Walking Friction Tester (WFT) [19]

and the center distance between the two rear wheels is 580 mm. The front wheels are solid smooth rubber tires with a width of 60 mm. The density of the tested tire rubber is 1.15 g/cm3 and the shore hardness is 75. A constant vertical load of 196 N is maintained on the test wheel, and the corresponding contact area is 19.775 cm2. The contact pressure is about 99.2 kPa. WFT is equipped with a speed sensor to measure the test speed in real-time, and a torque sensor is used to record the torque of the test wheel. According to the measurement principle of WFT and the friction coefficient calculated theoretically, the calibration is carried out. The friction coefficient measured by WFT is calculated as follows:

$$\mu = \frac{M}{R \times N}. \tag{1}$$

where M is the measured torque, R is the radius of the test wheel, and N is the vertical load on the test wheel. The data recorder of WFT calculates the friction coefficient at 0.2 s intervals to provide continuous recording of the friction coefficient values at close range. The test speed could be maintained at the normal walking speed.

3 Results and discussion

3.1 *Ice Breaking Performance*

As shown in Fig. 5, it shows the ice-breaking effect picture and the binary refinement image of the elastomer and groove level specimen, 2 mm ice thickness, and −5 °C groove filling. After binary refinement image processing, different types of specimens are obtained, and the calculation results of ice fragmentation rate of specimens with different ice thickness are shown in Table 1 and Fig. 6. When the thickness of ice layer is 1 mm, both the elastomer and groove level and the elastomer lower than groove specimen have a good ice-breaking effect, and the ice fragmentation rate of the elastomer and groove level specimen is significantly higher than that of the elastomer lower than groove specimen. This is because when the thickness of the ice layer on the pavement is 1 mm, the actual thickness of the ice layer on the upper part of the elastomer lower than the groove specimen could be close to 2 mm, and the actual strength of the ice layer at this time is far greater than that of the 1 mm thick ice layer of the elastomer and groove level specimen. Similarly, when the thickness of the ice layer on the pavement surface is 2 mm, the ice fragmentation rate of the elastomer higher than the groove is also less than that of the elastomer and groove level specimen. The reason is that when the elastomer is covered with a 1 mm thick ice layer, the actual road surface ice thickness is 3 mm. The integrity of the road surface ice layer is good, and the wheel load is difficult to damage its structure. Similarly, when the surface of the bituminous pavement is covered with 1 mm ice, the thickness of the ice inside the groove has reached 10 mm, which makes the ice

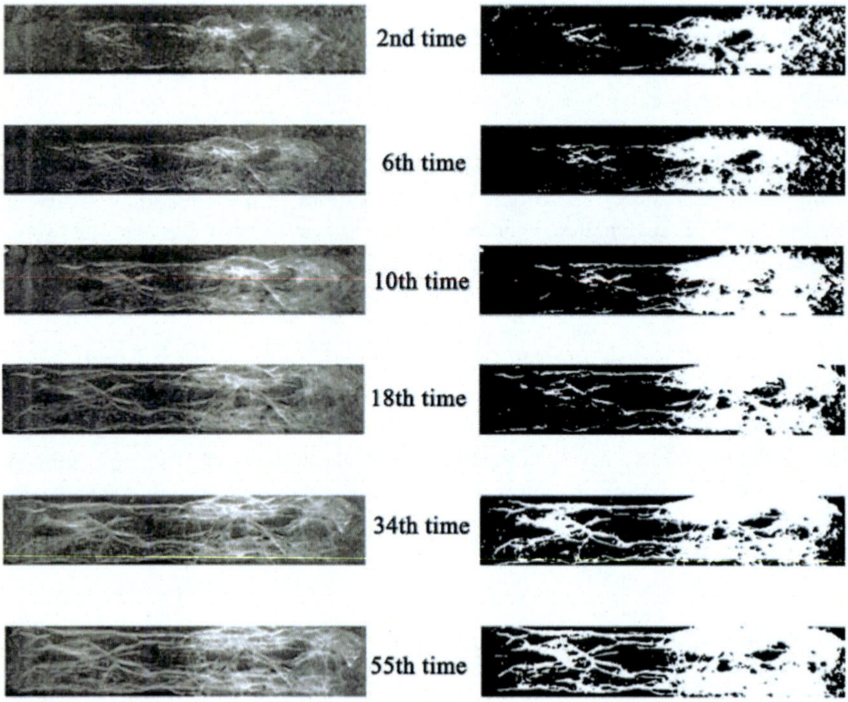

Fig. 5 The ice-breaking effect picture and binary processing image of the groove at different rolling times

Table 1 Results of ice fragmentation rate of different pavement types under different ice thickness

Pavement type	Ice thickness/mm			
	1 (%)	2 (%)	3 (%)	4 (%)
Grooved pavement (unfilled)	35.4	20.3	9.1	5.8
The elastomer and groove level	69.8	57.7	40.5	15.3
The elastomer lower than groove	65.3	55.6	32.8	7.2
The elastomer higher than groove	40.2	25.4	10.2	5.1

layer firmly solidified with the asphalt pavement. Therefore, the ice-breaking effect of the groove asphalt pavement is the worst. The results show that the ice-breaking effect of the four kinds of specimens is as follows: the elastomer and groove level > the elastomer lower than groove > the elastomer higher than groove > the grooved pavement (unfilled).

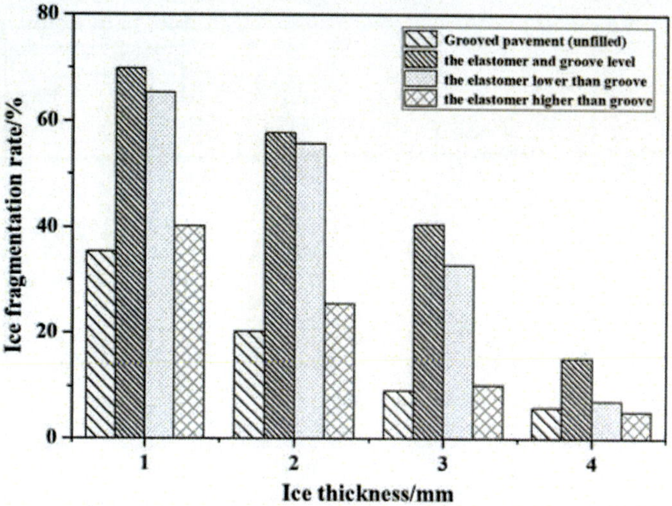

Fig. 6 Ice fragmentation rate of different pavement types under different ice thickness

3.2 *Skid Resistance Performance*

The test results are plotted as a histogram as shown in Fig. 7 and 8. Firstly, it could be seen that the BPN and friction coefficient data are consistent. In the dry state, the BPN and friction coefficient of the three pavement types have a few changes, and the groove-filled pavement slightly increases. Secondly, the skid resistance of AC-13 pavement decreases under the wet condition, while the skid resistance of

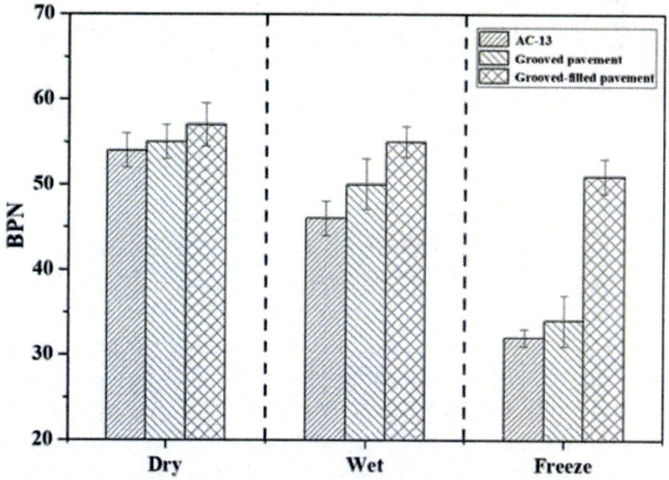

Fig. 7 BPN test results in different environments

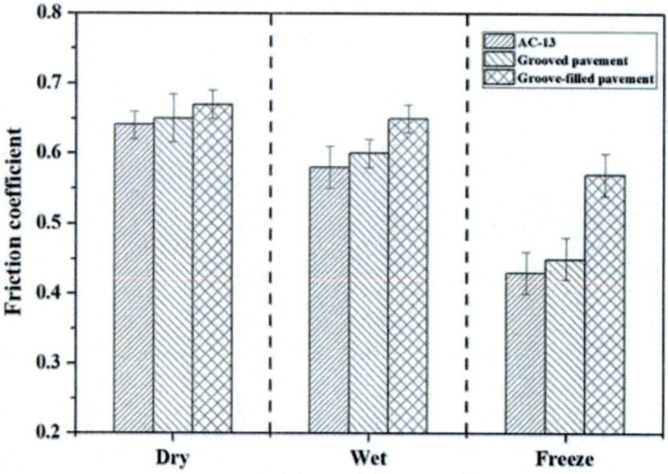

Fig. 8 Friction coefficient test results in different environments

grooved asphalt pavement and grooved filled pavement decreases slightly. Then, under the condition of icing, the skid resistance of AC-13 and grooved asphalt pavement decreases greatly, while the skid resistance of groove-filled asphalt pavement only slightly decreases, which indicates that the groove-filled asphalt pavement still has excellent skid resistance performance when the pavement is frozen. The main reason is that the groove-filled pavement could provide a large mechanical contrast with the asphalt pavement at the groove. Under the wheel rolling, the local area of the ice layer has a certain unbalanced stress, which makes the ice more easily broken so that the skid resistance performance of the pavement would not be greatly reduced. Compared with the common AC-13 asphalt mixture, the skid resistance of groove-filled pavement increases by 4.5%, 10.8% and 24.6% respectively under dry, wet and icing conditions. In the environment of ice and snow, the groove-filled pavement has obvious advantages.

4 Conclusions

In this study, a pavement surface treatment technology of groove-filled the elastomer was proposed. Through theoretical analysis, the ice-breaking mechanism of groove-filled the elastomer asphalt pavement was clarified. Then, through the indoor ice-breaking test, the ice-breaking performance of the asphalt pavement groove-filled with elastomeric materials under three filling forms of above, level and lower than the asphalt pavement surface and different ice thickness were studied. Finally, through BPN and friction coefficient test, the skid resistance performance of groove-filled elastomer asphalt pavement under dry, wet and icing conditions is verified.

It was verified the obvious advantages of the asphalt pavement of groove-filled the elastomer in ice-breaking and anti-skid. The ice-breaking performance of different pavement types is verified through the ice-breaking test. The results show that the ice-breaking effect of the four pavement types is as follows: the elastomer and groove level > the elastomer lower than groove > the elastomer higher than groove > the grooved pavement (unfilled). Finally, through the BPN and friction coefficient test, the skid resistance of the three types of pavement is verified under dry, wet and icing conditions, and the groove-filled asphalt pavement has the best skid resistance performance. Compared with the common AC-13 asphalt mixture, the skid resistance performance of groove-filled pavement increases by 4.5%, 10.8% and 24.6% respectively under dry, wet and icing conditions.

Although the the ice-breaking and skid resistance performance of groove-filled the elastomer asphalt pavement have been verified, the wear resistance of groove-filled asphalt pavement and determination of groove parameters still need to be further studied.

Acknowledgements This research was supported by the Shaanxi Provincial Department of Transportation Project(KY19-07,19-33K). These supports are gratefully acknowledged.

References

1. Petrova E, Shiryaeva AV Road accidents in Moscow: weather impact. AES Bioflux11 19–30
2. Tian J-J, Wang L-Q, He Y-L, Qu G-X, Liu L (2019) Driving safety assessment and management for hidden danger segments of low-grade roads under adverse weather. CICTP 2019:3845–3854
3. Wu LX, Cheng GZ (2011) Speed management on icy and snowy pavement of urban road. Adv Mater Res 225–226593–596
4. Malin F, Norros I, Innamaa S (2019) Accident risk of road and weather conditions on different road types. Accident Anal Prevention 122181–188
5. Autelitano F, Rinaldi M, Giuliani F (2019) Winter highway maintenance strategies: are all the sodium chloride salts the same. Constr Build Mater 226:945–952
6. Jiao W, Sha A, Liu Z, Jiang W, Hu L, Li X(2020) Utilization of steel slags to produce thermal conductive asphalt concretes for snow melting pavements. J Cleaner Prod 261:121197
7. Han S, Yao T, Yang X (2019) Preparation and anti-icing properties of a hydrophobic emulsified asphalt coating. Constr Build Mater 220:214–227
8. Zakerzadeh M, Abtahi SM, Allafchian A, Chamani MR (2019) Effectiveness of superhydrophobic material on the hydroplaning risk of asphalt pavements. Int J Pavement Eng 1–9
9. Wang X, Fan Z, Li L, Wang H, Huang M (2019) Durability evaluation study for crumb rubber-asphalt pavement. Appl Sci 9:3434
10. Tanaka S, Takeichi K, Masuyama Y, Takahashi N (2011) Study on properties of skid resistance on freezing pavements and quantitative evalution method of antifreezing effects. J Jpn Soc Civ Eng Ser. EI (Pavement Engineering) 67(1):53–64
11. Cong L, Yang F, Guo G, Ren M, Shi J, Tan L (2019) The use of polyurethane for asphalt pavement engineering applications: a state-of-the-art review. Constr Build Mater 225:1012–1025
12. Li M, Du M, Wang F, Xue B, Zhang C, Fang H(2020) Study on the mechanical properties of polyurethane (PU) grouting material of different geometric sizes under uniaxial compression. Construct Build Mater 259:119797

13. Fan C-J, Huang Z-C, Li B, Xiao W-X, Zheng E, Yang K-K, Wang Y-Z (2019) A robust self-healing polyurethane elastomer: from h-bonds and stacking interactions to well-defined microphase morphology. Sci China Mater 62:1188–1198
14. Sun M, Bi Y, Zheng M, Wang J, Wang L (2020) Performance of polyurethane mixtures with skeleton-interlocking structure. J Mater Civ Eng 32:04019358
15. Gao J, Wang H, Chen J, Meng X, You Z (2019) Laboratory evaluation on comprehensive performance of polyurethane rubber particle mixture. Constr Build Mater 224:29–39
16. Cong L, Guo G, Yang F, Ren M (2020) The effect of hard segment content of polyurethane on the performances of polyurethane porous mixture. Int J Transp Sci Technol
17. Wang D, Schacht A, Leng Z et al (2017) Effects of material composition on mechanical and acoustic performance of poroelastic road surface (PERS). Constr Build Mater 135:352–360
18. Jr James PW (2012) Evaluation of trapezoidal-shaped runway grooves. Geometric Configurations and Shapes
19. Han S, Liu M, Fwa TF (2020) Testing for low-speed skid resistance of road pavements. Road Mater Pavement Des 21:1312–1325

Improved Calibration Procedure for British Pendulum Tester

Weiwei Guo, L. Chu, and T. F. Fwa

Abstract Regular periodic calibration is important to ensure the accuracy of measurements of British pendulum test (BPT). The current standard calibration procedures of BS EN 13036 and ASTM E303 are widely adopted by pavement engineering community. Studies by different research groups have shown that test results of BPT devices complying with standard calibration requirements have good repeatability, but could have deviations of 10% or more. The present research presents a detailed examination of the impact of the limiting values of the main calibration parameters of BPT devices. Since it is difficult to study by means experimental measurements the impact of any pendulum parameter value within the allowable range on BPN (British pendulum number), a finite element model was adopted in this research based on the mechanics of the BPT test process. The simulation results showed that BPT devices satisfying the calibration requirements of any standard could produce measurements with uncertainty of more than 23% of their mean test value. By tightening the calibration parameter range according to the actual needs of the user, this study shows that the uncertainty of the BPT measurements can be controlled within an acceptable range.

Keywords British pendulum test · British pendulum number · Pendulum test value · Calibration procedure · Finite element model · Pavement skid resistance

1 Introduction

The British pendulum tester (BPT) is an easy-to-operate and robust piece of portable test device testing of surface friction of pavement materials. The accuracy and the reproducibility of the BPT measurements are critical factors for evaluating the friction coefficient of the test surface. More than 60 years of experience of laboratory and field

W. Guo · L. Chu (✉) · T. F. Fwa
School of Highway, Chang'an University, Xi'an, China
e-mail: longjiachu@chd.edu.sg

T. F. Fwa
National University of Singapore, Singapore, Singapore

© The Author(s), under exclusive license to Springer Nature Switzerland AG 2022

H. R. Pasindu et al. (eds.), *Road and Airfield Pavement Technology*, Lecture Notes in Civil Engineering 193, https://doi.org/10.1007/978-3-030-87379-0_15

BPT measurements by users in different parts of the world have shown practically no difficulty in achieving the precision required by standard specifications [1–4], such as the allowable error of 1.0 BPN (British pendulum number) unit at 95% confidence level specified by ASTM E303 [5].The calibration procedures and requirements specified by ASTM E303 [5] and BS EN 13036-4 [6] standard are generally adopted by most highway engineering practitioners and organizations in the world. Most of the verification requirements of the two calibration procedures are the same, but the calibration limit ranges are different.

Reproducibility studies on interlaboratory BPT measurements have shown some unsatisfactory results Strautins and Daniel [7] conducted an inter-laboratory study involving 5 laboratories and found that, for a reference specimen with a mean BPN of 27.4, the laboratory measurements varied from 17 to 44. The range of measured BPN values was from 49 to 65 for another reference specimen with a mean BPN of 56.6. Hiti and Ducman [3] tested a surface using four BPT devices. Eight BPN measurements were made per BPT device, each time setting the device to a different slider force-deflection curve within the allowable ASTM calibration limits. It was found that the BPN value measured on a referenced test surface by different BPT units varied from 44 to 60, resulting in a mean percentage difference higher than 28%. Such differences are undesirable for practical operations. These results show that the parameter calibration method cannot ensure satisfactory reproducibility in BPN measurements made by different BPT devices that follow the calibration standards of either ASTM or BS.

In order to reduce the variability of measured BPN values caused by calibration parameters, Hiti and Ducman [3] proposed strict requirements on the slider force-deflection curve. The research by Strautins [9] also showed that a reduction of tolerance limits for all calibration parameters can reduce the uncertainty of the measurements. By means of a finite element simulation model of the BPT test, Guo et al. [8] identified the limits of slider force-deflection curve as the single most influencing calibration requirement, and proposed a procedure to determine the limits to achieve a pre-specified level of measurement variability.

Given the current limitations of the calibration parameters specified by ASTM E303 and BS EN 13036-4, it is necessary to re-evaluate the impact of the uncertainty of the BPN value caused by the specified calibration parameter limits. In the impact evaluation, analyses are made on the effects of various influencing calibration parameters on the variability of BPT measurements. In view of the relatively large number of equipment parameters involved, and some parameters are unique for each equipment at the time of manufacture, it is practically impossible to experimentally perform the evaluation using physical pendulum equipment. Therefore, this research resorted to developing a computer simulation model to achieve the research goals. The BPT model developed by the finite element method provides a convenient tool that can conveniently vary the values of different parameters to provide a comprehensive evaluation of the impacts of different parameters on the measured BPT values.

2 Objectives and Method of Study

2.1 Research Objectives

The conventional approach of calibrating a given type of device is by setting control limits for selected device parameters. This is also the practice in use today by the pavement engineering community in the case of BPT. Currently, the two most commonly adopted BPT calibration methods are the ones published by ASTM and the British Standards Institute as standard procedures ASTM E303 [5] and BS EN 13036-4 [6] respectively. Although the two standards are largely similar, there are differences in the specific control limits of some calibration parameters. Table 1 highlights their main differences and their impacts on BPT measurements.

Comparing the entries in the second and third columns of Table 1, it can be seen that there are only slight differences between the two standards in the calibration requirements, except for item 5 concerning the slider force-deflection characteristic. Items (1) and (2) of Table 1 are fixed manufactured parameters that could not be corrected by users during calibration. For BPT with unworn rubber sliders, the impact of items (3) and (5) specified by the two standards are worthy of detailed analysis. This study employed a computer simulation model to evaluate the variations in the measured BPN values according to the parameters required by the two calibration standards.

Table 1 BPT device calibration requirements by ASTM E303-18 and BS EN 13036-4

Item	ASTM E303-18 Requirements	BS EN 13036-4 Requirements
1. Radius of pendulum swing	Sliding edge is 508 mm from the axis of rotation	Sliding edge is (514 ± 6) mm from the axis of rotation
2. Mass of pendulum arm with slider assembly	(1.50 ± 0.03) kg	(1.50 ± 0.03) kg
3. Center of gravity of pendulum arm	(411 ± 5) mm from axis of rotation	(410 ± 5) mm from axis of rotation
4. Width of rubber pad striking edge	Wear on striking edge shall not exceed 3.2 mm in slider plane or 1.6 mm vertical to it	Not more than 2.5 mm
5. Slider force-deflection requirements	Average vertical slider force of (2500 ± 100) g	Static force shall be (22.2 ± 0.5) N when deflected 4.5 mm. Change in static force shall be not greater than 0.2 N per mm deflection of the slider. Slider force at 0 to 8 mm deflection shall be within the envelope specified

2.2 Research Method

It is impractical to test a BPT device to cover all possible calibration ranges required for calibration by means of physical testing. However, the test mechanism of the pendulum instrument can be easily simulated by means of computer simulation. So far, researchers have developed two finite element simulation models of BPT. One is the NUS model developed by Liu et al. [10], and the other is an upgraded model based on the NUS model by Chu et al. [11]. The upgraded model shown in Fig. 1 more accurately expresses the friction performance of the rubber slider, and can completely simulate the entire test process of the pendulum.

In the finite element model, the aluminum beam is connected to the upper truss structure through hinge constraints and nonlinear springs to simulate the combined action of the spring and lever mechanism of the actual BPT pendulum assembly. In addition, the friction coefficient needs to be entered in advance to characterize the interaction between the slider and the pavement surface. The simulation model finally outputs the dwell height of the pendulum arm. The predicted BPN is obtained by using the conversion relationship between height and BPN. After calibration, the model can be applied to simulate a BPT test and calculate BPN value and other output information, such as contact pressure, sliding speed and friction at each point along the contact path. It can be used either to obtain the BPN value of a test surface with a known coefficient of friction, or to back-derive the coefficient of friction of

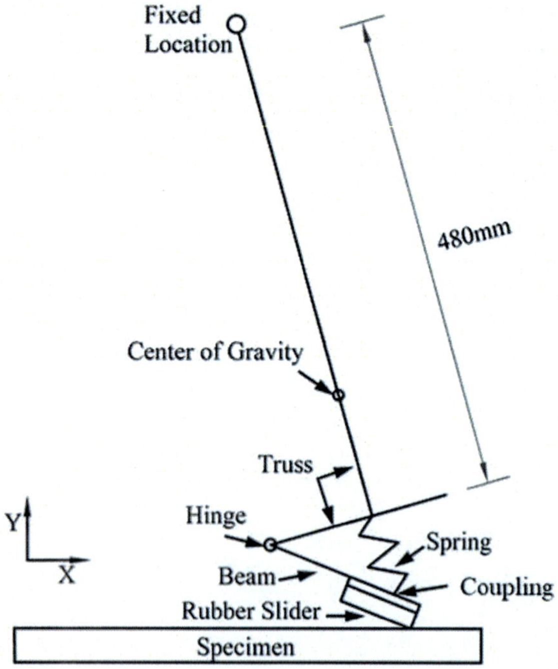

Fig. 1 Computer modeling of BPT pendulum-slider assembly [11]

a test surface with a known BPN value. The improved BPT finite element model was adopted in this study. Readers interested in the construction of 3D finite element simulation model can refer to Chu et al.'s work for more details [11].

3 Analysis of Impact of Improved Calibration Parameters

The focus of this research is to explore the variation range of the test results corresponding to the calibration requirements of ASTM E303 and BS EN 13036-4. Specifically, they are (i) the position of pendulum center of gravity, and (ii) the slider force-deflection characteristics. Improvement suggestions for these two parameters from different research will be used as the basis of comparison to illustrate the potential of the improved method to reduce the uncertainty of BPT measurements. The impact of the friction coefficient of the test surface in the range of 0.2–1.0 was considered for all the cases. This range covers all pavement friction coefficients that may be encountered in practice.

3.1 Center of Gravity of Pendulum Arm

The distances of the center of gravity (CG) from fixed location stipulated by ASTM E303 standard and BS EN 13036-4 standard are 411±5mm and 410±5mm, respectively. A narrower tolerance (410±3 mm) of the distance of center of gravity was proposed in Strautins [9] to reduce the uncertainty of BPT measurements. Therefore, in order to explore the impact of the change of the center of gravity on the measured BPN value, the simulation experiment is divided into three cases: the limit conditions of the two standards and narrower tolerances (410±3mm) are selected as parameters. The slider force-deflection curve is set as the lower limit of BS EN 13036-4.

The results of the analysis are shown in Figure 2. The curve shows the calculated BPN for all the cases analyzed. The vertical bars give the maximum differences $\Delta(BPN)_I$ in BPN value between BPT devices that meet the requirements of pendulum center of gravity position. The differences for ASTM E303 and BS EN 13036-4 standards were of the same order of magnitude. The absolute value of $\Delta(BPN)_I$ vary from 0.6 to 2.0 BPN units. The mean percentage of difference, i.e. percentage calculated based on the mean of the two BPN values, is less than 3% overall. As an improvement, the narrower tolerance method can reduce the uncertainty caused by the center of gravity to 1.8%. From the perspective of actual calibration, the position of the center of gravity is relatively easy to adjust, so it is worthwhile to adopt a stricter interval to reduce the uncertainty of the BPN value.

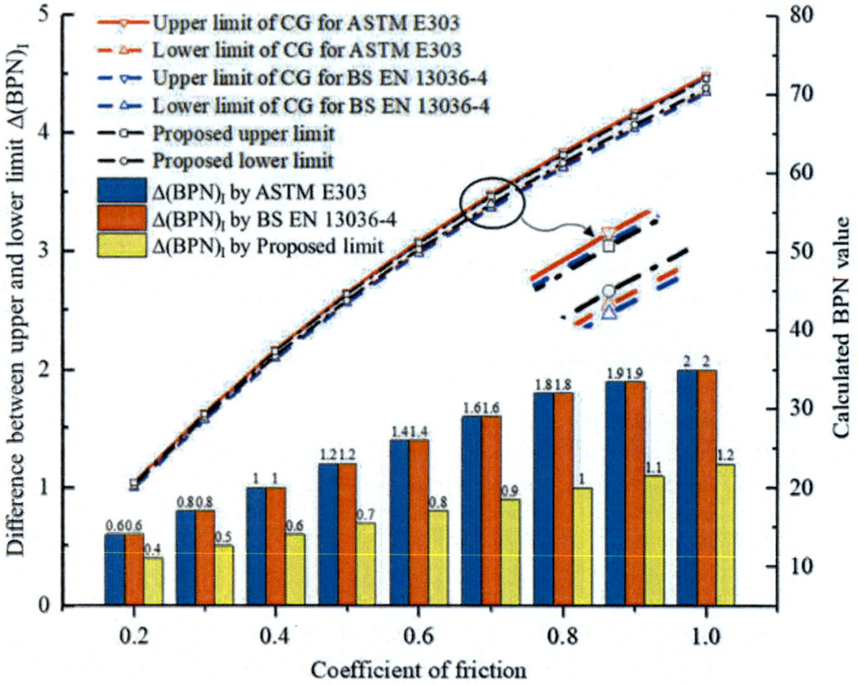

Fig. 2 Differences in BPT measurements due to different pendulum center of gravity (CG) positions

3.2 *Slider Force-Deflection Requirements*

The ASTM E303 standard only specifies the average slider force at the maximum slider displacement. The BS EN 130306-4 standard is stricter than the ASTM E303 standard. It clearly defines the upper and lower allowable limits for the calibration of the slide force-deflection curve, as shown in Fig. 3. For comparison, Fig. 3 also shows the implied upper and lower limits of the slider force-deflection relationship that meets the ASTM E303 maximum slider force-deflection calibration requirements. Both Guo et al. [8] and Strautins [9] recommended more stringent specific requirements for the slider force-deflection curve. The former stated a complete slider force-deflection calibration limit range in its recommendation. This section adopts the improvement recommendation of Guo et al. as the basis for comparison. Guo et al proposed to use the lower limit of BS EN 13036-4 standard as the lower limit of the improved calibration requirement. The lower limit of BS EN 13036-4 is valid, because any slide force-deflection curve below this limit will violate the slider force requirement with a maximum deflection of 3.9 mm. A trial-and-error analysis was performed using the computer BPT simulation model for the comparison study. Figure 3 also shows two proposed upper limits for calibration. In order to illustrate the impact of the slider force-deflection curve, the two sets of limits were analyzed for the ASTM E303 standard and the BS EN 130306-4 standard respectively.

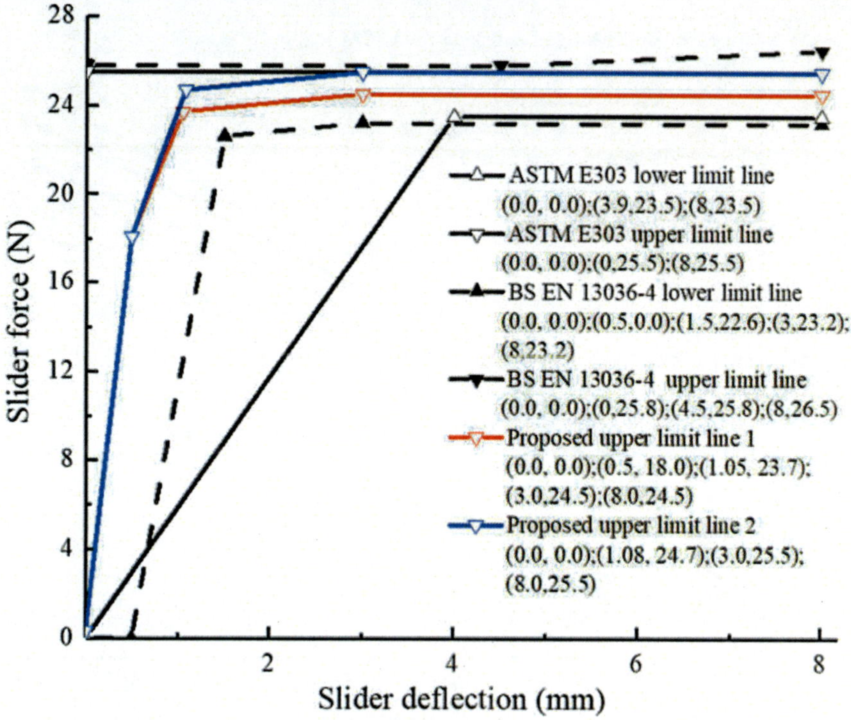

Fig. 3 Limits of calibration requirement for slider force-deflection relationships

Detailed measurement values calculated using the simulation model of the above four cases are plotted in Fig. 4. The maximum differences $\Delta(\mathrm{BPN})_{\mathrm{II}}$ in BPN values between the upper and lower limits of slider force-deflection curve for four cases are shown in Fig. 5. The absolute value $\Delta(\mathrm{BPN})_{\mathrm{II}}$ varies from 10.7 to 30.9 BPN units for the ASTM E303 standard, and 5.2–20.4 BPN units for BS EN 13036-4 standard. $\Delta(\mathrm{BPN})_{\mathrm{II}}$ of the proposed limits 1 and 2 can be reduced to about half of the BS EN 13036-4 standard, corresponding to 2.1 to 8.9 BPN units and 2.9 to 10 BPN units, respectively. In terms of the maximum percentage difference, the range varies from 56.2 % to 58.0 % for ASTM E303 standard and 21.9 and 25.1% for BS EN 13036-4 standard. Over the range of friction coefficient from 0.2 to 1.0, the maximum percentage difference are 10.5%, and 13.3% respectively. The magnitude of these differences is a significant improvement over the standard calibration procedures by ASTM E303 and BS EN 13036-4.

In the above analysis, it is obvious that the slider force-deflection curve has the largest impact on the BPN value. The possible absolute difference and maximum percentage difference based on the allowable range of the slider force-deflection curve of the current ASTM and BS standard are beyond the acceptable range for practical purposes. The position of the center of gravity is comparatively less important, but the adjustment of its value should also be taken into consideration to reduce measurement

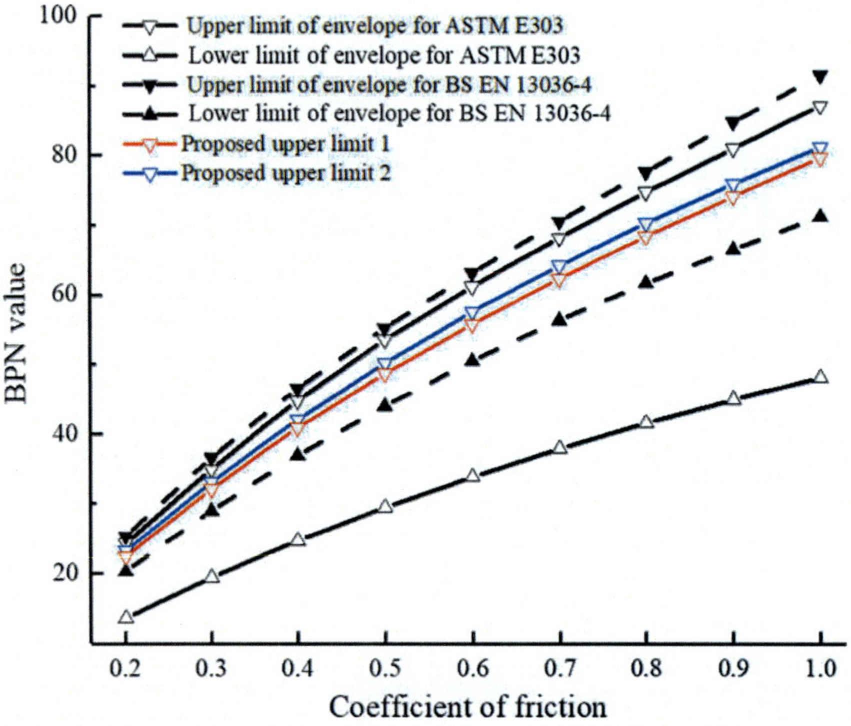

Fig. 4 Impacts of calibration limits of slider force-deflection relationship

uncertainty. Overall, reducing the allowable range of calibration parameters would reduce the uncertainty of measured BPN values, but it also means more stringent calibration limits. The allowable range of calibration parameters should be adjusted from the perspective of road maintenance and management.

4 Conclusion

This study adopts the finite element model to show that the uncertainty of the BPT test results within the allowable range of calibration parameter could reach at least 56.2 % for ASTM E303 standard and 21.9 % for BS EN 13036-4 standard. The uncertainty of the BPT measurements are much too large for effective pavement friction management. Tightening the allowable range of various calibration parameters has been shown to be feasible in this study to reduce the uncertainty of BPT measurements. The calibration limits can be determined according to the needs of a user in order to meet the user's operation requirements.

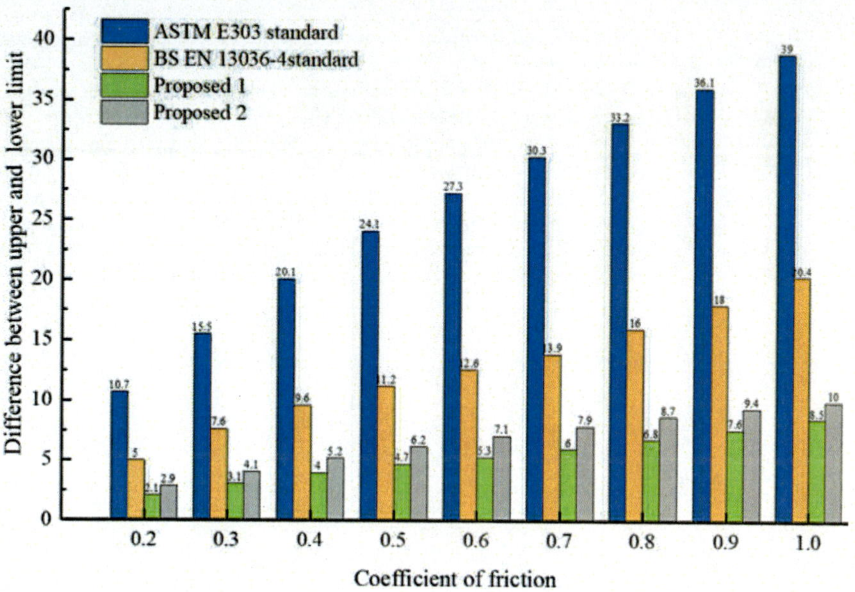

Fig. 5 Differences $\Delta(BPN)_{II}$ in BPN values between upper and lower limit range of slider force-deflection curve

Funding The authors gratefully acknowledge the financial support from Shaanxi Science and Technology Project 2020JQ-390.

References

1. Henry JJ (2000) Evaluation of pavement friction characteristics-a synthesis of highway practice. NCHRP Synthesis 291, National cooperative highway research program, Washington DC
2. Ahammed M, Tighe SL (2010) Effect of short-term and long-term weather on pavement surface friction. Int J Pavement Res Technol 3(6):295–302. https://doi.org/10.6135/ijprt.org.tw/2010.3(6).295
3. Hiti M, Ducman V (2014) Analysis of the slider force calibration procedure for the British pendulum skid resistance tester. Measure Sci Technol 25(2):025013. https://doi.org/10.1088/0957-0233/25/2/025013
4. Chu L, Cui X, Zhang K, Fwa TF, Han S (2019) Directional skid resistance characteristics of road pavement: implications for friction measurements by British pendulum tester and dynamic friction tester. Transp Res Record 2673(10):793–803. https://doi.org/10.1177/0361198119851453
5. ASTM E303-18 (2018) Standard test method for measuring surface frictional properties using the British pendulum tester. ASTM International, West Conshohocken, PA
6. British Standards Institute (2011) BS EN 13036-4 road and airfield surface characteristics. Test methods for measurement of slip/skid resistance of a surface. The Pendulum Test. London, UK
7. Strautins C, Daniel M (2013) Integration of slip resistance values within a risk management framework: a human gait based approach using reference samples. Proc of the international

conference on fall prevention and protection, National Institute of Occupational Safety and Health, Tokyo, Japan, pp 93-98
8. Guo W, Chu L, Fwa TF (2020) Evaluation of calibration procedures of British pendulum tester." J Test Eval 50(1). https://doi.org/10.1520/JTE20200288
9. Strautins C (2020) Pendulum calibration, metrological traceability and reference material. International slips trips and falls conference
10. Liu Y, Fwa TF, Choo YS (2003) Finite-element modeling of skid resistance test. J Transp Eng 129(3):316–321. https://doi.org/10.1061/(asce)0733-947x(2003)129:3(316)
11. Chu L, Guo W, Fwa TF (2020) Theoretical and practical engineering significance of British pendulum test. Int J Pavement Eng 1–8. https://doi.org/10.1080/10298436.2020.1726351

Runway Grooving Techniques and Exploratory Study of the Deterioration Model

Md. Tofail Miah, Erwin Oh, Gary Chai, and Phil Bell

Abstract The friction of runway pavement is critical for the safety of aircraft landing and movement on the runway. Tire hydroplaning may lead the aircraft to move off the runway and hinder the safe landing during wet weather conditions. Grooving on the runway is one way to develop frictional braking resistance and diminish hydroplaning's potential risk by improving runway surface drainage capacity during damp weather. According to the Federal Aviation Administration (FAA), groove construction must follow specific dimensions to maintain skid-resistant airport pavement surfaces. However, the groove area can be reduced for several reasons, and regrooving is essential if 40% of the runway groove of a substantial length decreased to 50% of its original dimension. Grooves initiate different potential distress mechanisms that are not found in an ungrooved pavement surface. Groove closure in different airports with hot weather is a frequent and prominent form of distress that substantially declines the grooves' effectiveness. Moreover, the degree of the declination of groove dimensions has not been quantified in a theoretical method. This paper discussed the current technique and importance of runway grooving. In addition to this, this paper reviews different potential distress mechanisms and issues related to groove deterioration. Finally, a brief of a predictive modeling requirement is illustrated, which is significant for the authority concern for maintenance and reinstate the grooving in the runway for friction development.

Keywords Runway · Groove techniques · Friction · Groove distress · Groove deterioration modeling

Md. T. Miah (✉) · E. Oh
School of Engineering and Built Environment, Griffith University, Brisbane, QLD, Australia
e-mail: mdtofail.miah@griffithuni.edu.au

G. Chai · P. Bell
Airport Consultancy Group, Gold Coast, QLD, Australia

© The Author(s), under exclusive license to Springer Nature Switzerland AG 2022
H. R. Pasindu et al. (eds.), *Road and Airfield Pavement Technology*, Lecture Notes in Civil Engineering 193, https://doi.org/10.1007/978-3-030-87379-0_16

1 Introduction

Pavement friction is a vital factor that administers the safe takeoff and landing of aircraft on runways. Runway conditions become worsen during wet weather by declining skid resistance diminishes significantly and responsible for runway expedition accidents [1].

Netherland Transport Safety Institute studied the overrun and veer-off accident factors and discovered that wet runways were vital components in both types of mishaps during landings and takeoff in Europe and worldwide. The outcome of the study revealed that wet runways caused about 40% of all landing overrun accidents in Europe and 60% worldwide. Hence, maintaining a required level of friction in all weather conditions has an immense consequence in the arena of airport pavement management. With the passage of time, research unveiled that some insightful measures are needed to accelerate the aircraft braking on the asphalt surface, especially in wet weather. Currently, one of the methods adopted worldwide is introducing transverse grooves on runway pavements to ensure adequate friction, especially during wet weather conditions [2].

NASA (National Aeronautics and Space Administration) studied first about grooves and introduced them at the landing tracks in 1962. Both NASA and FAA executed consecutive investigations to evaluate runway grooves' impacts and performance in hydroplaning aircraft tire braking. The findings reveal that grooving enhances runway surface drainage and thus declines hydroplaning risk and simultaneously improves aircraft braking capability and maneuvering the aircraft on the runway in wet weather conditions [2].

Grooves provide an exit of entrapped water from between aircraft tires and the pavement surface, which improve frictional braking resistance and mitigate hydroplaning's potential risk [1, 3]. Considering the advantage of hydroplaning and enhanced skid resistance of grooved runways over ungrooved ones, FAA (Federal Aviation Administration) recommended in Advisory Circular AC 150/5320-12C that all runways serve turbojet aircraft should install a standard saw-cut square groove [4]. However, several factors need to be considered before grooving a runway, for example, extreme hot or cold climates that are not appropriate for grooving. Despite some shortcomings, runways are deploying grooving as the most common method utilized to develop aircraft braking performance, especially on the wet asphalt runways [3].

Groove closure depends on various factors such as hot environments, aircraft speed, wheel loads magnitude, number of passing, and aircraft movements parallel to the grooves. The groove's familiar distresses are frequent rubber contamination, groove depth decrease due to surface erosion, edge break, and groove closure over time that were experienced in several international airports [3]. According to FAA's AC 150/5320-12C guidance, airport authorities should take immediate actions to reinstate the groove if 40% of the grooves in a certain length of the runway lost their dimension equal to or more than 50% from the original measurement [4]. Groove closure in different airports runway pavement is a common incident and a prominent form of distress. Groove closure definitely leads to a decline in the grooves' effectiveness, but the scale of the reduction of groove dimensions has not been quantified theoretically [5].

Condition prediction models are utilized to conduct analysis and forecasting the condition that is vital for maintenance and rehabilitation (M&R), budget planning, inspection scheduling, and work planning [6]. Condition prediction models are essential in a pavement management system. Condition prediction models mimic the function similar to that of a car engine [6].

This paper demonstrates an investigative review on specific aspects of pavement condition prediction modeling and techniques for developing prediction models to configure different models.

Wang and Larkin [7] discussed and evaluated the groove shape changes (Depth, width, and area) under a long-term loading period between July 2014 and June 2016 in National Airport Pavement Test Facility (NAPTF), where a series of full-scale tests of airport pavement grooves on flexible pavements were conducted to evaluate pavement groove's operational performance [7].

This paper exposes a brief of groove closure prediction modeling based on the NAPTF investigation outcome by Wang and Larkin [7]. GeneXProTools 5.0 software was implemented to generate the genetic programming (GP) model, including wheel pass number, load intensity, pavement layer thickness, and temperature as input parameters.

The paper renders useful information of modeling perception, variables selection, and a brief outcome of groove closure modeling derived from different input parameters. The generated model provides information on reducing groove areas to the airport runway pavement personnel and drawing attention to the gradual declination of surface friction. Finally, the groove closure prediction model significantly leads to timely planning and implementation of maintenance work.

2 Research Objectives

The objectives for performing the study are as follows:

1. Describes the significance of runway grooving, including current techniques and standards of groove installation.
2. Discuss possible groove distresses, relevant factors, and failure mechanisms.
3. Reviews various prospective prediction modeling tools.
4. A brief prediction model of groove area deterioration using GeneXProTools 5.0 software.

The model helps to predict the runway grooves' operational performance and associated financial and maintenance programs to the airport authority.

3 Runway Friction, Grooving Techniques, and Distresses

3.1 Runway Friction Deterioration and Resurgence

Runway friction is the critical element that a runway should have for aircraft's safe operation on runways. However, this friction can be decreased with time due to mechanical wear and polishing action by aircraft tires during rolling or braking and rubber deposition on the surface. These effects are connected with the volume and type of aircraft traffic. Moreover, variation in local weather conditions, type of pavement (HMA or PCC), type of materials used in the preliminary construction, and airport maintenance practices also persuade the skid resistance of runway pavement [4].

Furthermore, friction loss can be occurred due to pavement structural failures, such as rutting, cracking, raveling, joint failure, and settling. Besides, deposition of rubber, including other contaminants, such as oil spillage, dust particles, jet fuel, water, snow, ice, and slush, can trigger friction loss on runway pavement surfaces [4].

Runway grooving is an exceptional technique that now becomes admired to enhance aircraft braking on wet asphalt-surfaced runways. **Open Graded Friction Course** is useful for surface water drainage improvement but is turned blocked by debris. Conversely, **Stone Mastic Asphalt** is an open-graded mix that expresses improved surface texture without the risk of congestion or closure. Prior to the grooving, coarse asphalt mixes containing 20 mm aggregates were used to achieve better surface textures. Furthermore, **Sprayed** or **'chip' sealing** delivers significant surface texture and is commonly implemented in regional and remote airports [3, 5].

3.2 Runway Grooving Techniques and Requirements

In the runway, grooving has proven a suitable technique for providing sufficient skid resistance and avoiding hydroplaning during rain [4]. Approximately 40% in Europe and 60% of global accidents are closely connected to landing overrun. Hence, runways are required to be grooved to ensure satisfactory friction levels under all weather conditions [1].

FAA advisory circular AC-150/5320-12C [4] suggests that runways serving or are expected to serve turbojets shall be grooved. Existing runways should be considered for grooving considering annual rainfall, historical records related to hydroplaning and associated accidents, runway length, quality of surface texture under dry or wet conditions, improper seal coating, inadequate friction, and runway pavements strength. Moreover, Transverse and longitudinal grades, any drop-offs at the runway ends due to topographical constraints, and crosswind effects must be taken into

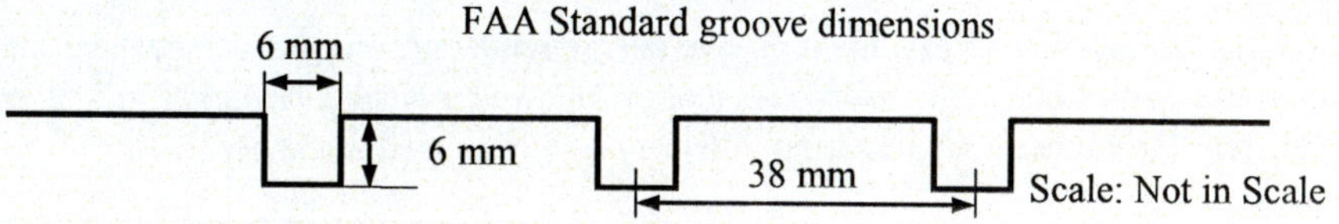

Fig. 1 FAA standard groove design section (Adapted from [4])

account. A reconnaissance survey containing bumps, bad or faulted joints, depressions, cracks, and the runway's structural conditions shall be performed as specified in ACs 150/5320–6 and 150/5370–10 before grooving [4].

FAA introduced a standard and groove configuration based on the tests and research. Present FAA standard square groove dimensions are depth 6 mm (1/4 in), width 6 mm (1/4 in), and 38 mm (1 ½ in.) center to center spacing, as depicted in Fig. 1 [4].

Moreover, a minimum of 90% of the grooves should have a depth of at least 3/16 in. (5 mm), at least 60% of the grooves should have depth no less than 1/4 in. (6 mm), and 10% of the grooves may not surpass a depth of 5/16 in. (8 mm) as per FAA advisory circular AC-150/5370-10G [8]. Hence, ensuring groove depth for adequate pavement friction during operation is critical [1]. However, groove depths can vary in construction and after the construction due to deterioration in different ways [9].

In General, grooves can be installed into the asphalt surface 4–8 weeks after the surface is constructed [3]. Grooves are normally fabricated across the runway surface, transversely to the runway length and perpendicular towards the runway's centre line [7]. Since the saw cutting techniques develop adequately, a new trapezoidal-shaped groove was proposed by Patterson [10].

Grooves are generally fabricated across the runway surface, transversely to the runway length and perpendicular towards the centerline of the runway [1].

Since the saw cutting techniques improve satisfactorily, a new trapezoidal-shaped groove was proposed by Patterson [10] to evaluate with the standard rectangular groove [1]. The trapezoidal groove dimensions consist 1/2 in. (12.5 mm) at the top, 1/4 in. (6 mm) at the bottom, and distributed 2 1/4 in. (56 mm) center to center as shown in Fig. 2 [10]. Newly designed trapezoidal grooves are shown in Fig. 2.

Recently, FAA constructed rectangular and trapezoidal grooves in both rigid pavement and asphalt pavement at the National Airport Pavement Test Facility (NAPTF) areas to compare the performance of both grooves. The results demonstrated that

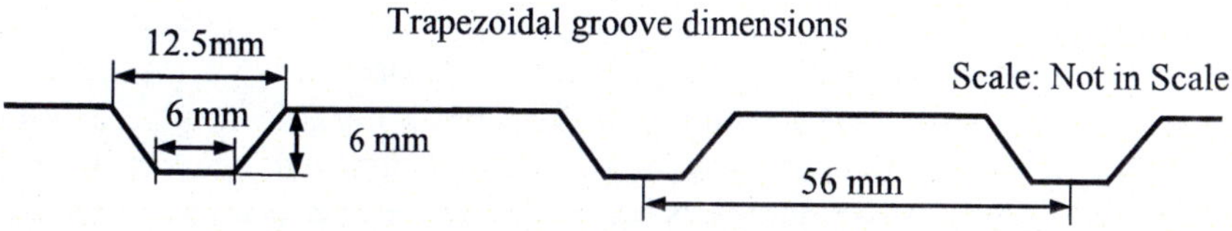

Fig. 2 Groove configurations of trapezoidal shapes (Adapted from [10])

the trapezoidal-shaped pavement groove design rendered some advantages over the current FAA standard square grooves that include better water evacuation, enhanced resistance to rubber contamination, integrity, and improved longevity [2].

3.3 Runway Groove Distress and Failure Mechanism

Grooves tend to produce various potential distress mechanisms that are not seen in an ungrooved pavement surface.

Runway Groove Distress. Some familiar distresses related to grooving are groove closure, rubber contamination, edge break, and groove depth diminishing due to surface erosion [3].

Groove Closure. Groove closure is an outstanding form of groove distress, which is quite common in hot environments and decreases its efficiency. In summer, it has successive high pavement temperatures, particularly in the asphalts made with relatively soft binders. It can also observe in places where aircraft travel slowly and parallel movement to the grooves, mainly in runway entry and exits. Moreover, comparatively new asphalt surfaces are more susceptible to groove closure along with high wheel loads. Nonetheless, the degree of groove area declination has not been computed in a theoretical process. However, friction testing is purposeful to measure the effect of partial groove closure [3].

Groove closure substantially reduces the effective volume of grooves and thus affects frictional performances of runways [11]. According to FAA's AC 150/5320-12C, if 40% of the grooves in the runway lost its shape equal to or less than 3 mm in depth and/or width in a consecutive length of 457 m, the efficiency of grooves for mitigating hydroplaning has been reduced significantly. Hence, the airport authority should reinstate the groove shape without delay [4]. Sometimes, re-sawing of closed grooves may create foreign object debris (FOD) hazards by breaking off weakly supported asphalt. This problem can be resolved by deploying a new asphalt overlay around 60 mm and grooving it once curing is completed [3].

Rubber Deposition. Rubber deposition can take place in touchdown zones in the runways, whether it is grooved or ungrooved. NASA investigated that transversely grooved runway pavements develop fewer rubber deposits during aircraft touchdown actions than non-grooved pavements. However, grooved pavements need to remove rubber contamination frequently to maintain groove effectiveness by ensuring friction [3].

Erosion. Groove depth can be decreased due to the fine particles' erosion from between the larger aggregates and settled inside the groove. Fortunately, this displacement of fines can improve the surface's general texture, which mitigates the surface's general texture effectiveness. Erosion of Asphalt surface primarily introduced by jet blast. Generally, aged binder promotes the dislocation of fine aggregate particles and binder from the surface. The rate of erosion can be increased for some reason, such as using unsound aggregates in asphalt production, using a specific binder that is more prone to rapid aging, improper aggregate gradation in the asphalt manufacturer, and unfavorable environmental conditions [3].

Edge Breaking. Edge breaking in the hot mixed asphalt (HMA) runway groove was caused due to the bitumen binder's cohesive failures rather than the aggregate-binder interface's adhesive failures. Hence, a high-temperature resistant binder is required to ensure cohesion and stiffness of pavement at high temperatures. On the contrary, this will reduce the aggregate loss associated with edge breaking and finally prevent groove closure [3].

Other Distress. Other distress, like cracking, spalling, wearing, and erosion, are innate to HMA pavements and can be observed in grooved and ungrooved asphalt surfaces. Furthermore, the movement of asphalt caused by binder flow may lead to a wavy groove shape. Most groove failures ended up with grove closure due to plastic flow and deeply concerned with asphalt characteristics containing binder and aggregates [12].

Mechanism of Groove Failure. Groove failure is closely connected with plastic flow resulting from the viscous flow, which is akin to the rutting mechanism in an ungrooved asphalt surface. Microscopic analysis of asphalt recommended that cohesive (or stiffness) shortage is more important than adhesive failure. Groove closure is deeply related to binder characteristics: temperature, loading time, and aging [13, 14]. Edge breaking is caused by the horizontal stresses induced by aircraft tires. Moreover, this recurring stress application leads to edge failure on the unsupported groove edges. The examination exposed that this edge beak is a cohesive failure and depends on the binder's viscosity (stiffness). Furthermore, asphalt becomes more brittle, and this groove edge breakage mechanism could be more serious in cold weather conditions than in hot environments. Also, age-induced hardening, freeze–thaw cycling, moisture effects, and chemicals' de-icing could enlarge the edge breakage [12].

4 Pavement Performance Prediction Models

Predicting pavement future performance and deterioration process is significant to understand among the concerned authorities. The emphasis is given to the precise estimation of pavement performance concerning time. Hence, M&R actions under PMS have become more dependent on this prediction of pavement performance, and necessity enhances than before.

Pavement prediction models are inevitable in the current pavement management system and play a vital role in many critical management decisions.

There is a various decision-making tool for adopting modeling such as dynamic programming, neural network, decision support system, the Genetic Algorithm, and System dynamic modeling [15], which have been exercised broadly in pavement management [16, 17]. Pavement performance prediction models can be categorized into three different types: empirical models, mechanistic models, and empirical-mechanistic models. Pavement performance prediction models are two types by some others which are deterministic and probabilistic [18].

The deterministic models predict every single number related to the level of distress or whatever parameter that needs to measure for describing the projected condition for a pavement's remaining life. Hence, it is considered an evaluation of pavement deterioration over time in the entire prediction process. Conversely, a probabilistic model predicts a distribution of such events that demonstrates different probable future conditions since the outcomes developed stochastically. In this process, the deterioration prediction is treated to some extent as ambiguous and does not reflect accurate predictions [18].

One of the limitations of the current deterioration prediction models is that they do not investigate the transition time of pavement deterioration and transition probabilities from one state condition to the next. Hence, the model developed excluding the transition probabilities between pavement conditions states is construct a deterministic model where the probabilistic characteristics of the pavement deterioration process remain absent [19].

4.1 Deterministic Models

Deterministic models are broadly used, prediction models. Two types of prediction models that include structural and functional performance prediction models mainly depend on the prediction type that needs to be determined [18].

Structural Performance Models. These models are utilized to predict all kinds of individual pavement distresses as a natural feature of pavement behavior. The prediction models can be empirical or mechanistic-empirical. Empirical models are based on experience or experimental results derived from several observations to achieve the correlation between the input variables and outcomes [20]. In the mechanistic-empirical model, the materials responses or accumulated deformation is regulated as per the experimental field data, which is why it is termed a mechanistic-empirical model.

These models indeed depend on the properties of the materials of pavement structure and the maximum allowable load, including the number of cycles of load applications before rupture occurs. Various distress such as fatigue cracking, predetermined rutting, and others are defined as failure criteria. Recently, materials characteristics, for instance, deflection calculation and future projected traffic, are considered for the prediction of the remaining structural life of the pavement and thus helps the agencies for future M&R planning and actions [18]. This type of prediction model has been developed for different pavement and utilized in several organizations such as the Asphalt Institute, the Portland Cement Association, and Shell International Petroleum company by the concerned engineers. Moreover, this structural prediction model is incorporated in APMS and associated software like PAVER and IAPMS [18].

Functional Performance Models. This type of model is frequently practiced in the PMS in highways that are utilized to determine the pavement surface friction (skid resistance) or to hydroplane potential during wet weather and present serviceability index (PSI). These models can be empirical or mechanistic types similar to the structural model. PMS of Denmark, the PARS system of the province of Ontario, and NCR models developed under this category [18].

In APMS, the only functional performance prediction models are utilized for PCI prediction models. The majority of them are introduced by the PAVER system but used by others as well, like AIRPAV and IAPMS. Nonetheless, in contrast to their highway counterparts, these models are empirical. Their analysis is not based upon the prediction of pavement deterioration related to loading and climate effect on the surface conditions, and rather they developed a correlation between PCI and other available data demonstrating the pavement structure. As a result, the prediction models that rely on previously observed data need to be calibrated through the procedure of statistical analysis [18].

In the beginning, the prediction under PAVER was executed by a straight-line extrapolation utilizing only the previous two PCI values. The projected PCI values were established by the straight line hitting these two PCI points plotted on a PCI vs. time graph. It is to be noted that no other variables were taken into consideration that way. It was too simple and comparative inaccurate due to a lack of modern techniques.

The most popular statistical analysis technique for prediction modeling is multiple regression analysis, which has been utilized in numerous situations to generate such models. Models evolved from this approach indicate that the projected PCI values are related to various explanatory variables such as pavement structure, time, load, and repetition of traffic to express the predictive mathematical equation.

PAVER software for pavement management was developed using such equations for both the flexible and rigid pavement by USACE, analyzing a substantial set of data collected from different US Air Force bases. The prediction performance was found satisfactory for the higher PCI values but declined significantly for the pavement surfaces with PCI less than 65 for rigid or 50 for flexible pavements correspondingly.

It was noticed that several universal models that were generated using the regional data not performed satisfactorily for the local ones. The model incorporated with the specific local climate, soil subgrade, and materials properties render comparatively better performance. Hence local models are more advantageous than universal and should be refined and updated as local and new information is accessible [18].

Regression analysis. Regression analysis is a handy tool for the development of a prediction model with due care. The equation derived from the modeling should be significant and useful in respect to the selected variables and not prioritized the best fit the available data only. This is inevitable to achieve a realistic model with a high level of confidence in its prediction. A large amount of data is required to get a precise model. However, accurate predictions may not be attained due to complex pavement characteristics. A substantial amount of data is utilized to develop the model. The regression models are defined within a range of data by which it is developed and hence cannot be fabricated extending far from the range. That is why the PAVER models are not precise for the PCI value less than 65 ± 50, and the predictive range was constrained to the upper part of the PCI scale [18].

4.2 *Probabilistic Models*

Probabilistic prediction models are three different types including, survivor curves and simulation models and Markovian models. In recent time Markovian models gain more recognition and acceptance as an effective prediction technique in the arena of the highway are now deploying towards airport pavement. The concept behind this type of model is that the pavement deterioration process is not deterministic in nature but uncertain. Hence, a probabilistic prediction model should depict the process in a stochastic way rather than assuming deterministic behavior the wrong way. Moreover, the Markovian probabilistic approach offers more rational models considering a different aspect of pavement characteristics [18]. Different methodologies have been generated to develop probabilistic pavement performance models with high prediction capabilities. Nonetheless, insufficient historical data generated preventive maintenance (PM) model often reveal erroneously predicted pavement condition and leads to non-optimum maintenance and rehabilitation (M&R) resolution [21].

Survivor Curves. Survivor curves are utilized by different agencies for planning, finding M&R alternatives on pavement networks. Authorities adopted previous construction, maintenance, and rehabilitation data to generate the curves containing probability vs. time. Generally, the probability declines with time from 1.0 to 0.0, which indicates serviceable pavement conditions without major maintenance or rehabilitation works [18].

Simulation Models. Simulation models are based on computer programs followed by mathematical models of pavement response against load and pavement behavior for a certain period. Different input parameters such as pavement layer interfaces and bitumen content are considered stochastic, and the model response is also stochastic accordingly. These programs can predict future pavement conditions. This modeling requires suitable computer resources and comprehensive repetitive calculations, thus not pragmatic in the planning framework [18].

Markovian Models. This type of model predicts pavements' functional condition and not aiming to analyze the deterioration process instead directs the model with readily available information regarding pavement characteristics.

The Markovian prediction model articulates the state by which it represents a pavement section's condition at any given time. The condition states could be designated with respect to the PCI of that sections. For instance, PCI ranging 100 ± 91 would be treated in state 1, and PCI ranges 81 to 90 would be considered in state 2. The pavement deterioration progression with time is modeled by shifting from one condition state to another with time advancement. This deterioration process's nature is probabilistic, and probabilities govern the evolution of pavement conditions in association with the different possible transitions. Each transition probability demonstrates the chances that a pavement section in the current condition will end up in a specific condition after a certain period. These probabilities are articulated in a matrix form (Markovian transition matrix), representing pavement sections with similar properties such as construction type, age, and traffic condition [18]. Under this approach, the transition probability matrices are derived from either engineering experience or analysis of historical pavement condition data. The analysis is done using several techniques, for instance, regression analysis or non-linear programming. Markov models were first implemented in pavement management under the Arizona highway PMS in 1980. Afterward, this prediction model was used in the airport pavement management. There are different benefits to adopting Markovian prediction models. These models offer better predictive accuracy than their counterparts when properly generated. A comparative study utilizing four PCI prediction models was carried out by Cook and Kazakov (1987), where two models were regression models (deterministic), and the other two were Markovian models. The average errors were observed higher in regression models than the Markovian models for the predicted PCI relative to the actual PCI values. Nonetheless, important historical data files are usually required to generate such Markovian models than with the regression models.

Another advantage of Markovian models is that they can make predictions far from the limit of the data and deliver typical outline deterioration conditions with respect to age, which regression models cannot give direct assurance. Consequently, these models can be integrated into most of the PMSs for the planning process [18].

4.3 System Dynamic Study

System dynamics is a standard method of modeling that identifies the correlation of a certain parameter with other variables and indicates the changes with respect to time. It must contain a flowchart or a conceptual model that expresses the entire process of the model. In the conceptual model, storages and flows are like building blocks. Storages act as accumulators in the system and facilitate describing the circumstances of the system. Conversely, flows specify the movement rate of possessions in and out of the system. Values and relationships among each storage and flow are designated in the form of constants and represented through equations or data tables. Once a system dynamics model is generated, it explains the cause-effect relations among the variables and maintains continuous interactions between its parameters [15].

4.4 Artificial Neural Network (ANN)

Artificial Neural Network (ANN) has been recognized as a powerful computational tool to resolve numerous engineering problems over many years. American Association of State Highway and Highway and Transportation Officials (AASHTO) utilized ANN to develop a new Mechanistic Empirical Pavement Design Guide (MEPDG) [22]. ANN is motivated by the biological neural network where the neuron consists of soma, dendrite, synapse, and axon and nucleus creates the input process. Dendrite is a tree-like fiber that acts as a receptor to receive the signal or input. Axon is the long single fiber cell that transfers the signal from a synapse to another recipient end of the neutron's synapse. The artificial neural network was developed by mimicking the basic structure and working procedure, including some significant attributes related to computing the model with pattern recognition tasks [23].

ANN is a plotting of input into the preferred outcome. It contains weighted inputs and transfers function into the output. A modest feed-forward neural network is generally adopted in ANN. The input at the first layer is supplied into the interconnecting layer called hidden layers, supports by the transfer function, which does not affect the feed-forward characteristics in the neural network and ultimately the output layer. The ANN model's performance is usually assessed by a different error, such as Mean Square Error (MSE) [23].

4.5 *Genetic Programming (GP)*

Gene expression programming (GEP) is like genetic algorithms (GAs) and genetic programming (GP), that customize populations of individuals utilizing genetic algorithm and decide as per fitness, and expresses genetic disparity using one or more genetic operators [24]. It is an emerging program that flourished from the Genetic Algorithm (GA). GP is an evolutionary algorithm-based system persuaded by biological evolution where independent input parameters are engaged to resolve the mathematical assignment, and an output parameter is generated utilizing linear or nonlinear equations [25]. Basically, it obeys the Darwinian principle of survival of fitness, where a number of solution candidates are set against a problem similar to the GA [23].

The GP's fundamental genetic operators mimic GA, for instance, mutation, reproduction, and crossover, except for the expression tree or syntax tree representing the GP rather than conventional codes. Furthermore, it can be uttered in linear mathematical formulae. In general, the leading operators in GP are Mutation and Crossover, and the preliminary population is randomly generated through individual computer programs. The selection method of individuals depends on their fitness for crossover.

In GP, the terminals are usually indicated as the input variables (i.e., x and denoted by d0, d1, etc.) and constants (symbolized by c0, c1, etc.) in the expression tree (ET). Likewise, the functions are interpreted through the internal nodes, such as different arithmetic functions (i.e., +, −, *, /, Sqrt), certain mathematical operations (i.e., ln, sin, cos, exp), including different conditional operations like if, greater, equal, then, and else. This arrangement of terminals and functions is determined as an intrinsic set in GP [23].

5 Predictive Modelling by Genetic Programming (GP)

5.1 Modeling Procedure

In this paper, GeneXProTools 5.0 was applied to generate the GP model. The software was employed to develop a relationship between the groove areas changing with several input parameters. GeneXproTools is a potent software package that can be utilized to execute a symbolic regression analysis based on Gene Expression Programming (GEP) [24, 26]. GeneXProTools is a simple to run comprising effective tool within the GEP technique and demonstrates the mathematical equation explaining the combined model to the users, including the intrinsic merits of Genetic Programming and Genetic Algorithms [27].

Generally, the task of linking input parameters is supported with a variety of functions in GP. This modeling utilized four basic operators (+, −, × , /) and some other functions, including × 2, $\sqrt{3}$, ln, and exp. Finally, a GP-based model delivered a mathematical equation evolved from the expression trees (ETs) that contain several genes denoted as sub-ETs. Every gene possesses a fixed length and constitutes a head that includes functions (for example, +, –, x2, and ln) and terminals (representing the input variables denoted by d), and a tail comprises only terminals. This head size indicates the complexity or maximum size of each sub-ET branch in the model. Though there is no standard way to run the model, the practice runs and monitors the results and adjusts the number of sub-ETs and chromosomes until attaining optimum accuracy. Distinct options of functions and original & derived variables and constants are developed in the terminals to model the data [28].

A model may not utilize the allocated maximum number of chromosomes and head sizes. A linking function is inevitable in linking the sub-ETs in the model when the number of genes becomes more than one. The addition function is deployed as a liking function in this research to link the sub-ETs.

In general, the experimental datasets are divided into training and validation parts. Modeling was continued until the coefficient of determination (R^2) become maximum for both training and validating phases. However, it is crucial to have a model trained and validated with about the same accuracy [23].

5.2 Importance of the Input Variables

The input variables considered for this modeling are Wheel pass number (n), Wheel load (L), Subbase Thickness (Tsb), Asphalt Thickness (Ta), and Maximum temperature (tmax), respectively.

Emery [13, 14] investigated the deterioration of grooves of asphalt on Australian runways and suggested that slow movement and heavy aircraft were accountable for most groove closure in Australia [13, 14]. Pavement layer thickness variations also create the differences in pavement performance substantially. Subgrade stiffness, granular subbase thickness, and asphalt thickness are the other parameters that affect the variability of anticipated deformation performance [29].

Groove failure and associated closure are considerably connected to asphalt rheological properties and successive plastic flow of the HMA. Aircraft loading in higher temperatures provides plastic flow and dislocated aggregates along the wheel tracks [12]. Asphalt pavements exposed to temperature variations daily or seasonally are more susceptible to fatigue damage than at a specific temperature [30]. As a result, groove closures are frequent in hot environments [3].

The input variables might significantly influence the output, and the correlations between the input variables and the output results could be meaningful for determining the groove area. However, insignificant variables could lessen the accuracy and overall performance of the model.

5.3 Data Collection, Input Variable, and Model Configuration

The wheel load, pass number, Subbase, and asphalt thickness data were extracted from asphalt pavement groove life and effect of aircraft traffic loading examined at FAA National Airport Pavement Test Facility (NAPTF) by [7]. The temperature during the loading period was taken from the website [31].

The input variables and response variables are denoted by X and Y. The inputs in this model were Wheel pass number (d0), load (d1), Subbase Thickness (d2), Asphalt Thickness (d3), and Maximum temperature (d4) correspondingly, and the response variable and output parameter were Groove Cross-Sectional Area (Y).

The experimental dataset (a total number of 81) was sub-divided into training (62 Nos, 76.54%) and validation (19 Nos, 23.46%). This GP model employed fifteen functions, including addition, subtraction, division, multiplication, ln, and exp. The addition was considered as the linking function, whereas the Root Mean Squared Error (RMSE) was utilized as the fitness function. The configuration of input settings and symbolic function set to develop the combined GP model in GeneXProTools: 5 are demonstrated in Tables 1 and 2.

6 Results and Discussions

This GP model's results were exposed through the expression tree (ET) associated with the gene is demonstrated in Fig. 3. Three sub-expression trees (ET's) with addition as a linking function were attained. In the ET's d0, d1, d2, d3, and d4 stands for Wheel pass number (n), Wheel load(L), Subbase Thickness (T_{sb}), Asphalt Thickness (T_a), and Maximum temperature (t_{max}), respectively. The value of constant

Table 1 General input in GEP Model structure

General input settings
Independent variables input:
• d(0): wheel pass number (Number)
• d(1): load (Kips)
• d(2): subbase thickness (inch)
• d(3): asphalt thickness (inch)
• d(4): maximum temperature (°C)
Dependent variable input:
• Groove cross-sectional area (Sq-inch)
Output: GEP model estimate (Sq-inch)
Number of training samples: 62
Number of validation samples: 19
Number of chromosomes: 30
Head size: 8
Number of genes: 3
Fitness function: RMSE (Root Mean Squared Error)
Linking function: addition
Stopping criterion: maximum fitness

Table 2 Symbolic function set for the GEP model

Symbolic function set			
Function	Symbol	Weight	Arity
Addition	+	4	2
Subtraction	-	4	2
Multiplication	*	4	2
Division	/	1	2
Exponential	Exp	1	1
Natural logarithm	Ln	1	1
x to the power of 2	X^2	1	1
Cube root	3Rt	1	1
Arctangent	Atan	1	1
Minimum of 2 inputs	Min2	1	2
Maximum of 2 inputs	Max2	1	2
Average of 2 inputs	Avg2	4	2
Hyperbolic tangent	Tanh	1	1
Complement	NOT	1	1
Inverse	Inv	1	1

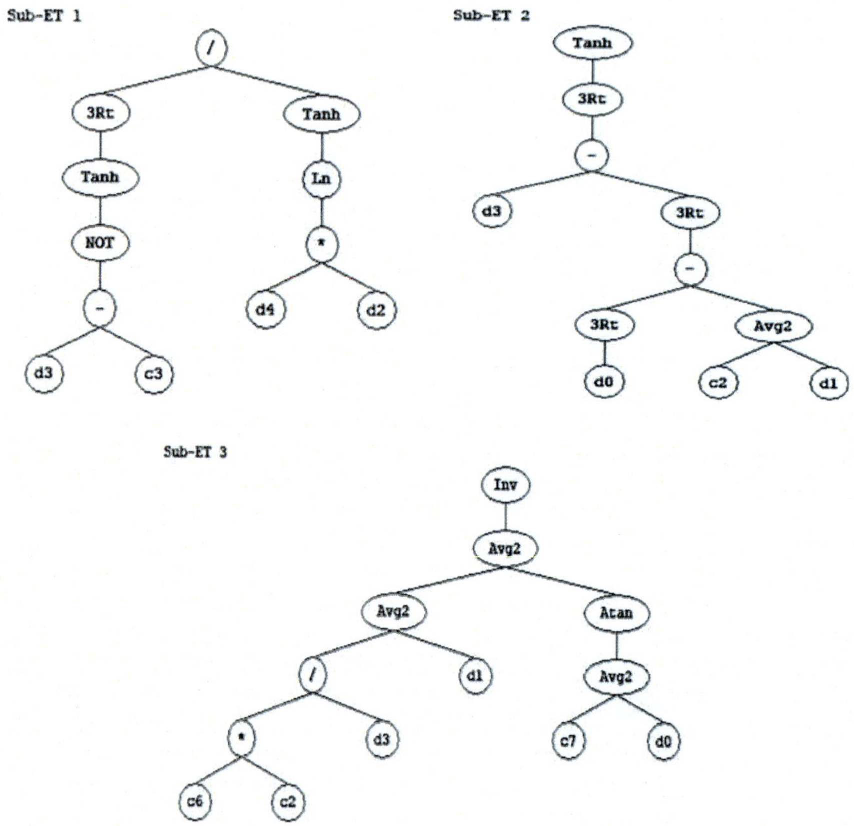

Fig. 3 GEP Expression Tree (ET)

c3 in sub-ET 1 is 6.410, c2 in sub-ET 2 is 0.90308. Likewise, the constant of c2, c6, and c7 in sub-ET 3 accounts for 6.625, 7.364, and -311.575, correspondingly.

The ultimate simplified equation articulates the Groove Cross-Sectional Area (f_{ga}) as indicates in Eq. (1), derived from these expression trees.

$$
\begin{aligned}
f_{ga} = {} & \frac{\sqrt[3]{\text{Tanh}(1-(\text{d3}-6.410))}}{\text{Tanh}(\ln(\text{d4}\times\text{d2})} \\
& + \text{Tanh}\sqrt[3]{\text{d3}-\sqrt[3]{\sqrt[3]{\text{d0}}-\frac{(0.9308+\text{d1})}{2}}} \\
& + \frac{1}{\frac{\frac{\frac{7.364\times6.625}{\text{d3}}+\text{d1}}{2}+\text{Atan}\left(\frac{-311.575+\text{d0}}{2}\right)}{2}}
\end{aligned}
\tag{1}
$$

The numbers of chromosomes and sub-ETs as genes are two important parameters that play a vital role in the preciseness of the model. However, a limited number of

Table 3 Parameters and outcomes of the model

Parameter	Value
Number of Genes (Sub ETs) evolved	3
Number of variables used	5
Maximum Number of Chromosomes used	13
Maximum Number Head size used	6
Lower bound	10
Upper bound	-10
Mutation	0.00138
Inversion	0.00546
Gene Transposition	0.00277
Constants per gene	10
Data Type	Floating-point
Coefficient of determination (R^2)	0.931 (Training), 0.946 (Validation)
RMSE (Root Mean Squared Error)	0.00156 (Training), 0.00278 (Validation)
Maximum fitness	998.42 (Training), 997.16 (Validation)

chromosomes and genes could lead to inferior accuracy and develop a complex and ineffective equation. Hence, there should be an optimum number of these essential parameters for any specific model. Input parameters were employed as observed in the prediction equation, which specifies the importance and relevance of all the input parameters into the output. The model outcome finally contributes to the precise prediction of the groove closure.

Other parameters and outcomes of the model are presented in Table 3. Maximum 13 chromosomes and 6 heads were used in sub-ET's. The model achieved an excellent fitness of 998.42 for training and 997.16 for validation phases in the maximum fitness scale of 1000. The coefficient of determination (R^2) for training and validation phases was obtained at a rate of 0.931 and 0.946, respectively. A minor error such as RMSE demonstrates that the GP model has been trained well, and it can forecast the groove areas with a high degree of accuracy and reliability. In addition to this, Table 4 delivers some statistical parameters of the variables.

Figure 4 depicted the experimental data in comparison with the predicted results for both the training and validation phases as determined in the GP model. These scatter plots expressed the correlation between input variables and derived variables.

The target sorted fitting curves articulates that the model followed the targeted values quite closely for both the training and validation phases, as illustrated in Figs. 5 and 6, respectively.

Table 4 Statistical parameters of variables

Variables	Statistical parameters					
	Minimum	Maximum	Median	Std deviation	Skewness	Kurtosis
Pass number (n)	0.00	37,290.00	11,338.00	10,289.19	0.88	−0.01
load (L)	55.00	65.00	55.00	2.64	3.31	9.21
Subbase Thickness (Tsb)	34.00	39.00	37.00	2.07	−0.24	−1.52
Asphalt Thickness (Ta)	10.00	15.00	12.00	2.07	0.24	−1.52
Maximum temperature (tmax)	9.00	34.00	27.00	7.52	−0.62	−0.69
Groove Cross-Sectional Area (Y)	0.034	0.065	0.049	0.006	−0.450	0.636

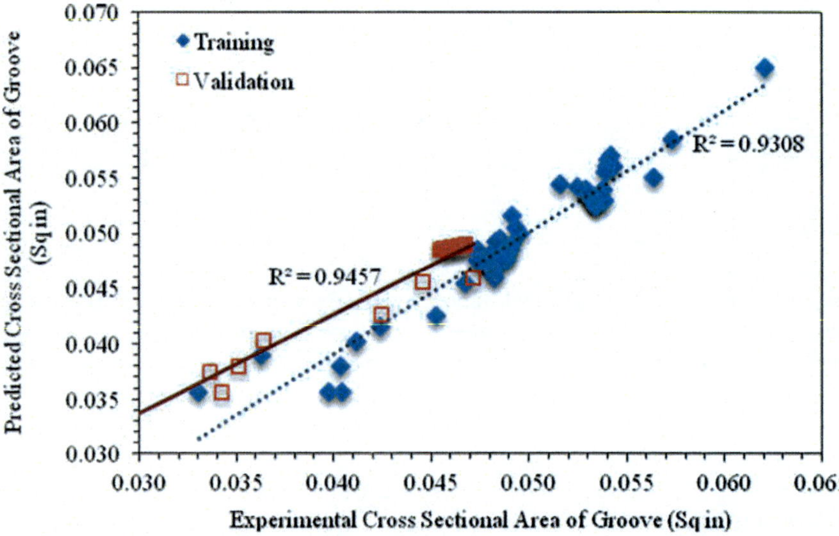

Fig. 4 Predicted and experimental Groove X-Sectional Area (Training & Validation Phase)

The results stated that the GP model could predict the groove area deterioration near the experimental results. Additionally, this GP model can be an effective and reliable model for groove closure prediction with a high degree of accuracy.

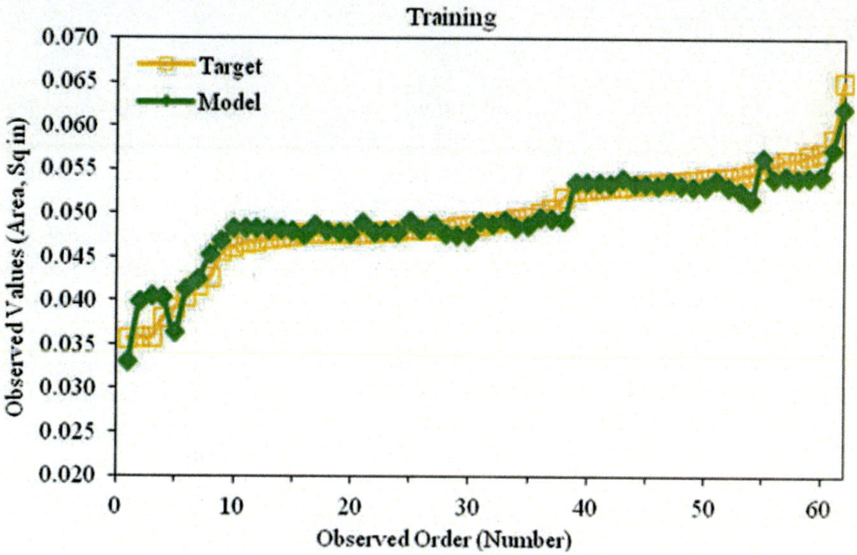

Fig. 5 Target sorted fitting of data (Training phase)

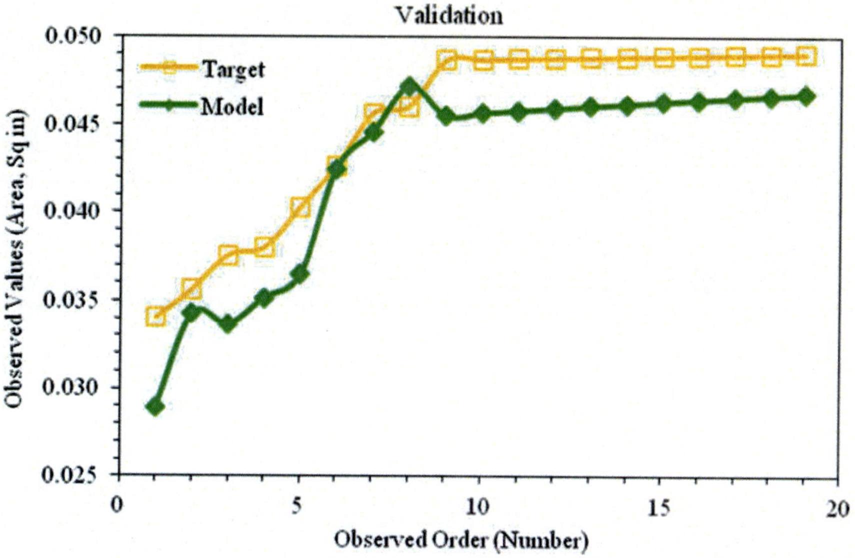

Fig. 6 Target sorted fitting of data (Validation phase)

7 Conclusions

This paper reviewed detail about runway grooving's contemporary practice, including potential distress mechanisms and factors involved with groove deterioration. A later different perspective of prediction modeling has been demonstrated. Finally, a GP Model has been developed successfully.

The coefficient of determination (R^2) obtained for training and validation are 0.931 and 0.945, correspondingly. The results indicate that the generated model fitted with the experimental data reasonably well, an expression of the prediction models' reliability.

It was tricky to predict the groove's service length on the runway in a theoretical process before. The prediction equation generated through this GP model renders useful information for the prediction of groove area deterioration derived from several input factors, including loads, pass number, the thickness of sub-base and Asphalt layer, and temperature. Airport authorities can refix and fit the model depending on their specific construction history, loading characteristics, and frequency. This will deliver a general prediction view of their groove life and timeframe of functional condition. Moreover, the model will shorten the evaluation time for groove area changes in a similar situation and calculate the rate of deterioration and the percent of groove closure regarding aircraft movement realistically.

Moreover, the authority can plan its budget for future repair and maintenance activities, including groove reinstallation to reinstate surface friction. However, some other factors like rubber contamination on runway grooves, aircraft wheel speed, and variation in materials configuration in the runway pavement layer, including fundamental characteristics like resilient modulus, could be considered for the modeling. More detailed examination facilities in the runway in operational conditions could provide realistic results and associated modeling.

Finally, this study has produced a deterioration model that will provide insight into the groove closure and groove life prediction for timely maintenance of runway pavements to ensure adequate surface friction.

References

1. Pasindu HR (2020) Analytical evaluation of impact of groove deterioration on runway frictional performance. Transp Res Procedia 48:3814–3823
2. Fwa TF, Pasindu HR, Ong GP, Zhang L (2014) Analytical evaluation of skid resistance performance of trapezoidal runway grooving. In Transportation research board 93th annual meeting, p 24
3. White G., Rodway B (2014) Distress and maintenance of grooved runway surfaces. Airfield Eng Maintenance Summit
4. Federal Aviation Administration (FAA) (1997) Measurement, construction and maintenance of skid-resistant airport pavement surfaces. FAA Advisory Circular 150/5320–12C, Washington DC
5. White, G.: Managing skid resistance and friction on asphalt runway surfaces. In World Conference on Pavement and Asset Management, Milan, Italy (2017).

6. Shahin MY (2007) Pavement management for airports, roads, and parking lots (2005). ISBN-10: 0–387–23464–0, ISBN-13, 978–0387, 2nd edn. Springer
7. Wang Q, Larkin A (2019) Asphalt pavement groove life analysis at the FAA national airport pavement test facility. In Airfield and highway pavements 2019: innovation and sustainability in highway and airfield pavement technology, Reston, VA: American Society of Civil Engineers, 418–426
8. Federal Aviation Administration (FAA) (2014) Standards for specifying construction of airports. FAA Advisory Circular 150/5370–10G, Washington DC
9. Washington DC Federal Aviation Administration (FAA) (2010) FAA grooving requirements. FAA worldwide technology transfer conference, New Jersey, USA
10. Patterson JW (2012) Evaluation of trapezoidal runway grooving. FAA-TC-TN12/7, Federal Aviation Administration, Washington, DC
11. White G (2018) State of the art: asphalt for airport pavement surfacing. Int J Pavement Res Technol 11(1):77–98
12. Apeagyei AK, Al-Qadi IL, Ozur H, Buttlar WG (2007) Performance of grooved bituminous runway pavement. Center Excel Airport Technol: Tech Rep 28:1–10
13. Emery SJ (2005) Asphalt on Australian airports. Australia asphalt paving association pavement industry conference, Surfers Paradise, Queensland
14. Emery SJ (2006) Bituminous surfacing for pavements on Australian airports. 24th Australia airports association convention, Hobart
15. Tarefder RA, Rahman MM (2016) Development of system dynamic approaches to airport pavements maintenance. J Transp Eng 142(8):04016027
16. Ansarilari Z, Golroo A (2020) Integrated airport pavement management using a hybrid approach of Markov Chain and supervised multi-objective genetic algorithms. Int J Pavement Eng 21(14):1864–1873. https://doi.org/10.1080/10298436.2019.1571208
17. de Moura IR, dos Santos Silva FJ, Costa LHG, Neto ED, Viana HRG (2020) Airport pavement evaluation systems for maintenance strategies development: a systematic literature review. Int J Pavement Res Technol 1–12
18. Gendreau M, Soriano P (1998) Airport pavement management systems: an appraisal of existing methodologies. Transp Res Part A: Policy Pract 32(3):197–214
19. Altarabsheh A, Altarabsheh R, Altarabsheh S, Asi I (2021) Prediction of pavement performance using multistate survival models. J Transp Eng Part B: Pavements 147(1):04020082
20. Pavementinteractive Homepage: Empirical Pavement Design retrieved from https://pavementinteractive.org/reference-desk/design/structural-design/empirical-pavement-design/. Last accessed 21 Jan 2021
21. Yamany MS, Abraham DM (2021) Hybrid approach to incorporate preventive maintenance effectiveness into probabilistic pavement performance models. J Transp Eng Part B: Pavements 147(1):04020077
22. Gopalakrishnan K, Ceylan H, Guclu A (2009) Airfield pavement deterioration assessment using stress-dependent neural network models. Struct Infrastruct Eng 5(6):487–496
23. Leong HY, Ong DEL, Sanjayan JG, Nazari A, Kueh SM (2018) Effects of significant variables on compressive strength of soil-fly ash geopolymer: variable analytical approach based on neural networks and genetic programming. J Mater Civ Eng 30(7):04018129
24. Ferreira C (2001) Gene expression programming: a new adaptive algorithm for solving problems. Complex Syst 13(2):87–129
25. Leong HY, Ong DEL, Sanjayan JG, Nazari A (2015) A genetic programming predictive model for parametric study of factors affecting strength of geopolymers. RSC Adv 5(104):85630–85639
26. Ferreira C (2006) Gene expression programming: mathematical modeling by an artificial intelligence. Germany, 2nd edn, Springer
27. Fernando AK, Shamseldin AY, Abrahart RJ (2009) Using gene expression programming to develop a combined runoff estimate model from conventional rainfall-runoff model outputs. In Proc 18th World IMACS congress and MODSIM09 international congress on modelling and simulation, 2377–2383

28. Gepsoft Homepage, https://www.gepsoft.com/tutorials/GettingStartedWithRegression.htm. Last accessed 26 Nov 2020
29. Valle PD, Thom N (2020) Pavement layer thickness variability evaluation and effect on performance life. Int J Pavement Eng 21(7):930–938. https://doi.org/10.1080/10298436.2018.1517873
30. Safaei F, Hintz C (2014) Investigation of the effect of temperature on asphalt binder fatigue. In International society for asphalt pavement conference, 1491–1500
31. TimeanddateHomepage, https://www.timeanddate.com/weather/@5101760/historic?month=6&year=2016. Last accessed 23 Nov 2020

Side Friction as a Cause of Poor Pavement Maintenance in India—A Study Using Terrestrial Laser Scanner Data

Shishodiya Ghanshyam Singh and S. Vasantha Kumar

Abstract Side friction is a peculiar characteristic that can be seen mainly in urban roads of India where the part of carriageway is utilized for activities like on-street parking, encroachment, etc. Though there are many studies on speed and capacity reduction due to side friction, however studies on carriageway width reduction are very limited. Hence an attempt has been made to measure the carriageway width loss due to side friction activities using the terrestrial laser scanner data and analyze the causes of such activities. Two locations with a stretch of about 50 m. each in Vellore, India was taken as study stretches and laser scanning was carried out using the Leica BLK 360 laser scanner. The total carriageway width and effective carriageway width were measured accurately using the point cloud data. It was found that, in one of the locations, almost one lane was occupied by on-street parking of vehicles. Aggregates were just poured over it along the lane and because of that vehicles were not moving on the lane. Hence it encourages illegal on-street parking of vehicles at that location as the lane was not utilized for vehicle movement. Similarly, at other location also, pavement maintenance was not carried out and thus a strip of sand particles along the shoulder-side lane was found which also eventually led to on-street parking as the lane was not utilized for vehicle movement. Thus, the results clearly says that lack of pavement maintenance is one of the major causes for side friction activities in India.

Keywords Side friction · Pavement maintenance · Terrestrial laser scanner · On-street parking · Effective carriageway

1 Introduction

Side friction basically refers to the interference caused to the moving vehicles on the carriage way due to various factors like on-street parking, roadside encroachment, presence of garbage bins, stopping of buses, pedestrian movement on carriageway,

S. G. Singh (✉) · S. V. Kumar
School of Civil Engineering, Vellore Institute of Technology (VIT), Vellore, Tamil Nadu 632014, India

© The Author(s), under exclusive license to Springer Nature Switzerland AG 2022

H. R. Pasindu et al. (eds.), *Road and Airfield Pavement Technology*, Lecture Notes in Civil Engineering 193, https://doi.org/10.1007/978-3-030-87379-0_17

entries and exists from approach roads, etc. These side friction elements not only obstruct the traffic flow and reduce the speed of vehicles but also endanger the life of drivers especially when hoardings and sand particles are present on the carriageway as shown in Fig. 1. For example, there are high chances of skidding of two-wheelers due to lack of proper tyre friction between the wheel and road. Though the chances of skidding are very less for other vehicle types like three wheelers and four wheelers, but they are not ready to pass over the sand particles due to the presence of banners/hoardings with sharp edges as seen in Fig. 1. Thus, the presence of side friction elements leads to under-utilized carriageway which eventually results in loss of efficiency and also raises safety concerns.

Many studies have attempted on the impact of these side frictions on travel time, speed, capacity and level of service (LOS) reduction. Salini and Ashalatha [1] attempted a study on four lane divided urban roads to assess the impact of side frictions on flow of traffic. Analysis revealed that the average speed of vehicles reduced considerably with bus stops, pedestrians and on-street parking by 21%, 27% and 15% respectively. Pallavi and Mehar [2] have studied the influence of side friction activities on vehicular speed. In their study they found that there is a 9% reduction in the capacity due to the presence of side friction elements. Ashish [3] developed a model to study the influence of crossing pedestrians at the undesignated locations on capacity of 4-lane urban midblock sections. Based on his study he concluded that the capacity reduces from 18 to 30% when the pedestrian cross-flow of 800 to 1500 pedestrians per hour is observed. Rao et al. [4] found that the average stream speed has reduced to 49–57% due to bus stops and bus bays with reduction in the capacity of 10–53%. The authors also reported that the on-street parking caused 45–67% of the

Fig. 1 Influence of side friction on traffic flow

stream speed reduction and 28–63% reduction in the capacity which thus adversely disturbing the functionality of the urban arterials. Pal and Roy [5] have studied the impact of roadside friction elements on travel speed and LOS of rural highways in India. It was found out that because of the impact of side friction in the market area, the design speed of the vehicles decreases and level of service gets altered from C to E. It can be seen that the most of the reported studies have focussed on the reduction in travel speed and/or capacity/LOS due to side friction elements. Studies on carriageway width reduction are very limited. Hence in the present study, an attempt has been made to measure the carriageway width loss due to side friction activities using the terrestrial laser scanner data and analyze the causes of such activities. The details of study area and data collection are explained in the following section.

2 Materials and Methods

2.1 Details of Study Area

The study area selected for the present work is located in Vellore city in the state of Tamil Nadu, India. Two study stretches with a distance of 50 m and 100 m were selected as shown in Fig. 2.

The study stretch-1 in Gandhi nagar is basically a sub-arterial road whereas the study stretch-2 near VIT is located on a state highway. The reason for selecting these two study stretches is at both the locations the side frictions were present. For example, in both the study stretches, illegal on-street parking was observed as seen in Fig. 2. Though both the study stretches are divided roads with median in between, but only one side of the road was considered as illegal parking was observed in one side only. In India, normally the single lane roads have 3.75 wide carriageway whereas the two-lane roads have 7 m carriageway with 3.5 m wide lanes. The roads that fall in between them are called intermediate lane roads as they have carriageways between 5.5 m and 6.0 m [6]. The study stretch in Gandhi nagar is basically of intermediate lane type as the width of the carriageway is 6 m with each lane of 3 m wide. The study stretch-2 near VIT is a 4-lane highway with two lanes in each direction and the carriageway width is 7 m in one direction.

2.2 LIDAR Data Collection

The Light Detection and Ranging (LIDAR) also called as laser scanning is a remote sensing method in which light is emitted continuously from a rapidly firing laser. This light after reflecting from earth surface features like buildings, tress, etc. returns to the LIDAR sensor where the distance to the target is calculated which is then used to make digital three-dimensional (3D) representations of the earth surface feature. The

Fig. 2 Map showing the location and photos of study stretches (map *Source* Google maps)

LIDAR data can be collected from either aerial or ground based terrestrial systems and in case of terrestrial systems, the laser scanner is normally kept over a tripod and then scanning is done. In the present study, terrestrial laser scanning was done using one of the popular instruments called Leica BLK 360 laser scanner (Fig. 3).

The scanner has a capability of emitting 3,60,000 laser pulses per second and thus can produce a point cloud of millions of data points of X, Y, Z values. The range of the scanner is 60 m. which means the objects that are within a distance of 60 m. can be covered in one scan. For distances of more than 60 m, the instrument needs to be shifted with sufficient overlapping between the subsequent scans. The scan time is generally 3 min/scan and the advantage of BLK 360 is it provides photos also taken with high-definition camera along with point cloud data. In the present study, a total of 4 scans were done with 2 scans in each location during the mid-day on January 20, 2021. The collected data were then transferred to the Cyclone Register 360 post processing software and the results are discussed in the following section.

Fig. 3 LIDAR survey using Leica BLK 360 at Study stretch-1

3 Results and Discussion

The results of point cloud data collected at the study locations using LIDAR were shown in Fig. 4. A total of 16.299 million points with X,Y,Z values were collected at study stretch-1. Similarly, for the study stretch-2, 15.601 million points were collected. Registration of scans were automatically carried out by the Cyclone Register 360 as the scans were taken with sufficient overlap between them. The bundle error is an important parameter to determine the accuracy of the LIDAR data collected and in the present study, the errors were calculated as 0.005 m and 0.004 m for the study stretches 1 and 2 respectively. An error of less than a 'cm' basically indicates that the collected LIDAR data is very accurate and hence can be used for measurement purposes. Using the registered point cloud data, measurements were taken in Cyclone Register 360 as shown in Fig. 5. The total carriageway width (in one direction) at both the study stretches was measured accurately. Then the width of carriageway covered by on-street parking & sand particles at stretches-1 and 2 respectively were measured. It was found that out of total carriageway width of 6 m, only 3.608 m is available as effective carriageway for the vehicles to go which means almost one lane width is occupied by on-street parking in study stretch-1 which is

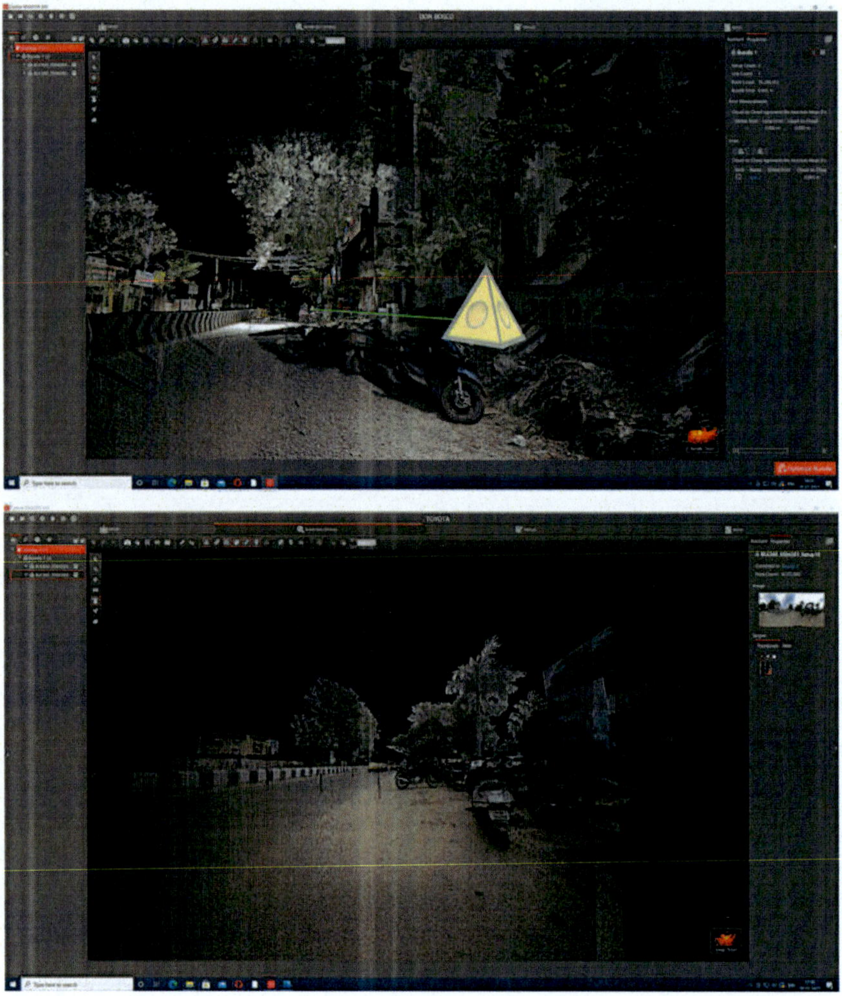

Fig. 4 Point cloud data of study stretch-1 (top) and 2 (bottom)

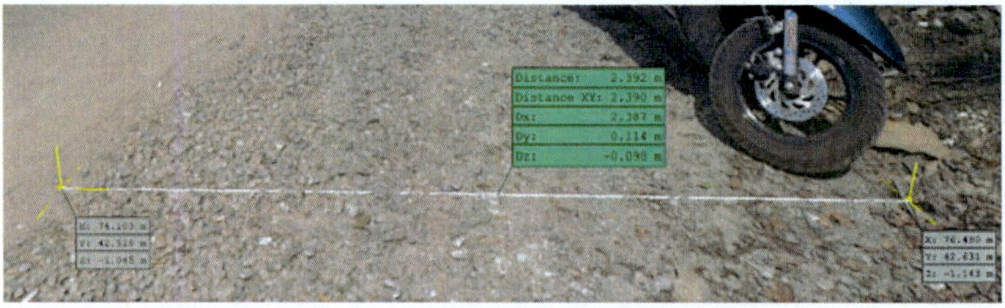

Fig. 5 A sample measurement using point cloud data

really unacceptable. The reason for this unauthorized parking on the carriageway is the shoulder-side lane was dug few months ago for pipe laying work and after that the pavement was not constructed (Fig. 3). Aggregates were just poured over it along the lane and because of that vehicles were not moving on the lane as sharp aggregates may puncture the tyres. Hence it encourages illegal on-street parking of vehicles at that location as the lane was not utilized for vehicle movement. In study stretch-2, out of total carriageway width of 7 m, almost one-third was covered with sand particles (1.259 m on shoulder side and 0.624 m on median side) and remaining portion only was available as effective carriageway width for vehicles to pass. This encourages vehicular parking as witnessed in Fig. 2. The sand particles not only reduce the carriageway width but also poses serious safety issues especially for two-wheelers and cycles as the chances of skidding is high when they drive over the pavement covered with sand particles. Thus, the results clearly says that lack of proper pavement maintenance is one of the major causes for side friction activities in India.

4 Conclusion

One of the important factors in India that affect the roadway capacity and LOS is the presence of side friction elements which not only hinder the traffic flow but also leads to safety issues especially when banners and sand particles are present on the carriageway. In the present study an attempt has been made to study how much carriageway is lost due to side friction elements using the terrestrial LIDAR data. The results showed that almost one-third to one-half of the carriageway was occupied with illegal on-street parking due to the poor pavement maintenance. The findings recommend that periodic maintenance of urban and rural roads is very essential in order to eliminate the side friction activities which act as a medium of obstruction to smooth flow of traffic.

References

1. Salini S, Ashalatha R (2020) Analysis of traffic characteristics of urban roads under the influence of roadside frictions. Case Studies on Transport Policy 8(1):94–100
2. Pallavi G, Arpan M (2018) Analysis of side friction on urban arterials. Transp Telecommun 19(1):21–30
3. Ashish D (2017) Influence of crossing pedestrians at undesignated locations on capacity of 4-lane urban midblock sections. J Traffic Logistics Eng 5(1):10–14
4. Rao AM, Velmurugan S, Lakshmi KMVN (2017) Evaluation of influence of roadside frictions on the capacity of roads in Delhi India. Transp Res Proc 25:4771–4782
5. Pal S, Roy S (2016) Impact of roadside friction on travel speed and LOS of rural highways in India. Transp Develop Econ 2(9):1–12
6. Indian Highway Capacity Manual (Indo-HCM) (2017) Council of scientific and industrial research (CSIR)—Central Road Research Institute (CRRI), New Delhi, India

Solar Energy Harvesting and Pavement Sensing

Kang-Won Wayne Lee, Michael Greenfield, Austin DeCotis, and Kevin Lapierre

Abstract An attempt was made to generate the required voltage in asphalt pavement to operate roadway sensors utilizing a temperature difference between two thermoelectric generators (TEGs). To enable output voltage by the TEGs below the asphalt surface layer, the harvester was installed with the copper plate 25 mm (1 in.) below the top surface layer. The copper plate is heated from the sun's rays penetrating the asphalt surface layer and transferring the energy into the harvester system. The power generated from the TEGs allows temperature difference readings, as well as maximum power output voltage. Optimizing the harvester for efficiency and sustainability were top priorities. Once the copper plate receives the required voltage from the heat generation, the Arduino can be turned on. To communicate with the Arduino board in the current set-up, a USB cord gets plugged into the Arduino with the other end into the computer. The software program *Arduino* should then be opened on the computer to read data from the apparatus. An SD card or Bluetooth receiver was implemented into the solar harvester unit. This allows data for retrieval to be stored without an external power source (i.e., computer), allowing the harvester to operate freely. The strain transducer was installed into the asphalt surface layer for strain monitoring of the roadway. The SD card/USB would be able to store the information from the pavement strain transducer. Data retrieval would be achieved simply by unplugging the card from the harvester unit and uploading it to a computer.

Keywords Solar energy harvesting · Thermoelectric generator · Copper plate · Arduino · Asphalt pavement sensing · Strain transducer

---

K.-W. W. Lee (✉) · M. Greenfield · A. DeCotis · K. Lapierre
University of Rhode Island, Kingston, RI 02881, USA
e-mail: leekw@uri.edu

© The Author(s), under exclusive license to Springer Nature Switzerland AG 2022

H. R. Pasindu et al. (eds.), *Road and Airfield Pavement Technology*, Lecture Notes in Civil Engineering 193, https://doi.org/10.1007/978-3-030-87379-0_18

1 Introduction and Background

1.1 Project Motivation

The properties of asphalt pavement allow the accumulation and dissipation of solar energy on a daily cycle. Heat is absorbed in pavements, causing many detrimental effects such as the degradation of pavement, heat island effect, and increased costs for cooling nearby structures [8]. Harvesting solar energy from pavement has the potential to provide many substantial benefits such as extending the service life of pavements, improving the air quality, lowering impacts to the climate, and producing energy [6]. The Rhode Island Transportation Research Center (RITRC) team at the University of Rhode Island (URI) investigated important aspects of energy harvesting via an embedded solar harvester system, which was created to reduce pavement distresses and to enhance evaluating pavement performance [7].

Typically, sun shining on pavement provides it with thermal energy throughout the day. Past studies have confirmed that pavement temperature rises and falls each day, and the magnitude of the daily fluctuations varies with albedo (fraction of reflected sunlight), light absorption by the pavement, and emission, all of which can vary with latitude, with season of the year, and with the pavement itself [4]. Pavement temperature falls at night as it radiates heat back into the atmosphere and as heat is conducted into underlying layers and subgrade soils.

Harvesting energy from pavement refers to injecting engineered processes into this daily thermal cycle. The goal is to capture part of the daily energy flow and to channel it to other uses. The benefits cited above for energy harvesting arise from the lower temperatures that can result from extracting energy from pavement more quickly than can be obtained in a typical daily cycle.

1.2 Objectives and Tasks

Efforts focused on building an apparatus that can harvest thermal energy from a pavement by a *thermoelectric* approach rather than relying on harvesting thermal energy directly. Opposing semiconductors in a "thermoelectric generator" create electrical voltage when they are brought to different temperatures. A ceramic barrier provides electrical insulation between the opposing surfaces and cuts down on direct heat transfer. Instead, energy flows from higher to lower electrical potential through a wire, transferring the energy electrically rather than thermally. This creates a possibility of harvesting energy by powering an external circuit with this voltage. The electrical power becomes available for use within the pavement itself.

2 Energy Harvesting and Advanced Technologies for Road Assessment

The URI research team performed experiments utilizing the Seebeck Effect [7, 13]. The Seebeck principle comprises two dissimilar electrical conductors or semiconductors that have a difference in temperature, which produces a voltage difference between the two substances [3, 10]. Recently, an experiment using the Seebeck principle was performed to develop a self-powered battery-less structural health monitoring (SHM) system for transportation infrastructure [5]. The system would be capable of processing analog voltage input from a variety of sensors, such as strain gauges, traffic counters and piezoelectric weighing strips [9]. An energy harvester driven by thermoelectric generators (TEGs) powered their system. TEGs function on the Seebeck/Peltier principle, allowing thermal differences between the upper and lower layers of asphalt concrete to be translated into electrical energy [11]. The surface heat of the pavement would be transferred from the surface to the lower layers through insulated copper plates, and the lower part of the harvester was kept cool through a heat sink [12]. Thermoelectric generators can power SHM systems when enough of a temperature differential exists to power the TEGs. This SHM system developed by the Texas A&M University (TAMU) team appeared to be successful. The system accepts analog voltage input from a variety of sensors, is readily programmable, functions without a storage battery, and can continuously retrieve wireless data. The URI research team has been conducting similar studies independently. Literature review utilizing the Seebeck/Peltier Effect (thermoelectric effect) was further researched and reviewed. An effort was made to harvest solar energy for developing a road performance assessment tool, as described below.

2.1 Methodology

The setup has two broad functions. The first setup harvests some of the thermal energy that is available by converting it into electric voltage. This provides a capability to do useful work in real time. The second setup uses the electrical power provided by the energy harvesting system to power a strain sensor. The sensor monitors pavement deformation or strains of passing traffic. This second setup involves conversion of the generated voltage into a steady flow that can operate the sensor and its electronics.

2.2 Apparatus to Harvest Solar Energy

2.2.1 Thermoelectric Generators

TEGs power the system by translating a thermal difference in temperature between their upper and lower layers into electrical energy. They convert the environmental energy that exists in the form of a thermal gradient into electrical energy. Normally heat would flow down a temperature gradient in the form of thermal conduction. Instead, the TEG replaces that energy flow with a form of electrical energy that flows from a higher voltage to a lower voltage. This voltage that drives that electrical current can be directed through a circuit, which provides for extracting useful electrical work from the pavement.

The layers within a TEG consist of two different semi-conductor materials (p-type and n-type elements) between two ceramic substrates to generate voltage. As one side of the TEG becomes hot due to heat transfer from the upper pavement layer, the other same remains a colder temperature, allowing a voltage to be produced. A cartoon depicts the overall structure (Fig. 1a) and a photograph depicts the system

Fig. 1 Solar energy harvesting system. **a** Cartoon of the harvesting system depicted next to an asphalt pavement. **b** Photograph of the TEGs between the copper plate and heat sink during benchtop testing. Pairs of black and red wires indicate the parallel physical and electrical geometries

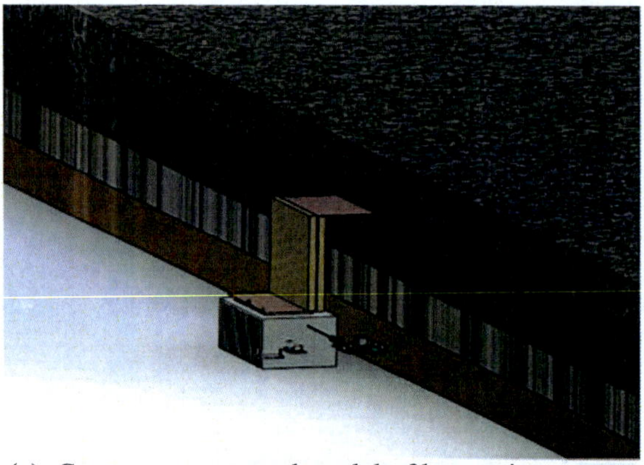

(a) Computer generated model of harvesting system

(b) Copper plate, TEGs, aluminium heat sink

during benchtop testing (Fig. 1b). A TXL-287–03 TEG from TXL Group Inc was used in this study. It employs a bismuth telluride thermoelectric material. Further information is available at TXL Group, Inc. Two TEGs in parallel were used to increase the available current.

2.2.2 Copper Plate Apparatus

A first step is to deliver heat from the hot pavement surface to the TEGs that extract the useful energy. Normally this heat flows from surface to interior directly through the pavement. This creates a temperature gradient within the pavement that is incapable of allowing energy to be extracted. To circumvent this typical equilibrium, an easier path was created for channeling heat flow to the embedded TEGs. This took the form of a long, thin plate. The main experimental setup consists of an "L" shape design 470 mm (18.5 in.) long × 150 mm (6 in.) wide × 1.59 mm (0.0625 in.) thick copper sheet that absorbs heat from the asphalt pavement. The size of the copper plate was determined in correspondence with the size of the TEGs. The lower end of the copper has a 63 mm (2.5 in.), 90-degree bend, where it meets the TEGs. Copper was selected because it is an excellent heat conductor and has a high thermal conductivity. Copper is also very durable and has a specific heat of 385 J/kg C at 25 °C [1]. Multiple on-line sources cite a density of 8.96 g/cm^3 and a thermal conductivity of 401 W/m–K at 0 °C. This copper sheet connects to two TEGs, which are electrically connected in parallel, by a thin layer of thermal paste (Fig. 1b).

The role of the copper plate is to deliver heat to the TEG, so the upper surface of the TEG becomes much warmer than its bottom surface. To decrease losses along the copper, Styrofoam insulation was added to the solar harvester. The R value of the insulation was approximately 2.8. The insulation was added to the 150 mm (6 in.) bent part of the copper plate on both sides. The two insulation pieces were wrapped in electrical tape to mimic a seal around the insulation. An adhesive was used to hold the insulation in place. The insulation enables heat transfer to stay within the vicinity of the copper.

2.2.3 Heat Sink

Attached to the bottom side of the TEGs was an aluminum heat sink, which is used to dissipate the heat away from the TEGs (Fig. 1). Dissipating the heat away from the TEG allows for a greater temperature difference between the two sides of the TEG, producing an electrical current for sustainable power generation. The heat sink is attached to the colder (bottom) side of the TEG. The heat sink is 180.3 mm (7.1in.) long × 99 mm (3.9in.) wide × 45.7 mm (1.8in.) tall.

2.3 *Apparatus to Use Solar Energy—Strain Sensing as an Example*

Electrical energy must be used in real time unless a reservoir such as a rechargeable battery is present to utilize the electrical voltage and current. As a demonstration that energy harvesting can work, and as an added benefit, the energy harvesting apparatus was used to power a strain sensor that can monitor pavement durability. The system to power and activate the sensor requires a steady voltage within a narrow range. Thus, electrical circuitry and hardware was implemented to employ the TEGs as the power source. Programming of the strain sensor controller (the "Arduino" board) was also required.

2.3.1 Boost Converter

A boost converter is used to amplify the voltage of the solar apparatus and to keep the available voltage around 7 V. It can take an input voltage of 2–24 V and make an output of 5–28 V. Without the boost converter, the TEGs would only be able to produce a certain temperature-dependent voltage that is not a constant power to supply the Arduino. The TEGs are then connected to a boost converter to amplify the voltage produced by the TEGs. A boost converter allows a low voltage to be converted to a higher voltage easily while stepping down the current. The TEGs must maintain an output voltage of at least 2 V to power the boost converter. This boost converter is connected directly to an Arduino Uno board to allow programming of a road performance monitoring sensor.

2.3.2 Analog Digital Converter

An HX711 analog digital converter is used to power the Wheatstone bridge connection sensor, whose characteristics and specifications are listed in Table 1. The converter easily converts the analog signal of the strain sensor into a digital signal, to allow electrical signals for data processing purposes. Without the digital converter, the signal from the sensor would not be easily readable.

2.3.3 Arduino Uno Board

The Arduino Uno is the control behind the solar apparatus. It is a microcontroller-based board that can be easily programmed to control sensors via an Arduino code written in a form of Java. The Arduino requires an input voltage of 7 to 12 V, so the boost converter is required to maintain the minimum voltage of 7 V. The Arduino Uno is then connected to an HX711 Analog Digital Converter to power the roadway sensor.

Table 1 Characteristics of dynatest PAST sensor

Specifications				
Type	PAST II (For AC or PCC)		PCCST (for PCC only)	
Range	Up to 1500 μstrain		Do.	
Configuration	Single strain gage (1/4 bridge)		Do.	
Cell material	Epoxy—Fiberglass		Do.	
Coatings	Epoxy—Silicon—PFT—Titanium		Epoxy—Silicon—PFT	
Resistance	120 Ω ±1.0%; GF=2.0		Do.	
Voltage	Up to 12 V (full bridge)		Do.	
Temperature	−30° to 150 °C	−22° to 300 °F	−30° to 150 °C	−22° to 300 °F
ΣE-modulus	≈2200 MPa	≈320 ksi	≈14,000 MPa	≈2000 ksi
Cross section	≈0.5 sq. cm	≈0.078 sq. in	≈0.25 sq. cm	≈0.04 sq. in
Cell force	0.110 N/μstrain	≈0.024 Ibf/μstrain	0.35 N/μstrain	≈0.08 Ibf/μstrain
Fatigue life	Theoretically up to 10^8 cycles		Do.	
Service life	Typically >36 months		Do.	

2.3.4 Wheatstone Bridge Sensor—5 kg Load Cell (Strain Sensor)

The Wheatstone bridge circuit is attached to the Arduino board. This sensor chosen was a 5 kg load cell (Fig. 2). It contains a Wheatstone bridge connection, which is accurate in measuring very low resistances without requiring a lot of power (Table 1).

Fig. 2 Dynatest wheatstone bridge connection sensor

2.4 Procedure to Build Solar Harvester

The two TEGs were connected in parallel by connecting the positive and negative wires with solder. Thermal epoxy was applied to the bottom end of the copper plate to act as an adhesive for the TEGs. The 2-bond epoxy was mixed with a stirring stick to ensure a proper homogenous mixture. The epoxy was then placed onto the hot side of the TEGs and attached onto the lip of the copper plate. After drying of the epoxy, shrink tubes were added to the negative and positive leads of the TEGs to secure the connection. The heat gun was applied to the shrink tubing which allowed the tubing to connect to the wiring. The shrink tubing replaced the black electrical tape and would ensure proper a moisture barrier from temperature changes. The epoxy was set for 24 h to allow proper drying time [9, 11].

Next, the thermal heat sink was attached to the cold side of the TEG. Another 24 h was required to allow the heat sink to dry completely. Styrofoam insulation with an R value of 2.8 was then cut out and attached to both sides of the 150 mm (6 in.) tall section copper plate. Electrical tape bonded the copper between the two pieces of insulation. Another section of Styrofoam was also attached to the bottom side of the 470 mm (18.5 in.) section. The insulation allows the heat loss from the sides and bottom the copper plate to be minimized, transferring the heat to the TEGs rather than to the surrounding pavement. An adhesive was used to hold the insulation in place.

The positive and negative ends of the TEGs were then soldered to the input side of the DC-DC boost converter. The boost converters set screw was turned to maintain a correct input voltage of 10 V for the Arduino board. The Arduino board can be powered from 7 to 12 V of power supply. The boost converters output pins are then soldered to a DC power cable (18 AWG), connecting into the DC barrel jack, allowing power to the Arduino Uno. The Arduino then connects to HX711 analog digital converter and the output connects to the 5 kg load cell sensor [5]. Code for the load cell was written in Java using the Arduino software. This code can be changed and altered depending on the type of sensor used.

2.5 Procedure to Install Solar Harvester and Sensor in the Field

To install the solar harvester and strain sensor, a cross section of asphalt pavement structure was developed as shown in Fig. 3. There was a preliminary survey with URI Facility officers at the Plains Road Extension on May 8, 2020. When a contractor cut out pavement sections, solar harvester and strain sensor were installed on June 29, 2020.

Figure 4 shows connections among the TEGs, boost converter, Arduino Uno, amplifier, and load cell sensor. Together, these units constitute the electronics of the

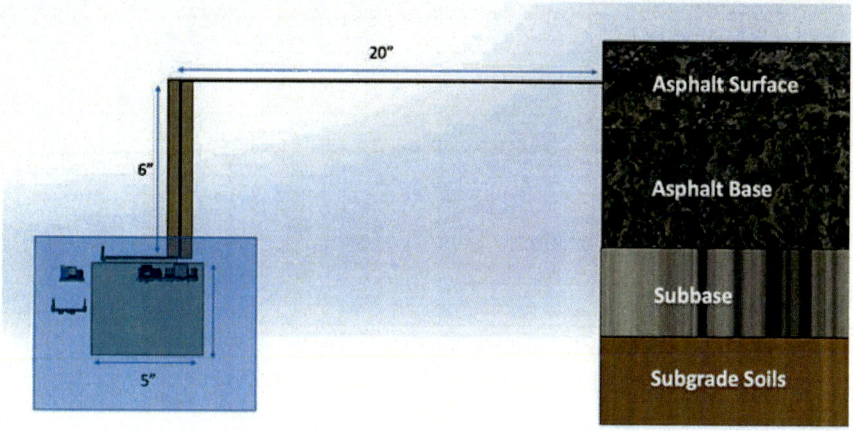

Fig. 3 Cross section of Solar Harvester, embedded in shoulder of road 25 mm (1in) below asphalt surface

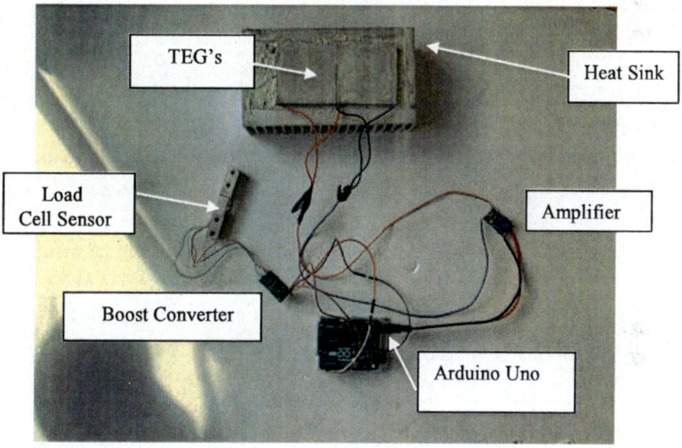

Fig. 4 Connections between TEGs, Boost converter, Arduino Uno, Amplifier, and load cell sensor

energy harvesting system. They were encased within an irrigation protection box for use in the field to protect them from moisture, rain, snow, and groundwater.

2.6 *Verification Before Field Installation*

In a laboratory-scale test of the electronics for powering the sensor, two AA batteries were connected in series as an initial test for a power supply to the Arduino Uno. The

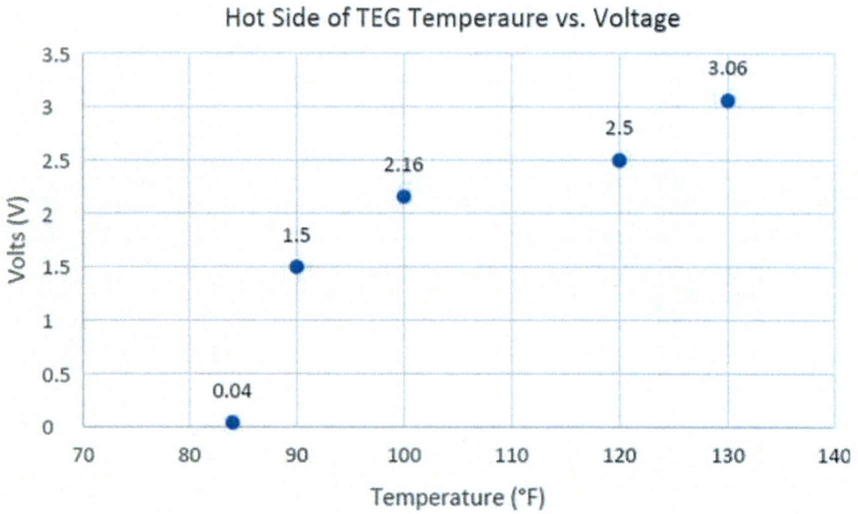

Fig. 5 Temperature of the hot side of TEG with its corresponding voltage; ambient air temperature 23 °C (74°F)

batteries were soldered to the boost converter and then the DC cable jack produced a current of approximately 200 mA, which successfully powered the Arduino.

The solar energy harvester was tested in the laboratory by applying a heat gun to the top layer of the copper. The copper was heated to approximately 54 °C (130° F). The copper was able to effectively transfer heat to the top layer of the TEG, resulting in a power generation of 3 V. This voltage exceeds the minimum two volts required for the boost converter, resulting in a powered Arduino. Varying the heating rate altered the temperature on the upper surface of the TEGs, which subsequently altered the voltage generated (Fig. 5). Even a hot-side temperature of 32.2 °C (90°F) was enough to generate some voltage when the cold-side temperature was 23.3 °C (74°F).

The load cell and Arduino were tested by using the five-volt USB jack on the Arduino and connecting it to a computer. The load cell was calibrated using Arduino software. The Arduino code allowed the Arduino to property read any loads (in grams) the sensor experiences; with collecting data continuously. Examples of these data are in Fig. 5.

2.7 *Installation*

The solar harvester was installed along Plains Road near the URI campus on July 29, 2020. The site was chosen to satisfy requests from the URI facilities office, who kindly incorporated the construction work into the scope of work of an existing project elsewhere at the university. Plans for the installation are shown in Fig. 6. A

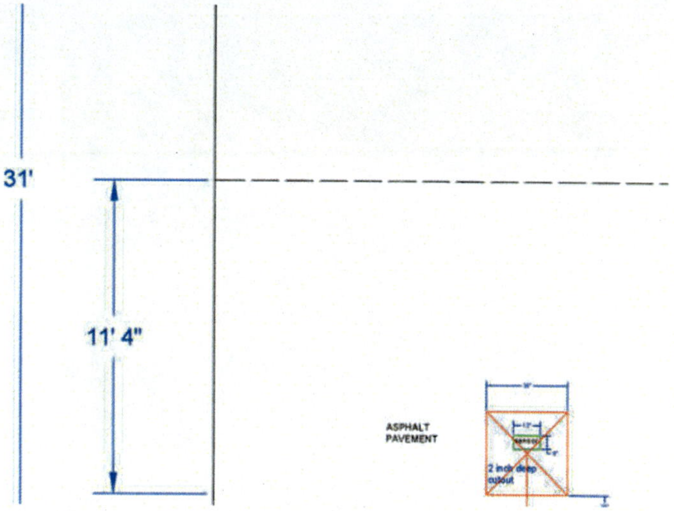

Fig. 6 Plans for the location, layout, and geometry of the installed energy harvester and strain sensor on Plains Road in Kingston, RI

realization of the plans on the road and shoulder, prior to construction, is shown in Fig. 7.

The first step was to dig a hole (Fig. 8) for the irrigation protection box to be inserted, which holds the electronics of the harvester. The hole was dug to the height of the irrigation box to make it flush with grade. The dirt was then back filled and compacted to hold the irrigation box securely. A 900 × 900 × 106 mm (36 × 36 × 4¼ in.) section of pavement was saw cut and removed from the road for the strain sensor. Next, the copper plate was inserted into the box with temperature gauges along the top, middle, and end section of the top surface. This allows for data collection of the varying temperatures along the copper heat plate.

The Dynatest strain sensor was placed into the base of the pavement cutout with the middle section of its H bar parallel with the roadway. The sensor was placed within the average passenger-side wheel path. The wiring of the sensor was placed in a 6¼ mm (¼ in.) sawcut line and sealed with silicon epoxy. Asphalt pavement was carefully placed over the sensor and hand tamped due to the sensitivity of the sensor. The remainder of the asphalt cutout was filled and compacted with a compacting machine, keeping away from the sensor. The asphalt temperature reading just before placement was recorded to be 110 °C (230°F). A 50 mm (2 in.) thickness of asphalt was then placed over the copper apparatus. It was compacted with the hand tamp and then properly dried. The next day, a hole was cut into the irrigation box to allow the end of the sensor wiring to be accessible in the irrigation box. The irrigation box has two bolts and two caps to make the top cover air/watertight.

Fig. 7 Markings for the saw cuts for strain sensor (in the road) and energy harvester (in the shoulder) on Plains Road

Fig. 8 Digging a hole for the irrigation box in the shoulder

2.8 Strain Measurement

The installed energy harvester and load cell strain sensor are now ready for in-field experiments. As a start, the sensor can be tested by using a P-3500 Strain indicator provided for the project by Kevin Broccolo of URI Civil Engineering. The wires connect into the wire jacks and the strain can be monitored.

2.9 Summary and Future Works

Further testing is needed to evaluate this implementation of the energy harvesting device into the roadway. Now that the solar harvester has been installed into the shoulder of a roadway, the ability of heat transfer to generate electrical energy needs to be tested in this real-world application. The temperature difference between the two TEGs will generate the required voltage to operate the roadway sensors. To enable output voltage by the TEGs below the asphalt surface layer, the harvester was installed with the copper plate 25 mm (1 in.) below the top layer. This allows temperature difference readings as well as maximum power output voltage. This copper plate will be heated from the sun heating the asphalt surface layer and transferring the energy into the harvester system.

During installation of the solar harvester apparatus, multiple K type thermocouples were placed among the solar harvester in the following areas:

1. Top part of road surface
2. Top of copper plate along two ends and middle
3. Hot temperature side of TEG junction
4. Cold temperature side of TEG junction
5. Cold temperature sink portion
6. Intermediate points along copper that transfers heat to/from the semiconductor, but not near it
7. Any sort of measurement in duplicate or triplicate for reliability.

These thermocouples will allow data to be collected. Subsequent analysis will enable estimating where thermal losses occur. Optimizing the harvester for efficiency and sustainability are top priorities.

RITRC recommends that a SD card or Bluetooth receiver be implemented into the solar harvester unit. This will allow data for retrieval to be stored without an external power source (i.e., computer), allowing the harvester to operate freely. The SD card/USB will be able to store the information from the pavement strain transducer or other structural health monitoring sensors. Data retrieval would be achieved simply by unplugging the card from the harvester unit and uploading it to a computer.

RITRC also recommends that the Dynatest Pavement Strain Transducer gets installed into the asphalt surface layer for strain monitoring of the roadway. This sensor is capable of 100,000,000 cycles and has a service life that exceeds 36 months.

Another recommendation is a box for the Arduino itself so it can be mounted inside the irrigation box without any movement of the electronic components. A box can be easily printed using the 3D printers if an appropriate size is not found in a commercial source.

This solar apparatus will allow any type of roadway sensor/road monitoring device to be installed. Any sensor can be programmed and configured into the Arduino board.

The solar apparatus can be further developed by creating a "manhole" for the electrical configuration of the harvester. A manhole will enable access to the electrical connection to allow different sensors to be tested from the Arduino board. It also prevents outside parameters from damaging the electrical connections. The recommended box configuration should be like an irrigation system box because it forms a watertight seal. Water cannot be allowed to enter the box because the electrical components would be compromised from the environment. The box will have to be approximately 300 mm by 300 mm (one foot wide by one foot) deep to allow proper room for the configuration. A small hole will have to be drilled through the box for the sensor wires going into the roadway. This hole can then be watertight by corking the remaining gaps in the hole.

It is also recommended adding an energy storage device such as a capacitor to the TEG's output wires to store voltage that is produced. The TEGs send current to the boost converter, but if there is not a continuous supply of two volts, the boost converter will not amplify the voltage to power on the Arduino Uno.

The Arduino can be turned on by a powered button that gets pressed/reset. Once the copper plate receives the required voltage from the heat generation, the Arduino can be turned on. To communicate with the Arduino board in the current set-up, a USB cord gets plugged into the Arduino with the other end into the computer. The software program *Arduino* should then be opened on the computer to read data from the apparatus. Code must be written in the software to run the sensor properly.

3 Thermodynamic Analysis of Solar Energy Harvesting Processes

The goals of thermodynamic modeling work are to assess the viability of the energy harvesting strategies that were considered. Most calculations were done for the thermoelectric energy generation approach.

3.1 Example of Heat Flow Condition During Thermoelectric Energy Harvesting

The energy harvesting system that uses a thermoelectric generator (TEG) was described earlier. It relies on bringing heat from a hot pavement surface down to the top surface of a TEG. Simultaneously the opposite side of the TEG is in thermal contact with cooler temperatures below the road. The temperature difference causes a voltage across a semiconductor. Connecting the opposite ends through a circuit allows for electrical work to be performed.

The heat is brought to the TEG through a copper plate. A wide upper surface warms the copper with an intent of it reaching the pavement temperature. Insulation along its sides is intended to allow heat to transfer *through* the copper and thus down from the surface while simultaneously blocking heat flow from the copper *into* the surrounding asphalt that is less hot.

A first calculation illustrates the rate that thermal energy can be carried through the copper to the TEG. Applying the First Law of thermodynamics within a thin section of the copper (the "system") indicates that heat flow changes the internal energy of the copper,

$$\mathrm{dQ} = \mathrm{Md}\,\hat{\mathrm{U}} \approx \mathrm{MC}_p\mathrm{dT} \tag{1}$$

The driving force for heat flow through an insulated portion of the copper is that there is a temperature difference between the asphalt near the surface and the buried location of the TEG. Fourier's Law of heat conduction expresses this driving force by using the thermal conductivity of copper,

$$\frac{\dot{Q}}{A} = k\nabla T = k\frac{\mathrm{dT}}{\mathrm{dx}} \tag{2}$$

Multiple sources cite a thermal conductivity k = 401 W/mK near room temperature for copper. Its heat capacity is Cp = 0.385 J/g°C at 25 °C [1, 2].

An estimate of the heat transfer rate can be made by replacing the derivative in Eq. (2) with finite differences in temperature and position. Here we assume a linear temperature gradient between 140°F (60 °C) and 68°F (20 °C) for simplicity along the 6″ length of copper. Inputting the dimensions of the copper plate and employing unit conversions as needed indicates

$$\dot{Q} = \mathrm{kA}\frac{\mathrm{dT}}{\mathrm{dx}} = (6.5 \times 0.0625\,\mathrm{in}^2)(401\ \mathrm{W/mK})\left(\frac{40\,\mathrm{K}}{6\,\mathrm{in}}\right)\left(\frac{0.0254\,\mathrm{m}}{\mathrm{in}}\right) = 27.6\,\mathrm{W} \tag{3}$$

We note that the temperature change of (60–20 °C) = 40 °C is identical in degrees Kelvin. The first factor on the right side provides the cross-sectional area of copper normal to the flow direction.

This calculation indicates that heat flow can proceed as quickly as 27.6 W between hot pavement and a cool TEG. The experiments of Chap. 3 involve work production rates that are closer to a range of 1–3 W. It can be concluded that sufficient energy is potentially available through heat transfer through the copper.

3.2 *Example of Heat Flow Dynamics During Energy Harvesting*

A more detailed example of the heat flow dynamics within the vertical section of the copper plate during energy harvesting can be obtained by combining Eqs. (1) and (2) into Fourier's second law of heat conduction,

$$\frac{\partial T}{\partial t} = \frac{k}{\rho C_p} \nabla^2 T \tag{4}$$

which can then be solved for the time- and position-dependence of temperature in the copper between positions $x = 0$ at the upper part with a hotter temperature (such as 140 °F or 60 °C) and $x = L$ at the lower part with a cooler temperature, such as 68°F or 20 °C. Solving this differential equation requires two boundary conditions in position and one initial condition in time. A possible initial condition specifies this lower temperature throughout the copper at $t = 0$. At $x = 0$, assume the hot temperature is maintained. To track the rate that the copper can heat up, neglect heat transfers out of the copper at $x = L$. From Eq. (2), this sets the temperature position derivative to zero.

When the thermal conductivity, mass density, and heat capacity are sufficiently independent of temperature, Eq. (4) can be solved analytically by using the separation of variables method. Briefly, it first is rewritten in terms of a dimensionless change in temperature compared to the warm and cool extremes,

$$\theta = \frac{T(x,t) - T_0}{T_1 - T_0} \tag{5}$$

Equation (4) remains the same with this substitution, other than T being replaced by θ. The boundary conditions become $\theta = 0$ at $x = 0$ at all times, $\theta\,' = 0$ at $x = L$ at all times; the initial condition is $\theta = 1$ everywhere at $t = 0$.

Next, the function $\theta(x,t)$ is written as a product of a time-dependent function $f(t)$ and a position-dependent function $g(x)$. Substituting these into Eq. (4), applying the derivatives, and rearranging leads to the differential equations

$$\frac{\rho C_p}{k}\frac{1}{f}\frac{\partial f}{\partial t} = \frac{1}{g}\frac{d^2 g}{dx^2} = -(\text{constant}) \tag{6}$$

The two equations for f and g must equal a constant because they are equal for various times and positions, with time affecting only the f equation and position affecting only the g equation. The negative sign in front of the constant is chosen for convenience later. Solving each side, applying the initial condition, and applying the boundary conditions leads to the following equation for the temperature change,

$$\theta = \frac{T(x,t) - T_0}{T_1 - T_0} = \sum_n \frac{2}{\pi} \frac{1}{(n - 1/2)} \exp\left(-\frac{k}{\rho C_p} \frac{\pi^2}{L^2} (n - 1/2)^2 t\right)$$
$$\sin\left(\pi (n - 1/2) \frac{x}{L}\right) \quad (7)$$

Theoretically, the integer n spans from 1 to infinity. In practice, the exponential term ensures that contributions become negligible for large enough n at long enough times. Here, a maximum of 200 terms was chosen. The ratio $(k/\rho C_p)$ is called thermal diffusivity. It is often notated as α and has units of (length squared per time).

As an example of using Eq. (7), the temperature profile along the length $L = 6$ in of the copper plate going into the ground can be calculated. Temperatures of $T_0 = 140°F = 60$ °C and $T_1 = 68°F = 20$ °C are specified. Thermal conductivity k and heat capacity C_p are listed near Eq. (2) above and in Sect. 2. The density $\rho = 8.96$ g/cm^3.

Figure 9 shows results from the model over times from 0 to 300 s. As a consequence of the high thermal diffusivity of copper, which is largely attributed to its high thermal conductivity, the temperature rises rapidly along the copper plate. The temperature is predicted to exceed 110 °F over the first inch after only 10 s. The entire plate is predicted to reach 100 °F after about 1 min and 120°F after just over 2 min. After 5 min, nearly the entire plate has reached its upper surface temperature of 140 °F. This suggests that the copper plate setup will be effectively at bringing heat from near the hot asphalt surface down to the TEG.

While this model calculation is encouraging, it is not expected to correspond precisely with experimental measurements. It assumes that the temperature near the asphalt surface instantaneously reaches 140°F (60 °C) at time zero, as does the horizontal copper plate. In reality, the asphalt will heat up over time due to incoming

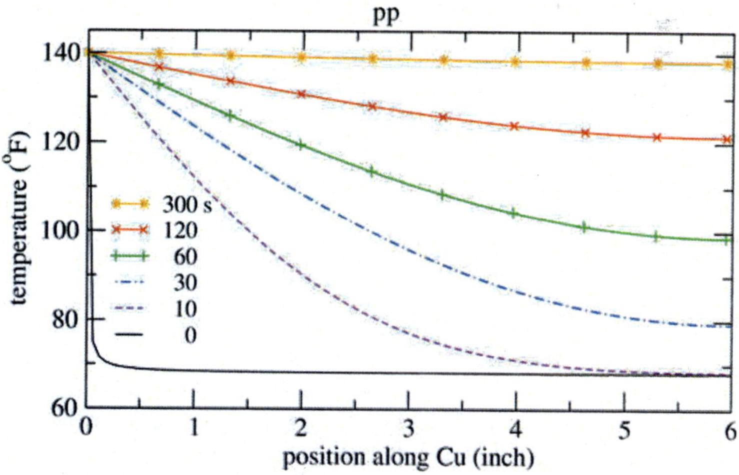

Fig. 9 Model results for temperature within the vertical copper plate, assuming perfect insulation along the sides and an instantaneous rise to 140°F at the top

solar radiation, and some time will be required to transfer that heat into the adjoining copper. This would provide a less sudden driving force for bringing heat to the TEG.

The calculation also neglects heat losses into the neighboring asphalt through the insulation. Such heat losses would cause a temperature gradient across the copper plate in its thin dimension at each time. The better the insulation, the smaller this effect. The rate of heat conduction into the insulation from the copper and into the asphalt from the insulation would each mediate these effects. Indeed, the intent of the Styrofoam insulation is to minimize the extent of this heat flow.

Calculations were not performed for the extraction of heat from the TEG via the heat sink. Fewer details about the geometry and the physical relationship between the heat sink and the road subbase were available for the calculations.

3.3 *Summary from Modeling*

A heat flow analysis of the TEG setup indicates that enough thermal energy flow is possible through the vertical copper plate to allow the TEG to generate electrical work at the desired voltages, though at low currents. A dynamic analysis shows that if the insulation works sufficiently well such that heat flow is channeled along the copper plate rather than through its sides, the temperature at the upper surface of the TEG can become very close to the upper pavement temperature. Thus, it is reasonable to consider a hot-side temperature driving force that is near the hot temperature of an asphalt pavement surface that receives solar heating.

4 Conclusions and Recommendations

An effort was made to create an efficient asphalt pavement solar collector using thermoelectric generators. Thorough testing was needed to evaluate the implementation of the energy harvesting device into the roadway. The temperature difference between the two TEGs will generate the required voltage to operate the roadway sensors. Now that the solar harvester has been installed into the shoulder of a roadway, the ability of heat transfer to generate electrical energy needs to be tested in this real-world application. To enable output voltage by the TEGs below the asphalt surface layer, the harvester was installed with the copper plate 25 mm (1 in.) below the top layer. This allows temperature difference readings as well as maximum power output voltage. This copper plate will be heated from the sun heating the asphalt surface layer and transfer the energy into the harvester system. Calculations show that this heat transfer is reasonable if heat flow along the plate into deeper asphalt layers can be neglected.

During installation of the solar harvester apparatus, multiple K-type thermocouples were placed among the solar harvester in the following areas: Top part of road surface, Top of copper plate along two ends and middle, Hot temperature side of TEG junction, Cold temperature side of TEG junction, Cold temperature sink portion,

Intermediate points along copper that transfers heat to/from the semiconductor, but not near it, and any sort of measurement in duplicate or triplicate for reliability. These thermocouples will allow time and temperature data to be collected. Subsequent analysis will enable estimating where thermal losses occur. Optimizing the harvester for efficiency and sustainability are top priorities. An SD card or Bluetooth receiver was implemented into the solar harvester unit. This will allow data for retrieval to be stored without an external power source (i.e., computer), allowing the harvester to operate freely. The SD card/USB will be able to store the information from the pavement strain transducer or other structural health monitoring sensors. Data retrieval would be achieved simply by unplugging the card from the harvester unit and uploading it to a computer.

Another recommendation is a box for the Arduino itself so it can be mounted inside the irrigation box without any movement of the electronic components. A box can be easily printed using the 3D printers if an appropriate size is not found in a commercial source. This solar apparatus will allow any type of roadway sensor/road monitoring device to be installed. Any sensor can be programmed and configured into the Arduino board.

The solar apparatus can be further developed by creating a "manhole" for the electrical configuration of the harvester. A manhole will enable access to the electrical connection to allow different sensors to be tested from the Arduino board. It also prevents outside parameters from damaging the electrical connections. The recommended box configuration should be like an irrigation system box because it forms a watertight seal. Water cannot be allowed to enter the box because the electrical components would be compromised from the environment. The box will have to be approximately 300 mm by 300 mm (one foot wide by one foot) deep to allow proper room for the configuration. A small hole will have to be drilled through the box for the sensor wires going into the roadway. This hole can then be watertight by corking the remaining gaps in the hole. It is also recommended adding an energy storage device such as a capacitor to the TEG's output wires to store voltage that is produced. The TEGs send current to the boost converter, but if there is not a continuous supply of two volts, the boost converter will not amplify the voltage to power on the Arduino Uno. The Arduino can be turned on by a powered button that gets pressed/reset. Once the copper plate receives the required voltage from the heat generation, the Arduino can be turned on. To communicate with the Arduino board in the current set-up, a USB cord gets plugged into the Arduino with the other end into the computer. The software program *Arduino* should then be opened on the computer to read data from the apparatus. Code must be written in the software to run the sensor properly.

The Dynatest Pavement Strain Transducer was installed into the asphalt surface layer for strain monitoring of the roadway. This sensor is capable of 100,000,000 cycles and has a service life that exceeds 36 months.

A combination of dynamic heat transfer and thermodynamics modeling indicates that the use of a copper plate should enable heat from near a pavement surface to reach a TEG that is further below the surface to enable a voltage-generating temperature difference to be obtained.

If the advanced technology approaches begun in this research are successful, they will enable pavement lifetime through extracting energy from asphalt pavement, enabling it to persist at a lower temperature. Sensing equipment that is powered through thermally generated electrical energy will enable for ongoing monitoring of road quality and stress–strain conditions without ongoing external power requirements. This in situ power generation enables technologies that are being pursued within other projects of the RITRC [7].

Acknowledgements The present study was sponsored by USDOT Region 1 University Transportation Center (UTC) Grant. The cooperation by Technical Champion from RIDOT, Mr. Steven Cascione was valuable to carry the research program. URI also provided partial support, and special thanks to Messrs. Kenneth Burke, Kevin Broccolo and Stephan Zaets.

References

1. Chase MW Jr (1998) NIST-JANAF thermochemical tables, 4th edn, J Phys Chem Ref Data, Monograph 9:1–1951
2. Dahm KD, Visco DP Jr (2015) Fundamentals of chemical engineering thermodynamics. Cengage Learning, Stamford, CT
3. Feng J, Ellis TW (2003) Feasibility study of conjugated polymer nanocomposites for thermoelectric. Synth Met 135–136:55–56
4. Han R, Jin X, Glover CJ (2011) Modeling pavement temperature for use in binder oxidation models and pavement performance prediction. J Mater Civ Eng 23:351–359
5. Karsilaya A, Dessouky S, Papagiannakis A (2018) "Development of a self-powered structural health monitoring system for transportation infrastructure" Publications 27
6. Lee KW, Correia A (2010) "A pilot study for investigation of novel methods to harvest solar energy from asphalt pavements," Final Report to Korea Institute of Civil Engineering and Building Technology (KICT)
7. Lee KW, Correia A, Park K (2012) "Solar energy harvesting from asphalt pavement to reduce dependence on finite fossil fuel supplies". Final Research Report to Expressway and Transportation Research Institute (ETRI), Korea Expressway Corporation (KEC), ETRI-2011–3
8. Lee KW, Kohm S (2014) "Cool pavements," Chap. 16, Springer's book on climate change, energy, sustainability and pavements, Springer, Inc., ISBN: 978–3–662–44719–2 _16. Gopalakrishnan, Kasthurirangan, Steyn, Wynand JvdM, Harvey, John (eds)
9. McCarthy P, Li W, Yang SC (2000) New water-borne electroactive polymers for coating applications. Polym Mater Sci Eng 83:315–316
10. Pinter E, Fekete ZA, Berkesi O, Makra P, Patzko A, Visy C (2007) "Characterization of poly 3-octylthiophene/silver nanocomposites prepared by solution doping." J Phys Chem C 111(32):11872–11878
11. Purohit K, Yang S, Lee KW (2013) "Polymer thermoelectric material for energy harvesting." Proc US-Korea Conference (UKC), East Rutherford, NJ
12. Racicot R, Brown R, Yang SC (1997) Corrosion protection of aluminum allows by double-strand polyaniline. Syn Met 85:1263
13. Yang SC, Brown R, Sinko J (2005) Anticorrosive coatings based on novel conductive polymers. Euro Coating Jrnl 11:48

Texture and Skid Resistance of Asphalt Concrete Surface Course Incorporating Geopolymer Artificial Aggregates

I Dewa Made Alit Karyawan, Iswandaru Widyatmoko, Januarti Jaya Ekaputri, and Ervina Ahyudanari

Abstract Skid resistance is required to prevent vehicle wheels from slipping on a road surface. The skid resistance is influenced by surface texture of the road, which is a function of aggregate properties and grading characteristics. This study aimed to examine the effect of using artificial aggregates on the texture depth and the skid resistance of two asphalt concrete surface course (ACSC) mixtures (dense and open graded). The artificial aggregate, which was made from a mixture of a locally sourced fly ash geopolymer paste and an alkaline activator, was used to partially substituting the coarse aggregate in these ACSC mixtures. ACSC slab samples, without and with 25% artificial aggregate, were produced in laboratory. The texture depth and skid resistance characteristics were subsequently determined. This study found that the differences in texture depth and skid resistance of the two ACSC mixtures were contributed by the difference in gradation of the mixtures and the properties of the coarse aggregate in the mixtures. The ACSC mixture containing 100% natural coarse aggregate (crushed stone) has a deeper and rougher texture depth than those containing 25% rounded artificial aggregate. Post-wheel track loading, the texture depth and skid resistance of open graded ACSC were found to be higher than those of dense graded ACSC. These trends were consistent regardless the type of aggregates used in these ACSC samples. Overall, all tested samples showed skid resistance values meeting the requirements for heavily trafficked roads in urban areas where the traffic speed does not exceed 95 km/h.

Keywords Artificial aggregate · Asphalt concrete · Geopolymer · Skid resistance · Texture depth

---

I D. M. A. Karyawan (✉)
Civil Engineering Department, University of Mataram, Mataram 83125, Indonesia
e-mail: dewaalit@unram.ac.id

I. Widyatmoko
Transportation and Infrastructure Materials Research, AECOM, Nottingham NG9 6RZ, UK

Civil Engineering Department, Universitas Pertamina, Jakarta 12220, Indonesia

J. J. Ekaputri · E. Ahyudanari
Civil Engineering Department, Institut Teknologi Sepuluh Nopember, Surabaya 60111, Indonesia

© The Author(s), under exclusive license to Springer Nature Switzerland AG 2022

H. R. Pasindu et al. (eds.), *Road and Airfield Pavement Technology*, Lecture Notes in Civil Engineering 193, https://doi.org/10.1007/978-3-030-87379-0_19

1 Introduction

Surface textures can represent different levels of roughness and/or smoothness of pavement surfacing. One of them is macrotexture that is broadly formed by graded aggregate mixtures, bonded and compacted to produce a relatively smooth running surface but with some irregularities. In this context, larger irregularities can be associated with voids between aggregate particles, and these are often contributed by size and physical characteristics of aggregate particles [1]. Macrotexture is also essential in providing escape channels to water during the dynamic tyre-surface interactions, and therefore reducing risk to hydroplaning [2]. The macrotexture parameter is often represented as either texture depth or mean profile depth [3].

Skid resistance is the most important safety parameter in preventing incidents on road and airfield runway pavements. Therefore, careful consideration should be taken when selecting the right material, design, and construction of these pavements [4, 5]. Skid resistance can prevent the wheels from slipping on the pavement surface [6]. External factors such as traffic speed, wheel load, surface water film thickness, and depth of wheel tyre-tread, can also affect skid resistance. Low skid resistance increases the stopping distance of running vehicles and can increase the risk of safety incidents [7–9]. Variations in vehicle speed and thickness of surface water are known to compromise skid resistance [8]. In the context of carrying out mix design for asphalt concrete surface course (ACSC), excessive bitumen content must be avoided to prevent bleeding which can cause slippery on the pavement surface. Aggregate characteristics such as micro texture, shape, size and resistance to polishing, are known to influence skid resistance of the compacted asphalt mixture [5, 10].

Previous research has reported development of artificial aggregate, manufactured from by-product of manufacturing industry [11–13]. The resulting construction materials can help reduce dependence on non-renewable resource materials [5]. In addition, it will help the environment with regard to waste and reduce high energy consumption in the procurement of these materials [11]. However, the manufacturing process resulted in aggregate having relatively porous but rounded shape. The apparent shape and microstructure (under Scanning Electron Microscopy, SEM) of the artificial aggregate are illustrated in Fig. 1.

Figure 1a shows the form of artificial aggregate produced using granulator. The size of artificial aggregate depends on the addition of alkali activator, the angle of granulator and the rotation speed of granulator [12]. Figure 1b presents the microstructure of the artificial aggregate [13].

Further study was considered necessary to assess the impact from incorporating the relatively porous but rounded artificial aggregate on the macrotexture and skid resistance of ACSC. This paper presents findings from the study, aiming to examine the impact from using the artificial aggregate on the macrotexture (texture depth) and skid resistance of dense and open graded ACSC mixtures.

Fig. 1 Artificial aggregate: **a** apparent shape [12] and **b** microstructure [13]

2 Material

2.1 Component Materials

Based on the recommendation from the previous studies [11, 13], artificial aggregate was manufactured from a mixture of fly ash geopolymer paste and activator alkali at a ratio of 75:25%. The alkali activator was a mixture of sodium hydroxide and sodium silicate with a ratio of 2.5 [11–13]. In addition, natural material is mineral quarried aggregate which have been crushed into different fractions, namely: coarse aggregate, fine aggregate and filler. The physical properties of the natural and the artificial aggregates, as well as the organoleptic check of the apparent shape and surface texture (microtexture) of these aggregates, are summarized in Table 1.

The binder used to produce asphalt mixture was paving grade bitumen 60/70. The component materials used for producing the asphalt mixture met the requirements of the Indonesia National Specification (SNI) [14]. These component materials are illustrated in Fig. 2.

Table 1 Properties of artificial and natural aggregates

Properties	Unit	Test method	Artificial aggregate	Natural aggregate
Specific gravity	gr/cm^3	SNI 1969:2008	1.9	2.5
Water absorption	%	SNI 1969:2008	6.1	2.3
Durability (soundness)	%	SNI 3407:2008	5.2	9.7
Toughness (Los Angeles)	%	SNI 2417:2008	22.8	30.8
Apparent shape	–	–	Rounded	Angular
Apparent microtexture	–	–	Porous	Dense

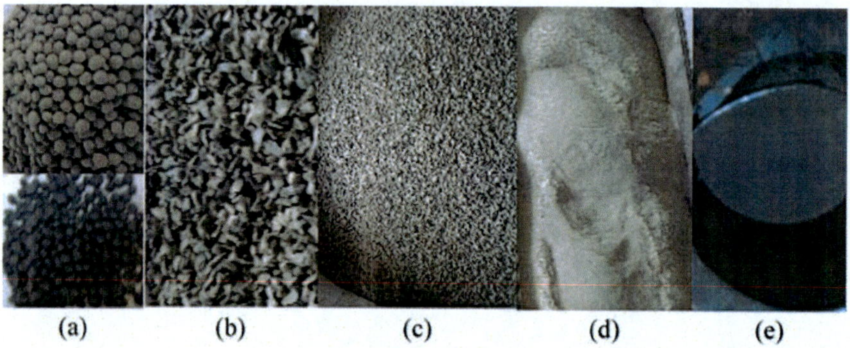

Fig. 2 **a** Artificial aggregate, **b** coarse aggregate, **c** fine aggregate, **d** filler, and **e** bitumen

2.2 *Target ACSC Mixtures*

Two sets of ACSC mixtures were produced, namely open and dense graded ACSC mixtures. These two mixture gradations are illustrated in Fig. 3.

Each sample set comprises either 100% natural aggregate (therefore without artificial aggregate), or, a mixture of 25% artificial aggregate and 75% natural aggregate. The proportions between artificial aggregate, natural aggregate and bitumen in each sample set are summarized in Table 2.

3 Methodology

This section presents the methodology for sample manufacturing and testing.

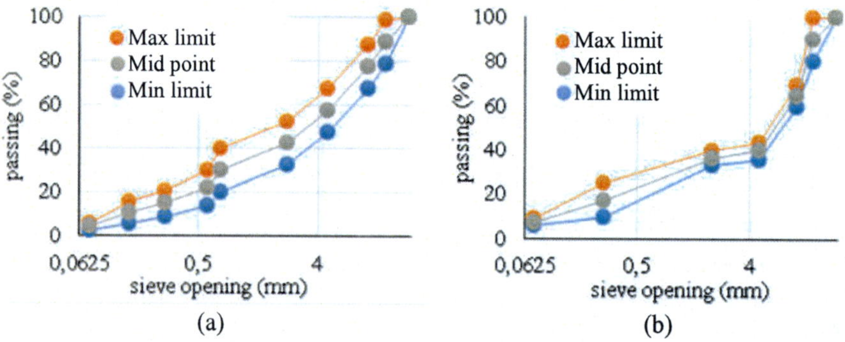

Fig. 3 Combined aggregate gradations: **a** dense graded and **b** open graded ACSC mixtures

Table 2 The proportions of material for each set of asphalt mixture

Sample set	Density (g/cm^3)	Total sample weight (g)	Artificial aggregate weight (g)	Natural aggregate (g)	Bitumen weight (g)
Dense graded without artificial aggregates	2.330	10,490	0	9871	619
Dense graded with 25% artificial aggregate	2.180	9810	2328	6982	500
Open graded without artificial aggregates	2.320	10,440	0	9814	626
Open graded with 25% artificial aggregate	2.220	9990	2345	7036	609

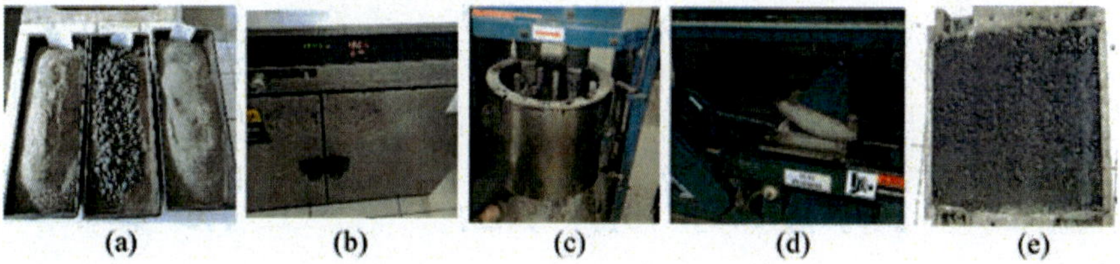

Fig. 4 Asphalt slab manufacturing process: **a** material preparation, **b** heating of the material, **c** mixing process, **d** compaction process, **e** samples ready to be tested

3.1 Sample Manufacturing

The four sets of ACSC samples were manufactured by using a laboratory roller compactor. Each loose asphalt mixture was compacted in a mould with a dimension of 30 cm × 30 cm × 5 cm (see Fig. 4). The four sets of compacted ACSC samples are illustrated in Fig. 5.

3.2 Macrotexture and Skid Resistance Measurements of Samples

Surface characteristics of the ACSC sample sets were assessed under two conditions: (1) pre-wheel track loading and (2) post-wheel-track loading. Wheel track loading was carried out using a laboratory small scale wheel track tester, running at 21 passes

(a) (b) (c) (d)

Fig. 5 Four asphalt mixture sample sets: **a** dense graded without artificial aggregate, **b** dense graded with 25% artificial aggregate, **c** open graded without artificial aggregate, **d** open graded with 25% artificial aggregate

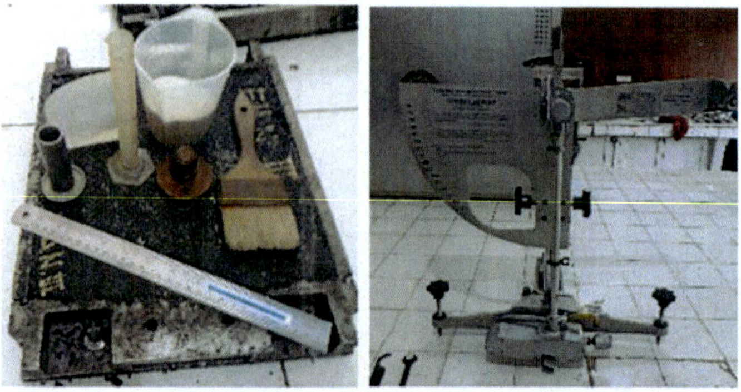

(a) (b)

Fig. 6 **a** Sand patch to measure texture depth, **b** BPT to measure skid resistance

per minute for 60 min at 60 °C. Measurement of surface characteristics includes: (a) determining texture depth using the sand patch method, and (b) assessing skid resistance using a British Pendulum Tester (BPT). These methods are illustrated in Fig. 6. Test results and discussions are presented in Sect. 4.

3.2.1 Macrotexture

The sand patch tests were carried out on the test section to determine an equivalent mean texture depth (MTD) [15]. The texture depth was calculated based on the amount of fine sand that is required to cover irregularity and pores of the sample surface. This process involves pouring fine sand with a certain volume (Vs) over the sample surface and then spread evenly with a circular motion in a circle until the sand is levelled with the surface (ASTM E965-06) [16]. The final diameter of the circled sand (d) is measured in 4 directions (horizontal, vertical, and diagonal) and the results are averaged. The area of the circle ($As = 1/4\pi d^2$), then the mean texture depth (MTD), can be calculated using "Eq. (1)".

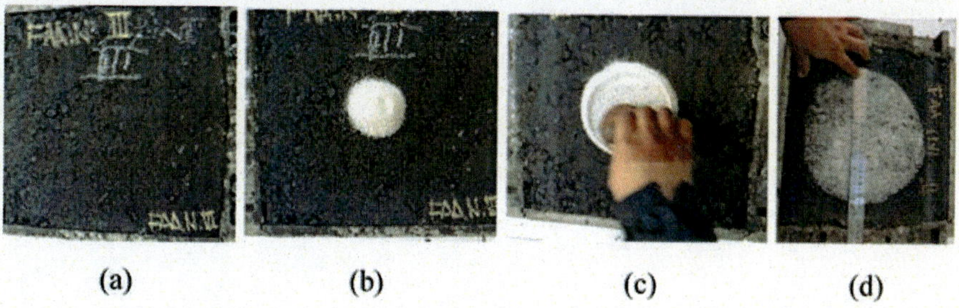

Fig. 7 Texture measurements on the ACSC sample (pre-wheel track loading): **a** the sample, **b** the sand is poured onto the sample surface, **c** the sand is leveled using a circular motion, **d** measurement of the diameter of the sand circle in four positions

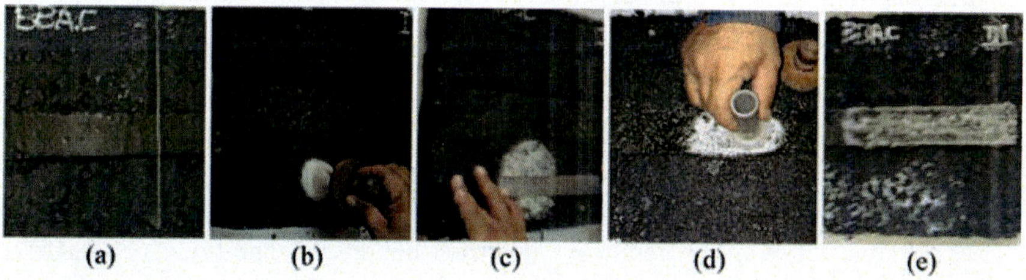

Fig. 8 Texture measurements on the ACSC sample (post-wheel track loading): **a** the sample under the wheel track load conditions, **b** the sand is poured onto the sample surface (outside the track), then leveled, **c** measurements in the four directions of the circle diameter, **d** the sand is poured and smoothed on the track, **e** measurement of the length of the sand distribution on the track

$$MTD = Vs/As = Vs/(1/4\pi d^2) = 4Vs/\pi d^2 \quad (1)$$

The texture depth was measured in two conditions: pre- and post-wheel track loading. For the post-wheel track loading condition, two set of measurements were made, namely outside and inside the wheel track. The pre loading test process is shown in Fig. 7, and post loading in Fig. 8.

3.2.2 Skid Resistance

Skid resistance was determined using a BPT. For this test, a small rubber foot (75 mm × 25 mm) was used. The pendulum was released, swinging over the test surface and the skid resistance value was determined as the British Pendulum Number (BPN). The final BPN value is the average of the five readings after temperature correction was applied to normalize the test condition to 20 °C, according BS EN 13036-4; see Table 3 [17].

Table 3 Temperature correction for BPN when tested at other than 20 °C [17]

Measured slider temperature (°C)	Correction to measured value
36–40	+3
30–35	+2
23–29	+1
19–22	0
16–18	−1
11–15	−2
8–10	−3
5–7	−4

Note The temperature correction can be affected by the texture of the surface

The samples used for skid resistance testing were the same sample as those subjected to the texture depth assessment. With a note that the skid resistance assessment was carried out before measuring the texture, so that the test surface remained clean without contamination from traces of sand on the surface.

The skid resistance assessment was carried out under two conditions: pre and post wheel track loading. In the post-loading condition, the skid resistance test was only carried out on the outside of the wheel track. The section within the wheel track was not tested, because the pendulum head is wider than the track width, which prevented it from reaching the tracked surface. The pre loading test process is shown in Fig. 9, and post loading in Fig. 10.

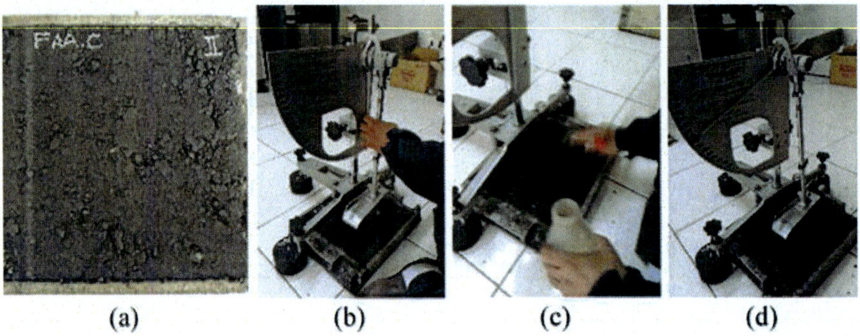

Fig. 9 Measure the skid resistance on the ACSC plate using a BPT (pre-wheel track loading): **a** the sample, **b** setting the BPT, **c** wetting the sample surface, then swinging the pendulum, **d** reading the dial

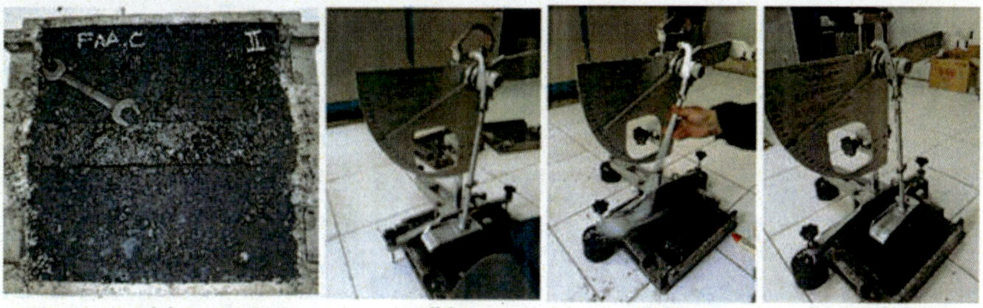

(a) (b) (c) (d)

Fig. 10 Measuring the slip resistance on the ACSC plate using a BPT (post-wheel track loading): **a** a sample plate that has been given a wheel track load, **b** setting the BPT, **c** wetting the sample surface, then swinging the pendulum, **d** reading the dial

4 Results and Discussion

The properties of artificial aggregates show advantages over the natural aggregates, specifically in terms of durability and toughness, as shown in Table 1. The durability by soundness test showed that the artificial aggregate has better resistance to degradation than the natural aggregate. Also, the toughness by Los Angeles abrasion test showed that the artificial aggregate has better wearing resistance than the natural aggregate. The organoleptic check of the apparent shape and microtexture of these aggregates suggested that the more rounded and porous artificial aggregate has surface characteristics which are less favourable than those of the more angular and dense natural aggregate. These can be seen as potential risks to compromising the surface characteristics of the asphalt mixtures, therefore, subsequent assessment of texture depth and skid resistance on the ACSC samples containing these aggregates were carried out.

Figure 11 shows that the inclusion of artificial aggregate did not have practical

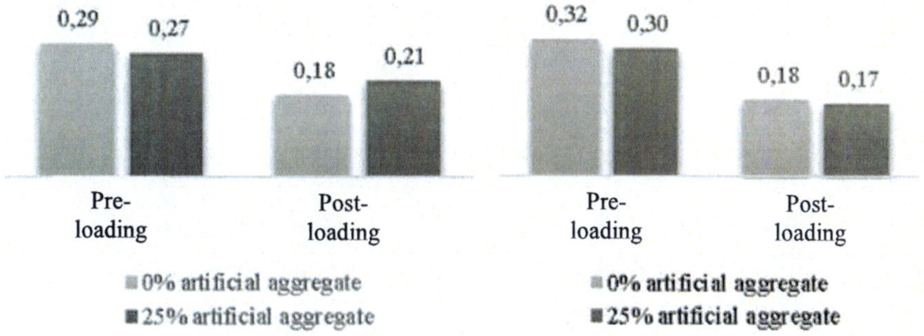

(a) (b)

Fig. 11 Comparison of the results of ACSC texture depth measurements without (0%) and with 25% artificial aggregate **a** dense graded, **b** open graded, pre- and post-wheel track loading impact to the texture depths of both sets of ACSC samples, dense and open graded. The texture depth values appear to be comparable for both samples sets. Similar trend is also observed for the texture depth values obtained before and after wheel track loading where the observed reduction in texture depths appear to be of a similar magnitude for both sample sets. Note that the reporting requirement for texture depth values is typically to the nearest 0.1 mm, however the test results are presented to the nearest 0.01 mm for the purpose demonstrating the slight differences in the values.

Higher skid resistance value is known to be related to rougher surface. In the BPT test, the measurement scale starts from 0 to 120. A scale of 0 means that the surface is very slippery (zero resistance against skidding). Conversely, the value 120 is the maximum value at which the pendulum does not swing further, after touching on the test surface. The test results show that for dense graded ACSC samples, the skid resistance value for those without artificial aggregate is higher than those with artificial aggregate. In contrast, for open graded ACSC samples, the skid resistance of those with artificial aggregates is slightly higher than those without artificial aggregate, after being loaded by the wheel tracking test (see Fig. 12).

Figure 12 explains the impact from apparent aggregate shape on skid resistance. For the ACSC mixture without artificial aggregate (therefore containing 100% angular natural aggregate), the skid resistance value seems to be higher than those containing 25% artificial aggregate. As stated previously, the apparent shape of the artificial aggregates was rounded. These results suggest that the rounded artificial coarse aggregate used in the mixture has contributed to the skid resistant values being lower than those containing 100% crushed natural aggregate. Overall however the skid resistant values for these samples are not less than BPN 65, indicating good skid resistance characteristics under the adopted test condition. When compared against the site category for the minimum values of 'skid resistance' as measured with the portable tester suggested by Road Note 27 [18], the BPN results for these samples

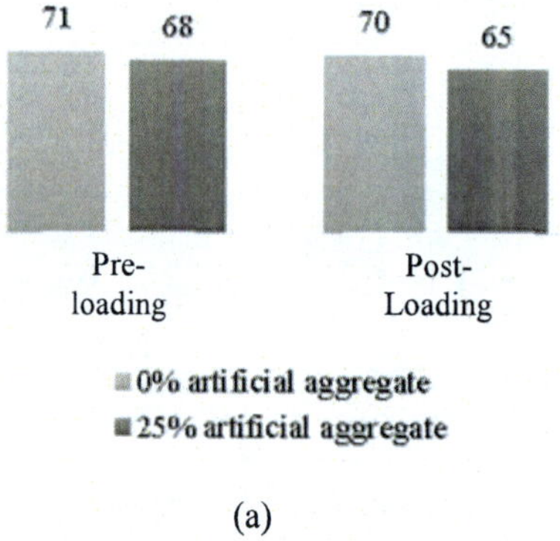

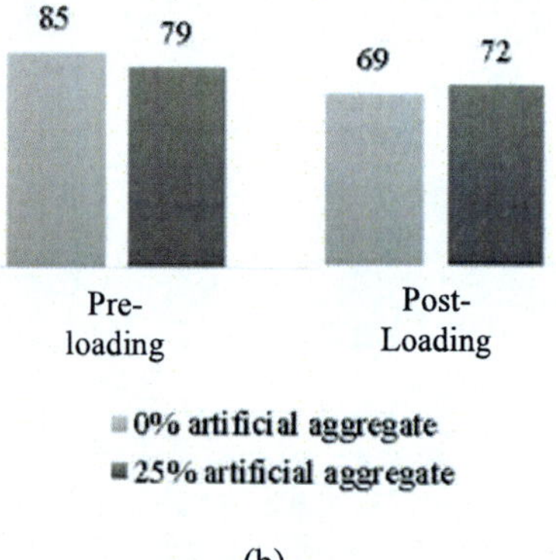

Fig. 12 Comparison of the results of ACSC skid resistance measurements without (0%) and with 25% artificial aggregate **a** dense graded, **b** open graded, pre- and post-wheel track loading

Table 4 Suggested minimum values of 'skid resistance' [18]

Category	Type of site	Minimum 'skid-resistance' (wet surface)
A	Difficult sites such as: (a) Roundabouts (b) Bends with radius less than 150 m on unrestricted roads (c) Gradients, 1 in 20 or steeper, of lengths greater than 100 m (d) Approaches to traffic lights on unrestricted roads	65
B	Motorways, trunk and class 1 roads and heavily trafficked roads in urban areas (carrying more than 2000 vehicles per day)	55
C	All other sites	45

Note For category A and B sites where speed of traffic is high (in excess of 95 km/h) an additional requirement is a minimum 'texture depth' of 0.65 mm

meet the requirements for category B, for site where traffic speed does not exceed 95 km/h (reproduced in Table 4).

5 Conclusion and Recommendation

Tests on aggregate properties demonstrated durability and toughness values of the artificial aggregate being better than those of the natural aggregate. The macrotexture of the ASCS produced with a blend of artificial and natural aggregates was found to be comparable to that produced with 100% natural aggregate. The apparent aggregate shape and surface texture seems to have influenced the skid resistance. The results showed that the use of 25% artificial aggregate resulted a decrease in skid resistance of the ACSC surface. Dense graded ACSC showed slightly lower texture depth and skid resistance values than the open graded ACSC. Post-wheel track loading, the texture depth and skid resistance of open graded ACSC were found to be higher than those of dense graded ACSC, regardless the type of aggregates used in these ACSC samples.

It is necessary to explore in the next research how to increase the skid resistance on the ACSC surface. The scope of the study on the shape of the aggregate, the strength of the aggregate, its impact on the area of contact with tyre and the grading of the aggregate still needs to be developed. Therefore, further research is recommended to explore these and other influencing factors such as finding alternatives to assess the sensitivity of the dimension of the contact area between the rubber wheel and the ACSC surface on the skid resistance value and carrying out field validation testing.

References

1. Bhadane KR, Naktode DPL (2020) A literature review: surface texture characteristics and frictional noise of the pavement. Aegaeum J 8(5):745–751
2. Kienle R, Ressel W, Götz T, Weise M (2018) The influence of road surface texture on the skid resistance under wet conditions. February 6, 2018
3. Li QJ, Zhan Y, Yang G, Wang KCP (2020) Pavement skid resistance as a function of pavement surface and aggregate texture properties. Int J Pavement Eng 21(10):1159–1169
4. Mataei B, Zakeri H, Zahedi M, Nejad FM (2016) Pavement friction and skid resistance measurement methods: a literature review. Open J Civ Eng 6:537–565
5. Wang GC (2016) Slag use in asphalt pavingin; the utilization of slag in civil infrastructure construction, 2016. https://www.sciencedirect.com/topics/engineering/skid-resistance
6. Dunlap DF, Fancher PS, Scott RE, MacAdam CC, Segel L (1976) Transp Res Rec 584:15–21
7. Yager TJ (2013) Retired NASA research engineer consultant on tire/pavement friction issues. Presented at Tenth ALACPA Airport Pavement Seminar Mexico City, Mexico; 30 September–4 October 2013
8. Pasindu HR, Fwa TF, Ong GP (2012) Analysis of skid resistance variation on a runway during an aircraft landing operation. In: Seventh international conference on maintenance and rehabilitation of pavements and technological control, Auckland, January 2012. New Zealand
9. Sekartadji R, Karyawan IDMA (2021) Analysis of the causes and prevention of runway excursions. In: ICST conference, December 14th 2020, Proceeding ICST (2021) vol 2, June 2021. e-ISSN: 2722-7375
10. Sengoz B, Onsori A, Topal A (2014) Effect of aggregate shape on the surface properties of flexible pavement; properties of flexible pavement. KSCE J Civ Eng 18(5):1–8
11. Karyawan IDMA, Ekaputri JJ, Widyatmoko I, Ahyudanari E (2019) The effects of Na_2SiO_3/NaOH ratios on the volumetric properties of fly ash geopolymer artificial aggregates. Mater Sci Forum 967:228–235
12. Yuliana HA, Karyawan IDMA, Murtiadi S, Ekaputri JJ, Ahyudanari E (2019) The Effect of slope granulator on the characteristic of artificial geopolymer aggregate. J Eng Sci Technol 14(3):1466–1481
13. Karyawan IDMA, Ekaputri JJ, Widyatmoko I, Ahyudanari E (2020) The effect of various Na_2SiO_3/NaOH ratios on the physical properties and microstructure of artificial aggregates. J Eng Sci Technol 15(2):1139–1154
14. Spesifikasi Umum Bina Marga Konstruksi Jalan dan Jembatan. Ditjen Bina Marga. Kementerian PUPR. 2018
15. Yaacob H, Bituminous pavement surface texture and skid resistance testing: the investigation of surface roughness of bituminous pavements
16. ASTM E965-06 (2006) Standard test method for measuring pavement macro texture depth using a volumetric technique. ASTM
17. BS EN 13036-4 (2011) Road and airfield surface characteristics—test methods part 4: method for measurement of slip/skid resistance of a surface: the pendulum test. BSI
18. Road Note 27 (1969) Instructions for using the portable skid-resistance tester. Road Research Laboratory. HMSO London

Pavement Design and Geotechnics

Analysis and Numerical Simulation of Semi-circle Bending Fatigue Crack Propagation of Asphalt Mixture

Likui Huang, Pan Zhang, Wei Song, Xiaodong Feng, and Shu Yang

Abstract In order to study the fatigue crack propagation characteristics of asphalt mixture, a model of fatigue crack dynamically propagating was established based on the Paris law modified with fracture energy release rate. Four different types of asphalt mixture were tested using semi-circular bending fatigue test (SCB fatigue test) to obtain the corresponding N-a curves. The extended finite element method (XFEM) was adopted to simulate the SCB fatigue tests for 4 types of asphalt mixture. The results show that the N-a curve obtained from the finite element model is highly consistent with the experimental results, which verifies the feasibility of fatigue crack propagation analysis based on modified Paris Law characterized by fracture energy release rate for asphalt mixtures. The extended finite element method was used to establish the fatigue crack dynamic propagation model of the asphalt mixtures. The fatigue crack propagation process of the specimen was simulated to verify the effectiveness of the method.

Keywords Asphalt mixture · Semi-circular bending test · Extended finite element method · Crack propagation · Modified Paris law

1 Introduction

Fatigue cracking is one of the common distresses of asphalt pavement. The fatigue cracking propagation characteristics is essential and critical for either new building or maintenance design of asphalt pavement. Many scholars have studied the fatigue damage and fracture characteristics of asphalt mixture [1–3]. The researches on the fatigue crack propagation of asphalt mixture can be divided into two categories: experimental research and numerical simulation. The most popular test methods for fatigue of asphalt mixture are indirect tensile test, mid-point loading beam test, Overlay test (OT) and semi-circular bending test (SCB) [4, 5]. Among them, the semi-circular bending test has been more and more adopted due to the simple loading

L. Huang (✉) · P. Zhang · W. Song · X. Feng · S. Yang
College of Civil Engineering, Hunan University, Changsha 410082, Hunan, China
e-mail: huanglikui@hnu.edu.cn

© The Author(s), under exclusive license to Springer Nature Switzerland AG 2022

H. R. Pasindu et al. (eds.), *Road and Airfield Pavement Technology*, Lecture Notes in Civil Engineering 193, https://doi.org/10.1007/978-3-030-87379-0_21

condition and the consistence of test data. However, the fatigue test has poor repeatability and high requirements for the accuracy of test equipment. What's more, the test is very time-consuming. Therefore, a numerical analysis method to simulate the process of crack propagation will greatly help to evaluate the fatigue properties of asphalt mixture. The existing numerical researches on the fatigue cracking of asphalt mixture mostly analyze the crack expansion trend based on the stress intensity factor of the static crack tip [6–8], and but they are difficult to directly describe the dynamic expansion process of crack. Using the recently developed extended finite element method (XFEM), it is unnecessary to re-divide the mesh and define the crack propagation path in advance when simulating crack propagation, and the crack propagation process can be simulated automatically [9, 10].

In this paper, the SCB test was used to evaluate the fracture characteristics of 4 types of different asphalt mixture, including AC-25, AC-20, AC-13 and AC-13 (with SBS polymer modified bitumen) by fracture toughness test and fatigue test. Based on the modified Paris law, the fatigue crack propagation processes of those 4 types of asphalt mixture were analyzed. The extended finite element method was used to establish the fatigue crack dynamically propagating model of the asphalt mixtures, and the fatigue crack propagation process of the specimen was simulated to verify the effectiveness of the method.

2 Materials and Specimen Preparing

2.1 Material

The asphalt binders used in these tests are Pen 60–80 bitumen and SBS polymer modified bitumen (PMB). The coarse aggregate, fine aggregate and mineral filler used are produced from limestone. For all the asphalt mixture, the gradation of AC-25, AC-20 and AC-13 are shown in Table 1. For AC-25 and AC-20, the Pen 60–80 bitumen are used, and for AC-13 both Pen 60–80 bitumen and PMB are used. The asphalt-aggregate ratio of the 4 types asphalt mixtures are 4.0%, 4.3% and 4.8% respectively. The Marshall Test parameters of the asphalt mixture are shown in Table 2.

Table 1 Gradation of asphalt mixtures

Material	Mass percent passing the following sieve(mm)/%												
	31.5	26.5	19	16	13.2	9.5	4.75	2.36	1.18	0.6	0.3	0.15	0.075
AC-25	100	95.0	82.5	74.0	66.5	55.0	38.0	27.3	19.1	14.1	10.0	7.5	4.4
AC-20		100	95.0	85.0	71.0	61.0	41.0	29.2	20.1	14.7	11.0	7.6	4.5
AC-13				100	95.0	76.5	52.8	36.0	24.3	17.4	12.4	8.9	5.4

Table 2 Marshall Test parameters of asphalt mixtures

Material	Marshall stability/kN	Flow value/mm	Void ratio/%	VFA/%	VMA/%
AC-25 (PEN)	11.5	3.98	5.5	60.7	14.2
AC-20 (PEN)	12.8	3.94	5.2	65.3	15.1
AC-13 (PEN)	10.3	3.37	4.9	67.7	15.3
AC-13(PMB)	11.2	3.26	5.0	66.6	15.6

2.2 *Specimen Preparing and Test Device*

The cylindrical specimen with a diameter of 150 mm and a height of 150 mm are prepared by a rotary compactor, and the cylindrical specimen are cut into the semi-circular specimen with a diameter of 150 mm and a thickness of 50 mm.

UTM-30 (Universal Testing Machine) and semi-circular bending testing kit are used in this research. The test model is shown in Fig. 1, in which the distance of the supporting points is 120 mm, and the load and displacement are automatically collected by the test system.

3 Fracture Toughness Test

The J-integral can reflect the intensity of the stress–strain field in the crack tip region, which is unrelated to the loop path of crack tip. Based on a large number of experiments, Bagley and Landes [11] established the J-integral criterion. When the J-integral of the crack tip reaches the critical value J_{IC}, the crack begins to expand.

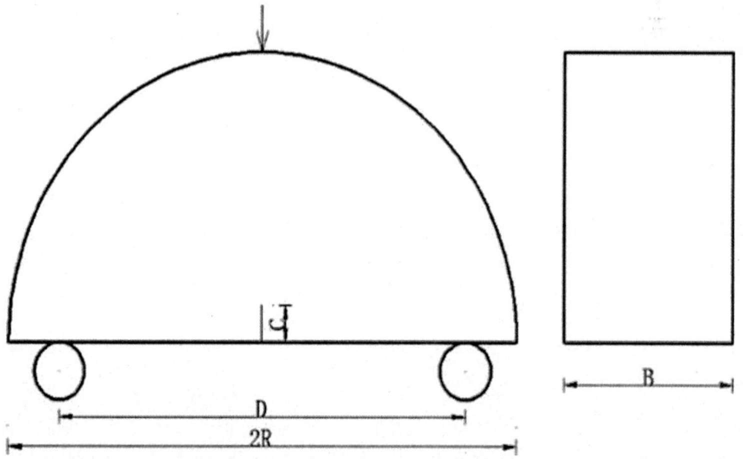

Fig. 1 Semi-circle bending experiments devices

The J_{IC} represent the integral fracture toughness and characterizes the anti-cracking performance of the material, which can be determined by laboratory tests. The greater the fracture toughness, the better is the anti-cracking performance of the material. The fracture toughness J_{IC} can be calculated by Eq. (1).

$$J_{IC} = \left(\frac{U_1}{b_1} - \frac{U_2}{b_2} \right) \frac{1}{a_2 - a_1} \tag{1}$$

where, U_1 and U_2 are the load work, N mm; b_1 and b_2 are the thickness of the specimen, mm; and a_1 and a_2 are the depth of notches, mm.

In order to obtain the fracture toughness of AC-13 (PEN), AC-20(PEN), AC-25(PEN), and AC-13 (PMB), the semi-circular bending tests are carried out on the three specimens with pre-cut depths (notch length) of 15 mm, 25 mm, and 35 mm respectively. The loading speed is set 0.5 mm/min and the circumambient temperature maintains at 15 °C. The load–displacement curves are obtained by UTM-30 test system for all specimens. The area enclosed by the curve is calculated when the maximum load (peak value) is reached in the load–displacement curve, and the energy required when the three types of notched specimen reach a state of failure respectively. The data of fracture energy and the depth of the specimens are fitted with straight lines, and the absolute value of the slope is the fracture toughness J_{IC} of the asphalt mixture at this temperature. It is shown that the fracture toughness J_{IC} of AC-13 (PEN), AC-20 (PEN), AC-25 (PEN), and AC-13 (PMB) specimens are 0.925, 0.890, 0.885, and 1.010 kJ/m^2, respectively. The fracture toughness of AC-13 (PMB) is the largest, and the one of AC-25 (PEN) is the smallest, but the values of J_{IC} of 4 types of asphalt mixtures are not much different. The AC-13 (PMB) manifests the best resistance to cracking.

4 Fatigue Test

4.1 Test Method

In the SCB fatigue test, a notch with a depth of 5 mm is precut at the center of the semicircle for each specimen to simulate the initial crack in the asphalt mixture. The loading schema is shown in Fig. 1, and the stress-controlled loading pattern is adopted. The loading form of a half sine wave with a frequency of 10 Hz is set, and the test temperature maintains at 15 °C. The load is in range of 340–1700 N for AC-13(PEN), AC-20(PEN) and AC-25(PEN), and 400–2000 N for AC-13(PMB). Considering the extension paths of fatigue crack are complicated, the projection length in the vertical direction are selected as the propagation length of the crack to calculate the stress intensity factor and fracture energy release rate of the crack tips. The crack extension area is painted with white paint, a calibration line is drawn in the crack extension area with a red pen, and the size of the calibration line is recorded for each specimen, as shown in Fig. 2.

The relationship between crack length and loading times is recorded by taking pictures of the testing process during the crack extension from 5 to 40 mm, as shown in Fig. 3.

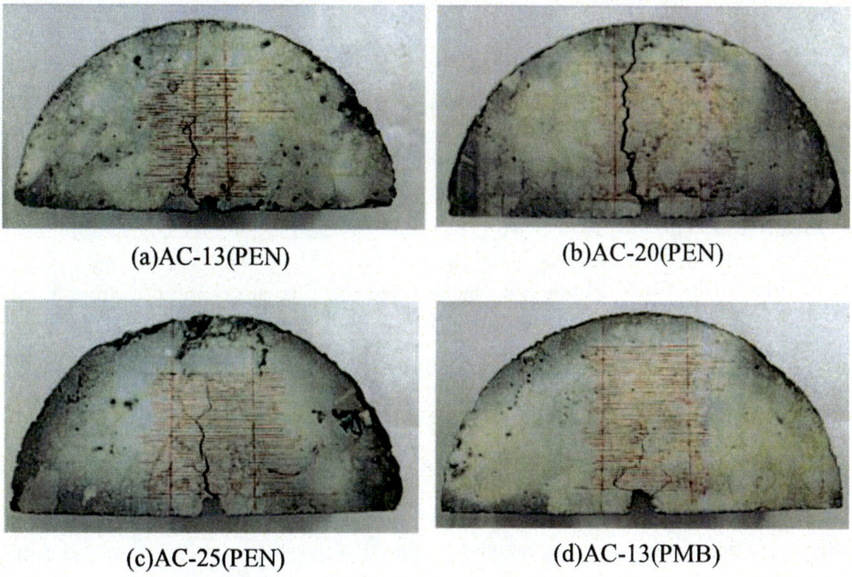

Fig. 2 Fatigue crack propagation of asphalt mixture specimens

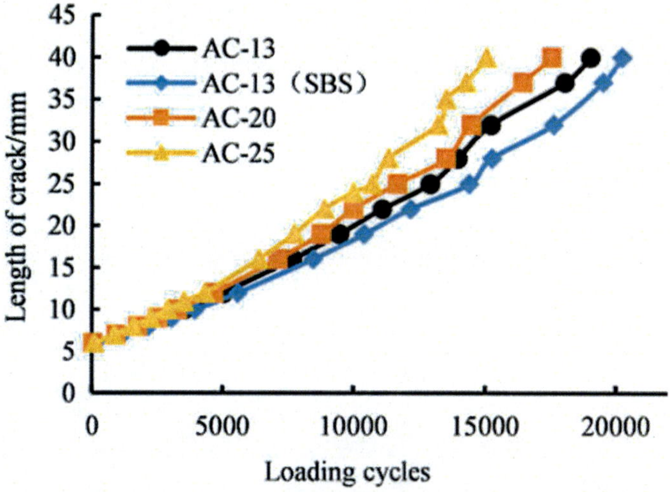

Fig. 3 *N-a* curve of different types of specimen

According to the results in Fig. 3, when the cracks extend to the same maximum length (40 mm), the times of cyclic load acted on AC-13 (PMB) is the most, and the applied load is also the largest, while the loading cycle is the least for AC-25 (PEN). That is, the AC-13 (PMB) has the longest fatigue life, and the AC-25 (PEN) has the shortest fatigue life.

4.2 Analysis of Crack Propagation Process

The ways used to study the fatigue cracking mechanism of asphalt mixtures mainly include the phenomenological and mechanical ways. The phenomenological way is mainly based on the fatigue test of asphalt mixtures to understand the fatigue characteristics. The mechanical way is to analyze the fatigue crack growth process of asphalt mixtures by using the principles of linear elastic fracture mechanics, and predict the fatigue life, which can characterize the formation and development of the fatigue crack more intuitively. It is generally believed that the Paris equation can be used to describe the crack propagation of the asphalt mixture well, as shown in Eq. (2).

$$\frac{\mathrm{d}a}{\mathrm{d}N} = A(\Delta K)^n \tag{2}$$

where, a is the crack extension length; N is the loading times; A and n are material-related parameters; ΔK is the difference between the maximum and the minimum stress intensity factor under cyclic loading, namely, $\Delta K = K_{\max} - K_{\min}$.

The traditional Paris equation ignores the fatigue life at the initial stage of crack propagation, and it is applicable only after the crack extends to a certain length. Therefore, in the process of simulating the fatigue crack propagation of asphalt mixtures using the extended finite element method, the traditional Paris equation needs to be modified. The difference of the stress intensity factor in Eq. (2) should be replaced by the difference of the fracture energy release rate. The stress intensity factor means the cracking tendency of the crack tip under load, and it is a quantitative indicator to reflect whether the crack will expand. The energy release rate signifies the energy released per unit area of the crack when the crack extends one unit length forward. In linear elastic mechanics, there is a conversion relationship, as shown in Eqs. (3)–(5).

$$\frac{\mathrm{d}a}{\mathrm{d}N} = C(\Delta G)^m \tag{3}$$

$$G_I = \frac{K_I^2}{E}(1 - \nu^2) \quad \text{(for plane strain)} \tag{4}$$

$$G_I = \frac{K_I^2}{E} \quad \text{(for plane stress)} \tag{5}$$

where, C and m are material-related parameters; E is the elastic modulus; v is the Poisson's ratio. ΔG is the difference between the maximum and the minimum fracture energy release rate under cyclic loading, namely $\Delta G = G_{\max} - G_{\min}$.

The curve of modified Paris equation in fracture mechanics can completely simulate the expansion process of fatigue crack of asphalt mixtures, which can be divided into three stages, namely stage I—low-speed stable expansion, stage II—medium-speed stable expansion, and stage III—rapid expansion, as shown in Fig. 4. G_{thresh} is the threshold of fracture energy release rate, and the fatigue crack does not expand when the ΔG of crack tip is smaller than G_{thresh}. G_{pl} is the upper linear limit of the fracture energy release rate, the fatigue crack propagation speed will accelerate when the ΔG exceeds G_{pl}, and the remaining fatigue life is small and negligible. G_{equivc} is the ultimate fracture energy release rate when the material is broken. According to relevant literatures, the value of G_{thresh} is taken as 0.01 times of G_{equivc}, and the value of G_{pl} is taken as 0.85 times of G_{equivc} in this paper.

A phenomenological model [12] is used to describe the first stage of crack low-speed stable expansion, defined as Eq. (6).

$$N_0 = c_1 \Delta G^{c_2} \tag{6}$$

where, N_0 is the loading times; c_1 and c_2 are the parameters related to the material properties, particularly, the value of c_1 is related to the unit system used in the calculation model.

In the second stage of crack propagation, the slope of the curve is relatively stable. The crack propagation length under each cyclic load is controlled by the modified Paris equation, shown in Eq. (7).

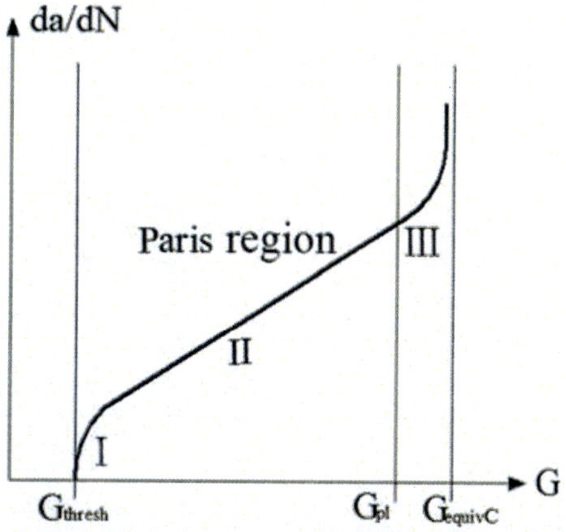

Fig. 4 Fatigue crack propagation process

$$\frac{da}{dN} = c_3(\Delta G)^{c_4} \tag{7}$$

where, c_3 and c_4 are material-related parameters, particularly, the value of c_3 is related to the unit system used in the calculation model.

It is necessary to accurately fit the *N-a* curve from the test, so as to obtain the parameters required for fatigue crack propagation analysis. The *N-a* curve is fitted by a second-order exponential growth function, in pattern of Eq. (8).

$$y = ae^{bx} + ce^{dx} \tag{8}$$

The relationship between the fatigue crack growth rate d*a*/d*N* and the fracture energy release rate ΔG can be obtained by taking the logarithm of Eqs. (6) and (7), and deriving Eq. (8), as shown in Fig. 5.

According to the slope of fitted curve in Fig. 5, each curve can be divided into three segments, which respectively represent the low-speed stable stage I, medium-speed stable stage II, and rapid expansion stage III. The first and second stages are fitted, and the results are shown in Figs. 6 and 7.

Considering the unit system adopted by the extended finite element model, from the results shown in Figs. 6 and 7, the parameters of the first and second stages of fatigue crack propagation for AC-13(PEN), AC-20(PEN), AC-25(PEN) and AC-13 (PMB) are obtained, as listed in Table 3.

For 4 types of asphalt mixtures, the fatigue crack propagation life ranks as AC-13(PMB) > AC-13(PEN) > AC-20(PEN) > AC-25(PEN), which implies that the longer is the fatigue life of the asphalt mixture, the larger are the parameters c_1 and c_4, and the smaller are the parameters c_2 and c_3.

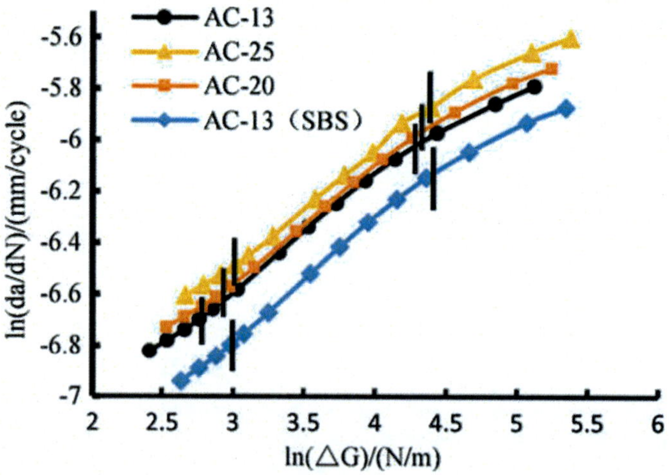

Fig. 5 Crack propagation rate of asphalt mixtures

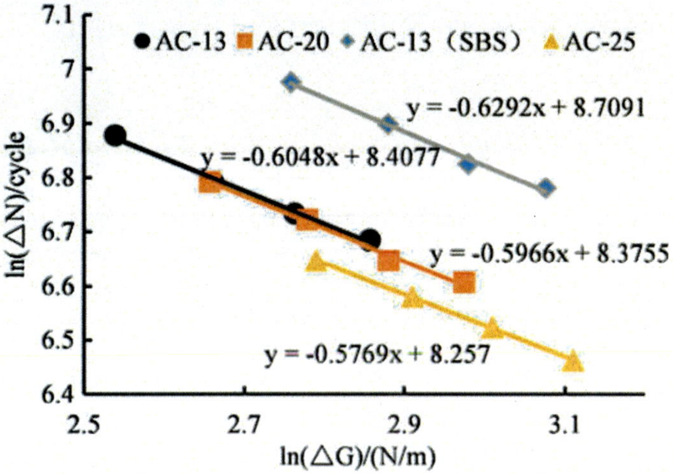

Fig. 6 First stage of fatigue crack growth

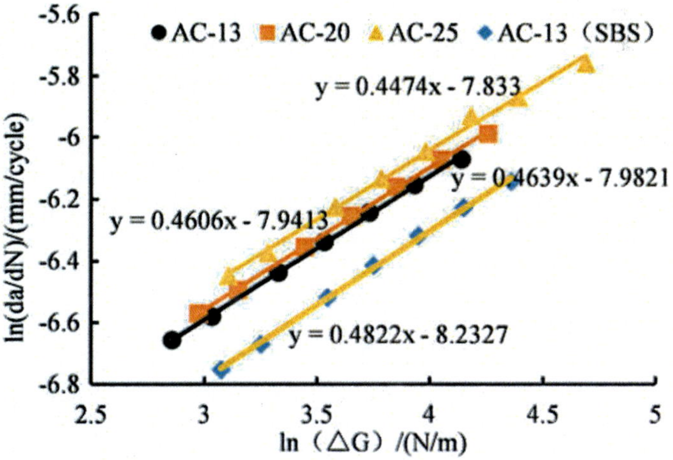

Fig. 7 Second stage of fatigue crack growth

Table 3 Model parameters of SCB crack propagation

Material	c_1	c_2	c_3	c_4
AC-25(PEN)	3854	−0.5769	3.9643×10^{-7}	0.4474
AC-20(PEN)	4339	−0.5966	3.5574×10^{-7}	0.4606
AC-13(PEN)	4481	−0.6048	3.4152×10^{-7}	0.4639
AC-13(PMB)	6058	−0.6292	2.658×10^{-7}	0.4822

5 Comparison Between Numerical and Experimental Results

5.1 *The Basic Principle of Extended Finite Element Method*

The extended finite element method is based on the theory of unit decomposition. A discontinuous function is added to the conventional displacement function of the node to characterize the asymptotic displacement field at the crack tip, so that the crack calculation is independent of the finite element mesh. The level set method is used to realize the pass of crack through the inside of the mesh, so that the automatic expansion of the crack can be realized without the need to define the crack propagation path in advance.

5.2 *Parameters of Numerical Model and Analysis Results*

A two-dimensional model to the SCB fatigue test is established refer to the method used in literature [13]. The material parameters for 4 types of asphalt mixtures are shown in Table 4.

Based on the modified Paris equation, the two-dimensional model established using the extended finite element method can simulate the dynamic expansion of fatigue cracks in asphalt mixture specimen. The relationships between crack length and cyclic loading times are recorded when the fatigue crack of each SCB specimen expands from 5 to 40 mm. The crack propagation processes of AC-13(PEN), AC-20(PEN), AC-25(PEN) and AC-13(PMB) specimens under cyclic loading are simulated with the model parameters in Tables 3 and 4, and the stress cloud diagrams are plotted in Fig. 8.

In actual situation, the expansion paths of fatigue cracking in asphalt mixture are complicated. In order to calculate the stress intensity factor and fracture energy release rate at the crack tip, the crack extension length is taken as the projection length of the crack in the vertical direction. Therefore, only the open cracks (type I) propagation are studied in tests and simulation analysis in this paper.

According to the fatigue crack expansion processes shown in Fig. 8, the relationship between the crack propagation length and loading times is obtained for each

Table 4 Material parameters of SCB specimen

Material	Elastic modulus/MPa	Poisson ratio	Ultimate fracture energy release rate/m
AC-25(PEN)	1400	0.25	900
AC-20(PEN)	1600	0.25	900
AC-13(PEN)	1800	0.25	900
AC-13(PMB)	2000	0.25	1000

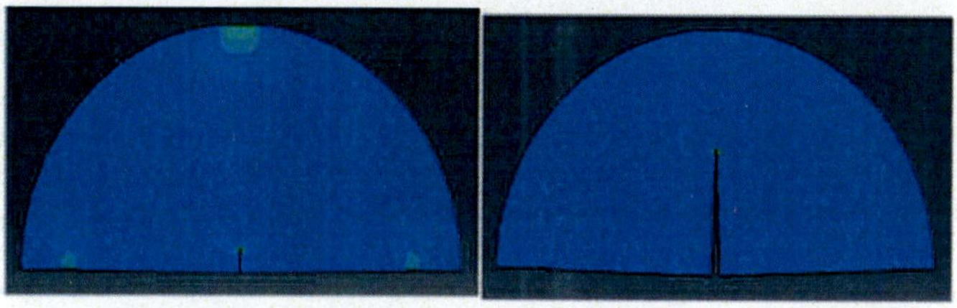

$a = 8$ mm, $N = 2012$ $a = 40$ mm, $N = 18588$

(a) AC-13(PEN)

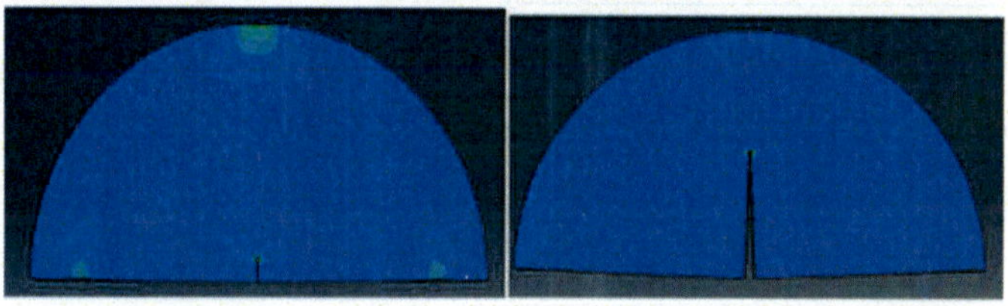

$a = 8$ mm, $N = 1848$ $a = 40$ mm, $N = 17076$

(b) AC-20(PEN)

$a = 8$ mm, $N = 1604$ $a = 40$ mm, $N = 15126$

(c) AC-25(PEN)

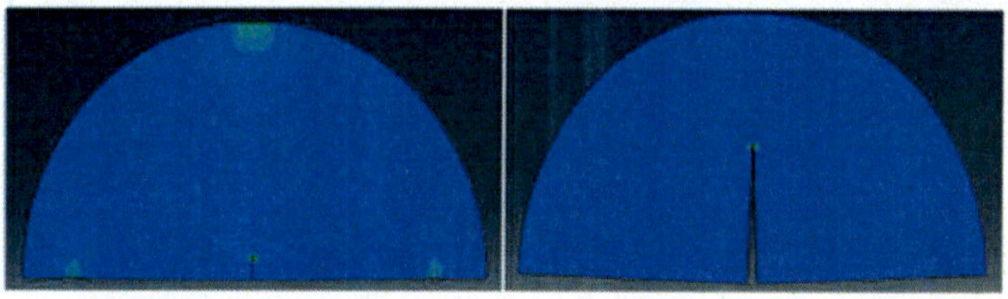

$a = 8$ mm, $N = 2204$ $a = 40$ mm, $N = 20156$

(d) AC-13(PMB)

Fig. 8 Stress nephogram of fatigue crack growth

SCB specimen. The comparison between simulation and test results are shown in Fig. 9.

Fig. 9 Relation diagram between crack propagation length and load cycle number

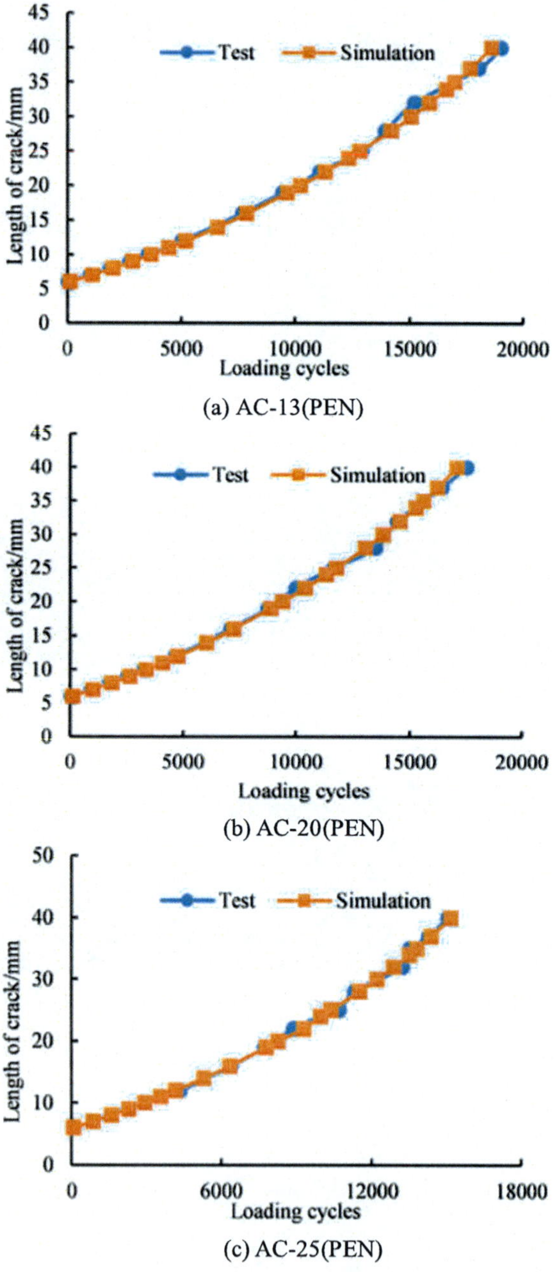

Fig. 9 (continued)

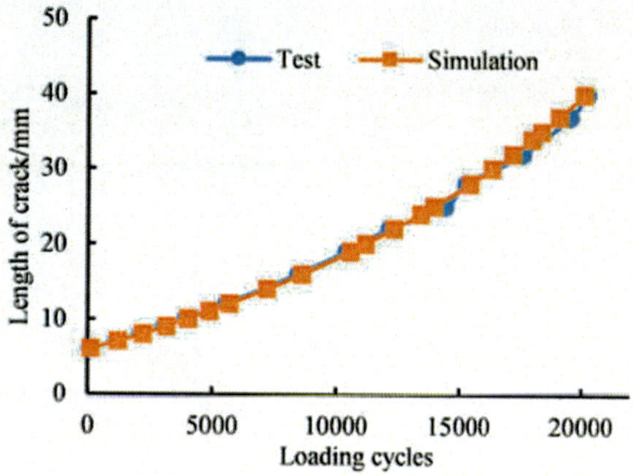

(d) AC-13(PMB)

According to Fig. 9, the simulation results of crack expansion are highly consistent with the test results. It is believable that the expansion process of fatigue crack for asphalt mixture SCB specimen can be reasonably simulate by using the extended finite element method based on the modified Paris equation.

5.3 *Analysis of the Fatigue Crack Propagation Process*

During the crack propagation process, the fracture energy release rate at the crack tip changes with the crack propagation length. The relationships between the fracture energy release rate and the crack length for AC-13(PEN), AC-20(PEN), AC-25(PEN), and AC-13 (PMB) are shown in Fig. 10.

As shown in Fig. 10, the fracture energy release rate of the asphalt mixture increases with the crack propagation length, and the longer is the crack propagation length, the faster the fracture energy release rate increases. The growth rate of the fatigue crack also increases with the crack propagation length, which may be due to the reduced area and ability of the specimens to resist loading deformation.

The fatigue life ranks as AC-13(PEN) > AC-20(PEN) > AC-25(PEN) for the same loading condition. When the crack extends to the same length, fracture energy release rate for AC-13(PEN) is the lowest, which means that the fatigue life of asphalt mixture can be increased, and the expansion of fatigue crack can be delayed by reducing the fracture energy release rate at the crack tip.

It is worth pointing out that the loads applied for AC-13(PEN), AC-20(PEN) and AC-25(PEN) are the same in SCB fatigue tests, and the load applied for AC-13(PMB) is slightly larger than other asphalt mixtures to shorten the test time.

Fig. 10 Relationship between the fracture energy release rate and the crack length

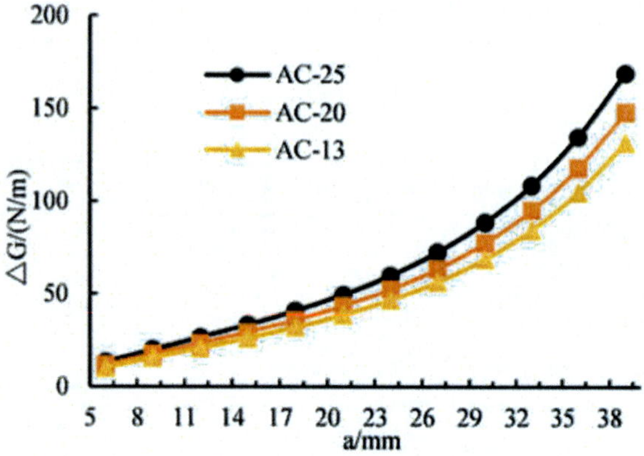

(a) AC-13(PEN), AC-20(PEN), AC-25(PEN)

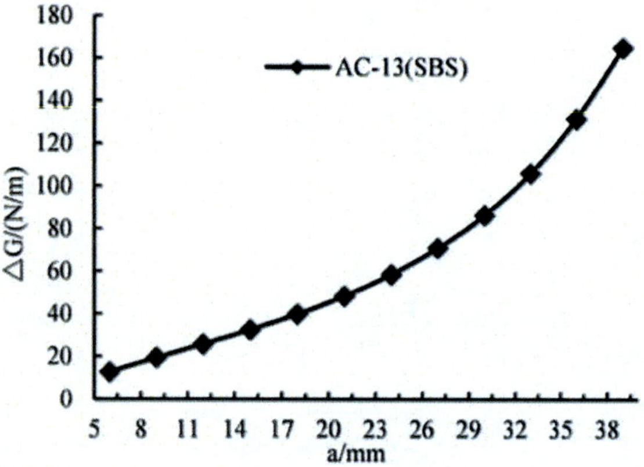

(b) AC-13 (PMB)

6 Conclusions

(1) The fatigue crack growth process of SCB specimen for asphalt mixtures can be completely described using the Paris law modified with fracture energy release rate. The *N-a* curve obtained from the SCB fatigue test can be fitted by the second-order exponential growth function. The model parameters related to the crack low-speed and medium-speed stable expansion stages are calculated with the *N-a* curve functions. The parameters are favorably related to the fatigue life of asphalt mixtures. The longer is the fatigue life, the larger are the parameters c_1 and c_4, and the smaller are the parameters c_2 and c_3.

(2) The dynamic expansion process of fatigue crack in asphalt mixtures can be simulated using extended finite element model, and the simulated *N-a* curve are highly consistent with the test results, which means the relationship between the

fatigue crack propagation length and the cyclic loading times can be accurately estimated by the numerical experiment.

(3) The fracture energy release rate at the crack tip of asphalt mixture specimen increases with the crack propagation length. As the crack propagation length increases, the growth rate of the fracture energy release rate also increases. The fatigue life changes reversely with the fracture energy release rate at the crack tip.

Acknowledgements This work was supported by the National Natural Science Foundation of China (No. 51678232) and the Science and Technology Project of Department of Transportation of Zhejiang Province (No. 2018019). The authors gratefully acknowledge their financial support.

References

1. Luo X, Zhang Y, Lytton RL (2016) Implementation of pseudo *J*-integral based Paris' law for fatigue cracking in asphalt mixtures and pavements. Mater Struct 49(9):3713–3732
2. Hu S, Zhou F, Scullion T et al (2012) Calibrating and validating overlay tester-based fatigue cracking model with data from national center for asphalt technology. Transp Res Rec 2296(2296):57–68
3. Yang J, Wang B, Chen S (2007) Estimating method of fatigue life during cracking initiation of asphalt pavement based on modified Neuber equation. J Traffic Transp Eng 7(2):50–54 (in Chinese)
4. Zhou F, Hu S, Scullion et al (2007) Transportation Development and verification of the overlay tester based fatigue cracking prediction approach (with discussion). J Assoc Asphalt Paving Technol (1):10~21
5. Liu Y (2009) Research on dynamic response and fracture performance of asphalt mixture based on semi-circular bending test. Harbin Institute of Technology, Harbin, Heilongjiang (in Chinese)
6. Zhou Z, Yuan X, Tan H (2013) Fatigue damage analysis of asphalt mixture specimen with pre-made gaps in direct tension test. China J Highw Transp 26(1):30–37 (in Chinese)
7. Luo H, Zhu H, Chen C (2009). The fatigue test and numerical study on pre-sawed bituminous mixture beams. China Civ Eng J 42(6):126–132 (in Chinese)
8. Zhao Y, Wang G, Wang Z et al (2010) Cracking analysis of asphalt pavement based on dynamic modulus. J Hunan Univ (Nat Sci)37(7):7–11(2010) (in Chinese)
9. Jin G, Huang X, Liang Y (2014) A numerical analysis of the fracture behavior of asphalt concrete under two different modes. J Hunan Univ (Nat Sci) 41(6):120–126 (in Chinese)
10. Ng K, Dai Q (2011) Investigation of fracture behavior of heterogeneous infrastructure materials with extended-finite-element method and image analysis. J Mater Civ Eng 23(12):1662–1671
11. Huang B, Hang Z, Kingery W (2004) Fatigue crack characteristics of HMA mixtures containing RAP. In: 5th RILEM conference on reflective cranking in pavements, France, pp 631–638
12. Pirondi A, Bonora N (2003) Modeling ductile damage under fully reversed cycling. Comput Mater Sci 26(1):129–141
13. Shen Q (2018) Research on reflective crack fatigue propagation of asphalt pavements with semi rigid base course based on XFEM. Hunan University, Changsha, Hunan (in Chinese)

Applications of Molecular Dynamics in Asphalt Pavement Mixture Studies

Huiping Wan, L. Chu, and T. F. Fwa

Abstract The rapid development of high-performance computing facilities and techniques in recent years has made molecular dynamics (MD) a feasible tool for studying the behaviors of asphalt mixtures under various operating conditions. MD is capable of building relationships between chemical structures of material and its macroscopic properties. This study first conducts a literature review of the various applications of MD in asphalt mixture studies. A summary is made on the techniques and approaches adopted by researchers, and the main application areas of MD in asphalt mixture studies so far. Next, based on the basic concepts of MD, an analysis is performed to evaluate the strengths of the different techniques and approaches adopted, including the types of molecular models employed. Also evaluated are their relative abilities in explaining the behaviors of asphalt mixtures, such as mechanical properties, self-healing and rejuvenator diffusion behaviors, and the stability of asphalt-aggregate interfacial bonding under the influence of several factors. Based on the results of these evaluations, the limitations of MD applications in asphalt mixture studies would be identified. Finally, taking into consideration the strengths and limitations of MD, possible new techniques and potential new application areas are proposed and explained.

Keywords Molecular dynamics · Asphalt pavement mixtures · Mechanical properties · Diffusion behavior · Interfacial behavior

1 Introduction

The majority of pavements throughout the world are surfaced with asphalt mixtures [1]. Asphalt pavements are exposed to the external environment from the beginning of its use, and have to bear the repeated loading effects of vehicles, leading to various

H. Wan · L. Chu (✉) · T. F. Fwa
School of Highway, Chang'an University, Xi'an, China
e-mail: longjiachu@chd.edu.cn

T. F. Fwa
National University of Singapore, Singapore, Singapore

© The Author(s), under exclusive license to Springer Nature Switzerland AG 2022

H. R. Pasindu et al. (eds.), *Road and Airfield Pavement Technology*, Lecture Notes in Civil Engineering 193, https://doi.org/10.1007/978-3-030-87379-0_22

physical distresses [2]. Ensuring the performance of asphalt pavements during long-term service and reducing the possibility of pavement diseases have been a focus of pavement engineering researchers.

Numerous laboratory tests have been developed to identify the durability of asphalt mixtures, and quantify their performance in rutting resistance [3–5] and moisture resistance [6, 7] etc. However, these experiments are performed at the cost of considerable time and money. Once any variance that may influence the properties of asphalt mixture is adjusted, the whole experiment needs to be conducted again. Sample preparation and testing condition are also restricted. Experiments at the macro scale can only provide a final result and phenomenon to researchers, while deeper exploration to mechanism at smaller scale cannot be demonstrated. Nevertheless, macroscopic properties of asphalt mixture are closely related to its microstructures [8–10]. Understanding the relationships among changes in the microstructure, composition of an asphalt mixture and its macroscopic properties is of great importance to modifying asphalt mixture design and improving field performance of asphalt pavement. Molecular Dynamic (MD) simulation provides an efficient approach to fulfill this need.

As an effective computational simulation tool, MD has been applied in many researches to investigate interaction between chemical microstructures of materials and its reflection on macroscopic properties [11–13]. MD can perform simulations for extreme conditions that are difficult to realize in experiment, such as high temperature and pressure. MD also does not need to prepare samples and conduct experiment, which greatly reduces time and cost. In addition, it can predict changes of macro properties for modified asphalt mixtures by changing their microstructure in the simulation process [14, 15].

To fully understand the MD applications in investigating asphalt mixtures properties, this paper firstly explains the concept of MD. Then asphalt models developed by past researchers are introduced, followed by their applications in studying the mechanical properties, diffusion behaviors, and interfacial behaviors of asphalt mixtures. Next, two case studies are exemplified to show the typical technique and process of utilizing MD to exploring adhesive properties at asphalt-aggregate interface considering anisotropic mineral surfaces and moisture effect. Finally, the limitations of MD are presented and its potential application areas in asphalt pavement mixtures are pointed out.

2 Literature Review

2.1 Concept of MD

Molecular dynamics (MD) is a computational technique used to simulate molecule behaviors under different conditions. Each atom in the materials is considered to follow the law of classical Newtonian mechanics. By predicting atom motions characterized by their positions, velocities and accelerations, one can understand how macroscopic properties are influenced by the collective behaviors of atoms [16].

The general simulation process starts with constructing a molecular model, then force field is assigned to every atom in the system to describe their interactions. Dynamic simulations are conducted to an ensemble of molecules with given velocities and positions until reaching an equilibrium state. Finally, desired properties are obtained through statistical analysis to the trajectory of system [17].

2.2 *Asphalt Molecular Model*

Asphalt Binder Model Zhang et al. first put forward two kinds of three-component asphalt binder models composed of asphaltene, naphthene aromatic and saturate [18]. Figure 1 shows the structures of these components. This model realized qualitatively but not quantitatively simulation of experimental data in SHRP program.

To overcome the shortcomings in former asphalt binder model, Zhang et al. proposed a new six-component model targeted toward core asphalt AAA-1 in SHRP program [19]. Its polarity was increased by adding polar aromatics molecules. Later, Li and Greenfield increased molecular sizes and types for previous six-component model [19] to create twelve-component models for AAA-1, AAK-1 and AAM-1 asphalts in SHRP program [20]. Figure 2 shows the detailed molecular structures of asphalt components. Among three asphalt models, model for AAA-1 asphalt was found to have much closer density value to experimental data than previous asphalt systems, resulting in its wide applications in molecular dynamics. There were also

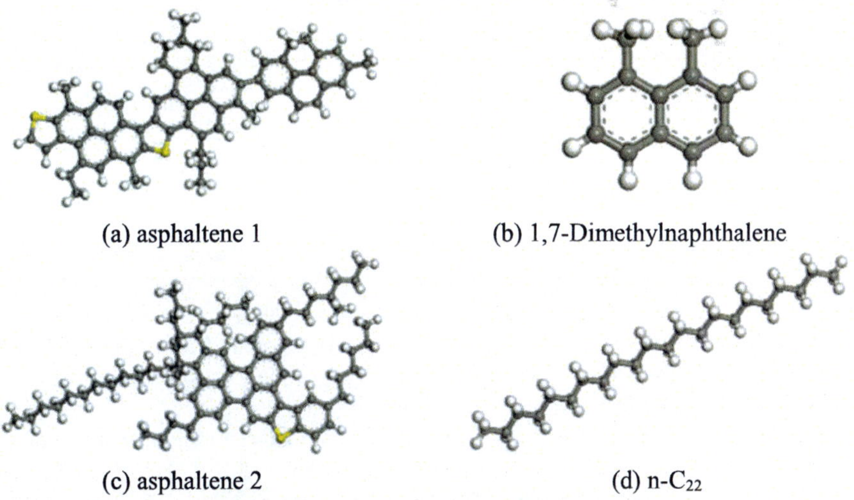

Fig. 1 Components for three-component asphalt model [18]

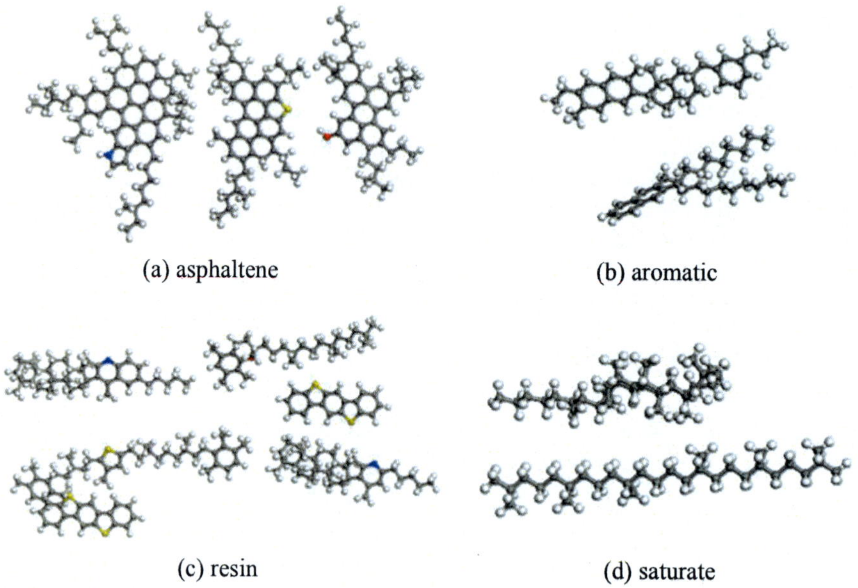

Fig. 2 Components for twelve-component asphalt model [20]

some asphalt models whose element ratios and fraction proportions were controlled to be close to other real asphalts [21, 22].

Aged Asphalt Binder Model Aging effect on asphalt mixture properties has attracted many researches' interests. When asphalt molecules interact with oxygen, they will be oxidized to form new functional groups, mainly ketones formed at the benzylic carbon atoms and the sulfoxides formed in the sulfide functional groups [23].

Based on this mechanism, an aged asphalt binder model was developed by replacing possible functional groups susceptible to oxygen with ketones and sulfoxides in 12-component AAA-1 virgin asphalt [24]. Considering that asphalt microstructure varies with aging degree [25], asphalt binder models with different aging levels were generated by adding different numbers of oxygen atoms to virgin asphalt [26]. In addition to molecule oxidation, the changes in saturate-aromatic-resin-asphaltene fractions are also observed in experiments [27–29]. Based on this principle, an aged three-component asphalt model was proposed by increasing the mass percentage of asphaltene [30]. There are also aged asphalt models considering both the change of molecular structure and component content. Molecular structures of aged asphalt binder were chosen from above researches, while SARA fractions were mainly determined by experiments [11, 31, 32].

Modified Asphalt Binder and Asphalt Mastic Models Modified asphalt binder models were generally constructed by mixing modifiers with asphalt components according to certain mass fractions. The modifiers used in MD researches include styrene–butadiene–styrene (SBS), carbon nanotube (CNT), graphene, exfoliated graphite nanoplatelets and rubber [22, 33–36]. Rubber was represented by natural rubber (NR), cis-polybutadiene rubber (BR) and styrene butadiene rubber (SBR). Figure 3 shows structures of all modifiers except for SBR which is a random copolymer.

Silica particle model (Fig. 4) was used to represent filler by Zhu et al. since silica accounted for the majority of most aggregates [37]. Then it was packed with asphalt components to construct asphalt mastic model.

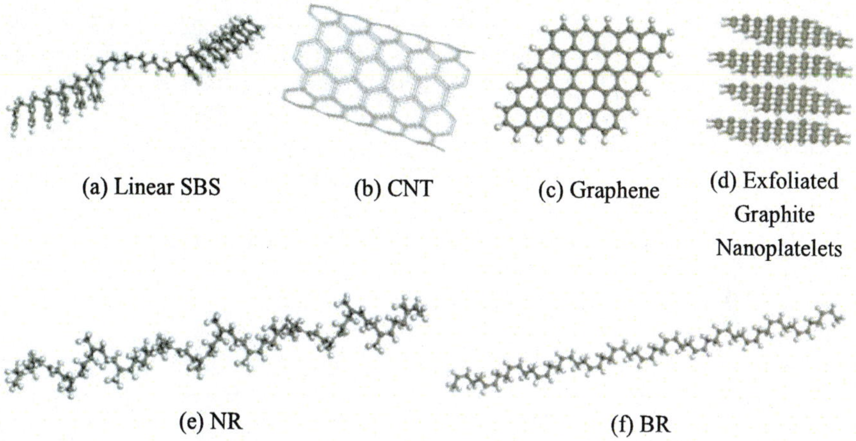

Fig. 3 Structures of SBS, CNT, graphene, exfoliated graphite nanoplatelets, NR and BR [22, 33–36]

Fig. 4 Silica particle [37]

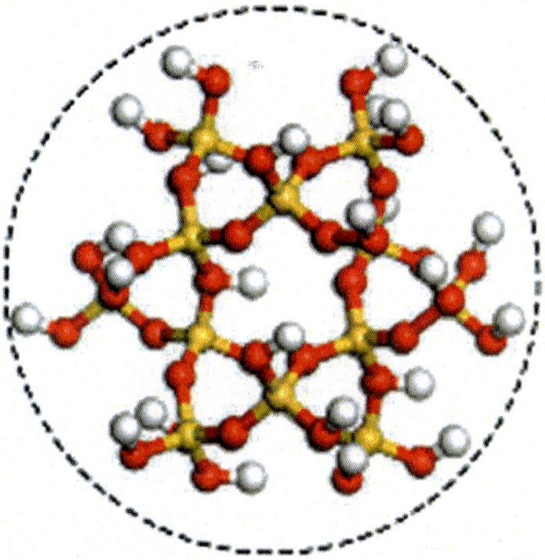

2.3 Applications of MD in Asphalt Mixtures

Once asphalt model is constructed, molecular dynamics can be applied to investigate some engineering properties for asphalt mixtures, such as their mechanical properties, diffusion behaviors and interface properties.

Mechanical Properties Micro-mechanical study has attained interests. MD was applied to study modifier effect on mechanical properties of asphalt under static condition through parameters such as Poisson's ratio, bulk modulus, shear modulus and Young's modulus. Modifiers including SBS [38], graphene and carbon nanotube (CNT) [34] and rubber [36]. It was found that they enhanced mechanical properties within a certain dosage. Filler also had a positive impact on mechanical properties of asphalt mastic probably due to its agglomeration [37].

Mechanical failure behaviors of virgin and aged asphalt were investigated under applied tensile and compressive stresses with different stress rates. The results showed that aged asphalt performed better in deformation resistance under both tensile and compressive stresses [24]. Hou et al. set vertical velocities to asphalt molecule groups in order to simulate crack initiation under tensile loading [39]. They found that during stretching process, gaps occurred among molecules in weak region might be the cause of crack.

Diffusion Behaviors MD has been applied to study two kinds of diffusion behaviors in asphalt mixtures, namely self-healing behavior and rejuvenator diffusion behavior.

Self-healing Behaviors For asphalt mixtures, self-healing refers to a process that gradually closes microcracks and partially recovers its mechanical properties (i.e. stiffness and strength) during rest periods [40].

Bhasin et al. first introduced molecular dynamics into asphalt self-healing by establishing self-healing model (Fig. 5) and applying diffusion coefficient, one-sixth of mean square displacement (MSD) versus time, to evaluate healing ability [41]. Based on Bhasin et al.'s work, Shen et al. found that asphalt healing was triggered by the diffusion of asphalt molecules and had high correlation with temperature and crack width [42]. The same conclusion was drawn by Qu et al. [22]. Sun et al. found asphalt binder had the strongest healing ability within phase transition (PT) temperature range while it healed slower outside this temperature range [43].

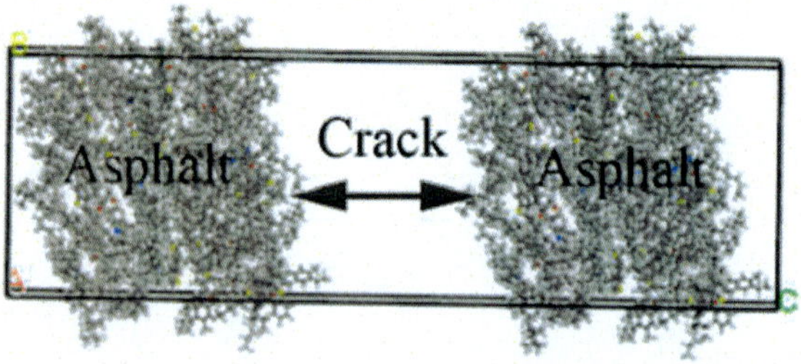

Fig. 5 Self-healing model for asphalt [41]

Temperature dependency of asphalt self-healing was further expressed by activation energy and pre-exponential factor derived from Arrhenius law [44]. The former represented the starting energy required for self-healing while the latter indicated instant healing after crack wetting. These two indicators were used to study the effects of modifier and oxidative aging on asphalt self-healing capability by Sun et al. [44] and Xu et al. [45]. The results showed that SBS modifier improved healing ability of asphalt binder while oxidative aging weakened it. Another MD demonstrated by He et al. also supported this conclusion [46]. Besides, He et al. observed that among SARA fractions, asphaltene moved slowest and saturates diffused fastest. Moreover, Sun et al. simulated healing test to prove the feasibility of MD to investigate healing phenomenon in macroscale tests from a micro perspective [47].

Rejuvenator Diffusion For energy saving purpose, reclaimed asphalt pavement (RAP) has gained popularity around the world and rejuvenator has been added into aged asphalt binder to improve recycling efficiency [48]. Through MD, the blending of rejuvenator and asphalt binder was found to be a diffusion process and related to temperature and adding sequence of rejuvenator [30]. Xu et al. found that rejuvenator diffused faster into virgin asphalt binder than aged asphalt binder by simulating rejuvenator diffusion into virgin and aged asphalt binder simultaneously [49]. Xu et al. hold opposite opinions [50]. They were also skeptical about the rationality of using Fick's law to calculate the diffusion coefficient of rejuvenator. For the effect of severity level of asphalt aging on rejuvenator diffusion, Xiao et al. found that rejuvenator diffused faster in severer aged asphalt binder, contrary to Sun et al.'s simulation results [26, 51]. The reason might be differences of model and calculation method of rejuvenator diffusion coefficient. However, they both agreed on that molecular structure of rejuvenator influenced its diffusion efficiency. Furthermore, rejuvenator dosage was also found to affect adhesion ability of rejuvenated asphalt binder [52].

Interface Properties Asphalt-aggregate interface property has a strong influence on overall performance and durability of asphalt pavement [53]. Researchers have also used molecular dynamics to do work in interface properties for asphalt mixtures.

Mechanical Property at Asphalt-aggregate Interface Efforts have been made on simulating asphalt-aggregate interface failure behaviors at the atomistic scale. Lu et al. conducted a micro uniaxial test at asphalt-aggregate interface and found that it was the rapid growth of nanovoids in asphalt model that caused final interface failure [13]. However, another tensile test simulated by Xu et al. demonstrated different result in which interface failure was mainly adhesive while cohesive failure existed if loading rate decreased to a certain level [54]. This tensile failure model was further applied to study asphalt-aggregate interface failure behavior under wet conditions. It was found that water intruded into asphalt-aggregate interface reduced adhesive strength but its influence level varied with asphalts with different compositions [55]. The simulated AFM test results of Xu et al. indicated that chemical structures of asphalt compositions also had an important impact on its adhesive properties [56].

Interfacial Bonding Property For interfacial bonding properties, Yao et al. found that it was aggregate's hydrophilicity that caused the displacement of asphalt from aggregate by building asphalt-aggregate and aggregate-water interface molecular models [57]. The asphalt-aggregate interface models under dry and wet conditions were first proposed by Xu et al. [58]. They also put forward three parameters, namely adhesion energy, work of debonding and energy ratio (ER) to characterize interfacial bonding strength. Equations (1–4) define these parameters.

$$\Delta E_{a-ag} = E_{total} - (E_a - E_{ag}) \quad (1)$$

$$E_{adhesion} = \Delta E_{a-ag}/A \quad (2)$$

$$W_{debonding} = (\Delta E_{a-w} + \Delta E_{ag-w} - \Delta E_{a-ag})/A \quad (3)$$

$$ER = \left|E_{adhesion}/W_{debonding}\right| \quad (4)$$

where ΔE_{a-ag} refers to interaction energy between asphalt and aggregate in dry conditions and E_{total} is total potential energy of asphalt-aggregate interface system while E_a and E_{ag} are potential energies for asphalt layer and aggregate layer respectively. Besides, ΔE_{a-w} is interaction energy between asphalt and water and ΔE_{ag-w} is interaction energy between aggregate and water. A refers to interface area of interface systems.

The two-layered and three-layered interface models and three parameters were widely used to investigate interfacial bonding properties considering the effects of aggregate mineral, moisture and oxidative aging.

Similar phenomenon was obtained for the effect of aggregate mineral type that more alkaline the mineral was, stronger adhesion it had with asphalt in dry and moisture conditions and more susceptible it was to moisture on asphalt-aggregate interface [59–63]. Furthermore, surface anisotropy of aggregate minerals also affected adhesion and debonding properties of asphalt-aggregate interface [62–64].

Same as Yao et al.'s findings [57], Xu et al. observed that water adversely affected adhesion between asphalt and aggregate by separating asphalt from aggregate spontaneously [58]. Interfacial water was also found to change asphaltene self-aggregation and distribution characteristics of SARA fractions near aggregate [65, 66].

Oxidative aging was found to improve adhesion ability between asphalt binder and silica under dry condition [32, 57]. However, there hasn't reached a unified conclusion on the effect of oxidative aging on asphalt-aggregate interfacial bonding properties when water existed. Xu et al. found that aged asphalt binder was more susceptible to moisture damage no matter water was at asphalt-aggregate interface or penetrated into asphalt [45]. The MD results of Gao et al. showed that the coupling effects of aging and moisture determined adhesion of interface systems [67]. Sun et al. found that interfacial water content had relations with adhesion between aged asphalt binder and aggregate [65].

Diffusion Behavior at Asphalt-aggregate In addition to adhesion, diffusion characteristics of asphalt components at asphalt-aggregate interface were also investigated by some researchers. Guo et al. found that temperature, molecular weight and absorbed medium were all influencing factors to the mobility of asphalt molecule at asphalt-aggregate interface [68], in agreement with the opinions of Du et al. [69]. Interfacial diffusion mechanism was investigated from the perspective of energy and the result indicated that asphalt diffusion at asphalt-aggregate interface was driven by Van der Waals forces between molecules [70].

3 Case Studies

Two case studies are presented to introduce the typical process and techniques of MD simulations to study the effects of aggregate mineral surface anisotropy and moisture on bonding strength at asphalt-aggregate interface.

3.1 Effect of Aggregate Mineral Surface Anisotropy on Bonding Strength at Asphalt-Aggregate Interface [62]

This research first studied how aggregate mineral surface anisotropy affected asphalt-aggregate interfacial bonding characteristics. The whole research was divided into three parts including model construction, molecule dynamics simulation and result analysis.

The asphalt-aggregate interface model was composed of asphalt referred from the work of Guo et al. [71] and acidic and alkaline aggregates represented by silica and calcite crystals from Cambridge Structural Database (CSD). To obtain a configuration with global minimum energy, annealing was performed on asphalt model. The close values of simulated density and glass transition temperature to experimental data validated the accuracy of asphalt model. For aggregates, silica crystal was cleaved into {0 0 1} {1 0 0} and {1 0 1} surfaces respectively while calcite crystal was exposed into {1 0 4}, {2 1 4} and {0 1 8} surfaces separately. The cleaved mineral crystals were then expanded to create supercells in order to match with asphalt model's geometric size. Asphalt-aggregate interface models were constructed by combining asphalt model and aggregate mineral supercells with 50 Å vacuum slab on the top to avoid the influence of periodic boundary conditions.

The interface models went through geometric optimization and dynamics simulations under canonical ensemble successively in order to reach equilibrium. Frames were extracted from trajectory for further analysis.

In result analysis stage, adhesion energy calculated by Eqs. (1) and (2) was used to characterize interfacial bonding strength between asphalt and aggregate. Through analysis, it was found that alkaline aggregate had higher bonding strength with asphalt than acidic aggregate, same as previous experimental findings [72]. Van der Waals interaction played a more important role than electrostatic interaction in adhesion energy for acidic aggregate while for alkaline aggregate, both of them made significant contributions to interfacial bonding strength. It was also found that aggregate surface anisotropy had an effect on interfacial bonding characteristics. For α-quartz-asphalt interface models, the ranking of adhesion energy was {1 0 1} > {1 0 0} > {0 0 1}, while that was {0 1 8} > {2 1 4} > {1 0 4} for calcite-asphalt interface models. The former could be explained by relatively high surface atom density on surface {1 0 1}. For the latter, the differences in Van der Waals energy also resulted from surface atom density while the number of dangling bonds for each Ca^{2+} ion at calcite surface was the reason for electrostatic energy difference.

3.2 *Effects of Aggregate Mineral Surface Anisotropy and Moisture on Bonding Strength at Asphalt-Aggregate Interface [63]*

This study considered the effect of both aggregate mineral surface anisotropy and moisture effect on asphalt-aggregate interface bonding properties.

Asphalt model applied in this research was derived from Li and Greenfield [20]. Acidic aggregate was represented by α-quartz crystals with exposed surfaces {0 0 1} and {1 0 1}. Their hydroxylated surfaces were also considered. {1 0 4}, {2 1 4} and {0 1 8} surfaces of calcite crystals were chosen to represent alkaline aggregates with different surface anisotropy. The procedures of asphalt-aggregate interface model construction under dry condition were same to case study one [62]. When developing interface models under wet condition, aggregate-water interface models with different water layer thicknesses were firstly generated to simulate the process of aggregate wetting. After a period of dynamics equilibration, asphalt model was added at the top of wetted aggregate to build the asphalt-water-aggregate interface system. Two-layered and three-layered interface models then underwent geometry optimization and dynamics equilibration under canonical ensemble respectively. The dynamics simulation results were exported for energy analysis.

From the results, it was found that under dry condition, the effect of aggregate surface anisotropy on interfacial bonding strength for calcite-asphalt interface model was the same as in case study one [62]. Hydroxylation on α-quartz surfaces reduced surface atom density but increased the number of hydrogen bonds formed by hydroxyl groups and functional groups in asphalt, resulting in its weaker Van der Waals interaction but stronger electrostatic interaction with asphalt.

An improved energy ratio (ER) considering residual adhesion between asphalt and aggregate was proposed to evaluate the moisture susceptibility of asphalt-water-aggregate interface models. Through calculation, ER values of three-layered interface system were found to decrease at first then remain stable at a certain value with the increase of interfacial moisture content. This might stem from continuous reduction of residual adhesion between asphalt and aggregate until it was zero. When comparing stabilized ER values among asphalt-water-aggregate interface models with anisotropic mineral surfaces, the performance of calcite surfaces for moisture susceptibility varied due to their surface dangling bond densities while for α-quartz surfaces, that was related to their surface atom densities and hydroxyl groups.

4 Limitations and Further Study

Although molecular dynamics has many advantages in the study of asphalt materials, there are also limitations in its applications. First of all, due to the limitations in time and length scales, only small size models (several nanometers) were generated to conduct MD simulations for a short period of time (hundreds of picoseconds). Secondly, MD simulation is able to investigate molecular interactions at the atomistic scale, but cannot simulate chemical reactions. However, during the use of asphalt materials, the change of macroscopic properties may also be related to some chemical reactions. Finally, though the mechanism of material macroscopic properties can be explained in MD simulation, it is hard to make a direct comparison between micro performance parameters and experimental values because of large size difference between micro models and a real material, and non-homogeneity of asphalt materials at the atomic scale.

There are still some other potential aspects that MD can be applied to asphalt mixtures. For example, currently aging model is only developed for virgin asphalt binder, but lack in modified ones such as SBS modified asphalt binder. More work is needed to analyze aging effect on mass fractions and molecular structures of modified asphalt binder components based on experiments and generate aged modified asphalt binder model according to these changes. Laboratory tests have proved that graphene can improve performance of asphalt binder [73, 74]. But there are only a small number of MD researches on graphene modified asphalt related to its thermodynamic properties, self-healing behavior, mechanical behaviors and interfacial interaction [22, 34, 35, 75]. The mechanism difference between graphene and other modifiers with asphalt components is not clear. The influence of graphene on other properties such as interfacial bonding properties still requires further study. Furthermore, when simulating dynamic mechanical behaviors of asphalt mixtures, tensile stress is applied to asphalt binder model or asphalt-aggregate interface model. Limited researches on other mechanical behaviors such as shear and torsion are conducted. Other potential fields like warm asphalt mixtures also deserve MD researches.

5 Conclusion

The applications of Molecular Dynamics (MD) in asphalt pavement mixtures were reviewed in this paper. The following conclusions can be drawn:

(1) Asphalt molecular model is the foundation of MD simulations. Improved molecule representation would make asphalt models more accurate and closer

to real asphalt materials. In this way, MD models can be better connected with real asphalt materials so that the mechanism of their macroscopic properties can be explained from an atomistic scale through MD simulations. Furthermore, the construction of more kinds of asphalt models considering aging and modifier effect would enhance MD applications in the study of more asphalt materials.

(2) MD acts effectively in investigating mechanical behavior, diffusion behavior and interfacial interaction for asphalt mixtures and evaluating the effects that related influencing factors can bring.

(3) MD has the inherent limitations in time and length scale, rendering it difficult to simulate long-term behaviors and quantitatively compare with experimental values. Besides, MD cannot reflect chemical reactions. The improvement in computational technology and multiscale investigations can help to overcome these limitations.

(4) By selecting an accurate force field type, MD can provide a feasible approach to study various properties for asphalt mixtures. There still exist potential aspects in asphalt mixtures that MD can be applied to, such as aged model construction of modified asphalt binder and investigations for graphene-modified asphalt binder and warm asphalt mixture.

Acknowledgements The authors gratefully acknowledge financial support by the Fundamental Research Funds for the Central Universities, China. Project No. 300102219307 and Shannxi Science and Technology Project 2021JQ-261.

References

1. Blackman BRK, Cui S, Kinloch AJ et al (2013) The development of a novel test method to assess the durability of asphalt road-pavement materials. Int J Adhes Adhes 42:1–10
2. Choudhary J, Kumar B, Gupta A (2020) Effect of filler on the bitumen-aggregate adhesion in asphalt mix. Int J Pavement Eng 21(12):1482–1490
3. Tayfur S, Ozen H, Aksoy A (2007) Investigation of rutting performance of asphalt mixtures containing polymer modifiers. Constr Build Mater 21(2):328–337
4. Li Q, Ni FJ, Gao L et al (2014) Evaluating the rutting resistance of asphalt mixtures using an advanced repeated load permanent deformation test under field conditions. Constr Build Mater 61:241–251
5. Shaheen M, Al-Mayah A, Tighe S (2017) Effects of high friction aggregate and PG Plus binder on rutting resistance of hot mix asphalt mixtures. Int J Pavement Eng 18(4):292–302
6. Moraes R, Velasquez R, Bahia H (2011) Measuring the effect of moisture on asphalt-aggregate bond with the bitumen bond strength test. Transp Res Rec 2209:70–81
7. Kim Y-R, Pinto I, Park S-W (2012) Experimental evaluation of anti-stripping additives in bituminous mixtures through multiple scale laboratory test results. Constr Build Mater 29:386–393
8. Ji X, Li J, Zou H et al (2020) Multi scale investigation on the failure mechanism of adhesion between asphalt and aggregate caused by aging. Constr Build Mater 265
9. Dai Z, Shen J, Shi P et al (2020) Multi-scaled properties of asphalt binders extracted from weathered asphalt mixtures. Int J Pavement Eng 21(13):1651–1661

10. Ji X, Sun E, Zou H et al (2020) Study on the multiscale adhesive properties between asphalt and aggregate. Constr Build Mater 249
11. Qu X, Liu Q, Guo M et al (2018) Study on the effect of aging on physical properties of asphalt binder from a microscale perspective. Constr Build Mater 187:718–729
12. Dong ZJ, Liu ZY, Wang P et al (2017) Nanostructure characterization of asphalt-aggregate interface through molecular dynamics simulation and atomic force microscopy. Fuel 189:155–163
13. Lu Y, Wang LB (2011) Nano-mechanics modelling of deformation and failure behaviours at asphalt-aggregate interfaces. Int J Pavement Eng 12(4):311–323
14. Su MM, Si CD, Zhang ZP et al (2020) Molecular dynamics study on influence of Nano-ZnO/SBS on physical properties and molecular structure of asphalt binder. Fuel 263
15. Wang P, Dong ZJ, Liu ZY (2017) Influence of carbon nanotubes on morphology of asphalts modified with styrene butadiene-styrene. Transp Res Rec 2632:130–139
16. Buehler MJ (2008) Atomistic modeling of materials failure. Springer, New York
17. Chen ZX, Pei JZ, Li R et al (2018) Performance characteristics of asphalt materials based on molecular dynamics simulation—a review. Constr Build Mater 189:695–710
18. Zhang L, Greenfield ML (2007) Analyzing properties of model asphalts using molecular simulation. Energy Fuels 21(3):1712–1716
19. Zhang LQ, Greenfield ML (2008) Effects of polymer modification on properties and microstructure of model asphalt systems. Energy Fuels 22(5):3363–3375
20. Li DD, Greenfield ML (2014) Chemical compositions of improved model asphalt systems for molecular simulations. Fuel 115:347–356
21. Xu M, Yi JY, Qi P et al (2019) Improved chemical system for molecular simulations of asphalt. Energy Fuels 33(4):3187–3198
22. Qu X, Wang DW, Hou Y et al (2019) Investigation on self-healing behavior of asphalt binder using a six-fraction molecular model. J Mater Civ Eng 31(5)
23. Petersen JC, Glaser R (2011) Asphalt oxidation mechanisms and the role of oxidation products on age hardening revisited. Road Mater Pavement Des 12(4):795–819
24. Pan JL, Tarefder RA (2016) Investigation of asphalt aging behaviour due to oxidation using molecular dynamics simulation. Mol Simul 42(8):667–678
25. Chen A, Liu G, Zhao Y et al (2018) Research on the aging and rejuvenation mechanisms of asphalt using atomic force microscopy. Constr Build Mater 167:177–184
26. Sun W, Wang H (2020) Molecular dynamics simulation of diffusion coefficients between different types of rejuvenator and aged asphalt binder. Int J Pavement Eng 21(8):966–976
27. Siddiqui MN, Ali MF (1999) Studies on the aging behavior of the Arabian asphalts. Fuel 78(9):1005–1015
28. Lesueur D (2009) The colloidal structure of bitumen: consequences on the rheology and on the mechanisms of bitumen modification. Adv Coll Interface Sci 145(1):42–82
29. Lin J, Hong J, Liu J et al (2016) Investigation on physical and chemical parameters to predict long-term aging of asphalt binder. Constr Build Mater 122:753–759
30. Ding YJ, Huang BS, Shu X et al (2016) Use of molecular dynamics to investigate diffusion between virgin and aged asphalt binders. Fuel 174:267–273
31. Fallah F, Khabaz F, Kim Y-R et al (2019) Molecular dynamics modeling and simulation of bituminous binder chemical aging due to variation of oxidation level and saturate-aromatic-resin-asphaltene fraction. Fuel 237:71–80
32. Luo L, Chu L, Fwa TF (2020) Molecular dynamics analysis of oxidative aging effects on thermodynamic and interfacial bonding properties of asphalt mixtures. Constr Build Mater 121299
33. Ding Y, Tang B, Zhang Y et al (2015) Molecular dynamics simulation to investigate the influence of SBS on molecular agglomeration behavior of asphalt. J Mater Civ Eng 27(8)
34. Zhou XX, Zhang X, Xu S et al (2017) Evaluation of thermo-mechanical properties of graphene/carbon-nanotubes modified asphalt with molecular simulation. Mol Simul 43(4):312–319

35. Yao H, Dai QL, You ZP et al (2017) Property analysis of exfoliated graphite nanoplatelets modified asphalt model using molecular dynamics (MD) method. Appl Sci-Basel 7(1)
36. Guo FC, Zhang JP, Pei JZ et al (2019) Study on the mechanical properties of rubber asphalt by molecular dynamics simulation. J Mol Model 25(12)
37. Zhu XY, Du Z, Ling HW et al (2020) Effect of filler on thermodynamic and mechanical behaviour of asphalt mastic: a MD simulation study. Int J Pavement Eng 21(10):1248–1262
38. Qu X, Wang DW, Hou Y et al (2018) Influence of paraffin on the microproperties of asphalt binder using MD simulation. J Mater Civ Eng 30(8)
39. Hou Y, Wang LB, Wang DW et al (2017) Using a molecular dynamics simulation to investigate asphalt nano-cracking under external loading conditions. Appl Sci-Basel 7(8)
40. Ayar P, Moreno-Navarro F, Rubio-Gamez MC (2016) The healing capability of asphalt pavements: a state of the art review. J Clean Prod 113:28–40
41. Bhasin A, Bommavaram R, Greenfield ML et al (2011) Use of molecular dynamics to investigate self-healing mechanisms in asphalt binders. J Mater Civ Eng 23(4):485–492
42. Shen S, Lu X, Liu L et al (2016) Investigation of the influence of crack width on healing properties of asphalt binders at multi-scale levels. Constr Build Mater 126:197–205
43. Sun D, Sun G, Zhu X et al (2018) Intrinsic temperature sensitive self-healing character of asphalt binders based on molecular dynamics simulations. Fuel 211:609–620
44. Sun D, Lin T, Zhu X et al (2016) Indices for self-healing performance assessments based on molecular dynamics simulation of asphalt binders. Comput Mater Sci 114:86–93
45. Xu GJ, Wang H (2017) Molecular dynamics study of oxidative aging effect on asphalt binder properties. Fuel 188:1–10
46. He L, Li G, Lv S et al (2020) Self-healing behavior of asphalt system based on molecular dynamics simulation. Constr Build Mater 254:119225
47. Sun W, Wang H (2020) Self-healing of asphalt binder with cohesive failure: insights from molecular dynamics simulation. Constr Build Mater 262
48. Behnood A (2019) Application of rejuvenators to improve the rheological and mechanical properties of asphalt binders and mixtures: a review. J Clean Prod 231:171–182
49. Xu GJ, Wang H (2018) Diffusion and interaction mechanism of rejuvenating agent with virgin and recycled asphalt binder: a molecular dynamics study. Mol Simul 44(17):1433–1443
50. Xu M, Yi JY, Feng DC et al (2019) Diffusion characteristics of asphalt rejuvenators based on molecular dynamics simulation. Int J Pavement Eng 20(5):615–627
51. Xiao Y, Li C, Wan M et al (2017) Study of the diffusion of rejuvenators and its effect on aged bitumen binder. Appl Sci-Basel 7(4)
52. Cui B, Gu X, Hu D et al (2020) A multiphysics evaluation of the rejuvenator effects on aged asphalt using molecular dynamics simulations. J Clean Prod 259:120629
53. Canestrari F, Cardone F, Graziani A et al (2010) Adhesive and cohesive properties of asphalt-aggregate systems subjected to moisture damage. Road Mater Pavement Des 11:11–32
54. Xu GJ, Wang H (2016) Molecular dynamics study of interfacial mechanical behavior between asphalt binder and mineral aggregate. Constr Build Mater 121:246–254
55. Wang H, Lin EQ, Xu GJ (2017) Molecular dynamics simulation of asphalt-aggregate interface adhesion strength with moisture effect. Int J Pavement Eng 18(5):414–423
56. Xu M, Yi JY, Feng DC et al (2016) Analysis of adhesive characteristics of asphalt based on atomic force microscopy and molecular dynamics simulation. ACS Appl Mater Interfaces 8(19):12393–12403
57. Yao H, Dai QL, You ZP (2015) Chemo-physical analysis and molecular dynamics (MD) simulation of moisture susceptibility of nano hydrated lime modified asphalt mixtures. Constr Build Mater 101:536–547
58. Xu GJ, Wang H (2016) Study of cohesion and adhesion properties of asphalt concrete with molecular dynamics simulation. Comput Mater Sci 112:161–169
59. Li R, Du H, Fan ZP et al (2016) Molecular dynamics simulation to investigate the interaction of asphaltene and oxide in aggregate. Adv Mater Sci Eng 2016
60. Gao Y, Zhang Y, Gu F et al (2018) Impact of minerals and water on bitumen-mineral adhesion and debonding behaviours using molecular dynamics simulations. Constr Build Mater 171:214–222

61. Liu JZ, Yu B, Hong QZ (2020) Molecular dynamics simulation of distribution and adhesion of asphalt components on steel slag. Constr Build Mater 255
62. Chu L, Luo L, Fwa TF (2019) Effects of aggregate mineral surface anisotropy on asphalt-aggregate interfacial bonding using molecular dynamics (MD) simulation. Constr Build Mater 225:1–12
63. Luo L, Chu L, Fwa TF (2020) Molecular dynamics analysis of moisture effect on asphalt-aggregate adhesion considering anisotropic mineral surfaces. Appl Surf Sci 527:146830
64. Feng PN, Wang HN, Ding HY et al (2020) Effects of surface texture and its mineral composition on interfacial behavior between asphalt binder and coarse aggregate. Constr Build Mater 262
65. Sun W, Wang H (2020) Moisture effect on nanostructure and adhesion energy of asphalt on aggregate surface: a molecular dynamics study. Appl Surf Sci 510:11
66. Cui WT, Huang WK, Xiao ZC et al (2020) The effect of moisture on the adhesion energy and nanostructure of asphalt-aggregate interface system using molecular dynamics simulation. Molecules 25:(18)
67. Gao YM, Zhang YQ, Yang Y et al (2019) Molecular dynamics investigation of interfacial adhesion between oxidised bitumen and mineral surfaces. Appl Surf Sci 479:449–462
68. Guo M, Tan YQ, Wang LB et al (2017) Diffusion of asphaltene, resin, aromatic and saturate components of asphalt on mineral aggregates surface: molecular dynamics simulation. Road Mater Pavement Des 18:149–158
69. Du Z, Zhu X (2019) Molecular dynamics simulation to investigate the adhesion and diffusion of asphalt binder on aggregate surfaces. Transp Res Rec 2673(4):500–512
70. Huang M, Zhang HL, Gao Y et al (2019) Study of diffusion characteristics of asphalt-aggregate interface with molecular dynamics simulation. Int J Pavement Eng
71. Guo M, Tan YQ, Bhasin A et al (2016) Using molecular dynamics to investigate interfacial adhesion between asphalt binder and mineral aggregate. Funct Pavement Des 60
72. Abo-Qudais S, Al-Shweily H (2007) Effect of aggregate properties on asphalt mixtures stripping and creep behavior. Constr Build Mater 21(9):1886–1898
73. Yao H, Dai Q, You Z et al (2016) Rheological properties, low-temperature cracking resistance, and optical performance of exfoliated graphite nanoplatelets modified asphalt binder. Constr Build Mater 113:988–996
74. Moreno-Navarro F, Sol-Sanchez M, Gamiz F et al (2018) Mechanical and thermal properties of graphene modified asphalt binders. Constr Build Mater 180:265–274
75. Ramezani MG, Rickgauer J (2020) Understanding the adhesion properties of carbon nanotube, asphalt binder, and mineral aggregates at the nanoscale: a molecular dynamics study. Pet Sci Technol 38(1):28–35

Assessment of Laboratory and Field Compaction of Dense Graded Aggregate Bases (DGAB)

A. R. Bambarandage, W. R. A. N. Jayantha, and W. K. Mampearachchi

Abstract Dense Graded Aggregate Base (DGAB) construction is a major contributing component in flexible pavement construction in terms of pavement structural capacity and the project cost. However, the production processes related to DGAB construction impose more significant financial, environmental, and social concerns, which stresses the effective compaction in DGAB construction. DGAB compaction process is controlled by three governing factors: moisture content (MC), compaction effort, and layer thickness, where compaction effort can be minimized when field compaction takes place at MC levels closer to the estimated Optimum Moisture Content (OMC). Hence, the relationship among maximum dry density (MDD), OMC, and compaction effort should be well established for a given layer thickness in such a laboratory method that best interprets the field practice. Information on the current industry compaction practices was gathered by conducting a questionnaire survey. Simultaneously, laboratory and field tests were carried out to compare the compaction behavior of DGAB at different MCs and energy levels. Moisture density plots of different laboratory and field compaction tests were compared to determine the most suitable laboratory compaction method to simulate the field compaction of DGAB. The study revealed that the vibratory hammer test resulted in the highest MDD and the lowest OMC, where moisture-density curves of the field compaction tests were approaching the vibratory hammer test curve with the increase of compaction effort, emphasizing that the vibratory hammer test best simulates field compaction, when compared to other laboratory impact compaction tests.

Keywords Compaction effort · Moisture content · Dry density · Vibratory hammer · Field compaction · Energy optimization

A. R. Bambarandage · W. R. A. N. Jayantha (✉) · W. K. Mampearachchi
Department of Civil Engineering, Faculty of Engineering, University of Moratuwa, Moratuwa, Sri Lanka
e-mail: jayanthawran.20@uom.lk

W. K. Mampearachchi
e-mail: wasanthak@uom.lk

© The Author(s), under exclusive license to Springer Nature Switzerland AG 2022
H. R. Pasindu et al. (eds.), *Road and Airfield Pavement Technology*, Lecture Notes in Civil Engineering 193, https://doi.org/10.1007/978-3-030-87379-0_23

1 Introduction

Aggregate Base Course (ABC) is a significant component of the pavement structure that facilitates the stress attenuation of the traffic loads, providing the desirable structural support for the asphalt surface layer and protection of the subbase and subgrade layers [1, 2]. Thus, DGAB is a major contributor to the pavement's stability, functionality, and durability. Performance of the DGAB relies on the material characteristics of the constituent aggregates and their interlocking behavior within the matrix [3–6].

Compaction of aggregate bases is defined as the reduction of void spaces among aggregate particles by densifying the aggregate materials through the application of mechanical energy to achieve the desired engineering performance [4, 7]. Many properties of aggregates can be enhanced through compaction, such as; increase in shear strength, reduction in permeability, decrease in compatibility, decrease in void ratio, and increased unit weight [8, 9]. Dense Graded Aggregate Base (DGAB) is a mechanically crushed rock aggregate mixture where the gradation has been engineered to minimize void spaces between aggregate particles when compacted [4, 10]. Three major governing factors control the DGAB compaction process: moisture content (MC), compaction effort, and layer thickness. However, as stated before, individual characteristics of aggregate such as shape, angularity, and texture will also affect the interlocking behavior in a packed matrix, like DGAB [3, 4]. The relationship between maximum dry density (MDD), optimum moisture content (OMC), and compaction effort should be well established for a given layer thickness before commencing the field operation. It will lead to an effective compaction process by saving energy, cost as well as the environment.

Standard Specification for Construction and Maintenance of Road & Bridges (SSCM) published by the Institute for Construction Training and Development (ICTAD) is the main guideline used for aggregate base course (ABC) compaction in this field study. However, these specifications vary from country to country [11–13].

Usually, for compaction of crushed rock material type, vibrating smooth drum equipment is used [12, 14]. Clause 405.3(b) of SSCM clearly states that "The required amount of water and MC shall be determined by carrying out field trails, but shall normally be within 2% of the predetermined OMC at the time of compaction" [11, 13]. However, practicing this norm is a question when the ABC construction process at present is examined. Variations in MC affect the material properties and pavement performance significantly [15]. Compaction of DGAB in the field is done under higher moisture content than stipulated in SSCM. As a result, segregation occurs, and fine particles migrate to the top surface of the layer by changing the gradation of the mix, which leads to an improper compaction process during the construction stage, resulting in the long term total permanent deformation (rutting accumulation) in highway pavements during the service life [1, 16, 17]. Consequently, the DGAB layer will not provide the expected structural capacity to the pavement due to this poor construction practice. Segregation control is an important task required to ensure proper compaction of DGAB [18]. The above poor construction practice has been reported worldwide [19].

Laboratory aggregate compaction tests used to establish target density for field evaluation of compaction are based on impact/dropped loading [8]. AASHTO T 99, AASHTO T 180, ASTM D698 and ASTM D1557 are such laboratory compaction test standards [4]. Moreover, according to the guideline given by SSCM, maximum dry density and optimum moisture content are determined by the Modified Procter Compaction Test (AASHTO T180). However, due to the use of much heavier vibrating rollers in compaction, field densities reach higher levels than that of laboratory density [20–23]. Although modified proctor compaction results in improved properties of the laboratory characterization of roadway base materials for some types of base course layers, the precision and compatibility of the laboratory characterization of the base layer to that in the field is questionable, especially for DGAB layers [23]. Besides, many studies have concluded that the impact compaction results in a finer gradation due to the fracturing of the aggregate particles under the heavy crushing load, which is not exactly resembling the actual scenario at the field [4, 24]. Laboratory compaction test methods such as the Vibratory Compaction Test and Gyratory Compaction Test have been engaged to overcome these problems in the global context [25, 26].

In this research, a questionnaire survey was carried out to obtain the current practices of DGAB compaction in the local context. According to the gathered data, field and laboratory compaction tests were done to examine the correlation between actual site compaction characteristics and the laboratory compaction characteristics.

2 Research Objectives

The research's ultimate goal is to evaluate the current practice in compaction of dense graded aggregate bases and identify the differences between field and laboratory compactions. Following tasks were carried out to achieve the objectives of the research.

1. Evaluate the current field practices used for ABC compaction
2. Identify the effects of compaction at the high moisture content (Segregation and change of gradation of DGAB)
3. Verification of the relationship between moisture content, dry density, and compaction efforts at different laboratory and field conditions.

3 Experimental Study

3.1 Survey Results

According to the questionnaire responses, the moisture level of DGAB is mainly controlled on four occasions—At the stockpile, at the site while laying, at the site just before the compaction begins and at the site while compacting. Table 1 shows the field practices of adding water on DGAB to control the MC of the mix. Figure 1 shows the survey results of the construction practice for the addition of water in DGAB.

Table 1 Stages of controlling MC

At stockpile	 • Water is added to graded DGAB at the stockpile • ABC is well mixed at its OMC and transported to site • Uniform MC can be maintained throughout the layer during compaction
While laying	 • Dry ABC is transported to the site • Water is added to dry ABC, mixed, and laid by using a motor grader, before compaction begins • Uniform MC can be maintained throughout the layer while compacting

(continued)

Table 1 (continued)

Just before compacting	 • Dry ABC is transported to the site • Uniformly lay ABC on a flat ground surface • Water sprayed on top of ABC layer and compaction is commenced MC will not be uniform within the layer
While compacting	 • Dry ABC is transported to the site • Uniformly lay ABC on a flat ground surface • Water sprayed on top of ABC layer during compaction • Only the top surface gets wet and excessive water flows on the surface, forming a segregated ABC layer

3.2 *Laboratory Experiments*

Five different tests were carried out at the laboratory to compare the moisture content (MC), and dry density (DD) relationship at different energy levels: Modified proctor compaction test (AASHTO T180, Method D), Standard proctor compaction test (AASHTO T99, Method C), Modified proctor compaction test with 25 blows per layer, Standard proctor compaction test with 25 blows per layer and Vibratory hammer test. Figure 2 shows the MC versus DD relationship for different compaction efforts under laboratory conditions. 98% compaction line is drawn for the test results

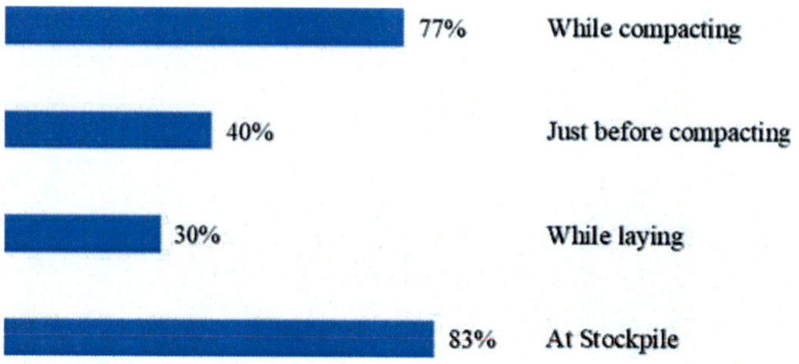

Fig. 1 Tendency to add water at different stages as a percentage of total number of survey locations

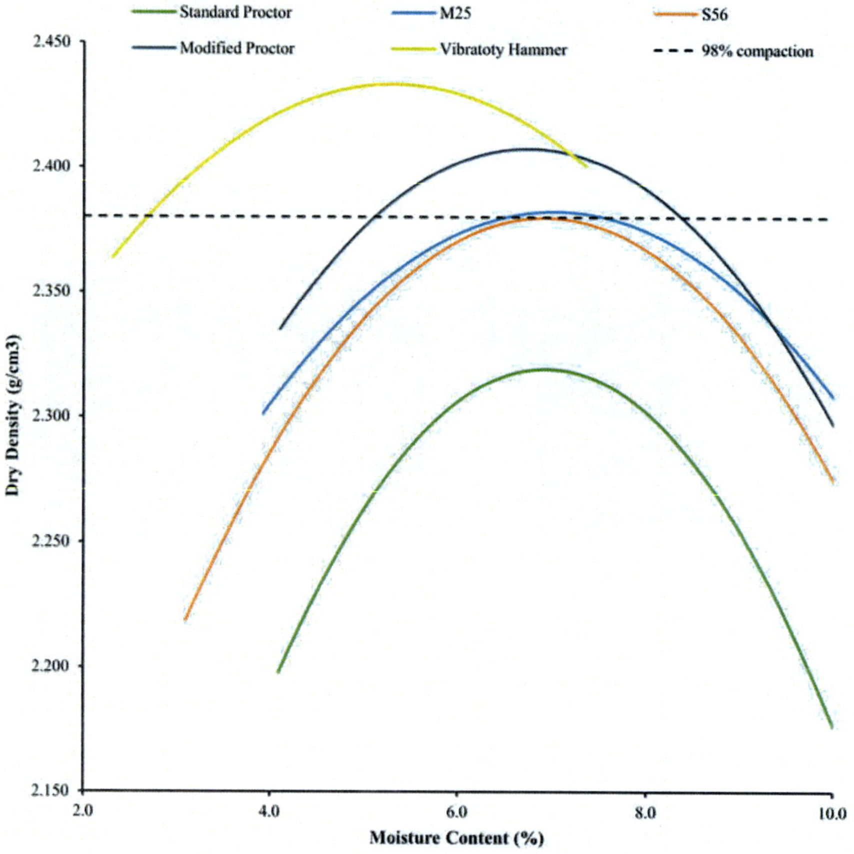

Fig. 2 MC versus DD in laboratory conditions over different compaction efforts

Table 2 MC ranges obtained for 98% compaction

Test method	Range of moisture for 98% MDD (%)	OMC (%)	Tolerance (%)
Standard proctor compaction test	–	6.9	–
Modified proctor compaction test	5.1–8.4	6.75	−1.65 to +1.65
Standard proctor compaction test method with 56 blows per layer	6.72–7.12	6.9	−0.2 to +0.2
Modified proctor compaction test method with 25 blows per layer	6.49–7.57	7.0	−0.54 to +0.54
Vibratory hammer compaction test	2.7–7.9	5.3	−2.6 to +2.6

obtained for the Vibratory hammer test, which gave the lowest OMC and highest MDD.

Table 2 shows the MC ranges obtained from different laboratory compaction tests for 98% MDD compaction.

3.3 Field Experiments—Compaction Tests

Welioya–Kalthota section (Chainage 43 + 500 − 43 + 900) of Bodagama-Hambegamuwa-Kalthota (B528) Road was selected as the case study location for field trials. The test section was a flat-straight section, and the cross-section of the test location is represented in Fig. 3. For the verification of the DD-MC relationship under field conditions, five no. of field trial strips were conducted under five different moisture levels. Six no. of field density tests were carried out at different energy levels (2, 4, 6, 8, 10 & 12 no. of roller passes) in each test strip. The in-situ density of compacted layers was obtained from the sand cone test, which is considered to be the common practice of density measurement in the field (Umashankar et al. 2016). Uniform layer thickness was maintained as 200 mm, while an 11-ton single drum vibratory roller was used as the compactor (Speed = 2.5 km/h, Frequency = 31 Hz, Amplitude = 1.97 mm, Drum width = 2100 mm) for all the trial strips.

DGAB thickness of the section is 200 mm, which was laid over the subbase layer. Six field trials were carried out at different strip locations for different moisture levels, as shown in Fig. 4. The dimensions of the strip are approximately 100 m in length and 3.1 m in width. The thickness of the DGAB layer was 200 mm for the entire case study. Figure 5 shows the variation of DD of field strips with MC under field conditions.

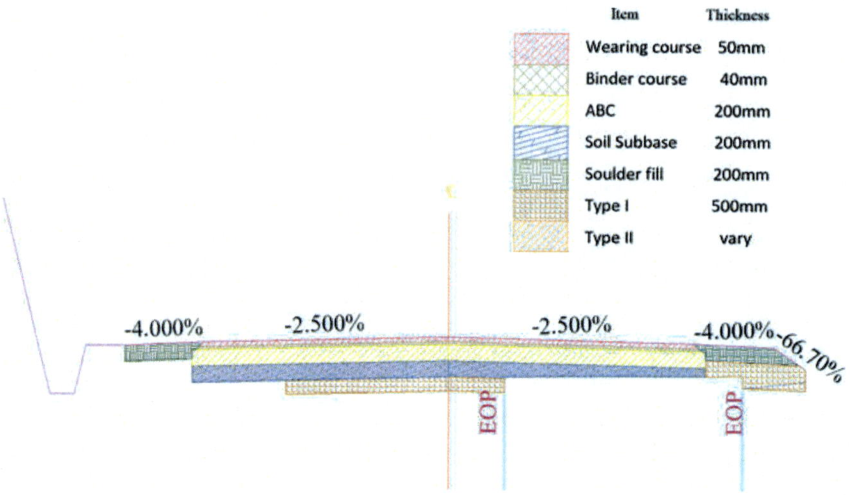

Fig. 3 Road cross-section of the test location

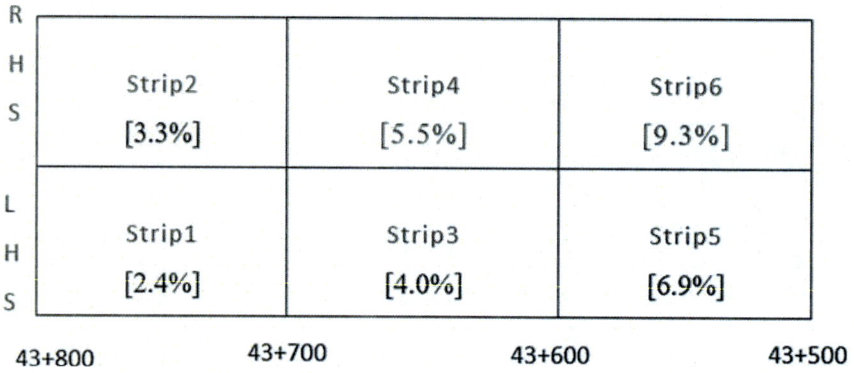

Fig. 4 Field trial strip plan [MC%]

3.4 Field Experiments—Gradation Change of DGAB Under High Moisture Content

When ABC is compacted at higher MC, fines will pump out with water, causing variations to the layer's designed gradation. Therefore, the gradation change that occurred is measured for 9.3% MC. Four such scenarios were chosen for conducting the sieve analysis test, such as the laid sample (before compaction) and after 4, 8 & 12 no. of roller passes. Two test samples were extracted for gradation test in each case from the top 100 mm layer & bottom 100 mm layer. Tables 3 and 4 represent the results of sieve analysis in the top 100 mm and bottom 100 mm layers, respectively. A graphical representation of the particle distribution is shown in Figs. 6 and 7.

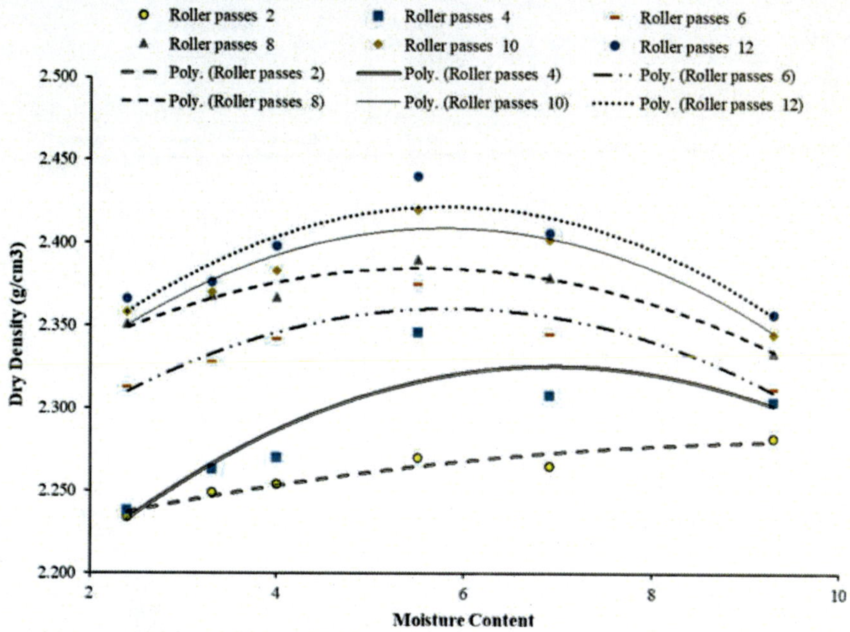

Fig. 5 DD versus MC under field conditions

Table 3 Sieve analysis test results of top 100 mm of DGAB

Sieve size (mm)	Spec limit		Design gradation	Top 100 mm layer			
	SSCM table 1701.5			Laid	4 Roller passes	8 Roller passes	12 Roller passes
	Min	Max					
50	100	100	100	100	100	100	100
37.5	95	100	100	100	100	100	100
20	60	80	74	64.6	60.1	62.8	67.1
10	40	60	53.3	51.3	46.3	47.1	48.9
5	25	40	34.7	32	33.2	37.3	38.5
2.36	15	30	25.2	22.7	27.3	29.1	33.3
0.425	7	19	15.3	12.1	16.3	17.8	21.1
0.075	5	12	8.7	4.5	4.2	4.9	4.8

When DGAB is compacted under high MC than optimum, fine particles in the mixture will move to the surface. In Tables 3 and 4, this fine particle displacement can be clearly notified. In the bottom layer, fine particle percentage decreases with the number of roller passes. Meanwhile, in the top 100 mm, the fine percentage increased with the number of roller passes.

Table 4 Sieve analysis test results of bottom 100 mm of DGAB

Sieve size (mm)	Spec limit SSCM table 1701.5		Design gradation	Bottom 100 mm layer			
	Min	Max		Laid	4 Roller passes	8 Roller passes	12 Roller passes
50	100	100	100	100	100	100	100
37.5	95	100	100	100	100	100	100
20	60	80	74	75.1	65.9	68.3	67.9
10	40	60	53.3	52.6	53.3	51	50.8
5	25	40	34.7	39.6	40.8	40.3	40.3
2.36	15	30	25.2	33.1	34.4	28	30.2
0.425	7	19	15.3	21.7	21.9	12.8	13.1
0.075	5	12	8.7	4.5	5.4	5	5.6

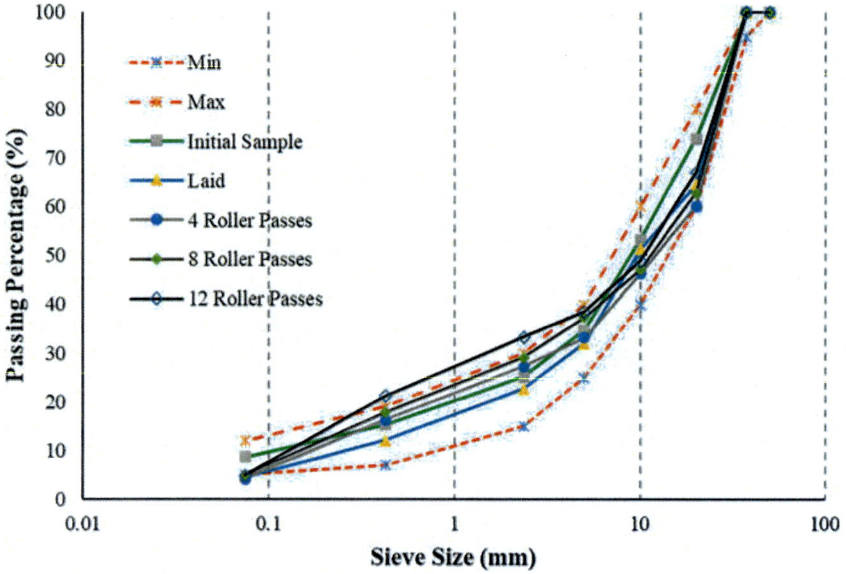

Fig. 6 Gradation change in top 100 mm of DGAB at 9.3% MC

4 Results and Discussion

4.1 Field Experiments

Laboratory test results verified that the increase of compaction effort would increase the MDD and decrease the OMC requirement. Field trial tests revealed that the

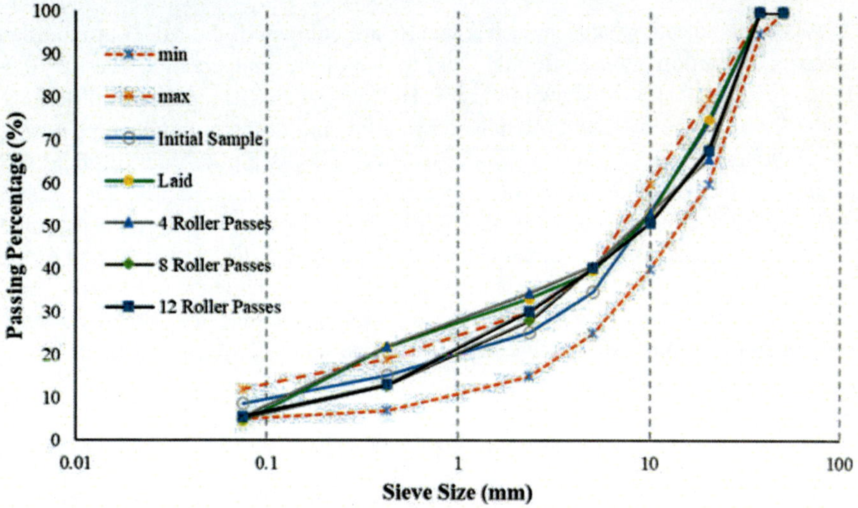

Fig. 7 Gradation change in bottom 100 mm of DGAB at 9.3% MC

required density could be achieved with lesser no. of roller passes if compaction occurs close to the estimated OMC (Fig. 8).

At 6.9% moisture level close to OMC, less than eight roller passes can accomplish 98% compaction. If the MC has slightly deviated from OMC, the required compaction should be achieved with more compaction effort. If the MC is vastly different from OMC (at 2.4 & 9.3 moisture levels), 98% compaction cannot be achieved even after 12 roller passes, as shown in Fig. 8.

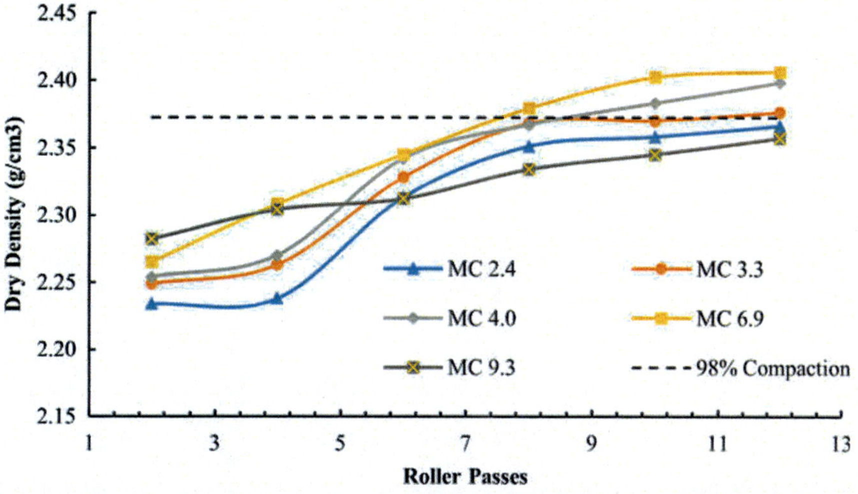

Fig. 8 Dry density versus roller passes

When the laboratory and field test results are compared, the MDD at modified proctor compaction curve is slightly close to that of the compaction curves of 10 & 12 no. of roller passes, as shown in Fig. 9. However, there is a significant difference of the OMC in the modified proctor compcation and the field compaction curves. The Vibratory Hammer test's moisture-density curve is higher than the field test results, yet closer to 12 no. of roller passes curve. Most importantly, the vibrator hammer test's OMC is closer to the field compaction test results than modified proctor test results. OMC of impact compaction tests has deviated from that of the field compaction tests. OMC is one of the most critical parameters to be maintained with more significant concern in the construction. Therefore, out of the laboratory compaction methods used in this research, the vibratory hammer test gave closer results to field compaction.

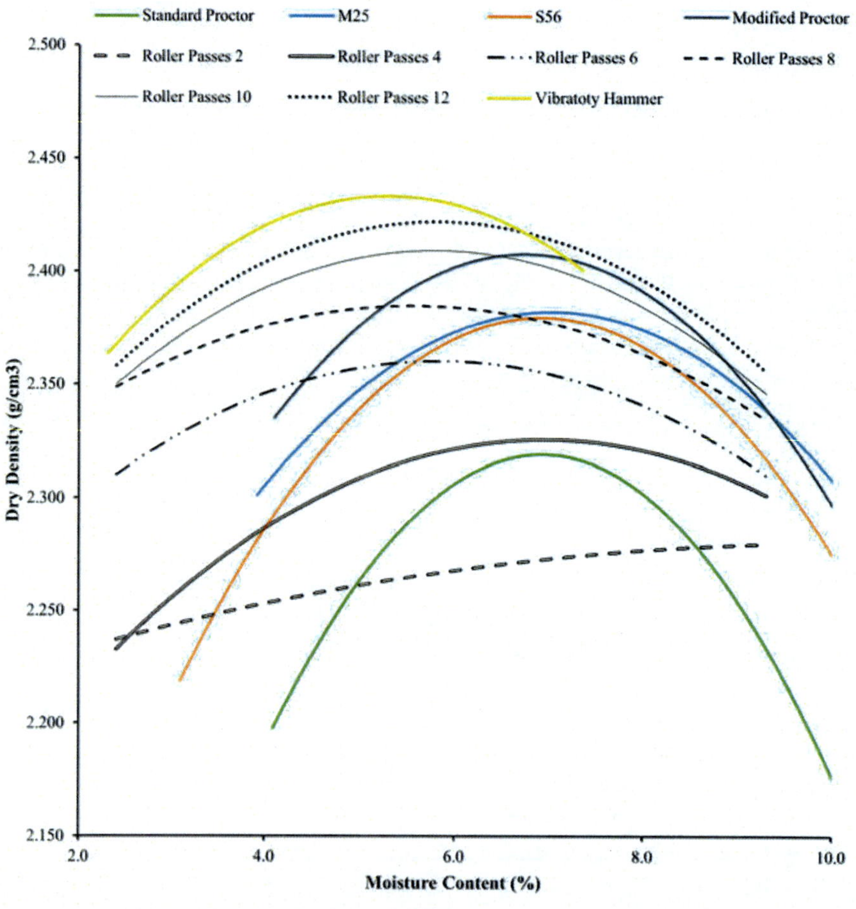

Fig. 9 Lab and field moisture-density plot comparison

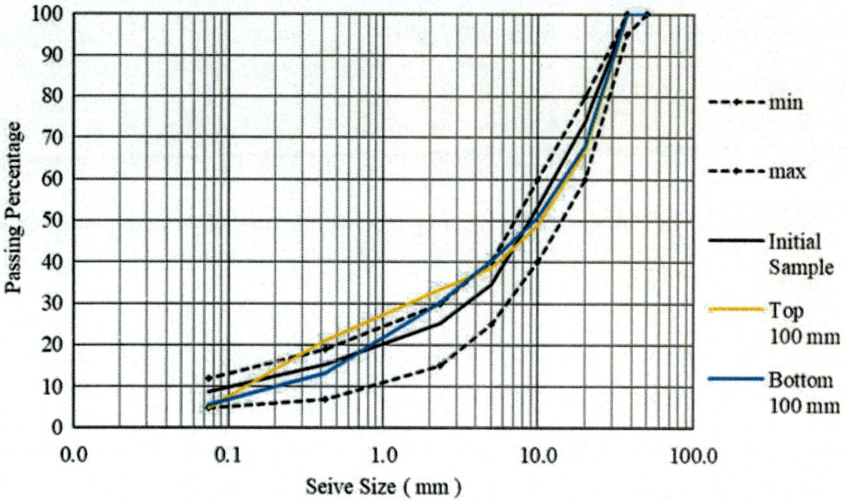

Fig. 10 Gradation curve of 9.3% MC test strip after 12 no of roller coverages

As determined from the questionnaire survey, moisture control is not properly conducted in the field. Even though OMC obtained from impact compaction tests (axial compaction in hammer compaction) is higher than that of field compaction tests, the moisture levels are maintained even above the impact compact test's OMC levels in usual field practice. Therefore, sieve analysis tests were carried out to determine the gradation change due to the presence of higher MC. Following observations were made with the use of field test results.

1. Fine particles move from the top surface to the bottom when an excessive amount of water is added to the laid sample.
2. After four no. of roller passes, fines tend to move upward again.
3. At the end of 8 roller passes, more fines move upward. However, in that instance, both top & bottom layers grading are within the specified limits.
4. But after 12 no. of roller passes, an excessive amount of fines move upward & accumulate on the top surface, as shown in Fig. 10, eventually leading to deviations of the gradation of the entire layer from the specified limits.

5 Conclusion

From the questionnaire survey results, it is revealed that the ABC compaction is carried out at higher moisture levels in most cases, leading to gradation deviations of top and bottom layers of DGAB.

It is shown that during the case study ABC cannot retain MC more than 10%. When MC is higher than 10%, then the excess amount expels through sides, top, or underneath the ABC layer.

98% compaction can be achieved by eight no. of roller passes of the 11-ton roller if the MC is around OMC (6.9%). However, if MC differs from OMC, proportionately more compaction effort should be deployed.

Dry Density—Moisture Content relationship of Modified compaction test and Vibratory Hammer test is similar to field compaction done by 10 & 12 roller passes of the 11-ton roller.

MDD obtained from vibratory hammer test results was higher than that of field compaction test results. However, OMC of field compaction is closer to OMC obtained from the vibratory hammer test.

References

1. Cetin A, Kaya Z, Cetin B, Aydilek AH (2014) Influence of laboratory compaction method on mechanical and hydraulic characteristics of unbound granular base materials. Road Mater Pavement Des 15(1):220–235. https://doi.org/10.1080/14680629.2013.869505
2. Xiao Y, Tutumluer E, Siekmeier J (2011) Resilient modulus behavior estimated from aggregate source properties, pp 4843–4852. https://doi.org/10.1061/41165(397)495
3. Barksdale RD (1972) The aggregate handbook, vol 1
4. Titi HH et al (2012) Base compaction specification feasibility analysis. December, 222p [Online]. Available: http://wisdotresearch.wi.gov/wp-content/uploads/WisDOT-WHRP-project-0092-11-02-final-report.pdf%5Cn, https://trid.trb.org/view/1249190
5. Rittenhouse G (1943) Relation of Shape to the passage of grains through sieves. Ind Eng Chem—Anal Ed 15(2):153–155. https://doi.org/10.1021/i560114a026
6. Barksdale RD, Itani SY Influence of aggregate shape on base behavior. Transp Res Rec 1227:173–182.
7. Choi YJ, Ahn D, Nguyen TH, Ahn J (2018) Assessment of field compaction of aggregate base materials for permeable pavements based on plate load tests. Sustain 10(10). https://doi.org/10.3390/su10103817
8. Shahin AW, Wilson D (2010) Investigation of the variability in the results of the NZ vibrating hammer compaction test
9. Miranda HMB, Batista FA, de Lurdes Antunes M, Neves J (2020) Influence of laboratory aggregate compaction method on the particle packing of stone mastic asphalt. Constr Build Mater 259:10. https://doi.org/10.1016/j.conbuildmat.2020.119699
10. Spencer WT (1961) Dense-graded aggregate base construction—ISHC specifications and practice
11. Institute for Construction Training and Development (ICTAD) (2009) Standard specification for construction and maintenance of road & bridges
12. RC (Vic. M. T. Department (1998) Guide to general requirements for unbound pavement materials. VicRoads
13. Unified Facilities Guide Specifications (UFGS) (2017) UFGS 32 11 23 aggregate base course
14. Vuong B, Newman G, Denham P, Fraser A, Field trials of constructibility and performance of recycled granular base materials with glass additive. In: 24th ARRB Conference – Building on 50 Years of Road and Transport Research, Melbourne, 2010, pp 1–18. Accessed: Mar. 11, 2021. [Online]. Available: https://sci-hub.do/https://trid.trb.org/view/1096887
15. Tamrakar P, Nazarian S, Paso E (2018) Assessment of moisture effects on road bases. Transp Res Board, (January):1–6
16. Lekarp F, Isacsson U, Dawson A (2000) State of the Art. II: permanent strain response of unbound aggregates. J Transp Eng 126(1):76–83. https://doi.org/10.1061/(ASCE)0733-947X(2000)126:1(76)

17. Ahmad K, Smalley IJ (1973) Observation of particle segregation in vibrated granular systems. Powder Technol 8(1–2):69–75. https://doi.org/10.1016/0032-5910(73)80064-6
18. Peng Z (2013) The base material segregation control technology research of strong interlocked skeleton dense graded aggregate. In: Proceedings—2013 4th international conference on intelligent systems design and engineering applications, ISDEA 2013, 2013, pp 617–620. https://doi.org/10.1109/ISDEA.2013.547
19. Frempong EM (1995) Field compaction control studies on road bases in a new settlement area in Ghana. Geotech Geol Eng 13(4):227–241. https://doi.org/10.1007/BF00422212
20. Karan P, Larkin T, Wilson DJ (2017) Vibratory compaction of base course aggregates. Transp Res Rec 2655(1):45–53. https://doi.org/10.3141/2655-07
21. Ping WV, Yang Z, Leonard M, Putcha S (2002) Laboratory simulation of field compaction characteristics on sandy soils. Transp Res Rec (1808):84–95. https://doi.org/10.3141/1808-10
22. Ping WV, Xing G, Leonard M, Yang Z (2003) Evaluation of laboratory compaction techniques for simulating field soil compaction (Phase II) (March)
23. Sebesta S, Harris P, Liu W (2006) Improving lab compaction methods for roadway base materials. Texas Department of Transportation and the Federal Highway Administration
24. Kaya Z, Cetin A, Cetin B, Aydilek A (2012) Effect of Compaction Method on Mechanical Behavior of Graded Aggregate Base Materials. pp 1486–1494. https://doi.org/10.1061/9780784412121.153
25. Adu-Osei A, Little DN, Tutumluer E, Lytton RL (2000) Effect of compaction on orthotropic properties of aggregates
26. Bozorgi A, Fried A, Montoya BM, Castorena C (2020) The effect of laboratory compaction method on the resilient behaviour and fabric of aggregate base course materials. Road Mater Pavement Des 21(7):1955–1967. https://doi.org/10.1080/14680629.2019.1580606

Areca Fiber Reinforced Alkali-Activated Black Cotton Soil Using Class F Fly Ash and Limestone Powder for Pavements

B. A. Chethan and A. U. Ravi Shankar

Abstract Alkali activation has gained importance in place of cement treatment in construction due to reduced CO_2 emissions. The precursors that are rich in silica, alumina, and calcium can be used for soil stabilization with a suitable alkali solution. In this investigation, 0–45% class F fly ash with a constant 5% limestone powder was used to stabilize black cotton soil. These mixes were reinforced with 0.5% areca fibers and stabilized using the alkali solution. Alkali solution was prepared using 8 molar NaOH solution and Na_2SiO_3 solution with Na_2SiO_3/NaOH of 1.5. The use of limestone powder has favoured the quick UCS gain on 3 days of room temperature curing. Fiber reinforcement has shown a significant influence on flexural strength and fatigue life improvement. Areca fibers reinforcement has resulted in enormous resistance to plunger penetration during the unsoaked CBR test. However, on further 4 days of soaking, samples lost the bonding and exhibited low CBR. The SEM images showed the compact microstructure of the set mix. The formation of cementitious products is evident from the XRD micrograms due to the dissolution of silica, alumina, calcium, and other compounds by the alkali solution. When subjected to wetting–drying and freezing–thawing durability tests, the set mixes were failed due to leaching of mineral constituents and further breaking of soil structure. Even though stabilized specimens exhibited significant strength improvement in dry conditions, they are unsuitable in wet conditions.

Keywords Alkali activation · Black cotton soil · Class F fly ash · Limestone powder · Areca fiber

1 Introduction

Black cotton soils are characterized by swell-shrink behaviour due to clay minerals like montmorillonite, kaolinite, illite, etc.; hence, they are considered a poor foundation material. Expansive soils will shrink during the draught due to moisture loss; thus, large cracks will form and propagate to the different pavement layers.

B. A. Chethan (✉) · A. U. Ravi Shankar
National Institute of Technology Karnataka, Surathkal, India

© The Author(s), under exclusive license to Springer Nature Switzerland AG 2022

H. R. Pasindu et al. (eds.), *Road and Airfield Pavement Technology*, Lecture Notes in Civil Engineering 193, https://doi.org/10.1007/978-3-030-87379-0_24

In contrast, due to floods during the rainy season, the material wets and loses its shear strength substantially [1]. The moisture imbalance leads to its high volumetric instability, causing severe foundation problems [2]. Various treatment methods were adopted to improve the weak soil properties, like preloading, replacing with good quality earth material, stone columns, stabilization with cementitious additives using industrial byproducts, etc. [3]. Cement is found to be the most common binder adopted [4]. Alkali activated binders were investigated nowadays due to their environmental friendliness. The alkali-activated binders reduce the CO_2 emission by utilizing various industrial byproducts, which otherwise lead to disposal problems [5]. The addition of lime has gained importance for soil stabilization, and for every 10% clay, it was suggested to use 1% lime [6]. With shallow mixing of highly compressible clay using 5% lime, the swelling was reduced by 99.8% [7]. Fine-grained soil (CL) treated with 5% lime showed excellent load-bearing capacity due to the formation of a flocculated and reticulated well-developed cemented crystalline structure [8].

Extensive ongoing research has shown that fly ash has gained importance as a soil stabilizer [9]. Precursor class F fly ash, a rich aluminosilicate material mixed with calcium-rich compounds, can become a cementitious binder with a suitable alkali activator [10, 11]. Waste materials containing silica, alumina, and ferrous oxide can be effectively utilized to form cementitious materials. A compact microstructure, good volume stability, higher mechanical strength, and low shrinkage were observed for alkali-activated clayey soil using NaOH and Na_2SiO_3 solutions [12]. The combination of NaOH and Na_2SiO_3 solutions can act as the best activator [13–16]. The NaOH solution dissolves Na_2SiO_3 (which provides Na^+ and Si^{4+}) and aluminosilicates during polymerization reaction [17]. An alkaline solution prepared with 8 molar NaOH solution concentration and Na_2SiO_3 solution, even though increased the UCS of loose sands, the leaching of the fluid motor at the wet side of OMC resulted in strength loss due to diminished reaction [3]. It was recommended not to use SS/SH ratio >1.5, as it may result in strength reduction of geopolymer treated black cotton soil [18].

Stabilization using natural fiber can be more economical due to its abundance [19–21]. The use of fiber reinforcement in the soil can enhance tensile strength. Coir fiber reinforced soil can exhibit better resilience when used for subgrades [19]. As the natural fibers are biodegradable, different chemical treatments were adopted in order to increase their lifespan when used for reinforcing the soil [22, 23]. Linseed oil treatment effectively improved the interface friction and tensile strength of the treated fibers [24]. The use of arecanut coir of length 20–30 mm up to 0.5% increased OMC, UCS, CBR, and reduced MDD of lateritic soil, and at higher dosages, distribution of fibers becomes difficult [25].

Aim of the investigation: Black cotton soil taken for the study has low strength and is highly susceptible to volume change due to moisture variation. A huge quantity of class F fly ash ends up in landfills. Limestone powder was used to increase the CaO content of the mix. In order to reduce shrinkage cracks on drying, natural areca fibers were used. Conventional activators NaOH and Na_2SiO_3 were blended to prepare alkali solution and used with precursors to evaluate stabilized soil performance concerning strength and durability criteria for use in pavements.

2 Materials and Experimental Program

2.1 *Black Cotton Soil*

Black cotton soil spread over the Chikmaglur region of Karnataka, India, is unsuitable as a foundation material for constructing any structure. A major portion of this soil has silt size particles in the IS sieve range of 75–2 μ. As per ISCS, it is classified as silty soil of high compressibility (MH) by considering liquid limit and plasticity index (Table 1). This soil has considerable swell potential and exhibits low strength when it comes in contact with water. The chemical properties of black cotton soil are tabulated in Table 2.

Table 1 Physical properties of black cotton soil

Test properties		Unit	Results
Specific gravity, Gs [26]		–	2.56
Modified Proctor compaction [27]	γ_d	kN/m^3	18.1
	OMC	%	16.0
Consistency limits [28]	Liquid limit	%	60
	Plastic limit	%	33
	Plasticity index	%	27
Shrinkage limit [29]		%	23
Grain size distribution [30]	Gravel	%	2
	Sand	%	26
	Silt	%	68
	Clay	%	4
Free swell index [31]		%	58
Organic content [32]		%	2.25
Indian Standard Soil Classification System [32]		–	MH

Table 2 Chemical properties of the materials

Test properties	Unit	Black cotton soil	Class F fly ash	Limestone powder
pH	–	8.24	10.7	11.4
SiO_2	%	75.4	70.5	6.6
Al_2O_3	%	7.06	10.98	0.15
Fe_2O_3	%	2.64	1.84	0.05
CaO	%	0.007	0.003	72.1
SO_3	%	0.16	0.15	0.65
LOI	%	2.30	1.70	42.9

2.2 Fly Ash

Fly ash collected from M/s Udupi Thermal Corporation Ltd. was found to be rich in silica and alumina (Table 2). The total percentage of silica, alumina, and iron oxides is >70%. It has a low CaO percentage (<18%). As per ASTM C618-19 [33], fly ash satisfies class F requirements. For the black cotton soil, the precursors with higher CaO content were found to be beneficial. However, due to the production of a large quantity of low calcium fly ash creating disposal problems, it was utilized in the investigation considering the environmental benefit.

2.3 Limestone Powder

In order to increase the calcium content in the mix, crystalline whitish limestone powder was used. It is mainly composed of calcite (93.85% $CaCO_3$) and other minerals like silica, aluminium, iron, magnesium, etc. The collected limestone powder has rich calcium content that may favour the development of calcium-based hydration products in an alkaline environment. A 5% limestone powder dosage was found to be suitable depending on the literature, as it considerably increases the mix pH. The chemical properties of limestone powder are tabulated in Table 2.

2.4 Areca Fiber

Areca fibers are natural, hence, considered environment friendly. Arecanut husk consists of fibers of various diameters. Fibers of smaller diameters break easily under the application of tensile force of smaller magnitude. Therefore, only strong fibers of an average diameter of 0.35 mm were extracted from the mature arecanut shells. These fibers were subjected to linseed oil treatment to help in the improvement of surface friction. Such treatment can provide enhanced bonding at the interface of stabilized soil and fiber. An optimized dosage of 0.5% fibers with 25 mm length is obtained from various trail mixes. A low fiber dosage of 0.25% was not significant in strength improvement. Whereas, at higher fiber dosages of 0.75 and 1%, the proper distribution of fibers becomes difficult. Also, at these higher dosages, the grouping, overlapping of fibers can not be avoided. But, adequate and homogeneous distribution is possible by carefully spreading soil-fiber mix in layers at optimized dosage. Due to this, compact packing of the mix is observed, contributing to a significant improvement in strength.

2.5 *Alkali Solution*

An integrated analysis of strength and cost was made by Rios et al. to evaluate the best mixture, shown that a solution of 7.5 molar concentration gives better strength at a low cost. The lower concentration of 5 molars gives low strength, and the higher concentrations of 10 and 12.5 molar increase the cost. A concentration >10 molar produces high pH, delaying polymerization and reducing initial strength [34]. Therefore, a 8 molar NaOH solution blended with Na_2SiO_3 solution to achieve a Na_2SiO_3/NaOH (or SS/SH) ratio of 1.5 is considered suitable for treatment. 1 L of 8 molar NaOH solution was prepared by dissolving 320 gm of sodium hydroxide flakes of 97% purity in distilled water. Due to the exothermic reaction of NaOH flakes with water, enormous heat was produced. Therefore, the NaOH solution is allowed to cool down. 3-parts of Na_2SiO_3 solution, which is available in the form of a viscous liquid, was mixed with 2-parts of NaOH solution to achieve the 1.5 SS/SH ratio. The prepared alkali solution was leftover for 24 h for proper dissolution of minerals contents.

2.6 *Sample Preparation and Curing*

Fiber treatment: Commercial linseed oil was used for pretreating areca fibers. Initially, the arecanut shells were immersed in water for 15 days. Linseed oil of quantity equal to 0.6 times the dry weight of fibers was added and mixed homogeneously. It was allowed to dry for 30 min at room temperature of 29 ± 1 °C. Then fibers were oven-dried for 2 h at 120 °C. This process gives a rougher morphology to the fiber surface, which improves the interaction between fiber surface and stabilized soil. Also, it helps to minimize the decomposition under seasonal variations when used for soil reinforcement, as found in earlier studies [24].

Mix preparation: The oven-dried (at 105–115 °C for 24 h) soil sample is pulverized. The soil passing through 4.75 mm IS sieve is initially blended with limestone powder and treated areca fibers on a mass basis. This mix is blended with water and compacted in the mould using a 4.5 kg rammer by 25 blows/layer in 5 layers for the compaction test. This process induces compaction energy of 2682 kJ/m^3 to the soil. A calculated quantity of alkali solution equivalent to OMC was added for alkali-activated mix preparation and mixed until a homogeneous mixture is achieved. The mix is visually inspected to avoid unblended lump formations and to ensure homogeneity. The moisture content of each mix is verified using the oven drying method. This mix is immediately used for moulding of samples at modified Proctor density.

UCS test: UCS samples of 38 mm diameter and 76 mm height were compacted in a stainless steel mould using a hydraulic plunger. After extraction weight and dimensions of these samples were measured and kept for curing at room temperature for 3, 7, and 28 days. Cured samples were loaded at a rate of 1.2 mm/min until failure. UCS is calculated as follows:

$$\text{UCS} = \text{Flailure load/corrected area}$$

Average UCS showing the variation within ±3% obtained from 3 specimens are represented.

Flexure test: This test performed as per ASTM D1635/D1635M-19 [35]. Prism specimens of 76 by 76 by 290 mm dimensions were cast in a beam mould to achieve the target density. The samples were extracted immediately, their weight and dimensions measured, and cured for 3, 7, and 28 days at room temperature. The third point loading method was used for determining the flexure strength of specimens. Bearing blocks shall ensure that forces applied to the beam will be vertical only without any eccentricity. While testing, the specimens were turned on their side, considering their moulded position, and centered on the lower half-round steel supports, which were placed apart a distance of three times the beam depth. Average flexural strength showing the variation within ±3% obtained from 5 specimens are represented.

Flexural strength (R) is calculated as follows:

If the fracture occurs in the middle third of span length, $\text{R} = l * (P + 0.75\ W)b * d^2$.

If the fracture occurs outside of the middle span, not more than 5% of the span length, $\text{R} = 3 * p * ab * d^2$.

where,

R = flexural strength (modulus of rupture) in kg/cm^2

l = length of span in cm

W = weight of the beam after curing in kg

P = maximum applied load in kg

b = width of the beam in cm

d = depth of the beam in cm

a = distance between the line of fracture and nearest support measured along the centerline of the bottom surface of the beams in cm.

CBR test: CBR specimens cured at room temperature for 28 days were tested for plunger penetration resistance. Further, on 4 days of soaking, CBR depletion was evaluated. All samples were cured at room temperature (35–40 °C). A strain rate of 1.2 mm/min was adopted for the test [36]. After the specified curing period, the change in weight and dimensions was noted to evaluate the achieved density and visually inspected for the signs of shrinkage cracks. The average CBR of specimens is calculated as,

CBR = Load sustained by stabilized soil at 2.5 mm penetration/load sustained by standard aggregates at 2.5 mm penetration.

Permissible variation in CBR values are ±1%, ±2%, ±5% for CBR test values of <5%, 5–10%, >31% obtained from 3 tests on each mix.

Durability tests: Seven days cured UCS samples are used for durability tests. In the wetting–drying test [37], two sets of specimens were submerged in water for 5 h. Further, they were transferred to an oven and kept for 42 h at a temperature of 71 ± 3 °C to complete one test cycle. In the freezing–thawing test [38], two sets of specimens were frozen for 24 h at <−23 °C. Further, they were thawed in a chamber for 23 h at 23 ± 2 °C to complete one test cycle. One set of specimens were unbrushed. Another set of specimens were brushed after each drying and thawing cycle. The diameter, height, and weight of specimens were recorded after curing and completing each durability cycle. Tests were continued until failure or completion of 12 durability cycles.

Fatigue test: In order to evaluate the fatigue phenomenon, repeated loads tests were conducted on the stabilized specimens using a fatigue machine. UCS samples (38 mm diameter and 76 mm height) cured for 28 days were used for testing. The repeated loading is applied at a frequency of 1 Hz and a rest period of 0.1 s on the unconfined sample at room temperature (35–40 °C). The applied stress is taken as a fraction of 28 days of unconfined compressive strength of the sample.

where, Fraction of load stress = Applied stress/28 days UCS.

Chemical and microstructural analysis: Powder samples of various materials and stabilized specimens are used for chemical analysis (using titration methods as per IS codes), SEM, and XRD analysis.

3 Results and Analysis

3.1 Compaction

When the alkali solution is added to the mix, the quick setting process initiates, making it difficult to continue the compaction process. The high alkali solution concentration on reacting with precursors releases more heat during the mixing. Also, as observed from trial mixes, compaction using alkali solution and water has not altered compaction parameters. Therefore, for obtaining modified compaction parameters, all the ingredients were compacted using water. It is observed that at 10% precursor dosage (5% fly ash +5% limestone powder calculated on a mass basis), compact packing of the mix was achieved due to the filling up of pore spaces. Beyond 10% precursor dosage, the soil particles were replaced by fly ash and limestone powder, which reduced the density of mixes. Dropdown in MDD may be related to the increased amount of low specific gravity of fly ash usage. The fly ash and limestone powder include finer particles, and due to their higher specific surface area, there was an increase in the OMC values at higher precursor dosages. Also, the water affinity of fibers leads to a slight increase in the OMC. Up to a dosage of 15% precursors, OMC was reduced due to lowered plasticity of the mix. The mix has become more workable with the increment in the precursor dosages (Fig. 1).

3.2 Unconfined Compressive Strength

The untreated BC soil has exhibited a UCS of 1.1 MPa at modified Proctor compaction. UCS values were significantly increased for the alkali-activated soil

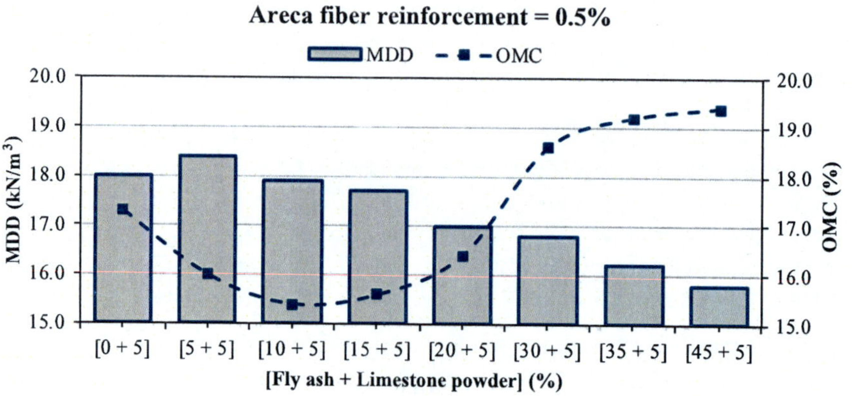

Fig. 1 Modified Proctor compaction characteristics of fly ash, limestone powder, and areca fiber added black cotton soil

with the addition of fly ash, cement, and areca fibers. A high dosage of fly ash is more significant in improving strength. Whereas, at low fly ash dosages, the UCS improvements were less due to the lesser dissolution of silica, alumina, and other constituents from the alkali reaction. Early age strength development of the mixes was excellent. In comparison, as the curing period was increased, a slight enhancement of strength was observed. This confirms the quick reaction of precursor materials in the alkaline environment. The plastic failure was observed for untreated black cotton soil. In contrast, the stabilized soil showed visible major cracks, which were strongly bonded by the areca fibers. As a result of tensile reinforcement, the crack propagation was delayed, and gradual splitting damage was observed. The stabilized specimens, due to the presence of tensile reinforcement, became more ductile. The sample with 45% fly ash and 5% limestone powder has achieved the highest UCS of 6.7 MPa on 28 days of curing (Fig. 2).

3.3 *Flexural Strength*

The flexural strength of the mixes was significantly improved with the incorporation of areca fiber reinforcement. Black cotton soil at modified Proctor density has shown a low flexural strength value of <0.02 MPa. Samples on 3 days of low curing duration has achieved considerable flexural strength for all combinations. The further curing period has helped to improve the strength slightly. The alkali solution could be able to react quickly to provide this early age strength. At 45% fly ash and 5% limestone powder, the flexural strength of 1.2 MPa was observed on 28 days of curing. Therefore, a higher dosage of fly ash with 5% limestone powder procured a higher amount of dissolvable silica, alumina, calcium, and other reactive elements for favouring the strength gain (Fig. 3).

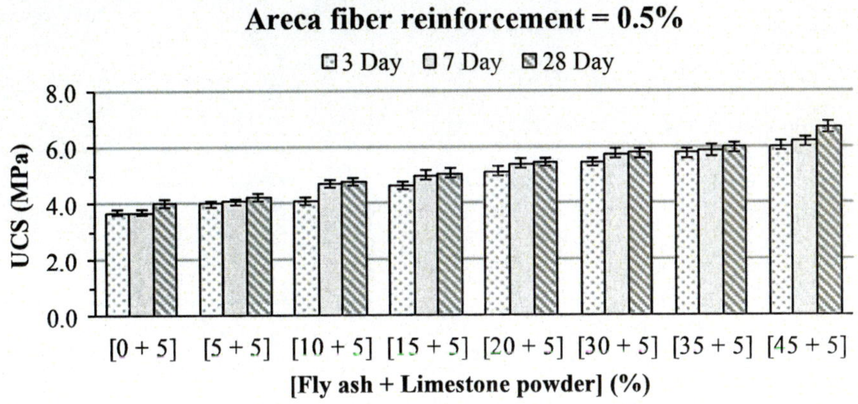

Fig. 2 Unconfined compressive strength variation of stabilized black cotton soil

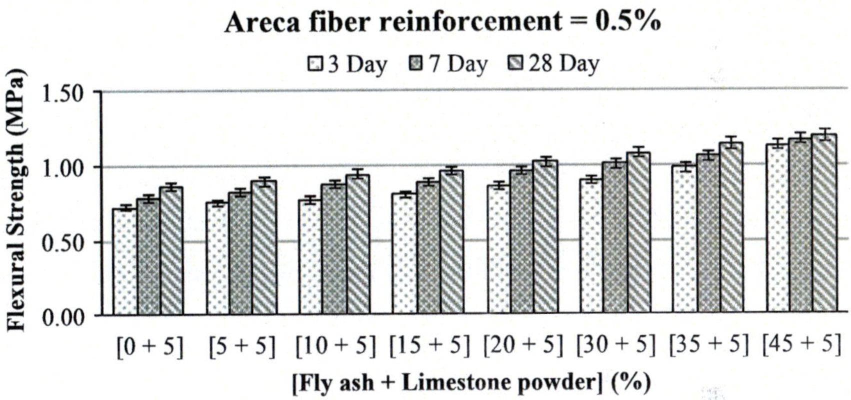

Fig. 3 Flexural strength variation of stabilized black cotton soil

3.4 California Bearing Ratio

The effect of increased dosage of fly ash was clearly reflected in the unsoaked CBR gain of stabilized samples. The natural areca fiber inclusion in the soil mixture enhanced the strength due to enhanced frictional resistance between the soil and treated fibers. The high tensile capacity fibers firmly held the soil mixture, and during plunger penetration, the set soil offered exorbitant resistance. On soaking due to absorption of water, the samples were softened, lost all their strength, and showed low resistance to the plunger penetration. Also, high expansion was observed due to the higher water affinity of the mixes. The slight improvement observed in soaked CBR values was due to the presence of low soil-fiber interaction (Fig. 4).

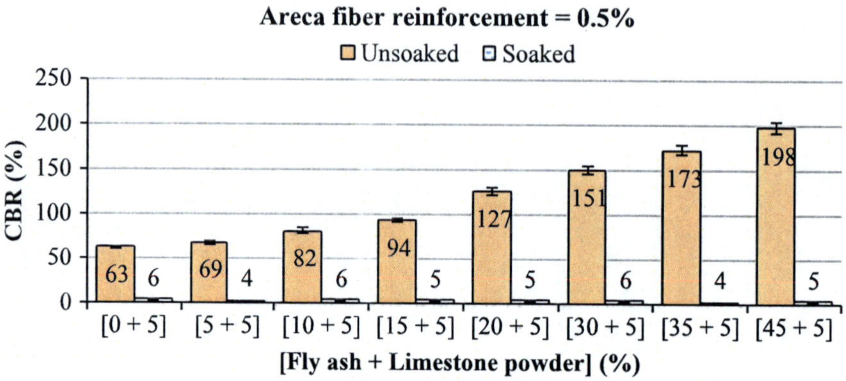

Fig. 4 California bearing ratio variation of stabilized black cotton soil

3.5 *Fatigue*

Due to the repeated application of wheel loads, the pavement layers will be continuously stressed. The repetition will also accumulate the strain in the pavement layer, and eventually, its lifespan will reduce. After a certain number of repeated load applications, the pavement layer may develop microcracks. These cracks will develop in length and width and convert to macrocracks. Such a cracked pavement layer will undergo catastrophic failure. In this investigation, a fraction of load calculated from 28 days UCS of each specimen was applied repeatedly until failure. The testing is conducted on different unconfined cylindrical specimens at different fractions of the load stresses of 0.5, 0.35, and 0.2 at room temperature. It is observed that the samples did not fail till 2×10^5 load repetitions at a fraction of load stress of 0.2. As the fraction of load stress was increased, the samples were failed at less number of repetitions of the load, as depicted in Fig. 5.

3.6 *Durability*

Cylindrical specimens of UCS cured at room temperature for 7 days were subjected to wet-dry (as per ASTM D 559) [37] and freeze–thaw (as per ASTM D 560) [38] durability tests. In the wet-dry test, the samples were soaked in water after the specified curing period. As the specimens came in contact with water, the bonding between the cemented particles was completely lost and collapsed within 5 min. The colour of the soaked liquid changed to blackish, which indicates the leaching of alkali solution with weak elements. When tested for freeze–thaw, samples did not exhibit any signs of failure during the first freezing cycle. During the following thaw cycle, samples absorbed water from felt pads and bulged. These samples could not resist brushing.

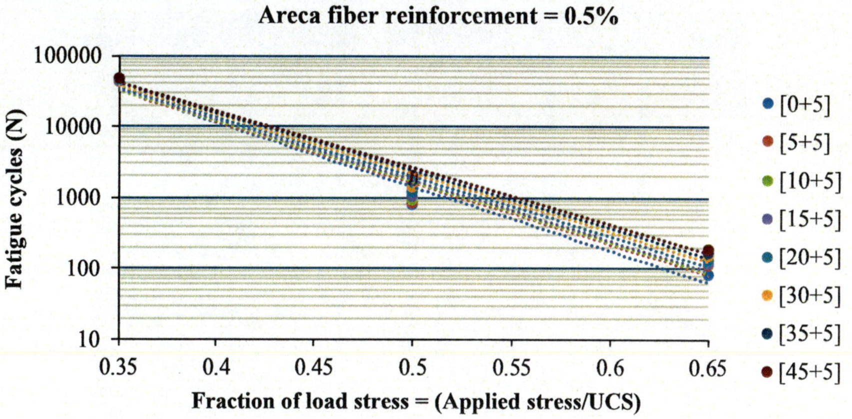

Fig. 5 Fatigue life of stabilized black cotton soil

Therefore, for all Na_2SiO_3/NaOH ratios and irrespective of the dosage of ingredients, the selected mix combinations could not resist the deterioration under the severe weather changes.

3.7 *SEM and XRD Analysis*

Bc soil without any treatment has developed many shrinkage cracks on drying (Fig. 6). These cracks were propagated during load application, thereby tending to fail the specimen at a low-stress value. The addition of fly ash and 5% limestone powder has resulted in a denser microstructure with no shrinkage cracks on curing. Limestone powder addition with fly ash results in the geopolymer formation of calcium-aluminosilicate gel in conjunction with amorphous aluminosilicate gel, which led to the change in the microstructure of composites. The calcium dissolution from limestone powder is responsible for both early and late age strength gain. Strength development and reaction rate are faster when fly ash is activated by an alkali activator of high concentration at ambient curing conditions. Due to the dissolution of calcium, the geopolymerization reaction continues for a longer duration. Areca fibers were strongly bonded in the alkali-activated set soil with precursors (Fig. 7). The interfacial friction between the soil and fiber surface was enhanced due to the deposition of hydrated products. During strength tests, the treated samples were failed at higher stress values due to the enhanced friction. The pullout of fibers was observed due to their high tensile capacity. The linseed oil treatment may be contributing to improving the roughness of the fiber surface. An enlarged view shows the deposits of hydrated gels on the fiber surface (Fig. 8). The alkali treatment could make the improved soil a compact mass with reduced void spaces. Also, shrinkage cracks were disappeared due to strong fiber bonding.

Fig. 6 Dry black cotton soil with shrinkage cracks

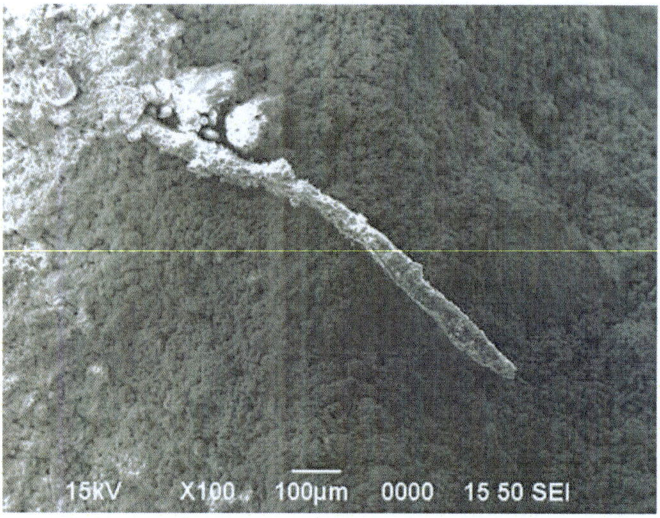

Fig. 7 Areca fiber firmly bound in alkali activated mix

Crystallographic information provided the formation of the following common products due to the alkali reaction during curing (Fig. 9). Gismondine ($CaAl_2Si_2O_8{\cdot}4H_2O$), Calcite ($CaCO_3$), Calcium Sulfate Hydrate ($CaSO_4{\cdot}0.15H_2O$), Aluminum Oxide (Al_2O_3), Millosevichite ($Al_2(SO_4)_3$), Gorgeyite ($K_2Ca_5(SO_4)_6{\cdot}H_2O$), Sodium Hydrogen Sulfate ($NaHSO_4$), Gypsum ($CaSO_4{\cdot}2H_2O$), Spurrite ($Ca_5(SiO_4)_2CO_3$), Xonotlite ($Ca_6Si_6O_{17}$ $(OH)_2$), Foshagite

Fig. 8 Enlarged view of hydration product deposition on the areca fiber surface

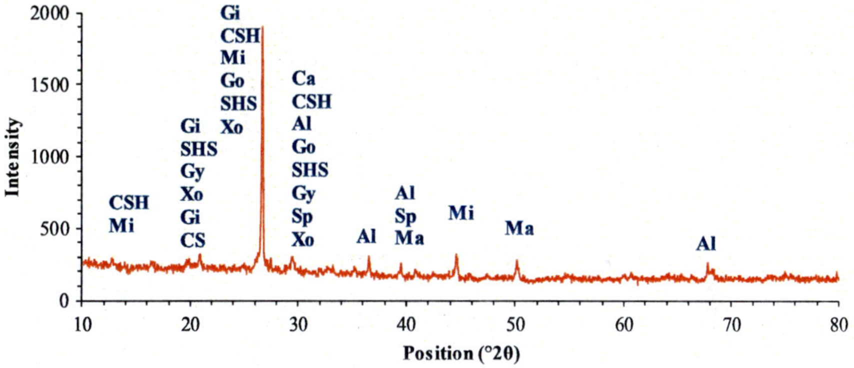

Fig. 9 XRD showing common hydration product formed in the alkali-activated soil with limestone powder and class F fly ash

($Ca_4(SiO_3)_3(OH)_2$), Magnesite ($MgCO_3$), Gibbsite ($Al(OH)_3$), and Calcium Sulfate ($CaSO_4$) are indicated in the XRD graph as per the maximum % of formation. However, these compounds are spread over different 2θ angles. It can be observed that these formations include dissolved silica, alumina, calcium, sulfate, potassium, etc., either obtained from alkali solution or from the soil and precursor materials. These hydration products are highly capable of retaining the bond between particles of the mix in the dry state.

4 Concluding Remarks

Following conclusions were drawn from the experimental studies conducted on the effectiveness of alkali activation for black cotton soil stabilization from the field application point of view.

1. The use of 5% limestone powder along with fly ash in the alkali-activated mix helps to gain early age (on 3 days) strength. On further curing, a slight improvement in strength was observed.
2. The fiber interaction with the soil enhances flexural strength and unsoaked CBR values significantly. Low soaked CBR values were observed due to mix softening and swelling due to absorption of water even in the presence of areca fibers.
3. The samples subjected to different weathering cycles are highly susceptible to degradation due to water sensitivity. The leaching of mineral constituents and breaking of soil structure leads to the nondurability of mixes.
4. The alkali activation of selected precursor materials is significantly effective in the development of high-strength mixes in a dry state. The hydration products avoid the formation of shrinkage cracks in the alkali-activated mix. The fiber soil boding is also enhanced by hydration product deposition on the fiber surface.

Strength, fatigue, and durability are the main parameters considered to suggest any stabilized material for pavement application. Investigations on stabilized black cotton soil have shown appreciable improvements in strength and fatigue characteristics. Whereas, when it comes to the field application point of view, the treated mixes are irresistible to the moisture and temperature variations. Therefore, these mixes can not be preferred for use in pavements.

References

1. Heitor A, Parkinson J, Kotzur T (2021) The Role of soil stabilisation in mitigating the impact of climate change in transport infrastructure with reference to wetting processes. Appl Sci 11(3):1080
2. Goodarzi AR, Akbari HR, Salimi M (2016) Enhanced stabilization of highly expansive clays by mixing cement and silica fume. Appl Clay Sci 132:675–684
3. Disfani MM, Mohammadinia A, Arulrajah A, Horpibulsuk S, Leong M (2021) Lightly stabilized loose sands with alkali-activated fly ash in deep mixing applications. Int J Geomech 21(3):4021011
4. Kurdowski W (2014) Cement and concrete chemistry. Springer Science & Business
5. Rakhimova NR, Rakhimov RZ (2019) Literature review of advances in materials used in development of alkali-activated mortars, concretes, and composites. J Mater Civ Eng 31(11):3119002
6. Ingles O (1987) Soil stabilization, Chapter 38, Ground engineer's reference book'. Butterworths, London, vol 38, pp 1–38
7. Hozatlıoğlu DT, Yılmaz I (2021) Shallow mixing and column performances of lime, fly ash and gypsum on the stabilization of swelling soils. Eng Geol 280:105931

8. Dash SK, Hussain M (2012) Lime stabilization of soils: reappraisal. J Mater Civ Eng 24(6):707–714
9. Rivera J, Coelho J, Silva R, Miranda T, Castro F, Cristelo N (2021) Compressed earth blocks stabilized with glass waste and fly ash activated with a recycled alkaline cleaning solution. J Clean Prod 284:124783
10. Siddiqua S, Barreto PNM (2018) Chemical stabilization of rammed earth using calcium carbide residue and fly ash. Constr Build Mater 169:364–371
11. Horpibulsuk S, Rachan R, Raksachon Y (2009) Role of fly ash on strength and microstructure development in blended cement stabilized silty clay. Soils Found 49(1):85–98
12. Zhang M, Guo H, El-Korchi T, Zhang G, Tao M (2013) Experimental feasibility study of geopolymer as the next-generation soil stabilizer. Constr Build Mater 47:1468–1478
13. Komnitsas K, Zaharaki D (2007) Geopolymerisation: a review and prospects for the minerals industry. Miner Eng 20(14):1261–1277
14. Ma C-K, Awang AZ, Omar W (2018) Structural and material performance of geopolymer concrete: a review. Constr Build Mater 186:90–102
15. Pacheco-Torgal F, Castro-Gomes J, Jalali S (2008) Alkali-activated binders: a review: Part 1. Historical background, terminology, reaction mechanisms and hydration products. Constr Build Mater 22(7):1305–1314
16. Singh B, Ishwarya G, Gupta M, Bhattacharyya SK (2015) Geopolymer concrete: a review of some recent developments. Constr Build Mater 85:78–90
17. Wang S, Xue Q, Zhu Y, Li G, Wu Z, Zhao K (2021) Experimental study on material ratio and strength performance of geopolymer-improved soil. Constr Build Mater 267:120469
18. Murmu AL, Jain A, Patel A (2019) Mechanical properties of alkali activated fly ash geopolymer stabilized expansive clay. KSCE J Civ Eng 23(9):3875–3888
19. Chauhan MS, Mittal S, Mohanty B (2008) Performance evaluation of silty sand subgrade reinforced with fly ash and fibre. Geotext Geomembr 26(5):429–435
20. Lekha BM, Goutham S, Shankar AUR (2015) Evaluation of lateritic soil stabilized with Arecanut coir for low volume pavements. Transp. Geotech. 2:20–29
21. Kumar A, Gupta D (2016) Behavior of cement-stabilized fiber-reinforced pond ash, rice husk ash–soil mixtures. Geotext Geomembr 44(3):466–474
22. Anggraini V, Asadi A, Farzadnia N, Jahangirian H, Huat BBK (2016) Reinforcement benefits of nanomodified coir fiber in lime-treated marine clay. J Mater Civ Eng 28(6):6016005
23. Sudhakaran SP, Sharma AK, Kolathayar S (2018) Soil stabilization using bottom ash and areca fiber: experimental investigations and reliability analysis. J Mater Civ Eng 30(8):4018169
24. Tan T, Huat BBK, Anggraini V, Shukla SK, Nahazanan H (2019) Strength behavior of fly ash-stabilized soil reinforced with coir fibers in Alkaline environment. J Nat Fibers pp 1–14
25. Das N, Singh SK (2019) Geotechnical behaviour of lateritic soil reinforced with brown waste and synthetic fibre. Int J Geotech Eng 13(3):287–297
26. IS: 2720: Part 3/Section-1 (2016) Methods of test for soils: determination of specific gravity—fine grained soils. Bur. Indian Stand. New Delhi
27. IS 2720: Part 8 (2015) Methods of test for soils: determination of water content-dry density relation using heavy compaction. Bur. Indian Stand. New Delhi
28. IS 2720: Part 5 (2015) Method of test for soils: determination of liquid and plastic limit. Bur. Indian Stand. New Delhi
29. IS 2720: Part 6 (2016) Methods of test for soils: determination of shrinkage factors. Bur. Indian Stand. New Delhi
30. IS 2720: Part 4 (2015) Methods of test for soils—grain size analysis. Bur. Indian Stand. New Delhi
31. IS 2720: Part 40 (2016) Methods of test for soils: Determination of free swell index of soils. Bur. Indian Stand. New Delhi
32. IS 1498: 1970 (2016) Classification and identification of soils for general engineering purposes. Bur. Indian Stand. New Delhi
33. ASTM C618-19 (2019) Standard specification for coal fly ash and raw or calcined natural pozzolan for use in concrete. ASTM Int. West Conshohocken, PA

34. Rios S, Ramos C, Viana da Fonseca A, Cruz N, Rodrigues C (2019) Mechanical and durability properties of a soil stabilised with an alkali-activated cement. Eur J Environ Civ Eng 23(2):245–267
35. ASTM D1635/D1635M-19 (2012) Standard test method for flexural strength of soil-cement using simple beam with third-point loading. ASTM Int. West Conshohocken, PA
36. IS 2720: Part 16 (2016) Methods of test for soil: laboratory determination of CBR. Bur. Indian Stand. New Delhi
37. ASTM D559/D559M-15 (2015) Standard test methods for wetting and drying compacted soil-cement mixtures. ASTM Int. West Conshohocken, PA
38. ASTM D560/D560M-16 (2016) Standard test methods for freezing and thawing compacted soil-cement mixtures. ASTM Int. West Conshohocken, PA

Clogging Resistance of High Strength Pervious Concrete

Ming-Gin Lee, Yung-Chih Wang, Wei-Chien Wang, E. A. Yatsenko, and Shou-Zjan Wu

Abstract This clogging resistant study includes high strength pervious concrete (HSP) and normal pervious concrete (NSP). The HSP concrete aims to carry out 28-day strength above 42 MPa and porosity as close to 10% as possible to achieve the technical specifications. Some of the key tests were porosity, compressive strength, and permeability test. Experimental simulation of sand and soil clogging tests of HSP pavement specimens were under evaluation. The test results indicate that the 28-day compressive strength of HSP higher than 42 MPa can be achieved with specific mix design and compaction energy by Proctor tamper, and their porosity ranged from 7.3 to 11.6%. The coefficient of permeability for HSP and NSP concretes meets Taiwan code requirements for structural concrete. The sand and soil clogging results show that all HSP specimens had better maintenance by vacuum sweeping and restored 60–75% of the overall permeability, while the high-pressure washing only improved the permeability by 4–6%. The reason is that the maximum water pressure of the test equipment is seriously insufficient. It is presumed that the pressure washing maintenance caused sand and soil to penetrate into the pores of the HSP specimen. Furthermore, a predictive model of the clogging potential of porous asphalt mixture has been found to fit well to HSP pavement.

Keywords Clogging · Pervious concrete · High strength · Vacuum sweeping · Permeability

M.-G. Lee (✉)
Chaoyang University of Technology, Taichung, Taiwan, Republic of China
e-mail: mglee@cyut.edu.tw

Y.-C. Wang · W.-C. Wang · S.-Z. Wu
National Central University, Taoyuan, Taiwan, Republic of China

E. A. Yatsenko
South Russian State Polytechnic University, Novocherkassk, Russia

© The Author(s), under exclusive license to Springer Nature Switzerland AG 2022

H. R. Pasindu et al. (eds.), *Road and Airfield Pavement Technology*, Lecture Notes in Civil Engineering 193, https://doi.org/10.1007/978-3-030-87379-0_25

1 Introduction

Permeable pavement provides a solution for architects or landscape designers who wish to use greenery in urban zones and parking lots [1]. The permeable concrete pavement possesses many advantages that improve urban and suburban environments as follows by the U.S. Environmental Protection Agency [1, 2]:

(1) Attenuation of runoff.
(2) Ground water recharge.
(3) Improved road safety because of better skid resistance.
(4) Retention of natural drainage patterns and vegetation.
(5) Decreasing need for curbing and storm sewers.
(6) Water treatment by pollutant removal.
(7) Improved erosion control.

Many plants have difficulty growing in areas covered by impermeable urban pavements, landscaping, and sidewalks, because both water and air have difficulty getting to the roots. Permeable pavement allows adjacent trees to receive more water and air into the roots and branches. Pervious concrete pavement is ideal for protecting trees in urban paved environments [3]. Pervious concrete is a special type of concrete with a high porosity used for concrete flatwork applications that allows water from precipitation and other sources to pass through it, thereby reducing the runoff from a site and recharging ground water levels. The void content can range from 18 to 35% with compressive strengths of 400–4000 psi (2.8–28 MPa) and the drainage rate of pervious concrete pavement will fall into the range of 2–18 gal./min/ft^2 (81–730 L/min/m^2) on pervious concrete reported by ACI Committee 522 [2, 4]. Pervious concrete traditionally has little to no fine aggregate and has just enough cement paste to coat the coarse aggregate particles while maintaining the interconnectivity of the voids [5]. Pervious concrete is typically used in parking areas, areas with light traffic, residential streets, pedestrian walkways, and greenhouses. Pervious Concrete is an important application for sustainable development and construction [5, 6]. The results of the pervious concrete study by Lee et al. [7] on parking lot in Taiwan composed of 8-in surface layer (pervious concrete) and a 4-in base layer (gravel) showed that the compression strength of the surface layer exceeds the ordinary concrete structure specification. The in site permeability test for the parking lot is about 1000 ml/15 s. The water penetration of the above pervious concrete is very good and no cracks were present on surface for over ten years.

Pervious concrete pavements have also been selected as an integral solution to the problem of hot pavements in the Cool Communities program. The hot pavements bring the phenomenon of heat island effects in the big city [8, 9]. Generally, the air temperature over pervious concrete parking lots is cooler than the temperature over asphalt parking lots. Pervious concrete parking lots also reduce snow and ice build-up. In addition, pervious concrete is considered a non-pollutant to the environment [4, 9–11]. Atmospheric deposition or suspended solids, heavy metals, and hydrocarbons are distributed over pavement surfaces by automobile exhaust and crankcase leakages

[8]. Motor vehicle pollutants accumulate on road surfaces until a precipitation event carries these pollutants away in runoff that finally enters waterways or groundwater [9, 10]. The above pollutants can potentially damage sensitive ecosystems in nearby bodies of water or enter the water supply through wells and surface water withdrawals, which can be harmful to human health. Various water quality studies of permeable concrete pavement systems report pollutant removal from water runoff [12].

The study tried to produce a high strength pervious concrete and it aims to carry out 28-day strength above 42 MPa and porosity as close to 10% as possible to achieve the technical specifications. The research tried to produce a high strength pervious concrete and use carbon dioxide for curing to find out whether it could enhance the compressive strength. The results show that the compressive strength of the control group increases rapidly and its 90-day compressive strength closed to 60 MPa [13]. However, the small pore structure of high strength pervious concrete might become clogged by sediment particles and its ability to drain storm-water runoff gradually decreases presented a recent review of this topic [14–16].

This high strength clog-resistant pervious concrete (CRP) was prepared by introducing straight pore channels of varying size and number into self-compacting mortar. The CRP can be engineered with high strength (>50 MPa), low porosity and high permeability, but does not clog despite extensive cyclic exposure to flow containing sand and clay [14]. The pervious concrete has merits in water retention, cooling, improving drainage and water purification, however, with the increase in traffic and age, the pervious pavement will gradually reduce its drainage function after the pores clogged, which is one of the major problems.

2 Materials and Methods

2.1 *Pervious Concrete Material and Concrete Mix*

The pervious concrete used in this study was type Π Portland cement conforming to ASTM C 150 which was manufactured by the Asia cement corporation. The coarse aggregate used was a commonly-used crushed stone with maximum diameter of 6 mm and fine aggregates are nonexistent in pervious concrete mixture. The specific gravity of superplasticizer (SP) is 1.05 and its solid content is 30%. The above materials including cement, crushed gravel, SP, and water were used to make the normal strength pervious concrete (NSP) mixture with zero-inch slump. High strength pervious concrete (HSP) mixtures with silica fume and SP have zero inch slump and water to binder ratio less than 0.2. Table 1 show three kinds of pervious concrete mixture designs used in this study. The first mix (NSP) is normal strength pervious concrete mixture. The second one (HSP1) and the third one (HSP1.2) are high strength pervious concrete mixture with different aggregate size of 3.0 and 3.6 mm individually. The HSP aims to carry out 28-day compressive strength above 42 MPa and porosity as close to 10% as possible to achieve the technical specifications.

Table 1 Mix proportions of normal and high strength pervious concrete

Mix	Target porosity (%)	Batch weight (kg/m^3)					A/C	F/A
		C	W	A	SF	SP		
NSP	20	340	100	1530	0	2	4.5	0
HSP1	10	475	68	1484	119	11	3.1	0
HSP1.2	10	475	68	1484	119	11	3.1	0

C—cement, W—water, A—coarse aggregate, F—fine aggregate, SF—silicate fume, SP—superplasticizer, W/B—water-binder ratio

2.2 Clogging Treatment

The porous pavement has merits in water retention, cooling, and improving drainage, however, with the increase in falling dust, sand, waste soil, or traffic and age, the porous pavement will gradually reduce its drainage function after the pores clogged, which is one of the main problems [12]. The major source of clogging materials on pervious concrete pavements has been the sand and clay deposited from vehicles carrying waste soils or construction materials. The study would simulate clogging effect induced by falling sand and soil on high strength pervious concrete pavement and explore the changes of permeability after the cycles of clogging treatment by vacuum cleaner or high pressure washer.

The component of the waste soil with sizes smaller than 1.18 mm was employed as the clogging material in the clogging treatment developed by Fwa et al. [16]. At the beginning of each cycle, an amount of 35.2 g clogging material was spread uniformly on the top face of the test unit of 0.31 m × 0.37 m. The clogging treatment followed by filling water flow to bring the clogging material into the test specimen. The next cycle of clogging treatment followed immediately with the application of another 35.2 g of clogging material. The clogging treatment was continued until there were negligible changes in the measured permeability in successive cycles of the clogging treatment [16]. The clogging treatment procedure including vacuum cleaner and high pressure washer was shown schematically in Fig. 1. The HSP clogging treatment tests and variables used in the experiment are shown in Table 2. Three different thickness of HSP test unit were 10 cm, 15 cm and 20 cm respectively.

3 Results and Discussion

3.1 Mechanical Properties of Pervious Concrete

The water cement ratio, unit weight, compressive strength, porosity and water permeability test results of NSP and HSP concretes are shown in Table 3. The water permeability of NSP and HSP concretes can be measured by a simple falling head

(a) Clogging material addition (b) Water injection (c) Outlet water

(d) High pressure washing (e) Vacuum cleaning

Fig. 1 Schematic diagram of procedure for laboratory clogging tests

Table 2 HSP clogging test variables used in the experiment

Experimental content	Variables
Concrete thickness (cm)	10, 15, 20
Pervious concrete specimen	HSP1, HSP1.2

Table 3 Properties of normal and high strength pervious concrete

Specimen code	HSP1	HSP1.2	NSP	ACI 522
Water cement ratio	0.17	0.17	0.30	0.25–0.45
Unit weight (kg/cm^2)	2157	2157	1972	1600–2000
Compressive strength (MPa)	55.52	41.46	18.26	2.8–28
Coefficient of permeability (cm/s)	0.06	0.06	0.25	0.16–1.22
Porosity (%)	6.9	11.6	16.8	18–35

permeameter. The test results indicate that the 28-day compressive strength of HSP concretes higher than 42 MPa can be achieved with specific mix design, and their porosity ranged from 6.9 to 11.6%. The coefficient of water permeability for HSP and NSP concretes is 0.06 and 0.25 cm/s respectively, and the unit weight is about 2157 and 1972 kg/cm^2 respectively. The coefficient of permeability for HSP and NSP concretes meets Taiwan code requirements for structural concrete.

3.2 *Clogging Simulation Test on High-Strength Pervious Concrete*

Pervious concrete pavements have good drainage capabilities; however, the clogging is more prone to occur in permeable concrete specimens with small porosity during storm water runoff with the clogging materials. Figures 2, 3 and 4 display the change in water permeability with increasing number of clogging cycles subjected to the vacuum cleaner and high pressure washer treatment for all HSP specimens with 10 cm, 15 cm and 20 cm thickness respectively. The sand and soil clogging results show that all HSP specimens had better maintenance by vacuum sweeping and restored 60–75% of the overall permeability, while the high pressure washing only improved the permeability by 4–6%. Pressure washing of a clogged pervious concrete pavement can restore 80–90% of the permeability in some comparison cases, and it is found that the maximum water pressure of our test is only 10 MPa, which is normally approximately 20 MPa. For most of the cases, the vacuum sweeping is faster than but not as effective as pressuring washing. And the highest permeability recovery was achieved by using vacuum sweeping followed by pressure washing [17]. These results show that the water permeability deterioration trend of a HSP concrete caused by clogging simulation test could be very closely represented by a decreasing exponential function of clogging cycles with very high coefficient of

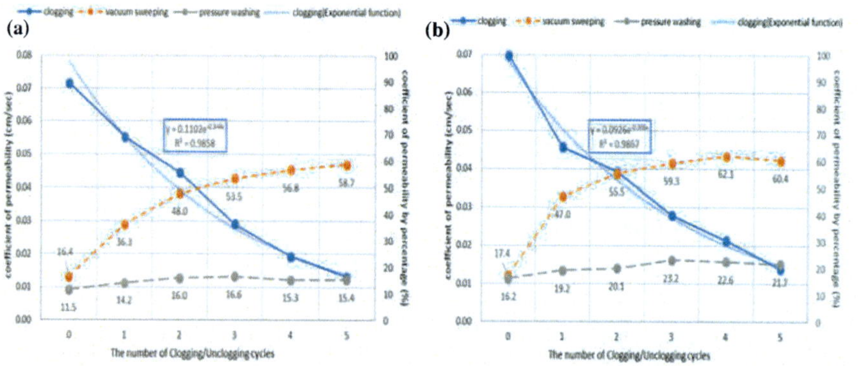

Fig. 2 Relationship between permeability and clogging cycle **a** 10 cm HSP1.2 specimen and **b** 10 cm HSP1 specimen

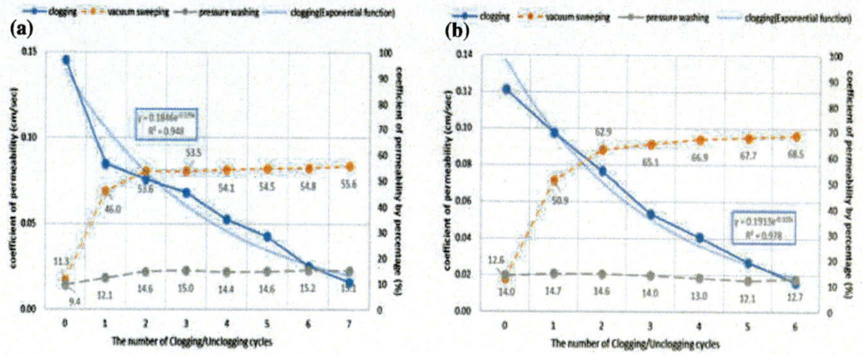

Fig. 3 Relationship between permeability and clogging cycle **a** 15 cm HSP1.2 specimen and **b** 15 cm HSP1 specimen

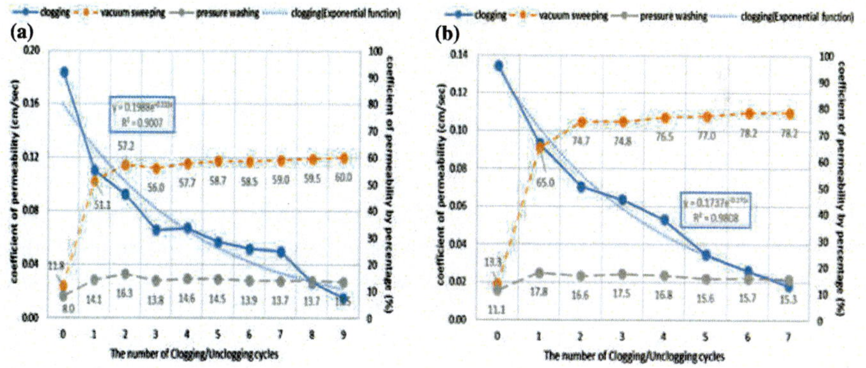

Fig. 4 Relationship between permeability and clogging cycle **a** 20 cm HSP1.2 specimen and **b** 20 cm HSP1 specimen

determination R^2 values (about 0.96) obtained from regression analyses. Table 4 lists the measured values of P_0, N_C, k_0, D_T, β and the number of clogging cycles (N_C) to reach a fully clogged state is greater than 5.

Table 4 Results of HSP clogging and permeability test

Variable/factor	β	Nc	k	D_T	P
HSP1.2–10 cm	0.344	5	0.071	3.636	10.1
HSP1–10 cm	0.305	5	0.070	3.03	9.6
HSP1.2–15 cm	0.279	7	0.145	3.636	9.2
HSP1–15 cm	0.333	6	0.121	3.03	11.1
HSP1.2–20 cm	0.222	9	0.184	3.636	11.6
HSP1–20 cm	0.271	7	0.123	3.03	13.7

Table 5 Regression models of HSP clogging induced permeability deterioration [16]

Specimen code	Regression model	R^2
HSP1.2 (10 cm)	$K_N = 0.110e^{-0.344N}$	0.986
HSP1 (10 cm)	$K_N = 0.093e^{-0.305N}$	0.987
HSP1.2 (15 cm)	$K_N = 0.185e^{-0.279N}$	0.948
HSP1 (15 cm)	$K_N = 0.191e^{-0.333N}$	0.978
HSP1.2 (20 cm)	$K_N = 0.199e^{-0.222N}$	0.900
HSP1.2 (10 cm)	$K_N = 0.110e^{-0.344N}$	0.986
Inductive equation	$K_N = k_0e^{-\beta N}$	–

3.3 Modeling of Clogging Performance and Clogging Prediction

Table 5 shows these regression models of clogging induced permeability deterioration for all the HSP test units with 10 cm, 15 cm and 20 cm thickness respectively. The clogging simulation test could be very closely represented by a decreasing exponential function of clogging cycles with very high coefficient of determination R-square value 0.90–0.99 obtained from regression analyses. It can be seen that the permeability deterioration in clogging cycles might be described by the inductive equation ($K_N = k_0e^{-\beta N}$). Where β is a positive regression constant, N the number of clogging cycles, K_N the permeability coefficient after N clogging cycles, and k_0 the initial permeability coefficient before clogging treatment [16]. The measured values of P_0, N_C, k_0, D_T, and β are considered to be factors that could have an effect on the clogging potential of a HSP pavement. The relationships between clogging factor and six predictor variables are shown in Fig. 5. From Figs. 5a, b, e, f, it is found that their factors all have very low R-square value 0.07–0.19. It can be inferred that the initial porosity (P_0) and aggregate size (D_T) have low effect on the logging potential and number of clogging cycle. According to Fig. 5c, d, their factors are found to have moderate and very strength respectively. The predictive models for the two clogging potential indicators established, namely the clogging potential, β and the number of clogging cycles to reach a fully clogged state, N_C. The predictive regression models obtained are [16]:

$$\beta = -0.004P_0 - 0.766k_0 - 0.006D_T + 0.448, R^2 = 0.673 \quad (1)$$

$$Nc = 0.193P_0 + 2.9452k_0 + 0.624D_T - 1.183, R^2 = 0.962 \quad (2)$$

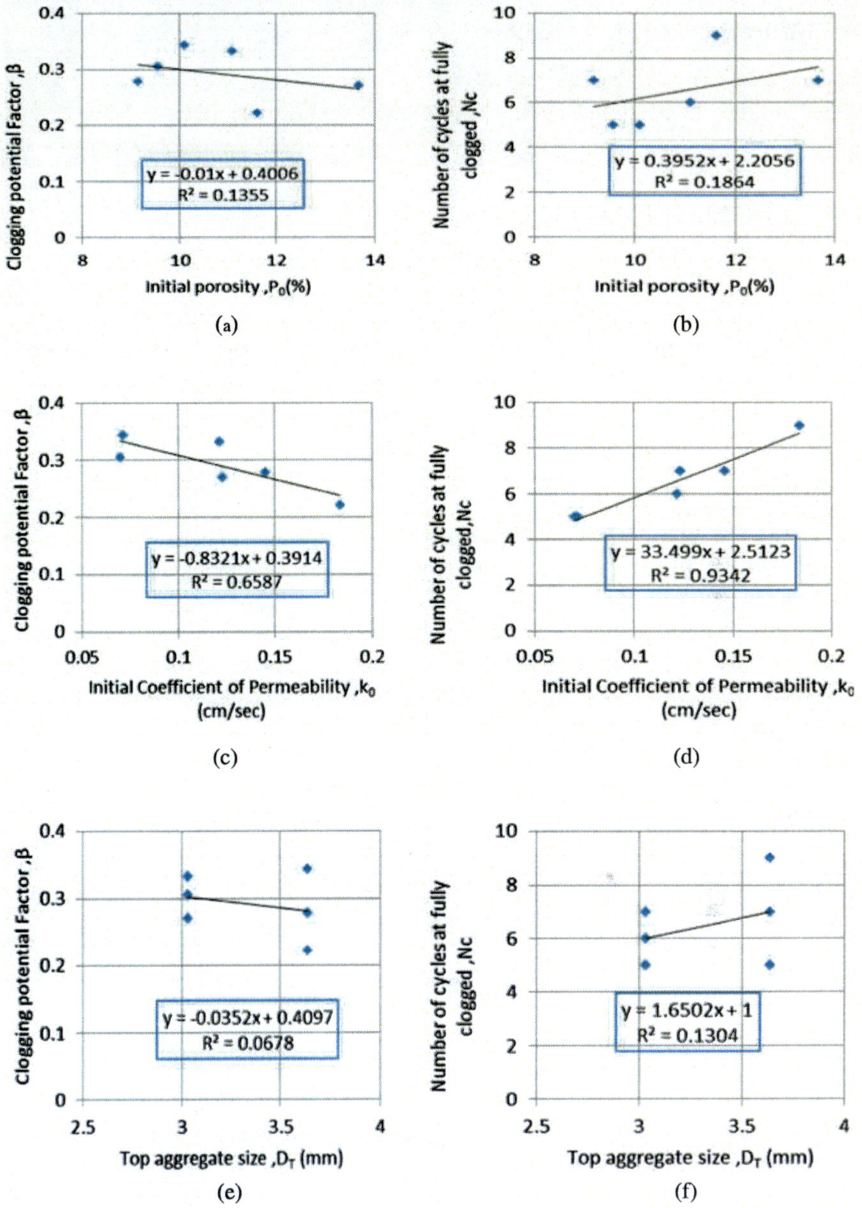

Fig. 5 Relationships between clogging factor and six predictor variables **a** β versus P_0 plot, **b** Nc versus P_0 plot, **c** β versus k_0 plot, **d** Nc versus k_0 plot, **e** β versus D_T plot and **f** Nc versus D_T plot

4 Summary

The main findings from this study are summarized as follows:

1. The results indicate that the 28-day compressive strength of HSP concretes higher than 42 MPa can be achieved with specific mix design, and their porosity ranged from 6.9 to 11.6%.
2. The clogging simulation test results show that high strength pervious concrete specimens had better maintenance by vacuum sweeping and restored 60–75% of the overall permeability, while the high pressure washing only improved the permeability by 4–6%. Pressure washing of a clogged pervious concrete pavement can restore 80–90% of the permeability in some comparison cases, and it is found that the maximum water pressure of our test is only 10 MPa, which is normally approximately 20 MPa.
3. The permeability deterioration in clogging cycles might be described by the inductive equation ($K_N = k_0e^{-\beta N}$) with very high coefficient of determination R^2 values (about 0.96) obtained from regression analyses. The results indicate that the number of clogging cycles (NC) to reach a fully clogged state is greater than 5.
4. The in-site permeability test for asphalt pavements has been found to work and perform well to high strength pervious concrete pavement. The test might be used before and after clogging treatment to quantify the effectiveness of a field permeability technique.

Acknowledgements This study was supported by the Ministry of Transportation and Communications for the financial support under the grant MOST-107-2221-E-324-010-MY2.

References

1. Lee MJ, Lee MG, Huang Y, Chiang CL (2013) Purification study of pervious concrete pavement. Int J Eng Technol 5(5):532–535
2. Tennis P, Leming ML, Akers DJ (2004) Pervious concrete pavements. EB 302, Portland Cement Association, Skokie, Illinois, pp 1–25
3. Yang J, Jiang G (2003) Experimental study on properties of pervious concrete pavement materials. Cem Concr Res 33(6):381–386
4. ACI 522R-06 (2006) Pervious Concrete American Concrete Institute. Farmington Hills, MI, pp 1–25
5. National Ready Mixed Concrete Association (2004) Freeze Thaw Resistance of Pervious Cocrete. Silver Spring, MD, pp 1–17
6. Obla K (2007) Pervious concrete for sustainable development. In: Proceedings, 1st international conference on recent advances in concrete technology, Washington DC
7. Lee MG, Chiu CT, Kan Y, Yen T (2009) Experimental study of pervious concrete on parking Lot. ASCE Geotech Spec Publ 193:125–131
8. Park SB, Lee BJ, Lee J, Young Il J (2010) A study on the seawater purification characteristics of water-permeable concrete using recycled aggregate. Resour, Conserv Recycl 54(10):658–665

9. Thomle JN (2010) The declining pH of waters exposed to pervious concrete. Master thesis, Washington State University, Department of Civil and Environmental Engineering, pp 1–160
10. Newman AP, Pratt CJ, Cresswell N (2002) Oil bio-degradation in permeable pavements by microbial communities. Water Sci Technol 45(7):51–56
11. Lee MG, Chiu CT, Yen T (2008) Application of pervious concrete on transportation engineering (2/2). Ministry of Transportation and Communications, Taiwan, pp 1–188
12. Lee MG, Tia M, Chuang S, Huang Y, Chiang C (2014) Pollution and purification study of the pervious concrete pavement. ASCE J Mater Civ Eng 26(8). https://doi.org/10.1061/(ASCE)MT.1943-5533.0000916
13. Lee MG, Tia M, Chuang S, Huang Y, Chiang C (2020) Effect of CO_2 curing on the strength of high strength pervious concrete. Key Eng Mater 846:207–212
14. Kia A, Hong SW, Christopher RC (2019) High-strength clogging resistant permeable pavement. Int J Pavement Eng. https://doi.org/10.1080/10298436.2019.1600693
15. Coughlin JP, Campbell CD, Mays DC (2012) Infiltration and clogging by sand and clay in a pervious concrete pavement system. J Hydrol Eng 17(1):68–73. https://doi.org/10.1061/(ASCE)HE.1943-5584.0000424
16. Chu L, Fwa TF (2018) Laboratory characterization of clogging potential of porous asphalt mixtures. TRR J Transp Res Board 2672:12–22
17. Tong B (2011) Clogging effects of Portland cement pervious concrete. Master thesis, Iowa State University, Department of Civil Engineering, pp 57–61

Comparison of Flexible Airfield Pavement Designs Using FAARFIELD v1.42 and APSDS 5.0

Gary Chai, Phil Bell, Ken McNabb, Leigh Wardle, and Erwin Oh

Abstract A case study has been carried out to compare the design of airfield pavements for a major airport using FAARFIELD v1.32 and APSDS 5.0. For the aircraft departure data used in the study, the pavement analysis shows that the APSDS 5.0 design method yielded pavement structure thicknesses that are nearly the same as FAARFIELD v1.32 for CBR greater than 10%. An adjustment factor k_c is required for APSDS design thickness to produce designs that are consistent with FAARFIELD for CBR less than 10%. In May 2017, FAA developed new subgrade failure models for flexible pavements in FAARFIELD v1.41 using the full-scale traffic test data collected at the NAPTF for Test Sections in Construction Cycles CC3 and CC5. FAARFIELD v1.41 was subsequently updated and evolved to version 1.42 in September 2017. In this paper, a comparison is carried out using the latest version of FAARFIELD v1.42 to examine if the new subgrade deformation models compute the design thicknesses that are compatible with that generated by APSDS 5. For the Boeing 737-800 (Code C) and 777-300ER (Code E) aircrafts spectrum and 100,000 movements analyzed in the study, the new subgrade failure models developed for the latest version of FAARFIELD generate the flexible pavement thicknesses that are not significantly difference from that of APSDS 5.0 for subgrade CBR $\geq$5%. The new failure model in FAARFIELD v1.42 produces flexible design thicknesses that differ less from APSDS 5.0 than FAARFIELD v1.32. The design thicknesses are more consistent for B737-800 with 2 wheels configuration. However, the differences are observed to be larger for CBR $\leq$5% when modelled with B777-300ER having 6 wheels configuration. The differences in the design thickness are attributed to the different coefficients adopted in the subgrade failure models in the design software.

G. Chai (✉) · E. Oh
Griffith University, Gold Coast, QLD, Australia
e-mail: g.chai@griffith.edu.au

G. Chai · P. Bell
Airport Consultancy Group, Gold Coast, QLD, Australia

K. McNabb · L. Wardle
Pavement Sciences, Richmond, VIC, Australia

© The Author(s), under exclusive license to Springer Nature Switzerland AG 2022

H. R. Pasindu et al. (eds.), *Road and Airfield Pavement Technology*, Lecture Notes in Civil Engineering 193, https://doi.org/10.1007/978-3-030-87379-0_26

Keywords Airfield pavement design · Subgrade failure models · APSDS 5.0 · FAARFIELD v1.42

1 Introduction

Road and airfield flexible pavement design methods are similar in that load-induced strains are estimated using layered elastic methods. Mechanistic analysis is used to predict state of stress beneath the wheel load and empirically relate stresses to performance. The empirical aircraft pavement performance data is obtained from full-scale test pavements each of which has been loaded with an actual undercarriage of a particular aircraft. The major differences between airfield and road pavements are the magnitude of load and the number of load repetition on the structures. Airfield pavements are subjected to a much higher magnitude of tyre load than that of road pavements.

The new generation and larger airplanes resulted in higher gross maximum take-off weight (MTOW), tyre load and contact pressure. Boeing 777-9, the latest series of long-range, wide-body twin-engine jet aircraft has two six-wheel main landing gears to support a gross taxi weight of up to 352,442 kg and contact tyre pressure of 1503 kPa [1]. In Airbus 350 family of aircraft, the gross taxi weight of the A350-900 is 272,900 kg and with the tyre contact pressure of 1682 kPa. The 575,000 kg Airbus A380 (with tyre contact pressure of 1503 kPa) has two six-wheel body gears in addition to two four-wheel wing gears, for a total of 20 wheels in the main gear assembly. The complex gear loads applied to airport pavements by these new aircraft types are quite different from the loads applied by the older generation of commercial airplanes. Complex wheel load interactions within pavement structures contribute to premature failure of the pavement structures and must therefore be considered in pavement design analyses [2].

A case study was carried out to compare the design of airfield pavements for a major airport using FAARFIELD v1.32 and APSDS 5.0 [3]. For the aircraft departure data used in the study, the pavement analysis shows that the APSDS5.0 design method yielded pavement structure thicknesses that are nearly the same as FAARFIELD v1.32 for CBR greater than 10%. An adjustment factor k_c is required for APSDS design thickness in order to produce designs that are consistent with FAARFIELD for CBR less than 10%. In May 2017, FAA developed new subgrade failure models for flexible pavements in FAARFIED v1.41 using the full-scale traffic test data collected at the NAPTF for Test Sections in Construction Cycles CC3 and CC5 [4]. FAARFIELD was subsequently updated and evolved to version 1.42 in September 2017. In view of the development of the new subgrade failure models, it is necessary to compare the design thicknesses generated by FAARFIELD v1.42 and to examine if the thicknesses are compatible with that computed by APSDS 5. To meet the objective, Boeing 737-800 (Code C) and B777-300ER (Code E) aircrafts and 100,000 movements were analyzed at various subgrade CBR using the two software programs. The same wander characteristics with a standard deviation of 773 mm is used in the pavement analysis.

2 Literature Review

The design philosophy for airport pavements in Australia has been discussed by Emery [5, 6] and Rodway [7]. The use of surfacing is inter-related to the pavement design philosophy. The Australian approach is to design lower cost pavements, with lighter surfacing, thinner layers, thinner surfacing, and less capable materials. This is common with South African and New Zealand practice, and differs from USA practice. Seals are used instead asphalt where possible. Thin asphalt is used instead of thick asphalt, if possible. And thin asphalt is used instead of concrete. The design philosophy for lower cost pavements has been successful, and is a reflection of the relative benign climate, a willingness to stretch designs (reduce reliability), a high local capability for inspection and maintenance (to repair failures), and less intense trafficking [5]. The American Federal Aviation Administration (FAA) developed FAARFIELD and the Australian APSDS (Airport Pavement Structural Design System) are the two commonly used for airport pavement designs. FAARFIELD has various default parameters with regard to materials and thicknesses, and some caution is needed in using it in Australia [6].

Following their full-scale tests on stabilized and un-stabilized pavements in 1948, the US Army Corp for Engineers concluded that at elevated temperatures bituminous bound pavement layers were not superior in load distributing capability to excellent quality (CBR 100%) base materials. The Corps specifically stated that their tests indicated that 50 mm asphalt surfacing was adequate over a crushed rock basecourse for 90 tonne wheel loads and 1 MPa tyre pressures. That is, high quality crushed rock basecourses had sufficient shear strength to withstand the high stresses produced in the zone directly beneath aircraft wheels and did not require thick asphalt surface layers to protect them. The Corps early preference for heavily-compacted, proof-rolled unbound crushed rock basecourses and relatively thin asphalt surfacings continues. From 1946 Australia adopted the Corps' approach and continues today to build flexible aircraft pavements that consist of 60 mm of asphalt supported by heavily compacted, proof-rolled, unbound fine crushed rock basecourse and sub-bases [7].

The development of APSDS in Australia was from a flexible pavement design program, CIRCLY [8]. Based on layered elastic analysis, APSDS has two unique features. The first is that it computes subgrade strains for all points across the pavement to capture all damage contributed by all the aircraft wheels. This approach contrasts with other methods of pavement thickness design that often only compute a single (maximum) strain value. Using the pattern of strains computed allows the development of equations to relate load repetitions to pavement rut-depth by calibrating against full-scale test data. The second unique feature is that, in order to adequately reflect the test data, different calibration parameters are used for each wheel configuration.

APSDS also uses the Barker and Brabston [9] approach to model unbound base and subbase layers, standard granular materials designated by FAA as P209 and P154 respectively. Both APSDS and FAARFIELD v1.42 use sub-layering techniques for these unbound layers to take account of stress-dependence of the materials.

APSDS also considers aircraft wander. Aviation traffic loads differ from road traffic loads as aircraft wheel loads are spread more across the width of the pavement. This is partly due to a lower degree of channelization and partly due to the wide variation in spacing of wheels and groups of aircraft wheels compared with the standardized wheel configuration on road vehicles. Field observations have shown that successive passes of aircraft along a runway or taxiway pavement follow a bell-curve distribution about the pavement center-line and follows a normal distribution, so the degree of aircraft wander can be characterized by the standard deviation. The standard deviation of wander is significantly different for runways, taxiways and aircraft docking bays. This affects the required pavement thickness at each of these locations.

Data collected in the 1970s indicates wander widths of 1778 mm for taxiways and 3556 mm for runways. The standard deviation for a taxiway is 775 mm and for a runway is 1524 mm [10]. The wander width is the zone containing 75% of the aircraft center lines (1.15 standard deviations on either side of the mean value with a normal distribution).

In 2001, Wardle et al. [11] published a calibration of APSDS 4.0 against S77-1 designs, known as the *Chicago Calibration.* The performance parameters obtained in the calibration is shown in Table 1.

Validation of APSDS against the FAA methodology was performed by comparing thicknesses calculated from this software to those calculated from the S77-1 empirical design curve. The comparison covered a range of 108 design scenarios, including medium to large passenger jet aircrafts in Australia and a wide range of subgrade moduli. The analysis showed that good general agreement between APSDS and S77-1 with a medium difference of 36 mm (or 6.7% of the S77-1 thickness) [12]. White [13] confirmed that better agreement between APSDS 4.0 and COMFAA pavement thicknesses could be obtained by using different calibration parameters for each wheel configuration. The performance parameters reported in Table 1 were subsequently recalibrated in 2010 and the update performance parameters are referred to *Melbourne calibration* [14].

Table 1 Performance parameters obtained from *Chicago calibration* [11]

Subgrade CBR (%)	Subgrade modulus, E (MPa)	k	B
3	30	0.0032	9.5
6	60	0.0030	10.9
10	100	0.0024	15.0
15	150	0.0020	23.6

3 Subgrade Performance Models

3.1 *Apsds 5.0*

In airport pavement design, layered elastic models are used to compute values of chosen damage indicators, most commonly subgrade strain, which are then related to pavement life (strain repetitions). The strain value can be converted to damage using a performance relationship of the form:

$$N = \left(\frac{k}{\varepsilon}\right)^b \tag{1}$$

In this relationship, N is the predicted strain repetitions to cause failure, k is a material constant determined by calibration, b is the damage exponent for the material determined by calibration and ε is the static load-induced strain.

The introduction of new generation aircraft, such as the Boeing 777 and the Airbus A380, both of which have 6-wheel configurations was the major impetus for FAA to conduct new full-scale tests the US National Airport Pavement Test Facility (NAPTF) to improve the accuracy of pavement thickness designs for such aircrafts. Using this calibration data, APSDS 5.0 produced pavement thickness designs that more accurately reflects the performance of the full-scale test pavements than did the 2001 calibration [14]. APSDS 5.0 correlation with the S77-1 designs was improved by using different calibration parameters for each wheel configuration (see Table 2). The recalibration of the APSDS 5.0 was referred to *Melbourne calibration.*

The Damage Factor for the i-th loading is defined as the number of repetitions (n_i) of a given damage indicator divided by the 'allowable' repetitions (N_i) of the damage indicator that would cause failure. The Cumulative Damage Factor (*CDF*) is given by summing the damage factors for all loadings in the traffic spectrum using Miner's hypothesis:

$$CDF = \sum \frac{n_i}{N_i} \tag{2}$$

where, *CDF* is the cumulative damage factor, n_i is the number of repetitions and N_i is the allowable repetitions.

3.2 *FAARFIELD v1.42*

Prior to introduction of the layered elastic method, flexible aircraft pavements were usually designed using the US Army Corps of Engineers CBR pavement design method detailed in Instruction Report S-77-1 [15]. Aircraft induced deflections at

Table 2 Performance parameters obtained from *Melbourne calibration* [14]

Subgrade CBR (%)	Wheel configuration	Weighted error (mm)	Weighted error (%)	k	b
3	1 wheel	17.10	1.7	0.00382	7.8
	2 wheels	20.14	1.9	0.00254	12.4
	4 wheels	28.03	2.0	0.00204	17.8
	6 wheels	35.67	2.3	0.00200	27.1
	All wheel groups	143.14	11.0	0.00180	25.3
6	1 wheel	9.66	1.6	0.00382	9.3
	2 wheels	11.56	1.9	0.00297	12.5
	4 wheels	31.72	4.2	0.00216	18.7
	6 wheels	29.96	3.6	0.00188	27.1
	All wheel groups	50.65	7.0	0.00204	21.7
10	1 wheel	12.41	3.1	0.00382	10.4
	2 wheels	7.46	2.0	0.00300	13.1
	4 wheels	17.30	3.9	0.00225	19.0
	6 wheels	9.85	2.0	0.00192	27.1
	All wheel groups	25.14	5.8	0.00254	16.2
15	1 wheel	8.71	3.2	0.00382	11.0
	2 wheels	7.04	3.0	0.00280	15.1
	4 wheels	16.58	6.1	0.00252	18.3
	6 wheels	10.60	3.5	0.00217	27.1
	All wheel groups	19.57	7.3	0.00275	16.3

subgrade level were calculated using Boussinesq's single-layer equations which were correlated with the load repetitions observed to cause rutting failure in full-scale tests. The method was adapted from highway design practice in 1942, modified and extrapolated to cater for higher loads, multiple wheel undercarriages and aircraft wander. The FAA computerized its S-77-1 design method in its COMFAA software program.

COMFAA is a general-purpose computer program that operates in two computational modes: Aircraft Classification Number (ACN) Computation Mode and Pavement Design Mode. In ACN Computation Mode, COMFAA calculates the ACN number for aircraft on flexible pavements. In Pavement Design Mode, COMFAA calculates flexible pavement thickness based on the CBR method in AC 150/5320-6D for a CBR value specified by the user [16].

In 1995, the FAA introduced its LEDFAA software design to predict wheel-load interactions and to provide the airport community with a pavement design methodology addressing the needs of the B-777 aircraft [2]. The FAA issued an upgraded version (LEDFAA 1.3) in 2004. It features an expanded aircraft library with design procedures implemented as a computer program. LEDFAA represented a significant departure from earlier FAA design philosophies. Apart from design procedures implemented as a computer program, instead of nomographs, the main change from the user perspective was the replacement of the "design aircraft" concept design for fatigue failure now expressed by a "cumulative damage factor" (CDF) using Miner's Rule. The major material property of the pavement layers was now uniformly expressed as an elastic modulus instead of the previous CBR (California Bearing Ratio) for flexible pavements or k-value for rigid pavements [2].

FAARFIELD has now replaced LEDFAA. FAARFIELD software uses elastic layer theory and finite element methods for flexible and rigid pavement design respectively. The core of the program is a structural response model that consists of two programs (LEAF and NIKE3D). LEAF is a multilayer elastic computational program and NIKE3D is a program based on finite element methods. In May 2017, FAA developed new subgrade failure models for flexible pavements in FAARFIELD v1.41. FAARFIELD v1.41 was subsequently updated and had evolved to version 1.42 in September 2017. The subgrade failure models FAARFIELD v1.42 are shown in Eqs. 4 and 5.

FAA [4] used a commercial analysis program, CurveExpert© Basic 1.4 to identify the best regression of the lower bound of vertical strain to the common logarithm (base 10) of coverages. From the generated possible models by CurveExpert, the Bleasdale model was selected as showing the best fit. The format of the Bleasdale model is shown in Eqs. (3) and (4). For coverages equal or greater than 1000, the Bleasdale model is used directly. For coverages less than 1000, a straight line model was adopted, tangent to the Bleasdale curve and parallel to the FAARFIELD v1.3 failure model. The failure models used in FAARFIELD v1.42 to find the number of coverages to failure for a given vertical strain at the top of the subgrade are summarized below:

$$\log_{10} Coverages = \left(\frac{1}{a + b * \varepsilon}\right)^{\left(\frac{1}{c}\right)} \quad (3)$$

$$\log_{10} Coverages = \left(\frac{1}{-0.1638 + 185.19 * \varepsilon}\right)^{0.60586} \quad (4)$$

when C > 1000 coverages; and

$$Coverages = \left(\frac{0.00414131}{\varepsilon}\right)^{8.1} \quad (5)$$

when C $\leq$ 1000 coverages

where:

ε_v = maximum vertical strain at the top of the subgrade

4 Previous Study

A case study has been carried out to compare the design of airfield pavements for a major airport using FAARFIELD v1.32 and APSDS5 [3]. The aircraft spectrum and movements adopted in the study is shown in Table 3. With a growth rate of 2%, the total aircraft movements for wide body aircrafts are 212,640 for a design period of 40 years.

For the aircraft departure data, the pavement analysis shows that the APSDS 5.0 design method yielded pavement structure thicknesses that are nearly the same as FAARFIELD v1.32 for CBR greater than 10%. Figure 1 shows the design thicknesses produced by the two design programs. It shows that FAARFIELD produces design thicknesses that are significantly thicker than designs produced by APSDS for subgrade CBR value less than 10%. An adjustment factor k_c is required for APSDS design thickness to produce designs that are consistent with FAARFIELD for CBR less than 10%. The design thicknesses produced by FAARFIELD correlates to the thicknesses generated by APSDS by a polynomial equation:

$$t_{FAARFIELD} = (k_c)(t_{APSDS}) \tag{6}$$

Table 3 Aircraft departure data

Aircraft	Name	Tire pressure (kPa)	Total movements
1	B747-400ER	1572	47,280
2	B777-300 ER	1524	147,120
3	A380-800	1338	13,920
4	A340-300	1420	4320

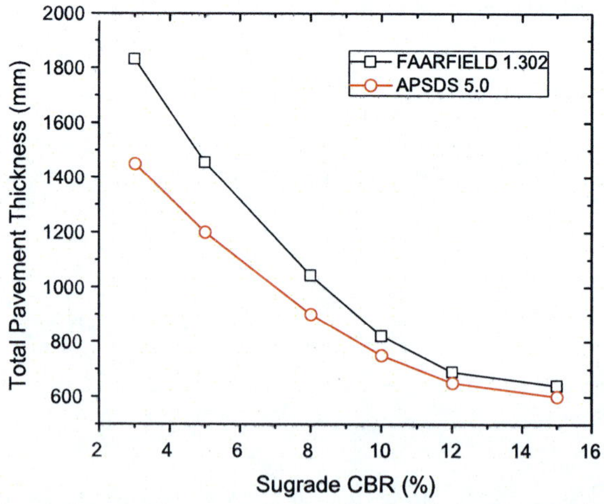

Fig. 1 Pavement thickness produced by APSDS and FAARFIELD v1.32

Table 4 Aircraft traffic movements

Aircraft	Name	Tire pressure (kPa)	Total movements
1	B737-800	1407	100,000
2	B777-300 ER	1524	100,000

$$k_c = \left(1.3748 - 0.0399CBR + 0.00123CBR^2\right) \quad (7)$$

where, $t_{FAARFIELD}$ is the design thickness produced by FAARFIELD, t_{APSDS} is the design thickness produced by APSDS, k_c is the adjustment factor and CBR is the subgrade California Bearing Ratio (%).

5 Current Study

5.1 Aircraft Traffic Spectrum

The development of the new subgrade failure models in 2017 prompted the need to compare the design thicknesses generated by FAARFIELD v1.42 and to examine if the thicknesses are compatible with that computed by APSDS 5.0. To meet the objective, Boeing 737-800 (Code C) and Boeing 777-300ER (Code E) aircrafts and with 100,000 aircraft movements (see Table 4) were analyzed at various subgrade CBR values using the two design software.

5.2 Methodology

Using the traffic spectrum and movements, the pavement models outlined in Table 5 was analysed using subgrade CBR value varying from 3%, 6%, 8%, 12% and 15%. The minimum thickness requirements specified in AC 150/5320-6F for the P-401/P-403 Hot Mix Asphalt Surface and P-401/P-403 St (flexible) are adopted. To model

Table 5 Pavement Models for B737-800 & B777-300ER aircrafts

Layer	Material type	Thickness (mm)	Modulus (MPa)
1	P-401/ P-403 HMA surface	100	1,378.95
2	P-401/ P-403 St (flexible)	125	2,757.90
3	P-209 crushed Aggregate	H_1	Variable
4	P-154 uncrushed Aggregate	H_2	Variable
5	Subgrade CBR (%)	3, 5, 8, 12 & 15	

the B737-800 and B777-300ER loadings, the pavement structures consist of 100 mm of P-401/P-403 (HMA) Surface, 125 mm of P-401/P-403 St (flex) which is a plant mix bituminous material and variable P-209 Crushed Aggregate base course and P-154 Uncrushed Aggregate subbase layers. Once the models have been developed, the pavement structures are analyzed iteratively with varying thickness of P-209 and P-154 layers until the CDF is close to 1.0.

6 Results and Discussions

Tables 6, 7 and 8 present the pavement thicknesses computed by FAARFIELD v1.42 and APSDS 5.0 respectively for B737-800 at subgrade CBR values of 3%, 5%, 8%, 12% and 15%. Tables 9, 10 and 11 are the design thicknesses for B777-300ER at the range of subgrade CBR specified in the design.

Figure 2 depicts the design thicknesses produced by the two design programs for B737-800 and Fig. 4 shows the design thicknesses for B777-300ER. For the B737-800 and B777-300ER aircrafts spectrum and 100,000 movements analyzed in the study, the new subgrade failure models developed for the latest version of FAARFIED v1.42 generate the flexible pavement thicknesses that are not significantly difference

Table 6 Pavement structures for subgrade CBR of 3% & 5%

No.	Aircraft	Thickness (mm)		Thickness (mm)	
	Boeing 737-800	FAARFIELD	APSDS	FAARFIELD	APSDS
1	P-401/P-403 HMA surface	100	100	100	100
2	P-401/P-403 St (flexible)	125	125	125	125
3	P-209 crushed Aggregate	300	300	300	300
4	P-154 uncrushed Aggregate	615	665	360	335
	Subgrade CBR (%)	3		5	
	Total pavement thickness (mm)	1140	1190	885	860

Table 7 Pavement structures for subgrade CBR of 8% & 12%

No.	Aircraft	Thickness (mm)		Thickness (mm)	
	Boeing 737-800	FAARFIELD	APSDS	FAARFIELD	APSDS
1	P-401/P-403 HMA surface	100	100	100	100
2	P-401/P-403 St (flexible)	125	125	125	125
3	P-209 crushed Aggregate	300	270	165	150[a]
4	P-154 uncrushed Aggregate	110	100[a]	101	100[a]
	Subgrade CBR (%)	8		12	
	Total pavement thickness (mm)	635	595	491	475

[a]Minimum FAA thickness

Table 8 Pavement structures for subgrade CBR of 15%

No.	Aircraft	Thickness (mm)	
	Boeing 737-800	FAARFIELD	APSDS
1	P-401/ P-403 HMA surface	100	100
2	P-401/ P-403 St (flexible)	125	125
3	P-209 crushed Aggregate	150[a]	150[a]
4	P-154 uncrushed Aggregate	100[a]	100[a]
	Subgrade CBR (%)	15	
	Total pavement thickness (mm)	475	475

[a]Minimum FAA thickness

Table 9 Pavement structures for subgrade CBR of 3% & 5%

No.	Aircraft	Thickness (mm)		Thickness (mm)	
	Boeing 777-300ER	FAARFIELD	APSDS	FAARFIELD	APSDS
1	P-401/ P-403 HMA surface	100	100	100	100
2	P-401/ P-403 St (flexible)	125	125	125	125
3	P-209 crushed Aggregate	450	450	450	450
4	P-154 uncrushed Aggregate	1485	1110	825	615
	Subgrade CBR (%)	3		5	
	Total pavement thickness (mm)	2160	1785	1500	1290

Table 10 Pavement structures for subgrade CBR of 8% & 12%

No.	Type	Thickness (mm)		Thickness (mm)	
	Boeing 777-300ER	FAARFIELD	APSDS	FAARFIELD	APSDS
1	P-401/ P-403 HMA surface	100	100	100	100
2	P-401/ P-403 St (flexible)	125	125	125	125
3	P-209 crushed Aggregate	400	400	300	205
4	P-154 uncrushed Aggregate	250	145	101	100[a]
	Subgrade CBR (%)	8		12	
	Total pavement thickness (mm)	875	770	626	530

[a]Minimum FAA thickness

from that of APSDS 5.0 for subgrade CBR $\geq$5%. The differences in the design thickness are observed to be marginal (Figs. 3, 4 and 5).

For B737-800 loading, APSDS 5.0 produces design thicknesses that are slightly thicker than designs produced by FAARFIELD v1.42 at subgrade CBR value of 3%. The thickness difference is 50 mm in the P-154 Uncrushed Aggregate subbase layer. This indicates that the new subgrade failure model in FAARFIELD generates design

Table 11 Pavement structures for subgrade CBR of 15%

No.	Aircraft	Thickness (mm)	
	Boeing 777-300ER	FAARFIELD	APSDS
1	P-401/P-403 HMA surface	100	100
2	P-401/P-403 St (flexible)	125	125
3	P-209 crushed Aggregate	220	150[a]
4	P-154 uncrushed Aggregate	101	100[a]
	Subgrade CBR (%)	15	
	Total pavement thickness (mm)	546	475

[a]Minimum FAA thickness

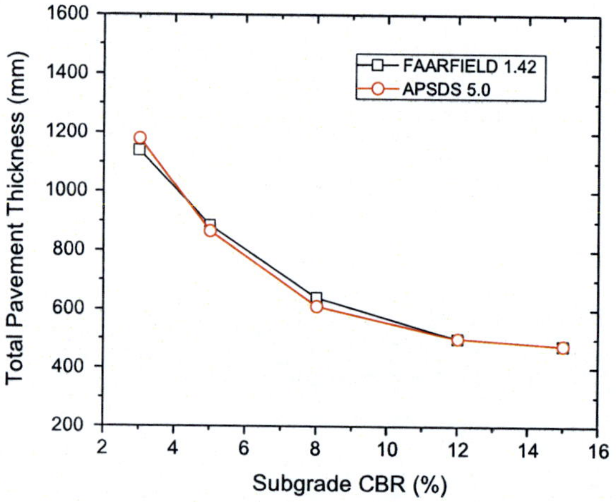

Fig. 2 Design thicknesses produced by APSDS 5.0 and FAARFIELD v1.42 (B737-800 aircraft)

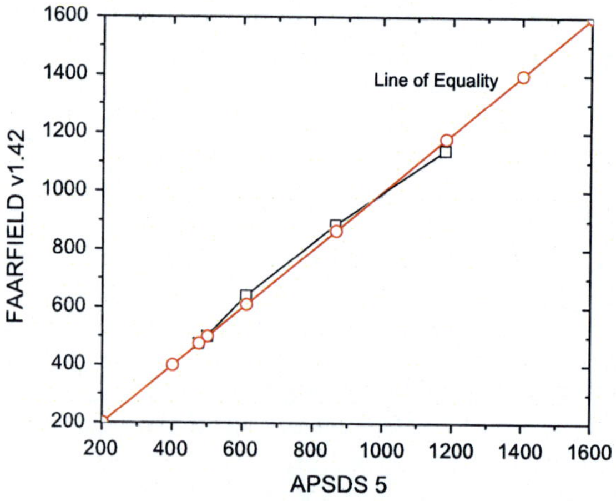

Fig. 3 APSDS 5.0 versus FAARFIELD v1.42 for CBR 3, 5, 8, 12 & 15% (B737-800 aircraft)

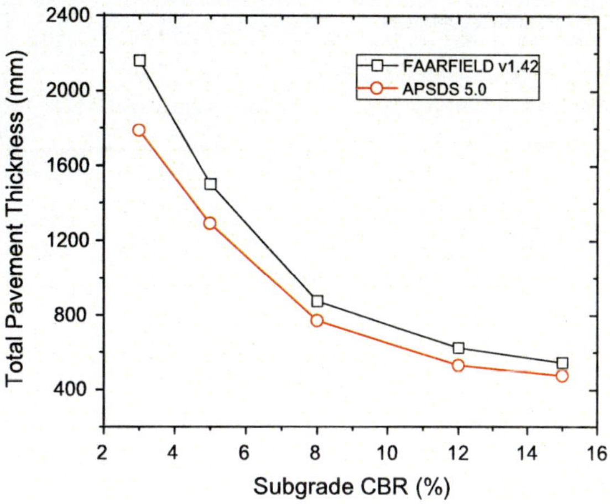

Fig. 4 Design thicknesses produced by APSDS 5.0 and FAARFIELD v1.42 (B777-300ER aircraft)

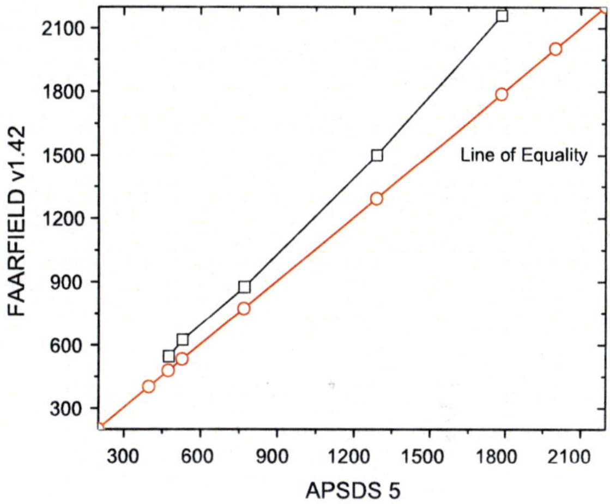

Fig. 5 APSDS 5.0 versus FAARFIELD v1.42 for CBR 3, 5, 8, 12 & 15% (B777-300ER)

thicknesses that is consistent with APSDS for the B737-800 aircraft with 2 wheels configuration.

On the other hand, for B777-300ER aircraft, FAARFIELD v1.42 produces design thicknesses that are significantly thicker than designs produced by APSDS for subgrade CBR value less than 5%. The thickness differences are observed to be significance for CBR ≤5% when modelled with B777-300ER having 6 wheels configuration. The P-154 Uncrushed Aggregate layer is 375 mm thicker in the FAARFIELD model than that of APSDS at CBR of 3%. Overall, the thicknesses differences are observed to be less for CBR ≥5%. The differences in the design thickness are attributed to the different coefficients adopted in the subgrade failure models in the design software.

7 Conclusions

A brief literature review of the subgrade failure models used in APSDS 5.0 and FAARFIELD v1.42 has been conducted. The key differences are that in APSDS 5.0 different calibration parameters for the subgrade failure model are adopted for each wheel configuration, whereas in FAARFIELD v1.42 different failure models are used for coverages equal or greater than 1000 and for coverages less than 1000.

The development of the new subgrade failure models in FAARFIELD in 2017 prompted the need to compare to the design thicknesses generated by FAARFIELD v1.42 and to examine if the thicknesses are compatible with that computed by APSDS5.0. To make an assessment, Boeing 737-800 (Code C) and B777-300ER (Code E) aircrafts and 100,000 aircraft movements were analyzed at various subgrade CBR values using the two software programs. The conclusions of the study are as follows:

- For the aircraft spectrum and movements analyzed in the study, the new subgrade failure models developed for the FAARFIELD v1.42 generate the flexible pavement thicknesses that are not significantly difference from that of APSDS 5.0 for subgrade CBR $\geq$5%.
- For CBR $\leq$3% and with B737-800 aircraft loading, APSDS5 produces design thickness that are marginally thicker than that generated from FAARFIELD v1.42. On the contrary, FAARFIELD v1.42 generates thicker pavement than APSDS 5.0 for B777-300ER aircraft on subgrade CBR $\leq$3%.
- Overall, it is observed that the new subgrade failure model in FAARFIELD v1.42 produces flexible design thicknesses that differ less from APSDS 5.0 than FAARFIELD v1.32. The design thicknesses are more consistent for B737-800 aircraft with 2 wheels configuration. However, the differences are observed to be larger for CBR $\leq$5% when modelled with B777-300ER having 6 wheels configuration on the main landing gear. The differences in the design thickness are attributed to the different coefficients adopted in the subgrade failure models in the design software.

At the time of writing, FAARFIELD 2.0 [1] was still in the development phase. It is recommended that a similar comparison be carried out using a wide spectrum of aircrafts using the latest version of FAARFIELD in next stage of the study. This will include the 4 wheels configuration on the main landing gear such as the A350-900 and B787-9 aircrafts.

References

1. FAARFIELD 2.0.0.i RC (2020) The evaluation version of FAARFIELD 2.0 airport pavement thickness design software accompanying DRAFT AC 150/5320-6G, Airport Pavement Design and Evaluation
2. Federal Aviation Administration (FAA) (2012) Layered elastic design (LEDFAA), Advanced pavement design, Federal Aviation Administration, US Department of Transportation, Washington, D.C., USA
3. Chai G, Wardle L, Haydon M (2016) Airfield pavement design for a major airport using FAARFIELD and APSDS. In: Eight international conference on maintenance and rehabilitation of pavements, Singapore
4. Kawa, I (2017) Development of new subgrade failure model for flexible pavements in FAARFIELD, Report No. DOT/FAA/TC-17/28
5. Emery S (2005) Bituminous surfacings for pavements on Australian Arports, 24th Australia Airport Association Convention, Hobart
6. Emery S (2015) Pavements and surfacings on WA Airports. Aust Geomech 50(1)
7. Rodway B (2016) Flexible aircraft pavement surfacing-Australian Practice. In: Eight international conference on maintenance and rehabilitation of pavements, Singapore
8. Wardle L (1999) CIRCLY 4 user manual, MINCAD Systems Pty Ltd., Richmond, Victoria, Australia
9. Barker W, Brabston W (1975) Development of a Structural design Procedure for Flexible Airport Pavements, Report No. S-75-17, US Army Corps of Engineers, Waterways Experiment Station, Vicksburg, Mississippi
10. Ho Sang VA (1975) Field survey and analysis of aircraft distribution on airport pavements, Report No. FAA-RD-74-36. U.S. Federal Aviation Administration
11. Wardle L, Rodway B, Rickards I (2001) Calibration of advanced flexible aircraft pavement design method to S77-1 method. Advancing Airfield Pavements. In: Buttlar WG and Naughton JE (eds) American Society of Civil Engineers. 2001 airfield pavement specialty conference, Chicago, Illinois
12. White G (2007) An investigation of the australian layered elastic tool for flexible aircraft pavement thickness design. Master of Engineering Thesis, Queensland University of Technology
13. White G (2008) Recalibration of airport pavement structural design system. In: Roesler JR, Bahia U, Al-Qadi I (eds) Airfield and highway pavements: efficient pavements supporting transportation's future: proceedings of the 2008 airfield and highway pavements conference
14. Wardle L, Rodway B (2010) Advanced design of flexible aircraft pavements. In: Proceedings of the 24th ARRB conference-building on 50 years of road and transport research, Melbourne, Australia
15. Pereira AT (1977) Procedures for development of CBR design curves, Instruction Report S-77-1, US Army Corps of Engineers, Waterways Experiment Station, Vicksburg, Mississippi, USA
16. Kawa I, Hayhoe GF (2002) Development of a computer program to computer pavement thickness and strength. In: Proceeding of the 2002 Federal Aviation administration airport technology transfer conference, Atlantic City, New Jersey

Development of Design Guidelines for Composite-Geogrid Reinforced Unpaved Pavements

Chamara Prasad Gunasekara Jayalath, Chaminda Gallage, and Kasun Wimalasena

Abstract It is well-known that the inclusion of composite geogrids (CGGs) into the pavement structure assists to maximise the benefits of geosynthetic-reinforcement in flexible pavements. However, the use of CGGs in practice is limited in Queensland, Australia regardless of its potential benefits, mainly due to the absence of rational design guidelines. Therefore, this study was conducted to develop guidelines to design unpaved granular pavements/working platforms constructed with local pavement materials and reinforced with CGGs at the base-subgrade interface. Firstly, six laboratory-scale pavement model tests were conducted varying the granular base thickness as 200, 300 and 400 mm, and with and without the CGG reinforcement at the interface of the subgrade-base layer. In all these tests, a 500 mm thick subgrade was prepared to achieve a CBR value of 2.5%, and then the granular layer was compacted on top of the subgrade to achieve 91% of its maximum dry density, in a steel box with length, width and height of 1 m, 1 m and 1.2 m respectively. The pavement models were subjected to repeated loading (up to 150, 000 cycles) at the centre using a 200 mm diameter plate to simulate the maximum tyre pressure of 550 kPa. By following a step-by-step procedure, design guidelines were developed based on the rut depths of both the reinforced and unreinforced pavement models obtained after 5000 load cycles and every 25,000 load cycles up to 150,000. Both economic and environmental benefits can be obtained by adopting the proposed guidelines.

Keywords Composite geogrids · Unpaved pavements · Design guidelines · Design traffic · Rut depth

1 Introduction

It is well-known that weak subgrade materials are commonly found in most areas of Australia. This is mainly because the majority of Australian soils are expansive in nature [1–9]. As a result, thick granular layers have to be placed as the subbase or base of flexible pavements to cater to design traffic. However, in most countries including

C. P. G. Jayalath (✉) · C. Gallage · K. Wimalasena
Queensland University of Technology, Brisbane, QLD 4000, Australia

© The Author(s), under exclusive license to Springer Nature Switzerland AG 2022

H. R. Pasindu et al. (eds.), *Road and Airfield Pavement Technology*, Lecture Notes in Civil Engineering 193, https://doi.org/10.1007/978-3-030-87379-0_28

Australia, the availability of required quality aggregate materials for road construction is limited. It is even rarer for these materials to be available within a shorter haulage distance [10]. The use of recycled materials in granular base construction is an alternative option [11–13]; however, fulfilling the demand has become problematic. Consequently, the abovementioned circumstances have become challenges for transportation professionals, especially when infrastructure systems are built and maintained under shrinking budgets [14]. It, therefore, demands a solution to significantly reduce the required quantity of aggregate materials for the construction and rehabilitation of roadways.

Geosynthetics, since their introduction in late 1970, have been successfully used as a pavement-reinforcement material to increase performance. The primary applications of geosynthetics in the geotechnical-engineering discipline are reinforcing earth embankments and slopes [15], improving the performance and the service life of pavements and railways [16, 17], and coastal and landfill protection [18–22]. Many researchers [14, 23–32] agree that the inclusion of geogrids in pavement systems with the purpose of subgrade reinforcement is a viable option for reducing the granular-base thickness, extending the service life of pavements and reducing costs. The inclusion of both geotextiles and geogrids in a pavement structure has been identified as the most effective geosynthetic application [26, 29]. It can therefore be concluded that the inclusion of composite geogrids (CGGs) which are the geogrids combined with a nonwoven geotextile component, into the pavement structure helps to maximise the benefits of geosynthetic-reinforcement in granular flexible pavements.

CGGs are the next generation of geogrids for roadway and pavement systems. This type of geogrid provides both reinforcement and separation functions because it is combined with a nonwoven geotextile component, and therefore has been identified as the most effective geosynthetic application in pavement systems [29]. CGGs allow for direct cost savings when designing a roadway structure as they reduce the required amount of base material and excavation. Therefore, CGGs are generally considered to be a solution to the excessive demand for natural crushed rocks [1, 2, 11–13, 33, 34]. Furthermore, CGGs do not require two separate installation processes, as is the case with traditional materials (i.e. geogrids and geotextiles). Therefore, the one-step installation process associated with CGGs increases overall productivity, which also results in cost savings. Consequently, CGGs are commonly used by pavement engineers in the construction and rehabilitation of pavements on weak subgrades such as expansive soils.

Several design methods are available in designing of the geosynthetic-reinforced unpaved pavements. According to the literature survey, the 'Steward et al.' method [35], 'Giroud and Noiray' method [36] and 'Giroud and Han' method [37] can be identified as the most used design methods in the world to design geogrid-reinforced unpaved pavements. However, most of the currently available geogrid-reinforced pavement-design methodologies are both specific to the product and the conditions under which the testing had been performed. Hence, those design procedures are not applicable to design geogrid-reinforced granular pavements/working platforms with different geogrid materials, pavement materials and testing conditions. At the same time, design guidelines which can be adapted to design the geogrid-reinforced granular pavements/working platforms in Queensland, Australia, are also absent. Therefore, this study was conducted to develop guidelines for pavement-engineering practitioners to estimate the unbound granular base thickness over a poor subgrade (i.e. California bearing ratio (CBR) value of 3% or below) reinforced with the CGG at the base-subgrade interface, complying with testing conditions and the properties of selected locally available materials. Firstly, the performance of CGG reinforced unpaved granular pavements was investigated through laboratory-scale pavement models that were subjected to cyclic loading of 550 kPa tire pressure with a 0.33 Hz frequency. The subgrade CBR of 2.5% and granular base compaction of 91% of its maximum dry density were adopted in all model tests (length = 1.0 m, width = 1.0 m and height = 1.2 m). Finally, by following a step-by step procedure, design guidelines were developed based on the rut depths of both the reinforced and unreinforced pavement models obtained after 5000 load cycles and every 25,000 load cycles up to 150,000.

2 Methodology

2.1 Materials

Geogrid

A CGG (see Fig. 1) made of polypropylene was used in this experimental study. Table 1 demonstrates the manufacturer-specified properties of the CGG for both the machine direction (MD) and cross-machine direction (XMD). The nominal strength of the CGG is 40 kN/m in both directions.

Fig. 1 Composite geogrid

Table 1 Properties of composite geogrid

Property	Units	MD/XMD
Geogrid		
Maximum tensile strength	kN/m	≥40/≥40
Elongation at nominal strength	%	≤8/≤8
Tensile strength at 2% elongation	kN/m	16/16
Tensile strength at 5% elongation	kN/m	32/32
Aperture size	mm	31/31
Geotextile		
Maximum tensile strength	kN/m	7.5/11
Elongation at maximum tensile strength	%	40/30

Subgrade Material

Black Soil which is one of the common soil types in Queensland was used as the subgrade material. The soil has a liquid limit of 74%, plastic limit of 54% and linear shrinkage limit of 13.5%. According to the Unified soil classification system (USCS), this soil is classified as silt of high plasticity (MH). The specific gravity of the subgrade material is 2.62. Based on the standard Proctor compaction test, the maximum dry density and the optimum moisture content of subgrade material are 1.32 g/cm^3 and 32% respectively.

Base Material

Type 2.3 unbound granular material (As per the technical specifications in "Transport and Main Roads Specifications MRTS05-Unbound Pavements" [38]) was used as the base material in this study. According to the Standard Proctor compaction test, the maximum dry density and the optimum moisture content of granular material are 2.21 g/m^3 and 7.5% respectively.

2.2 *Pavement Model Testing*

Three base course thicknesses, namely 200, 300 and 400 mm, were adopted in this experimental study. For each base thickness, two identical model pavements: one without CGG reinforcement and the other with CGG reinforcement at the interface of the subgrade-base layer were prepared in a steel box with length, width and height of 1 m, 1 m and 1.2 m, respectively. The 500 mm thick subgrade was prepared to achieve a CBR value of 2.5%, and then the granular layer was compacted on top of the subgrade to achieve 91% of its maximum dry density. All pavement models were subjected to repeated loading at the centre using a 200 mm diameter plate to simulate the maximum tyre pressure of 550 kPa. The loading was continued until a 75 mm permanent deformation was accumulated at the centre of the surface or 150,000 loading cycles, whichever occurred first. More details of the methodology

Table 2 Summary of rut depths

Number of cycles	Rut depth (mm)					
	200 mm base		300 mm base		400 mm base	
	URF[a]	RF[a]	URF	RF	URF	RF
5000	27.9	21.7	22.2	20.8	16.6	17.0
25,000	56.5	37.3	38.5	40.0	29.5	32.0
50,000	70.3	44.5	46.5	51.5	36.5	45.0
75,000	82.0	49.0	51.5	56.5	43.0	59.0
100,000	90.0	51.6	58.0	60.6	48.0	63.6
125,000	-	53.3	63.0	64.2	50.6	65.8
150,000	-	54.6	67.0	68.0	52.8	68.0

[a]URF—unreinforced model; RF—reinforced model

of pavement model tests can be found in the published research article on these experiments [39].

The reduction of permanent deformation with increasing base thickness was evident when comparing the rut depth development in unreinforced-pavement models. In the case of 200 mm granular base, the reinforced model has exhibited a better rutting resistance compared to the unreinforced model indicating that the CGG is effective as a subgrade reinforcement. However, in the long run, the performance of the reinforced-pavement models of 300 and 400 mm granular bases became worse compared to the performance of the corresponding unreinforced pavements, as shown in Table 2. This permanent-deformation behaviour may be due to the 'rigid boundary effect'.

2.3 *Development of Design Guidelines*

The outputs of the pavement model tests were used to develop the guidelines to design and construct granular pavements/working platforms reinforced with the CGG at the base-subgrade interface. The guidelines were developed using the following steps.

Step 1: Determine the rut depth (mm) for both the reinforced and unreinforced pavement models, after 5000 load cycles and every 25,000 load cycles up to 150,000 cycles.

The rut depths after 5000 load cycles and every 25,000 load cycles up to 150,000 cycles were obtained separately for both the reinforced and unreinforced pavement models of 200, 300 and 400 mm granular bases. Figure 2 demonstrates the way of obtaining rut depths after 5000 load applications. Rut depths obtained after each considered number of load cycles have been summarised in Table 2.

Step 2: Plot the variation of the rut depth with the granular-base thickness for both the reinforced and unreinforced pavement models, separately for each considered

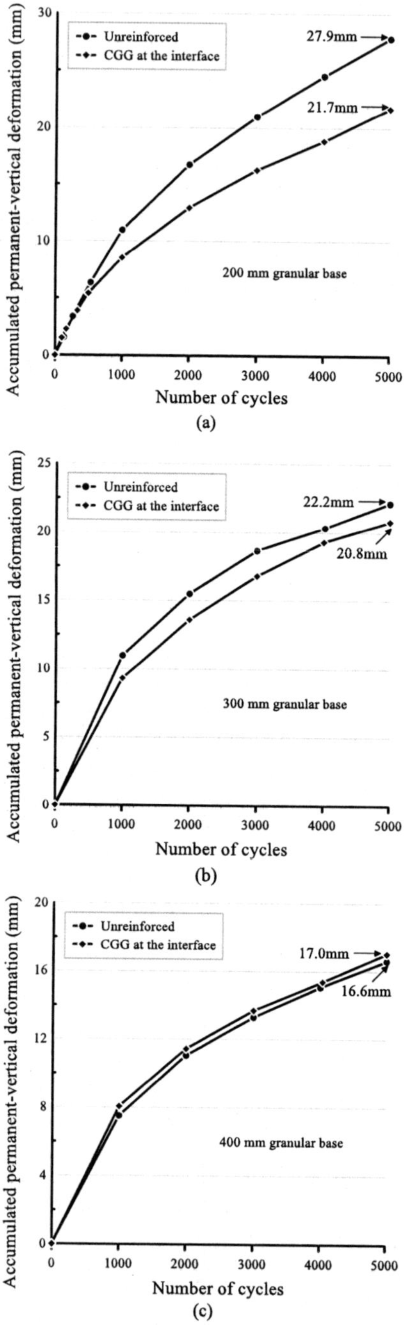

Fig. 2 Determining the accumulated-permanent-vertical deformation after 5000 cycles for the reinforced and unreinforced pavement models of the granular-base thickness **a** 200 mm **b** 300 mm **c** 400 mm

loading cycle (i.e. 5000 load cycles and every 25,000 load cycles up to 150,000 cycles).

Step 3: Determine the equivalent base thickness after each considered loading cycles.

The equivalent base thickness is defined as the base thickness of the unreinforced pavement that equals the rut depth development of the CGG-reinforced pavement of the 200 mm granular base. The determination of the equivalent base thickness for 5000 loading cycles is demonstrated in Fig. 3. The equivalent base thicknesses for the remaining considered load cycles were determined, and the obtained values were summarised in Table 3.

Step 4: Refer to the design curve of the structural base thickness for the California bearing ratio (CBR) of the subgrade equal to or less than 3%, given in 'Guide to Pavement Technology Part 6: Unsealed Pavements' [40].

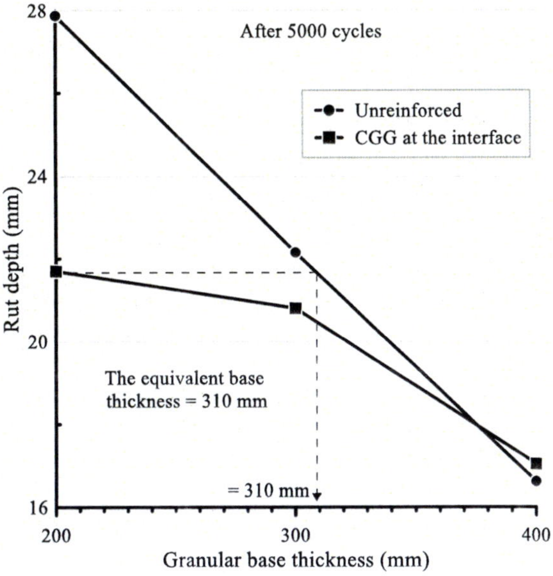

Fig. 3 Determination of the equivalent base thickness after 5000 cycles

Table 3 Summary of the equivalent base thicknesses for different loading cycles

Number of cycles	Equivalent base thickness
5000	310
25,000	310
50,000	320
75,000	330
100,000	365
125,000	380
150,000	390

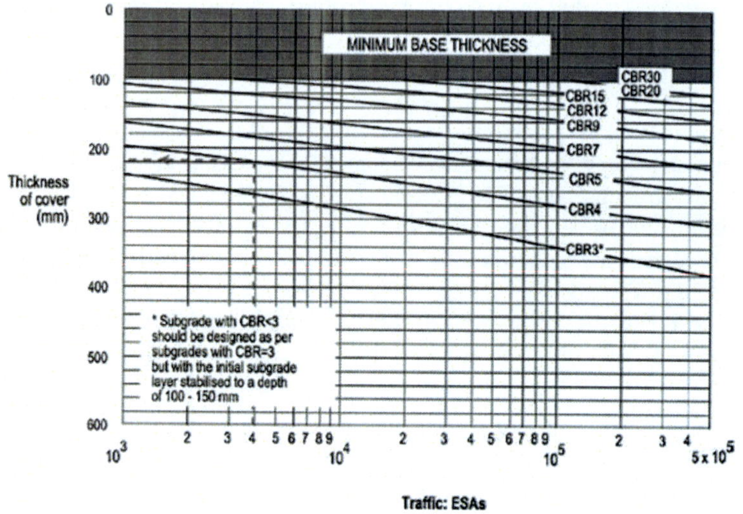

Fig. 4 Design curves for granular pavements, given in Austroads Standard [40]

As shown in Fig. 4, Fig. 4.3 of the 'Guide to Pavement Technology Part 6: Unsealed Pavements' [40] represents a minimum structural thickness to protect the subgrade from deformation (rutting) under traffic load during its design life. The design curve given for CBR = 3% was considered in developing the design guidelines of this chapter, and it fulfils the requirement of the minimum structural thickness for subgrades with CBR ≥3%. However, the Austroads Standard [40] recommends that subgrades with CBR <3% should be designed as per subgrades with CBR = 3% but with the initial subgrade layer stabilised to a depth of 100–150 mm.

Step 5: Compare the design curve of the structural base thickness (for the California bearing ratio (CBR) of the subgrade equal to or less than 3%) to the variation of the equivalent base thickness for the unreinforced condition with the design traffic.

In Austroads Standard [40], the design tyre-pavement contact stresses for pavement analysis are taken as 750 kPa, referring to the 'Guide to Pavement Technology Part 2: Pavement Structural Design' [41]. In order to develop the proposed design guide, it was assumed that the variation of equivalent base thickness with the number of loading cycles is similar for the tyre-pavement contact stress of 550 kPa and 750 kPa. Based on this assumption, the design curve of the structural base thickness (for the California bearing ratio (CBR) of the subgrade equals or less than 3%) was compared to the variation of the equivalent base thickness for the unreinforced condition with the design traffic, as illustrated in Fig. 5.

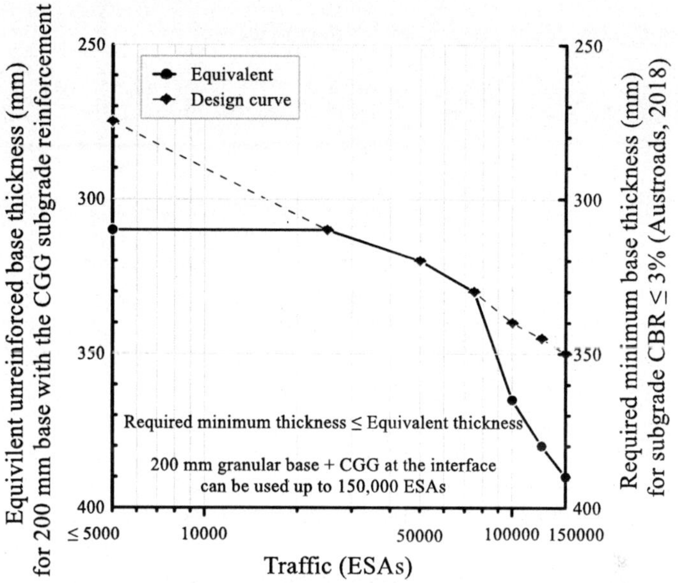

Fig. 5 Comparison of the curves for the equivalent base thickness and design curve

3 Discussion

Figure 5 shows that the curve for equivalent base thickness, which has been developed up to always 150,000 ESAs (equivalent standard axles), lies on or below the design curve. Therefore, the required minimum base thickness is always lower than or equal to the equivalent base thickness. Thus, as a guideline, it can be recommended that a 200 mm granular base with the CGG at the base-subgrade interface can be used to cater to the design traffic up to 150,000 ESAs for subgrades with CBR ≥2.5% of unsealed granular pavements, lightly trafficked granular pavements with thin bituminous surfacing and working platforms. Subgrades with CBR <2.5% should be designed as per subgrades with CBR = 2.5% but with the initial subgrade layer stabilised to a depth of 100–150 mm. The proposed guidelines can be used globally on the condition that the physical properties and mechanical behaviour of pavement material, and properties of material layers of the designed pavements comply with the testing conditions of the current study.

The cost saving when the CGG reinforcement is included in the granular pavement can be demonstrated through the below example.

Considering,

The design traffic is 25,000 ESAs
The subgrade CBR is 2.5%
The average in-place cost of Type 2.3 standard granular material is 140 AUD/m^3
The average in-place cost of CGG is 6 AUD/m^2

Two-lane granular pavement with 3.5 m lane width.

Calculation

In order to cater to 25,000 ESAs,
The required granular base thickness for the unreinforced pavement condition = 310 mm (see Fig. 6).
The required granular base thickness for the CGG-reinforced pavement condition = 200 mm.
The reduction of the granular layer thickness due to the inclusion of CGG reinforcement is 110 mm [310–200 mm].

Considering a 1 m long pavement section:

Cost saving due to the inclusion of CGG = $(0.110\ \text{m} \times 140\ \text{AUD/m}^3 - 6\ \text{AUD/m}^2) \times 3.5\ \text{m} \times 2 = 65.8\ \text{AUD/m}$
Material saving due to the inclusion of CGG = $(0.110 \times 3.5 \times 2 \times 1000)\ \text{m}^3 = 770\ \text{m}^3/\text{km}$
Cost saving for 1 km long pavement section = 65.8 AUD/m × 1000 m = 65,800 AUD/km.

Figure 6 indicates that for the designed traffic of 25,000 ESAs, an unreinforced pavement model with a 310 mm granular base can be replaced by the 200 mm granular base with the CGG at the base-subgrade interface, achieving an approximately

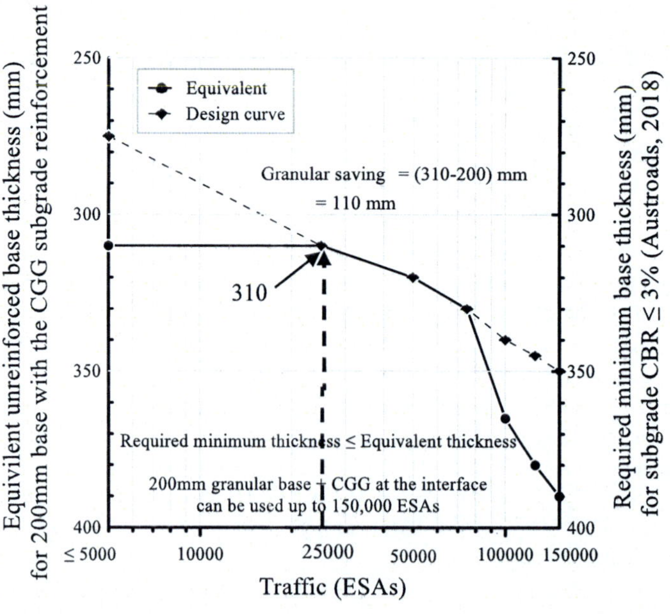

Fig. 6 Design calculation

35% reduction of the base thickness. This implies that 770 m^3 of granular materials can be saved for a one-kilometre road strip that has two 3.5 m-wide lanes, which is equivalent to 65,800 AUD in terms of savings in material cost. In addition, the associated costs for material transportation and labour are also reduced. Thus, this reduction of the granular-base thickness results in a significant saving of the construction cost of Australian roads/working platforms. Further to the aforementioned economic benefits, there are substantial environmental benefits related to aggregate savings. For example, less transportation of base materials by trucks results in less air pollution (gasoline vapours, dust), less carbon emission, less noise and less energy consumption.

4 Summary

Based on the outputs of the pavement model tests carried out in this research, the use of a 200 mm granular base with the CGG at the base-subgrade interface can be recommended to cater to the design traffic up to 150,000 ESAs for subgrades with CBR $\geq$2.5% of unsealed granular pavements, lightly trafficked granular pavements with thin bituminous surfacing and working platforms. Subgrade with CBR <2.5% should be designed as per subgrades with CBR = 2.5% but with the initial subgrade layer stabilised to a depth of 100–150 mm. Adaptation of the proposed guidelines results in both economic and environmental benefits.

References

1. Udukumburage RS, Gallage C, Dawes L (2019) Oedometer based estimation of vertical shrinkage of expansive soil in a large instrumeted soil column. Heliyon
2. Udukumburage RS, Gallage C, Dawes L (2020) An instrumented large soil column to investigate climatic ground interaction. Int J Phys Model Geotech
3. Udukumburage RS, Gallage C, Dawes L, Gui Y (2020) Determination of the hydraulic conductivity function of grey Vertosol with soil column test. Heliyon
4. Chan D, Gallage CPK, Rajeev P, Kodikara J (2015) Field performance of in-service cast iron water reticulation pipe buried in reactive clay. Can Geotech J
5. Chan D, Rajeev P, Kodikara J, Gallage C (2016) Field performance of in-service cast iron gas reticulation pipe buried in reactive clay. J Pipeline Syst Eng Pract
6. Gallage CPK, Chan D, Kodikara J (2012) Response of a plastic pipe buried in expansive clay. Proc Inst Civ Eng Geotech Eng
7. Gallage CPK, Chan D, Kodikara J, Ng PCF (2013) Discussion: response of a plastic pipe buried in expansive clay. Proc Inst Civ Eng Geotech Eng
8. Gallage C, Cochrane M, Ramanujam J (2012) Effects of lime content and amelioration period in double lime application on the strength of lime treated expansive sub-grade soils. In: Advances in transportation geotechnics II—proceedings of the 2nd international conference on transportation geotechnics, ICTG 2012
9. Gallage CPK, Tehrani K, Williams D (2017) Instrumented large soil-column to investigate climate-induced ground deformation in expansive soil. In: Proceedings of the 19th international conference on soil mechanics and geotechnical engineering, pp 1147–50

10. Duncan-Williams E, Attoh-Okine NO (2008) Effect of geogrid in granular base strength—an experimental investigation. Constr Build Mater
11. Jayakody S, Gallage C, Kumar A (2014) Assessment of recycled concrete aggregates as a pavement material. Geomech Eng. 6(3):235–248
12. Jayakody S, Gallage C, Ramanujam J (2019) Effects of reclaimed asphalt materials on geotechnical characteristics of recycled concrete aggregates as a pavement material. Road Mater Pavement Des
13. Jayakody S, Gallage C, Ramanujam J (2019) Performance characteristics of recycled concrete aggregate as an unbound pavement material. Heliyon
14. Jersey S, Tingle J, Norwood G, Kwon J, Wayne M (2012) Full-scale evaluation of geogrid-reinforced thin flexible pavements. Transp Res Rec
15. Garcia EF, Gallage CPK, Uchimura T (2007) Function of permeable geosynthetics in unsaturated embankments subjected to rainfall infiltration. Geosynth Int
16. Askarinejad H, Barati P, Dhanasekar M, Gallage C (2018) Field studies on sleeper deflection and ballast pressure in heavy haul track. Aust J Struct Eng
17. Gallage C, Eom T, Barker D, Ramanujam J (2015) Falling Weight Deflectometer (FWD) tests on granular pavement reinforced with geogrids—case study. In: Proceedings of the international conference on geotechnical engineering (ICGE—Colombo—2015): Geotechnics for sustainable development
18. Cheah C, Gallage C, Dawes L, Kendall P (2017) Measuring hydraulic properties of geotextiles after installation damage. Geotext Geomembr
19. Cheah C, Gallage C, Dawes L, Kendall P (2016) Impact resistance and evaluation of retained strength on geotextiles. Geotext Geomembranes
20. Weerasinghe IA, Gallage C, Dawes L (2019) Optimising geosynthetic clay liner overlaps: implications on hydraulic performance. Environ Geotech
21. Weerasinghe IA, Gallage C, Dawes L (2021) Effect of overburden confining stress on hydraulic performance of geosynthetic clay liners (GCLs). Heliyon
22. Weerasinghe IA, Gallage C, Dawes L, Kendall P (2020) Factors affecting the hydraulic performance of a geosynthetic clay liner overlap. J Environ Manage
23. Abu-Farsakh MY, Gu J, Voyiadjis GZ, Chen Q (2014) Mechanistic-empirical analysis of the results of finite element analysis on flexible pavement with geogrid base reinforcement. Int J Pavement Eng
24. Tang X, Stoffels S, Palomino AM (2014) Mechanistic-empirical performance prediction of geogrid-modified soft soil subgrade. In
25. Al-Qadi IL, Dessouky S, Tutumluer E, Kwon J (2011) Geogrid mechanism in low-volume flexible pavements: accelerated testing of full-scale heavily instrumented pavement sections. Int J Pavement Eng
26. Bräu G, Vogt S (2018) Field and laboratory tests on the bearing behaviour of unpaved roads reinforced by different geosynthetics Geotech Eng
27. Ferrotti G, Canestrari F, Virgili A, Grilli A (2011) A strategic laboratory approach for the performance investigation of geogrids in flexible pavements. Constr Build Mater
28. Ghafoori N, Sharbaf M (2015) Evaluation of triaxial geogrids for reduction of base thickness in flexible pavements. In: Bituminous mixtures and pavements vi—proceedings of the 6th international conference on bituminous mixtures and pavements, ICONFBMP 2015
29. Kwon J, Tutumluer E (2009) Geogrid base reinforcement with aggregate interlock and modeling of associated stiffness enhancement in mechanistic pavement analysis. Transp Res Rec
30. Leonardi G, Bosco D Lo, Palamara R, Suraci F (2020) Finite element analysis of geogrid-stabilized unpaved roads. Sustain
31. Sayida MK, Evangeline SY, Girish MS (2020) Coir geotextiles for paved roads: a laboratory and field study using non-plastic soil as subgrade. J Nat Fibers
32. Schuettpelz C, Fratta D, Edil TB (2009) Evaluation of the zone of influence and stiffness improvement from geogrid reinforcement in granular materials. Transp Res Rec
33. Clark BR, Gallage C, Yeaman J (2020) Temperature variation through deep multigrade asphalt pavements and proposed method for accounting for fluctuations. J Mater Civ Eng

34. Clark BR, Gallage C (2020) Superior performance benefits of multigrade bitumen asphalt with recycled asphalt pavement additive. Constr Build Mater
35. Steward J, Williamson R, Mohney J (1977) Guidelines for use of fabrics in construction and maintenance of low-volume roads. Transp Res Board
36. Giroud JP, Noiray L (1981) Geotextile-reinforced unpaved road design. J Geotech Eng Div
37. Giroud JP, Han J (2004) Design method for geogrid-reinforced unpaved roads. I. Development of design method. J Geotech Geoenviron Eng
38. TMR (Transport and Main Roads) (2017) Technical specification, MRTS05-Unbound Pavements. Queensland Department of Main Roads, Brisbane
39. Wimalasena K, Jayalath CPG (2020) Effect of geogrid reinforcement in weak subgrades. Int J Geomate
40. Austroads (2009) Guide to pavement technology part 6: unsealed pavements. Austroads Inc., Sydney, Australia
41. Austroads (2019) Austroads guide to pavement technology part 2: pavement structural design. Austroads Inc., Sydney, Australia

Development of Performance Grade Specification for Polymer Modified Bitumen in SEA Region

Lu Jia, Manpreet Sethi, Kluttz Bob, and Fitts Gary

Abstract With the increase of traffic volume and traffic loads, polymer modified bitumen has been widely used in Southeast Asia (SEA) region under Superpave Performance grading specification. Because of the tropical climatic in the SEA region no freezing air temperature, the used PG grade specification in the SEA region only included the high temperature grade requirement without the low temperature grade. Consequently, only the DSR (Dynamic Shear Rheometer) and RTFO (Rolling Thin Film Oven) are used by most of the local PMB producers and test companies without BBR (Bending Beam Rheometer) and PAV (Pressure Aging Vessel). This paper reviews the development of Superpave Performance Grading system in the US and its application. The constraints and limitations of the existing Performance Grading system used in SEA region have been investigated and are explained. Combined with the development of latest research for polymer modified bitumen and its test method, a PG-based polymer modified bitumen specification is proposed at the end of this paper, which is more suitable for the climatic conditions and existing equipment setups in the SEA region.

Keywords Polymer modified bitumen · Performance grade specification · Superpave · Multi-stress creep recovery (MSCR)

1 Background

With the increase of traffic volume in Southeast Asia (SEA), asphalt pavement such as rutting, cracking, shoving and potholes is becoming more common. As one of the most practical solutions, polymer modified bitumen has become more and

L. Jia (✉)
Kraton Polymer Singapore Pte. Ltd, Singapore 308900, Singapore
e-mail: Lu.jia@Kraton.com

M. Sethi
Kraton Polymers India Private Limited, Mumbai 400076, India

K. Bob · F. Gary
Kraton Polymers LLC, 16400 Park Row, Houston, TX 77084, USA

© The Author(s), under exclusive license to Springer Nature Switzerland AG 2022

H. R. Pasindu et al. (eds.), *Road and Airfield Pavement Technology*, Lecture Notes in Civil Engineering 193, https://doi.org/10.1007/978-3-030-87379-0_29

more popular since the 2000s. Meanwhile for quality control and supply contract requirements, different specifications for polymer modified bitumen have also been developed and applied in SEA which can be divided into 3 categories as listed below:

(1) Superpave Performance Grade (PG) system, such as in Singapore, Malaysia and Hong Kong;
(2) Superpave Performance Grade (PG) system plus special requirements, such as Indonesia. In the Indonesian PMB specification, the ductility after RTFOT aging is required in addition to the normal Superpave PG requirements.
(3) Empirical grading systems based on Pen and softening point, such as Vietnam and Thailand. PMB products in Vietnam have been classified as PMB I, II and III depending on the penetration value and softening point value, and elastic recovery tested at 25 °C. In engineering practice, for important projects, some of the performance tests in PG are also run to make sure that the PMB products can meet at least PG76 or PG82 requirements. For the PMB specification in Thailand, its classification is based on penetration and softening point value with extra empirical requirements such as toughness and tenacity, elastic recovery at 25 °C and ductility at 13 °C.

Meanwhile, in some SEA countries, there are several highway projects under directly foreign financial investment, such as JIIC or China's Belt and Road Initiative. Most of the project design and requirements are directly referenced from or influenced by those investing countries. As a result, the specification for PMB also uses the PMB specification either from Japan or China. Such complex scenarios are beyond the scope of this paper.

Since almost all of the SEA region has a tropical climate, the Performance Grade (PG) specifications used in SEA often only includes the high temperature grade requirements without low PG grade. So formal PG grades such as PG 70-16 or PG 76-16 are simplified to PG70, PG76 or PG82.

For most countries in SEA only Dynamic Shear Rheometer (DSR) for the original sample and Rolling Thin Film Oven Test (RTFOT) aged sample are required. Some countries will also include requirements for DSR after RTFOT aging plus Pressure Aging Vessel (PAV) aging tested at 31 °C for PG70 or 34 °C for PG76 (Indonesia specification) based on the intermediate temperature formula for PG70-16 and PG76-16.

Consequently, most PMB producers and 3rd party test laboratories only use the DSR (Dynamic Shear Rheometer) with 25 mm plates for testing and RTFO (Rolling Thin Film Oven) for conditioning. 8 mm DSR plates BBR (Bending Beam Rheometer) and PAV (Pressure Aging Vessel) are not used.

In order to check the reasonableness behind using PG specifications in SEA in this paper, the development of the Performance Grade specification system in North America is reviewed, and the rationale behind some aspects of the PG system are also explained.

Based on the early learnings from implementing PG specifications in North America, the tropical climate and pavement distresses in SEA are considered and a PMB specification system suitable for SEA region is proposed.

2 Superpave PG Specification in North America

2.1 Development of Superpave Specification in US

From 1987 to 1993, the Strategic Highway Research Program (SHRP) conducted a $50 million research to develop a new system named SUperior PERforming PAVEments, aka. "Superpave". And there were 3 main parts in Superpave [1]:

(1) Performance based asphalt binder specification (PG asphalt binder specification) and the supporting test methods and equipment.
(2) A design and analysis system based on volumetric properties of asphalt mix. This is Superpave mix design method.
(3) Mix analysis tests and performance prediction models.

By 2001, the majority of US states had adopted the Superpave PG asphalt binder specification.

2.2 Superpave Binder PG Grade Selection

Superpave performance grades are represented by two numbers. The first is the pavement high temperature at a depth of 2 cm and the second is the pavement low temperature. The original high temperature model was based on the highest average seven-day maximum pavement temperature while the low temperature is the actual pavement temperature. More recently, the high temperature model was changed to a cumulative damage model which more accurately represents damage from infrequent high temperatures versus long term high temperatures. The models include reliability which is normally selected as 95 or 98%.

The original LTPPBind model used weather station data from across North America. The most recent model uses National Oceanic and Atmospheric Association (NOAA) satellite data and thus can be applied anywhere in the world [2].

A "base" PG binder grade is determined by the climatic high and low pavement temperatures. To account for heavy and/or slow moving traffic, "grade bumping," increasing the PG high temperature grade, was implemented. The difference between the PG high and low temperatures is the Useful Temperature Interval (UTI). As a general rule-of-thumb, for a PG binder if the UTI is more than 90 °C, then the modification is required.

Table 1 Performance tests in PG specification and related pavement performance/distress

Bitumen samples	Test methods	Related pavement performance/distress
Original sample	Rotational viscosity	Workability & compactability
	DSR	Rutting or perpetual deformation at high temperature
Short-term aging sample	RTFO aging	Aging during mix production construction
	DSR	Rutting or permanent deformation at high temperature
Long-term aging sample	RTFO + PAV aging	Aging during asphalt pavement service life
	DSR	Fatigue cracking at intermediate temperature
	Bending beam Rheometer (BBR)	Low temperature cracking

2.3 Test Methods in PG Specification

The Superpave specifications are designed to be performance-related to address specific pavement distresses, rutting, fatigue cracking and thermal cracking.

And it is based on the idea that an asphalt binder's properties should be related to the conditions under which it is used. For asphalt binders, this involves expected climatic conditions as well as aging considerations. Therefore, the PG system uses a series of tests but specifies that a particular bitumen binder must pass these tests at specific temperatures that are dependent upon the specific climatic conditions in the area of use.

Superpave performance grading used bitumen binder tests and their link with pavement performance are listed in Table 1.

2.4 Behind Superpave Specification

Climatic limitation. Although the concept of the Superpave PG specification has become common around the world, there is one key factor that can easily be neglected in tropical or maritime climates. That is the existing Superpave PG specification was specifically designed or developed for the continental climatic condition in the U.S. and Canada. Its suitability for other climatic conditions has never been closely evaluated, especially for the SEA region where the ambient temperature never drops below freezing.

According to the database from LTPPBind, the typical low temperature PG grade by weather stations with 98% reliability is illustrated below (Pic. 1) below [3]. From the picture, it can be found that the lowest pavement temperature at southern states in NA is only −10 °C (Fig. 1).

Theoretically, the appropriate base binder grades along the U.S. Gulf Coast should be PG XX-10. But in reality, most southern states selected −22 as the lowest PG grading requirement as listed in Table 2 [4]. The only exceptional case are PG 64-10 in California, where a special mixture design method are adopted and different with all the other states in NA. The reason behind those are ensure that for basic binders used in asphalt pavement, it should have minimal UTI more than 80 °C.

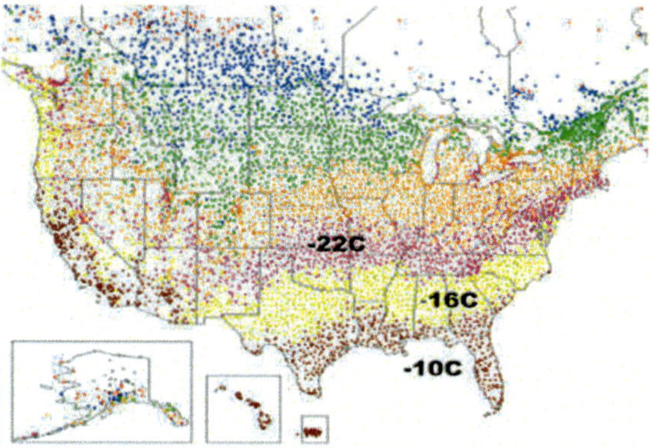

Fig. 1 Low temperature grade from LTPPBinder database, emphasis on Southern US

Table 2 General performance grade of base bitumen in southern states of US

State	Minimal binder grade	PG range[a] °C	Common binder grade	PG range[a] °C
HI	PG 64-16	80	PG 64-16	80
CA	PG 64-10	<u>74</u>	PG 64-16	80
	PG 64-16	80		
AZ	PG 70-10	80	PG 70-10	80
	PG 76-16	92	PG 76-16	92
	PG 64-16	80	PG 64-16	80
NM	PG 64-22	86	PG 64-22	86
TX	PG 64-16	80	PG 64-22	86
LA	PG 64-22	86	PG 67-22	89
MS	PG 64-22	86	PG 67-22	89
AL	PG 58-22	80	PG 67-22	89
GA	PG 58-22	80	PG 67-22	89
FL	PG 58-10	80	PG 67-22[a]	89
SC	PG 64-22	86	PG 64-22	86

[a]For unmodified grades, Florida DOT requires different grades depending on how much reclaimed asphalt pavement is used in the mixture. For 0–20% RAP, PG67-22 is required

Table 3 Statistical summary of basic Performance Grade used in US

PG grades		Low temperature							
		−10	−16	−22	−28	−34	−40	−46	−52
High temperature	**40**	2	11	38	43	35	23	32	37
	46	14	36	35	98	143	97	55	28
	52	28	53	119	313	524	436	100	18
	58	32	74	321	810	585	232	1	0
	64	201	471	726	580	136	27	0	0
	70	441	335	49	4	0	0	0	0
	76	71	1	0	0	0	0	0	0

PG ranges	56	62	68	74	80	86	92	98	104
Subtotal	25	102	163	527	1712	2595	1743	500	46
Percentages (%)	0.3	1.4	2.2	7.1	23.1	35.0	23.5	6.7	0.6

For all the locations in both US and Canada, including the southern states that have a subtropical to semitropical climate, most of the adopted PG grades for binder have a UTI of more than 80 °C, as illustrated in Table 3 [4] above.

Some researchers have pointed out that the higher UTI inherently means a lower slope on the master curve of binder, which can be regarded as a much larger penetration index or low temperature susceptibility. Conversely, a smaller UTI means greater temperature susceptibility. A UTI of less than 80 °C may allow the use of bitumen that previously would be considered unacceptable due to too low penetration index [4].

2.5 *Challenges of Superpave PG Specification.*

From the performance control side, grade bumping based on the design traffic volume and the slow moving or static traffic loads can provide extra performance to resist permanent deformation deformation, as listed in Table 4. But it also brings a problem for fatigue damage control at intermediate temperature. In Superpave PG grading, the intermediate test temperature of bitumen binder is calculated from the high

Table 4 Effect of traffic amount and traffic conditons on grade bumping

	Desgin traffic volume	Traffic coditions
No Bumping	<10 × 10^6	Fast
Bump 1 Grade	10–30 × 10^6	Slow
Bump 2 Grades	>30 × 10^6	Standing or intersection

Table 5 Intermediate temperature for PG grade with and without high temperature grade bump

	Superpave PG	High temp. (°C)	Low temp. (°C)	Intermediate temp. (°C)
No bumping	PG 58-22	58	−22	22
Bump 1 Grade	PG 64-22	64	−22	25
Bump 2 Grades	PG 70-22	70	−22	28
Bump 3 Grades	PG 76-22	76	−22	31

performance grade and low performance grade as below:

$$\text{Intermediate temp} = (\text{high PG grade temp} + \text{low PG grade temp})/2 + 4 \quad (1)$$

A problem arises with this formula when grade bumping. A higher test temperature is more lenient and so the intermediate test temperature should be based on the climate grade, not the bumped grade. If applied incorrectly for grade bumping, as shown in Table 5, the intermediate test temperature also increases. Therefore an unmodified bitumen such as PG 64-22, which is roughtly equivalent to Pen 60–70, has a stricter figure resistance requirement (tested at 25 °C) than polymer modified bitumen such as PG 76-22, whose fatigue performance was tested at 31 °C.

In order to address the problem, some southern states in US, such as Alabama, Florida and Louisiana, have required that for modified binders, the PAV DSR should be tested at same temperature as the base grade bitumen used locally. For example, in Florida, the base bitumen used is PG 67-22, whose intermediate test temperature is 26.5 °C. And the PMBs used there are PG 76-22 and "High Polymer" (HiMA™) bitumen, which are also required to be tested at 26.5 °C for fatigue performance [5–9].

Development of Superpave PG Specification. Although the Superpave PG specification is the basis for state DOT specifications in the US, many agencies have added empirical binder test methods, "PG Plus," to serve the specific situation in each state. Currently, 90% of the states in the US have some sort of elasticity requirement such as elastic recovery, force ductility, toughness and tenacity, phase angle, MSCR recovery and sometimes prescriptive requierments [4].

Since 1993, the PG specifications have proved adequate for permanent deformation for unmodified binders. However, ample evidence showed that G*/sinδ underpredicts the resistance to permanent deformation of elastomer-modified binders. This is partly due to the low strain applied, staying within the linear viscoelastic range, and partly to the oscillatory nature of the test which dampens delayed elastic recovery.

The Multiple Stress Creep and Recovery (MSCR) test published in T350 and ASTM D7405) was developed ten years ago. It uses the well-established creep and recovery test concepts to evaluate the asphalt binder's potential for permanent deformation overcoming the limitations of oscillatory DSR [10]. Using the DSR on RTFO-aged binder, a one-second creep load is applied followed by a 9 s recovery period. This is applied 10 times at a creep load of 0.1 kPa to equilibrate the specimen followed by another 10 cycles at 0.1 kPa for data collection and 10 more cycles at 3.2 kPa.

Table 6 Effect of traffic amount and traffic conditons on grade bumping

Traffic	M 320 test temp (°C)	M 332 J_{nr} limit at (kPa^{-1})
Standard (S)	64	4.5
Heavy (H)	70	2.0
Very heavy (V)	76	1.0
Extremely heavy (E)	82	0.5

This protocol captures both the non-recoverable creep compliance of the specimen, Jnr and the stress sensitivity of the binder. According to related research, MSCR test results correlate better with field rutting performance of asphalt pavements for various bituminous materials [11].

The MSCR test goes well outside the linear viscoelastic regime, so time temperature superposition which is the basis for PG grade bumping is no longer valid. For that reason, MSCR testing should be performed only at the climatic high temperature for conventional grades. For increasingly high traffic loads, decreasingly low values of Jnr are required. A comparison of traffic load handling between M 320 and M 332 is shown in Table 6.

Since the binder will be tested to a constant high and low grade temperature, the issue of selecting the proper test temperature for intermediate temperature PAV DSR due to grade bumping becomes a moot point.

3 Superpave PG Specification in SEA Region

3.1 Climatic Conditions in SEA Region

The climate of Southeast Asia is mainly tropical and so the weather is hot and humid most of the year. There is a lot of rainfall during the wet, monsoon season, due to seasonal shift in winds and the effect of the tropical rain belt, but the temperature is still warm.

Historical weather data shown in Fig. 2 [12] indicates that the PG climatic low temperature for SEA is +4 °C or higher.

3.2 Typical Asphalt Pavement Distress

The most common distresses of asphalt pavement in SEA region are:

(1) Permanent deformation at high temperature.
(2) Fatigue cracking which mainly happens in the temperature range of 15–25 °C.

Fig. 2 Typical temperature grade requirement in SEA

(3) Potholes are typically pavement failures that result from deteriorated fatigue cracks and/or pavement structural failure. Consequently, they are less driven by binder selection than by inadequate pavement structural design and/or inadequate drainage.

3.3 Principles to Develop PG Specification in SEA Region

The main principles for PG specifications used in the SEA region should follow the same principles as developed in the United States:

(1) For the high temperature grade: Use the high temperature grade limit as determined by LTPPBind using the desired level or reliability.
(2) For low temperature grade: As the actual low pavement temperatures do not adequately describe a high quality bitumen, it is recommended to use a grade temperature resulting in a Useful Temperature Interval of at least 80 °C, This will result −22 °C or −16 °C. In any case, assure that:

   a. For the base bitumen: the range of Performance grading is greater than 80 °C to ensure that, under conventional gradeing, the bitumen will have an adequate penetration index.
   b. For polymer modified bitumen: the PG grade range is greater than 92 °C to ensure an adequate level of modification.

(3) Fatigue resistance of binders in SEA region is more important since low temperature cracking is not a concern due to the climate The appropriate intermediate

Table 7 Recommended performance grade for base bitumen and PMB used in SEA region

Old specification	Minimal superpave PG specification	General superpave PG specifcation
Pen 80–100	PG 58-22	PG 58-28
Pen 60–70	PG 64-16	PG 64-22
PMB PG70	PG 70-22	PG 70-22
PMB PG76	PG 76-16	PG 76-22
PMB PG82	PG 82-10	PG 82-16

temperature should be selected carefully to avoid the confusion in existing PG specification in NA caused by high temperature bumping:

a. For most commonly used base bitumen Pen 60–70 in SEA region, the most possible closed PG should be PG 64-16, which has an intermediate temperature at 28 °C.
b. For SBS modified bitumen it should have better anti-cracking performance if compared with normal Pen 60–70, therefore, the recommended anti-fatigue cracking performance check should be also carried out 25 °C which is a higher requirements compared with PG 64-16 tested at 28 °C.

3.4 Recommended Superpave PG Grade Binders in SEA

Therefore, after considering all above factor, the list binder PG specification in Tables 7 and 8 are recommended for the general use of asphalt pavement in SEA region.

3.5 Application of Superpave PG Grade Binders in SEA

As for most of the commercial lab and/or quality control/assurance lab in contractor or PMB producer side, some of the critical equipment in this recommend specification have not been equipped or installed, such as PAV, BBR or 8 mm loading head of DSR.

Considering the reality in SEA region, the author of this paper would recommend paving industry to consider use listed tests as replacements for fatigue control in the proposed specification. The proposed requirements for alternative test methods are referred from the literature, which cover the relationship between RTFOT aging and PAV aging and pointed out that the G*Sinδ value of binder with RTFOT aging only is around half of the value of binder after RTFOT and PAV aging together [13]. Therefore, the specification requirement has been reduced from 5000 kPa down to 2500 kPa (the Authors would also suggest to strengthen the requirement as 2000 kPa to assure better performance). Beside that, for users who only have 25 mm loading plates with DSR equipment, the author suggest to adjust the gap size of DSR test

Table 8 Recommended Minimal PG specification for base bitumen and PMB in SEA region

Test property	Method	Unit	Requirements			
High temperature PG grades[a]	AASHTO M 320 or ASTM D6373		PG 64	PG 70	PG 76	PG 82
Minimal low temp PG grade[b]			−16	−22	−16	−10
Penetration at 25 °C	ASTM D5	1/10 mm	60–70	Report	Report	Report
R&B softening point	ASTM D36	°C	≥48	Report	Report	Report
Elastic recovery at 25 °C[c]	ASTM D6084	%	N.A	≥75	≥85	≥90
Rotational Viscosity@135 °C	AASHTO T 316 ASTM D4402	Pa·S	≤3.0			
Dynamic shear, G*/Sin δ @10 rad/s ≥ 1.0 kPa, Test Temp,	AASHTO T315 ASTM D7175	°C	64	70	76	82
Flash point	ASTM D92	°C	≥230			
RTFOT aged	AASHTO T 240 ASTM D2872					
Dynamic shear, G*/Sinδ @10 rad/s ≥ 2.20 kPa, Test temp	AASHTO T 315 ASTM D7175	°C	64	70	76	82
Mass loss	AASHTO T240 ASTM D2872	%	≤1.0			
Elastic recovery at 25 °C[c]	ASTM D 6084	%	N.A	≥70	≥80	≥95
RTOFT + PAV aged	AASHTO T240 + R28 ASTM D2872 + D6521					
Dynamic shear, G*Sinδ @10 rad/s ≤ 5000 kPa, test temp	AASHTO T 315 ASTM D7175	°C	28	25[d]	25[d]	25[d]
Bending beam Rheology	AASHTO T 313 ASTM D6648					
Creep Stiffness ≤ 300 MPa, @test temp		°C	−6	−12	−6	0
M-Value ≥ 0.300 MPa, @ test temp		°C	−6	−12	−6	0
Storage stability	EN 13399					
Softening point difference	ASTM D36	°C	N.A	≤2		

(continued)

Table 8 (continued)

Test property	Method	Unit	Requirements		
Penetration difference	ASTM D5	1/10 mm	N.A	≤5	
Premium performance requirements[e]					
MSCR, %R @ 3.2 kPa, test at 76 °C	AASHTO T350 or ASTM D 7405	%	N.A	N.A	≥90
MSCR, Jnr @ 3.2 kPa, test at 76 °C		kPa^{-1}	N.A	N.A	≤0.10
Modifiers/additives in PMB[f]			N.A	Additive used in formulation, such as PPA, Wax, SBS or SBR, etc.	

[a]As this specification recommended here are still based on the Superpave performance grade which still allow to use the high temperature bumping according to traffic vol or speed
[b]For the negative grade listed in this table, the recommend value should be the lowest requirement or minimal requirements, project can select lower negative grade to enhance the performance requirement. And PG 76-22 and PG 82-16 are recommended as general selection in SEA region
[c]For the fatigue performance requirement, for the base bitumen, recommend to use 28 °C as minial requirement and 25 °C as general test temperature and for all the other PMB grade, a higher grade requirement is adopted as 25 °C as minimal requirements and 22 °C as general test temperature. Here a constant temperature requirement for PMB was used to solve the intermediate temperature confusion as explained in Sect. 3.3
[d]The elastic recovery test are included in this specification to ensure that adequate level of modification are employed rather than blowed soft bitume is used as wholely or partially replacement
[e]The premium performance requirements were recommended for specific project which expected for high performance, "High Polymer" PMB products or facing challenging heavy duty traffic, such as port, runway or junction, which are reference from Florida DOT specification [9]. This specification requires performing MSCR testing at higher temperature than the climate to develop a response to loading at the stress level applied during the test
[f]As in SEA region, there are many PMB producers are using lots of additives rather than SBS, SBR, etc., which cause lots of uncertainty for the long term performance of asphalt binder, mixture or pavement. The statements for modifier used in PMB products will help local authorities or highway engineers to collect more information to verify their long term performance in SEA region, which will benefit the industry for future specification development

from 2 to 1 mm to ensure the test results will not beyond the measure limitation of DSR equipment. But unfortunately, the proposed requirements for that loading condition are still under development now in our lab (Table 9).

4 Conclusion and Future Direction for PG Specification in SEA Region

After reviewing the development of Superpave PG specification in US or North America, the basic principles and consideration behind the PG specification had been disclosed and summarized in this paper. And the latest development of PG specification in NA and US had also been updated. Combined that basic principles

Table 9 Proposed quality controls test for fatigue performance test

Test method	Sample aging conditions		Loading plante		gap sizes	Requirements
	RTFOT	PAV	8 mm	25 mm		
Standard PG specification test	YES	YES	YES		2 mm	Dynamic shear, G*Sinδ @ 10 rad/s < 5000 kPa @ test temp
Alternative method 1	Yes	No	YES		2 mm	Dynamic shear, G*Sinδ @ 10 rad/s < 2500 kPa* (2000 kPa) @ test temp
Alternative method 2	Yes	No	No	Yes	1 mm	Under development

for PG specification with the climatic conditions in SEA region and the typical pavement damage, it has been found that existing PG specification used in SEA region had many shortcomings such as the negative grade, intermediate temperature for fatigue test, etc. This compromised the ultimate purpose of PG specification in SEA region to control the pavement damage under typical local climatic conditions. Based on those consideration, a new Performance Grade Specification in SEA region has been proposed and recommended. Besides that, the author also provided some alternative methods for local users with test equipment limitations or constrains.

As for the future development of Performance grade in SEA region, the author belive that with the wide utilization and implementation of MSCR and other performance test methods in paving industry in SEA region, the future direction for PG specification could become:

(1) Performance grade specification combined with the MSCR requirements based on the true temperature condition without considering the grade bumping for traffic volume or speed.
(2) Test equipment for low temperature performance evaluation such as BBR or PAV will become available and popular in SEA region, which will help the end user to determine the negative grade of binder. It will also help the paving industry in SEA region to catch the latest development of binder performance evaluation method based on BBR test, such as Delta Tc [14] etc.
(3) Combined with latest test method to evaluate the fatigue performance of binder. There are already some researches has been carried out in Euro and US to find out other test methods rather DSR to evaluate the fatigue performance of binder, such as Linear Amplitude Sweep (LAS) [15], etc.

Acknowledgements The Authors believe the information set forth above to be true and accurate, but any findings, recommendations or suggestions that may be made in the foregoing text are without any warranty or guarantee whatsoever, and shall establish no legal duty or responsibility on the part of the Authors or their employer(s). Furthermore, nothing set forth above shall be construed as a recommendation to use any product in conflict with any existing patent rights. We expressly disclaim any and all liability for any damages or injuries arising out of activities related to use of information herein.

References

1. McGennis RB, Shuler S, Bahia Federal HU (1994) Background of superpave asphalt binder test methods. Asphalt Institute. FHWA-SA-94-069
2. https://infopave.fhwa.dot.gov/Tools/LTPPBindOnline. Last access 2021/04/20
3. Blow MD (2016) Asphalt binder basics specifications, history and future. In: North Dakota asphalt conference (2016)
4. Bob Kluttz (2014) Latest developments in SBS polymer-modified bitumen technology. In: Society of asphalt technology workshops
5. US State Binder Specifications homepage, http://www.asphaltinstitute.org/engineering/specification-databases/us-state-binder-specifications/. Last access 2021/03/30
6. Alabama State DoT Binder Specification. Asphalt institute. http://www.asphaltinstitute.org/engineering/specification-databases/us-state-binder-specifications/. Last access 2021/03/30
7. Louisiana State DoT Binder Specification. http://www.asphaltinstitute.org/engineering/specification-databases/us-state-binder-specifications/. Last access 2021/03/30
8. California State DoT Binder Specification. http://www.asphaltinstitute.org/engineering/specification-databases/us-state-binder-specifications/. Last access 2021/03/30
9. Florida State DoT Binder Specification. http://www.asphaltinstitute.org/engineering/specification-databases/us-state-binder-specifications/. Last access 2021/03/30
10. The MSCR test in plain language, Technical Bulletin TB 03.16.17 P1, McASPHALT Industries Limited
11. The multiple stress creep recovery (MSCR) procedure, Asphalt institute, TechBrief FHWA-HIF-11-038. http://www.asphaltinstitute.org/engineering/mscr-information/. Last access 2021/03/30
12. Nurkamal Y (2018) Application of high performance PMBs in Asia Pacific Airport Projects. In: 11th Asia Pacific Conference on Transportation and the Environment
13. Phromsorn CJ, Kennedy TW (1995) Evaluation of laboratory methods simulating aging effects of asphalt binder. Transp Res Rec, p 13–20. ISSN: 0361-1981
14. Asphalt Institute Technical Advisory Committee (2019) State-of-the-knowledge: use of the delta Tc parameter to characterize asphalt binder behavior. Asphalt Institute Technical Document. ISBN: 978-1-934154-77-9
15. Hintz C, Velasquez R, Tabatabaee H, Bahia H (2019) Fatigue characterization of asphalt binders with the linear amplitude sweep (LAS). Modified Asphalt Research Center

Factors Affecting the High Early Strength Development and the Methods for Testing High Early Compressive Strength of the Rigid Pavement

Wei-Chien Wang, Shuei-Lien Fang, Tien-Yu Chen, Tai-Chiang Kao, Chang-Chu Liu, Yu-Yang Li, and Hoang Trung Hieu Duong

Abstract Maintenance highway rigid pavement can cause a significant impact on traffic congestion due to the lack of the traffic lane. Typically, the general maintenance plans are arranged from midnight to early morning before the morning peak traffic periods to mitigate the possibility of traffic congestion. As a result, to open to traffic, the maintenance highway rigid pavement usually requires the concrete to meet a compressive strength requirement of 210 kgf/cm^2 after 4–6 h. The high early strength mix design (S-mix design) is examined and published by the Taiwan Freeway Bureau, MOTC. However, this research shows that the S-mix design's early compressive strength is significantly affected by the exterior temperature. Therefore, it is practically important to control the temperature at the early ages of the concrete. In addition, the maturity method and rebound hammer test results can be used to evaluate high early compressive strength of the construction site. Due to the rigid pavement requires high early compressive strength in a short period of time, taking the specimen to the laboratory to conducting the compressive strength test is challenging. Thus, using the rebound hammer result to simulate the hydraulic compressive testing value is one of the suitable method for evaluating high early compressive strength

W.-C. Wang · Y.-Y. Li (✉) · H. T. H. Duong (✉)
National Central University, Taoyuan City 320, Taiwan
e-mail: yuyang111422@g.ncu.edu.tw

H. T. H. Duong
e-mail: trhieu.ncu@g.ncu.edu.tw

W.-C. Wang
e-mail: weichien@ncu.edu.tw

S.-L. Fang · T.-Y. Chen · T.-C. Kao · C.-C. Liu
Taiwan Freeway Bureau, Northern Region Branch Office, Taipei City 114, Taiwan
e-mail: fang@freeway.gov.tw

T.-Y. Chen
e-mail: tianyu@freeway.gov.tw

T.-C. Kao
e-mail: tck0909@freeway.gov.tw

C.-C. Liu
e-mail: ccl0111@freeway.gov.tw

© The Author(s), under exclusive license to Springer Nature Switzerland AG 2022

H. R. Pasindu et al. (eds.), *Road and Airfield Pavement Technology*, Lecture Notes in Civil Engineering 193, https://doi.org/10.1007/978-3-030-87379-0_30

of the construction site. The method for testing the high early compressive strength of the rigid pavement is analysis of the correlation between rebound hammer test results and the hydraulic compression testing value, creating the S-mix design, then establish the maturity index value of the S-mix design.

Keywords High early compressive strength · Rigid pavement · Temperature · Maturity method · Rebound hammer

1 Introduction

Nowadays, rigid pavement is generally used in highway and airport. For long-term application, the rigid pavement usually occurs in corner breaks, transverse cracking, scaling and other damage. When the damage reached a critical condition, the pavement can be repaired by a variety of methods such as partial depth patching, full-depth patching with concrete pavement to repair the damaged pavement [1]. To avoid the problem of traffic congestion, on-site rigid pavement's maintenance is usually constructed during midnight and open to traffic the next morning. The open traffic requirement conditions are usually evaluated by the compressive strength of the rigid pavement. According to the International Building Codes (IBC), the minimum compressive strength requirements allowed for open traffic is 175 kgf/cm^2, 210kgf/cm^2 is the minimum design compressive strength requirements in roads, bridges and buildings, while 280 kgf/cm^2 is the minimum design compressive strength requirements in airport rigid pavement [2]. To achieve the requirement for quick roads repair, bridges, and high-rise building's slab, a new standard procedure for early strength concrete is developed, which is defined as High Early Strength Concrete (HSC). The compressive strength requirement for the High Early Strength Concrete is about 210 kgf/cm^2 after 4 to 6 h of curing [3].

In a normal rigid pavement project, the compressive strength testing method uses the same method as the building construction project. The specimen is brought to the laboratory and testing the compressive strength by a hydraulic compressor machine. However, due to the short construction time of the rigid pavement maintenance project, which is only from midnight to early morning, using the traditional testing method is extremely difficult. In normal circumstances, after on-site construction finished, the specimens will be brought to the laboratory for compressive strength testing and verification. However, due to the short curing duration requirement of the rigid pavement repair, various researchs show that the concrete characteristics are significantly affected by air temperature and moisture conditions, such as the low temperature in the winter reduces the early strength of concrete. Thus, laboratory maintenance cannot evaluate the accurate quality of on-site concrete, especially at an early age. This research develops a non-destructive method for on-site compressive strength testing by combining the maturity method with the rebound hammer.

The maturity method is a non-destructive method that can predict the compressive strength of the on-site concrete which is developed by Saul in 1951. Establishing a

Maturity-Strength relationship of the concrete in the laboratory base on the different air temperature then use the Maturity-Strength relationship to evaluate the compressive strength in the construction site. However, there is not much research showing the accuracy of maturity method when applied for high early strength concrete adding accelerator and quick setting admixtures as well as using the different curing methods. Besides, the rebound hammer is a testing instrument that is extremely fast and easy to measure the compressive strength on-site. However, the result is affected by the steel bar and the subbase of the rigid pavement. Therefore, this study establishes a method that uses the maturity method to predict the compressive strength of concrete based on concrete's core temperature and using the rebound hammer to verify the accuracy of strength prediction in the construction site.

2 Experimental Methods

To simulate the compressive strength of concrete in the site using different curing methods, this research established a Maturity-Strength relationship of the concrete using the air curing method. Curing the specimen in the laboratory at a constant air temperature of 23, 26, 30, 32 and 35 °C, the specimen core's temperatures are collected every 30 min in order to develop the Maturity-Strength relationship in 4 h and 6 h of curing. Then, the specimens will be tested the compressive strength at 4 h and 6 h by both the hydraulic compressor machine and the rebound hammer to verify the accuracy of the predicted strength of the Maturity's formula.

2.1 Material Properties and Preparation

The concrete mix design is developed base on the Freeway Bureau (MOTC) [4] that has the cement contained is 500 kg/m^3, and 0.33 W/B ratio. To mitigate the amount of cement used [5], the fine aggregate and coarse aggregate size distribution are designed as in Figs. 1 and 2.

The concrete admixture is Type E high range water-reducing with the ratio of 7% of the cement weight. The specimens are labeled according to the rule Cement ratio (kg/m^3) x Air temperature (°C) x Curing method (OA-Air curing, OB-Heat insulation, OC-Electric blanket, OD-Hot water blanket curing) which is shown in Table 1.

2.2 Curing Method

The curing methods used in the research are shown in Fig. 3. In the construction site, the rigid pavement has a large surface that is surrounded by other concrete

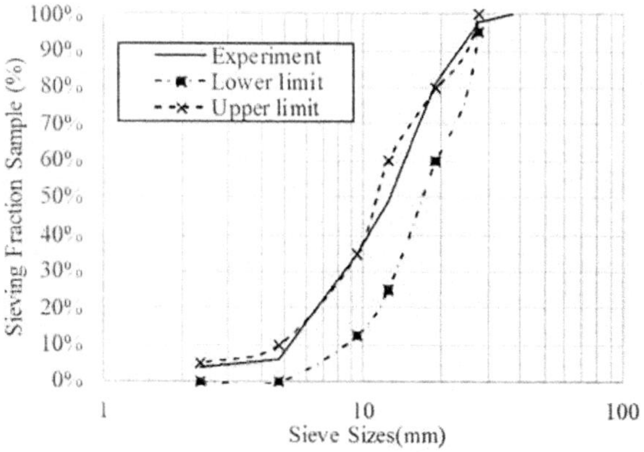

Fig. 1 Corse aggregate gradation

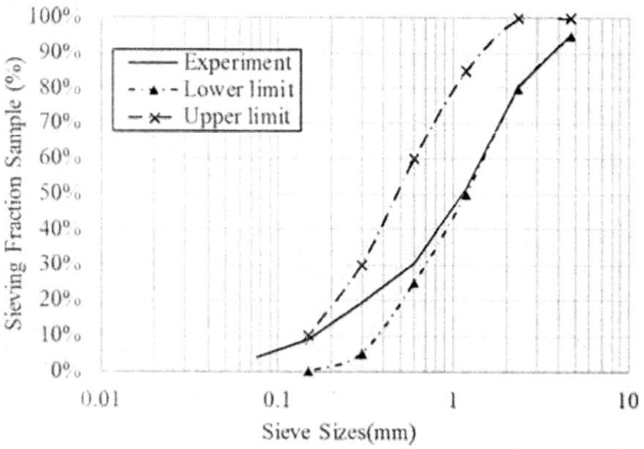

Fig. 2 Fine aggregate gradation

Table 1 Concrete mix design

Curing Equipment	No.	Water	Cement	Coarse Agg.	Fine Agg.	Admixture	W/C
		kg/m^3					
Air curing	C500-T23-OA	131	500	994	741	35	0.33
	C500-T26-OA						
	C500-T30-OA						
	C500-T32-OA						
	C500-T35-OA						
Heat insulation	C500-T23-OB						
Electric blanket	C500-T23-OC						
Hot water blanket	C500-T23-OD						

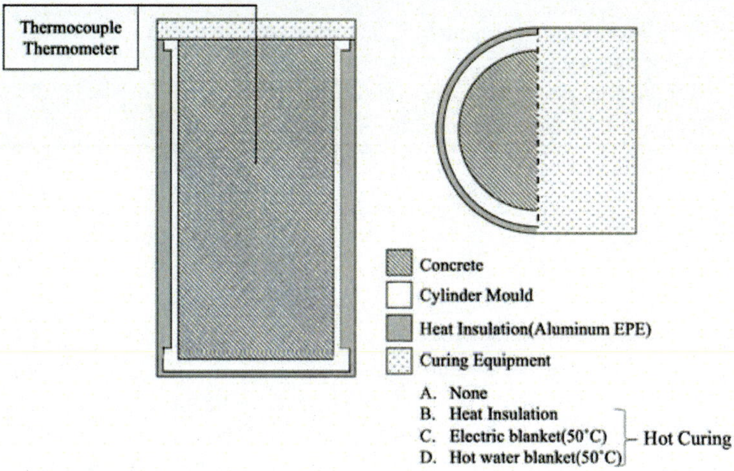

Fig. 3 The curing method design

panels. Thus, the heat of the concrete is mostly dissipated by the surface. Aiming to simulate the on-site concrete curing situation, after mixing, the mold will be covered by the 0.5 cm heat insulation blanket (Aluminum EPE) to prevent losing the temperature in the edges and bottom. The top of the specimen is covered differently base on the designed curing methods such as (A) Air curing, (B) Heat Insulation, (C) Electric blanket, (D) Hot water blanket. In addition, the electric blanket is heated to 50 °C, and the hot water blanket curing the concrete with 50 °C hot water every 5 min.

2.3 *Compressive Strength Testing Method*

According to ASTM C39, the specimens are cylinders with 15 cm diameter and 30 cm height. The thermometer device is put in the middle of the specimens, which is shown in Fig. 3. Each curing method is used separately and measure the core's temperature every 30 min during 6 h of curing. Besides, the specimens are tested for the compressive strength by both hydraulic compressor machine and rebound hammer at 4 h and 6 h of curing. Based on ASTM C1074, using the core temperature's value every certain time, the maturity value is calculated using Eq. 1, where Datum temperature is assumed to be −10 °C [6]. By developing the logarithm equation between the maturity value (M) and the compressive strength obtained by the hydraulic compressor machine, the Maturity-Strength Index is developed. Then using the rebound hammer value to compare with the Maturity value as a verification method.

$$\sqrt{M(t) = \sum (T - T_0) \cdot \Delta t} \quad (1)$$

M(t) The Maturity value at time t (°C.hours).
Δt A time interval (hour).
T_a Average concrete temperature during Δt (°C).
T_0 Datum temperature (assumed to be −10 °C).

2.4 Testing Specimen

Portland cement type I is used in this research is produced by Taiwan Cement Company, with a surface area of 3150–3550 cm^2/g and the Calcium-Silica (CaO/SiO_2) ratio of 2.92. The admixture is provided by Bonddex that contained plasticizers and C-S–H nanocrystals (C-S–H seeds). The admixture has a specific gravity of 1166 g/cm^3, with a solid composition of 23%, which can accelerate the early strength of concrete significantly. The fine aggregate and coarse aggregate are designed according to CSN 1240 (concrete particle specifications). The coarse aggregate has specific gravity of 2.58, water absorption ratio of 1.96%, particle size 8–16 mm, dry density of 1550 kg/m^3. The fine aggregate has specific gravity of 2.64, water absorption ratio of 2.04%, fineness module of 2.9.

3 Results and Discussions

3.1 Strength-Maturity Relationship

Aiming to simulation the on-site concrete curing, the specimens of the (A)—Air curing group are exposed to air and measured the core's temperature every 30 min. Based on Eq. (1), the value of maturity (M) is calculated and are shown in Table 2. According to the maturity value (M) and the compressive strength value at 4 h and 6 h curing, the logarithm equations between the maturity and the compressive strength at 4 h and 6 h are developed as formula (2) and (3). The logarithm regression analysis of 4 h and 6 h of maturity and compressive strength show a positive trend as well as a high correlation coefficient. The 4-h regression analysis correlation coefficient R^2 is 0.9207, and 6-h regression analysis correlation coefficient R^2 is 0.9858, which can indicate that maturity value is highly related to the compressive strength value and the accuracy of the compressive strength prediction using the logarithm equation and Maturity value can achieve high accuracy (Figs. 4and5 and 3).

$$PFc_4 = 624.7 \times \log(M_4) - 1223 \quad (2)$$

$$PFc_6 = 736.9 \times \log(M_6) - 1520 \quad (3)$$

$PFc_{(t)}$ Predicted value of compressive strength at time t (kgf/cm^2).

Table.2 Specimen's core temperature and maturity value using air curing method

No.	C500-T23-OA		C500-T26-OA		C500-T30-OA		C500-T32-OA		C500-T35-OA	
Age (hr)	T	M	T	M	T	M	T	M	T	M
0	24.0	0	27.6	0	28.0	0	30.1	0	30.8	0
0.5	24.2	17	28.4	19	28.8	19	32.4	21	33.0	22
1	24.0	34	27.4	38	29.1	39	33.8	43	38.1	46
1.5	23.6	51	27.6	57	30.1	59	35.6	66	36.1	69
2	23.6	68	27.5	75	31.2	80	38.4	90	39.6	93
2.5	24.0	85	29.2	95	33.4	101	42.8	117	43.8	120
3	25.2	102	31.6	116	34.6	124	43.2	143	44.8	148
3.5	25.0	120	31.6	137	35.4	146	44.2	170	45.2	175
4	25.6	138	32.4	158	34.8	169	44.9	198	43.6	202
4.5	27.2	156	32.4	179	34.2	191	46.1	226	44.2	229
5	27.2	175	32.8	200	34.2	213	46.1	254	44.2	256
5.5	28.4	194	32.8	222	34.3	235	46.2	282	44.3	283
6	28.4	213	33.0	243	34.8	257	46.8	310	44.8	311

T Concrete core's temperature (°C)
M Maturity value (°C.hr), T_0 is assumed to be −10 °C

Fig. 4 Logarithm regression analysis of 4-h maturity and compressive strength

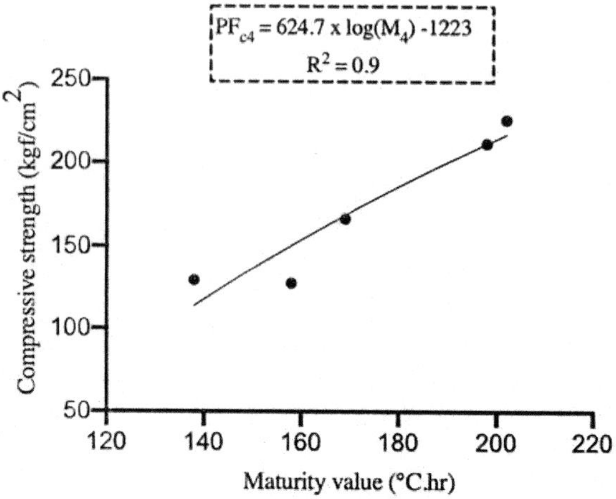

Fig. 5 Logarithm regression analysis of 6-h maturity and compressive strength

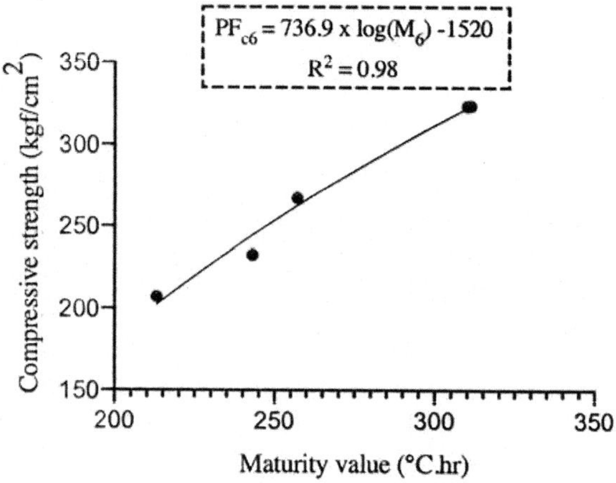

Table.3 Compressive strength and maturity value at 4 h and 6 h curing

No.	4 h curing		6 h curing	
	Fc_4	M_4	Fc_6	M_6
C500-T23-OA	129.1	138	207.4	213
C500-T26-OA	127.3	158	232.7	243
C500-T30-OA	166.5	169	267.3	257
C500-T32-OA	211.6	198	323.6	310
C500-T35-OA	225.8	202	323.8	311

Fc Compressive strength at 4 h and 6 h (kgf/cm^2)
M Maturity value (°C.hr)

$M_{(t)}$ Maturity value at time t (°C.hour).

3.2 *Thermal Curing Methods and Prediction of Strength-Maturity Relationship*

To simulate the thermal curing method such as (B) Heat insulation,(C) Electric blanket, and (D) Hot water blanket, the specimens are treated differently depending on the different curing methods. The specimen's core temperatures are taken every 30 min to obtain the maturity value (M) according to Eq. (1), which is shown in Table 4. After that, using the hydraulic compressor machine to test the compressive strength of the specimens at 4 h and 6 h. Base on the maturity value (M), using Eq. (2) to predict the compressive strength at 4 h and using Eq. (3) to predict the compressive strength at 6 h. The comparison between the prediction compressive strength and the testing compressive strength is shown in Table 5. In addition, Figs. 6 and 7 also show that the maximum deviation of prediction compressive strength and the testing compressive strength at 4 h and 6 h is ± 8%. This result indicates that using the same maturity equation for different curing methods is efficiency.

Table.4 Temperature and maturity value of the different curing methods

No.	C500-T23-OB		C500-T23-OC		C500-T23-OD	
Age (hr)	T	M	T	M	T	M
0	29.2	0	30.0	0	30.0	0
0.5	30.2	20	32.2	21	32.8	21
1	31.2	41	34.4	43	34.2	44
1.5	33.4	62	38.3	67	36.9	67
2	34.7	85	42.2	94	39.5	92
2.5	36.0	108	45.9	122	41.8	118
3	37.1	131	46.0	150	44.0	145
3.5	38.1	155	46.1	178	45.8	172
4	39.3	180	46.1	206	47.5	201
4.5	40.6	205	44.9	233	52.0	232
5	42.0	231	44.4	260	53.0	264
5.5	42.5	258	44.1	287	55.1	296
6	42.7	284	43.8	314	56.1	329

T Concrete core temperature (°C)
M Maturity value (°C.hr), T_0 assumed to be −10 °C

Table.5 The predicted compressive strength and testing compressive strength at 4 h and 6 h

No.	4 h curing				6 h curing			
	Fc_4	M_4	PFc_4	Δ_4	F_{c6}	M_6	PFc_6	Δ_6
C500-T23-OB	177	180	186	-5%	267	284	288	-8%
C500-T23-OC	225	206	222	1%	332	314	320	4%
C500-T23-OD	231	201	216	7%	353	329	335	5%

Fc Tesing compressive strength (kgf/cm^2)

M Maturity value (°C. hour)

$PFc_{(t)}$ Predicted strength with ages in *t* (kgf/cm^2)

$\Delta_{(t)}$ The deviation: $(Fc - PFc_{(t)})/Fc$

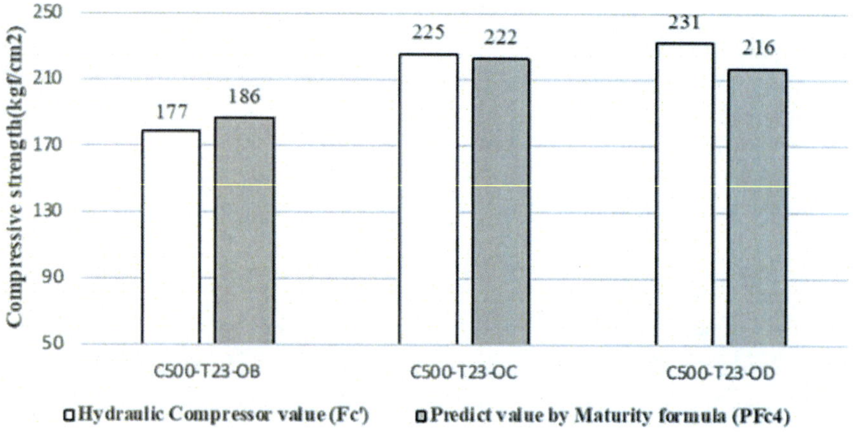

Fig. 6 The predicted compressive strength and testing compressive strength at 4 h

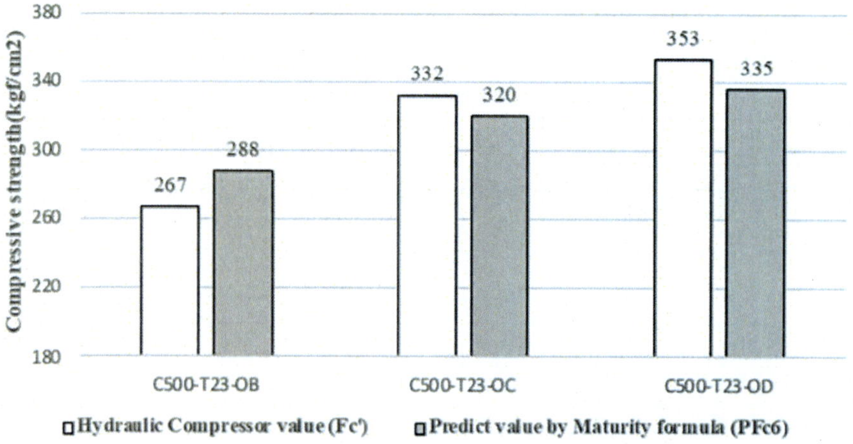

Fig. 7 The predicted compressive strength and testing compressive strength at 6 h

3.3 Estimating In-Place Strength

On the pavement's construction site, the pavements are divided into pavement 1 and pavement 2 which are cured by the hot water blanket curing method. After concrete reaches the initial setting, cover the concrete's surface with the insulation plastic sheet. Then, using the 50^{0}C hot water flows on the top of the insulation plastic sheet to maintain the curing temperature of the concrete. Collect the core's temperature of pavement 1 and pavement 2 every 30 min and use a rebound hammer to test the compressive strength of the pavements after 4 h and 6 h of curing. The compressive strength result of the rebound hammer and the prediction value by the maturity method are displayed in Table 6. Besides, Figs. 8 and 9 also show that the maximum deviation of prediction compressive strength by the maturity method and the rebound hammer's compressive strength at 4 h and 6 h is ± 9%. This result indicates that it is possible to use the maturity method to predict the compressive strength on-site and verify the result again by using the rebound hammer.

Table.6 The maturity prediction compressive strength and rebound hammer compressive strength at 4 h and 6 h

No.	4 h curing				6 h curing			
	Fc'_4	M_4	PFc_4	Δ_4	Fc'_6	M	PFc_6	Δ_6
Pavement 1	210	208	225	−7%	312	325	331	−6%
Pavement 2	258	215	234	9%	322	333	339	−5%

Fc' Compressive strength value using rebound hammer (kgf/cm^2).
M Maturity value (°C.hr).
$PFc_{(t)}$ Predicted strength by maturity method at time t (kgf/cm^2).
$\Delta_{(t)}$ The deviation: $(Fc'-PF_{c(t)})/Fc$

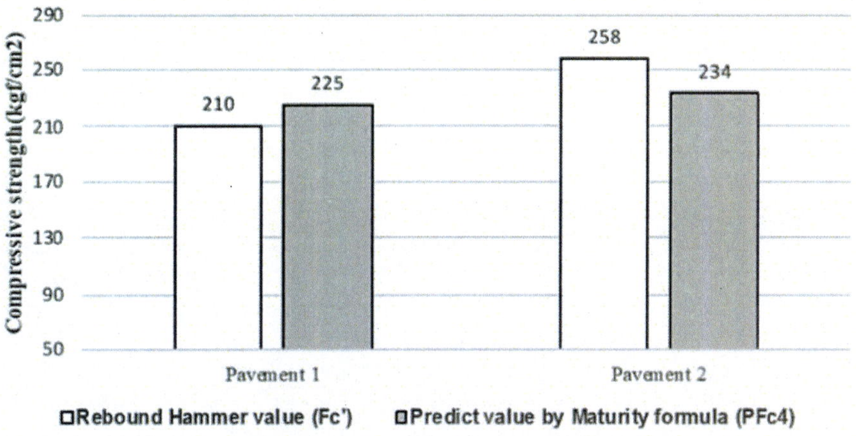

Fig. 8 The maturity prediction compressive strength and rebound hammer compressive strength at 4 h

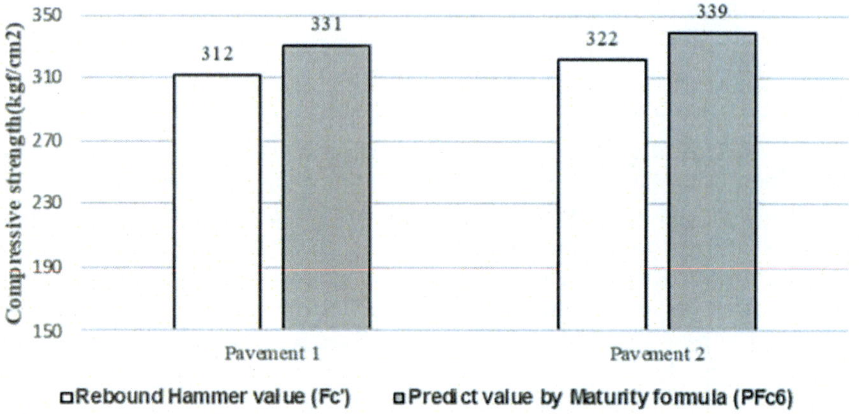

Fig. 9 The maturity prediction compressive strength and rebound hammer compressive strength at 4 h

4 Conclusion

Aiming to develop a non-destructive method for measuring the high early compressive strength on the construction site, this research establishes the maturity formula that uses for prediction of the compressive strength of concrete. Under various stable air temperatures such as 23, 26, 30, 32, and 35 °C, the strength-maturity relationship after 4 h and 6 h of curing is developed in the laboratory. Then use the developed maturity's formula to predict the compressive strength of the different curing methods and compare with the hydraulic compressor machine's result in the laboratory and the rebound hammer's result in the construction site. The results indicate that:

1. Using different constant air temperature (23, 26, 30, 32, and 35 ^{0}C) in the laboratory to predict the compressive strength development of the same mix design but different curing methods. The developed maturity equation shows that the correlation coefficient R^2 for 4 h of curing is 0.9 and the correlation coefficient R^2 for 6 h of curing is 0.98. The overall correlation coefficients reach values higher than 0.9, which can state that the maturity value and compressive strength are essentially related and using the different curing methods do not affect the accuracy of the maturity method.
2. According to CNS 1232, the maximum acceptable tolerance of the compressive strength value is 9.5%. Using the developed strength-maturity formula shows an acceptance tolerance, which achieves the maximum tolerance ± 9% for different environmental conditions as well as different curing methods.
3. In concrete subjected to real environmental conditions, the pavement is cured by the 50^{0}C hot water blanket due to the winter. It can be observed that the developed strength-maturity achieves high accuracy. The predicted compressive strength by the maturity method and the rebound hammer show a similar result that has a maximum deviation of ± 9%. Thus, combining the maturity method which

is developed in the laboratory and the rebound hammer on the construction site can be the potential method for high early strength rigid pavement.

References

1. Ansari F, Luke A (1996) "High early strength concrete forfast-track construction and repair," Report to NJDOT
2. Federal Highway Administration(FHWA) (1999) "Materials and procedures for rapid repair of partial-depth spalls in concrete pavements." Manual of Practice
3. Mitchell M, Link R, Punurai S, Punurai W, Hsu CT (2007) A very early strength concrete for highway construction. J Test Eval 35:152
4. Chen TY, Fang SL (2018) "Rigid pavement repair mixing design in 4 hour-concrete". Freeway Bureau, MOTC, ICSII 2018
5. Shakhmenko G, Birsh J (1998) Concrete mix design and optimization. In: 2nd international Ph.D. symposium in civil engineering, Budapest
6. Jeffery M, Anton K, Robert W, Samuel A (2008) "Evaluation of the maturity method to estimate concrete strength in field applications". ALDOT Res Project 930–590:8–16

Influence of Coarse Aggregate Size and Type on the Design Thickness of Rigid Pavements for Indian Conditions

Ashik Bellary and S. N. Suresha

Abstract The pavement quality concrete (PQC) is used in the construction of rigid pavements, commonly called concrete pavements. The coarse aggregates (CA) are the major constituents of the PQC mix, which influence the strength of the PQC mix. The strength of the PQC mix used in the construction of rigid pavements is one of the major factors that govern the design thickness of these pavements. In this study, an attempt has been made to characterize the influence of CA size and type on strength characteristics of the PQC mix experimentally. The crushed basalt and granitic CA are used in the preparation of the PQC mix. Also, three different nominal maximum aggregate size (NMAS) 19, 26.5 and 31.5 mm, are selected based on the IRC 44: 2017 guidelines. The experimental results are given as input in EverFE 2.24 software for determining the safe design thickness of rigid pavement over the design period. It is concluded that the safe design thickness of the PQC mix prepared using basaltic CA is comparatively lower than the PQC mix prepared using granitic CA for the same NMAS. Also, the safe design thickness of concrete pavement increased with an increase in NMAS irrespective of CA type. Thus, it is suggested to use stronger CA with lower NMAS based on the availability at the site.

Keywords Coarse aggregate (CA) · Design thickness · EverFE 2.24 · Nominal maximum aggregate size (NMAS) · Pavement quality concrete (PQC) · Rigid pavements

1 Introduction

The concrete pavement design depends on many factors such as design period, axle load spectrum, the strength of the foundation, climatic conditions and other factors. The strength properties of concrete used in pavement construction is one of the major factors that govern pavement design thickness. The concrete used in the construction

A. Bellary (✉) · S. N. Suresha
Department of Civil Engineering, National Institute of Technology Karnataka, Surathkal, India

S. N. Suresha
e-mail: sureshasn@nitk.edu.in

© The Author(s), under exclusive license to Springer Nature Switzerland AG 2022

H. R. Pasindu et al. (eds.), *Road and Airfield Pavement Technology*, Lecture Notes in Civil Engineering 193, https://doi.org/10.1007/978-3-030-87379-0_31

of rigid pavements is called Pavement quality concrete (PQC). The strength properties and other mechanical properties of the PQC are influenced by the ingredients used for the preparation of PQC mix. The ingredients used in the concrete mix preparation are coarse aggregates, fine aggregates or sand, binder material (Cement and cementitious materials) and admixtures if necessary. Coarse aggregates (CA) are the stones retained on the 4.75 mm IS sieve obtained from the quarry. These are the major constituent of pavement quality concrete (PQC). The CA used in the concrete mix act as a skeleton of concrete matrix and are responsible for the strength characteristics of the concrete mix, whereas the fine aggregates/sand contribute to the workability of concrete. The CA used in concrete pavements must be strong, durable and possess low porosity.

There are different types of CA available and are used in the preparation of PQC. Granite, limestone, dolomite, basalt are the naturally available CA. In the recent past, recycled aggregates have also been used, and efforts are being made to manufacture the CA. The different types of CA have resulted in different mechanical properties of PQC when used in the preparation. In the past, researchers have studied the influence of CA type on the mechanical properties of concrete. Ozturan and Cecan [1] studied the effect of basalt, limestone and gravel CA on the concrete with a target strength of 30, 60 and 90 MPa. They found that the concrete produced using granite CA resulted in higher compressive strengths than the concrete produced using the other two aggregates. Also, the authors concluded that the difference is the strengths of concrete made using those CA is the lowest for concrete with a target strength of 30 MPa. Similar findings were reported by Wu et al. [2].

Besides the CA type, the nominal aggregate size of CA plays a significant impact on the mechanical properties of concrete. Researchers have studied the influence of the different type of aggregates and nominal maximum aggregates size (NMAS) on the mechanical properties of concrete. Kasu et al. [3] conducted flexural strength tests on PQC specimens prepared using 10 mm, and 20 mm NMAS and reported that smaller-sized aggregates improved the static flexural strength and fatigue repetitions. Wolinski et al. [4] reported that the increase in aggregate size resulted in the reduced compressive strength of concrete specimens. Ćosić et al. [5] studied the influence of aggregate size on PQC used in pervious concrete and reported that the smaller NMAS results in better mechanical properties.

In India, different types of aggregates are available in different parts of the country. The granitic aggregates and basalt aggregates are largely used in the construction of PQC. Three different NMAS 19, 26.5 and 31.5 mm are suggested in IRC 44: 2017 [6] to prepare PQC. In the design of concrete pavements as per IRC 58: 2015 [7], the flexural strength and elastic modulus of the PQC are considered. From the literature, it is found that type of CA and NMAS of CA have a significant influence on the concrete strength properties. Thus, in this study, an attempt has been made to determine the mechanical properties of PQC prepared with granitic and basalt aggregates of three different NMAS experimentally. The obtained results are used to determine the safe design thickness for each PQC mix. The main objective of this work is to study the influence of CA size and type on the design thickness of rigid pavements.

2 Experimental Investigation

2.1 Materials

Cementitious materials

The Ordinary Portland Cement (OPC-43 grade) conforming to IS 269–2015 [8] and Ground Granulated Blast Furnace Slag (GGBS) conforming to IS 12089–1987 [9] are used in the preparation of PQC mixes. The percentage of GGBS is limited to 50% as per IRC 44–2017 [6]. The physical properties of binder materials, cement and GGBS, are tabulated in Table 1a, b, respectively.

Coarse aggregates (CA)

Two types of CA, granite and basalt are used to prepare the PQC mix in the present study. Granitic aggregates are derived from intrusive igneous rocks and are composed predominantly of silica. The grain or texture varies from fine to coarse. Granitic aggregates are more abrasion resistant and hard to break. The Basaltic aggregates are derived from the extrusive igneous rocks that are of a very dark colour (green or black) with low silica content [10]. The physical properties of both granitic and basalt coarse aggregates are evaluated as per IS: 2386 [11–14], and the results are tabulated in Table 2.

Fine aggregates

In the present research, clean, dry river sand obtained from local supplier conforming to IS: 383–2016 [15] is adopted. The physical properties of the fine aggregates are presented in Table 3.

Table 1 Physical properties of binder materials

a Physical properties of cement (OPC-43 grade)		
Tests	Test results	Requirements as per code
Initial and final Setting time	52 min and 242 min, respectively	Minimum 30 min and Maximum 600 min, respectively
Specific surface of cement	3240 cm^2/gm	Minimum 2250 cm^2/gm
Specific gravity test	3.15	-
b. Physical properties of GGBS		
Physical Properties		
Loss of ignition		0.04%
Specific gravity		2.8
Insoluble residue		0.3
Bulk density (kg/m^3)		1200
Specific surface area (m^2/kg)		370

Table 2 Physical properties of aggregates

Properties	Obtained results		Requirement as per MoRT&H (5th revision) [16] specifications (%)
	Crushed granite aggregates (%)	Crushed basalt aggregates (%)	
Aggregate Los Angeles abrasion value	24	21	Maximum 35
Water absorption	0.3	0.2	Maximum 2
Combined flakiness and elongation index	25	28	Maximum 35
Specific gravity test	2.69	2.89	–
Aggregate crushing value	21	19	–
Aggregate impact value	22	18	–

Table 3 Physical properties of fine aggregates

Tests	Test procedure as per IS-code	Results
Specific gravity test	IS: 2386–1963 (Part 3) [13]	2.54
Water absorption	IS: 2386–1963 (Part 3) [13]	1%
Sieve analysis	IS: 383–2016 [15]	Zone-II

Three gradations with different nominal sized aggregates (NMAS) are used in this research in order to study the effect of nominal maximum aggregate size (NMAS) on the Mechanical properties of PQC mixes. The adopted gradations for three different NMAS is shown in Fig. 1.

Chemical admixtures

A commercially available Chloride free Naphthalene Sulfonate (Fosroc-Conplast SP 430) having a specific gravity of 1.05 and conforming to IS: 9103–1999 [17] has been added to improve the workability of the mixture, without affecting the other concrete parameters such as segregation, ultimate strength, and concrete permeability. In order to achieve the desired slump of 25–50 mm, the amount of superplasticizer was varied up to 1% by weight of the binder.

Water

Potable water conforming to IS: 456–2000 [18] is used for mixing and curing of PQC samples.

2.2 *Mix Design of PQC*

The mix design is carried out for the M40 grade of concrete as per IRC 44–2017 [6]. The concrete mixtures with NMAS 19 mm, 26.5 mm and 31.5 mm, are designated

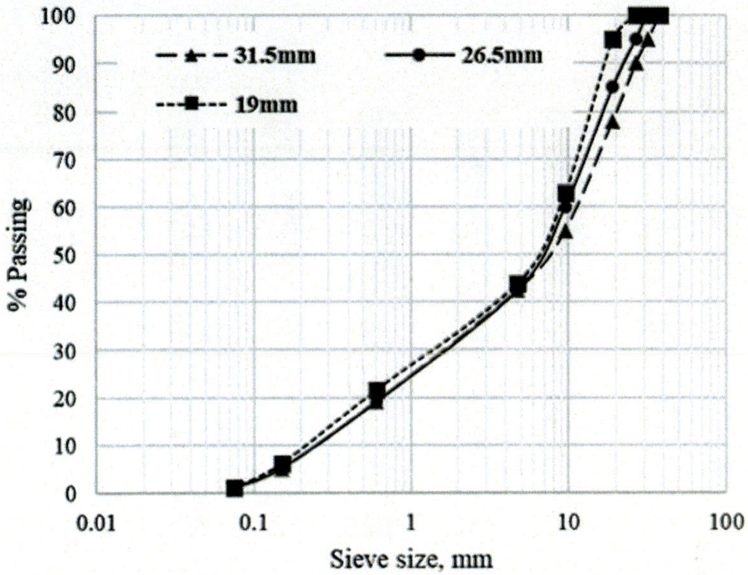

Fig. 1 Combined aggregate gradation curves for NMAS 19, 26.5 and 31.5 mm [6]

with G19, G26.5 and G31.5 for concrete mix prepared using crushed granite as coarse aggregates; B19, B26.5 and B31.5 for concrete mixes prepared using crushed basalt coarse aggregates, respectively. The proportions of concrete ingredients made using these two different aggregates and three different NMAS for M40 grade concrete are presented in Fig. 2.

The aggregates (CA and Fine aggregates) and the cementitious materials (Cement and GGBS) are dry mixed in the ribbon mixer of 125 kg capacity for about three minutes. Later water and admixture are added to the dry mix, and wet mixing is continued for another four minutes. After proper mixing, the ease of compaction is determined using the slump cone tests in conformation to IS: 1199:1959 [19]. The test is conducted immediately after mixing the ingredients. Once the desired slump is achieved, the fresh concrete is poured into moulds for preparation of the specimen in three layers for proper compaction. The wet mix is thoroughly compacted on a vibrator and then allowed to stand in the humid and cool place at ambient room temperature for about 24 h. After 24 h, the specimens are demoulded. After demoulding, the PQC specimens are cured in the water tank at room temperature. In order to evaluate the strength and other characteristics, the various specimens are cast. The details of the samples used for various tests are given in Table 4.

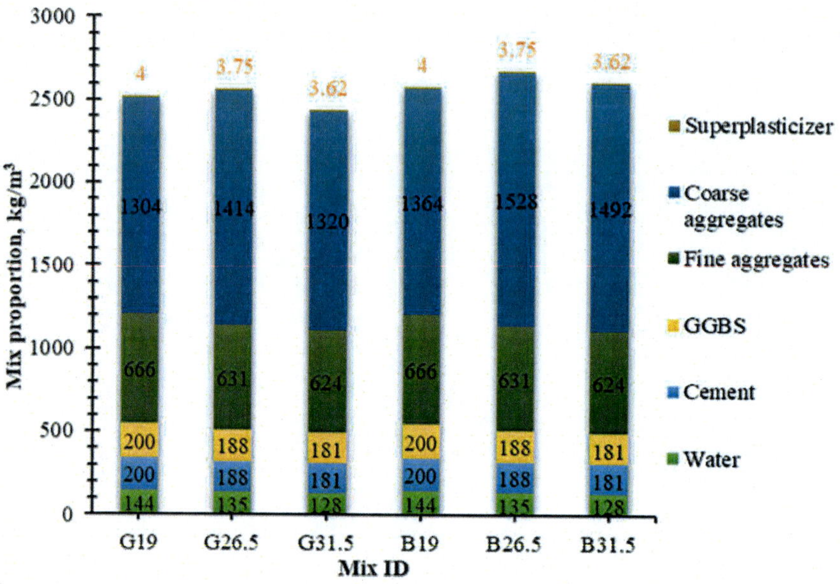

Fig. 2 Mix proportions for PQC in kg/m^3 as per IRC: 44–2017

Table 4 Different strength tests for PQC mix and specimen details

Tests	Specimen size	Relevant code
Compressive Strength	150 mm x 150 mm x 150 mm	IS: 516–1959 [20]
Flexural Strength	100 mm x 100 mm x 500 mm	IS: 516–1959 [20]
Modulus of Elasticity	Diameter -150 mm, Height-300 mm	IS: 516–1959 [21]
Split Tensile Strength Test	Diameter–150 mm Height-300 mm	IS: 5816–1970 [22]

3 Analysis of Experimental Results

3.1 *Fresh Concrete Properties (Slump)*

The slump test is conducted to measure the workability of concrete. The slump test is conducted on the concrete mixes prepared with different aggregate types and with different NMAS using the slump cone. The slump values of concrete mixes are tabulated in Table 5.

Table 5 Slump values of different concrete mixes obtained using slump cone test

Mix ID	Slump Value, mm
G19	16
G26.5	17
G31.5	18
B19	18
B26.5	19
B31.5	22

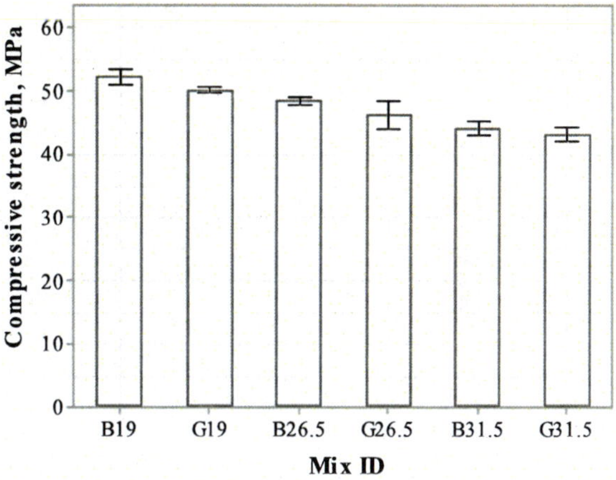

Fig. 3 Compressive strength results for different concrete mixes (after 28 days of curing)

3.2 *Hardened Properties of Concrete*

The tests are conducted according to the Indian standards (IS), and a minimum of three samples are tested for each mix and for each test. The test results (with 95% confidence interval) of the different concrete mixes along with their Mix IDs is shown in Figs. 3, 4, 5 and 6.

4 Design of Concrete Pavement Using EverFE 2.24 for Concrete Prepared Using Different CA for Indian Conditions

In India, the design thickness of concrete pavement is determined as per IRC 58–2015. The code of practice provides regression equations and design charts to determine the safe design thickness for the design period. The design procedure considers the cumulative damage that is incurred due to both traffic and temperature. However, the regression equations and the design charts developed to determine the flexural stress

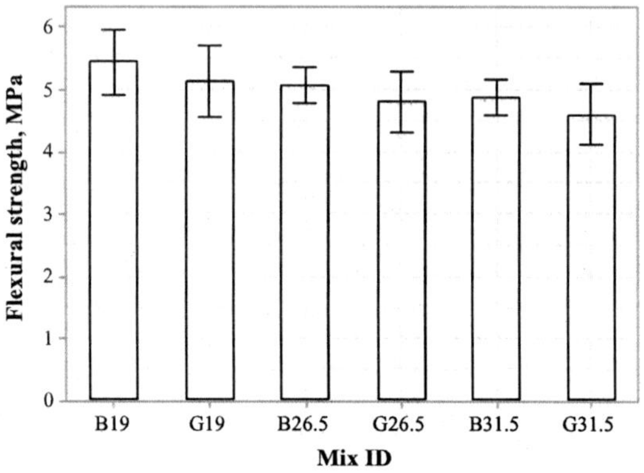

Fig. 4 Flexural strength results for different concrete mixes (after 28 days of curing)

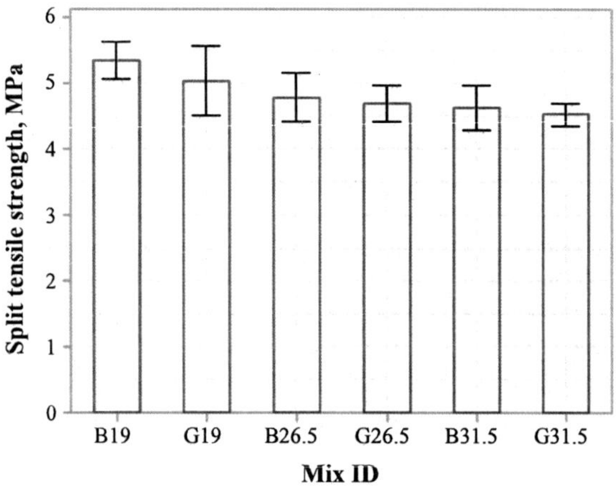

Fig. 5 Split tensile strength results for different concrete mixes (after 28 days of curing)

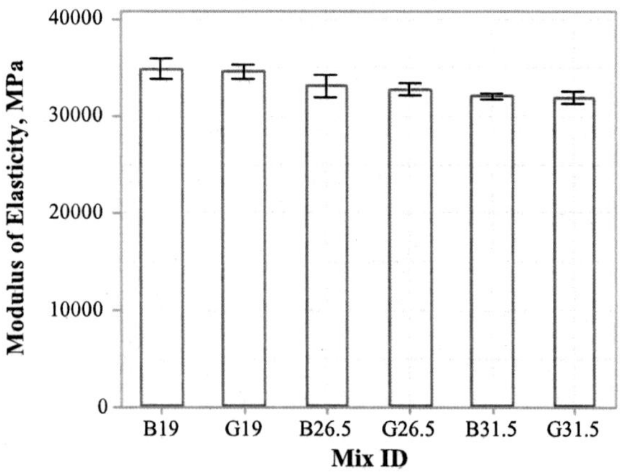

Fig. 6 Modulus of Elasticity (MOE) results for different concrete mixes (after 28 days of curing)

are based on the single slab results in FEM software. These are applicable for fixed concrete strength properties (for example, E = 30,000 MPa is a fixed parameter). Thus in this study, EverFE 2.24 software is used to determine the stresses. This software is free and open-source software developed by Davids et al. [23].

For determining the safe design thickness using EverFE software, first, the trial thickness is assumed, and the flexural stresses are calculated for each loading class and load group both during day and night. The corresponding stress ratio (SR) for each loading class and load group is then determined by dividing the flexural stress by the flexural strength of the mix. Using the SR, the allowable number of repetitions is calculated using the following equations [7].

Fatigue life (allowable repetitions),

$$\mathrm{N} = \text{unlimited/infinite for SR} < 0.45 \tag{1}$$

$$\mathrm{N} = \left[\frac{4 \cdot 2577}{SR - 0.4325}\right]^{3\cdot 268} \text{for } 0.45 \le \text{SR} \le 0.55 \tag{2}$$

$$\mathrm{N} = \left[\frac{0.9718 - SR}{0.0828}\right]^{3\cdot 268} \text{for SR} > 0.55 \tag{3}$$

The damage accumulated for the design period of 30 years is calculated. The axle load spectrum is given in Appendix—VII of IRC 58 2015 is considered to determine the design number of axle load repetitions for both bottom-up cracking (BUC) and top-down cracking (TDC) analysis. The input parameters used in EverFe 2.24 software are given in Table 6. The flow chart of the design process followed in this work is shown in Fig. 7.

The safe design thickness computed by following the above procedure. The safe design thickness of concrete pavement for each of the mix is shown in Fig. 8.

It can be seen from Fig. 8, the safe design thickness of concrete pavement varies between 28 to 33 cm. The B19 concrete mix demands the lowest safe design thickness of 28 cm, and the G31.5 concrete mix requires 33 cm of safe design thickness. The safe design thickness of concrete mix prepared using granitic CA are higher than the safe design thickness of concrete mix prepared using basaltic CA irrespective of the NMAS. The safe design thickness of concrete pavement is lower for the concrete mix with lower NMAS irrespective of the aggregate type.

The safe design thickness computed for the B31.5 concrete mix is about 12.5% higher than that of the B19 concrete mix and about 12.12% higher for G31.5 when compared to the G19 concrete mix. There is a marginal change in safe design thickness concrete when aggregate is varied, keeping the NMAS constant. There is about a 6 to 6.5% increase with an increase in NMAS from 19 to 26.5 mm and 26.5 to 31.5 mm.

Table 6 Dimension and material properties used as input in EverFE software

Dimensions and material properties	Input value
Concrete slab	
Length	4.5 m
Width	3.5 m
Thickness	Varied for safe design thickness (28 to 35 cm)
Modulus of Elasticity	From Fig. 2.4 for different mixes
Poisson's ratio (constant irrespective of mix)	0.15
Temperature differential	16.8 °C during day and 13.4 °C during night (assumed)
Coefficient of thermal expansion, α	1×10^{-5} /°C (constant irrespective of mix)
Dry lean concrete (DLC) Granular sub-base (GSB)	
Length	4.5 m
Width	3.5 m
Thickness	150 mm
Modulus of Elasticity	20,000 MPa
Poisson's ratio	0.2
Granular sub-base (GSB)	
Length	4.5 m
Width	3.5 m
Thickness	150 mm
Modulus of Elasticity	200 MPa
Poisson's ratio	0.2
Subgrade	
Length	4.5 m
Width	3.5 m
Thickness	500 mm
Modulus of Elasticity	100 MPa
Poisson's ratio	0.2
Dowels	
No. of dowels	11
Diameter	32 mm
Length	450 mm
Embedded length	225 mm
Modulus of Elasticity	200,000 MPa
Poisson's ratio	0.3
Ties	
Spacing	1000 mm c/c

(continued)

Table 6 (continued)

Dimensions and material properties	Input value
Diameter	13 mm
Length	1000 mm
Embedded length	500 mm
Modulus of Elasticity	200,000 MPa
Poisson's ratio	0.3

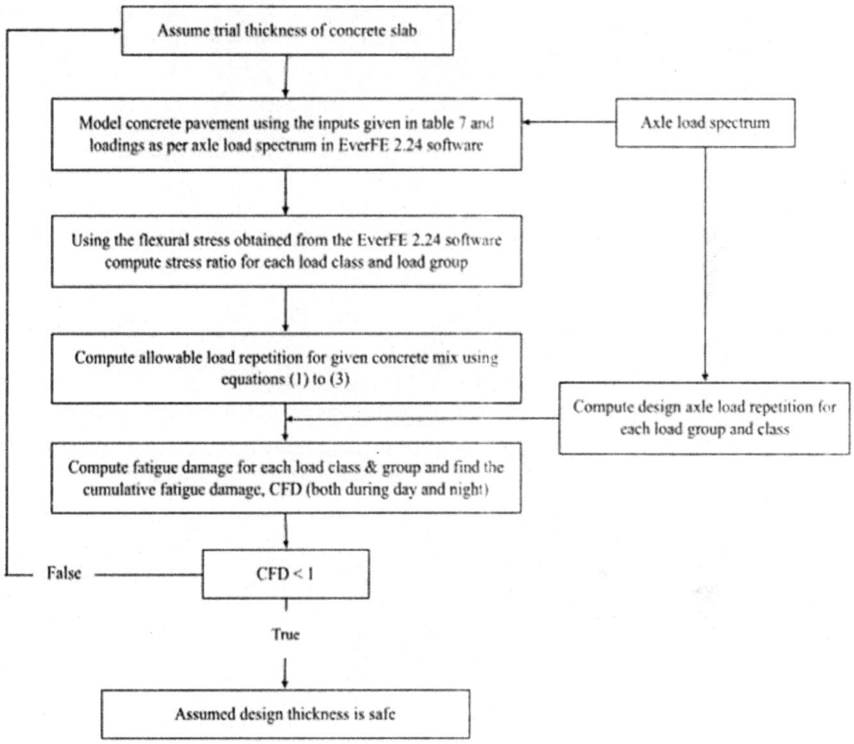

Fig. 7 Flow chart of the procedure for determining design thickness of the concrete pavement

5 Summary and Conclusions

In this study, the influence of type and size of coarse aggregates (CA) on the safe design thickness of concrete pavement is studied. The experimental investigation has been carried out to determine the concrete strength properties. The obtained experimental results are used to determine the safe design thickness of concrete pavement for Indian conditions using EverFE 2.24 software. The experimental studies found that all the concrete mix used in this study satisfy the minimum requirements to use

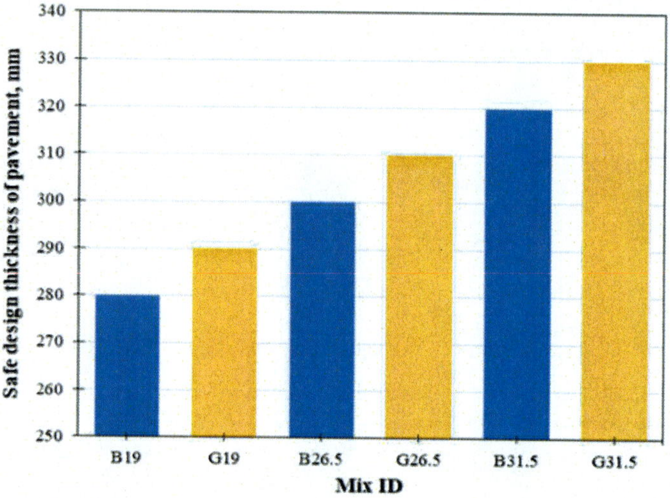

Fig. 8 Safe design thickness for various concrete mix

them as PQC. However, the mechanical properties of the PQC mix improved with the use of CA of lower NMAS. The mechanical properties of the PQC mix prepared using the basaltic CA showed slightly better results than the one prepared using granitic CA. The safe design thickness of concrete pavement depends to a larger extent on the NMAS of CA used in the mix than the CA type. It is concluded that CA with lower NMAS would result in lower safe design thickness and may save the material cost incurred in the concrete pavement projects. It is suggested to use stronger CA with lower NMAS based on the availability at the site.

References

1. Özturan T, Çeçen C (1997) Effect of coarse aggregate type on mechanical properties of concretes with different strengths. Cem Concr Res 27:165–170. https://doi.org/10.1016/S0008-8846(97)00006-9
2. Wu KR, Chen B, Yao W, Zhang D (2001) Effect of coarse aggregate type on mechanical properties of high-performance concrete. Cem Concr Res 31:1421–1425. https://doi.org/10.1016/S0008-8846(01)00588-9
3. Kasu SR, Deb S, Mitra N, Muppireddy AR, Kusam SR (2019) Influence of aggregate size on flexural fatigue response of concrete. Constr Build Mater 229. https://doi.org/10.1016/j.conbuildmat.2019.116922.

4. Wolinski S, Hordijk DA, Reinhardt HW, Cornelissen HAW (1987) Influence of aggregate size on fracture mechanics parameters of concrete. Int J Cem Compos Light Concr 9:95–103. https://doi.org/10.1016/0262-5075(87)90025-X
5. Ćosić K, Korat L, Ducman V, Netinger I (2015) Influence of aggregate type and size on properties of pervious concrete. Constr Build Mater 78:69–76. https://doi.org/10.1016/j.conbuildmat.2014.12.073
6. IRC 44 (2017) Guidelines for cement concrete mix design for pavements, Indian Road Congr. New Delhi
7. IRC 58 (2017) Guidelines for cement concrete mix design for pavements, Indian Road Congr. New Delhi
8. Indian Standard Code of Practice for Ordinary Portland Cement—Specifications, IS 516: 1959, Bureau of Indian Standards, New Delhi
9. Indian Standard Code of Practice for Specification for granulated slag for the manufacture of portland slag cement, IS 12089:1987, Bureau of Indian Standards, New Delhi
10. Taylor P, Kosmatka SH, Voigt G (2007) Integrated materials and construction practices for concrete pavement: a state of the practice manual, FHWA HI—07—004
11. Indian Standard Code of Practice for Methods of test for aggregates for concrete: particle size and shape, IS-2386:1963, Bureau of Indian Standards, New Delhi
12. Indian Standard Code of Practice for Methods of test for aggregates for concrete: estimation of deleterious materials and organic impurities, IS-2386: 1974, Bureau of Indian Standards, New Delhi
13. Indian Standard Code of Practice for Methods of test for aggregates for concrete: specific gravity, density, voids, absorption and bulking, IS-2386:1974, Bureau of Indian Standards, New Delhi
14. Indian Standard Code of Practice for IS-2386:974, Methods of test for aggregates for concrete: mechanical properties, Bureau of Indian Standards, New Delhi.
15. Indian Standard Code of Practice for Coarse and Fine Aggregate for Concrete—Specification, IS-383: 2016. Bureau of Indian Standards, New Delhi
16. MoRTH—2013, Specifications for road and bridge works, Indian Roads Congr 1(2013):1–883
17. Indian Standard Code of Practice for Concrete admixtures-Specifications, IS-9103:1999, Bureau of Indian Standards, New Delhi
18. Indian Standard Code of Practice for Plain and reinforced concrete - code of practice, IS-456 : 2000, Bureau of Indian Standards, New Delhi
19. Indian Standard Code of Practice for Methods of sampling and analysis of Concrete, IS-1199 :1959, Bureau of Indian Standards, New Delhi
20. Indian Standard Code of Practice for Methods of Tests for Strength of Concrete, IS-516:1959, Bureau of Indian Standards, New Delhi
21. Indian Standard Code of Practice for Methods of Tests for strength of concrete: Determination of Modulus of Elasticity, IS-516:2020 Bureau of Indian Standards, New Delhi
22. Indian Standard Code of Practice for Splitting tensile strength of concrete- method of test, IS-5816:1999, Bureau of Indian Standards, New Delhi
23. Davids WG, Turkiyyah G, Mahoney JP (1998) EverFE rigid pavement three-dimensional finite element analysis tool. Transp Res Rec. https://doi.org/10.3141/1629-06

Investigating Impact of Boundary Layer in Pervious Concrete

M. Sajeevan, N. Ahilash, and D. N. Subramaniam

Abstract Pervious concrete is a green concrete because of its environmental benefits including reduced carbon footprint. However, industrial application of pervious concrete is limited by uncertainty of its mechanical properties. While porosity plays an important role in mechanical properties and understanding on porosity distribution and characteristics is essential in optimising design of pervious concrete. Image analysis is a modern tool which could employ different analyses techniques and significantly reduce error than in conventional methods of analysing porosity. This study aims to analyse distribution of porosity in pervious concrete, using image analyses tools. Cubes were cast, cores were obtained, and images of the surfaces were painted and photographed. Image was converted to binary and analysed and total porosity of cubes were obtained. A small layer in the top and bottom have high porosity than total porosity and between those layers the porosity, which is lower than total porosity, is called effective porosity. Therefore, the top and bottom boundary layers were analysed with design parameters. Top boundary layer of a concrete cube is 4.7 mm and it does not depend on design parameters. The bottom boundary layer of the concrete cube increases with the aggregate to cement ratio and decreases with applied compaction energy. For blows more than 45, the bottom boundary layer does not significantly change. Minimum required number of blows for the 1st layer of the cube to get optimum bottom boundary layer is 10 for all Aggregate to cement ratio (A/C) except for design with A/C of 2.5.

Keywords Pervious concrete · Boundary layer · Porosity distribution · Effective porosity · Image analysis

1 Introduction

From the beginning of the industrial revolution the environment got polluted significantly by anthropogenic activities with the intention of improving their quality of life [4, 15]. However, when nature started to respond to those activities, humans

M. Sajeevan (✉) · N. Ahilash · D. N. Subramaniam
Department of Civil Engineering, Kilinochchi Premises, University of Jaffna, Jaffna, Sri Lanka

© The Author(s), under exclusive license to Springer Nature Switzerland AG 2022

H. R. Pasindu et al. (eds.), *Road and Airfield Pavement Technology*, Lecture Notes in Civil Engineering 193, https://doi.org/10.1007/978-3-030-87379-0_32

realized the importance of the environment and concept of sustainable development [20, 22, 24, 25]. After this every field focused on eco-friendly approach whereas in constructions the concept of Green Buildings is a major element of sustainable development. One of the approaches of building green is the usage of green materials that make comparatively lower impact to the environment than conventional materials [2, 6, 17].

Pervious concrete is a structural element of Water Sensitive Urban Design, which is based on the green urban development concept [10, 16]. This type of concrete is considered a green material for various reasons including, (a) elimination of usage of fine aggregates, sand, (b) less material consumptions, (c) usage of more environmental friendly cements with less carbon footprints, (d) usage of waste material such as fly ash and bottom ash as fillers, (e) increase in permeability that facilitate percolation of rainwater to recharge aquifers in urban spaces, (f) reduced heat transfer makes it a good substitute for walls to manage heat and energy losses related to indoor thermal comfort in buildings are to name a few [1, 7, 14].

Pervious concrete is often considered as a type of light weight concrete (although the higher compaction can lead to higher density), typically change in the density range of 1500—2400 kg/m^3, primarily depending on aggregate to cement ratio and compaction of fresh concrete [1, 8]. Absence of fine aggregates, and a designed aggregate to cement ratio thereby enhance the porosity of the concrete [23]. Increase in porosity in turn affects the performance of the resulting concrete, predominantly as a decrease in compressive and flexural strength [10, 11, 19]. Porosity of conventional concrete is generally less than 1%, where the concrete matrix would ensure that the coarse aggregate particles are completely surrounded by the binder and fine aggregates. On the other hand, porosity of pervious concrete may vary from as low as 5% to as high as 50% depending on the application sought [5, 9]. Higher porosity also indicates that the coarse aggregates may not be covered completely while the binding of the aggregates would also be incomplete. Aggregate shape and size distribution would therefore play a vital role in packing of the cube and hence the performance of hardened concrete and consistency of the same [3, 13]. Aggregate size distribution and shape in turn would also affect the pore size distribution and shape. This will affect the connectivity of pores, and effective conductivity of the pervious concrete.

Contemporary literature on pervious concrete research mainly focusses on laboratory scale experiments and data analysis using correlation techniques to recognize correlating concrete design and performance parameters [12, 18]. Thereby, mathematical models were developed to predict concrete performance based on the design parameters [5, 19]. The characteristics, importantly the pore characteristics are often not considered in such models, and this leads to the need of experimental studies prior to each application. A few studies have analysed some pore characteristics including porosity spatial distribution and pore size distribution with very limited analyses relating to design parameters and performance parameters [26]. Image analysis is a potential tool that is being used primarily in the analysis of pore characteristics in concrete [5, 21]. However, the potential of this tool has not been developed to be employed improve current understanding on pervious concrete.

There is boundary layer effect in both top and bottom of the pervious concrete which has high porosity compared to the interior of concrete. Therefore, the top and the bottom layers can bear less load compared to inner concrete. Then the observed compressive strength test values are not equal to the real bearing capacity of the design often underestimated. Therefore, the boundary layers should be analysed to better understand properties of pervious concrete. This study analyses the impact of design parameters on top and bottom boundary layers and thereby on properties of pervious concrete.

2 Material and Methods

Ordinary Portland cement (OPC) and crushed aggregates were obtained from local distributers in Northern Province, Sri Lanka. Required water for the experiments was collected from the pipe line of the Faculty of Engineering, University of Jaffna and tested for negligible Chlorine content.

A consistent concrete mixing protocol and duration was adopted in this current study that lasted for 20 min in total. Optimum water to cement ratio 0.28 was used throughout the study to maintain zero slump (0–15 mm). In the beginning the aggregates were added with surface dry condition and started to mix with cement in the drum. After 3 min half of the measured water was sprinkled and after 6th minute balance water also sprinkled to the mixture and stopped in 12th minute. Chunks and lumps of materials which was in the interior wall of the drum was removed and dropped in the concrete mixture. The mixing was started after 3 min of rest and stopped after 5 min. Then, slump of the mix was tested and the transferred to the moulds by three layers. The compaction energy was given to the cubes according to the mix design by three layers with a standard proctor Rammer which has 2.5 kg weight. A Total number of 15, 30, 45, 60 and 75 blows were given to cubes for experiments. The compaction energy was distributed equally for all three layers.

The particle size distribution of the aggregates was determined using image analysis. Sample of coarse aggregates were taken and soaked in black color oil paint. After drying, the particles were arranged in a white sheet which has known area. An image was taken by covering the complete white sheet. Arranged particles were mixed, re-arranged and images were taken. The images were cropped and converted to binary scale of black and white. The image was then processed, and area of each black zone was obtained by scaling the image for white sheet as reference. The image processing and analysis were done using ImageJ™ which is an open-source software. The equivalent diameter of each black zone was computed. The cumulative area percentage against particle size distribution was plotted as shown in Fig. 1. The distribution of particles in the graph is between 12 to 25 mm and mean diameter is 18.7 mm.

The aggregate to cement (A/C) ratios 2.5, 3.0, 3.5, 4.0, 4.5, 5.0, 6.0 and 7.0 were used throughout the experiment. By changing the total number of blows and the A/C ratio, different types of mix design were completed as shown in Table 1.

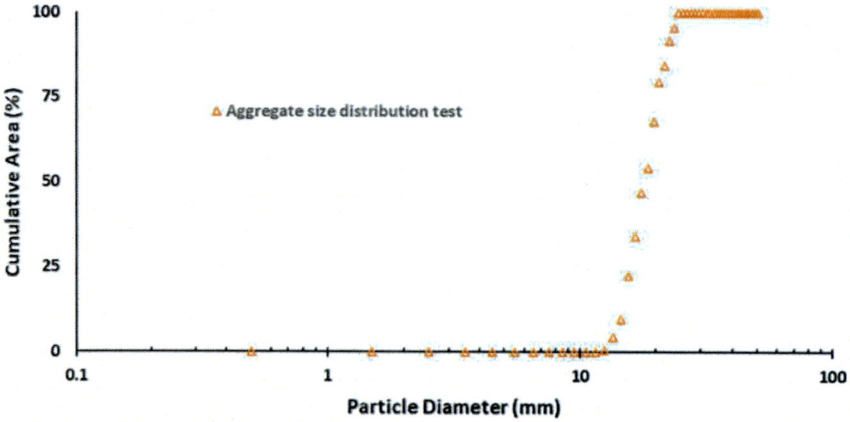

Fig. 1 Particle size distribution of aggregates used in this study

Table 1 Details of cubes casted with difference mix designs in experiment

A/C Ratio	Compaction (number of blows)				
	15	30	45	60	75
2.5	√	√	√	√	√
3.0	√	√	√	√	√
3.5	√	√	√	√	√
4.0	√	√	√	√	√
4.5	√	√	√	√	√
5.0	√	√	√	√	√
6.0	√	√	√	√	√
7.0	√	√	√	√	√

In total, 40 different mix designs were fabricated, and twelve cubes were cast in each design mix amounting to a total of 480 concrete cubes. Each cube cast were of dimensions 150 mm × 150 mm × 150 mm (standard cube size). The weight of each concrete mix in each cube were measured using a digital balance with a least count of 1 g. The slump of each design was also measured to ensure that it was zero. The cubes were then air-dried for 24 h in room temperature and then transferred to water curing tank and cured for 28 days (total of 29 days). Then 6 cubes of each design were tested for compressive strength and the core of the balance 6 cubes were separated for analysis. The core was dipped in the water filled graduated glass tank (vernier calliper with a least count of 0.05 mm) of dimensions 150 mm × 150 mm × 250 mm (width, length and depth, respectively). The porosity of the core sample was measured from the volume of the replaced water by solid of the concrete.

The core samples were then oven dried at 104 °C for 24 h. The curved surface of the core samples was then sprayed with black aerosol spray paint and air-dried. Black colour covered all pore, paste and aggregate surface of the core. The cut surface (of the cross-section) of paste and aggregates were painted with white paint carefully using a brush. It allows to easily differentiate the pores and cut surface of paste aggregate. The dried core sample was placed in a revolving table which can be controlled by a stepper motor. The image of curved surface was taken in panoramic view using Sony A7 III camera with 50 mm/f 1.8 prime lens from a distance of 350 mm.

2.1 Quality Control

The whole batch of aggregates were obtained in one purchase and stored in room temperature and at controlled humidity. The quality of the aggregate was tested using standard testing methods. Cement for the experiments were purchased from same brand and same batch in one purchase and stored in a dry place wrapped in polythene covers. The moisture content of aggregates was measured prior to each experiment to ensure the initial moisture content was similar across all experiments. Good laboratory practices were followed at all times during the experiments.

Before mixing each batch of concrete, the mixing drum was washed with water to eliminate dust and other particles from the last session and dried. On every casting day another three control cubes were cast to monitor the impact of other factors on the properties of concrete. The control cubes were cast for the same A/C ratio (3.5), and a constant W/C ratio of 0.28 and a compaction of 30 blows. Same mixing method used for experimental cubes were applicable for control cubes as well. The weight of fresh concrete, dry concrete and compressive strength were monitored for the control cubes as well.

Instead of natural light, ring light was used to maintain same light condition for all cubes. Aperture was used above f 9 to reduce the blur effect in pores when the camera is focused in the cut surface. ISO below 2000 was used to reduce the noise. All other camera settings had not been changed throughout the study to maintain same image features. Camera was fixed at 350 mm distance from the centre of the rotating stage. These precautions were taken to reduce the error in the image and improve the quality of the image.

2.2 Data Analysis

The images were converted to binary format and analysed using Matlab, which is a software containing tools for image analysis. In the binary image a black pixel represents pore surface and white pixel represents aggregate or binder cut surface.

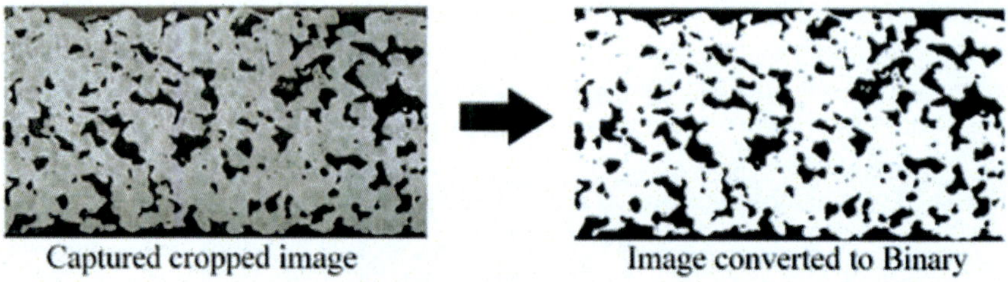

Fig. 2 Cropped image and processed binary image of A/C 3.0 and 30 blows mix design cube 1

When a binary image loaded in Matlab, it is stored in matrix format. As the image is binary, the matrix will be two dimensional which contains 0 and 255. It means 0 represents black colour and 255 represents white. In the image or piece of image total number of black pixels by total number of pixels gives the porosity. The process was done by algorithms of Matlab (Fig. 2).

3 Results and Discussion

When the core of the concrete cube was dipped in the water tub, the water enters the pores of the core. If the water entered into all pores we can take, the replaced volume of water by the core of the concrete cube is equal to solid volume of the core. However, when some pores are not connected to the surface, the water wouldn't enter them. In turn, such pores would be counted as solids in the concrete and the replaced water will be equal to the addition of unconnected pores and solids. When the core of the cube was separated and the image of the curve surface was analyzed, all pores, including closed pores will be measured. This value will therefore reflect the true porosity of pervious concrete. The porosity computed from water replacement is referred to as Measured Porosity (MP) while the porosity compute from image analysis is referred to as Actual Porosity (AcP). The variation of MP and AcP with Aggregate to Cement (A/C) ratio is shown in Fig. 3

For all designs the AcP is higher than MP as expected from the discussion above. In addition, the difference between AcP and MP gradually increases with A/C ratio. On a closer analysis, it could be observed that MP is a fraction of AcP, which remains constant (approximately 0.6) across all A/C ratio. This observation shows that using comparison of performance of pervious concrete across different design may employ MP while it is not a correct indicator of porosity of pervious concrete.

The AcP is computed by analysing the curved surface of the core sample obtained from the cube concrete samples. The pore area of the surface divided by the total area of the picture was AcP. The image was then split into 10 layers horizontally (the width of each layer was 15 mm, whereas the total height of the image was 150 mm the height of the core sample). Figure 4 shows the variation of porosity with depth from top and A/C ratios for pervious concrete samples cast with 30 blows of compaction.

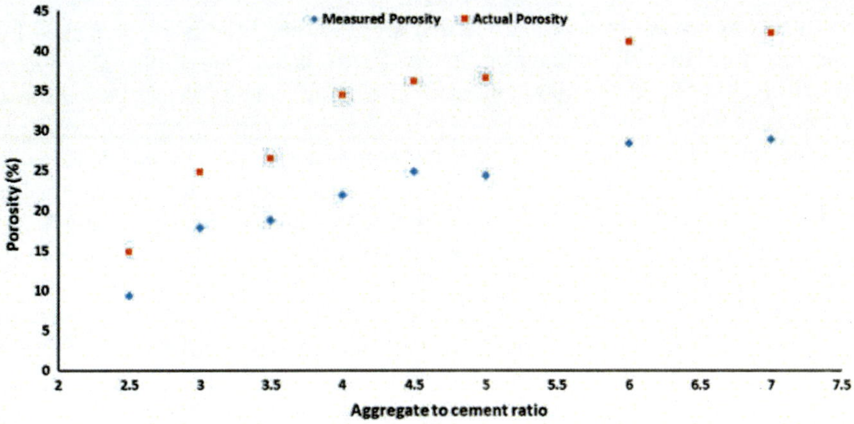

Fig. 3 Aggregate to Cement ratio versus Apparent Porosity and Actual Porosity of 30 blow compaction designs

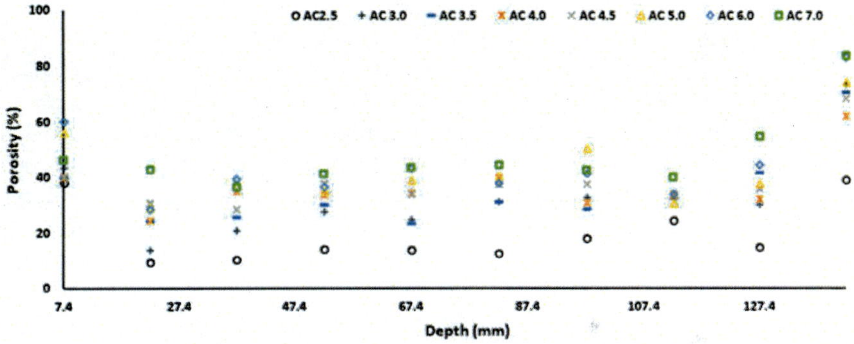

Fig. 4 Porosity versus depth graph for all aggregate to cement ratio designs with 30 blow compactions

It is apparent from Fig. 4 that top and bottom strips of all samples have higher porosity than other strips (middle strips). In addition, it also could be observed that all eight strips in the middle have same porosity indicating uniform distribution of porosity in the inner layers. These two observations were also consistent throughout samples of all compaction levels. The uniform porosity of the inner layers of a sample is termed as Effective Porosity (EP) in this study. Because the application of the pervious concrete depends primarily on the permeability, EP would be a better representation than AcP or MP. It could also be observed that the boundary layer was situated in 10% from top and bottom boundaries. The impact of A/C ratio and compaction on the extent of boundary layers in the sample was therefore analysed subsequently.

Cube was divided by five and the first layer and the last layer was picked for this part of the analysis on boundary layers. Picked layers were further sliced in to 10, 20, 30, 40, 50, 75 and 100 number of strips. The width of the strips were then varied from 0.15 mm to 1.5 mm (4–40 pixels). The porosity of each strip was then computed based on the area of the pore and total area. This process was done to one cube per design for all A/C ratios in 30 blows compaction and for all total number of blows compaction in 3.0 A/C ratio in. From preliminary analysis the number of strips were optimized to 50 as a strip was representative element for the analysis of boundary layers.

For all cubes the chosen top and bottom layers were divided in to 50 strips and the porosity of each strip were computed. From six cubes of each design the average porosity for each strip was calculated. A graph was plotted for depth against porosity for both top and bottom layer of both experiments, separately. The graph obtained was in the shape of a mirror image of 'S' curve.

$$y = \frac{a}{b + c{*}e^{-d{*}x}} \tag{1}$$

In Eq. (1), a, b, c and d are arbitrary constants. Equation 1 indicated an 'S' curve which starts in zero and ends in a/b. The maximum value of porosity is 1 (fraction) and the curve starts from its highest value and ends at EP of the particular concrete sample. The function was modified as in Eq. 2, for the mirror image of Eq. 1, so that the function starts at the maximum porosity and ends at EP. A constant was added to end the curve in effective porosity of the cube, where the constant 'c' is EP of the sample. In Eq. (2), a, b and c are arbitrary constants.

$$y = \frac{1 - c}{1 + a * e^{b*x}} + c \tag{2}$$

Curves of all samples with varying A/C ratio and compaction were fitted for the function given in Eq. 2, for variation of porosity (y, in fraction) and depth (x). The point at which the porosity of the strip approaches EP was mathematically computed from the function and noted.

Figure 5 shows the variation of thickness of boundary layers with A/C ratio and compaction levels.

Pervious concrete samples cast with 15 blows of compaction showed a sudden increase from A/C 3.0 to 3.5, decreased till 5.0 and then plateaued at 6 and 7. In comparison to a definite pattern observed for samples with higher compaction, samples with 15 blows of compaction showed a random variation with A/C ratio. This observation could be partially attributed to the fact that the bottom layer of the cubes that received 15 blows in total had received only 5 blows which had not met the minimum required compression to effect uniform compaction. Samples with 30 blows or higher of compaction in contrast, had received 10 blows or more for bottom layer. In addition, the thickness of the boundary layer did not show significant impact of compaction, however, a gradual increase was observed with A/C ratio for all

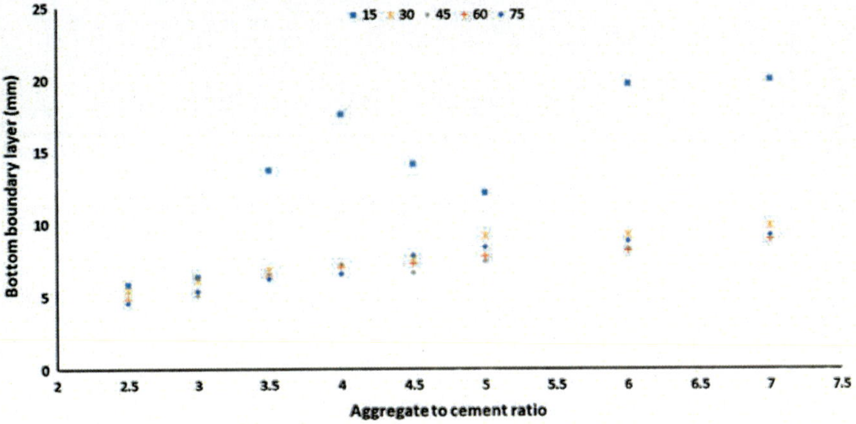

Fig. 5 Bottom boundary layer versus aggregate to cement ratio for all compaction mix designs

samples with compaction higher than 30 blows in total. This observation in contrast to that observed for samples with 15 blows indicate that the bottom layers receive optimum compaction between 5 and 10 blows on the bottom layer to attain uniform compaction.

The boundary layer thickness of samples that received 15 blows of compaction had fluctuated up to 20 mm, which is approximately 15% of the volume of the cube cast. On the other hand, the boundary layer thickness of the samples that received 30 blows or higher of compaction had increased from approximately 5 mm to 10 mm accounting for a maximum of 7% of the volume of the cube. When A/C increases, the workability decreases. So, the minimum required compaction energy increases. The bottom boundary layer increases with aggregate to cement ratio slightly when the cube received compaction more than minimum requirement.

When the compaction increases the bottom boundary layer decreases, but after 45 blows the bottom boundary layer doesn't have significant difference. We can observe it from Fig. 6 and the one way ANOVA test was done for the obtained data within designs and between designs (Fig. 7).

Top boundary layer doesn't have any relationship with any parameters. In this case when the compaction was applied to the cube, there would be a space in the top of the mold because of the compressed concrete. Therefore, the space will be filled with fresh concrete and it doesn't receive the compaction of the mix design. The top boundary layer depends on the added extra fresh concrete after compaction. From the analyzed data, the top boundary layer is 4.74 mm (95% CI 4.46–5.01 mm).

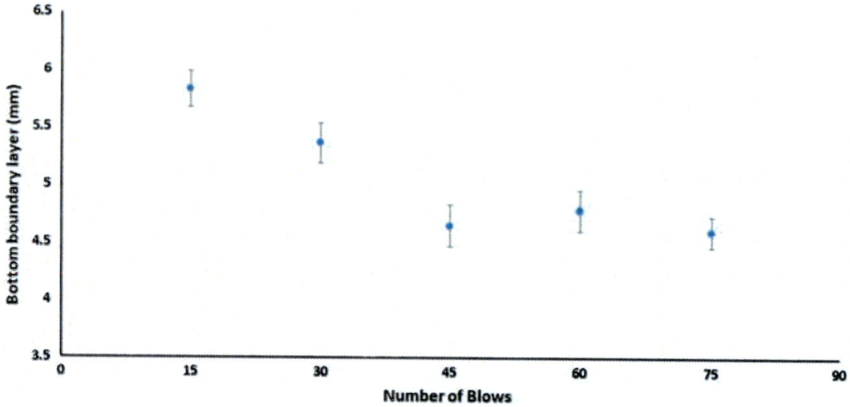

Fig. 6 Bottom boundary layer versus number of blows for samples with A/C ratio of 2.5

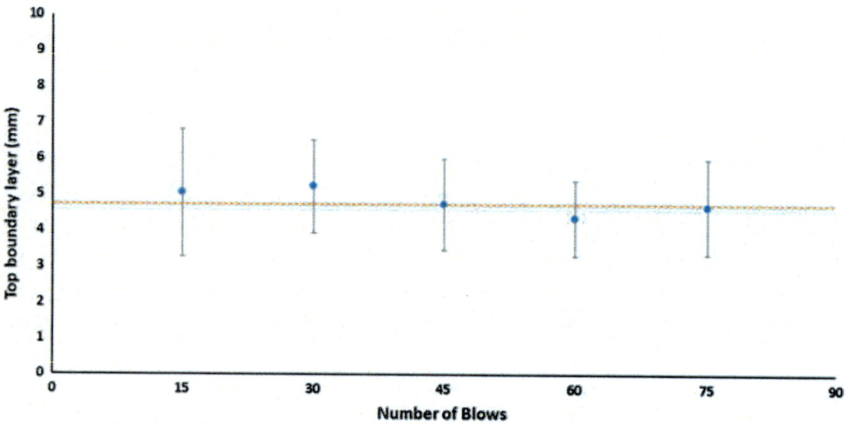

Fig. 7 Top boundary layer versus Number of Blows Error bar graph

4 Conclusion

Measured Porosity (MP) estimated from water replacement experimental method is only a fraction of Actual Porosity (AcP) of pervious concrete, due to occurrence of inaccessible pores. AcP assessed through image analysis includes the impact of porosity distribution in the boundary layers (top and bottom). Porosity is distributed evenly across the inner layers of pervious concrete (in between top and bottom boundary layers) for samples that had more than 30 blows of compaction, irrespective of A/C ratios used. This is termed as Effective Porosity (EP). EP increases with decrease in compaction and increase A/C ratio.

The pervious concrete has boundary layers in the top and bottom which have higher porosity than EP (even more than AcP). The thickness of bottom boundary layer decreases with compaction and increases with A/C ratio. An optimum thickness of less than 10 mm for bottom boundary layer was obtained for all samples that had more than 30 blows of compaction, irrespective of A/C ratio. However, lower A/C ratio had even lower boundary layer thickness at the bottom of approximately 5 mm, compared to approximately 10 mm for higher A/C ratio (7.0). Thickness of the top boundary layer on the contrary, did not depend on neither A/C ratio nor compaction, and was observed to be almost constant at 4.74 ± 0.28 mm (95% confidence interval) and depends primarily on the filling the cubes.

References

1. Aamer Rafique Bhutta M, Hasanah N, Farhayu N, Hussin MW, Tahir MBM, Mirza J (2013) Properties of porous concrete from waste crushed concrete (recycled aggregate). Construct Build Mater 47:1243–1248
2. Adabre MA, Chan APC, Darko A, Osei-Kyei R, Abidoye R, Adjei-Kumi T (2020) Critical barriers to sustainability attainment in affordable housing: international construction professionals' perspective. J Cleaner Prod 253:119995
3. Agar Ozbek AS, Pedersen RR, Weerheijm J, van Breugel K (2019) Mesoscopic modeling of the impact behavior and fragmentation of porous concrete. Cement Concrete Composites 102:116–133
4. Alvarez-Larrauri R, Fogel I (2008) Environmental audits as a policy of state: 10 years of experience in Mexico. J Cleaner Prod 16(1):66–74
5. Ba M-F, Qian C-X, Guo X-J, Han X-Y (2011) Effects of steam curing on strength and porous structure of concrete with low water/binder ratio. Constr Build Mater 25(1):123–128
6. Brodie IM (2006) Prediction of stormwater particle loads from impervious urban surfaces based on a rainfall detachment index, in Deletic A, Fletcher TD (eds) 7th international conference on urban drainage modelling and the 4th international conference on water sensitive urban design, Melbourne, Australia
7. Brown RA, Borst M (2015) Nutrient infiltrate concentrations from three permeable pavement types. J Environ Manage 164:74–85
8. Carsana M, Tittarelli F, Bertolini L (2013) Use of no-fines concrete as a building material: Strength, durability properties and corrosion protection of embedded steel. Cement Concrete Res 48:64–73
9. Chen J, Wang H, Xie P, Najm H (2019) Analysis of thermal conductivity of porous concrete using laboratory measurements and microstructure models. Constr Build Mater 218:90–98
10. Chindaprasirt P, Hatanaka S, Chareerat T, Mishima N, Yuasa Y (2008) Cement paste characteristics and porous concrete properties. Constr Build Mater 22(5):894–901
11. Chindaprasirt P, Hatanaka S, Mishima N, Yuasa Y, Chareerat T (2009) Effects of binder strength and aggregate size on the compressive strength and void ratio of porous concrete. Int J Minerals Metall Mater 16(6):714–719
12. Debnath B, Sarkar PP (2019) Permeability prediction and pore structure feature of pervious concrete using brick as aggregate. Constr Build Mater 213:643–651
13. Deo O, Neithalath N (2011) Compressive response of pervious concretes proportioned for desired porosities. Constr Build Mater 25(11):4181–4189
14. Djerbi A (2018) Effect of recycled coarse aggregate on the new interfacial transition zone concrete. Constr Build Mater 190:1023–1033
15. Evans CD, Monteith DT, Cooper DM (2005) Long-term increases in surface water dissolved organic carbon: observations, possible causes and environmental impacts. Environ Pollut 197:55–71

16. Golroo A, Tighe SL (2011) Alternative modeling framework for pervious concrete pavement condition analysis. Constr Build Mater 25(10):4043–4051
17. Kazemi F, Hill K (2015) Effect of permeable pavement basecourse aggregates on stormwater quality for irrigation reuse. Ecol Eng 77:189–195
18. Khan MI (2002) Factors affecting the thermal properties of concrete and applicability of its prediction models. Build Environ 37(6):607–614
19. Lian C, Zhuge Y, Beecham S (2011) The relationship between porosity and strength for porous concrete. Constr Build Mater 25(11):4294–4298
20. Lucke T, Nichols PWB (2015) The pollution removal and stormwater reduction performance of street-side bioretention basins after ten years in operation. Sci Total Environ 536:784–792
21. Lyu K, Garboczi EJ, She W, Miao C (2019) The effect of rough vs. smooth aggregate surfaces on the characteristics of the interfacial transition zone. Cement Concrete Composites 99:49–61
22. Pepper IL, Gerba CP, Brusseau ML (2006) Environmental and pollution science. Elsevier Inc., United States of America
23. Shen W, Shan L, Zhang T, Ma H, Cai Z, Shi H (2013) Investigation on polymer–rubber aggregate modified porous concrete. Constr Build Mater 38:667–674
24. Subramaniam DN, Egodawatta P, Gallage CPK, Mather P, Rajapakse J (2014) Significance of drying periods on nitrate removal in experimental biofilters. J Water Manage Model, Toronto, Canada, CHI Water
25. Subramaniam DN, Logeswaran T, Tharshikka V, Nilakshan B (2018) Dynamics of clay particles in non-vegetated stormwater biofilters. Water Air Soil Pollut 229(9):302
26. Sumanasooriya MS, Neithalath N (2011) Pore structure features of pervious concretes proportioned for desired porosities and their performance prediction. Cement Concr Compos 33(8):778–787

Laboratory Characterization of Cement-Treated Rock Rubble as Airfield Pavement Base Layer

J. H. Tan, Q. Khan, C. H. Ng Yannick, and G. P. Ong

Abstract As a means to ensure economic and environmental sustainability in pavement works, the use of locally obtained waste materials (rocks) in base layer design as well as exploring energy-efficient construction methods is highly encouraged. This study explores the use of cement-treated recycled rocks of up to 1.5 in. (37 mm) in size as an alternative base layer material for the construction of airfield pavements. The Preplaced Aggregate Concreting Method (PAC) was implemented to construct the base layer without any vibration or compaction effort. This is an unconventional airfield pavement base construction method, which solely relies on the flowability of the grout mix to ensure the voids between rocks are filled up by means of gravity. The experimental program explores the compressive strength, modulus of elasticity and tensile strength of the material. The preliminary results demonstrate the potential of CTRR as a viable base layer alternative to existing stabilized bases.

Keywords Preplaced aggregate concrete · Rock rubble · Cement-treated · Stabilized base · Airfield pavement

1 Introduction

Preplaced Aggregate Concreting (PAC) is a concreting technique whereby coarse aggregates are preplaced and grout subsequently inserted to fill up the voids between the aggregates [1–3]. Unlike conventional concreting methods, which require high

J. H. Tan (✉) · Q. Khan · C. H. N. Yannick · G. P. Ong
Department of Civil and Environmental Engineering, National University of Singapore, Singapore, Singapore
e-mail: ceejht@nus.edu.sg

Q. Khan
e-mail: ceeqk@nus.edu.sg

C. H. N. Yannick
e-mail: yannick_ng@nus.edu.sg

G. P. Ong
e-mail: ceeongr@nus.edu.sg

© The Author(s), under exclusive license to Springer Nature Switzerland AG 2022

H. R. Pasindu et al. (eds.), *Road and Airfield Pavement Technology*, Lecture Notes in Civil Engineering 193, https://doi.org/10.1007/978-3-030-87379-0_33

energy consumption for vibration and compaction, PAC utilizes a grout mix that is flowable enough to fill the voids by self-weight. No vibration or compaction effort is required [1]. PAC can carry up to 60% of coarse aggregates compared to about 40% in conventional concrete [1–3]. As a result, PAC requires a lower proportion of grout, making it an economical option to utilise. In contrast to normal concrete, where coarse aggregates are relatively dispersed, the coarse aggregates in the PAC method rest atop one another, creating more point-to-point contact for improved stress distribution [2]. Adopting the PAC method over the conventional concreting technique reaps eco-efficient and mechanical performance benefits, such as using lower cement content, preventing segregation of coarse aggregates and reducing concrete shrinkage [3]. The application of the PAC technique has been well-established for repair and reinforcement of structures, most notably the rehabilitation of Hoover Dam [4]. However, its application in the aviation industry has yet to be explored.

Construction of stabilized bases for airfield pavement works often requires vibration/compaction as well as having smaller aggregates, which requires additional crushing at the plant. These existing preparation techniques are considered to be energy-intensive and costly. Moreover, most construction materials for stabilized base construction in Singapore are usually imported. To promote sustainability in pavement works, exploration of locally obtained materials as substitution is highly encouraged. This study aims to utilize waste rocks from rock cavern excavation to construct a cement-treated rock rubble (CTRR) as an alternative sustainable material for stabilized base layer in airfield pavement design construction. As an initial step, this study characterizes the mechanical properties of CTRR in terms of compressive strength, modulus of elasticity, and tensile strength, to evaluate its suitability as an alternative to existing stabilized bases.

2 Experimental Program

2.1 Material Properties

Ordinary Portland Cement (OPC) Type I was used. Its minerals are shown in Table 2. Locally obtained granite rock aggregates with particle sizes ranging from 0.75 in. to 1.5 in. (19–37 mm) were used as coarse aggregate. These rock aggregates originate from the Bukit Timah granite formation in the northern region of Singapore. The rocks have a specific gravity of 2.67 and a water absorption capacity of 0.85%. Table 1 summarizes the aggregate performance and quality based on the Coarse Aggregate Material Requirement of the Federal Aviation Administration Advisory Circular FAA-10H [5]. The gradation of the rock rubble is shown in Fig. 1a. A mixture of quarry sand (QS), a by-product from the rock cavern excavation process, and commercially obtained river sands (W7 and W9), was used as fine aggregates. As shown in Fig. 1b, the overall gradation of fine aggregates (14% of W7, 30%

Table 1 Test results on coarse aggregates according to FAA requirements

Material test	Result	Requirement	Standard
Resistance to degradation	16% by weight	Loss: 40% maximum	ASTM C131
Soundness of aggregates by use of magnesium sulfate	1.8%	Loss after 5 cycles: 15% maximum	ASTM C88
Flat particles, elongated particles or both	0%	10% maximum by weight	ASTM D4791
Percentage of fractured particles	100%, by weight, of particles with at least two fractured faces	Minimum 60% by weight of particles with at least two fractured faces	ASTM D5821
Clay lumps and friable particles	0.45%	Less than or equal to 3 percent	ASTM C142

Table 2 Mineralogy of OPC

Minerals	Composition (%)
SiO_2	22
Al_2O_3	5
MgO	2.5
Fe_2O_3	3.1
SO_2	2
K_2O	0.6
Na_2O	0.6
Mn_2O	0.3
CaO	65.4

of QS and 56% of W9) falls within the recommended American Concrete Institute Committee 304 (ACI-304.1) Gradation 2 limit [4].

2.2 *Optimal Grout Mix Proportion*

In the PAC technique, rheological properties of grout mix significantly affect the bond between rock rubble. The fluidity and consistency of the grout mix determine its ability to fill up the void spaces without segregation. To reach an optimum mix design, it is crucial to determine the appropriate combination of cement, water, and fine aggregates. The principal factor affecting strength is the water/cement (*w/c*) ratio, while fines/cement (*f/c*) ratio influences the consistency of the grout mix [6, 7]. Hence, several mix proportions were investigated to optimize both consistency and fluidity. Grout mixes at varying *w/c* and *f/c* were first subjected to bleeding tests

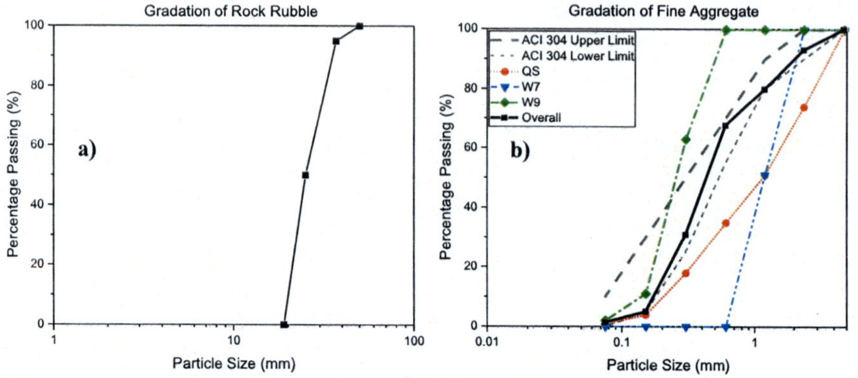

Fig. 1 a Rock rubble gradation; **b** Fine aggregate gradation

following ASTM C940 [8]. The target criteria for bleeding limit is less than 5% after 120 min. As Fig. 2a shows, while the *f/c* ratio influences bleeding more significantly than the *w/c* ratio, excess fines also result in a more viscous and less workable grout.

The effect of mineral additives was also explored to enhance the rheological properties of the grout mix without significantly increasing the *f/c*. Pozzolanic materials such as bentonite have been widely used as an admixture during grout slurry process. Its superior water-absorbent capability, as well as the slow pozzolanic reaction with cement hydration, makes bentonite a suitable candidate to improve the rheological properties of the mix without accelerating the rate of strength gain in grout [9, 10]. Luo et al. [10] recommend the optimum amount of bentonite added to be between 5 to 15% of cement weight. In this study, the optimum dosage of bentonite is determined using the bleeding test mentioned above. To evaluate the influence of bentonite on the reduction in bleeding, preliminary experiments were conducted at $f/c = 1$. Bentonite was added at increasing dosage, in proportion to cement weight, across

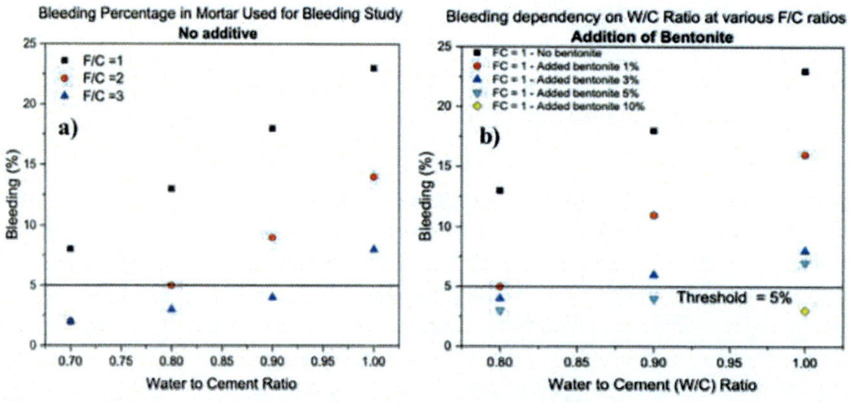

Fig. 2 Bleeding limit: **a** Without bentonite; **b** with bentonite

Table 3 Optimal grout mix proportion

w/c	f/c	Bentonite (%) by cement wt	Superplasticizer (%) by cement wt
0.6	2	5	1.5
0.7	2	5	1.0
0.8	2	5	1.0
0.9	2	5	0.5
1.0	2	5	0

varying *w/c* ratio as shown in Fig. 2b. From the preliminary study, it is concluded that increasing bentonite content significantly reduces the rate of bleeding at a fixed *f/c* ratio. However, an excessive amount of bentonite results in a viscous grout mix. Therefore, bentonite dosage of 5% was found to be optimal without compromising on the workability of the grout. In order to further ensure the bleeding criteria is met across all *w/c* ratios during specimen preparation, the *f/c* was increased to 2 while maintaining bentonite dosage of 5% by cement weight.

As for grout fluidity, it is determined using the flow cone method according to ASTM C939 [11]. Depending on the mix ratio, superplasticizer was added to ensure a homogeneous and flowable mix. The addition of superplasticizer increases the grout fluidity at low *w/c* ratio, which is beneficial for filling up the voids between aggregate particles. Abdelgader [6] proposed a superplasticizer dosage of 1.2–2% of cement weight for PAC specimens. According to ACI 304.1, the recommended time of efflux ranging from 18 to 26 s is deemed ideal for PAC specimens [4]. However, this is proposed for high strength applications such as structural beams and columns, and a more fluid grout mix is thus preferred for low strength applications like CTRR. The final grout mixes are shown in Table 3.

2.3 Specimen Preparation

CTRR cylindrical specimens were prepared in moulds of 6 in. diameter × 6 in. height (150 mm × 150 mm). Rock rubble was initially laid in the mould, and the grout mixture was separately mixed via a shear mixer for three minutes. The grout was subsequently poured into the mould and allowed to flow without vibration. In addition, cylindrical moulds of dimensions 2 in. diameter × 4 in. height (50 mm × 100 mm) were also used to prepare grout specimens with varying *w/c* ratio. After casting, the specimens were sealed to prevent moisture loss and cured for 7–28 days. After one-day of curing, the specimens were de-moulded to cure in an air-tight container until testing age. Figure 3 shows a schematic diagram of the cylindrical specimen preparation.

Fig. 3 Schematic procedure of cylindrical preparation

2.4 Testing Procedure

Material characterization is a crucial step for pavement analysis because fundamental engineering parameters are essential for mechanical-empirical design to better predict its pavement behaviour and distress response [12]. The mechanical properties of CTRR were examined in terms of compressive strength, modulus of elasticity and splitting tensile strength. Following ASTM C39 [13], the unconfined compression test was conducted at a displacement-controlled rate at 0.04 in./min (1 mm/min). External Linear Variable Displacement Transducers (LVDT) in contact with the top loading platen measure the axial strain. The modulus of elasticity test was conducted as per ASTM C469 [14]. Using a Virtual Infinity Stiffness (VIS) load frame by GDS Instruments, loading and unloading cycles could be configured within the software; the test consists of four load-unload cycles up to 40% of the peak strength during loading phase and subsequently back to near zero load during the unload cycle. The loading rate of the VIS load frame was kept constant at 0.04 in./min (1 mm/min). The modulus was calculated as shown in Eq. 1, where P_{40} represents 40% of peak strength; $P_{0.00005}$ is the strength at 50-microstrain, and ε_{40} is the longitudinal strain produced at P_{40}. Local LVDTs with a gauge length of 4.33 in. (110 mm) were installed on the specimens to measure the axial displacement during load-unload cycles. The splitting tensile test was performed in accordance with ASTM C496 [15]. The test was carried out at 11.24 kips/min (50kN/min). The tensile strength of the specimen was calculated as shown in Eq. 2, where P is the peak load, and l and d are the length and diameter of the specimen respectively. Figure 4 illustrates the test setup of the

Fig. 4 Testing setup: **a** unconfined compression test; **b** modulus of elasticity test; **c** splitting tensile test

three tests mentioned above.

$$E = \frac{P_{40} - P_{0.00005}}{\varepsilon_{40} - 0.00005} \quad (1)$$

$$T = \frac{2P}{\pi l d} \quad (2)$$

3 Results and Discussion

3.1 Unconfined Compressive Strength

Although compressive failure rarely occurs in pavement structures, the compressive strength of cementitious materials is one of the most widely used measure of quality assurance during design and construction because of its convenience [16]. FAA-10H standard imposes an upper limit of 800 psi (5.516 MPa) on the 7-day strength of stabilized bases to mitigate the risks of reflective cracking [5]. Table 4 presents the 7-day and 28-days compressive strength of CTRR at *w/c* ratio ranging from 0.6 to 1.0. As expected of cementitious materials, the strength of CTRR is inversely proportional to its *w/c* ratio. For a given *w/c* ratio, the 7-day strength appears to be a reasonable estimation of the long-term strength as it is approximately 70% of the 28-day strength. Figure 5a shows the 7-day strength of CTRR clearly exceeds FAA's strength criteria, which may result in significant reflective cracking. This can be alleviated by constructing a layer of unbound granular base that acts as a buffer, thus protecting the overlying surface course [17]. Given the presence of a granular interlayer, higher strength gain can be allowed within the CTRR layer to allow for heavier aircraft loading.

Figure 5b compares the strength of CTRR with that of pure grout at 7-day. The grout strength is generally higher than CTRR. This agrees with past studies on PAC [18]. In addition, the difference in compressive strength between CTRR and the grout

Table 4 Compressive strength results

w/c ratio	Cement content (lb/ft^3)	CTRR 28-day mean compressive strength (psi)	CTRR 7-day mean compressive strength (psi)	Grout 7-day mean compressive strength (psi)
0.6	34.3	3646.7	2217.3	2610.7
0.7	32.5	3091.5	1179.2	2593.3
0.8	30.6	2283.9	1627.8	2364.1
0.9	29.0	1312.3	1123.0	1435.9
1.0	28.1	1072.6	845.4	847.2

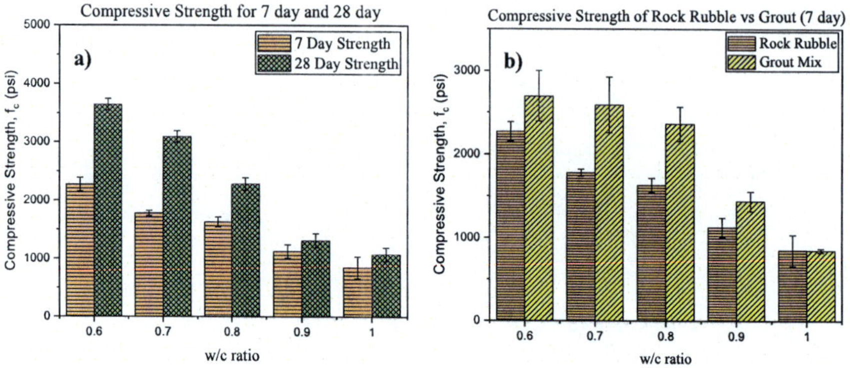

Fig. 5 Compressive strength: **a** 7 versus 28 days; **b** CTRR versus grout

mix reduces with increasing *w/c* ratio. As the rock aggregates are very likely to have a higher compressive strength than the grout, the strength of the CTRR is governed by the strength of the grout mix and the adhesion bonding at the interfacial transition zone (ITZ). Xie et al. [19] reported that a higher *w/c* ratio results in weaker ITZ interface between the cement paste and rock aggregates. As shown in Fig. 5b, this phenomenon explains the comparable strength between grout and CTRR at $w/c =$ 1.0, thereby suggesting that bond strength at ITZ is relatively as weak as the strength of grout at higher *w/c* ratios.

Figure 6 shows the failure pattern of a CTRR specimen, wherein fracturing is observed at the ITZ. This is attributed to the weak adhesive interface between the rock rubble and the grout. Since these large discontinuities are not present in pure grout specimens, the latter tend to return a higher strength.

Fig. 6 Cross-section of CTRR after failure

3.2 Modulus of Elasticity

The modulus of elasticity is a crucial input parameter in pavement design. Cementitious stabilized bases such as lean concrete and cement-treated bases are typically used as structural layers in a pavement system and behave similarly to an elastic slab [20]. A base layer with adequate stiffness is preferred to control the warping stresses induced by temperature and shrinkage differential [12]. However, a high modulus of elasticity may result in brittle behaviour leading to reflective cracking propagating up to surface courses, resulting in premature failure to the pavement. Xuan et al. [21] reported common modulus values of 0.15 to 2.90 million psi (1–20 GPa) for cement-treated aggregate material. Scimemi performed a series of heavy-weight deflectometer (HWD) tests on cementitious bound base layer on an airfield pavement and found that the modulus of the cement-treated base varies from 0.16 to 0.71 million (1.1–4.9 GPa) [22].

Figure 7a shows the stress–strain curves obtained from load-unload cycles of CTRR for different *w/c* ratios. The mean 7-day modulus values are 0.95, 1.7 and 2.0 million psi for a *w/c* ratio of 0.9, 0.7 and 0.6 respectively. These results are consistent with the reported values in the literature. Similarly, stiffness is derived from stress–strain curve obtained from compressive strength tests, using a secant modulus at 50% of the peak stress. From Fig. 7b, the moduli are 0.075, 0.17 and 0.24 million psi for *w/c* ratio of 0.9, 0.7 and 0.6 respectively. The moduli from compressive strength tests are almost one order lower than the moduli obtained from the load-unload tests. The measurement of axial deformation based on local LVDTs from load-unload tests yields more accurate moduli values. On the other hand, measurements of displacement recorded by external LVDTs from compression testing significantly overestimates the strain values, due to extraneous movements induced by the loading platen and/or specimen bedding errors.

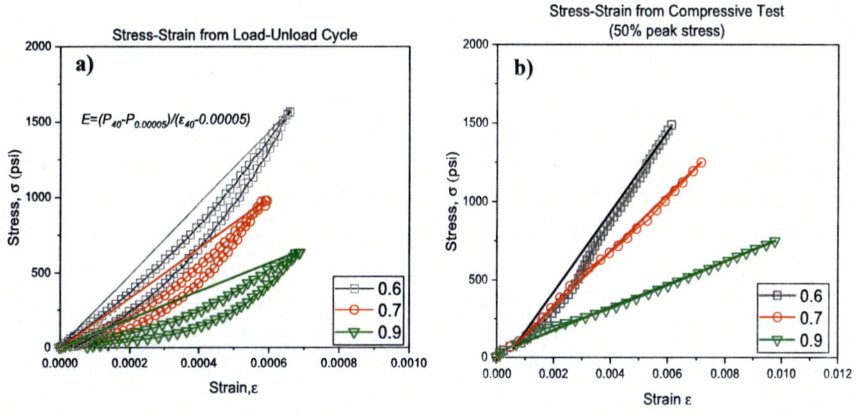

Fig. 7 Stress–strain curves from **a** Load-unload cycles; **b** compression test

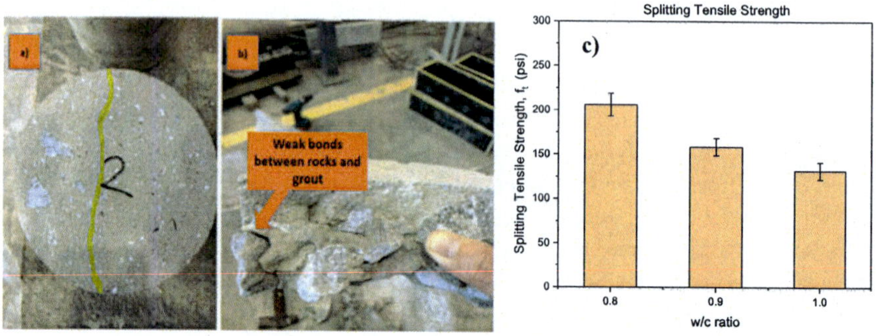

Fig. 8 **a** Line of split (highlighted in yellow), **b** Failure plane of CTRR, **c** split tensile strength

3.3 *Splitting Tensile Strength*

Splitting tensile strength is an essential parameter in understanding CTRR as a potential base layer in a pavement structure. The major pavement distresses commonly associated with cement-treated bases are transverse cracking (caused by warping stresses) and fatigue cracking. Cementitious materials are known to be weaker in their tensile capacity rather than compressive strength. Hence, its tensile properties reflect the cracking potential as it simulates tensile stresses at the bottom of the base course when subjected to loading and environmental conditions.

Figure 8c shows that the mean 7-day splitting tensile strength at varying *w/c* ratio, ranges from 131.7 to 206 psi (0.91–1.42 MPa). However, a significant variation of 14% in strength was observed for each *w/c* ratio. This was also observed by Guinea et al. [23] for conventional concrete and they attributed this effect to the influence of aggregate-mortar interface on the cracking mechanism for concrete. In the context of CTRR, the larger maximum rock sizes, as compared to the maximum aggregate sizes used for conventional concrete, would result in a smaller overall specific surface contact area with the grout, thereby reducing the effectiveness of the interfacial bond on the tensile behaviour of CTRR specimens. Figure 8b depicts the weak adhesion between the grout and rocks of a post-test specimen after tensile failure.

4 Conclusion

This study has explored the usage of recycled rocks (of up to 1.5 in. or 37 mm in diameter) to develop an alternative and more sustainable stabilized base layer for future airfield pavement applications. The main findings of the study can be summarized as follows:

- The mechanical properties of CTRR, namely compressive strength, modulus of elasticity and splitting tensile strength, are comparable to that of existing base materials.
- Despite exceeding the FAA's 7-day compressive strength criteria, the concern of reflective cracking can be mitigated by constructing a layer of unbound granular base that acts as a buffer to protect the overlying surface course.
- Compressive and tensile failure of CTRR material was observed to be caused by the weak rock-grout adhesion bond. The mode of failure across different *w/c* ratios is dependent either on the strength of grout mix or the interfacial bond relative to grout strength.

From this preliminary study, it is shown that CTRR has potential to be a more sustainable substitute for existing stabilized bases because it uses a lower cement content and incorporates waste materials such as rocks and quarry sand. However, more advanced tests have to be carried out to understand its flexural and fatigue properties under long term loading.

References

1. Das KK, Lam ESS (2020) Feasibility of producing two-stage (preplaced aggregate) concrete by gravity process. Struct Concr 21(3):1157–1163. https://doi.org/10.1002/suco.201900356
2. Najjar MF, Soliman AM, Nehdi ML (Jul. 2014) Critical overview of two-stage concrete: properties and applications. Constr Build Mater 62:47–58. https://doi.org/10.1016/j.conbuildmat.2014.03.021
3. Lv J, Zhou T, Li K (Jan. 2020) Development and investigation of a new low-cement-consumption concrete—preplaced aggregate concrete. Sustainability 12(3):1080. https://doi.org/10.3390/su12031080
4. ACI 304.1 R92 (1997) "Guide for the use of preplaced aggregate concrete for structural and mass concrete applications." ACI Committee 304
5. FAA AC150/5370–10H (2018) "Standard specifications for construction of airports. "Federal Aviation Admin
6. Abdelgader HS, Górski J (Aug. 2002) Influence of grout proportions on modulus of elasticity of two-stage concrete. Mag Concr Res 54(4):251–255. https://doi.org/10.1680/macr.2002.54.4.251
7. Chairunnisa N, Fardheny AF (2019) The study of flowability and the compressive strength of grout/mortar proportions for preplaced concrete aggregate (PAC). MATEC Web Conf 280:04010. https://doi.org/10.1051/matecconf/201928004010
8. ASTM Standard C940 (2016) "Standard test method for expansion and bleeding of freshly mixed grouts for preplaced-aggregate concrete in the laboratory. "ASTM International, West Conshohocken, PA
9. Ahmad S, Barbhuiya SA, Elahi A, Iqbal J (Mar. 2011) Effect of Pakistani bentonite on properties of mortar and concrete. Clay Miner 46(1):85–92. https://doi.org/10.1180/claymin.2011.046.1.85
10. Luo J, Li C, Ma Y, Wang L (2019) "Bentonite replacing part of cement concrete for resistance to chloride ion attack." E3S Web Conf 136:03011. https://doi.org/10.1051/e3sconf/201913603011.
11. ASTM Standard C939 (2010) "Standard test method for flow of grout for preplaced aggregate concrete. "ASTM International, West Conshohocken, PA

12. Rao C, Titus-Glover L, Bhattacharya B, Darter MI, Stanley M, Von Quintus HL (Aug 2012) "Estimation of key pcc, base, subbase, and pavement engineering properties from routine tests and physical characteristics," [Online]. Available: https://trid.trb.org/view/1216257
13. ASTM Standard C39 (2015) "Standard test method for compressive strength of cylindrical concrete specimens. "ASTM International, West Conshohocken, PA
14. AATM Standard C469 (2014) "Standard test method for static modulus of elasticity and poisson ratio of concrete in compression. "ASTM International, West Conshohocken, PA
15. ASTM Standard C496 (2011) "Standard test method for splitting tensile strength of cylindrical concrete specimens. "AATM International, West Conshohocken, PA
16. Lv S et al (Jul 2019) "Strength and fatigue performance for cement-treated aggregate base materials." Int J Pavement Eng. https://doi.org/10.1080/10298436.2019.1634808
17. Brown SF (1979) "Design of pavements with lean-concrete bases." Presented at the 58th annual meeting of the transportation research board. [Online]. Available: https://trid.trb.org/view/148093
18. Cheng Y, Liu S, Zhu B, Liu R, Wang Y (2019) Preparation of preplaced aggregate concrete and experimental study on its strength. Constr Build Mater 229:116847. https://doi.org/10.1016/j.conbuildmat.2019.116847
19. "Experimental study of the interfacial transition zone (ITZ) of model rock-filled concrete (RFC)," (Jan 2015) Cement and Concrete Composites 55:223–231. https://doi.org/10.1016/j.cemconcomp.2014.09.002
20. Lim S, Zollinger DG (Jan. 2003) Estimation of the compressive strength and modulus of elasticity of cement-treated aggregate base materials. Transp Res Rec 1837(1):30–38. https://doi.org/10.3141/1837-04
21. Xuan DX, Houben LJM, Molenaar AAA, Shui ZH (Jan. 2012) Mechanical properties of cement-treated aggregate material—a review. Mater Des 33:496–502. https://doi.org/10.1016/j.matdes.2011.04.055
22. Fileccia Scimemi G, Turetta T, Celauro C (Aug 2016) "Backcalculation of airport pavement moduli and thickness using the lévy ant colony optimization algorithm." Construct Build Mater 119:288–295. https://doi.org/10.1016/j.conbuildmat.2016.05.072
23. Guinea GV, El-Sayed K, Rocco CG, Elices M, Planas J (2002) The effect of the bond between the matrix and the aggregates on the cracking mechanism and fracture parameters of concrete. Cem Concr Res 32(12):1961–1970. https://doi.org/10.1016/S0008-8846(02)00902-X

Laboratory Investigations on Lateritic Soil Stabilized with RBI Grade 81, Coconut Fiber and Aggregates

B. A. Chethan and A. U. Ravi Shankar

Abstract Soil stabilization is an excellent old technique adopted to improve the properties of any weak soil. In the recent past, many chemical stabilizers came to the market. In this study, a commercial stabilizer-RBI Grade 81's (dosage 2–8%) potential to improve lateritic soil properties was investigated. At 6% RBI Grade 81 dosage, the effect of reinforcement was evaluated by using 0.5 and 1.0% coconut fibers. Strength improvement of 6% RBI Grade 81 stabilized soil admixed with <12.5 mm size aggregates (5, 10, and 15%) was studied to know the effect of granular layer demolition waste incorporation. CBR and UCS tests were used for strength evaluation. Considerable improvement in mix strength was observed above 6% RBI Grade 81 dosage on 28 curing days for the soil incorporated with fibers and aggregates. A linear fit was established between UCS and soaked CBR, which shows a high correlation (>0.9), and Prob > F is <0.06 depicting high reliability for the estimation of soaked CBR from UCS. Further, fatigue and durability tests were carried out to understand stabilized soil's behavior under dynamic load application and performance during monsoon to simulate the submerged condition of the pavement in dense rainfall areas (>3500 mm). Stabilized lateritic soil mixes sustained freeze–thaw cycles effectively with weight loss of <14%. However, even at an 8% RBI Grade 81 dosage, the mix has not satisfied the wetting–drying test requirement. Stabilized soil blends showed excellent fatigue performance under repeated loads. The 6% RBI Grade 81 treated soil mixes can be used for modified subgrade.

Keywords Lateritic soil · RBI Grade 81 · Coconut fiber · Aggregate · Stabilization

1 Introduction

Over recent years, sustainable highway construction practices used stabilization techniques for improving unsuitable soil by enhancing shear strength and bearing capacity [1]. Subgrade gives ultimate support to upper pavement layers [2]. Its stability and

B. A. Chethan (✉) · A. U. R. Shankar
National Institute of Technology Karnataka, Surathkal, India

© The Author(s), under exclusive license to Springer Nature Switzerland AG 2022

H. R. Pasindu et al. (eds.), *Road and Airfield Pavement Technology*, Lecture Notes in Civil Engineering 193, https://doi.org/10.1007/978-3-030-87379-0_34

durability should be improved economically to give better pavement performance under repeated wheel loads.

RBI Grade 81 is a commercial soil stabilizer. A fair to excellent volume stability was observed when soil containing fine silt to coarse sand particles was treated using this stabilizer [3]. When used with expansive soil, the mix density has increased [1]. A high content of this stabilizer resulted in the increase of CBR and UCS of soils [4], [5]. For kaolinite, red soil, and lateritic soil with 2–8% RBI Grade 81, a significant CBR improvement was observed [6]; and for expansive soil, CBR improvement was >400% [7]. For soil with >6% stabilizer, a moderate increase in UCS and CBR was achieved [8]. Expansive soil has shown maximum UCS at 4%, and CBR (19%) at 5% stabilizer dosages [1]. Peak UCS and CBR values were obtained at 6% RBI Grade 81 for BC soil [9], plastic clay [10], and highly compressible clay [11].

Cement kiln dust along with RBI Grade 81 has resulted in improved UCS and CBR of clayey soil due to CSH and CAH gel formations [12]. Micro-level structural changes that occur due to hydration of stabilizer contributed to the reduction of soil pores and strength gain [11]. The CaO, Al_2O_3, and SiO_2 contents predominated in the stabilization process [9]. The CaO/SiO_2 indicates the abundance of CaO and SiO_2, depicting the potential of a binder for pozzolanic reaction; at a higher ratio, it is found to be more active [13]. Binders with higher CaO/SiO_2 or $CaO/(SiO_2 + Al_2O_3)$ ratios may still be inefficient due to lesser CaO, SiO_2, and Al_2O_3 contents [14].

The eco-friendly agricultural product, coconut fiber, has a density of 670–1000 kg/m^3 [15]. Its infield service life is 4–10 years [16]. The strength and long-term durability of stabilized soil should not discourage fiber usage [17]. Wet coconut fiber exhibit better tensile strength with high breaking stress compared to other natural fibers. A high lignin content reduces its degradation rate [18]. Reinforcements used for earthen embankments and foundations [19] will form a mesh-like configuration [20]. The high water absorption rate of light palm fibers increased optimum moisture content (OMC) and lowered the mix density [21]. Red sandy soil with a fiber length of 15–25 mm takes high stress due to increased stiffness, low settlement [22]. Sandy soil–cement incorporated with 12.8 mm long 0–3% coconut fiber became ductile from brittle and exhibited high strength. Fiber reinforcement was also found adequate for uncemented soil [23].

Illinois Department of Transportation (IDOT 2016) used stabilized weak soil with large granular materials for low frost susceptible 'Aggregate subgrade'. IDOT Bureau of Design and Environment has given gradation bands of large-sized virgin and recycled material for subgrade (Division of Highways 2013). However, the structural contribution of such subgrade was not acknowledged in IDOT mechanistic pavement design [24]. These aggregates were satisfactory, but less data is available for use in weak subgrade [25], [26].

RBI Grade 81 was used for stabilizing diverse soil types with marginal materials. However, much information is not available on the strength, durability, and fatigue performance of lateritic soil treated with coconut fibers and aggregates with RBI Grade 81. Therefore, investigations were done to appreciate the influence of these materials for use in pavements.

2 Materials

Geotechnical properties of lateritic soil (LS) collected from Surathkal, India, are tabulated in Table 1. The particle size distribution curve (Fig. 2) is obtained from wet sieve analysis [27]. It has a majority of sand and silt size particles; therefore, classified as a silty-sand (SM) as per IS 1498–1970 [28]. The main constituent of LS is SiO_2, with a considerable amount of Al_2O_3 and CaO contents. A commercial stabilizer developed by Road Building International, South Africa, and accredited in India by Central Road Research Institute, New Delhi, was used for soil stabilization. Road Building International Grade 81 (RBI Grade 81) is an odourless, light brown coloured powder, with a specific gravity of 2.5, with 98% particles in the silt–clay size range. It is insoluble in water, non-degradable to ultraviolet radiation exposure, inert, and chemically stable. Its saturated paste is highly alkaline, with a pH of 12.5. India is one of the major countries producing coconut fiber (CF). Commercially available brown CF's were extracted from the exterior shell of mature coconuts. It is a prevalent natural fiber used for soil reinforcement [29]. CF's exhibiting good tensile strength were procured from Mangalore, India (Table 2). Long fibers (bristle) were cut short for obtaining an aspect ratio of 20. Good, angular aggregates of <12.5 mm size

Table 1 Lateritic soil properties

Property, [Code of practice followed]	Value
Specific gravity, G, [31]	2.66
Consistency limits, [32]	
Liquid limit, W_L, %	36
Plastic limit, W_P, %	27
Plasticity index, I_P, %	9
Standard Proctor compaction, [33]	
OMC, %	12.9
γ_d, kN/m^3	19.1
Modified Proctor compaction, [34]	
OMC, %	13.1
γ_d, kN/m^3	20.5
UCS, (@ Modified Proctor compaction), kN/m^2, [35]	
@ OMC-2%	227
@ OMC	589
@ OMC + 2%	306
CBR, %, (@ Modified Proctor compaction), [36]	
Unsoaked	49
Soaked	18
WD durability, [37]	Failed in 1st cycle
FT durability, [38]	< 6.4% weight loss

Fig. 1 Images of **a** lateritic soil, **b** coconut fibers, and **c** aggregates

used belong to the igneous rock—granite group. Flaky (thickness <60% of mean dimension) and elongated (length >180% mean dimension) aggregates that break down quickly under loading were discarded. The selected aggregate has a specific gravity (G) of 2.66, impact value of 17.0%, crushing value of 22.4%, abrasion value of 18.0%, flakiness index of 12.0%, elongation index of 14.1%, and water absorption of 1.2% as per IS 383–1970 [30] (Fig. 1).

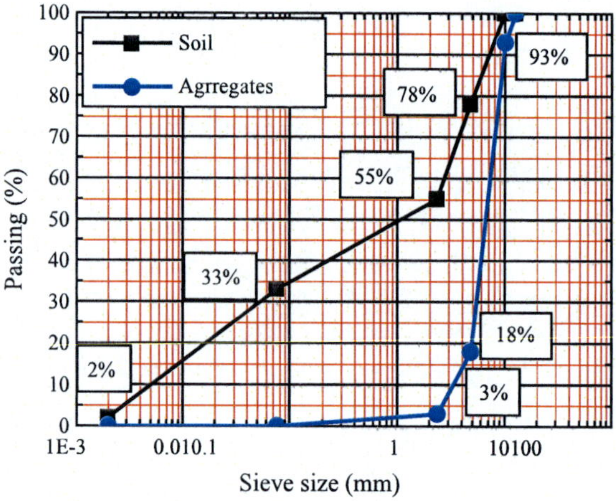

Fig. 2 Particle size distribution curves for lateritic soil and aggregates

Table 2 Properties of coconut fibers [39]

Length, m	Diameter, mm	Unit weight, kN/m^3	Lignin, %	Cellulose, %	Breaking elongation, %	Tensile strength, MN/m^2
0.1–0.3	0.25–0.65	12	45	43.5	32	60–130

3 Experimental Methodology

Air dehydrated, pulverized LS, CF's, and aggregates (Fig. 2) are oven-dried at 105–115 °C temperature for 24 h. Mixes of LS with 2–8% RBI Grade 81 were prepared and tested to identify compaction parameters (OMC and γd). By considering high volume road applications, all tests were conducted at Modified Proctor density. UCS specimens of 38 mm diameter and 76 mm height were prepared in a stainless steel tube mould by filling in a single layer and compacted by static compression. The compressive load is applied slowly using a hydraulic plunger and retained for 1 min to ensure the compact packing of particles. These specimens were cured in a desiccator for 7, 14, and 28 days (at temperature 25 ± 1 °C, relative humidity 100%), and loaded at a strain rate of 1.2 mm/min to determine strength. Three specimens exhibiting UCS within ± 5% of the average were considered for the analysis. As mix M3 has exhibited the highest γd and UCS, this mix (94% LS and 6% RBI Grade 81) was selected as the optimum combination. Optimized mix M3 was reinforced with 0.5 and 1.0% CF's to evaluate the effect of natural fiber inclusion. Also, mix M3 was partially replaced with 5, 10, and 15% aggregates to evaluate the effect of aggregates addition. Fiber-reinforced and aggregate blended mixes were tested for UCS and soaked CBR. Initially, LS and RBI Grade 81 were blended with a trowel using water equal to 50% of OMC. The excess fiber concentration only at certain portions of the specimen hampers the soil reinforcement objective. Therefore, CF's were distributed properly to avoid lump formation due to jumbling (grouping). Then the remaining 50% OMC was added and mixed to achieve homogeneity. Higher CF's dosages (>1%) resulted in jumbling with a nonuniform distribution, therefore, considered unsuitable. Alike, LS-RBI Grade 81-aggregate blends were prepared. Mix details used for the study are tabulated in Table 3.

Soaked CBR tests were conducted on 7, 14, and 28 days cured (covered in plastic wraps) and 4 days soaked samples. Weight loss of 7 days cured UCS samples of LS

Table 3 Mixture details

Mix ID	Lateritic soil, %	RBI Grade 81, %	Coconut fiber, %	12.5 mm downsize aggregates, %
M1	98	2	–	–
M2	96	4	–	–
M3	94	6	–	–
M4	92	8	–	–
M5	94	6	0.5	–
M6	94	6	1.0	–
M7	94	6	–	5
M8	94	6	–	10
M9	94	6	–	15

and M1-M4 mixes were evaluated by performing durability tests. In the wetting–drying (WD) test, the specimens are water-soaked for 5 h and then oven-dried at 71 ± 3 °C for 42 h. In the freezing–thawing (FT) test, water-saturated felt pad covered specimens were frozen for 24 h at −23 °C and then thawed in a moist chamber at 23 ± 2 °C for 23 h. A repeated load equivalent to 1/3, 1/2, and 2/3 of the UCS were applied by a load cell on 7 days cured M1–M4 samples until failure with 1 Hz frequency, 0.1 s rest period, under half sinusoidal waveform at 35–40 °C temperature using a repeated load testing machine. The fatigue life of the UCS samples was registered in the data acquisition software.

The finely powdered LS and stabilized LS samples were subjected to chemical analysis in the laboratory using titration methods, as per Indian standard codes, to understand the changes in mineral constituents composition and justify the influence on strength parameters.

4 Discussions of Test Results

4.1 *Chemical Analysis*

LS-RBI Grade 81 mix has enhanced silica, alumina, and calcium contents, which were effectively involved in the pozzolanic reactions (Table 4). The formation of cementitious products viz. Calcium aluminum silicate (CAS) bound the LS particles together, creating a crystalline matrix. The alkalinity of the mix has increased, promoting the liberation of silicon and aluminum ions from both LS and RBI Grade 81. These ions reacted with calcium ions forming irreversible CAS compounds. The CaO/SiO_2 and $CaO/(SiO_2 + Al_2O_3)$ ratios of the LS are 0.07 and 0.064, which were increased to 0.11 and 0.085 respectively for mix M3, depicting the favoured chemical changes. LS contains 76% of silt to coarse sand particles, hence, justified the suitability for RBI Grade 81 treatment.

Table 4 Chemical composition of lateritic soil, stabilized lateritic soil, and RBI Grade 81

Material	Chemical composition, %							Conductivity (ms)
	SiO_2	Al_2O_3	Fe_2O_3	CaO	MgO	SO_3	pH	
Lateritic soil	65.8	11.95	3.35	5.05	2.2	0.008	10.1	1.2
Lateritic soil stabilized with 6% RBI Grade 81	58.8	16.95	3.85	6.43	3.5	0.28	10.4	1.229
RBI Grade 81	15–19	5–7	0–2	52–56	0–1	9–11	12.5	-

4.2 Compaction

Indian Road Congress has suggested subgrade compaction by standard and modified Proctor methods for low (<2 msa) and high volume roads, respectively. A sudden decrease of OMC from 12.9% to 11.5% and 13.1% to 10.0% were observed at standard and modified Proctor compactions, respectively, for M1 mix with 2% RBI Grade 81 addition. Higher RBI Grade 81 dosages showed greater water demand for hydration, resulting in the rise of OMC. For M1, M2, and M3 mixes, γd was increased due to the filling up of LS's void spaces. However, the γd of the M4 mix has decreased due to excess lighter RBI Grade 81 particles occupying more massive LS particles space. Mix M3 has shown the highest γd of 19.1 and 20.5 kN/m^3 at standard and modified Proctor compaction, respectively (Fig. 3). In a previous study, the treatment of black cotton soil has exhibited considerable strength improvement at 6% RBI Grade 81 dosage and was found to be economical [40]. Therefore, 6% RBI Grade 81 was chosen as an optimum dosage for stabilized mixes with CF's and aggregates.

Mixes M5 and M6 showed lower OMC due to CF's low water absorption capacity compared to the higher water affinity of LS and RBI Grade 81 due to higher specific surface area particles. Good bonding of LS with 0.5% CF's has resulted in compact packing of M5 mix, leading to the highest γd of 21.6 kN/m^3 under modified Proctor compaction. Low specific gravity CF's has resulted in a slight reduction of the γd of M6 mix due to higher fiber concentration. Aggregates showed lower water absorption capacity; hence, they reduced OMC and γd values due to the less packing of M7, M8, and M9 mixes. Due to the increased demand for water to moisten aggregate surface, the OMC has increased with increased aggregate dosage. Uniform dispersion of aggregates in the mix M8 resulted in a noticeable increase in γd. Whereas, high aggregate content of 15% resulted in a decrease of γd due to the increased resistance of aggregates during compaction. The presence of large size nondeformable aggregate

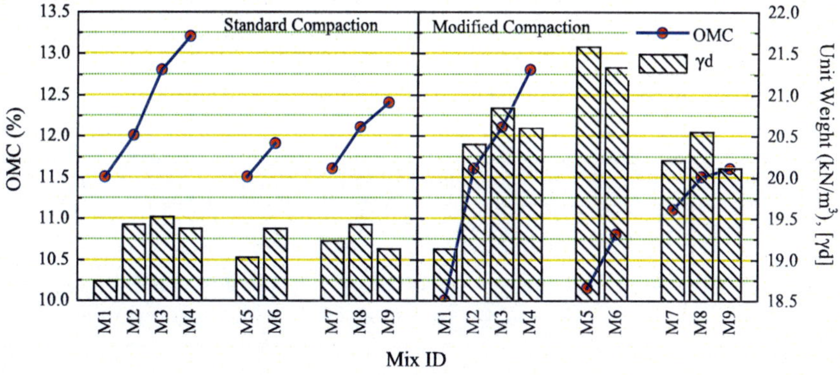

Fig. 3 Variation of standard and modified Proctor compaction characteristics

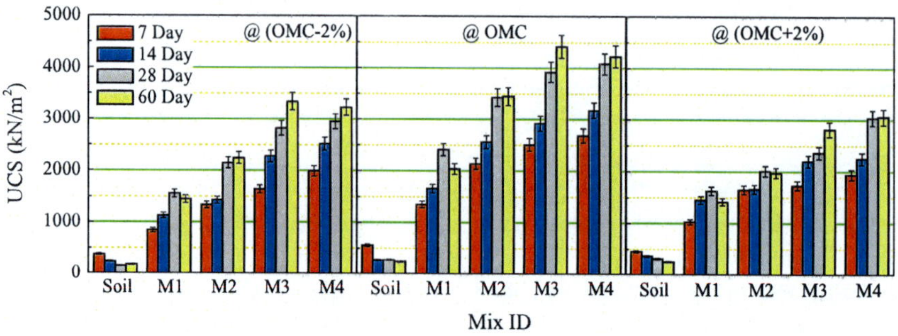

Fig. 4 Variation of UCS of different mixes at (OMC-2%), OMC, and (OMC + 2%)

particles under the applied compaction energy will resist the mix packing. Therefore, CF's has mainly led to form a densest composite matrix, whereas, aggregate augmentation has resulted in gradation change.

4.3 *Unconfined Compressive Strength*

The variation of UCS values of M1–M4 mixes at OMC and OMC ± 2%, on 7, 14, 28, and 60 days curing, are depicted in Fig. 4. The deviation of ± 2% from OMC was taken considering the probable field moisture fluctuations persisting during construction. Also, absorption of free moisture during curing (i.e. OMC + 2%) may reduce UCS due to the increased lubrication of particles leading to low frictional resistance. Due to an effective reaction of LS with RBI Grade 81, samples prepared at OMC exhibited the highest UCS. Specimens that lose water (i.e. OMC-2%) due to evaporation were incapable of developing higher strength due to insufficient water available for the hydration process. After 60 days of curing, the UCS of M1–M4 mixes prepared at OMC were increased by 3.5, 5.9, 7.5, and 7.2 times that of LS. Bernadette Cynthia et al. 2017 showed that 6% RBI Grade 81 amended expansive soil resulted in considerable UCS improvement [10].

4.4 *California Bearing Ratio*

Soaked CBR of M1–M4 mixes was increased by 3.4, 4.0, 4.9, and 5.2 times that of LS after 28 days of curing due to continuous binding and aggregation of particles by the formation of hydration products. Denser M5 and M6 mixes obtained by homogeneous interaction of CF's produced a stronger matrix. The high tensile strength of CF's and shear strength developed in cemented LS promoted the overall stability. The strength of fiber-reinforced LS was controlled by the pullout rather than their breakage behaviour [21]. Continuous cementation may overcome the strength lost with time due to the degradation of CF's. After 28 days of curing, the soaked CBR values of M5, M6, and M8 mixes were increased by 1.9, 2.1, and 1.2 times that of mix M3. The change of LS gradation due to aggregate augmentation has resulted in a marginal change in CBR. The RBI Grade 81 has more affinity towards LS for the chemical reaction due to its larger specific surface area than inert aggregates. Fiber-reinforced mixes M5 and M6 showed a soaked CBR > 100% on 14 and 28 days of curing. Also, mix M9 with 15% aggregate cured for 28 days has shown a soaked CBR of >100% (Fig. 5). In stabilized LS during the CBR test, the plunger could not penetrate due to higher mix stiffness imparted by hydration products, the tensile strength of CF's, and coarser aggregates. Therefore, IRC recommends WD, FT durability tests to evaluate such stabilized soils.

A linear fit model with 3 degrees of freedom developed between UCS and soaked CBR values of LS, M1, M2, M3, and M4 mixes on 3, 7, and 28 days curing shows a good statistical relationship (Fig. 6 and Table 5). Adj. R-square values are >0.9;

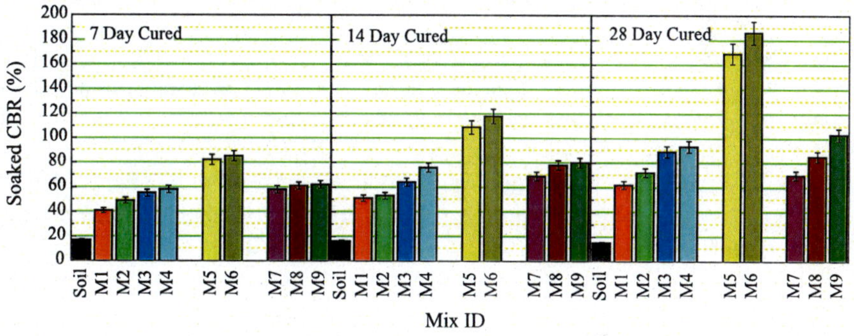

Fig. 5 Variation of soaked CBR of different mixes

Fig. 6 Linear fit between UCS and soaked CBR of RBI Grade 81 stabilized lateritic soil

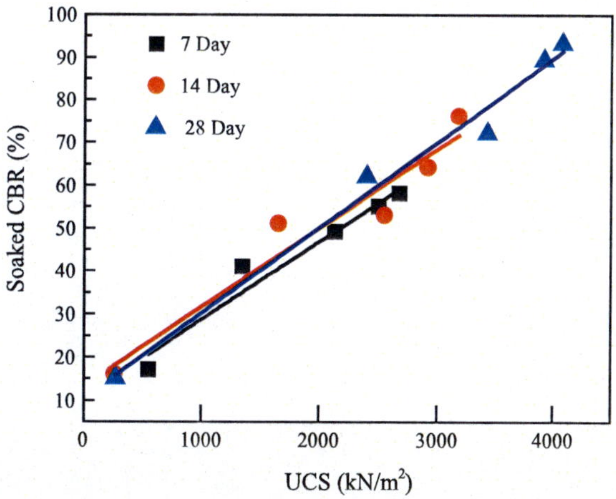

Table 5 Statistical relationship between UCS and Soaked CBR

Curing, days	Linear fit	Adj. R-Square	Prob > F
3	Soaked CBR = 10.55 + 0.018UCS	0.94	0.00448
7	Soaked CBR = 13.19 + 0.018UCS	0.91	0.00761
28	Soaked CBR = 10.33 + 0.020UCS	0.98	8.29E-04

hence, UCS and soaked CBR are strongly correlated. The Prob > F is <0.05; therefore, the soaked CBR values can be predicted from UCS values with high reliability.

4.5 *Fatigue*

The stabilized soil used in the pavement layer will be subjected to repeated application of wheel loads. The load repetitions will induce critical stress and strains in the pavement layer. In order to check the sustainability and long-term performance, the fatigue load test was conducted. Repeated axial loads of various magnitudes were applied to the UCS specimen under unconfined conditions. Even though the specimens were subjected to higher repeated load stresses when compared to actual wheel load stresses, all the specimens sustained $>1.2\times10^5$ fatigue cycles (Fig. 7). High dosages of RBI Grade 81 in LS significantly improved the resistance to fatigue degradation by strongly bonding the particles together. The formation of CAS compounds has proved the advancement in the performance of stabilized mixes, which may help in restricting the structural failure of the pavement due to the accumulation of distresses and propagation of cracks under repeated load application.

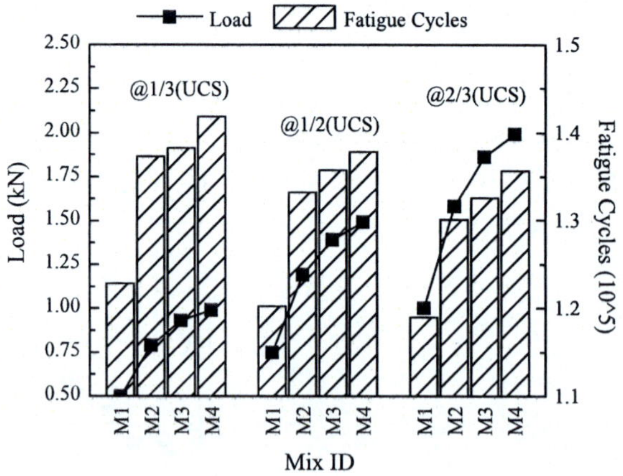

Fig. 7 Fatigue performance of RBI Grade 81 stabilized lateritic soil samples

4.6 Durability

LS specimens were failed during the first 5 h of wetting. Whereas, M1, M2, and M3 mixes failed after 1, 2, and 10 WD cycles. M4 mix was intact after 12 WD cycles, with weight loss >14% (Fig. 8). Hence, it is not an effective stabilizer for dense rainfall areas but can be suitable for low rainfall areas. Higher RBI Grade 81 dosage has a significant influence on cementation. The binding of LS makes it insensitive to moisture and temperature changes, resulting in less soil loss. M1–M4 mixes were resistant to FT cycles due to lowered voids and intactness with a weight

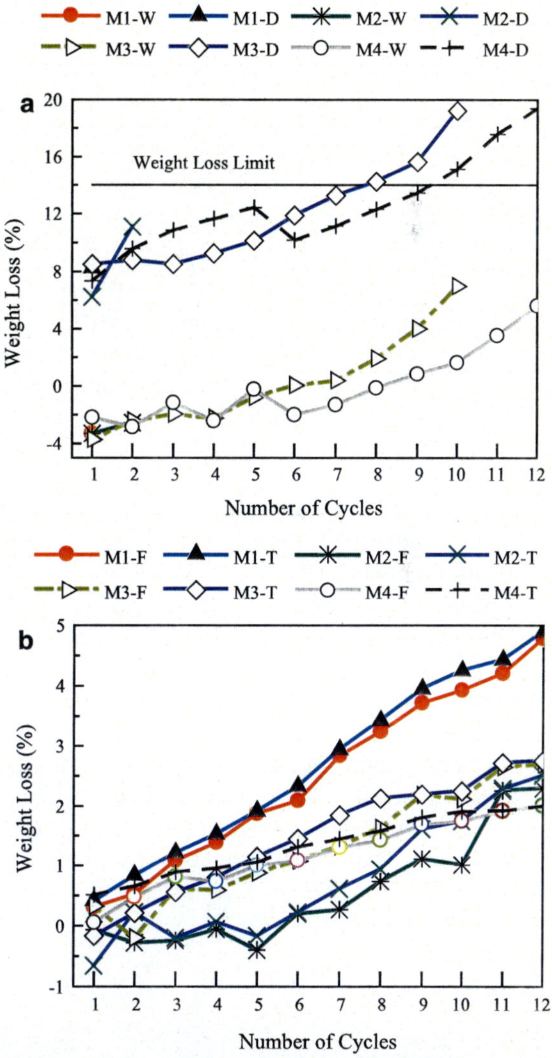

Fig. 8 Weight loss of RBI Grade 81 stabilized lateritic soil samples under **a** WD and **b** FT cycles. Notations: Prefixes M1, M2, M3 and M4 are Mix ID's, Suffixes W = wetting, D = drying, F = freezing, and T = thawing cycles

loss reduction from 4.9 to 2.0%. Freezing reduced the temperature of pore water and led to the formation of ice crystals, exerting bursting force on the surrounding LS. Consequent thawing softened the cemented LS. However, the hydration products formed showed magnified resistance to WD and FT effects leading to slight soil loss depicting more potent mixes against cyclic seasonal changes.

5 Conclusions

The following conclusions were drawn for describing the effectiveness of RBI Grade 81 as a lateritic soil stabilizer.

i. The strength improvement of the stabilized mix was contributed due to the presence of reactive components like silica, alumina, calcium oxide, etc. For stabilized LS, the CaO/SiO_2 and $CaO/(SiO_2 + Al_2O_3)$ ratios were increased from 0.07 to 0.11 and 0.064 to 0.085, respectively, which helped the formation of CAS compounds.

ii. Proper interaction of LS, RBI Grade 81, and CF's has resulted in dense matrix formation. Bonding due to the hydration products and aggregation of LS particles firmly held the CF of high tensile and shear strength, which led to the higher pullout resistance from the cured mix.

iii. RBI Grade 81 dosages >6% resulted in considerable improvement in UCS. At OMC, due to higher hydration capabilities, mixes showed higher UCS. Insufficient water available for hydration at OMC-2% and excess lubrication at OMC +2% lowered the friction between particles led to strength loss.

iv. Reinforcement has significantly contributed to a considerable increase in the soaked CBR. Appreciable tensile strength and tearing resistance of the CF's combined with RBI Grade 81 cementation promoted the resistance to mix softening on soaking. The linear fit between UCS and soaked CBR has shown a high correlation of >0.9. The Prob > F is <0.05, indicates that the prediction of soaked CBR from UCS is highly reliable.

v. Stabilized LS mixes showed high fatigue degradation resistance under 1/3, 1/2, and 2/3 of respective UCS loads. High RBI Grade 81 dosage resulted in enhanced resistance to WD and FT durability cycles. At 6% and 8% RBI Grade 81 dosages, the samples passed 8 and 9 WD durability cycles, with a specified weight loss of <14%. Hence, these mixes significantly improved the resistance to cyclic seasonal variations. However, the treated soil could not pass 12 WD durability cycles; therefore, it can be used as a modified subgrade.

References

1. Kumar K, Solanki AJ (2017) Evaluation of RBI grade 81 for stabilization of expansive soil as sub-grade material. Mater Today Proc 4(9):9737–9741
2. Yoder EJ, Witczak MW (1991) Principles of pavement design. Wiley
3. Kodicherla SPK, Nandyala DK (2017) Effect of RBI Grade 81 on strength characteristics of clayey subgrade. Int J Geo-Eng 8(1):1–11
4. Madurwar KV, Dahale PP, Burile AN (2013) Comparative study of black cotton soil stabilization with RBI Grade 81 and sodium silicate. Int J Innov Res Sci Eng Technol 2(2):493–499
5. Jayatheja M, Kumar ND, Kumar SPK (2016) "Influence of zycosoil on consolidation, shear strength characteristics and permeability of clayey subgrade." In 10th international conference on lowland technology, Manglore
6. Anitha KR, Ashalatha R, Johnson A (2009) "Effects of RBI grade 81 on different types of subgrade soil." In 10th national conference on technological trends (NCTT09), pp 6–7
7. Haricharan TS, Vinay Kumar KS, Durga Prashanth L, MR A, AU R (2013) "Laboratory investigation of expansive soil stabilized with natural inorganic stabilizer." Int J Res Eng Technol 2(13):201–204
8. Venugopal G, Chetan Fakkerappa B (2014) "Studies on black cotton soil stabilization using RBI Grade-81." Int J Innov Res Sci Eng Technol 3(2014):18205–18210
9. Lekha BM, Shankar AUR (2014) Laboratory performance of RBI 81 stabilized soil for pavements. Int J Civ Eng Res 5(2):105–110
10. Cynthia JB, Kamalambikai B, Kumar RP, Dharini K (2017) Laboratory study on subgrade soil stabilization using RBI grade 81. IOP Conf Series: Earth Environ Sci 80(1):12058
11. Gunturi M, Ravichandran PT, Annadurai R, Krishnan D (2014) Effect of RBI-81 on CBR and swell behavior of expansive soil. Int J Eng Res 3(5):336–339
12. Kumar JS, Janewoo U (2016) Stabilization of expansive soil with cement kiln dust and RBI grade 81 at subgrade level. Geotech Geol Eng 34(4):1037–1046
13. Janz M, Johansson SE (2002) "The function of different binding agents in deep stabilization." Swedish Deep Stab Res Centre Rep 9:1–35
14. Odadjima H, Noto S, Nishikawa J, Yamazaki T (1995) "Cement stabilization of peaty ground with consideration of organic matter." In Proc Int Workshop on Engineering Characteristics and Behavior of Peat
15. Ali M (2011) Coconut fibre: A versatile material and its applications in engineering. J. Civ. Eng. Constr. Technol. 2(9):189–197
16. Rowell RM (2000) "Characterization and factors effecting fiber properties." Nat Polym agrofibers based Compos
17. Shankar AU, Chandrasekhar A, Bhat H P (2012) "Experimental investigations on lithomargic clay stabilized with sand and coir." Indian Highw 40(2)
18. Rao GV, Balan K (2000) Coir geotextiles: emerging trends. Kerala State Coir Corporation Limited
19. Gray DH, Ohashi H (1983) Mechanics of fiber reinforcement in sand. J Geotech Eng 109(3):335–353
20. Nataraj MS, McManis KL (1997) Strength and deformation properties of soils reinforced with fibrillated fibers. Geosynth Int 4(1):65–79
21. Sarbaz H, Ghiassian H, Heshmati AA (2014) "CBR strength of reinforced soil with natural fibres and considering environmental conditions." Int J Pavement Eng 15(7), Taylor and Francis, pp 577–583
22. Sivakumar Babu GL, Vasudevan AK (2008) "Strength and stiffness response of coir fiber-reinforced tropical soil." J Mater Civ Eng 20(9):571–577
23. Consoli NC, Prietto PDM, Ulbrich LA (1998) Influence of fiber and cement addition on behavior of sandy soil. J Geotech Geoenviron Eng 124(12):1211–1214
24. Heckel G (2009) Aggregate subgrade thickness determination. Illinois Department of Transportation, Bureau of Materials and Physical Research

25. Mathis D (1991) "Rock cap: a true free draining base." Proc 42nd Annu Road Build Clin Coeur D'Alene, ID, pp 153–156
26. Uhlmeyer JS, Pierce LM, Lovejoy JS, Gribner MR, Mahoney JP, Olson GD (2003) Design and construction of rock cap roadways: case study in Northeast Washington State. Transp Res Rec 1821(1):39–46
27. IS 2720 : Part 4 (2015) "Methods of test for soils—grain size analysis." Bur Indian Stand, New Delhi
28. IS 1498 : 1970 (2016) "Classification and identification of soils for general engineering purposes." Bur Indian Stand New Delhi
29. Faruk O, Bledzki AK, Fink H-P, Sain M (2012) Biocomposites reinforced with natural fibers: 2000–2010. Prog Polym Sci 37(11):1552–1596
30. IS 383 (2016) "Coarse and fine aggregate for concrete—specification." Bur Indian Stand
31. IS: 2720 : Part 3/Section-1 (2016) "Methods of test for soils: determination of specific gravity—fine grained soils." Bur Indian Stand New Delhi
32. IS 2720 : Part 5 (2015) "Method of test for soils: determination of liquid and plastic limit." Bur Indian Stand New Delhi
33. IS 2720 : Part 7 (2016) "Methods of test for soils: determination of water content-dry density relation using light compaction." Bur Indian Stand New Delhi
34. IS 2720 : Part 8 (2015) "Methods of test for soils: determination of water content-dry density relation using heavy compaction." Bur Indian Stand New Delhi
35. IS 2720 : Part 10 (2015) "Methods of test for soils: determination of unconfined compressive strength." Bur Indian Stand New Delhi
36. IS 2720 : Part 16 (2016) "Methods of test for soil: laboratory determination of CBR." Bur Indian Stand New Delhi
37. ASTM D559 / D559M-15 (2015) "Standard test methods for wetting and drying compacted soil-cement mixtures." ASTM Int West Conshohocken, PA
38. ASTM D560 / D560M-16 (2016) "Standard test methods for freezing and thawing compacted soil-cement mixtures." ASTM Int West Conshohocken, PA
39. Pillai MS, Vasudev R (2001) "Applications of coir in agricultural textiles." In International Seminar on Technical Textiles, Mumbai, India, pp 2–5
40. Patil V, Joshi P, Kale S, Pingle T (2019) "Design of flexible pavement on black cotton soil stabilized with rbi grade 81." In Environmental geotechnology, Springer, pp 193–200

Monotonic Loading Test to Investigate the Benefits of Composite Geogrids for Subgrade Improvement

Kasun Wimalasena, Chaminda Gallage, Chamara Jayalath, and Jack Churchill

Abstract The presence of weak subgrades is one of the greatest challenges in constructing road pavements. Conventionally, techniques such as refiling with suitable material and soil stabilisation are considered to improve subgrade condition, ignoring the additional project cost. However, geogrids have gained popularity as economical, expedite and sustainable subgrade improvement techniques over recent times. Although many studies have been conducted to assess the suitability of biaxial geogrids, limited studies have been performed to check the suitability of composite geogrids for subgrade improvement, despite assuming that composite geogrids extend additional benefits. In this study, two model tests: one unreinforced and one composite geogrid reinforced, were constructed in a steel box with length, width, and height of 1 m, 1 m and 1.2 m, respectively. The subgrade was prepared to a thickness of 500 mm, achieving CBR 2.5% bearing capacity. A granular layer of 200 mm was constructed on top of the subgrade achieving 100% degree of compaction from maximum dry density. In the reinforced section, composite geogrid was placed at the base subgrade interface. Both model sections were subjected to a monotonic load at a rate of 1 mm/min, applied on the top surface of the granular layer through a circular plate of 200 mm diameter, until the occurrence of ultimate failure. Results demonstrate that reinforcing the weak subgrade by a composite geogrid has increased the ultimate bearing capacity by 53%. In addition, the overall section modulus of the composite geogrid reinforced section is higher than the modulus of the unreinforced section.

Keywords Pavement engineering · Pavement material · Composite geogrid · Geogrids · Weak subgrade · Bearing capacity

K. Wimalasena (✉) · C. Gallage · J. Churchill
Queensland University of Technology, Brisbane, Australia
e-mail: mataramb@qut.edu.au

C. Jayalath
Aurecon, Brisbane, Australia

© The Author(s), under exclusive license to Springer Nature Switzerland AG 2022

H. R. Pasindu et al. (eds.), *Road and Airfield Pavement Technology*, Lecture Notes in Civil Engineering 193, https://doi.org/10.1007/978-3-030-87379-0_35

1 Introduction

Road design engineers have to face numerous challenges to deliver high standard road networks constraining themselves to limited investments. The weak subgrade is one of the key challenges that could substantially increase the construction cost[1, 2]. The issue of the weak subgrade is extremely common in most of the road construction sites in the state of Queensland, Australia, owing to the presence of expansive clay type soil [3–10]. Therefore, soft subgrade treatment methods: such as backfilling with suitable materials, increase the granular cover thickness or soil stabilisation, are required to be considered to improve the subgrade condition [11–16]. In fact, finding suitable subgrade soil for backfilling is extremely difficult in the state of Queensland and therefore, backfilling, if requires, will mostly be done using granular material. Therefore, inevitably, the presence of a weak subgrade will consequentially increase the demand for natural gravel material [17, 18]. Although recycled aggregate has been recommended to fulfil the extensive demand for granular material in road construction [19–21], the ongoing rate of production would not fulfil a significant portion of the demand for granular material. Accordingly, the road construction industry is searching for an effective and sustainable solution for the issue of weak subgrades.

Geosynthetics are used in different geotechnical applications: such as for pavement construction [12, 22, 23], water-related works [24, 25] and in environmental geo-techniques [26–28]. Geosynthetics have become popular in pavement engineering applications due to economic and environmental benefits, convenient and expedite construction and durability [29]. The reinforcement function of geosynthetics can effectively be used to improve the condition of weak subgrades [30]. Although geosynthetics are available in different types, such as geogrids and geotextiles, researchers have confirmed that geogrids are the best type to achieve the reinforcement function of a road pavement [31]. Besides, geotextiles are also popular in road construction as they can provide layer separation and act as a filter layer between the subgrade and the granular layer [32]. The migration of soil particles from subgrade to the granular is common in weak subgrades, and therefore, the specifications of the Queensland Department of Transport and Main Roads states that a geotextile should be placed on subgrade before placing the geogrid [33]. As a result, composite geogrids, which has a geotextile layer attached under the geogrid, are popular in the local road construction industry as a hybrid product that extends the functions of reinforcement and separation simultaneously [2].

Geogrids can extend the pavement life, reducing the rutting depth and reducing the required base layer thickness [32, 34]. This performance depends on many factors such as type of geogrid, type of granular and the location of geogrid [35]. In order to assess the impact of different factors, researchers have conducted large scale pavement model testing under monotonic loading and cyclic loading [36, 37]. The cyclic loading tests can assess the long-term performance (rut depth and permanent deformation), while monotonic loading tests will assess the bearing capacity of the surface. Bearing capacity-based design methodologies are used mainly for unbound granular pavements and also to design working platforms [38]. This study focuses on assessing the bearing capacity of weak subgrade reinforced with a composite geogrid and a granular cover. The bearing capacity of the improved subgrade surface will be a direct input for the empirical design method [39] of unbound granular pavement to develop a rational pavement design that accounts the effect of geogrid reinforcement.

The main objective of this study is to examine the possibility of using composite geogrids for weak subgrade improvement under local conditions. Accordingly, two laboratory scale model tests were conducted using locally available materials. This study compares the stress versus deformation results of the two model tests to verify that composite geogrids are beneficial in reinforcing weak subgrades. Moreover, vertical stress distribution on weak subgrade surface was analysed.

2 Material Properties

2.1 Subgrade Soil

This project used a clay type black soil, collected from a road construction site in Toowoomba, Australia (Figure 4a). The collected subgrade soil was subjected to a series of standard geotechnical tests aiming to estimate the basic soil properties. These classification tests were conducted based on the standard procedures stipulated in the Material Testing Manual (MTM) of Queensland Department of Transportation and Main Roads [40]. Figure 1 illustrates the particle size distribution of the subgrade soil and the estimated subgrade soil properties are listed in Table 1. This soil was classified as high plastic silt based on the standard classification guidelines given in Unified Soil Classification System (USCS). Besides, the same soil stands in par with the category A-7–6 of AASHTO soil classification method.

2.2 Granular Material

A bulk of granular material, classified as type 2.1 based on MTRS05 specification of the Queensland Department of Transport and Main Roads [40] was received from the Logan City Council material storage for road construction (Figure 4b). The laboratory gradation test was performed, following the TMR specification, and verified that the selected granular material complies with the requirements for "Grading C' under granular material type 2 in MTRS05 specification (ref Figure 2). In addition, fine ratio, the ratio between the percentage of passing of 0.075 mm sieve to the 0.425 mm, was estimated at 0.51 and hence, it was confirmed that this granular material complies with TMR specification type 2.1. The standard proctor compaction test for this material confirmed that Maximum Dry Density and Optimum Moisture Content of the granular material as 2.3 g/cm^3 and 7% respectively.

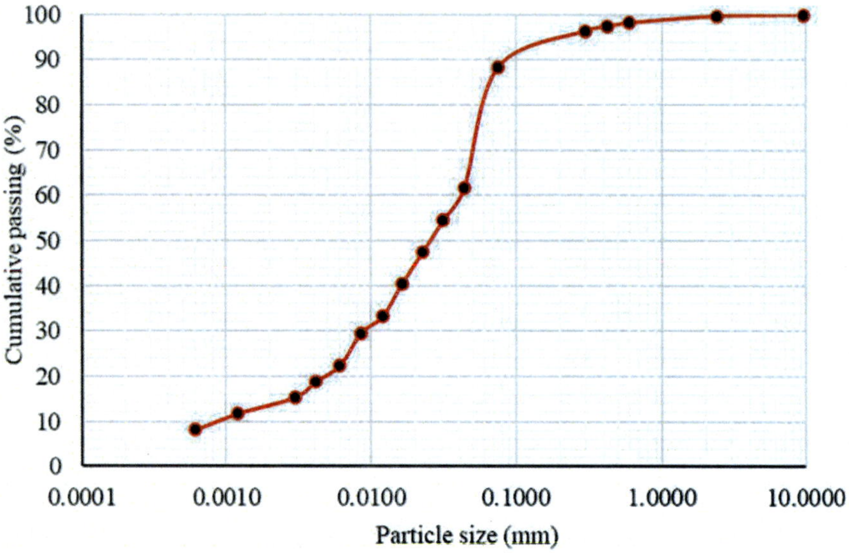

Fig. 1 Particle size distribution of subgrade soil

Table 1 Properties of subgrade soil

Soil property	Value
Soil Particle Density	2.62
Maximum Dry Density (g/cm^3)	1.316
Optimum Moisture Content (%)	32
Liquid Limit (%)	73
Plastic Limit (%)	53
Shrinkage Limit (%)	20
USCS Classification	MH

2.3 *Reinforced Material*

The reinforced material selected for this pavement model testing was a welded type composite geogrid made of a polypropylene with a non-woven geotextile as the bottom layer (ref. Figure 3). The aperture size of the geogrid was measured as 31 mm x 31 mm and the tensile strength in both machine direction and cross machine direction was specified as 40kN/m. Table 2 lists the manufacturer specifications of the selected type of composite geogrid.

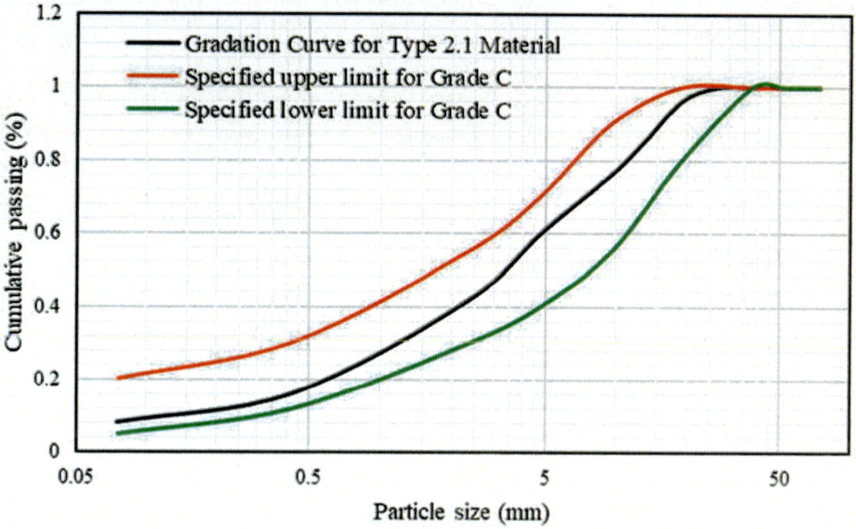

Fig. 2 Particle size distribution of type 2.1 material

Fig. 3 Composite geogrid

3 Large Scale Model Test

The pavement models were constructed in a steel test box with length width and height of 1.0 m, 1.0 m and 1.2 m, respectively. In this study, two test models: i.e. one unreinforced and one reinforced with composite geogrid, were prepared and subjected to monotonic loading. In each pavement model, a weaker subgrade of CBR 2.5% was constructed at the bottom of the model box to a thickness of 500 mm. The granular

Table 2 Technical specifications of composite geogrid

Property	Value	Unit
Geogrid		
Ultimate tensile strength (MD/CMD)	≥ 40/40	kN/m
Elongation at nominal strength	800%	%
Tensile strength at 2% elongation	≥ 16/16	kN/m
Tensile strength at 5% elongation	≥ 32/32	kN/m
Aperture size	31 × 31	mm
Geotextile		
Maximum tensile strength (MD/CMD)	7.5/11	kN/m
Elongation at maximum tensile strength (MD/CMD)	40/30	%

MD—Machine direction, CMD—Cross machine direction

(a) (b) (c)

Fig. 4 **a** Clay type black soil; **b** type 2.1 granular; **c** mechanical soil crusher

layer was constructed on top of the subgrade layer to a height of 200 mm for both reinforced and unreinforced tests. In some countries, unbound granular is mainly used to construct the granular base layer of pavement, while subbase is usually constructed by suitable soil. However, it is common to use granular material to construct a subbase layer in Queensland, Australia, as finding suitable soil is challenging in the vicinity. Therefore, this study used unbound granular material to construct a cover layer on the subgrade. In the reinforced test section, the composite geogrid was placed between subgrade and granular cover and hereafter will be referred as "interface".

3.1 Material Preparation

The black soil was air dried aiming to remove moisture to a possible extent. This has left larger lumps of clay that are rock hard and difficult to break for remixing with water to achieve the desired moisture content. Therefore, a mechanical crusher was used for crushing the clay lumps into small sizes (ref Fig. 4c). Afterwards, the subgrade soil was oven-dried at 60 degrees of Celsius for further removing of moisture. A series of trial CBR tests were conducted to establish the relationship between the unsoaked CBR of subgrade soil with moisture content and the degree of compaction. Accordingly, it was decided to mix dried subgrade soil with water to achieve 46.5% moisture content, which can be used to create a weaker subgrade of 2.5%. Further details of estimating CBR relationship with soil properties could be found in [41]. The prepared subgrade soil was stored in airtight containers and completely sealed and cured for minimum of 7 days to sequalise moisture throughout the soil.

As same as subgrade soil, type 2.1 gravel were oven-dried at 60 degrees of Celsius for at least 2 days to remove moisture. Thereafter, the gravel was mixed with water to achieve a gravimetric moisture content of 5.5%. In real road construction, water is added into granular to increase moisture content closer or above optimum moisture content, which demands less effort to achieve the desired level of compaction. Subsequently, the compacted granular layer is left to be dried for a certain time to reduce the moisture content. In this way, the premature shear failure of gravel layer can be eliminated. However, as it is practically difficult to follow the same method in model box testing, it was decided to maintain the moisture content of the granular layer 1.5% below the optimum moisture content to avoid premature shear failure.

3.2 *Preparation of Model Sections*

The schematic arrangement of a model section is illustrated in Fig. 5. At first, a 500 mm thick subgrade layer of CBR 2.5% was compacted at the bottom of the model box. This subgrade was compacted as 10 equivalent layers of 50 mm each to maintain uniformity across the subgrade layer. From the standard CBR trials, it was confirmed that a subgrade of CBR 2.5% could be created by compacting subgrade soil of 46.5% moisture to a density of 1.118 g/cm^3. Accordingly, the required amount of soil for a 50 mm layer was calculated and dumped into the model box. Thereafter, the soil was carefully levelled and manually compacted using a hand tamper that is 20 kg in weight and has a square shape bottom of 200 mm length of each side. The tamping hammer was dropped from a height of 150 mm approximately throughout the compaction process to ensure applying of equal compaction energy. After compacting the layer to 50 mm height, the top surface of the layer was scratched prior to compacting the next layer to ensure bonding between adjacent layers. The granular cover was constructed on the completed subgrade by following the same procedure as the subgrade layer. However, a mechanical compactor (see Fig. 6c) was used to compact granular layer instead of hand tamper owing to the fact that the achievement of the desired compaction of 100% (from MDD) in the unbound granular layer is difficult with a hand tamper. In addition, all efforts were taken to maintain equal duration to prepare each layer.

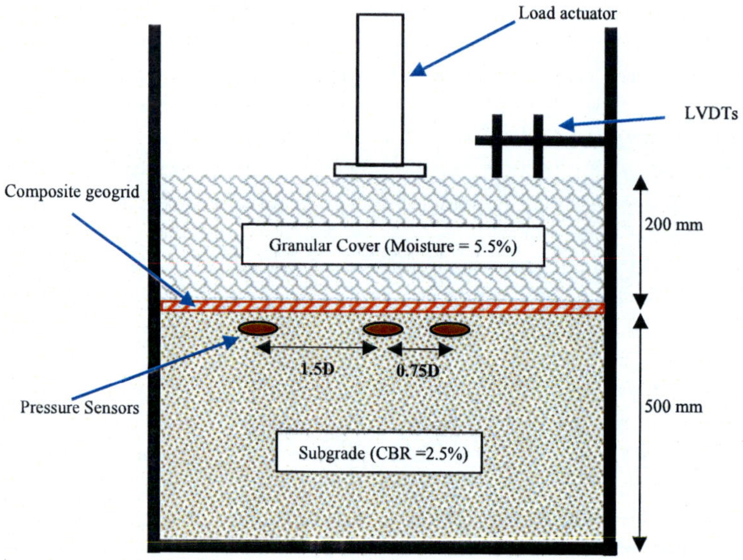

Fig. 5 Schematic arrangement of a test section

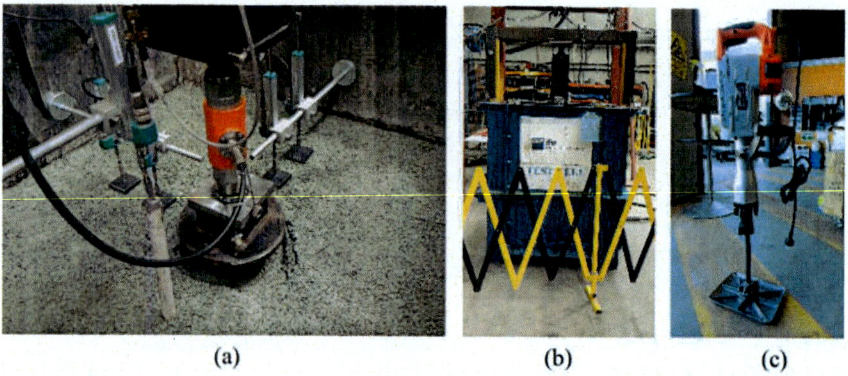

Fig. 6 **a** Loading with 200 mm plate; **b** model test box; **c** mechanical compactor

The composite geogrid was placed at the interface in the reinforced test section. The geogrid was trimmed to fit the internal dimensions of the steel test box and thereafter, was placed on the subgrade. The four corners were anchored with u shaped pins of 5 cm to ensure that the geogrid stays flat until the first 50 mm granular layer was constructed. Placing the geogrid flat is important, owing to the reason that an initial deformation can have a significant impact on the performance of geogrid. Both reinforced and unreinforced test pavement models were instrumented with pressure plates, moisture sensors and Linear Variable Differential Transducers (LVDTs) to obtain the necessary data to analyse the behaviour of two pavements under monotonic loading. Finally, a monotonic loading at a rate of 1 mm/min deformation was applied on the granular surface through a circular loading plate of 200 mm diameter until to the ultimate failure state (see Fig. 6a.).

4 Results and Discussion

4.1 Effect of Composite Geogrid Layer

The surface stress versus deformation graphs for unreinforced and geogrid reinforced model subgrade sections are illustrated in Fig. 7. It was observed that the stress required for a certain deflection in reinforced section is significantly higher than that of the unreinforced section. However, the stress to make a deflection up to 2 mm was found almost equal for both reinforced and unreinforced sections. Moreover, the ultimate stress of the reinforced section was observed as 1975 kPa while the unreinforced section has recorded ultimate stress of 1300 kPa. Hence, it can be seen that the composite geogrid has increased the ultimate stress of the improved subgrade by 1.52 times. Hence, it is evident that composite geogrids can contribute to a significant improvement of the bearing capacity of a weak subgrade. In fact, Abu-Farsak et al. [38] has also observed similar behaviour for biaxial and triaxial geogrids covered by a 305 mm thick granular layer.

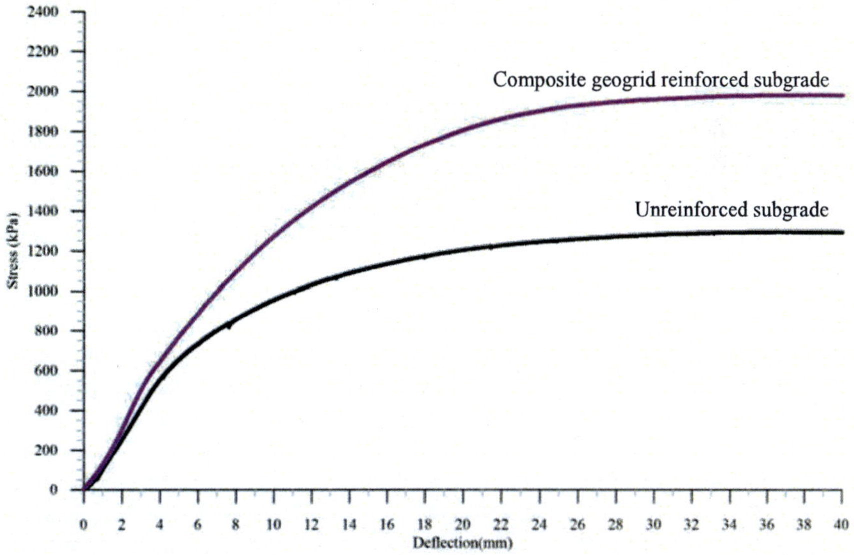

Fig. 7 Stress versus deformation of unbound granular test pavements under monotonic loading

The Plate Load Test results can be used to derive different types of elastic modulus such as initial tangent modulus, tangent modulus at given stress level, reloading modulus and secant modulus [38]. The elastic modulus from the plate load test could also be calculated at given stress by the following equation:

$$E_{PLT} = \frac{2P(1 - \upsilon^2)}{\pi R\delta} \tag{1}$$

Where P is the applied load; R is the radius of the loading plate, δ is the deflection of plate at load P, and υ is the Poisson ratio. Using Eq. 1, elastic modulus for both reinforced and unreinforced sections were calculated at 17.27 kN vertical load which is equal to 550 kPa stress on top of the granular surface, equal to the standard tyre load. The Poisson ratio was assumed as 0.5 and the elastic modulus of the unreinforced section was estimated as 10.31 MPa, while the elastic modulus of the reinforced section was estimated as 13.3 MPa. This provides clear evidence that composite geogrid reinforcement can contribute to the improvement of pavement modulus.

4.2 Vertical Stress at Base-Subgrade Interface

The vertical stress distributions at interface level for reinforced and unreinforced sections were measured by three pressure transducers placed 50 mm below the base subgrade interface. These pressure transducers were arranged as one at the centre of the loading plate and other two are 1.0D and 1.5D away from the center of the loading plate in the same line, where D is the diameter of loading plate. Figure 8 shows the vertical stress distribution measured along the centerline of the plates when the vertical stress applied on the granular surface is 1300 kPa. The geogrid reinforced section shows a clear reduction of transferred vertical stress at the centre compared to the unreinforced subgrade section. This is due to the increased vertical stress distribution angel and tension membrane effect of geogrid. Moreover, it is also evident that vertical stress is distributed at a maximum circular area of 1.5D radius at the interface level. In addition, vertical stress at the interface is negligible beyond 1.5D away from the centre of the interface along the centerline of pressure plates.

In past studies, researchers have reported a reduction of vertical stress at the centre of base subgrade interface in the reinforced section compared to the unreinforced model [36]. Furthermore, Abu-Farsakh et al. [38] has observed an increase of vertical stress at 1.0D and 2.0D away from the centre at the interface in the reinforced section compared the unreinforced section. Moreover, they have observed a vertical distribution of stress around an area of 2.0D radius. Hence, that study has concluded that geogrid reinforcement redistributes the applied load to a wider area. In contrast, this study has recorded a lower vertical stress at 0.75D away from the center of interface. A slight increase of vertical stress was observed at 1.5D away from the center of interface in the reinforced section than the unreinforced section.

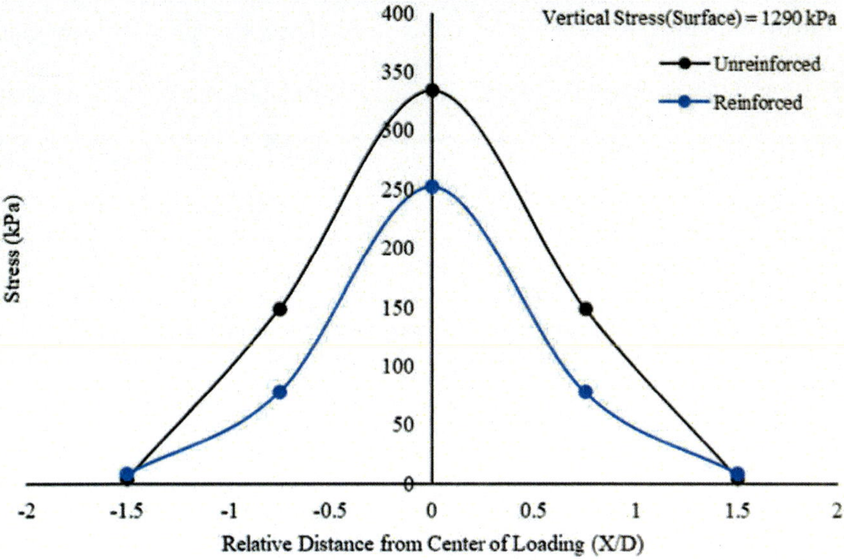

Fig. 8 Vertical stress distribution at interface level

5 Conclusion

This study has tested two instrumented laboratory scale model sections under a monotonic load. One of the subgrades was reinforced with a composite geogrid, placed at granular subgrade interface, while the other model was considered as the control test. Based on the observed test results, the study has drawn the following conclusions:

- A composite geogrid on the subgrade with a 200 mm granular cover improves the ultimate bearing capacity of a weak subgrade with the same granular cover by approximately 1.5 times.
- The composite geogrid at interface contributes to the improvement of overall pavement modulus.
- Composite geogrid reinforcement demonstrates a 25% reduction of vertical stress at the centre of the granular-subgrade interface.
- The vertical stress distribution spans across a circular area of 1.5D radius on the interface. Moreover, the measured vertical stress at a point beyond 1.5D away from the center is negligible compared to the applied vertical stress inside the circular area of 1.5D radius.
- Although a slight increase of the measured vertical stress at 1.5D away from the center of the granular-subgrade interface was measured in the reinforced section, further investigations are needed to conclude the capacity of composite geogrids to redistribute the stress bulb.

In general, the presented results in this study clearly emphasis that composite geogrids can be used effectively to improve the condition of weak subgrades. However, future studies are needed to investigate the suitability under different conditions such as: increased granular covers and type of gravel. Moreover, it is also recommended to conduct full scale field studies to link laboratory-scale monotonic results with the long-term performance of a geogrid reinforced pavement.

Acknowledgements This research work is part of a research project (Project No IH18.06.1) sponsored by the SPARC Hub at Department of Civil Eng, Monash University funded by the Australian Research Council (ARC) Industrial Transformation Research Hub (ITRH) Scheme (Project ID: IH180100010). The financial and in-kind support from Depart-ment of Transport and Main Roads (Queensland), Logan City Council, Global Syn-thetics (Australia), Polyfabrics (Australia) and Queensland University of Technolo-gy (QUT) is gratefully acknowledged. Also, the financial support from ARC is high-ly acknowledged.

References

1. Jayalath CPG, Wimalasena KD (2020) Measuring strains on geogrid specimens under the laboratory tensile test. Int J Geomate 18:86–93
2. Gunasekara Jayalath CP, Gallage C, Dhanasekar M, Dareeju B, Ramanujam J, Lee J (2018) Pavement model tests to investigate the effects of geogrid as subgrade reinforcement. In: Proceedings of the 12th Australian and New Zealand young geotechnical professionals conference, pp 1–8
3. Udukumburage RS, Gallage C, Dawes L (2019) Investigation of the effect of initial water content and surcharge on volume change behaviour of compacted grey vertosol. In 13th Australia New Zealand conference on geomechanics, pp 1029–1033
4. Udukumburage RS, Gallage C, Dawes L (2018) Loaded swell tests to estimate the heave of the expansive soil in instrumented soil column. Int J Geomate
5. Gallage C, Tehrani K, Williams D (2017) Instrumented large soil-column to investigate climate-induced ground deformation in expansive soil. In Proceedings of the 19th international conference on soil mechanics and geotechnical engineering, Seoul, pp 1147–1150
6. Kodikara J, Rajeev P, Chan D, Gallage C (2014) Soil moisture monitoring at the field scale using neutron probe. Can Geotech J 51:332–345
7. Udukumburage RS, Gallage C, Dawes L, Gui Y (2020) Determination of the hydraulic conductivity function of grey Vertosol with soil column test. Heliyon 6:e05399
8. Udukumburage RS, Gallage C, Dawes L (2019) Oedometer based estimation of vertical shrinkage of expansive soil in a large instrumeted soil column. Heliyon 5:e02380
9. Chan D, Rajeev P, Kodikara J, Gallage C (2016) Field performance of in-service cast iron gas reticulation pipe buried in reactive clay. J Pipeline Syst Eng 7:04015025
10. Gallage CPK, Chan D, Kodikara J (2012) Response of a plastic pipe buried in expansive clay. Proc Inst Civ Eng-Geotech Eng 165:45–57
11. Jayakody S, Gallage C, Ramanujam J (2019) Effects of reclaimed asphalt materials on geotechnical characteristics of recycled concrete aggregates as a pavement material. Road Mater Pavement Des 20:754–772
12. Gallage C, Eom T, Barker D, Ramanujam J (2015) Falling weight deflectometer (FWD) tests on granular pavement reinforced with geogrids–case study. In Proceedings of the international conference on geotechnical engineering, geotechnics for sustainable development, pp 597–600
13. Jayakody S, Gallage C, Kumar A (2014) Assessment of recycled concrete aggregates as a pavement material. Geomech Eng 6:235–248

14. Jayakody S, Gallage C, Ramanujam J (2014) Effects of moisture content on resilient properties of recycled concrete aggregates (RCAs). In Proceedings of fourth international conference–geomate, geotechnique, construction materials and environment, pp 394–399
15. Jayakody S, Gallage C, Kumar A (2012) Assessment of recycled concrete aggregate for road base and sub-base. In Proceedings of the second international conference on geotechnique, construction materials and environment, pp 575–579
16. Gallage C, Cochrane M, Ramanujam J (2012) Effects of lime content and amelioration period in double lime application on the strength of lime treated expansive sub-grade soils. In Proceedings of the 2nd international conference on transportation geotechniques, pp 99–104
17. Clark BR, Gallage C, Yeaman J (2020) Temperature variation through deep multigrade asphalt pavements and proposed method for accounting for fluctuations 32:04020005
18. Clark BR, Gallage C (2020) Superior performance benefits of multigrade bitumen asphalt with recycled asphalt pavement additive. Construct Build Mater 230:116963
19. Wimalasena K, Pradeep T, Rengarasu T (2019) Suitability of demolished concrete waste in hot mix asphalt concrete for local roads in Sri Lanka. J Eastern Asia Soc Transp Studies 13:1641–1647
20. Clark BR, Piacere L, Gallage C (2018) Effects of recycled asphalt pavement on the stiffness and fatigue performance of multigrade bitumen asphalt. J Mater Civ Eng 30:04017278
21. Jayakody S, Gallage C, Ramanujam J (2019) Performance characteristics of recycled concrete aggregate as an unbound pavement material. Heliyon 5:e02494
22. Wimalasena K, Jayalath CPG (2020) Effect of geogrid reinforcement in weak subgrades. Int J GEOMATE 18:140–146
23. Suku L, Prabhu SS, Babu GS (2017) Effect of geogrid-reinforcement in granular bases under repeated loading. Geotextiles Geomembranes 45:377–389
24. Cheah C, Gallage C, Dawes L, Kendall P (2017) Measuring hydraulic properties of geotextiles after installation damage. Geotext Geomembr 45:462–470
25. Cheah C, Gallage C, Dawes L, Kendall P (2016) Impact resistance and evaluation of retained strength on geotextiles. Geotext Geomembr 44:549–556
26. Weerasinghe I, Gallage C, Dawes L, Kendall P (2020) Factors affecting the hydraulic performance of a geosynthetic clay liner overlap. J Environ Manage 271:110978
27. Weerasinghe I, Gallage C, Dawes L, Kendall P (2019) Liquid flow mechanism at a geosynthetic clay-liner overlap
28. Weerasinghe I, Gallage C, Dawes L (2021) Effect of overburden confining stress on hydraulic performance of geosynthetic clay liners (GCLs), Heliyon 7:e05770
29. Nazeri A, Ziaie-Moayed R, Ghiasinejad H (2020) Evaluating the efficiency of a composite geosynthetic reinforcement in container yard pavements via laboratory plate load test. Int J Pavement Eng pp 1–12
30. Zornberg J (2015) Advances in the use of geosynthetics in pavement projects. In Keynote lecture, proceedings of the 2nd iberic conference on geosynthetics, Geosintec Iberia, pp 07–08
31. Ferrotti G, Canestrari F, Virgili A, Grilli A (2011) A strategic laboratory approach for the performance investigation of geogrids in flexible pavements. Construct Build Mater 25:2343–2348
32. Zornberg J, Gupta R (2010) Geosynthetics in pavements: North American contributions. In Theme speaker lecture, proceedings of the 9th international conference on geosynthetics, Guarujá, Brazil, (May 2010), pp 379–400
33. QD. o. T. a. M (2017) Roads: MRTS58 Subgrade reinforcement using pavement geosynthetics. Ed
34. Imjai T, Pilakoutas K, Guadagnini M (2019) Performance of geosynthetic-reinforced flexible pavements in full-scale field trials. Geotext Geomembr 47:217–229
35. Al-Qadi IL, Dessouky SH, Kwon J, Tutumluer E (2012) Geogrid-reinforced low-volume flexible pavements: pavement response and geogrid optimal location. J Transp Eng 138:1083–1090
36. Nazzal M, Abu-Farsakh M, Mohammad L (2007) Laboratory characterization of reinforced crushed limestone under monotonic and cyclic loading. J Mater Civ Eng 19:772–783

37. Abu-Farsakh MY, Chen Q (2011) Evaluation of geogrid base reinforcement in flexible pavement using cyclic plate load testing. Int J Pavement Eng 12:275–288
38. Abu-Farsakh MY, Akond I, Chen Q (2016) Evaluating the performance of geosynthetic-reinforced unpaved roads using plate load tests. Int J Pavement Eng 17:901–912
39. Austroads: Guide to Pavement Technology Part 2: Pavement structural design. Ed (2018)
40. Q. D. o. T. a. M. Roads: Materials testing manual. Ed, p. 725 (2018)
41. Wimalasena K, Gallage C (2021) Predicting california bearing ratio (CBR) value of a selected subgrade material. In 12th international conference on road and airfield pavement technology, Virtual Conference (Submitted)

Numerical Simulation of Differential Settlement Resistance for Reinforced Joint Between New and Existing Pavement with RoadMesh Steel Wire Mesh

Xu-Dong Zha, Hao-Jun Liu, and Jian-Hua Zhu

Abstract In order to analyze the ability of longitudinal cracking resistance due to differential settlement between new and existing subgrade in the joint between new and existing pavement reinforced by RoadMesh steel wire mesh, and relying on the existing road widening project, three kinds of layer location reinforcement schemes were selected, which are between surface layers (Scheme I), between surface layer and base (Scheme II) as well as between base and subbase (Scheme III). The maximum void value of 10 mm due to differential settlement was set on the surface of new subgrade under the outer edge of new shoulder. The three-dimensional finite element numerical models were established to carry out the comparative analysis of mechanical responses of widening asphalt pavement structure for different reinforced layer locations. The results show that the peak value of maximum tensile stress and the deflection value of double wheel gap center in pavement surface can be used as the main control indexes for differential settlement resistance. All the orders of simulation results of the two indexes under each scheme are Scheme I < Scheme II < Scheme III < Unreinforced. The simulation results of the two indexes in Scheme I are 53.4% and 60.6% lower than those of the unreinforced scheme respectively. It indicates that the joint between new and existing pavement reinforced by the RoadMesh can effectively improve the differential settlement resistance of widening asphalt pavement structure. The closer the reinforced layer location is to the pavement surface, the more significant the reinforcement effect is, so the optimum reinforcement layer location is between surface layers.

Keywords Steel wire mesh · Joint between new and existing pavement · Pavement reinforcement · Differential settlement resistance · Three-dimensional finite element method · Existing road widening

---

X.-D. Zha (✉) · H.-J. Liu
School of Traffic and Transportation Engineering, Changsha University of Science and Technology, Changsha 410114, Hunan, China

J.-H. Zhu
Quzhou Management Center of Highway, Port, Waterway and Transportation, Quzhou 324000, Zhejiang, China

© The Author(s), under exclusive license to Springer Nature Switzerland AG 2022
H. R. Pasindu et al. (eds.), *Road and Airfield Pavement Technology*, Lecture Notes in Civil Engineering 193, https://doi.org/10.1007/978-3-030-87379-0_36

1 Introduction

In the existing road widening project, because of the uncoordinated deformation between new and existing subgrade, there must be differential settlement [1], which makes the widening asphalt pavement structure bear additional stress, and it is easy to cause the longitudinal cracking at the joint between new and existing pavement under the action of traffic load [2]. Therefore, it is necessary to carry out step excavation and reinforcement treatment for the joint between new and existing subgrade pavement, in which geosynthetics reinforcement is a common treatment measure [3]. However, only in the case of large deformation can it give full play to the role of reinforcement. At this time, the pavement is very easy to produce more serious longitudinal cracking, and it is difficult to play the expected effect of cracking resistance.

For this purpose, a kind of pavement reinforcement technology with RoadMesh steel wire mesh has been more and more applied in the past decade, which is woven by hexagonal double-twisted steel wire mesh and transverse steel rods. Laying it between the structure layers of asphalt pavement along the longitudinal direction can form three-dimensional mesh hoop reinforcement together with the mixtures of structure layers [4]. The existing researches have showed that its excellent characteristics can effectively enhance the resistances to cracking, rutting and fatigue as well as other performances of asphalt pavement structure, and prolong the service life of pavement [5–8]. Therefore, it has been successfully applied in pavement reinforcement projects such as asphalt overlay on existing cement concrete pavement and joint between new and existing pavement in existing road widening, etc. [9–11]. Accordingly, relying on the existing road widening project, and through three-dimensional finite element modeling and numerical simulation, this paper will analyze the ability of differential settlement resistance for reinforcing joint between new and existing pavement with RoadMesh to provide theoretical basis for engineering applications.

2 Design of Widening Pavement Structure

The existing road of background project is a third-class highway with cement concrete pavement and width of 7 m, which will be widened and upgraded to a second-class highway with width of 10 m, bilateral symmetrical widening and asphalt pavement structure. The corresponding widening design scheme is shown in Fig. 1, in which AC is surface layer of Asphalt Concrete, CSM is base and subbase of Cement Stabilized Macadam, GM is cushion of Graded Macadam, CSGS is subgrade treated layer of Cement Stabilized Gravel Soil, ECCP is cracking layer of Existing Cement Concrete Pavement slab, and EMBM is base of Existing Mud Bound Macadam, respectively.

For the convenience of construction, the cracking layer of existing cement concrete pavement slab and the base of mud bound macadam in the existing pavement are equal in thickness and flush in top surface to the base of cement stabilized macadam and the cushion of graded macadam in the new pavement respectively, and then

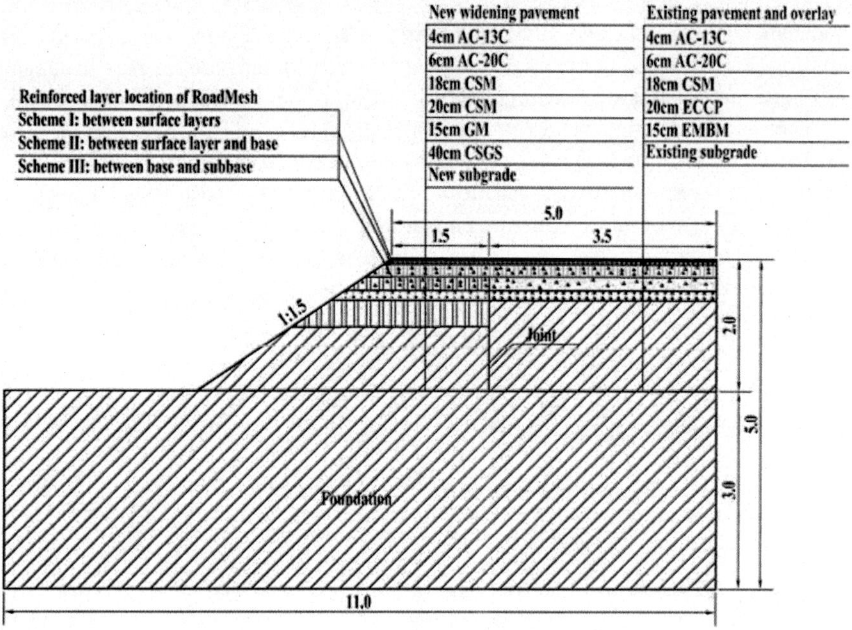

Fig. 1 Semi-range cross section of widening subgrade and pavement (unit: m)

the AC-13C upper surface layer, AC-20C lower surface layer and cement stabilized macadam base are uniformly overlaid.

3 Numerical Modeling of Widening Pavement Structure

3.1 Modeling of Three-Dimensional Finite Element

The assumed conditions of pavement structure analysis model are linear elastic homogeneous body, isotropy and regardless of self-weight, and the interface conditions are continuous contact. Considering the bilateral symmetrical widening, one half cross section was taken for three-dimensional modeling. The boundary conditions were set as horizontal displacement constraints on the vertical outer edge around the model and full constraints on the bottom surface of foundation. In order to compare the ability of differential settlement resistance with different RoadMesh reinforcement schemes between layers, and according to the joint reinforcement between new and existing pavement from up to down, three kinds of layer location reinforcement schemes were selected, which are between upper and lower surface layers of asphalt (Scheme I), between lower surface layer of asphalt and base of cement stabilized macadam (Scheme II) as well as between base and subbase of cement stabilized macadam (Scheme III) respectively, and the unreinforced scheme was also selected. Among them, for existing pavement, Scheme III is reinforced between base of cement stabilized macadam and cracking layer of existing cement concrete pavement slab.

When modeling with ANSYS, the SOLID45 elastic solid hexahedron element was used for each structure layer model of pavement, subgrade and foundation, and the BEAM4 three-dimensional elastic beam element was used for the RoadMesh. The whole steel wire mesh structure constituted a network structure layer in accordance with the actual size, which was embedded between the structure layers in the corresponding layer location reinforcement scheme, as shown in Fig. 2. Among

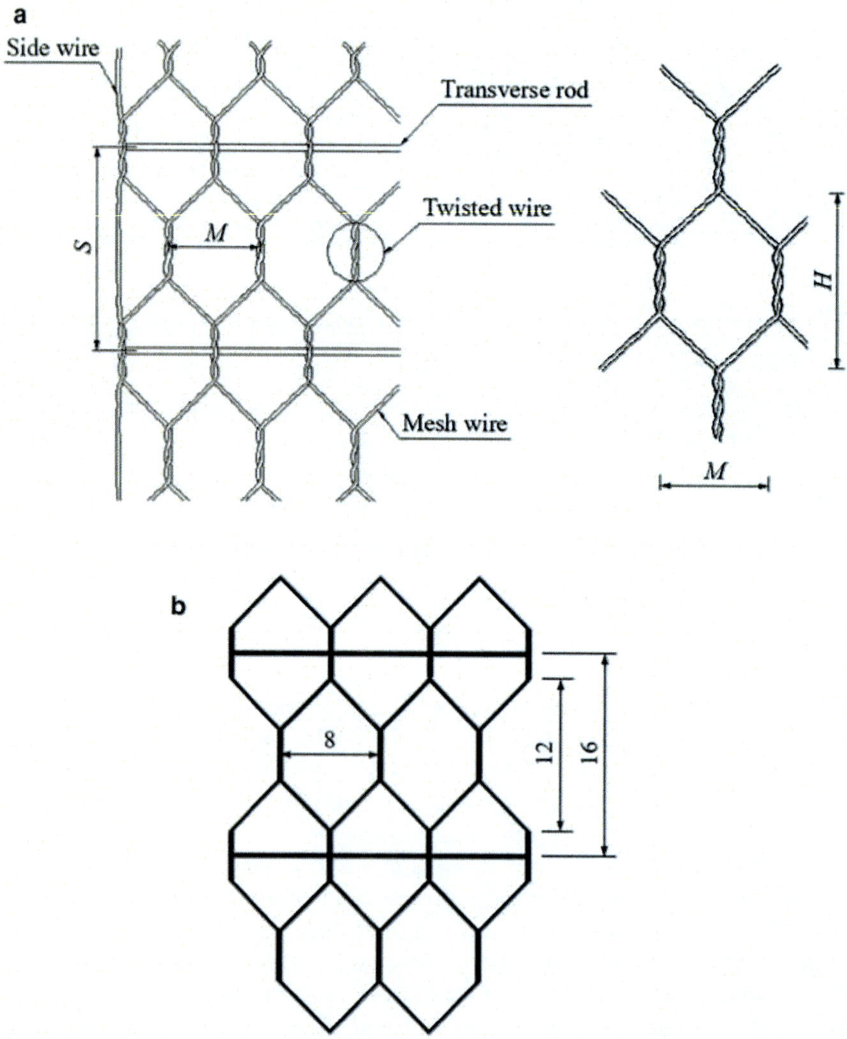

Fig. 2 RoadMesh structure and its finite element model

them, the mesh cell width $M = 8$ cm and its height $H = 12$ cm, the mesh surface wire diameter $d = 2.2$ mm, and the transverse steel rod spacing $S = 16$ cm and its diameter $D = 3.4$ mm, respectively.

Referring to the existing research [12], the surface differential settlement of symmetrically widened subgrade presents a "cosine curve" distribution along the cross section direction. That is, the settlement at the center of subgrade is the smallest. With the increase of distance from the center, the settlement gradually increases until the surface of new subgrade under the outer edge of new shoulder reaches the peak value, and then extends to the slope surface and gradually decreases. In order to simulate the differential settlement between the new and existing subgrade, the settlement of existing subgrade was ignored in the analysis. It was assumed that the surface settlement value on the side of existing subgrade at the joint is zero, the surface settlement value of new subgrade under the outer edge of new shoulder is largest and taken as 10 mm, and the settlement value between the two is in a linear distribution.

According to the semi-range cross section in Fig. 1, the existing pavement width $b = 7$ m, the new pavement width $B = 10$ m, the widening width of both sides $W = 1.5$ m, the filling height of subgrade pavement $F = 2$ m, the slope rate $n =$ 1:1.5, and the foundation thickness $h = 3$ m and its width $Z = 22$ m, respectively. The RoadMesh rolls have a fabric width of 3.5 m, which were spanned cross the joint along the transverse direction and spread over the full length of route along the longitudinal direction to lay between the corresponding pavement structure layers. According to the reinforced layer location schemes, 1.5 m and 2.0 m width of RoadMesh were respectively laid on the new and existing pavement underlayers on both sides of the joint along the transverse direction. Hereby, taking the longitudinal length of widening road as 6 m, the corresponding solid modeling and element meshing of three-dimensional finite element can be obtained as shown in Fig. 3. In order to ensure the calculation accuracy and improve the calculation efficiency, and through

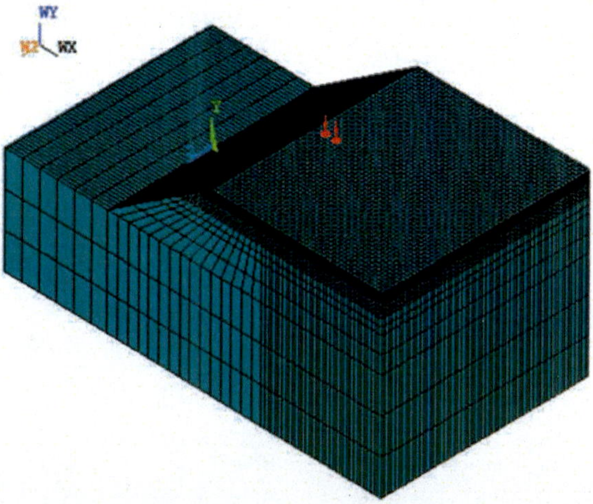

Fig. 3 Solid modeling and meshing for three-dimensional finite element

the convergence analysis and comparison, it was determined that the elements were divided by the mapping mesh. Among them, the element size is 0.1 m along the longitudinal Z-axis. The element size is 0.1 m on the road surface along the transverse X-axis, and gradually becomes sparse from the joint to 0.5 m of the foundation layer outside the embankment slop foot. Along the vertical Y-axis, the element size gradually becomes sparse from 0.02 m of the surface layer to 1 m of the foundation layer according to the thickness combination of structure layers.

3.2 *Taking Values of Calculation Parameters*

The calculation parameters of each structure layer were taken according to Table 1. The elastic modulus and Poisson's ratio of RoadMesh are $E_R = 200$ GPa and $\mu_R = 0.30$, respectively. The standard axle load BZZ-100 with double circular vertical uniform distribution was adopted for the driving load, in which the grounding pressure, load radius and center distance of double wheels are $p = 0.7$ MPa, $\delta = 10.65$ cm and 3δ, respectively. The numerical simulation shows that the worst loading position was the road surface of new shoulder at the maximum settlement position. As shown in Fig. 3, and considering the requirements of lane layout and traffic safe distance, the load acting position was taken as: the double wheels were arranged along the transverse direction, and the center of double wheel gap is 0.5 m from the outer edge of new shoulder along the transverse X-axis and located at 3 m of the model center along the longitudinal Z-axis.

Table 1 Calculation parameters of each structure layer

No.	Structure layer	Thickness h_i (cm)	Elastic modulus E_i (MPa)	Poisson's ratio μ_i
1	AC-13C upper surface layer	4	1400	0.25
2	AC-20C lower surface layer	6	1200	0.25
3	CSM base	18	1500	0.25
4	CSM subbase	20	1300	0.25
5	GM cushion	15	225	0.25
6	CSGS subgrade treated layer	40	95	0.25
7	New subgrade	97	30	0.35
8	ECCP cracking layer	20	1000	0.15
9	EMBM base	15	200	0.25
10	Existing subgrade	137	100	0.35
11	Foundation	300	20	0.35

4 Analysis of Numerical Simulation Results

In order to compare and analyze the ability of differential settlement resistance in the joint between new and existing pavement reinforced by RoadMesh, and according to the mechanical characteristics of widening asphalt pavement structure, five mechanical response values of the vertical displacement (l) and maximum tensile stress (σ_{s0}) of pavement surface, and the maximum tensile stresses at the bottom of upper surface layer (σ_{s1}), lower surface layer (σ_{s2}) and base (σ_{s3}) were selected as the evaluation indexes. Thus, according to the results of ANSYS numerical simulation, the transverse distribution curves of five indexes along the central connection line of double wheels for different layer location reinforcement schemes can be obtained as shown in Fig. 4, where L is the distance from the center of subgrade, and the dotted line is the joint between new and existing pavement. The peak values of each index under the corresponding schemes are shown in Table 2.

It can be seen from Fig. 4 that the overall change laws of five evaluation indexes along the transverse distribution under each scheme are basically the same, and there are significant changes at the joint. Among them, the shapes of the maximum tensile stress change curves are alike at the pavement surface, the bottom of upper surface layer and the bottom of lower surface layer, and there is an obvious reverse bending tensile state of the upper part of the pavement near the joint.

As can be seen from Table 2, and compared with the unreinforced scheme, the pavement surface deflections of double wheel gap center (vertical displacement at 4.5 m) of Scheme I, II and III are decreased by 60.6%, 52.3% and 27.5%, the peak values of maximum tensile stress of pavement surface are decreased by 53.4%, 46.4% and 25.4%, the peak values of maximum tensile stress at the bottom of upper surface layer are decreased by 52.5%, 45.4% and 24.0%, the peak values of maximum tensile stress at the bottom of lower surface layer are decreased by 52.3%, 45.3% and 24.4%, respectively, and the differences of peak values of maximum tensile stress at the bottom of base are very small. At the same time, among the four maximum tensile stress indexes, the peak value of pavement surface tensile stress is the largest, indicating that it is the key index to control differential settlement cracking.

Figure 4a shows that the existing pavement surface in Scheme I has slight counter-arch. Figure 4b–d show that all the surface layers of existing pavement are in tension state under each layer location reinforcement scheme, in which all the peak values of maximum tensile stress appear slightly near the side of existing pavement at the joint, and that of the unreinforced scheme appears at the joint. However, all the surface layers of new pavement under the wheel load are in compression state. Meanwhile, Fig. 4e shows that the base bottom of each scheme is in tension state, but the differences of peak value of maximum tensile stress under the load for each scheme are small. Therefore, it indicates that the differential settlement between new and existing subgrade results in the reverse tension of widening asphalt pavement structure, which is easy to cause the top-down longitudinal cracking at the joint between new and existing pavement. After reinforcement by the RoadMesh, it can play the role of dispersing load. When it is laid between the upper and lower surface layers, it

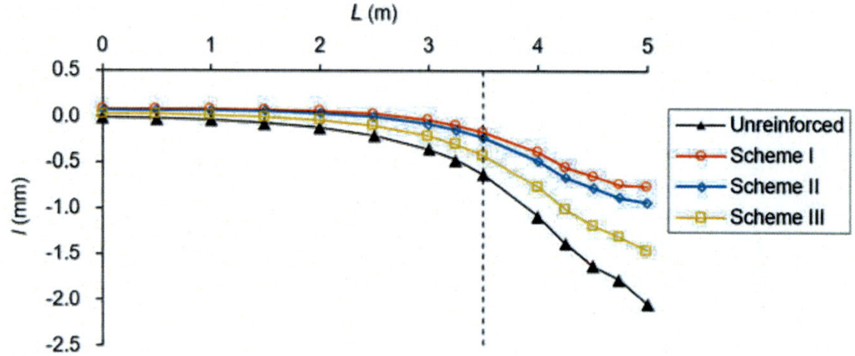

a. l~L change curves.

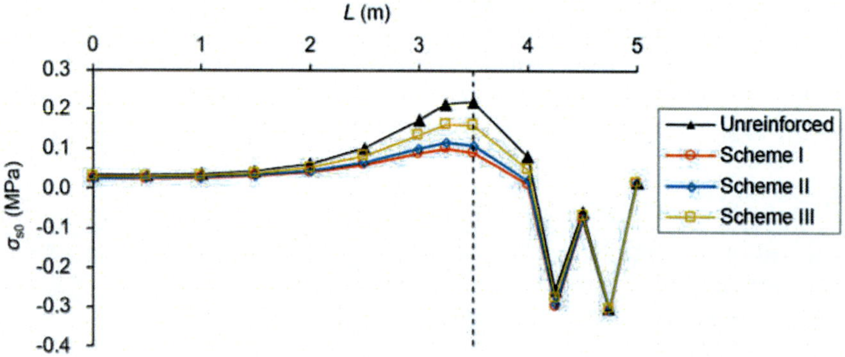

b. σ_{s0}~L change curves.

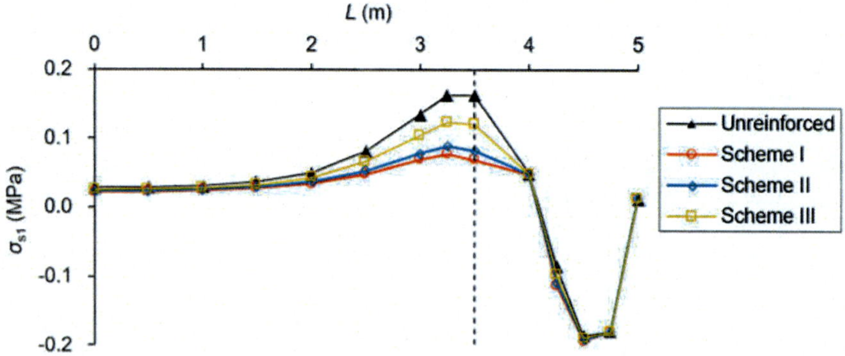

c. σ_{s1}~L change curves.

Fig. 4 Transverse distribution curves of each index under each reinforcement scheme

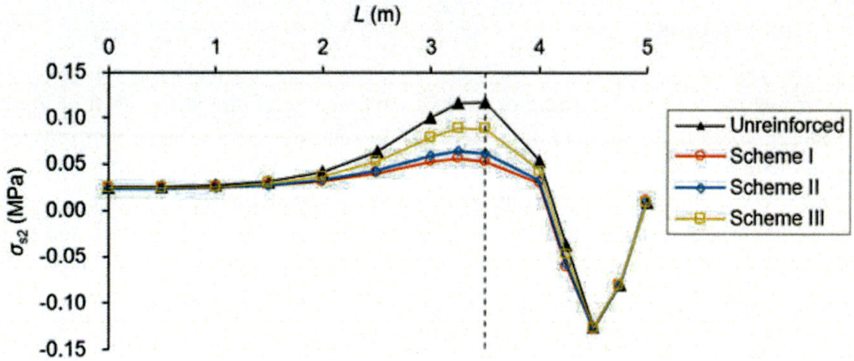

d. σ_{s2}~L change curves.

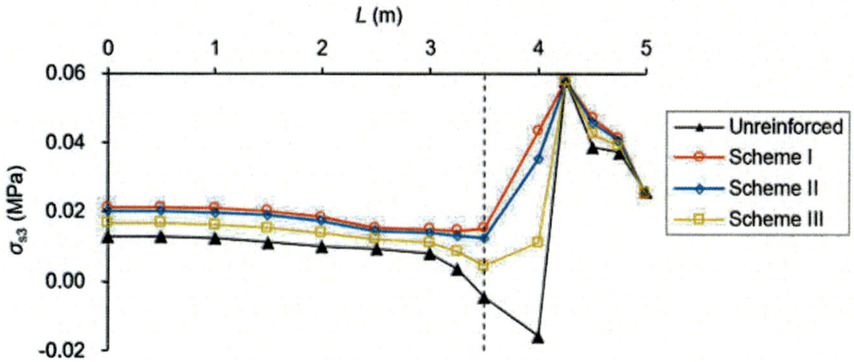

e. σ_{s3}~L change curves.

Fig. 4 (continued)

Table 2 Peak values of 5 evaluation indexes under each reinforcement scheme

Reinforcement scheme	Scheme I	Scheme II	Scheme III	Unreinforced
Deflection l_0 (mm)	0.6402	0.7742	1.1763	1.6230
$\sigma_{s0,max}$ (MPa)	0.1018	0.1170	0.1628	0.2183
$\sigma_{s1,max}$ (MPa)	0.0773	0.0889	0.1237	0.1628
$\sigma_{s2,max}$ (MPa)	0.0560	0.0642	0.0887	0.1174
$\sigma_{s3,max}$ (MPa)	0.0579	0.0579	0.0580	0.0580

can significantly make the stress redistribution in widening pavement structure, that is, the closer the RoadMesh reinforcement layer location is to the pavement surface, the better the reinforcement effect is, and the stronger the corresponding ability of differential settlement cracking resistance is.

5 Conclusions

According to the design scheme of asphalt pavement structure in the existing road widening project, three kinds of layer location reinforcement scheme and unreinforced scheme were considered. Through three-dimensional finite element modeling and numerical simulation, the ability of differential settlement resistance in the joint between new and existing pavement reinforced by the RoadMesh was compared and analyzed. The following conclusions can be obtained:

(1) After producing the differential settlement between new and existing subgrade, the widening asphalt pavement structure is subjected to reverse tension under the action of traffic load, which will easily lead to the top-down longitudinal cracking at the joint between new and existing pavement. Therefore, the peak value of maximum tensile stress and the deflection value of double wheel gap center in pavement surface can be used as the main control indexes for resistance to differential settlement.

(2) The reinforcement between layers with the RoadMesh can effectively play a role of dispersing load to reduce the stress and deformation of widening asphalt pavement structure, and the closer to the pavement surface, the more significant the reinforcement effect is. Therefore, the best reinforcement layer location for differential settlement resistance is between the surface layers, followed by between surface layer and base, and the worst is between the base layers.

To sum up, in the practical applications of existing road widening project, and under the conditions of satisfying the construction, the RoadMesh should be laid between the upper pavement structure layers as far as possible across the joint between new and existing pavement to give full play to its ability of differential settlement resistance.

Acknowledgements This paper is a part of the Project 51878077 supported by National Natural Science Foundation of China and the Project (2020 [12]) supported by China Railway No. 5 Engineering Group. Co., LTD.

References

1. Jia BX, Liu FP, Zhao L et al (2020) Influential factors analysis of the new and old roadbed on the differential settlements due to the highway reconstruction and extension. J Saf Environ 20(1):69–72
2. Weng XL, Zhang C, Li CL et al (2013) Large scale test on additional deformation of pavement widening with differential settlement. China J Highw Transport 26(4):1–6
3. Guan ZB, Yang T (2017) Pavement crack resistance effect analysis of geogrid treatment in old road widening. Chin J Undergr Space Eng 13(S1):263–267
4. Zha XD, Xie G, Lv CZ (2011) Reinforced technology of asphalt pavement with double twisted steel wire mesh. J China Foreign Highw 31(3):63–67
5. Cafiso S, Di Graziano A (2009) Monitoring and performance of AC pavements reinforced with steel mesh. Int J Pavement Res Technol 2(3):82–90
6. Zha XD, Zhu K, Tang T (2015) Mechanical analysis of AC+PCC composite pavement structure reinforced with RoadMesh steel mesh. J Changsha Univ Sci Technol: Nat Sci 12(1):1–9
7. Huang X, Zha XD, Wu H (2018) Numerical simulation on crack resistance of reinforced joint between new and existing pavement with RoadMesh steel mesh. J Changsha Univ Sci Technol: Nat Sci 15(1):1–7
8. Tang T, Zha XD, Xiao QM et al (2018) Laboratory characterization and field validation of RoadMesh-reinforced asphalt pavement in China. Int J Civ Eng 16(3):299–313
9. Victor U, Aleksander I (2019) Evaluation of performance of the RoadMesh reinforcement pavement at trial stretch on Chita-Khabarovsk Autoroad. Lect Notes Civ Eng 29:45–55
10. Suku L, Mahajan R, Sivakumar Babu GL (2020) Experimental and field investigation of road mesh reinforcement in flexible pavements. Indian Geotech J 50(2):184–195
11. Xiao QM, Zha XD, Hu HW et al (2021) Construction technology and treatment effect of RoadMesh reinforcement in joint section between new and existing asphalt pavement. In: 11th International conference on road and airfield pavement technologies 2019, ICPT, vol 1075(012029). IOP Publishing, Malaysia, pp 1–9
12. Zhang MK, Xu ZH, Huang XM (2010) Research on classification and controlling criterion of differential settlement after widening expressway. J Transp Eng Inf 8(1):109–113+119

Optimising Pervious Concrete Design with Partial Replacement of Cement with Fly Ash

N. Ahilash, M. Sajeevan, D. N. Subramaniam, and M. Rajakulendran

Abstract Pervious concrete is a light-weight concrete that does not contain fine aggregates (sand) and of enhanced porosity compared to conventional concrete. Partial replacement of cement paste by waste material, reduces environmental impact both by reducing carbon footprint of cement usage and usage of waste material. Partial replacement of cement by fly ash in conventional concrete is an accepted substitution that also significantly enhance mechanical properties. However, porous concrete being different from conventional concrete, studies are scarce in applicability of such replacements in porous concrete. This study aims to optimise partial replacement of cement by fly ash in porous concrete. By changing the aggregate to cement ratio (A/C) and water cement ratio (W/C) mix designs were formulated, cubes were cast and slump and compressive strength of fresh concrete were monitored. Having obtained highest strength of A/C of 2.5 cubes were cast with varying W/C ratio (0.3, 0.35, 0.4 and 0.45) and the percentage of replacement of cement by fly ash (5, 10, 15 and 20%). Replacement of 10% was observed to be optimum replacement while higher the W/C was advantageous for compressive strength. However, for porous concrete the slump of the fresh concrete should be below 15 mm. Therefore, the optimum W/C was 0.3. Finally, the mix design with 2.5 Aggregate to cement ratio, 0.3 water to cement ratio and 10% replacement of cement by fly ash gives 15.22 MPa (95% CI 14.8–15.6 MPa) strength. It is 13% extra strength than using 100% cement for binder.

Keywords Pervious concrete · Fly ash · Compressive strength · Water to cement ratio · Aggregate to cement ratio

---

N. Ahilash (✉) · M. Sajeevan · D. N. Subramaniam
Department of Civil Engineering, Kilinochchi Premises, University of Jaffna, Jaffna, Sri Lanka

M. Rajakulendran
Colombo International Nautical and Engineering College, Malabe, Sri Lanka

© The Author(s), under exclusive license to Springer Nature Switzerland AG 2022

H. R. Pasindu et al. (eds.), *Road and Airfield Pavement Technology*, Lecture Notes in Civil Engineering 193, https://doi.org/10.1007/978-3-030-87379-0_37

1 Introduction

Pollution of resources increases with time, mostly due human activities challenging the conducive environment for living organisms [17]. Of all anthropogenic activities, the impact on the environment is immensely felt from acquiring materials for production from non-renewable sources to mismanagement of waste materials produced during manufacturing. Reducing the environmental foot-print of a product entails increasing efficiency and reduction of waste production. Green structures were introduced to control pollution by infrastructure development of the world [15, 16]. Pervious concrete is one such green materials for construction of permeable structures by reducing the consumption of the fine aggregates (sand). When fly ash is replaced for cement at a certain percentage in pervious concrete, the greenness of the material is further enhanced. The carbon foot-print will decrease when the percentage of cement used is reduced and the waste material which should otherwise end in land fill, is used in concrete.

As fly ash has more than 35% silica content, it increases late strength of conventional concrete [12]. The workability of the concrete mix was higher due to the fact that the particles are more spherical and that they are approximately of the same size. Fineness of the particles reduce the size of voids in the paste and the additional binder produced by fly ash reaction with available lime allows fly ash concrete to continue to gain strength over time [7, 9]. Also because of this particle shape and size (Particle size is finer than OPC particles) total contact surface area increases, and in turn each particle binds with more particles. These phenomenon increases the ultimate compressive strength of concrete. Durability of pervious concrete is affected by intrusion of aggressive chemicals, water and other deleterious substances into the concrete. Fly ash reacts with alkali and sono-alkali, and eventually acts as a sulphate resistance agent, protecting from chlorine penetration into concrete [1, 2, 6] (Arpita Bhatt 2019). As fly ash replaced cement, the hydration of cement in turn reduces, and thermal cracks are reduced subsequently enhancing the durability of the concrete [13].

Pervious concrete allows free transfer of more moisture into the matrix so that the durability is compromised significantly compared to conventional concrete. Many researchers have investigated the durability issues of pervious concrete from various perspectives. Lo et al. [13] mentioned three conditions that cause freeze/thaw durability problems in pervious concrete: Initially, if pervious concrete is clogged, it influences the movement of water in pervious concrete. Second, if the average daily temperature is below freezing for a long period in specific regions, it causes contraction in the concrete. Third, if the ground water table reaches within three feet of the top of the concrete, pervious concrete can become saturated [4]. The freezing and thawing damage happens quickly if large voids of pervious concrete become saturated [14]. Also, the degree of saturation of paste layers coating the coarse aggregate particles affects the durability pervious concrete [5, 18]. Mixture materials containing sand and/or latex are more durable during alternating freezing and thawing conditions. Finally, compaction energy (which controls the macro pore structure) greatly influences the freeze–thaw durability of pervious concrete [11]. Low carbon content of the fly ash will controls the freeze and thawing event in concrete.

Anyhow, there are some disadvantages in using fly ash. The quality of fly ash should be ensured as poor quality will give the deteriorating results. The setting time is substantially reduced when the percentage of fly ash in concrete is increased [3, 4, 8]. High carbon content of fly ash adds air into cement paste by working as air entraining agent; it increases the freeze and thaw effect and durability reduces. At the same time high carbon content of fly ash adsorbs more water from the mix [10].

Although the benefits of using fly ash as replacement for cement in conventional concrete outweighs the disadvantages, the impact of fly ash content in pervious concrete is scarcely covered in contemporary literature. The increased perviousness and the increased penetration of moisture into the pervious concrete structure necessitates the analysis of the performance of fly ash replaced concrete. This study aims to analyse the impact of fly ash as replacement for cement and to optimize the content for performance of pervious concrete.

2 Methodology

2.1 Material and Selection

This part includes mix design of pervious concrete that used in this study. Water was collected from the water supply (Tap water) where the free chlorine was negligible in content. Aggregates were obtained from a local supplier and tested according to standards to ensure that it can be used in concrete production. The aggregates sizes between 5 and 10 mm were used for the mix throughout this study. Fly ash (FA) was obtained from Nuraicholai Coal Power Plant in Puttalam, Sri Lanka and sieved through 800 mm sieve, and tested for particle size distribution using hydrometer. Particle size distributions of cement, Aggregates and Fly ash are in Fig. 1.

Aggregate Impact Value (AIV) test was carried out to evaluate the property of aggregate that resist to impact (toughness). AIV indicates the results in percentage of fines passing through the 2.36 mm sieve of the total weight of aggregates used to test; after impact applied. Aggregate samples were dried in oven at 105 °C for 24 h and placed in the cup and 15 blows were provided. Then the weight of aggregates passing 2.36 mm sieve measured. Average of 5 repeated same procedure was obtained as 26.9% for the aggregates used in this study.

Absorption of aggregates signifies that amount of water that is absorbed by the aggregate because of gathering of water in the pores inside aggregates. Durability and quantity of water required for mix design is affected by absorption of aggregates. Absorption was measured according to ASTM-C127 and found to be 1.83%. Specific density of aggregates was observed to find specific density and it was 2728 kg/m^3. Ordinary Portland cement was used as the binder material for all experiments in this study.

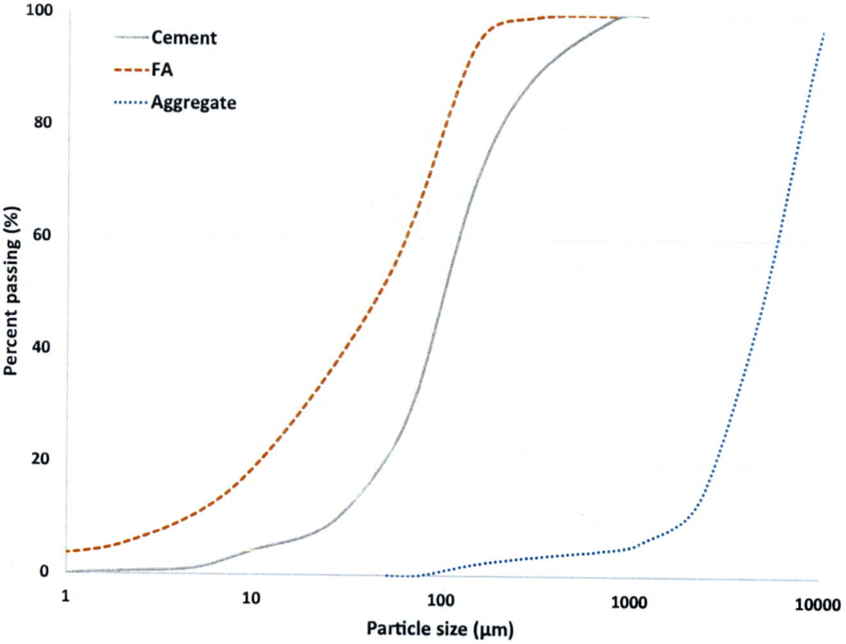

Fig. 1 Particle size distribution curve of cement, aggregates and fly ash

2.2 Concrete Casting

Concrete mixing protocol and duration of mixing play a decisive role in the compressive strength of the pervious concrete. In this study casting was carried out in two phases based on two experiments (Experiment 1 and 2).

Experiment 1

A reliable protocol and duration for mixing was implemented in this study that lasted for 20 min in total. An aggregate to cement ratio (A/C) of 2.5 and 3.0 were used in experiment 1. Water to cement ratio (W/C) was changed 0.25–0.35 with the increment of 0.05 (0.25, 0.3, 0.35). Initially aggregates and cement were mixed for 3 min and half of the computed water was sprinkled on the mixture (while mixing was continued). After mixing for another 3 min, the rest of the water added in similar way and kept mixing for another 6 min (total of 12 min). The mixture was then left (stopped mixing) for 3 min, while any claggy concrete on the walls of the drums were removed from the wall and dropped in the drum. Finally, the mixture was again mixed for 5 min before transferring to the concrete moulds.

The mould was filled in three layers, after reaching third layer of the mould the concrete was tamped by the trowel to spread the concrete evenly throughout the mould's cross-section. After filling the concrete in the mould was compacted using proctor of 2.5 kg with 5 blows from a drop of 300 mm height, one for each corner of the mould and one in the middle of the top surface and wiped the excess concrete from the surface. A sample was taken from the mix and the slump test was carried out. Six cubes of 15 cm × 15 cm × 15 cm were cast for each design and tested for compressive strength after 7 and 28 days of curing. Simultaneously six control cubes were also cast.

Experiment 2

In this part aggregate to cement ratio was used as 2.5 and water to cement ratio was varied from 0.3 to 0.45 with the increment of 0.05. The lower boundary of W/C ratio was decided from the results of Experiment 1. In Experiment 2, a percentage in weight of cement was replaced by fly ash and used as binder material. Replacement percentage was varied from 5 to 20% with an increment of 5%. Mix proportions carried out in this part is given in Table 1.

This mixing protocol was followed as in Experiment 1. In the beginning aggregates, cement and fly ash were mixed for 3 min and half of the calculated water was added to the mix while mixing. After mixing another 3 min rest of the calculated water was added and mixed for 6 min. Then mixing was stopped for 3 min while concrete sticked on the drum was scraped. Again, mixture was mixed for another 5 min and filled in the concrete moulds. Moulds were filled in three layers, after filling the third layer 5 blows of compaction was applied by the standard proctor (2.5 kg) to the top surface, one in each corner of the mould and fifth one in the middle of the top surface. Excess concrete was removed and surface was smoothed by trowel. Mean

Table 1 Mix proportions of Experiment 2

Group	FA %	Cement %	W/C
G1	05	95	0.30
			0.35
			0.40
			0.45
G2	10	90	0.30
			0.35
			0.40
			0.45
G3	15	85	0.30
			0.35
			0.40
			0.45
G4	20	80	0.30
			0.35
			0.40
			0.45

time slump test also done to the mixture. In this experiment replacement percentage of cement with Fly ash was limited to 20%, because of its low strength gain rate and long setting time.

2.3 Quality Control

Mixing drum was cleaned with the same water which was tested and selected to concrete mix and dried before each mix design batch (Even for the first batch) of concrete to neglect the errors cause by the excess water rather than calculated amount in the mixing drum. Moulds were cleaned properly and oiled before filled by fresh concrete. The whole batch of aggregates, Fly ash and Rice husk were obtained in one purchase and stored in one room which at room temperature and controlled humidity. Before each mixing moisture content of aggregates, FA and RHA were measured to ensure the parameters and conditions except study parameters were same across all experiments. Fully shield cement pack was used in each day to irradicate the moisture content issues of cement. Ordinary water from national water supply board was used in each experiment and before each mix water was check on chlorine content according to WHO standards. All mixing were done in the same room, where the effect of temperature and humidity change could be neglected.

Six control cubes were cast simultaneously for each batch for monitor the impact of environment in study. Control cubes were cast for the A/C ratio of 2.5 and the constant W/C ratio of 0.3 for all experiment. Same mixing protocol and duration, and same material were used to cast control cubes. Three of these control cubes were tested for compressive strength after 7 days curing and remaing were tested after 28 days curing period (same day when similar test cubes tested for compressive strength).

Results and Analysis

First experiment of this study was done to optimize aggregate to cement ratio (A/C ratio) and water to cement ratio (W/C ratio) to perform the next step that is replacing the percentage of cement by Fly Ash (FA). Figure 2 shows the compressive strength of casted cubes with various W/C ratio and A/C ratio of 2.5 and 3.0 after 7 days and 28 days. Both graphs show increasing trend of compressive strength of casted cubes with increasing W/C ratio. Because of pervious concrete must have approximately zero slump, W/C ratio above 0.3 were considered not appropriate, even though compressive strength was higher. 0.25 water to cement ratio indicates lower compressive strength than others. Because small amount of water is contributing for the hydration reaction of cement compared to higher W/C ratios. At the same time in 3.0 A/C ratio mix quantity of cement is less than 2.5 aggregate to cement ratio mix. Therefore, the need of water for hydration reaction is less in 3.0 A/C ratio mix. Hence the compressive strength of aggregate to cement ratio 3.0 is higher that is 7.831 MPa average in 28 days than 2.5 A/C ratio (5.35 MPa) when W/C ratio is 0.25.

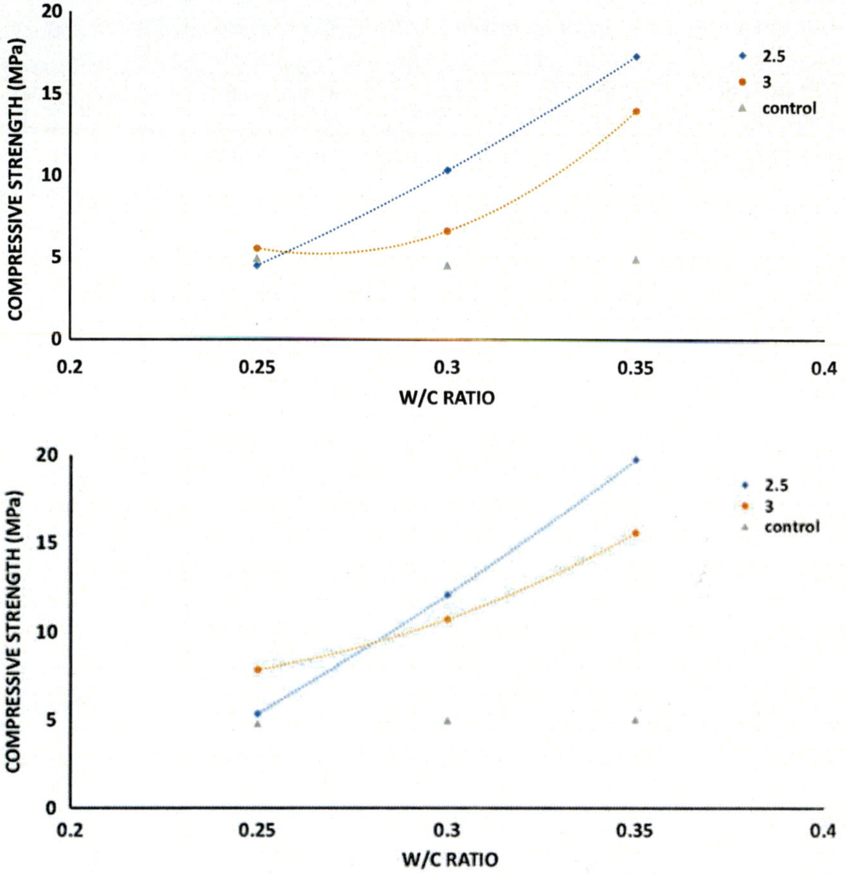

Fig. 2 Compressive strength of experimental cubes with different A/C ratio and control cubes after (1a) 7 days and (1b) 28 days of curing

When W/C ratio increased to 0.3 this scenario had changed to the contrary, because the available amount of water has now increased. Now there is enough water for the hydration reaction and therefore the compressive strength of test cubes was higher than 0.25 W/C ratio. Now, test cubes with 2.5 A/C ratio have enough water for the hydration reaction. So, these cubes had high compressive strength due to high cement content than 3.0 A/C ratio. The same phoenomenon was observed for W/C ratio 0.35. And the compressive strength of control cubes after 7 days and 28 days curing were very consistent, indicating the factors other than W/C ratio and A/C ratio such as weather conditions, water quality, mixing procedure, curing or material quality were not significantly affected.

In addition, since the same size of aggregates were used in all mix design that is 5–10 mm, the impact of aggregate packing is not a factor that affected compressive strength. Thus with the A/C ratio increment the required amount of the binder paste is also increasing. If the binder increased it will cover the aggregates with more thicker layer and aggregates will bind together more, consequently increasing compressive strength. This was observed for A/C ratio of 2.5 where available amount of cement is higher than that was available in A/C ratio 3.0. Furthermore, aggregates coat well in thick binder. So the surface of bonding increased, and the distribution of forces also increased due to this. In overall, with increasing water to cement ratio compressive strength of the cast test cubes were high for 2.5 A/C ratio compare to 3.0 A/C ratio. Mean time W/C ratio of 0.25 is not suitable for pervious concrete mix design.

Using the optimum ratios from Experiment 1, Experiment 2 was carried out with 2.5 aggregate to cement ratio and water to cement ratio with and greater than 0.3 that is 0.3, 0.35, 0.4 and 0.45. Although W/C ratios higher than 0.3 yielded higher slump which is unsuitable for pervious concrete, replacement of cement with ash and fillers may increase the requirement of water and hence a higher W/C ratio may result in lower slump. Aggregate size was maintained same as Experiment 1 (5–10 mm). Figure 3 shows the compressive strength gained in 28 days. Cement was replaced by fly ash (FA) up to 20% with the increment of 5% replacement. W/C ratio 0.4 and 0.45 yielded highest strength, that also corresponded to cement replacement by FA of 10%. For both W/C ratios, the trend clearly indicated increase and decrease in compressive strength before and beyond 10% replacement, respectively. This is because of FA properties, where particles of fly ash are spherical and approximately uniform in size and are of smooth in texture. Also, FA will not consume water from the mix. Thus, for all percentage of replacement of cement the availability of water for cement hydration process did increase. Water availability increases with increasing percentage replacement of FA and W/C ratio. Though the compressive strength reduces when excess water remains in the fresh concrete and the slump is vise versa. This phenomenon was observed at W/C ratio 0.4 and 0.45 above 10% replacement. 0.45 W/C ratio have more compressive strength than 0.4 W/C ratio till 5% FA replacement further proves this phenomenon. Likewise, if we take W/C ratio 0.35 and 0.4, since 15% replacement 0.4 is greater than 0.35. At 20% replacement compressive strength of W/C ratio 0.35 is surpassing the strength of W/C ratio 0.4.

However, because of high water to cement ratio at 0.4 and 0.45, compressive strengths are high than W/C ratio 0.3 and 0.35; but when the replacement of FA comes to 15%, compressive strengths are almost same except W/C ratio 0.3. After 15% that is at 20% W/C ratio 0.35 got strengthen than 0.4 and 0.45. So, if take W/C ratio 0.4 or 0.45 as optimum water to cement ratio 10% replacement of fly ash yielded high strength. But for pervious concrete, there must be existence of pores and pore must be connected to facilitate transportation of fluids. From Fig. 4 it could be observed that, there are no pores at the bottom of cubes. Its look like an ordinary conventional concrete. Therefore, design with W/C ratio above 0.35 were considered not suitable. Simultaneously, W/C less than 0.3 yields poor compressive strength. So, the optimum W/C ratio for the pervious concrete mix design with the replacement with fly ash lying between 0.3 and 0.4.

There is no significance difference in compressive strength after 10% replacement for W/C ratio of 0.3 and 0.35. Higher replacement of cement with FA only decreases the cost of concrete and increase environmental friendliness. However, the clump of

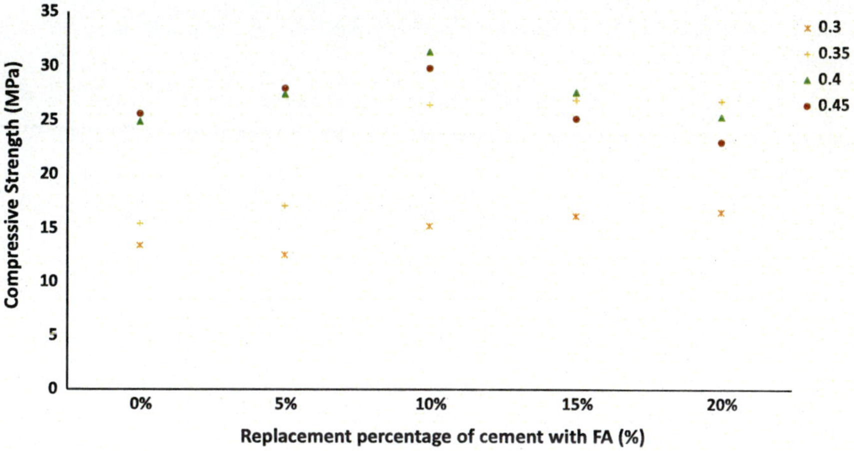

Fig. 3 Compressive strength of experimental cubes with fly ash for various W/C ratio

Fig. 4 Bottom of test cube with W/C ratio 0.4

fresh concrete is yet another crucial factor. Table 2 shows the average slump values of all mix design in this study. For pervious concrete slump value must be less than 15 mm. Table 2 clearly indicates that the slump value increases with increasing percentage of replacement of cement by FA. It's because of the particles of fly ash; its texture and shape friction between the particle is very low. When percentage replacement increases excess water will function as a lubricant between particle and increases the flow of the mortar. Those properties affect the slump value actually increases it. Considering the values in column of W/C ratio 0.35, it has 38 mm for 0% replacement. All other slump values (for replacement percentages—5%, 10%, 15% and 20%) are more than 38 mm for water to cement ratio 0.35. So, 0.35 W/C ratio also not suitable for pervious concrete mix with fly ash. For 0.3 W/C ratio slump value reaching the requirement of slump value for 10% of FA presence. Finally, the

Table 2 Slump values of mix designs with percentage of fly ash varies with W/C ratio

FA %	W/C ratio				
	0.25	0.30	0.35	0.40	0.45
00	0.0	00	38.00	108.00	193.00
05	0.0	05	45.00	117.00	206.00
10	0.0	12	55.67	130.50	220.83
15	3.5	20	70.00	144.83	240.00
20	7.0	35	92.50	175.00	270.50

optimum water to cement ratio for pervious concrete with fly ash is 0.3. And the optimum percentage replacement is 10% cement replaced by fly ash.

3 Conclusions

Fly ash being a waste product, and specifically due to the size of particles and chemistry, it has high potential of polluting the environment when not managed. Production of cement also imparts more detrimental effect on environment predominantly by releasing large amount of carbon-dioxide. So, replacing certain percentage of cement in concrete production with these Fly Ash irradicates both impact on environment.

This study highlights the importance of selecting optimum A/C ratio, W/C ratio and percentage of replacement the cement by FA to obtain high compressive strength for a pervious concrete without compromising on porosity. First higher compressive strength obtained for A/C ratio 2.5 for W/C ratio higher than 3.0. For mix which contain only FA as the replacement of cement 10% replacement is giving high strength. Compressive strength of 10% replacement mix design increasing with increment of W/C ratio. A slump requirement of less than 15 mm for casting pervious concrete was kept because even minor increment than 15 mm covers the bottom surface with binder paste and result continuous porosity became zero. W/C ratio as 0.3 exposed high compressive strength for 10% replacement with Fly Ash, with low slump value and it's 15.22 MPa.

References

1. Bhatt A, Priyadarshini S, Acharath Mohanakrishnan A, Abri A, Sattler M, Techapaphawit S (2019) Physical, chemical, and geotechnical properties of coal fly ash: a global review. Case Stud Constr Mater 11:e00263
2. Corinaldesi V, Moriconi G (2009) Influence of mineral additions on the performance of 100% recycled aggregate concrete. Constr Build Mater 23(8):2869–2876

3. Faella C, Lima C, Martinelli E, Pepe M, Realfonzo R (2016) Mechanical and durability performance of sustainable structural concretes: an experimental study. Cem Concr Compo 71:85–96
4. Faleschini F, Zanini MA, Brunelli K, Pellegrino C (2015) Valorization of co-combustion fly ash in concrete production. Mater Des 85:687–694
5. Gao P, Chen Y, Huang H, Qian Z, Schlangen E, Wei J, Yu Q (2020) Effect of relative humidity on drying-induced damage in concrete: a comparative study of digital image correlation and lattice modelling. Mater Des 196:109128
6. Golewski GL (2018) Evaluation of morphology and size of cracks of the interfacial transition zone (ITZ) in concrete containing fly ash (FA). J Hazard Mater 357:298–304
7. Golewski GL (2019) Estimation of the optimum content of fly ash in concrete composite based on the analysis of fracture toughness tests using various measuring systems. Constr Build Mater 213:142–155
8. Hemalatha T, Ram Sundar KR, Murthy AR, Iyer NR (2015) Influence of mixing protocol on fresh and hardened properties of self-compacting concrete. Constr Build Mater 98:119–127
9. Hwang SS, Moreno Cortés CM (2021) Properties of mortar and pervious concrete with co-utilization of coal fly ash and waste glass powder as partial cement replacements. Constr Build Mater 270:121415
10. Jo M, Soto L, Arocho M, St John J, Hwang S (2015) Optimum mix design of fly ash geopolymer paste and its use in pervious concrete for removal of fecal coliforms and phosphorus in water. Constr Build Mater 93:1097–1104
11. Kevern JT, Wang K, Schaefer VR (2010) Effect of coarse aggregate on the freeze-thaw durability of pervious concrete. J Mater Civ Eng 22(5):469–475
12. Limbachiya M, Meddah MS, Ouchagour Y (2012) Use of recycled concrete aggregate in fly-ash concrete. Constr Build Mater 27(1):439–449
13. Lo F-C, Lee M-G, Lo S-L (2021) Effect of coal ash and rice husk ash partial replacement in ordinary Portland cement on pervious concrete. Constr Build Mater 286:122947
14. Neithalath N, Sumanasooriya MS, Deo O (2010) Characterizing pore volume, sizes, and connectivity in pervious concretes for permeability prediction. Mater Charact 61(8):802–813
15. Subramaniam DN (2015) Dynamics of nitrogen and suspended solids removal in experimental stormwater biofilters under intermittent wetting and drying [Doctor of Philosophy: Queensland University of Technology
16. Subramaniam DN, Egodawatta P, Mather P, Rajapakse JP (2015) Stabilization of stormwater biofilters: impacts of wetting and drying phases and the addition of organic matter to filter media. Environ Manage 56(3):630–642
17. Subramaniam DN, Logeswaran T, Tharshikka V, Nilakshan B (2018) Dynamics of clay particles in non-vegetated stormwater biofilters. Water, Air, Soil Pollut 229(9):302
18. Yang L, Jiang X, Huang T, Jiang W (2015) Physicochemical characteristics and desulphurization activity of pyrolusite-blended activated coke. Environ Technol 36(22):2847–2854

Performance of OGFC Using West Kalimantan Aggregates and Polymer Modified Bitumen

S. Mayuni, S. P. R. Wardani, and B. H. Setiadji

Abstract Open Graded Friction Course (OGFC) has been used as a permeable pavement in many countries. OGFC introduces safe road conditions as a pavement material compared to normal asphalt while facilitating the lateral and vertical drainage functions. In Indonesia, one type of OGFC researched to date is porous asphalt. This mix has an advantage in terms of its drainage function, but it lacks the ability to withstand against traffic loads. This research aims to see how binder content affected the performance of porous asphalt mixtures in order to create porous asphalt pavements that met structural and functional requirements. To do so, local aggregates from the West Kalimantan region and several contents of PG-76 polymer modified bitumen were used in this study. Marshall and Cantabro Loss tests were conducted to observe the structural performance, while volumetric properties were used to evaluate the functional performance. The outcomes of this study presents the insight of the relationship between the different content of the binder and the stability, durability, and voids content of the mixture.

Keywords Porous asphalt · Binder content · Stability · Durability · Voids content

S. Mayuni (✉)
Postgraduate Program at Civil Engineering, Faculty of Engineering, Universitas Diponegoro, Jalan Hayam Wuruk No. 5-7, Semarang 50241, Indonesia
e-mail: sitimayuni@students.undip.ac.id

Department of Civil Engineering, Faculty of Engineering, Universitas Tanjungpura, Jalan Prof. Dr. Hadari Nawawi, Pontianak 78124, Indonesia

S. P. R. Wardani · B. H. Setiadji
Department of Civil Engineering, Faculty of Engineering, Universitas Diponegoro, Jalan Prof. Soedarto, SH, Semarang 50275, Indonesia
e-mail: wardani@live.undip.ac.id

B. H. Setiadji
e-mail: bhsetiadji@ft.undip.ac.id

© The Author(s), under exclusive license to Springer Nature Switzerland AG 2022

H. R. Pasindu et al. (eds.), *Road and Airfield Pavement Technology*, Lecture Notes in Civil Engineering 193, https://doi.org/10.1007/978-3-030-87379-0_38

1 Introduction

OGFC (Open Graded Friction Course) as an open-graded pavement has been widely implemented in many developed and developing countries [1]. In Indonesia, commonly referred to as Porous Asphalt, is a thin layer of asphalt mixture placed on top of a conventional asphalt layer [2]. In other words, the open-graded porous asphalt is placed on top of the dense graded asphalt, as a drainage system that functions to remove rainwater that falls on the road surface. This layer has a structure with a high void content, which is 18% [3]. This serves as drainage during the rainy season, and as a safe road that can reduce the chances of hydroplaning [4], as well as overcoming the effects of sprays and splashes that interfere with the driver's view [5]. Coarse surfaces are also advantages in favor of road safety, good skid resistance which can prevent slippage between the road surface and the tire [6]. Widhianto et al. [6], Poulikakos and Partl [7] stated that porous asphalt is an environmentally friendly road construction technology, which can reduce noise and control rainwater puddles.

One country in Asia, Malaysia, which has a climate similar to Indonesia, started pioneering the application of porous asphalt in 1991 and officially announced the national specification standard in 2008. Malaysia succeeded in reducing the number of accidents and promoting porous asphalt as a safe road in the rainy season [8].

In addition to having advantages, porous asphalt mixture also has its drawbacks [9]. In terms of strength, it is generally lower than conventional dense graded asphalt. It has low stability which opens up a greater chance for deformation.

Structural performance and functional performance are the two main factors in the application of porous asphalt layers. High void content in an open-graded mix is needed to form a structure that can drain the water. Instead, mix with a high void content tend to have lower structural strength when compared with a dense-graded mixture [4].

Many cities in Indonesia having problems with puddles on the road due to heavy rainfall for a long time [1]. Djakfar et al. [10] recommend that structurally, porous asphalt mixtures can be used on roads with light traffic loads. In terms of void content, which is one of the functional performance parameters, the present invention is not adequate, because the void content (VIM 8.1%) produced does not meet specifications (18–23%).

Research development on porous asphalt mixtures in Indonesia is still limited. Djakfar et al. [11] suggested developing research with the approach of the material used. Several recent studies have experimented to improve the stability of the mixture by adding a variety of additives, powder, and fiber [10, 12–14]. The use of additives increases the interlocking so that the stability increases, but on the other hand it disturbs the permeable function.

For a long time, efforts to improve the performance and quality of porous asphalt mixtures have been continuously carried out with many studies in various countries, including the use of polymer modified asphalt [15], other modified asphalt [16], adding additives, including the use of fiber [17], or using coarser aggregate [18]. Cellulose fiber is one of the additives that are often used in studies of porous asphalt.

Adding 0.3–0.5% cellulose fiber to the OGFC mixture provides a good stability value compare to polypropylene fiber and polyester fiber [19]. In the Cantabro test, cellulose fibers significantly improved the behaviour, especially for porous asphalt mixtures with high binder content [20]. Polymer modified bitumen improves the adhesion between aggregate and the binder [21]. Like PG-76, it can increase the durability of the mixture [8]. This study tries to find the performance value of a porous asphalt mixture that meets the requirements, where the aim of the study is to determine the effect of different binder content on the value of stability, durability and voids content.

2 Experimental Program

2.1 *Material Properties*

The mixture contained of dry sieved aggregates with asphalt binder (PG-76) and 0.3% cellulose fiber. Indonesia Bina Marga Standard 2018 recommends 0.3% cellulose fiber for asphalt mixture [22]. The mixing temperature was 186 °C and at the temperature of 177 °C, 50 blows per side were applied. The temperature was determined based on the result of bitumen viscosity test. The aggregates were obtained from a crushed stone quarry in Sungai Kunyit, Kalimantan Barat, which is commonly used in Kalimantan Barat for asphalt mixture. Tables 1 and 2 demonstrate the physical properties of the modified asphalt (PG-76) and aggregate used in this analysis, respectively. All the material properties followed material quality specifications from Indonesia Bina Marga Standard 2018 [22].

Gradation has an important role in porous asphalt performance [23]. This study applies Australia porous asphalt gradation [24], which was recommended because it would create a mixture with the most homogenous void distribution and high enough permeability [25], with a little modifications where a 6.7 mm sieve is replaced by a 6.3 mm sieve, according to sieve availability (Table 3).

The sample of porous asphalt mixture was prepared using a Marshall Compactor with 4, 4.5, 5, 5.5 and 6% design binder content. Three specimens were tested for each binder content. The sample was compacted with 50 blows per side. After that,

Table 1 Properties of PG-76

Properties	Result	Requirement
Penetration at 25 °C (mm)	41.80	–
Softening point (°C)	66.50	≥ 48
Flash point (°C)	275.50	≥ 230
Ductility	101.67	≥ 100
Viscosity at 135 °C (CSt)	1587.65	≤ 3000
Specific gravity	1.119	≥ 1

Table 2 Properties of aggregate

Test	Result	Requirement
Los Angeles abrasion (%)	15.2	≤ 30
Flakiness (%)	1.29	≤ 5
Elongation (%)	1.49	≤ 5
Soundness (%)	5.93	≤ 12
Aggregate impact value (%)	6.00	≤ 30
Specific gravity	2.67	≥ 2.5
Water absorption (%)	0.37	≤ 3
Affinity for bitumen (%)	99	≥ 95

Table 3 Porous asphalt gradations (AAPA 2004 Modifications)

Sieve Size		Percentage passing (%)
ASTM	mm	
½"	12.7	100
3/8"	9.5	85–100
¼"	6.3	35–70
No. 4	4.75	20–45
No. 8	2.36	10–20
No. 16	1.18	6–14
No. 30	0.60	5–10
No. 50	0.30	4–8
No. 100	0.150	3–7
No. 200	0.075	2–5

the compacted samples were tested for volumetric properties (air void content) and mechanical performance using Stability Marshall and Cantabro tests. According to ASTM D1559, the Stability Marshall Test is used to assess the stability value of porous asphalt mixtures, which shows the strength of the samples. According to ASTM D7064, the Cantabro Test is used to determine the particle loss of porous asphalt mixtures, which indicates sample durability. Figure 1 illustrates the specimens with 4–6% binder content.

2.1.1 Volumetric Properties

Volumetric parameters depend on the mixture's Bulk Specific Gravity (Gmb) and the Theoretical Maximum Specific Gravity (Gmm).

- *Theoretical Maximum Specific Gravity (Gmm).* According to ASTM 2041–11, maximum theoretical specific gravity is a virtual value reflecting a compressed sample without air voids.

Fig. 1 Specimen of porous asphalt

- *Bulk Specific Gravity of Mix (Gmb).* According to ASTM D3203-11, the bulk specific gravity of the compacted mixture is determined by the specimen's geometric measurements.

Air voids (%). Air voids content, commonly lies between 18 and 23%, is one of porous asphalt mixture's most significant parameters. In porous asphalt, Av (air voids) are calculated as [26]

$$\mathrm{Av} = \left(1 - \frac{\mathrm{Gmb}}{\mathrm{Gmm}}\right).100\% \tag{1}$$

2.1.2 Cantabro Loss Test

The durability of porous asphalt was assessed in terms of its resistance to abrasion loss. The Cantabro loss test is a special porous asphalt test commonly used to determine mixture resistance to particle loss from traffic impact and abrasion, as well as stripping resistance [27]. The procedures described by Jimenez and Perez (1990) [28] were used as the test method. In The Los Angeles Abrasion Machine, the compacted porous asphalt samples were subjected to 300 revolutions at 30 rpm, without the charge of steel spheres. Then, the mass loss percentage determined the sample's resistance to abrasion. Specimens and Loss Angeles Machine are shown in Fig. 2. The following equation determines the test result.

$$\mathrm{P} = \left[\frac{\mathrm{P1} - \mathrm{P2}}{\mathrm{P1}}\right].100\% \tag{2}$$

(a) Los Angeles Abrasion Machine

(b) Weighing specimen after Cantabro Loss Test

(c) Weighing specimen before Cantabro Loss Test

Fig. 2 Cantabro loss test

where: P = Cantabro abrasion loss (%).
P1 = initial weight of the sample (gram).
P2 = final weight of the sample (gram).

2.1.3 Marshall Test

Marshall test was used to determine the porous asphalt compacted sample's stability value. Stability indicates one of porous asphalt's most important structural performance and a measure of the asphalt layer's ability to resist deformation and distortion cause of traffic loads such as corrugation and rutting.

3 Result and Discussion

Table 4 and Fig. 3 display the values of the test, especially the value of stability

Table 4 Effect of binder content on porous asphalt performance

Binder content (%)	Stability (kg)	Cantabro loss (%)	Voids content (%)
4	418.81	46.1	22.29
4.5	512.54	24.09	21.77
5	567.81	14.19	21.07
5.5	525	5	18.6
6	466.07	3.58	17.65
Australian Standard	> 500	< 30	18–23

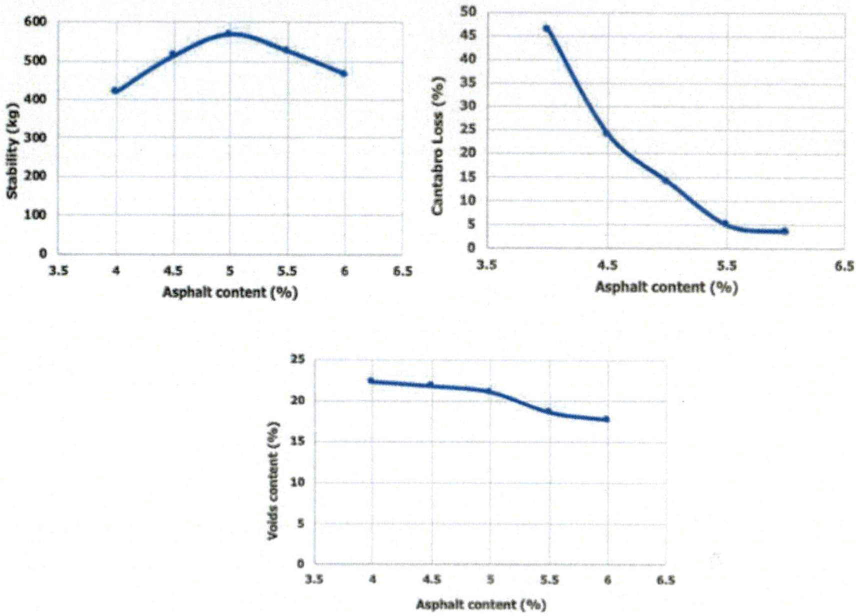

Fig. 3 Effect of binder content on porous asphalt performance

and Cantabro Loss as structural performance, and voids content as functional performance which follows Australia porous asphalt specifications [24]. It can be seen that the highest stability is achieved 567.81 kg in the mixture with 5% binder content. However, the 4.5 and 5.5% binder content mixture still meets the specifications.

The trend shows that as the binder content is increased, the durability is increased. The best durability is indicated by the smallest Cantabro Loss value, 3.58%; the result of a mixture with asphalt content of 6%, but in terms of stability and voids content do not meet the specification of porous asphalt mixture.

The trend shows that as the binder content is increased, the voids content is decreased. The mixture with 4–5.5% binder content has the voids content that meets the requirements (18–23%). Porous asphalt mixtures with open grading tend to have low strength and low durability as a result of relatively low interlocking due to high voids content, so the minimum voids content needs to be considered provided the structural performance records meet the specifications. The test results show that the voids content meet the requirements at the minimum value (18.6%) is produced by 5.5% binder content of porous asphalt mixtures, which the structural performance is eligible and meets the specifications; stability 525 kg and the Cantabro Loss 5% indicates that the mixture has high durability. The results of this study can be compared with studies in Malaysia [29] (Table 5). Although this study recommends a higher binder content than the Malaysian porous asphalt mixture, it has a better voids content which is closer to the minimum specifications, as well as better durability.

Table 5 Porous asphalt performance with polymer-modified bitumen PG-76

Year	Country	Additives	Binder content (%)	Stability	Voids content (%)	Cantabro loss (%)
2019	Malaysia	Mineral filler	5	–	21.7	6.1
2021	Indonesia	Cellulosa fiber	5.5	525 kg	18.6	5

4 Conclusions

This study evaluates porous asphalt's performance at varying binder content. Based on the findings, it can be concluded that different asphalt content affects the final product's volumetric and mechanical properties. As the binder content increases, porous asphalt's resistance against abrasion loss also increased. Decreased voids content due to the increasing of binder content, but can be seen best voids content was approaching the minimum requirements (18.6%). Then it can be recommended that the asphalt porous used aggregate of West Kalimantan with 5.5% polymer modified bitumen PG-76 and 0.3% cellulose fiber, which resulted in the performance of porous asphalt with 18.6% voids content and 5% Cantabro Loss, as well as the stability of 525 kg on the road with the light to the medium traffic load.

Acknowledgements The support provided by PT Strada Multi Perkasa Pontianak, PT Buntara Megah Inti Tangerang, and PT Enceha Pasific Jakarta for this study is very much appreciated.

References

1. Djakfar L, Hendi Bowoputro H, Zaika Y (2015) Evaluation of the use of steel slag and pyrophyllite materials in the porous asphalt mix. Int J Civ Struct Eng IJCSE 2(1):147–151
2. Hernandez-Saenz MA, Caro S, Arámbula-Mercado E, Epps Martin A (2016) Mix design, performance and maintenance of permeable friction courses (PFC) in the United States: state of the art. Constr Build Mater 111:358–367
3. Nashir M, Parung H, Ali N, Hariyanto T (2013) Kinerja Campuran Aspal Berpori dengan Menggunakan Aspal Polimer Starbit Jenis E-55. Semin. Nas. IX-2013 Tek. Sipil ITS
4. Xing M, Chen S, Wang B, Wei S (2010) Research on influence of aggregate gradation on the performance of porous asphalt pavement, pp 3046–3054
5. Zhang H, Anupam K, Scarpas A, Kasbergen C, Erkens S (2019) Effect of stone-on-stone contact on porous asphalt mixes: micromechanical analysis. Int J Pavement Eng 1–12
6. Widhianto B, Setyawan A, Sarwono D (2013) Desain Aspal Porus dengan Gradasi Seragam Sebagai Bahan Konstruksi Jalan yang Ramah Lingkungan. e-Jurnal Matriks Tek Sipil 1:165–170
7. Poulikakos LD, Partl MN (2010) Investigation of porous asphalt microstructure using optical and electron microscopy. J Microsc 240(2):145–154
8. Hamzah MO, Hasan MRM, Che Wan CN, Abdullah NH (2010) A comparative study on performance of Malaysian porous asphalt mixes incorporating conventional and modified binders. J Appl Sci 10(20):2403–2410
9. Mansour TTN, Putman BJB, Asce AM (2013) Influence of aggregate gradation on the performance properties of porous asphalt mixtures. J Mater Civ Eng 25(2):281–288

10. Djakfar L, Zaika Z, Harimurti (2017) Ealuation of porous hotmix asphalt performance under accelerated loading. Int J Geomate 13(39)
11. Djakfar L, Bowoputro H, Prawiro B, Tarigan N (2015) Performance of recycled porous hot mix asphalt with gilsonite additive. Adv Civ Eng 2015
12. Ayun Q, Mahardi P (2017) Pengaruh Penambahan Sulfur Terhadap Karakteristik Marshall dan Permeabilitas pada Aspal Berpori. J Ilm Tek Sipil UNS 1(1):117–122
13. Noris TG, Mahardi P (2017) Analisa Pemanfaatan Limbah Styrofoam Sebagai Bahan Substitusi ke Dalam Aspal Penetrasi 60/70 Terhadap Karakteristik Campuran Aspal Porus. J Rekayasa Tek Sipil UNS 01(01):65–70
14. Ramadhan RP, Mahardi P (2017) Pengaruh Penambahan Limbah Plastik (PET) Terhadap Karakteristik Marshall dan Permeabilitas pada Aspal Berpori. J Ilm Tek Sipil UNS 01(01):129–135
15. Chen J-S, Lee C-T, Lin Y-Y (2016) Influence of engineering properties of porous asphalt concrete on long-term performance. J Mater Civ Eng 29(4):04016246
16. Shirini B, Imaninasab R (2016) Performance evaluation of rubberized and SBS modified porous asphalt mixtures. Constr Build Mater 107
17. Punith VS, Suresha SN, Raju S, Bose S, Veeraragavan A (2011) Laboratory investigation of open-graded friction-course mixtures containing polymers and cellulose fibers. J Transp Eng 138(1):67–74
18. Hassan NA, Zul M, Mahmud H, Ramadhansyah PJ (2014) Air void characterisation in porous asphalt using X-ray computed tomography. Adv Mater Res 911:443–448
19. Ye Z, Jian L (2019) The effect of fiber on the performance of open graded friction course (an environmental survey). Ekoloji 28(107):4449–4453
20. Nielsen CB (2006) Durability of porous asphalt. Denmark
21. Gupta A, Rodriguez-Hernandez J, Castro-Fresno D (2019) Incorporation of additives and fibers in porous asphalt mixtures: a review. Materials (Basel) 12(19)
22. Marga B (2018) Spesifikasi Umum 2018, no. September
23. Król JB, Khan R, Collop AC (2018) The study of the effect of internal structure on permeability of porous asphalt. Road Mater Pavement Des 19(4):935–951
24. A. Australian Asphalt Pavement Association (2004) National asphalt specification, no. April
25. Hassan NA, Mahmud MZH, Ahmad KA, Hainin MR, Jaya RP, Mashros N (2016) Air voids characterisation and permeability of porous asphalt gradations used in different countries. ARPN J Eng Appl Sci 11(24):14043–14047
26. Alvarez AE, Mora JC, Espinosa LV (2018) Quantification of stone-on-stone contact in permeable friction course mixtures based on image analysis. Constr Build Mater 165:462–471
27. Poulikakos LD et al (2006) Mechanical properties of porous asphalt, recommendations for standardization. Fac. l'environnement Nat. Archit. construit, Lab. des voies Circ. - LAVOC, no. Dezember 2006, p 110
28. Aman MY, Shahadan Z, Tamin NRM (2015) A comparative study on properties of Malaysian porous asphalt mixes with different bitumen contents. Res J Appl Sci Eng Technol 9(10):797–806
29. Luxman NN et al (2019) Effect of compaction temperature on porous asphalt performance. IOP Conf Ser Earth Environ Sci 244(1)

Performance Study of Pervious Concrete as a Road Pavement Infrastructure System

Sujit Kumar Pradhan, Niranjan Behera, and Anil Palai

Abstract Pervious concrete is recognised as an environment friendly building material to meet the growing demands for pavement infrastructure. It is one of the best practices for storm water management by capturing rain water and allowing it to seep into the ground. In this study, influence of fine aggregate and coarse aggregate quantities on the properties of pervious concrete and performance characteristics of pervious concrete in terms of permeability, compressive strength, flexural strength, split tensile strength were carried out. Materials used are OPC Type-I of Grade 43, fine aggregate confirming gradation (Zone II) and coarse aggregate mix of 4.75–10 mm, 10–12.5 mm, 12.5–16 mm, 16–20 mm of 15, 40, 30 and 15% of total weight of coarse aggregate respectively. Mixes were prepared with the water-cement ratio of 0.34 and maintaining aggregate-cement ratio as 3.5:1. Here, total five nos. mixes were prepared, out of which one for Control concrete mixture having M35 grade and remaining four mixtures were prepared by partial replacement of fine aggregated with coarse aggregate in the range of 70–100% by weight. It was observed from the study that with the increase of fine aggregates, various mechanical properties such that compressive strength, flexural strength, split tensile strength increases, and coefficient of permeability decreases. Also, the relationship between the strength, permeability and total void present in graded aggregate based on angularity number was developed. From the study, we conclude that pervious concrete act as a suitable sustainable pavement infrastructure.

Keywords Pervious concrete · Permeability · Flexural strength · Split tensile strength

1 Introduction

Pervious concrete has high void content so it can be used to infiltrate water coming from different sources to pavement surface to other media (like ground water table,

---

S. K. Pradhan (✉) · N. Behera · A. Palai
Indira Gandhi Institute of Technology, Sarang, Dhenkanal, Odisha 759146, India
e-mail: sujitpradhan@igitsarang.ac.in

© The Author(s), under exclusive license to Springer Nature Switzerland AG 2022
H. R. Pasindu et al. (eds.), *Road and Airfield Pavement Technology*, Lecture Notes in Civil Engineering 193, https://doi.org/10.1007/978-3-030-87379-0_39

rain water harvesting system, underground water extraction channel, etc.) and by this process wastage of rain water and artificial flood condition can be reduced. There are certain ecological benefits of using pervious concrete for pavement structure because of its good potentiality to charge up groundwater table and reduce storm water runoff which leads to prevent the artificial flood in the urban area.

Pervious concrete come up with an extra flexibility for architect to employ greenery in parking lots. Pervious concrete is an eco-friendly construction material and EPA has also point out that it as a Best Management Practice (BMP) for storm water Management.

The basic composition of pervious concrete is same as of conventional concrete with a little fine or no fine aggregate which leads to formation of an interconnected network of voids that allow the water to percolate through the pavement. In terms of use pervious concrete can have many places in the field of construction. Some of the areas where we can use pervious concrete are residual roads, alleyways, low traffic roads, ford (crossings), footpath and cycle way, parkway, tennis courts, slope stabilization, walkway in park and garden etc.

2 Objective

The main objective of study includes carrying out the various tests on pervious concrete. To develop the relationship between the strength, permeability and percentage of fine aggregate.

3 Literature Review

Pervious concrete is a green material and it does not consume natural resources. It helps in storm water quality issue. It can be used as a paving material in road construction, efficiently if we increase the abrasion resistance and other strength characteristic of the pavement [1]. Pervious concrete have reasonable voids as compare to the nominal mix concrete so characteristics strength of the concrete is very less as compare to the nominal mix. So, with the addition of extra fine aggregate provide more resistance against abrasion and increase the strength characteristic property and decreases permeability coefficient [2]. PPC provides higher resistance to cracking and gives high flexural and compressive strength for that reason construction of road pavement should use PPC [3]. The nonlinear permeability coefficient k value of pervious concrete depends upon aggregate size and void ratio [4]. The coefficient of permeability of pervious concrete is inversely proportional to cement content and angularity number [5]. Keeping in account good workability of the mix and to avoid the clogging up void space the proportion of water should be carefully taken to prepare the pervious concrete mix. For better strength and permeability the optimum range of water-cement ratio should be 0.30–0.38 [6]. Cube strength of pervious concrete drops down as the particle sizes of coarse aggregate increase. The hydraulic conductivity increases as the particle sizes of coarse aggregate is increased [7]). Pervious concrete properties varies with mix proportion, times space between mixing and casting, method of casting, and environmental condition [8]. To increase the resistance against freeze thaw action, air entering admixture can be used [9]. Pervious concrete along with managing the surface runoff it also reduces the night time temperature and noise pollution. To maintain good permeability of the pavement void space should be at least 16% [10]. The permeability of pervious concrete is determined by considering all the pore structure features and permeability also depends upon gradation of aggregate [11]. With increase in void content, void ratio and void network, the strength characteristic of porous concrete decreases [12]. The system permeability depends upon the permeability of sub-base and sub-grade [13].

4 Materials

4.1 Cement

Cement used in all mixes conforming to [14], provides 43 MPa or 43, 00, 000 kg/sqm. And also provide initial strength gaining facility. Its initial setting times also less as compare to other grade of cement. This grade cement is mostly used in M30 and M35 grade concrete mix.

4.2 Coarse Aggregate

A mixture of coarse aggregate of different size is formed (Routhfourth's method of gradation). The percentage taken are 15, 40, 30 and 15% from 4.75 to 10 mm, 10 to 12.5 mm, 12.5 to 16 mm, 16 to 20 mm of the total weight of coarse aggregate respectively. Main functions of the coarse aggregate are to bear the load coming on the structure. Figure 1 and Fig. 2 presents the particle size distribution curves of coarse aggregate and fine aggregate, respectively.

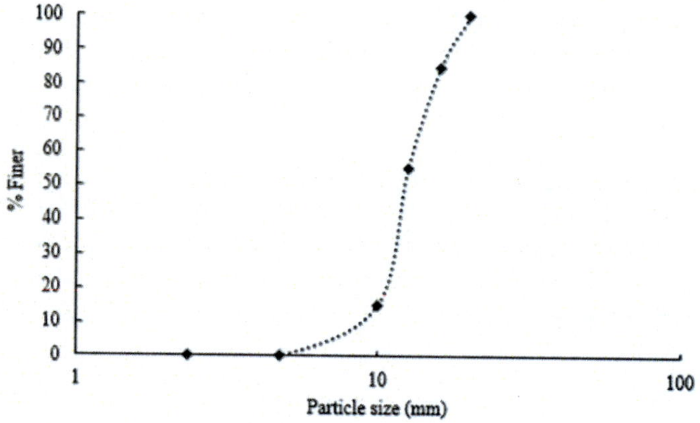

Fig. 1 Particle size distribution of coarse aggregate

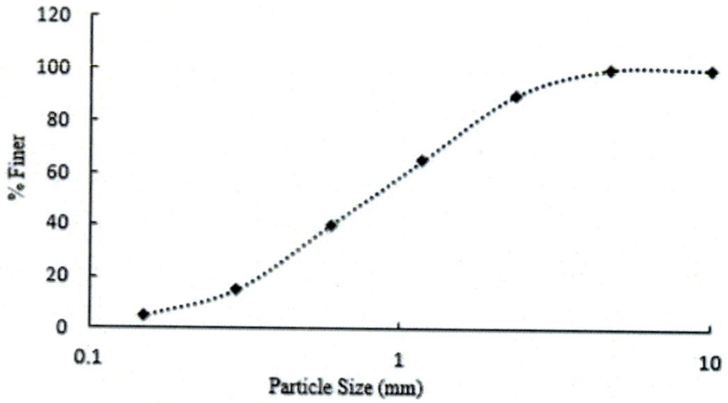

Fig. 2 Particle size distribution of fine aggregate

4.3 Fine Aggregate

Fine aggregate used falls under zone II. Zone I–Sand being very coarse and Zone IV sand is very fine. It is generally recommended by code to use sands of Zones I to III for Structural concrete works. Zone II sand known as normal sand, which has the total void content and interconnecting void in a good proportion as per [15].

Fine aggregate plays the main role of filling the void among the coarse aggregate which helps in increasing the strength of the concrete as a whole at the same time it decreases the permeability. Table 1 present the specific gravity of different materials.

Table 1 Material properties

Properties	Value
Specific gravity of course aggregate	2.7
Specific gravity of fine aggregate	2.68
Specific gravity of cement	3.1

4.4 Water

Water-cement ratio to be selected which generally varies from 0.26 to 0.40 as per [16].

5 Methodology

5.1 Concrete Mix

The concrete mix was designed based on the calculation obtained from [17]. And also follow the [18] for calculation of the quantities of Cement, Coarse Aggregate and Fine Aggregate in the previous concrete mix. Adopted target strength is M35 and water-cement ratio is 0.34. The ratio between the cement, fine aggregate and coarse aggregate was taken as 1:1.436:2.006.

Mixing of concrete was done by electrical concrete mixer machine. Total four mixes were prepared by varying the quantity of fine aggregate as 0, 10, 20 & 30% of fine aggregate while fine aggregate was replaced by coarse aggregate in the concrete mix.

5.2 Compressive Strength Test

Casting of cubes having inner dimension 150 × 150 × 150 mm is done as per [19] recommendations. The concrete was poured into the mould with three layers and each layer was compacted using 16 mm tamping rod. The specimens were kept in the mould for 24 h and after removal of all the specimens were left in the curing tank for a period of 28 days. After the curing of concrete cubes they were removed and placed in the compression testing machine which is having a least count of 1.00 tonne. The load was applied gradually over the specimen till to the failure of the specimen at a rate 140 kg/cm^2/min or 22.1 kN/mm^2/s confirming to the range given by IS code and the corresponding load was noted as ultimate load and the strength is calculated in MPa. Average of all the values was taken and reported as the compressive strength of that mix. Similarly test was conducted for all types of mixes and the corresponding compressive strength was obtained.

5.3 Flexural Strength Test

Casting of beams having inner dimension 100 × 100 × 300 mm is done as per [19] recommendations. The concrete was poured into the mould with three layers and each layer was compacted using 16 mm tamping rod. The specimens were kept in the mould for 24 h and then after removal of all the specimens were left in the curing tank for a period of 28 days. Concrete sample was placed in the machine after the curing period. Specimen was placed in the machine centred with the longitudinal axis of the specimen at right angles to the rollers and the load is applied gradually at a rate of 29.5 N/sec. Maximum load that was bear by the specimen was noted down and flexural strength was calculated by using the formula. The average of all the values was taken as the flexural strength of mix. Similarly test was conducted for all types of mixes and the corresponding flexural strength was obtained.

5.4 Split Tensile Strength Test

Casting of cylinders having inner dimension 100 × 300 mm is done as per [20] recommendations. The concrete was poured into the mould with three layers and each layer was compacted using 16 mm tamping rod. The specimens were kept in the mould for 24 h and then after removal of all the specimens were left in the curing tank for a period of 28 days. The cylinder specimen was placed in the machine after the curing period in such a way that the longitudinal axis was perpendicular to the bed. Two bearing strips were provided on the upper and lower side of the specimen and load is applied gradually at a rate of 17.25 kN/s. The maximum applied load indicated by the machine at failure was noted down and split tensile strength was calculated by using the formula. The average of all the values was taken as the split tensile strength of that mix. Similarly test was conducted for all types of mixes and the corresponding split tensile strength was obtained.

5.5 Angularity Number

30 kg of oven dried sample was taken by using a metal scoop and the metal container was filled in three layers. Each layer was compacted by 100 blows at a rate of 2 blows/second by lifting the tamping rod 5 cm above the aggregate surface. The last layer was filled with overflowing quantity and after compaction of the last layer the excess materials were removed by using tamping rod as a straight edge. Weight of the aggregates with cylinder measured and recorded to the nearest 10 g. Three different determinations were made and mean weight of the aggregates was calculated. Then by using the obtained data angularity number was calculated by the "Eq. (1)"

$$\text{Angularity Number} = 67 - \frac{100\ \text{W}}{\text{C}\ \text{G}_\text{A}} \quad (1)$$

where, W = mean weight in g of the aggregate in the cylinder,
C = weight of water in g required.
G_A = specific gravity of aggregate.

5.6 *Permeability*

Casting of slab having inner dimension 500 × 500 × 150 mm is done as per [21] recommendations. The concrete was poured into the mould with three layers and each layer was compacted using 16 mm tamping rod. The specimens were kept in the mould for 24 h and after removal of all the specimens were left in the curing tank for a period of 28 days. The concrete slab was cleaned by brooming Putty was applied around bottom edge of the infiltration ring and placed on the concrete slab. The applied putty was pressed in to the surrounding to avoid the leakage of water through the ring area. Then water was poured into the ring maintaining a head between two marked lines. The test was performed by constant head method and time required from pouring of water to when there was no free water on the surface is recorded and the permeability was calculated. The test was conducted at three different positions on the concrete slab and the average of all was recorded as the permeability of that respective mix. Similarly test was conducted for all types of mixes and the corresponding permeability was obtained.

6 Result and Discussion

6.1 *Total Void Content*

Total void content was calculated by using the angularity number as per "Eq. (2)".
Total void content = 33 + Angularity number (IS 2386 part (I):1963)

$$\text{Total Void Content} = 100 - \frac{100\ \text{W}}{\text{C}\ \text{G}_\text{A}} \quad (2)$$

where, W = mean weight in g of the aggregate in the cylinder.
C = weight of water in g required.
G_A = specific gravity of aggregate.

Generally, preferable angularity number in coarse aggregate in between 0 and 11. Total void in aggregate calculated concept based on angularity number. As increase in fine aggregate percentage, angularity number decreases as well as total void content also decreases (Fig. 3). Angularity number our sample varies between 5 and 12

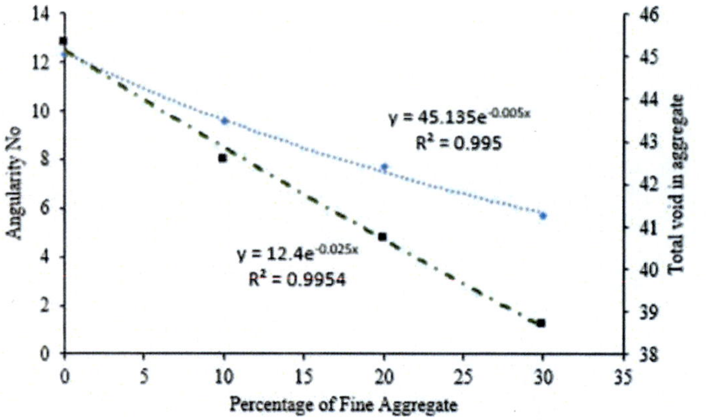

Fig. 3 Angularity number versus total void in aggregate versus percentage of fine aggregate

and total void of aggregate varies from 38.7 to 45.3 as increase in fine aggregate in aggregate sample. In road construction angularity number of 5–10 is generally preferred [22]. Adding of fine aggregate leads to filling the voids in coarse aggregate and decrease void content in the previous concrete mix.

6.2 Strength Characteristics

Figures 4, 5, and 6 represents the variation of compressive strength, split tensile strength and flexural strength with respect to the percentage of fine aggregate respectively. From the graph it is observed that with the increase in percentage of fine

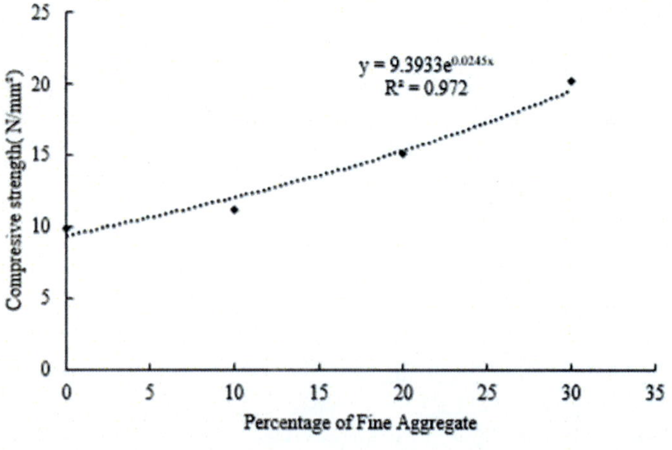

Fig. 4 Compressive strength versus percentage of fine aggregate

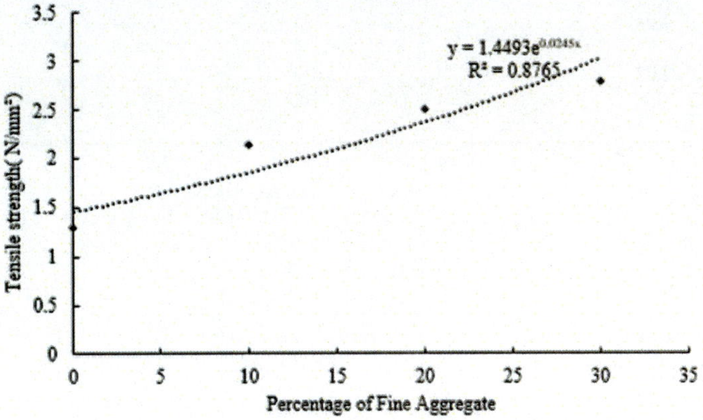

Fig. 5 Split tensile strength versus percentage of fine aggregate

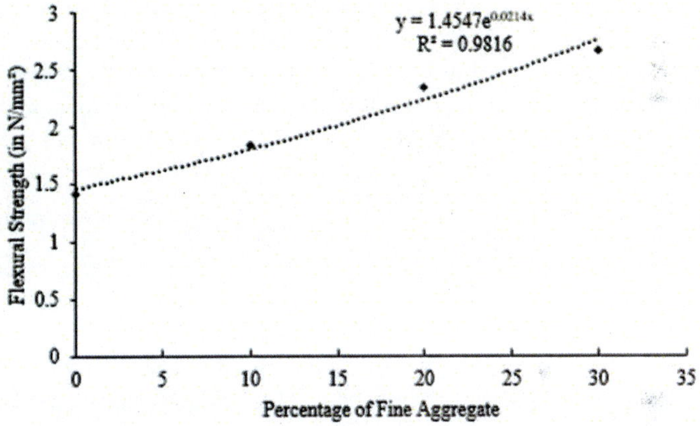

Fig. 6 Graph of flexural strength versus percentage of fine aggregate

aggregate the compressive strength, split tensile strength and flexural strength of the concrete exponentially increases. Compressive strength of pervious concrete varies from 9 to 20 MPa. Split tensile strength varies from 1.295 to 2.777 MPa. Flexural strength varies from 1.416 to 2.666 MPa. This happens due to increase of bond inside the structure due to addition of fine aggregate.

6.3 *Permeability*

Figure 7 shown represents the variation of permeability with respect to the percentage of fine aggregate. Co-efficient of Permeability of pervious concrete varies from 0.908 to 1.015 cm/min. From the figure it is observed that with the increase in percentage

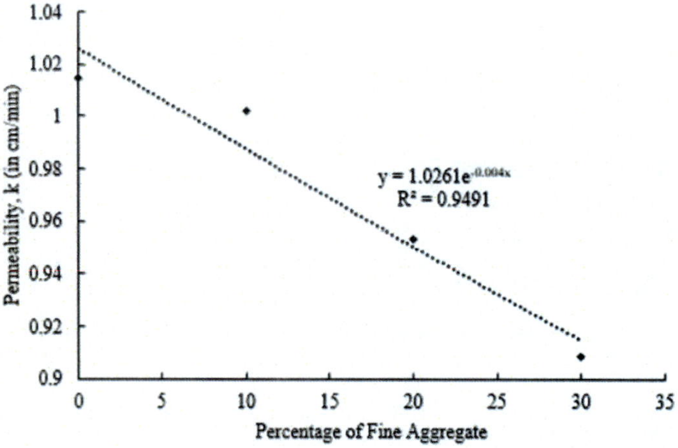

Fig. 7 Graph of permeability versus percentage of fine aggregate

of fine aggregate the permeability of the concrete exponentially decreases. This happens due to decrease in voids content inside the structure due to addition of fine aggregate. And addition fine aggregates blocks the interconnection of voids as well as void network.

6.4 Optimum Mix

By using the Fig. 8, it can prepare mix of desire strength and permeability. Angularity number our sample varies between 5 to 12 and total void of aggregate varies from 38.7 to 45.3 as increase in fine aggregate in aggregate sample. Compressive strength of

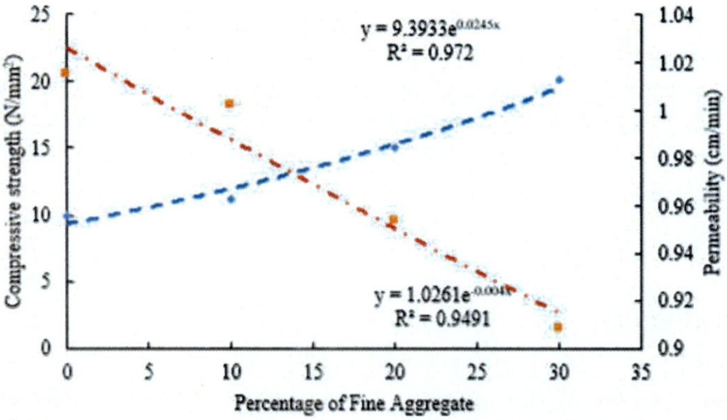

Fig. 8 Compressive strength versus permeability versus percentage of fine aggregate

pervious concrete varies from 9 to 20 MPa. Co-efficient of Permeability of pervious concrete varies from 0.908 to 1.015 cm/min. If our required strength and permeability are high, then we have to negotiate with one characteristic of pervious concrete.

7 Conclusion

From the experiment it is seen that the compressive strength of the previous concrete with 0, 10, 20, 30% fine aggregate is 9.88, 11.19, 15.11 and 20.20 MPa respectively and the permeability is 1.01, 1.00, 0.95 and 0.90 in cm/min respectively. This indicates that with increasing the percentage of fine aggregate the compressive strength of the previous concrete increases and permeability decreases and vice-versa. In order to bring a balance between the compressive strength and permeability of the previous concrete a relationship between them is obtained in the form of graph. By using this graph we can get a particular mixture composition for a specific value of permeability and strength. The angularity number falls in range of 5–10 which is suitable for pavement construction for light traffic area.

References

1. Lian C, Zhuge Y (2010) Optimum mix design of enhanced permeable concrete—an experimental investigation. Constr Build Mater 2:2664–2671
2. Bonicelli A, Arguelles GM, Pumarejo LGF (2016) Improving pervious concrete pavements for achieving more sustainable urban roads. Procedia Eng 161:158–1573
3. Chen Y, Wang K, Wang X, Zhou W (2013) Strength, fracture and fatigue of pervious concrete. Constr Build Mater 42:97–104
4. Hatanaka S, Kamalova Z, Harada M (2019) Construction of a non-linear permeability model of pervious concrete and drainage simulation of heavy rain in a residential area. Results Mater 3:100033
5. Maguesvari MU, Narasimha VL (2014) Study of pervious concrete with various cement content. Int J Adv Technol Eng 02(08)
6. Zhuge Y, Lian C (2009) Development of environmentally friendly and structural enhanced permeable concrete pavement material. In: Southern engineering conference
7. Deepashree S, Mohanraj N, Krishnaraj C (2016) An experimental study on the durability characteristics of pervious concrete. ARPN J Eng Appl Sci
8. Putman BJ, Neptune AI (2011) Comparison of test specimen preparation techniques for pervious concrete pavements. Constr Build Mater 25:3480–3485
9. Shu X, Huang B, Wu H, Dong Q, Burdette EG (2011) Performance comparison of laboratory and field produced pervious concrete mixtures. Constr Build Mater 25:3187–3192
10. Weiss PT, Kayhanian M, Khazanovich L, Gulliver JS (2015) Permeable pavements in cold climates: state of the art and cold climate case studies. Minnesota Department of Transportation, University of Minnesota, Research project final report 2015-30
11. Neithalath N, Sumanasooriya MS, Deo O (2010) Characterizing pore volume, sizes, and connectivity in pervious concretes for permeability prediction. Mater Charact 61:802–813
12. Deo O, Neithalath N (2010) Compressive behaviour of pervious concretes and a quantification of the influence of random pore structure features. Mater Sci Eng A 528:402–412

13. Haselbach LM, Valavala S, Montes F (2006) Permeability predictions for sand-clogged Portland cement pervious concrete pavement systems. J Environ Manage 81:42–49
14. IS 8112 (2013) Indian standard ordinary portland cement, 43 grade-specification, Second revision
15. IS 383 (2016) Coarse and fine aggregate for concrete-specification, Third revision
16. ASTM C94/C94M-21 (2021) Standard specification for ready-mixed concrete. ASTM International, West Conshohocken, PA
17. ASTM C1688/C1688M-14a (2014) Standard test method for density and void content of freshly mixed pervious concrete. ASTM International, West Conshohocken, PA
18. IS 10262: Indian standard concrete mix proportioning-guidelines, First revision (2009).
19. IS 516 (1959) Indian standard methods of tests for strength of concrete
20. IS 5816 (1999) Indian standard splitting tensile strength of concrete—method of test, First revision (1999, Reaffirmed 2004)
21. IS 3085 (1965) Indian standard method of test for permeability of cement mortar and concrete (1965, Reaffirmed 2002)
22. Kim Y-R, Souza LT (2009) Effects of aggregate angularity on mix design characteristics and pavement performance. Nebraska Department of Transportation Research Reports, 126

Performance-Based Design for Binary-Blended Filler Application in Dense-Graded Cold Mix Asphalt

Kabiru Rogo Usman, Mohd Rosli Hainin, Mohd Khairul Idham Mohd Satar, Muhammad Naqiuddin Mohd Warid, Ibrahim Aliyu Adamu, and Ashraf Abdalla Mohammed Radwan

Abstract The quest to developing a universally acceptable mix design procedure for cold mix asphalt (CMA) is in the spotlight of continuous research. Moreover, the performance improvement of CMA coupled with sustainable construction drives had witnessed the inclusion of industrial by-products and biomass waste alike in CMA. Nonetheless, various transportation departments in different countries tailor CMA's design to suit their geographic regions' peculiar challenges. Despite such efforts, mix designs need to produce laboratory results replicative of service conditions capable of ameliorating the high void content, weak early strength, and slow rate of strength gain of CMAs. This study proposed a performance-based mix design based on the combined result of indirect tensile stiffness modulus (ITSM) and Cantabro loss tests compared to the standard Marshall mix design. The mixtures contain binary blending of 1–4% palm oil fuel ash (POFA) with 3–6% granite filler by total aggregate weight in a fine dense-graded CMA using a polymer-modified cationic quick set (CQS-1h) and an unmodified rapid set (RS-1K) emulsified asphalt. A gradation with a nominal maximum aggregate size (NMAS) of 4.75 mm (FGCMA-4.75 mm) was employed. In addition to Cantabro and ITSM, the designed mix was tested for Indirect Tensile Strength (ITS), and modified Lottman's test. Results revealed a significant correlation between the usual Marshall design with the proposed design based on the established optimum emulsion contents (OEC). An optimal 3% POFA yielded moisture-resistant mixtures with enhanced mechanical and Marshall volumetric properties in void reduction, increased stability, and adequate early strength.

K. R. Usman · M. R. Hainin (✉) · M. K. I. M. Satar · M. N. M. Warid · I. A. Adamu · A. A. M. Radwan
School of Civil Engineering, Universiti Teknologi Malaysia, 81310 Johor Bahru, Johor, Malaysia
e-mail: mrosli@utm.my; roslihainin@ump.edu.my

M. R. Hainin
College of Engineering, University Malaysia Pahang, 26600 Pekan, Pahang, Malaysia

K. R. Usman
Department of Civil Engineering Technology, School of Engineering Technology, Nuhu Bamalli Polytechnic Zaria, P.M.B 1061, Zaria, Kaduna State, Nigeria

I. A. Adamu
Department of Civil Engineering Technology, Federal Polytechnic Bauchi, PMB 0231, Bauchi State, Nigeria

© The Author(s), under exclusive license to Springer Nature Switzerland AG 2022

H. R. Pasindu et al. (eds.), *Road and Airfield Pavement Technology*, Lecture Notes in Civil Engineering 193, https://doi.org/10.1007/978-3-030-87379-0_40

Keywords Cold mix asphalt · Fine dense-graded cold mix asphalt · Palm oil fuel ash · CQS-1h · RS-1K

1 Introduction

Cold mix asphalt (CMA) emerges as an environmentally friendly cost-effective alternative to hot mix asphalt (HMA) for varying applications. CMA comprises of well-graded aggregates, emulsified asphalt, water, and additives mixed and placed at ambient temperatures [1]. The most common application of CMAs is in pavement repair and rehabilitation works, and to a lesser extent as wearing courses of light-trafficked roads. Advances in material science and emulsified asphalt formulation have witnessed the use of CMA as a base course and wearing courses of heavily-trafficked roads [2]. However, absence of a universal design method has slowed the pace at which CMA is used globally. Most transportation departments or companies in Malaysia, the US, Europe, and Africa have different design approaches for CMA, often customised to suit specific needs [3]. Efforts were made to develop acceptable mix designs including volumetric, simplified, and performance-based mix designs amongst others [4–6]. Performance-based mix designs utilising mechanical performance tests reflects a closer replica of service conditions. Moreover, indirect tensile stiffness modulus (ITSM) or resilient modulus and Cantabro were identified as simplified yet, robust tests for designing CMA [7].

In recent times sustainable construction has encouraged the addition of waste material and industrial by-products in CMA to further improves its performance and reduce cost. This study introduces the addition of palm oil fuel ash (POFA) as supplemental to granite filler to produce a fine dense-graded CMA with a nominal maximum aggregate size (NMAS) of 4.75 mm formulated to suit reinstatement works of pipes laid in pavement cuts. POFA is the by-product of burning palm oil biomass, including kernels, fruit bunches, leaves, and fibres in power generating plants or palm oil mills. POFA is often dumped in landfills, causing environmental menace. POFA waste exponentially increases over the years in Malaysia, concomitant to increased palm plantation, which was reported to stand at 5.90 million hectares in 2019 [8]. The race in curbing the environmental threat posed by POFA's landfill disposal resulted in its application in hot mix asphalt (HMA) [9] and as cement replacement in concrete with great success and performance improvement [10]. The use of POFA as filler in HMA improves the Marshall volumetric properties, enhances stiffness, tensile strength, durability, and resistance to moisture damage [9, 11, 12]. Despite these achievements, there is the need to extend further the use of POFA in other construction material, including CMA.

Research efforts at improving CMA's performance have witnessed the addition of numerous industrial waste and by-products with attendant mixture improvement. An improvement in mechanical properties, thermal and moisture susceptibility of a CMA employed as a binder course was achieved by incorporating binary filler. The filler comprises high calcium fly ash and a fluid catalytic cracking catalyst residue [13]. CMA exhibited higher emulsion and water content demand in a related vein, despite enhancement in stiffness and fatigue cracking consisting of a complete aggregate replacement with construction and demolition waste (CDW). However, the complicated design procedure was highlighted as a great challenge for this mix [14]. Thanaya's research is among the pioneer studies target simplifying design methodology for CMA couple with the addition of wide-ranging waste materials as aggregate and filler replacements. Thanaya study's main objective was to simplify the CMA design procedure and make it practicable to site conditions. The approach for determining the optimum total liquid content at compaction was particularly impracticable for site applications [15]. Researchers tried several mix design approaches, including a volumetric mix design that takes account of residual water after compaction [5]. In recent times, optimisation techniques are employed to enhance the mix design of CMA [16]. Due to the high variability in the mixture's characteristics, the two notable design procedures for CMAs viz. Illinois method based on the modified Marshall method, the Asphalt institute method based on modified Hveem are often utilised. As a consequence, several transportation departments significantly vary these modified designs to suit peculiar needs. Consequently, CMA mix design improvement remained a grey area with recorded successes and opened for further enhancement.

Notwithstanding the successes recorded, the lingering challenge associated with CMA of absence for a simplified, universally accepted mix design remains far from being attained [17]. This study aimed to bridge the gap by proposing a performance-based mix design replicative of site conditions in designing a CMA blended with <0.75 μm sized POFA and granite as fillers.

2 Materials and Test Methods

The detailed characteristics of the materials used in this study and the methodology employed are elucidated in Sects. 2.1–2.3.

2.1 *Emulsified Asphalts*

Emulsified asphalt is a stable dispersion of typically two immiscible liquids called 'continuous' and 'discontinuous' phase. They are blended in proportions and varying viscosities, which allows low temperatures application as emulsified asphalt mixtures or as recycled mixtures [18]. The emulsion is gaining wider usage, especially for rehabilitation and maintenance work as CMA or slurries, for ease of application and minimal emission, with attendant cost and energy-saving [19]. In a drive to ensure a fast-curing mix, cationic quick (CQS-1h) and rapid (RS-1K) setting emulsions with medium viscosities were employed in this study. The emulsions' choice was intended to ensure material compatibility, as they are compatible with both alkaline and acid-nature aggregates [20]. Even though it is a well-established fact that alkaline aggregates like limestone bonds better with bitumen than acidic aggregates yet, acidic aggregate offers better abrasion and skid resistance. Moreover, the aforementioned fact applies to a neat bitumen in bitumen-aggregate mixes not emulsion-aggregate mixes [21, 22]. The reason is that, water present in emulsified asphalts may prevent proper adhesion between the bitumen's active components from reacting with the calcium (Ca+) from the limestone [21].

The polymer-modified cationic quick set (CQS-1h) emulsion is a unique formulation for micro-surfacing, enabling fast curing of the mixture and early opening to traffic. Thus, CQS-1h breaks and set faster than standard emulsions. The rapid setting (RS-1K) emulsion is also a medium viscosity fast curing emulsion typically used in Malaysia for spray applications [23]. The curing of emulsified asphalt is a determinant factor of their earlier life strength and the type of application intended. In high humidity, low wind speed or night time application, fast curing emulsions become desirable for adequate early strength development [24].

The emulsions and their residues were tested for mechanical and rheological properties according to the American Society for Testing and Materials (ASTM) and American Association of State Highway and Transport Officials (AASHTO) standards as the case may be. The result has satisfied the Jabatan Kerja Raya (JKR)—Malaysian Public Works Department requirements and presented in Table 1 [25].

Table 1 Mechanical, and physical, properties of the emulsion

Test on emulsified asphalt						
Test description	Test method		Result		Specification (CQS/RS)	Requirement
	ASTM	AASHTO	CQS-1h	RS-1K		
Physical appearance	–	–	Brownish	Brownish		JKR, 2008, MS 161, and ASTM D2397
Storage stability (%)	D6930	T 59	0.9	1.0	1% max	
Settlement test (%)	D6930	T 59	1.0	1.0	1% max	
Residue by evaporation (%)	D6934	T 59	65	63	57/50 min	
Particle charge	D7402	–	Positive	Positive		
Saybolt Furol viscosity (s)	D7496	–	20	22	15–50	
Test on emulsified asphalt residue						
Penetration at 25 °C (0.1 mm)	D5	T 49	40	68	40/60–90/200	
Softening point (°C)	D36	T 53	69	–	54–57 °C min	JKR
Solubility in Trichloroethylene (%)	D2042	–	95	100	97.5 min	

Table 2 Physical and mechanical properties of cement, granite filler, and POFA (Salam et al., 2018)

Property tested		Materials		
		POFA	Granite	Cement
Colour	–	Dark grey	Grey	Grey
Specific gravity	–	2.51	2.97	3.16
Fraction passing through a 75 μm sieve	(wt%)	97	90	100
Bulk density	(g/cm^3)	2.51	2.96	1.14
Solubility in water	–	Insoluble	Insoluble	Insoluble

2.2 Palm Oil Fuel Ash

Palm oil fuel ash (POFA) is the ash by-product of burning palm tree biomass, including empty palm fruit bunches, fibres, kernels/shells, palm leaves in palm oil mills, and power generating plants. POFA is ideally grey but turns black depending on the proportion of unburnt carbon. A higher percentage of unburnt carbon makes POFA black; hence further burning at 1000 °C reverts the POFA to grey. POFA's specific gravity ranges from 1.89 to 2.60, and it has a high silica and alumina content, thus pozzolanic [26]. POFA's characteristics and chemical constituent are distinctly different, primarily influenced by several factors comprising (i) the Biomass part that was used in its production, (ii) geographic location of the palm biomass, (iii) burning temperature, (iv) particle size, and (v) pre or post-production treatment applied [26].

Moreover, Malaysia is among the World's leading oil palm producers, with Indonesia topping the list [27]. The large quantity of POFA being disposed of as waste in Malaysia has reached alarming levels [26], quantity over 1000 tons are dumped in landfills [28]. Consequently, research in notable Asian palm oil-producing countries and the World on the sustainable application of POFA in construction sprung [10, 29, 30]. POFA's application varies from one study to another depending on the produced material—either concrete, modified bitumen, or asphalt concrete; the POFA's particle size required, which dictates the adopted drying and grinding method. The usual POFA's percentage replacement in asphalt or bitumen ranges between 1 and 7% [11, 30]. Evidence of the use of POFA in concrete at percentage replacement between 0 – 70% exist, and this produces compressive strength of up to 95 MPa [31, 32]. POFA's properties in comparison to cement and granite filler are given in Table 2.

2.3 Aggregate

The aggregate used in this study is a crushed granite with angular features inherent of an impact cone crusher. This crusher shares the dual advantage of both impact and cone crushers, as such granite aggregates obtained from such crushers is desirable for asphalt production as it ensures good stone-on-stone interlock.

Table 3 Mechanical Properties of fine and Coarse aggregate

Aggregate	Properties		Granite	Limit	Relevant specification
Fine aggregate (≤ 4.75 mm)	Specific gravity	–	2.89	-	ASTM C 128
	Sand equivalent	(%)	96.0	>45	ASTM D 2419
	Water absorption	(%)	0.48	≤2	ASTM C 128/MS 30
	Bulk density	(Kg/m^3)	1815	–	ASTM C 128

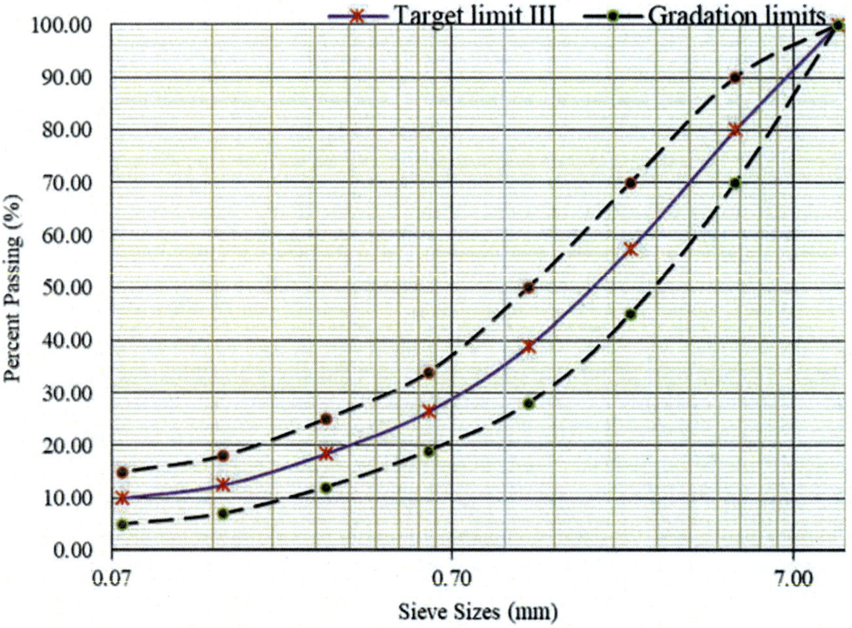

Fig. 1 Aggregate gradation plot

The nominal maximum aggregate size used for this study is 4.75 mm with a filler of <0.075 mm in size. Result of the physio-mechanical testing on the aggregate is presented in Table 3.

The plot of the aggregate gradation used in this study is presented in Fig. 1.

3 Experimental Program

The experimental program entails testing the individual aggregate for physical and mechanical properties. The properties of individual constituents in CMA are pertinent to ensuring material compatibility, hence, good workability and performance. Among the crucial parameters tested is the specific gravity, which translates to the materials' volume/mass measures. The specific gravity was evaluated with an air-displacement pycnometer. The reason for choosing this device is because the gravimetric test is not suitable for testing the specific gravity of granite filler and POFA as aggregates need to be submerged.

Subsequently, material preparation, mixture design, and performance testing were conducted. The laboratory tests are highlighted in the subsequent sections.

3.1 Specimen Preparation

The traditional water addition to CMA for ease of workability and coating was not used in this study particularly due to coating problems associated with the RS-1K emulsion. Mixing was conducted mechanically in a Hobart steel bowl mechanical mixer at ambient temperature. Mixtures are manually tilled in the mixer to enhance aggregate coating. Mixing was limited to within a maximum of three (3) minutes as typically obtainable in the literatures. Nonetheless, higher or lower mixing times are also possible depending on the gradation, emulsion type, and perhaps water content [33, 34]. Thereafter, the mixture was transferred into oiled moulds equipped with circular plastics at the bottom and top of the moulds having diameters same with the Marshall mould and tamped in three (3) layers. Each layer was tamped 15 blows by the mould's periphery and ten (10) blows at the centre, as spelt out in ASTM D6926-10 [35].

The standard compaction level for CMA is typically 50 blows as specified by the Malaysia's public works department—Jabatan Kerja Raya (JKR) and India's ministry of road, transport and highways (MoRTH) [25, 36]. Nonetheless, this study adopted 75 blows per face for all samples in order to keep void levels to a practical minimum and to achieve maximum density. However, some studies highlighted that 3–5% air void is difficult to achieve even at compaction levels higher than 75 blows [37]. Arguably, 3–5% VTM can be achieved with CMA depending on the mixture gradation. Some studies adopt compactive effort even higher than 75 blows, though higher compaction levels lead to aggregate crushing for coarser mixes, hence mixture weakening [38]. As such the Asphalt Institute (AI) and MoRTH CMA mix designs specified 50 blows per face of the Marshall sample. This is so because most CMA are used as pavement maintenance mixtures and for light-trafficked road construction. It is worth mentioning that water was observed to seep under the mould, especially with the RS-1K emulsion, but minimal amount of water was observed with CQS-1h. This is perhaps attributable to the higher water content of RS-1K, which stood at 37%, while CQS-1h has 35% water content.

After compaction, samples were oven cured at 40 °C in their moulds for 24 h after which they are further cured for additional 72 h at the same temperature [39]. A similar curing protocol was reported, only that the curing temperature was set at 60 °C [40].

Sample testing. The designed samples were subjected to mechanical performance testing. Test conducted on the designed mixtures included ITSM, indirect tensile strength (ITS), and moisture damage (modified Lottman test). Samples are conditioned before testing according to the requirement for the respective individual tests.

3.2 *Mix Design*

A performance-based mix design base on result of indirect tensile stiffness modulus (ITSM) and Cantabro was used to design the mixtures. Result were compared with that obtained by the modified Marshall method adopted by Illinois design method as outlined in asphalt institute (AI) manual series MS-19 as elucidated in MoRTH specification [36]. The design procedure is as outlined in Sects. 3.2.1 and 3.2.2.

The mixture acronyms used in this study are presented in Table 4. FGCMA stands for fine dense-graded cold mix asphalt, the 'C' and 'P' suffices in front of the acronyms stands for the control and POFA-modified mixes respectively. The size of the grounded POFA is <0.075 mm while the nominal maximum aggregate size (NMAS) of the mixture is 4.75 mm as shown in Table 4.

Modified Marshall Stability and Flow Test. The procedure contained in the revised Marshall flow and stability specification was used to design all mixtures according to ASTM D6926-10 [35].

Stability and flow tests were performed in accordance with ASTM D 6927-06, with samples conditioned for forty minutes in a water bath at 60 °C prior to testing. The maximum failure load in Kilo Newton (kN) at a constant deformation rate of 50 mm/min was recorded as the stability values. Similarly, flow is the maximum deformation measured in millimetres (mm) at the failure load using a flow meter. Triplicate samples for each binder content were tested and the average recorded.

Confirmatory tests were conducted to re-assess the established optimum emulsion contents (OEC). Four key Marshall volumetric parameters were used for the

Table 4 Mix Design Acronym Description

S/N	Mixture acronym	Mixture description	Binder type	Nominal Maximum aggregate size (mm)
1	FGCMA-C	Control mixture (0% POFA)	CQS-1h & RS-1K	4.75 mm
2	FGCMA-P 1%	Different percentages of filler (pan material < 0.075 mm) replacement with POFA (1–4%) and (3–6%) granite filler		
3	FGCMA-P 2%			
4	FGCMA-P 3%			
5	FGCMA-P 4%			

confirmatory test which include stability, flow, voids in total mix (VTM), and voids filled with binder (VFB).

Performance-Based Mix Design Based on Indirect Tensile Stiffness Modulus and Cantabro. Concomitant to the absence of a universally acceptable mix design for CMA, this study further adopted ITSM and Cantabro test results to achieve the optimum residual asphalt emulsion content.

The highest values of resilient modulus with the least Cantabro losses in addition to some fundamental Marshall Volumetric properties were utilised to arrive at the optimum emulsion content. The utilisation of performance-based mix design for CMA was explored with success using foamed emulsion for CIPR mixtures (Saidi et al., 2019). Similarly, dry and immersion ITS performance test results at 15 °C were employed to arrive at the OEC of cold recycled emulsified asphalt mixtures (Zhu et al. 2019).

3.2.1 Mix Design Based on Indirect Tensile Stiffness Modulus Test

This study uses the peak value of the average values from triplicate samples tested within the range of emulsion contents. It is worth noting that the trend exhibited by Marshall Specimens was similar to that of the Resilient Modulus samples. Resilient modulus is a measure of a material's Modulus of Elasticity under the influence of rapidly applied loads similar to traffic loading on pavements.

This study conducted the resilient modulus test according to ASTM D 7369–11 as a re-approval from the initial ASTM D4123- 82/96 [41]. Laboratory fabricated samples were conditioned at 25 °C and 40 °C for three (3) hours before being tested under a repeated haversine/waveform indirect tension loading. Triplicate samples were prepared for each binder content, and each sample was tested at 0° and 90° orientation. Average values for each set of samples are used for plotting the graph. The maximum value in a range of binder contents (7–9%) taken from the curve's peak is taken to be the optimised value.

3.2.2 Mix Design Based on Cantabro Test

Cantabro test was utilised as part of the performance-based mix design protocol. The Texas Department of Transportation's specification (Tx-DOT-Tex-245) guided the conduct of the test [42]. It measures a compacted bituminous specimen's resistance to abrasion, which translates to an asphalt mixture's durability. The sample's crumbling is simulated by subjecting the Marshall specimen to 300 revolutions in a Los Angeles Abrasion (LAA) machine without the steel balls.

Triplicate samples for each individual mixture at their respective OEC determined from ITSM were cast and tested. Thus, at OEC, each individual mixture will produce a value of Cantabro loss commensurate to its abrasion resistance capability. The Cantabro loss in the mass loss is expressed as a percentage of the initial weight. The smaller the Cantabro loss the better the samples are in resisting abrasion loss thus better durability performance.

The best mixture in all the mix design process was chosen by utilising multi-criteria decision method to arrive at 3% POFA replacement.

4 Results and Discussion

4.1 Results of Optimum Emulsion Content for Various Mixtures

The combined optimum emulsion contents (OEC) for CQS-1h and RS-1K designed by Marshall, and ITSM coupled with Cantabro percentage losses are presented in Table 5. In terms of the Marshall design CQS-1h has the least OEC with control mix while RS-1K has similar OEC for both the control and POFA modified mixes. Also, for the ITSM design method, CQS-1h has the least OEC of 7.9%. Generally, there is increasing emulsion demand at higher POFA replacements. Similarly, there is increasing Cantabro loss with increasing POFA content except for the 3% POFA replacement. The least Cantabro loss with FGCMA-3% of both CQS-1h and RS-1K as compared to other mixtures signified the mixture's better performance than the others.

The Cantabro losses in Table 5 for CQS-1h and RS-1K represents the respective average percentage degradation of the individual mixtures fabricated with their respective OEC. It is not optimum emulsion percentage hence, lower than the OEC for Marshall and ITSM as evident in Table 5. From Table 5, it is discernable that the Cantabro losses of CQS-1h mixtures are lower compared to that of RS-1K mixtures.

All samples were verified for Marshall parameters of stability, flow, VTM, and VFB for CQS-1h and RS-1K and the result presented in Table 6. Tick marks indicated

Table 5 Combined optimum emulsion contents for Marshall, and ITSM tests with Cantabro losses

Percentage replacement	Mixture designation	Average ITSM OEC (%)		Average Marshall OEC (%)		Average Cantabro loss (%)	
		CQS-1h	RS-1K	CQS-1h	RS-1K	CQS-1h	RS-1K
Control	FGCMA-C	8.1	7.8	7.7	8.1	4.5	8.5
1% POFA	FGCMA-P 1%	8.1	8.3	8.5	8.3	4.2	9.6
2% POFA	FGCMA-P 2%	7.9	8.4	8.6	8.2	5.5	4.6
3% POFA	FGCMA-C 3%	7.7	8.0	8.1	8.4	3.8	4.0
4% POFA	FGCMA-C 4%	7.9	8.0	8.7	8.8	6.8	8.7

Table 6 Summary of Marshall verification of result for CQS-1h and RS-1K mixtures

Mix types	CMA mix requirements and Marshall result verification to MoRTH [36]							
	Stability		Flow		VTM		VFB	
	(Maximum)		(Minimum)		(Range)		(Minimum)	
	2.2 (kN)		2.0 (mm)		3.0–5.0 (%)		15.0 (%)	
CQS-1h								
FGCMA-C	9.70	✔	2.40	✔	10.50	×	60.20	✔
FGCMA-P 1%	10.20	✔	2.10	✔	4.70	✔	80.00	✔
FGCMA-P 2%	8.30	✔	2.30	✔	4.40	✔	81.20	✔
FGCMA-P 3%	15.35	✔	2.40	✔	4.00	✔	82.40	✔
FGCMA-P 4%	5.39	✔	2.51	✔	4.70	✔	80.10	✔
RS-1K								
FGCMA-C	10.60	✔	2.10	✔	10.20	×	62.10	✔
FGCMA-P 1%	10.45	✔	2.90	✔	4.40	✔	80.60	✔
FGCMA-P 2%	11.51	✔	2.42	✔	4.00	✔	83.00	✔
FGCMA-P 3%	13.28	✔	2.16	✔	3.80	✔	83.90	✔
FGCMA-P 4%	6.49	✔	2.80	✔	5.40	×	78.10	✔

that the sample has passed the requirement, while the crosses indicated the opposite. All the mixtures have satisfied the minimum criteria set by India's Ministry of Road Transport and highways (MoRTH) specifications for emulsified asphalt mixtures [36].

However, the control mixtures of both emulsions and the 4% POFA replacement for RS-1K emulsion did not met the criteria. All the control mixtures of both emulsions and the FGCMA-P 4% of RS-1K have high void content. High voids in CMA lead to greater susceptibility to moisture damage. From the result in Table 6, it can be deduced that FGCMA-P 3% of both emulsions has higher stability than either of control mixtures and other POFA percentages. Progressively higher POFA replacement percentages from 1 to 4% lead to corresponding increase in flow. The VFB for all POFA modified mixtures are higher than FGCMA-C for CQS-1h and RS-1K.

Sample ITSM values of the control mixtures are presented in Table 7. From Table 7, 7.5 and 8% emulsion of the control mixture have the highest ITSM values at both 25 °C and 40 °C for both emulsions. Nonetheless, only samples with 8% emulsion content at the two test temperatures for both emulsions has recorded all through improvement irrespective of test temperature or emulsion type. At 8% emulsion

Table 7 Average ITSM values for CQS-1h and RS-1K at 25 °C and 40 °C

Average ITSM for FGCMA-C (control mixture) (MPa)				
Emulsion	CQS-1h		RS-1K	
(%)	25 °C	40 °C	25 °C	40 °C
7.00	4192.0	2424.0	6947.0	4895.0
7.50	3665.0	2078.0	8706.0	6047.0
8.00	5178.0	2626.0	7648.0	5640.0
8.50	2617.0	1055.0	7334.0	857.0
9.00	2591.0	2449.0	7509.0	2178.0

content, the ITSM values for CQS-1h and RS-1K at 25 °C and 40 °C are 23.50%, 8.3%, and 10.10%, and 15.22%, respectively, in comparison with the control mixture. Moreover, 7.5% emulsion recorded the highest ITSM value at 25 °C among all other mixtures, though, higher ITSM values at 25 °C signifies a low cracking resistance. On the other hand, at 8.5% - 9.0% emulsion content CQS-1h mixtures present higher cracking resistance with lower ITSM than RS-1K mixtures. More so, RS-1K samples exhibited greater rutting resistance for their higher ITSM at 40 °C than CQS-1h between 7.0 and 8.0% emulsion.

The combined ITSM plots at 25 °C and 40 °C for both emulsions is plotted and presented in Figs. 2 and 3. From Figs. 2 and 3, the OEC is computed by adding the emulsion contents corresponding to the peak ITSM similar to the Marshall test. The data in Table 7 was plotted to obtain the OEC in Figs. 2 and 3. Similarly, the OECs for all the POFA modified mixtures were obtained by plotting their corresponding ITSM data as determined above for the control (FGCMA-C) mixture as detailed in Table 5.

The OEC obtained from combined ITSM and Cantabro are closely related, meaning that the two test produces OEC similar to that obtained by the Marshall method.

4.2 *Indirect Tensile Strength Result*

The indirect tensile strength test result is depicted in Fig. 4. There is improved fatigue cracking resistance by the addition of 3% POFA. This is linked to the pozzolanic attribute of POFA which aid adequate adhesion hence higher ITS values than the control mixtures for both CQS-1h and RS-1K. However, RS-1K FGCMA-P 3% recorded higher ITS than the corresponding CQS-1h mix, this could be connected to the higher elasticity and less stiffness of RS-1K revealed by the penetration result on its residue.

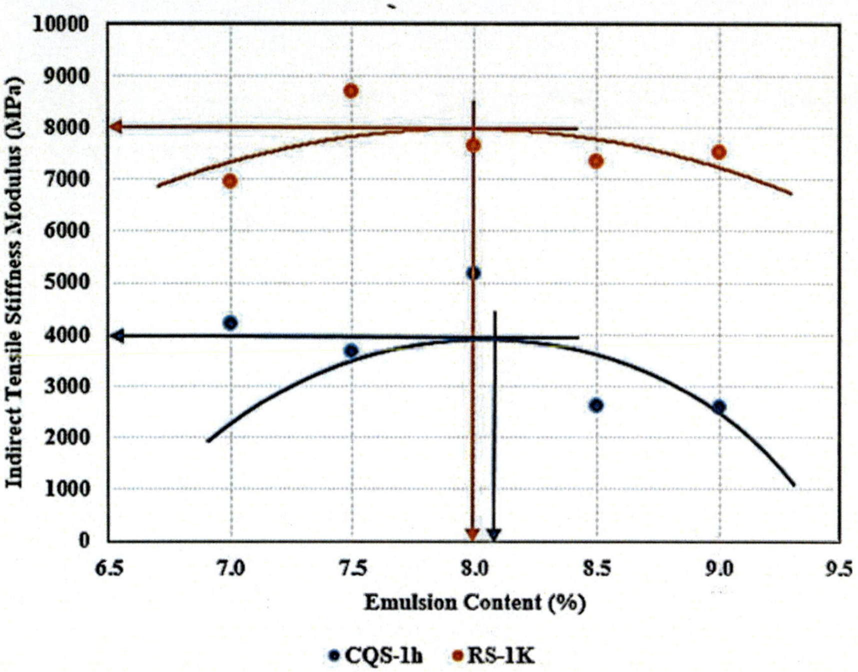

Fig. 2 Combined indirect tensile stiffness modulus at 25 °C—FGCMA-C

4.3 *Moisture Damage*

The ITS and tensile strength ratios (TSR) of FGCMA-P 3% is presented in Fig. 5. FGCMA-P 3% has higher than FGCMA-C mixtures for both dry and wet subset samples. The improved moisture damage resistance is linked to the enhanced adhesion of the POFA and that of added cement. Hence, FGCMA-P 3% could ameliorate moisture-induced damage in CMA, which is identified as the critical element affecting the performance of CMA.

5 Conclusion

The OEC obtained with ITSM and Cantabro are in close range to that obtained by the Marshall test, as such these tests could be employed for designing fine dense-graded CMA, especially those with NMAS of 4.75 mm. Furthermore, the incorporation of POFA in CMA improves Marshall volumetric properties and mechanical performance. Importantly, FGCMA-P 3% produces mixtures with enhanced Marshall volumetric properties in void reduction, increased stability, and adequate early strength.

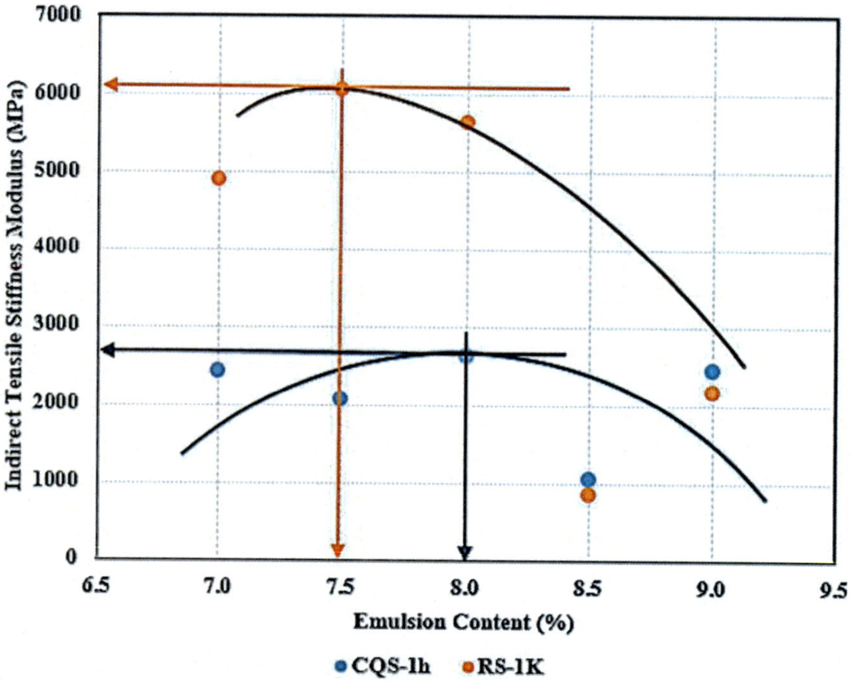

Fig. 3 Combined indirect tensile stiffness modulus at 40 °C—FGCMA-C

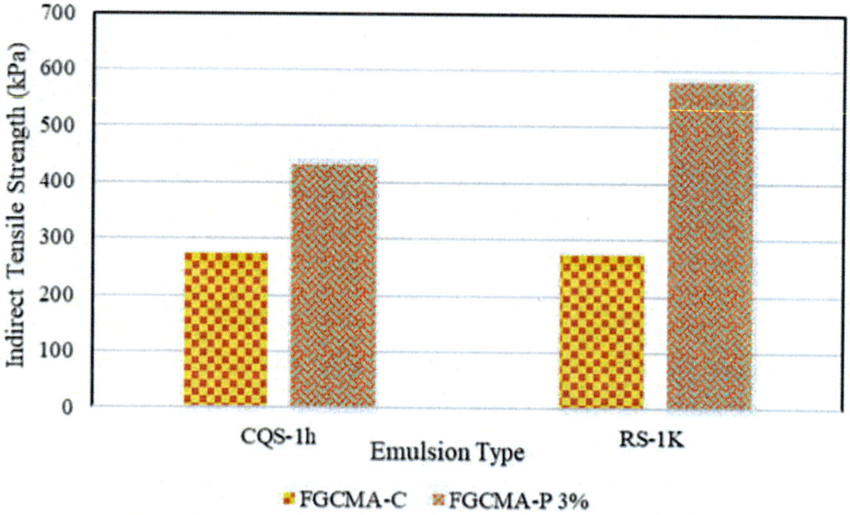

Fig. 4 Indirect tensile strength result for CQS-1h and RS-1K

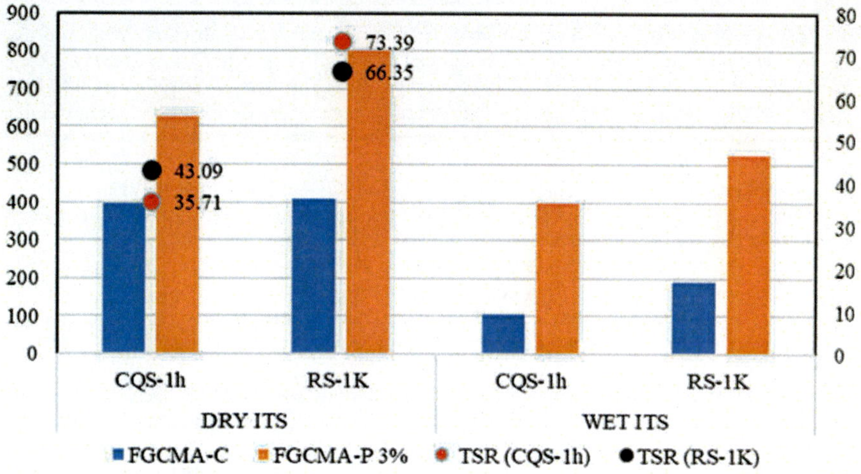

Fig. 5 Moisture damage (Modified Lottman's) test result for CQS-1h and RS-1K

More so, the optimal 3% POFA replacement yielded improved cracking and moisture damage resistance as revealed by the ITS and modified Lottman's tests respectively.

Acknowledgements This work was supported/funded by: Malaysia's Ministry of Higher Education under the Fundamental Research Grant Scheme (FRGS/1/2019/TK01/UTM/02/1). Nigerian Tertiary Education Trust (TET Fund) intervention/sponsorship through Nuhu Bamalli Polytechnic.

Conflict of Interest The authors declare no conflict of interest concerning this paper's publication.

References

1. Khan A, Redelius P, Kringos N (2016) Toward understanding breaking and coalescence of bitumen emulsions for cold mix asphalts. Prague, Czech Republic, pp 1–13
2. Dulaimi A, Al-Nageim H, Ruddock F, Seton L (2017) Performance analysis of a cold asphalt concrete binder course containing high-calcium fly ash utilizing waste material. J Mater Civ Eng 29:04017048
3. Waller HF (1980) Emulsion mix design methods: an overview. In: 59th Annual meeting. Transportation Research Board, pp 1–9
4. Gómez-Meijide B, Pérez I (2014) A proposed methodology for the global study of the mechanical properties of cold asphalt mixtures. Mater Des 57:520–527
5. Ling C, Bahia HU (2018) Development of a volumetric mix design protocol for dense-graded cold mix asphalt. J Transp Eng Part B Pavements 144:04018039
6. Lee KW, Brayton TE, Gress D, Harrington J (2010) Performance-based mix-design method for cold in-place recycling of bituminous pavements for maintenance management. In: Proceedings of the ninth maintenance management conference, pp 11–19
7. Singh DN, Patel A, Kulkarni MP, Gumaste SD, Bartake PP, Rao KVK (2011) A methodology for determination of resilient modulus of asphaltic concrete. Adv Civ Eng 1–6

8. Adebayo MK, Ab-Kadir MA, Ismail M (2020) A review on self-compacting concrete incorporating palm oil fuel ash as a cement replacement. Constr Build Mater 258:1195410
9. Ahmad JB, Nizam K, Hidayah N, Zainorabidin A (2012) The practical use of palm oil fuel ash as a filler in asphalt pavement. In: International conference on civil, environmental, engineering and sustainability, pp 3–5
10. Bamaga SO, Hussin MW, Ismail MA (2013) Palm oil fuel ash: promising supplementary cementing materials. Korean Society of Civil Engineers J Civ Eng 17:1708–1713
11. Maleka AM, Hamad AW, Jaya RP (2015) Effect of palm oil fuel ash (POFA) on the durability of asphaltic concrete. Appl Mech Mater 744–746:1560–1565
12. Maleka AM, Alkali IA, Jaya RP (2014) The indirect tensile strength of palm oil fuel ash (POFA) modified asphaltic concrete. Appl Mech Mater 587–589:1270–1275
13. Dulaimi A, Al-Nageim H, Ruddock F, Seton L (2017) High performance cold asphalt concrete mixture for binder course using alkali-activated binary blended cementitious filler. Constr Build Mater 141:160–170
14. Gómez-Meijide B, Pérez I, Airey G, Thom N (2015) Stiffness of cold asphalt mixtures with recycled aggregates from construction and demolition waste. Constr Build Mater **77:**168–178
15. Thanaya INA, Negara INW, Suarjana IP (2014) Properties of cold asphalt emulsion mixtures (CAEMs) using materials from old road pavement milling. Procedia Eng 95:479–488
16. Nassar AI, Thom N, Parry T (2016) Optimizing the mix design of cold bitumen emulsion mixtures using response surface methodology. Constr Build Mater 104:216–229
17. Al-Merzah S, Al-Busaltan S, Al-Nageim H (2019) Characterizing cold bituminous emulsion mixtures comprised of palm leaf ash. J Mater Civ Eng 31:04019069
18. Usman KR, Hainin MR, Satar MKI, Warid MNM, Giwangkara GG (2020) Rheological criteria assessment of a rapid setting emulsion as compared to quick set for emulsified asphalt cold mixes. IOP Conf Ser: Earth Environ Sci 476:2
19. Usman KR, Hainin MR, Idham MK, Warid MNM, Yaacob H, Hassan NA, Azman M, Che Puan OC (2019) Performance evaluation of asphalt micro surfacing—a review. IOP Conf Ser Mater Sci Eng 527:012052
20. Ronald M, Luis FP (2016) Asphalt emulsions formulation: State-of-the-art and dependency of formulation on emulsions properties. Constr Build Mater 123:162–173
21. Yin Y, Chen H, Kuang D, Song L, Wang L (2017) Effect of chemical composition of aggregate on interfacial adhesion property between aggregate and asphalt. Constr Build Mater 146:231–237
22. Yilmaz M, Kök BV, Kulolu N (2011) Effects of using asphaltite as filler on mechanical properties of hot mix asphalt. Constr Build Mater 25:4279–4286
23. Yaacob H, Hainin MR, Aziz MMA, Warid MNM, Chang FL, Ismail CR, Hassan NA (2013) Bitumen emulsion in Malaysia-a conspectus. Jurnal Teknology (Sci Eng) 65:97–104
24. Yaacob H, Chang FL, Hainin MR, Jaya RP (2015) Curing of asphalt emulsified tack coat subjected to Malaysian weather conditions. J Mater Civ Eng 27:1–10
25. JKR (2008) JKR/SPJ/2008-S4: standard specification for road works part 4 flexible pavement. JKR, pp 1–187
26. Hamada HM, Jokhio GA, Yahaya FM, Humada AM, Gul Y (2018) The present state of the use of palm oil fuel ash (POFA) in concrete. Constr Build Mater 175:26–40
27. Abdullah N, Sulaim F. The oil palm wastes in Malaysia. Biomass now—sustainable growth and use. InTech
28. Awal ASMA, Hussin MW (1997) The effectiveness of palm oil fuel ash in preventing expansion due to alkali-silica reaction. Cem Concr Compos 19:367–372
29. Altwair NM, Kabir S, Brameshuber W (2010) Palm oil fuel ash (POFA): an environmentally-friendly supplemental cementitious material for concrete production. In: International Rilem conference, material science (Matsci), vol III(77), pp 113–126
30. Borhan MN, Ismail A, Rahmat RA (2010) Evaluation of palm oil fuel ash (POFA) on asphalt mixtures. Aust J Basic Appl Sci 4:5456–5463
31. Megat Johari MA, Zeyad AM, Muhamad Bunnori N, Ariffin KS (2012) Engineering and transport properties of high-strength green concrete containing high volume of ultrafine palm oil fuel ash. Constr Build Mater 30:281–288

32. Momeen M, Islam U, Mo KH, Alengaram UJ (2016) Durability properties of sustainable concrete containing high volume palm oil waste materials. J Clean Prod 137:167–177
33. Thanaya NA (2007) Review and recommendation of cold asphalt emulsion mixtures Caems design. Civ Eng Dimension 9:49–56
34. Lundberg R, Jacobson T, Redelius P, Östlund JA (2016) Production and durability of cold mix asphalt. In: Proceedings of 6th Eurasphalt & Eurobitumen congress. Czech Technical University, Prague
35. American Society for Testing Materials ASTM D6926 (2015) Standard practice for preparation of bituminous specimens using Marshall. ASTM International, West Conshohocken, PA
36. MoRTH (2013) Specifications for road and bridges works. In: Indian roads congress on behalf of the Govt. of India, Ministry of Road Transport and Highway, vol 1, pp 234–683
37. Dash SS, Panda M (2018) Influence of mix parameters on design of cold bituminous mix. Constr Build Mater 191:376–385
38. Joni HH (2018) Studying the effect of emulsified asphalt type on cold emulsified asphalt mixtures properties. IOP Conf Ser Mater Sci Eng 433
39. Thanaya INA, Zoorob SE, Forth JP (2009) A laboratory study on cold-mix, cold-lay emulsion mixtures. Proc Inst Civ Eng Transp 162:47–55
40. Du S (2015) Performance characteristic of cold recycled mixture with asphalt emulsion and chemical additives. Adv Mater Sci Eng 2–8
41. American Society for Testing Materials ASTM D4123 (1995) Standard test method for indirect tension test for resilient modulus of bituminous. ASTM International, West Conshohocken, PA
42. Texas DoT M and T D 2014 Cantabro loss. TxDoT pp 3–5

Predicting California Bearing Ratio (CBR) Value of a Selected Subgrade Material

Kasun Wimalasena and Chaminda Gallage

Abstract The subgrade bearing capacity is an important parameter in flexible pavement design, and it is largely influenced by the variation of subgrade moisture. California Bearing Ratio (CBR) is the most popular method of assessing the subgrade bearing capacity. It compares the load required to make a particular penetration on a given subgrade with the load to make the same penetration on standard material. Although it would be beneficial to perform CBR tests at more frequent intervals on a road section to design, it would not be practical in certain instances owing to the laborious and time-consuming CBR test process. In that case, developing a method to predict the subgrade CBR based on the variation of moisture would be an advantage. Accordingly, this study aims to assess the influence of moisture content and the compacted density on the subgrade CBR value. Hence, the possibility of developing a model to predict subgrade CBR was investigated. A series of standard CBR tests were performed for different combinations of moisture and compaction densities, and standard CBR values were estimated. Thereafter, a statistical model was developed to predict the CBR value for the selected clay material and validated with laboratory test results.

Keywords Pavement Engineering · Pavement materials · Unsaturated soils · California Bearing Ratio · Subgrade · Subgrade moisture

1 Introduction

A well-connected transportation network plays a vital role in the economic development of a country [1–3]. In such networks, road and rail links establish connections between major cities to facilitate logistics and to commute. These networks should be constructed not to fail before the end of design life and, therefore, need a strong subgrade as the foundation of constructing resilient roads and rail lines [4]. Overestimation of subgrade condition would develop an inadequate design to

K. Wimalasena (✉) · C. Gallage
Queensland University of Technology, Brisbane, Australia
e-mail: mataramb@qut.edu.au

© The Author(s), under exclusive license to Springer Nature Switzerland AG 2022

H. R. Pasindu et al. (eds.), *Road and Airfield Pavement Technology*, Lecture Notes in Civil Engineering 193, https://doi.org/10.1007/978-3-030-87379-0_41

bear traffic loads and, as a result, will induce early pavement distresses. Underestimation of the subgrade condition would lead to the decision that the subgrade is weak and hence, would recommend weak subgrade improvements, such as geogrid reinforcement and increasing granular cover [5–8], which could create unnecessary demand for resources [9, 10]. Therefore, an accurate subgrade assessment will ensure a long-lasting pavements and railroads.

California Bearing Ratio (CBR) test is the most popular method being used to assess the bearing capacity of subgrades [11, 12]. This was first developed by the California Division of Highways in 1920 and subsequently adopted by most of the transportation agencies for subgrade assessment [13, 14]. In some instances, researchers have used the CBR test to assess the stability of recycled concrete aggregates [9] and stress–strain stage of railroad subgrades [15, 16]. Although CBR is not a fundamental material property that can be incorporated in mechanistic-empirical road design methods, its simplicity and inexpensiveness have made it a popular test for subgrade assessment [13]. Moreover, Austroads pavement design guidelines assume tenfold of subgrade CBR as the resilient modulus in the absence of experimentally estimated subgrade resilient modulus for mechanistic-empirical pavement design [17].

Usually, the CBR test is performed multiple times and obtain the average to eliminate test errors. Also, in order to assess the existing subgrade condition of a road section to be designed, the CBR test has to be performed at multiple locations with a reasonable gap. Despite doing CBR tests at more frequent intervals with repetitions would be beneficial, number of tests must be limited owing to the time and labour requirement. In that case, a prediction model could be helpful to assess the subgrade while reducing the number of CBR trials [18]. Moreover, prediction models could also be helpful to estimate the variation of CBR with changes to soil properties. Subgrade conditions drastically change with the variation of groundwater height and the presence of expansive clay type soil [19, 20]. This condition is prominent in Queensland, Australia [21–29], and as a result, weak subgrades have become a frequent challenge in road construction [30]. With the support of prediction models, variation of CBR can be pre-assessed for design processes while estimating the current condition with laboratory experiments [13].

In literature, researchers have found that subgrade CBR depends on soil properties: such as optimum moisture content, liquid limit, plastic limit, plastic index, shrinkage limit and density [13, 14, 31–33]. Even though the relationship between the CBR and the soil properties is complex to be defined, studies have reported that multiple linear regression method and the Artificial Neural Network (ANN) method can effectively be used to predict subgrade CBR from soil properties with reasonable accuracy [11, 13, 18, 34]. However, the developed models have their own limitations and needs to be calibrated for different conditions if required to be used.

This study has been conducted as a part of a large-scale laboratory plate load testing project to evaluate the performance of geogrid reinforced flexible pavements. In order to create a subgrade of a predetermined bearing capacity, it is required to know the relationship between the bearing capacity of the subgrade and its properties. Otherwise, a large number of standard CBR trials has to be performed to determine required subgrade properties for desired bearing capacity. Accordingly, this paper reports the development of a statistical model to predict unsoaked CBR by considering subgrade soil properties as input variables. The scope of the study is limited to one clay type soil. Moreover, only the moisture content of subgrade soil and the Degree of Compaction (DOC) of the subgrade have been selected as independent variables to develop a prediction model.

2 Materials and Testing

2.1 Subgrade Soil

The subgrade soil was collected from a road construction site in Queensland, Australia. Standard procedures specified in the Material Testing Manual (MTM) of Queensland Department of Transport and Main Roads [35] were followed to determine the basic soil properties listed in Table 1. This subgrade soil can be classified as high plastic silt as per the Unified Soil Classification System (USCS). Furthermore, it also matches with classification A-7-6 of the AASHTO soil classification method.

The standard proctor compaction test was conducted, following Q142A standard procedure stipulated in MTM [35], aiming to estimate the relationship between the dry density and the moisture content for the selected subgrade soil. Figure 1 illustrates the compaction curve of the subgrade soil. Moreover, Maximum Dry Density (MDD) and Optimum Moisture Content (OMC) of the subgrade soil was determined as 1.3 g/cm^3 and 33%, respectively.

Table 1 Properties of subgrade soil

Soil property	Value
Soil particle density	2.62
Maximum dry density (g/cm^3)	1.316
Optimum moisture content (%)	33
Liquid limit (%)	73
Plastic limit (%)	53
Shrinkage limit (%)	20
D_{10} (mm)	0.00081
D_{30} (mm)	0.0085
D_{60} (mm)	0.041
Coefficient of curvature (Cc)	50.62
Coefficient of uniformity (Cu)	2.18
USCS classification	**MH**

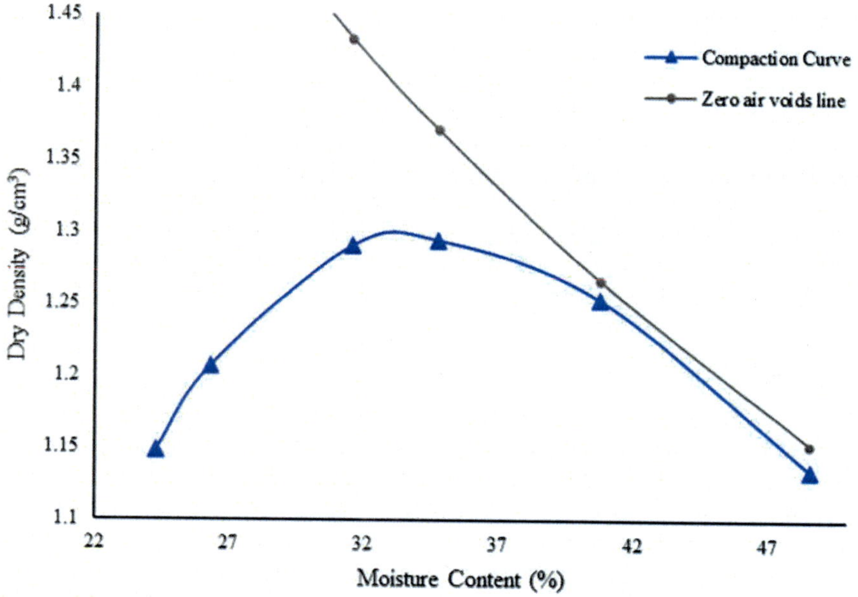

Fig. 1 Dry density versus moisture content of subgrade soil

2.2 *Standard CBR Testing*

The subgrade soil was oven-dried at 60 °C for three days to remove the moisture. Thereafter, subgrade soil was mixed with water in small batches, and water content for each batch was changed to achieve different soil moisture contents. In this study, the moisture content was varied between 30 and 50%. The mixed soil was sealed in polythene bags and cured for 7 days as considerable time is required for clay type soil to equalise moisture throughout the bulk.

The prepared clay soil was used to create subgrades, with known moisture content and density, inside the standard CBR mould. The DOC was maintained between 80 and 100%. First, as the moisture content of the subgrade soil and the volume of the standard CBR mould is known, the required weight of the mixed subgrade soil to achieve the desired degree of compaction was calculated. Then, the measured soil was compacted inside the CBR mould in three equally thick layers. Thereafter, annular weights were placed on the top surface, and the sample was tested in Instron machine as shown in Fig. 2. The CBR of each trial was calculated based on standard calculation procedure of Q113A in TMR specification [35] using the load-deformation data recorded from testing. The results of the 26 trials are listed in Table 2.

Fig. 2 Performing standard CBR test using Instron machine

3 Statistical Analysis

The results from CBR trials were analysed using the regression method to understand the correlations between selected subgrade soil properties and subgrade bearing capacity. Hence, a bearing capacity prediction model was estimated. Initially, simple linear regression was applied, selecting one subgrade property as an exogenous variable and the estimated CBR value as an endogenous variable. Thereafter, the multiple linear regression method was considered selecting both subgrade moisture and DOC as independent variables while subgrade bearing capacity (CBR) as dependent variable. All 26 observations in Table 2 were used for both simple and multiple linear regression analysis.

3.1 Simple Linear Regression Models

First, the moisture content was selected as the independent variable and the estimated subgrade CBR as the dependent variable. The model equation is expressed by Eq. 1 and also illustrated in Fig. 3. This model estimated a linear relationship

Table 2 Results of CBR trials

Test No.	Moisture (%)	Density (g/cm³)	DOC (%)	Measured CBR (%)
1	30.9	1.33	102	7.4
2	37.0	1.04	80	3.5
3	37.0	1.11	85	3.8
4	37.0	1.17	90	5.7
5	37.0	1.24	95	6
6	37.6	1.10	85	4
7	37.6	1.17	90	3.5
8	37.6	1.23	95	4.5
9	37.6	1.29	100	6.8
13	41.9	1.12	86	4.3
14	41.9	1.19	91	4.7
15	41.9	1.25	96	5.1
10	43.0	1.10	84	4.3
11	43.0	1.16	89	3.6
12	43.0	1.23	94	5.1
16	43.7	1.12	86	3.4
17	43.7	1.19	91	3.4
18	43.7	1.26	97	5
19	47.9	1.11	85	2.6
20	47.9	1.17	90	3.5
21	47.9	1.24	95	4.1
22	48.9	1.11	86	2.6
23	48.9	1.18	91	3
24	48.9	1.24	96	3.5
25	50.3	1.13	87	2
26	50.3	1.20	92	1.9

between subgrade CBR and the moisture content with Multiple R = 0.7404 and $R^2 = 0.5482$. Multiple R > 0.5 implies that moisture content has a strong relationship with subgrade bearing capacity. The *p*-values of all mode coefficients are less than 0.05, which indicates that the estimated linear model is statistically significant ($\alpha < 0.05$). However, R^2 value confirms that this model can only explain 54.8% of subgrade CBR variance only.

$$\text{CBR} = 12.23 - 0.19(\text{Moisture}) \tag{1}$$

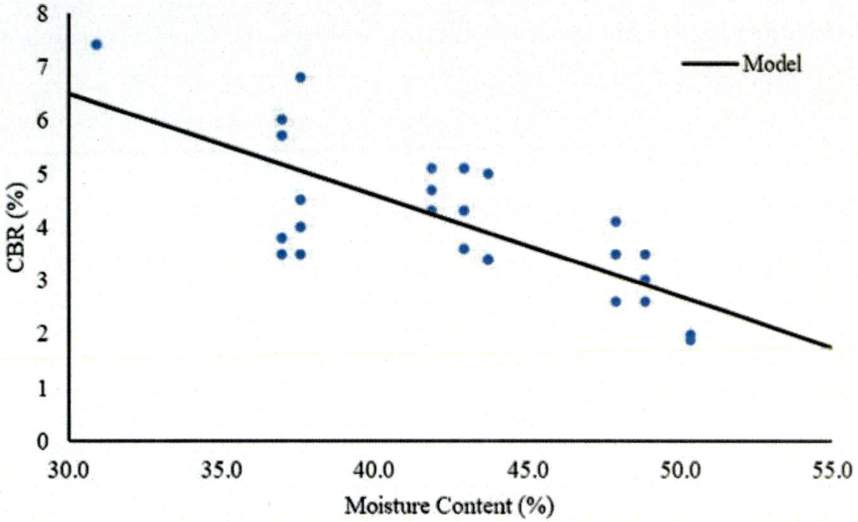

Fig. 3 CBR (%) versus moisture content (%)

In the second analysis, the DOC and subgrade bearing capacity were selected as independent and dependent variables, respectively. Equation 2 defines the model equation with Multiple R = 0.6329 and R^2 = 0.4006 and the linear relationship between two variables are illustrated in Fig. 4. All model coefficients are statistically significant with *p*-value < 0.05. Although DOC has a great correlation with subgrade

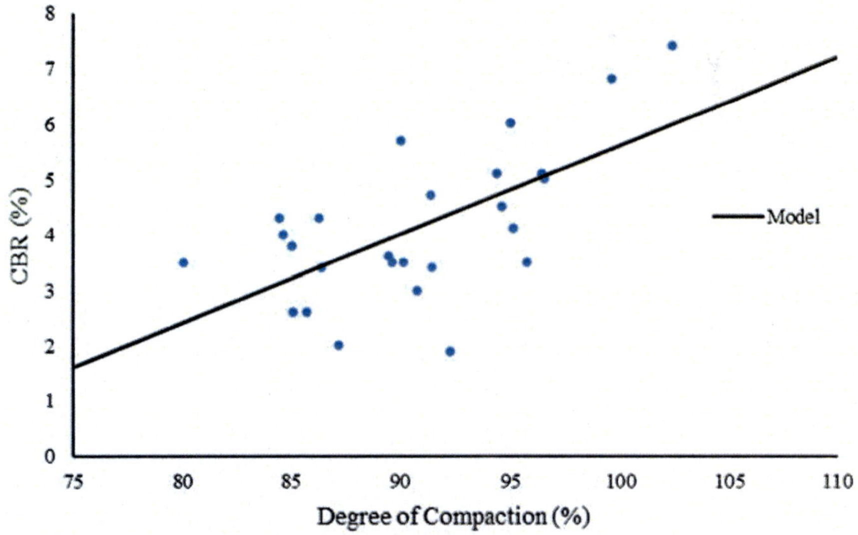

Fig. 4 CBR (%) versus Degree of Compaction(%)

bearing capacity (Multiple R > 0.5), the R^2 indicate that the DOC itself could explain 40% of the subgrade CBR only.

$$CBR = -10.42 + 0.16(DOC) \tag{2}$$

3.2 *Multiple Linear Regression Model*

The simple linear regression analysis in Sect. 3.1 confirms that all estimated models are statistically significant and both moisture content and DOC have greater correlation with subgrade CBR. However, aforementioned estimation considered only one subgrade property as exogenous variable and hence, does not result powerful statistical models that can predict the subgrade bearing capacity with greater accuracy. This is due to the fact that subgrade CBR is dependent on both moisture content and DOC. Accordingly, multiple linear regression was applied by considering both moisture content and the DOC as independent variables while considering subgrade CBR as the dependent variable. The model equation is given in Eq. 3 and the regression statistics are listed in Table 3. Accordingly, the MLM model can explain 82.4% of the CBR variation, which is high in accuracy compared to the case of simple linear regression models. Moreover, *p* values of the two independent variables estimated below 0.05 confirmed that both variables are statistically significant in MLM model.

$$CBR = 12.24 + 13.53(DOC) - 20.069(\text{Moisture Content})^2 \tag{3}$$

The estimated MLM model was further validated by performing Analysis of Variance (ANOVA) on regression results. The analysis compares the null hypothesis: subgrade CBR not related to the subgrade moisture content and the DOC, to the alternative hypothesis of subgrade CBR is dependent on these two variables. Table 4 summaries the results of ANOVA test. As the Significance F value records 2.11E−09, which is exceedingly lower than 0.01, the null hypothesis can be rejected with 99% confidence, supporting that the subgrade CBR is greatly dependent on moisture

Table 3 MLM model statistics

Regression statistics		
Multiple R	0.9077	
R square	0.8239	
Variable	Coefficients	*P*-value
Intercept	12.2369	5.08E−02
DOC (%)	13.5314	3.72E−06
Moisture content (%)	−20.0692	1.47E−07

Table 4 Results of ANOVA test

	df	SS	MS	F	Significance F
Regression	2	37.4654	18.7327	53.8177	2.11E–09
Residual	23	8.0058	0.3481		
Total	25	45.4712			

content and DOC. Hence, the developed multiple linear regression model is a valid prediction of subgrade CBR against moisture content and DOC.

4 Validation

The developed subgrade CBR prediction model was validated by conducting another CBR test series with 10 experimental trials. The combination of the moisture content and the DOC for those trials are listed in Table 5. Figure 5 illustrates the experimentally estimated CBR and the predicted CBR for the ten trials. The straight line represents the points where predicted CBR and experimental CBR are exactly same. Almost all points are closer to the straight line depicts that predicted CBR value is closer to the experimentally estimated value. Furthermore, the standard error of the predicted versus experimental CBR was estimated as 0.5141.

Table 5 CBR trials for validation

Test	Moisture content (%)	DOC (%)	Predicted CBR (y)	Estimated CBR (y′)	$(y - y')^2$
1	41.6	95.3	4.9	4.8	0.0205
2	38.4	89.8	4.7	4.4	0.1036
3	44.9	94.4	4.3	4.8	0.2901
4	46.0	78.9	2.0	2.7	0.5380
5	45.8	90.1	3.5	3.5	0.0003
6	47.3	90.5	3.3	3.5	0.0460
7	49.0	84.4	2.1	2.1	0.0016
8	52.5	83.7	1.3	2.6	1.6367
9	51.4	85.9	1.9	1.9	0.0021
10	49.2	81.7	1.7	1.8	0.0041
				Standard error	0.5141

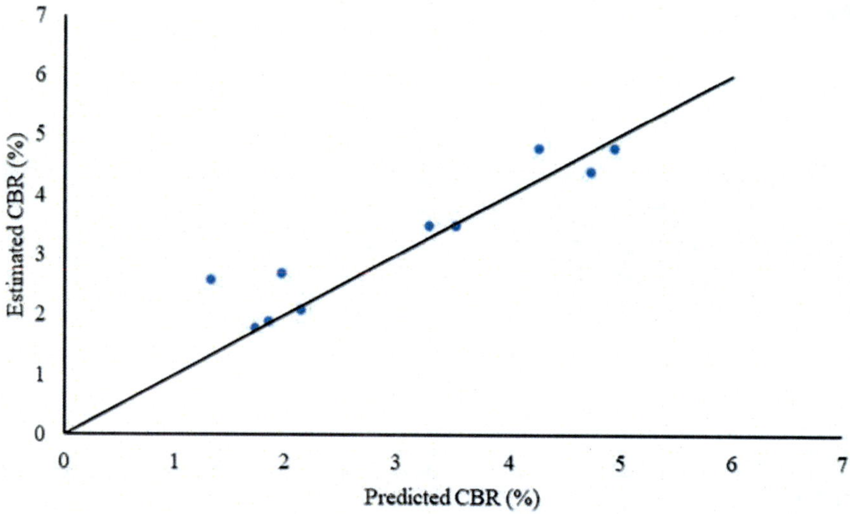

Fig. 5 Estimated versus predicted CBR

5 Conclusion

This study was aimed to develop a relationship between the subgrade moisture content and DOC with its unsoaked CBR value. The simple linear regression analysis indicated that moisture content and the DOC has a significant correlation with the unsoaked CBR of subgrade soil. A multiple linear regression model was developed with $R^2 = 0.8239$, selecting subgrade moisture and the DOC as independent variables to predict the unsoaked CBR of the subgrade. Additional 10 CBR trials were performed and obtained data to validate the proposed CBR prediction model, and it was confirmed that the proposed MLM model can predict CBR value with a standard error of 0.5141. Therefore, it can be concluded that MLM can successfully be used to develop statistical models to predict the unsoaked CBR of subgrades. Moreover, the proposed model could be used to estimate the unsoaked CBR of a subgrade, given that the subgrade soil moisture and the DOC are known. In addition, the prediction model could also be used to determine the required moisture content and the degree of compaction to create a subgrade of a known CBR value for laboratory testing purposes.

Acknowledgements This research work is part of a research project (Project No. IH18.06.1) sponsored by the SPARC Hub at Department of Civil Eng, Monash University funded by the Australian Research Council (ARC) Industrial Transformation Research Hub (ITRH) Scheme (Project ID: IH180100010). The financial and in-kind support from Department of Transport and Main Roads (Queensland), Logan City Council, Global Synthetics (Australia), Polyfabrics (Australia) and Queensland University of Technology (QUT) is gratefully acknowledged. Also, the financial support from ARC is highly acknowledged.

References

1. Clark BR, Gallage C (2020) Superior performance benefits of multigrade bitumen asphalt with recycled asphalt pavement additive. Constr Build Mater 230:116963
2. Clark BR, Piacere L, Gallage C (2018) Effects of recycled asphalt pavement on the stiffness and fatigue performance of multigrade bitumen asphalt. J Mater Civ Eng 30:04017278
3. Duncan-Williams E, Attoh-Okine NO (2008) Effect of geogrid in granular base strength—an experimental investigation. Constr Build Mater 22:2180–2184
4. Tenpe AR, Patel A (2020) Application of genetic expression programming and artificial neural network for prediction of CBR. Road Mater Pavement Des 21:1183–1200
5. Wimalasena K, Jayalath CPG (2020) Effect of geogrid reinforcement in weak subgrades. Int J GEOMATE 18:140–146
6. Gallage C, Gunasekara Jayalath CP (2019) Use of Particle Image Velocimetry (PIV) technique to measure strains in geogrids. In: Proceedings of the 7th international symposium on deformation characteristics of geomaterials, E3S Web of conferences, vol 92, pp 1–6
7. Gunasekara Jayalath CP, Gallage C, Dhanasekar M, Dareeju B, Ramanujam J, Lee J (2018) Pavement model tests to investigate the effects of geogrid as subgrade reinforcement. In: Proceedings of the 12th Australian and New Zealand Young geotechnical professionals conference, pp 1–8
8. Jayalath CPG, Wimalasena KD (2020) Measuring strains on geogrid specimens under the laboratory tensile test. Int J GEOMATE 18:86–93
9. Jayakody S, Gallage C, Ramanujam J (2019) Effects of reclaimed asphalt materials on geotechnical characteristics of recycled concrete aggregates as a pavement material. Road Mater Pavement Des 20:754–772
10. Jayakody S, Gallage C, Ramanujam J (2019) Performance characteristics of recycled concrete aggregate as an unbound pavement material. Heliyon 5:e02494
11. Taskiran T (2010) Prediction of California bearing ratio (CBR) of fine grained soils by AI methods. Adv Eng Softw 41:886–892
12. Sreelekshmypillai G, Vinod P (2019) Prediction of CBR value of fine grained soils at any rational compactive effort. Int J Geotech Eng 13:560–565
13. Alawi M, Rajab M (2013) Prediction of California Bearing Ratio of subbase layer using multiple linear regression models. Road Mater Pavement Des 14:211–219
14. Taha S, El-Badawy S, Gabr A, Azam A, Shahdah U (2015) Modeling of California Bearing Ratio using basic engineering properties. In: 8th international engineering conference, At Sharm Al-Sheikh, Egypt
15. Dareeju B, Gallage C, Ishikawa T, Dhanasekar M, Dawes L (2018) Effects of particle size distributions and principal stress axis rotation on cyclic plastic deformation characteristics of coarse materials. Heliyon 13:161–170
16. Dareeju B, Gallage C, Ishikawa T, Dhanasekar MJS (2017) Effects of principal stress axis rotation on cyclic deformation characteristics of rail track subgrade materials. J Soils Found 57:423–438
17. Austroads (2018) Guide to Pavement Technology Part 2: Pavement Structural Design
18. Katte VY, Mfoyet SM, Manefouet B, Wouatong ASL, Bezeng LA (2019) Correlation of California Bearing Ratio (CBR) value with soil properties of road subgrade soil. Geotech Geol Eng 37:217–234
19. Jayakody S, Gallage C, Ramanujam J (2014) Effects of moisture content on resilient properties of recycled concrete aggregates (RCAs). In: Proceedings of 4th international conference of GEOMATE, pp 394–399
20. Udukumburage RS, Gallage C, Dawes L (2019) Oedometer based estimation of vertical shrinkage of expansive soil in a large instrumented soil column. Heliyon 5:e02380
21. Cheah C, Gallage C, Dawes L, Kendall P (2017) Measuring hydraulic properties of geotextiles after installation damage. Geotext Geomembr 45:462–470
22. Abeykoon A, Udukumburage RS, Gallage C, Uchimura T (2017) Comparison of direct and indirect measured soil-water characteristic curves for a silty sand. Int J GEOMATE 13:9–16

23. Cheah C, Gallage C, Dawes L, Kendall P (2016) Impact resistance and evaluation of retained strength on geotextiles. Geotext Geomembr 44:549–556
24. Chan D, Rajeev P, Kodikara J, Gallage C (2016) Field performance of in-service cast iron gas reticulation pipe buried in reactive clay. J Pipeline Syst Eng 7:04015025
25. Jayakody S, Gallage C, Kumar A (2014) Assessment of recycled concrete aggregates as a pavement material. Geomech Eng 6:235–248
26. Gallage C, Chan D, Kodikara J (2012) Response of a plastic pipe buried in expansive clay. In: Proceedings of the Institution of Civil Engineers-Geotechnical Engineering, vol 165, pp 45–57
27. Gallage C, Cochrane M, Ramanujam J (2012) Effects of lime content and amelioration period in double lime application on the strength of lime treated expansive sub-grade soils. In: Proceedings of the 2nd international conference on transportation geotechniques, pp 99–104
28. Gallage C, Tehrani K, Williams D (2017) Instrumented large soil-column to investigate climate-induced ground deformation in expansive soil. In: Proceedings of the 19th international conference on soil mechanics and geotechnical engineering, Seoul 2017, pp 1147–1150
29. Chan D, Gallage CPK, Rajeev P, Kodikara J (2015) Field performance of in-service cast iron water reticulation pipe buried in reactive clay. Can Geotech J 52:1861–1873
30. Priddle J, Lacey D, Look B, Gallage C (2013) Residual soil properties of South East Queensland. Aust Geomechan J 48:67–76
31. Rehman Z, Khalid U, Farooq K, Mujtaba H (2017) Prediction of CBR value from index properties of different soils. Tech J 22:17–26
32. Sathawara J, Patel A (2013) Comparison between soaked and unsoaked CBR. Int J Adv Eng Res Stud (IJAERS) II(III):132–135
33. Zumrawi MM (2012) Prediction of CBR value from index properties of cohesive soils
34. Yildirim B, Gunaydin O (2011) Estimation of California Bearing Ratio by using soft computing systems. Expert Syst Appl 38:6381–6391
35. Q. D. o. T. a. M. Roads: Materials Testing Manual. edn 2018, p 725

Selection of Image Processing Algorithms for Evaluation of Pervious Pavement Pore Network Properties

Ajayshankar Jagadeesh, Ghim Ping Ong, and Yu-Min Su

Abstract Digital Image Processing (DIP) algorithms are often required as a precursor to measure the internal characteristics of pavement structures during X-ray computed tomography (XRCT) based non-destructive evaluation (NDE) of pavement materials. The improper use of DIP algorithms can result in the significant under- or over-estimation of internal pavement characteristics, thereby affecting pavement design and maintenance strategies. Past research studies highlighted the significance of threshold segmentation algorithms and binarization of greyscale images on the porosity and permeability characteristics of pervious pavement mixtures. In addition, the use of a watershed segmentation algorithm was introduced to separate interconnected pore network structure into multiple pores. However, isolated pores were not removed in past analyses found in the literature due to a lack of consideration in using ungrouping algorithm to segregate connected and isolated pores. The main objective of this study is to select the appropriate DIP algorithms that can be used to evaluate pervious pavement pore network properties from three-dimensional XRCT based images. In this paper, a key microstructural pore parameter was investigated using various DIP algorithms for different pervious pavement mixtures and recommendations are made. It is expected that the results presented in this paper can help researchers understand the importance of DIP algorithms on XRCT-based pavement evaluation studies.

Keywords Pervious concrete · X-ray computed tomography · Digital Image Processing · Air voids

A. Jagadeesh · G. P. Ong (✉)
National University of Singapore, Block E1A #07-03, 1 Engineering Drive 2, Singapore 117576, Singapore
e-mail: ceeongr@nus.edu.sg

Y.-M. Su
National Kaohsiung University of Science and Technology, 415 Chien-Kung Road, Sanmin, Kaohsiung 80778, Taiwan

© The Author(s), under exclusive license to Springer Nature Switzerland AG 2022

H. R. Pasindu et al. (eds.), *Road and Airfield Pavement Technology*, Lecture Notes in Civil Engineering 193, https://doi.org/10.1007/978-3-030-87379-0_42

1 Introduction

Pervious pavements are widely used because of their increased permeability, skid resistance and acoustic absorption coefficients resulting in reduced surface water runoffs, accidents and tire/road noise [1]. This necessitates an understanding of pore network properties and their relationships to structural and functional performance characteristics. In recent years, the usage of X-ray computed tomography (XRCT) gained momentum because of its non-destructive nature and enabling the accurate measurement of pervious concrete pore network parameters such as volume, surface area, size and shape parameters which are difficult to obtain experimentally. Digital Image Processing (DIP) algorithms are often required as a precursor to measuring the internal characteristics of pavement structures during X-ray computed tomography (XRCT) based non-destructive evaluation (NDE) of pavement materials.

Earlier studies in pavement XRCT processing were mainly limited to two-dimensional analysis of pores, mainly the total porosity profiles versus depth. Kayhanian et al. [2] and Manahiloh et al. [3] studied the effect of clogging in pervious concrete field samples with the porosity profiles obtained from CT scanning in the image processing software such as Matlab and Image-Pro Plus. With significant improvements in image processing techniques, many researchers started exploring the three-dimensional characteristics of pavement pore properties. Cosic et al. [4] evaluated the influence of aggregate type and size on the inter-connected and isolated pore volume and pore connectivity of pervious concrete mixtures in Avizo Fire 3D image analysis software. Kuang et al. [5] investigated the pore parameters of porous concrete samples such as pore size distribution, effective and total porosity, tortuosity, specific surface area using Bayes decision theory of pattern recognition. Chandrappa and Biligiri [6] investigated the effect of pervious concrete pore parameters such as surface area, porosity, pore radius and tortuosity on unconfined compressive strength and permeability. Although the application of digital image processing in pervious pavement mixtures started long ago, a brief literature review indicated that proper usage of image processing algorithms such as thresholding, watershed segmentation and ungrouping algorithm was not carried out, resulting in improper analysis and lack of experimental validation. The improper use of DIP algorithms can result in the significant under- or over-estimation of internal pavement characteristics, thereby affecting pavement design and maintenance strategies. Past research studies highlighted the significance of threshold segmentation algorithms and binarization of greyscale images on the porosity and permeability characteristics of pervious pavement mixtures [7, 8]. Jagadeesh et al. [9–11] investigated the effect of different thresholding algorithms and gradient functions of watershed segmentation algorithm on porosity, permeability, pore network and skid resistance properties. In addition, the use of watershed segmentation algorithm was introduced to separate interconnected pore network structure into multiple pores. However, isolated pores were not removed in past analyses found in the literature due to a lack of consideration in using ungrouping algorithm to segregate connected and isolated pores. Ong et al. [12] recently investigated the effect of pore network properties on non-Darcy permeability characteristics of pervious concrete samples using discharge thresholding, watershed and ungrouping algorithms. The current study is aimed at investigating the effect of image processing algorithms such as thresholding, watershed and ungrouping on pervious concrete pore network properties using XRCT scanning and DIP techniques.

2 Materials and Experiments

Three pervious concrete cylinders of 250 mm height and 150 mm diameter were cast using the siliceous aggregates obtained from southern Taiwan. The mix design parameters are same for all three samples: coarse aggregates = 1530 kg/m^3, cement = 340 kg/m^3, superplasticizer = 2 kg/m^3 and water-cement ratio of 0.30. The specific gravity and the percent absorption of the aggregates were same for all mixtures and are 2.64 and 1.35% respectively. The aggregate gradations are different for three samples, named P1 (nominal maximum aggregate size of 9.5 mm), P2 (nominal maximum aggregate size of 12.5 mm) and P3 (a combination of both). More details on the aggregate gradations can be found in Jagadeesh et al. [8]. The machine mixing was carried out followed by the compaction using tapping rods in three layers. The hardened samples were then extracted and cured for 28 days in a limewater tank. The XRCT scanning, effective porosity and permeability tests are performed for all samples to allow a comparison between specimen properties to functional properties of pervious concrete and pore network characteristics. The Somatom Emotion 16-channel medical XRCT scanner with an Ultra-Fast Ceramic detector, high-performance X-ray tube of DURA 422 MV was used to obtain the internal structure of the pavement samples at the resolution of 0.326 mm × 0.326 mm × 0.7 mm. The Simpleware ScanIP N-2018.03 software was used in the conversion of voxel absorption coefficients into 12-bit greyscale intensities ($2^{12} = 4096$, varying from 0 to 4095). The image pre-processing was optimized using the convolution kernel algorithm in the XRCT scanner and carried out in default. In addition to this, the effective porosities and permeabilities of the hardened pervious concrete samples was determined from the ASTM C1754 [13] standard (Drying Method B) and the falling head permeameter [14].

3 Evaluation of Pore Network Properties

This section explains in detail the evaluation of XRCT-based pore network properties using various Digital Image processing (DIP) algorithms. Figure 1 illustrates the steps in Digital Image processing (DIP) for the processing of XRCT images for pore network properties. The first step in the evaluation of pore network properties using pre-processed medical XRCT images is the image threshold segmentation algorithm, which is the process of dividing the greyscale images into two phases such as solids

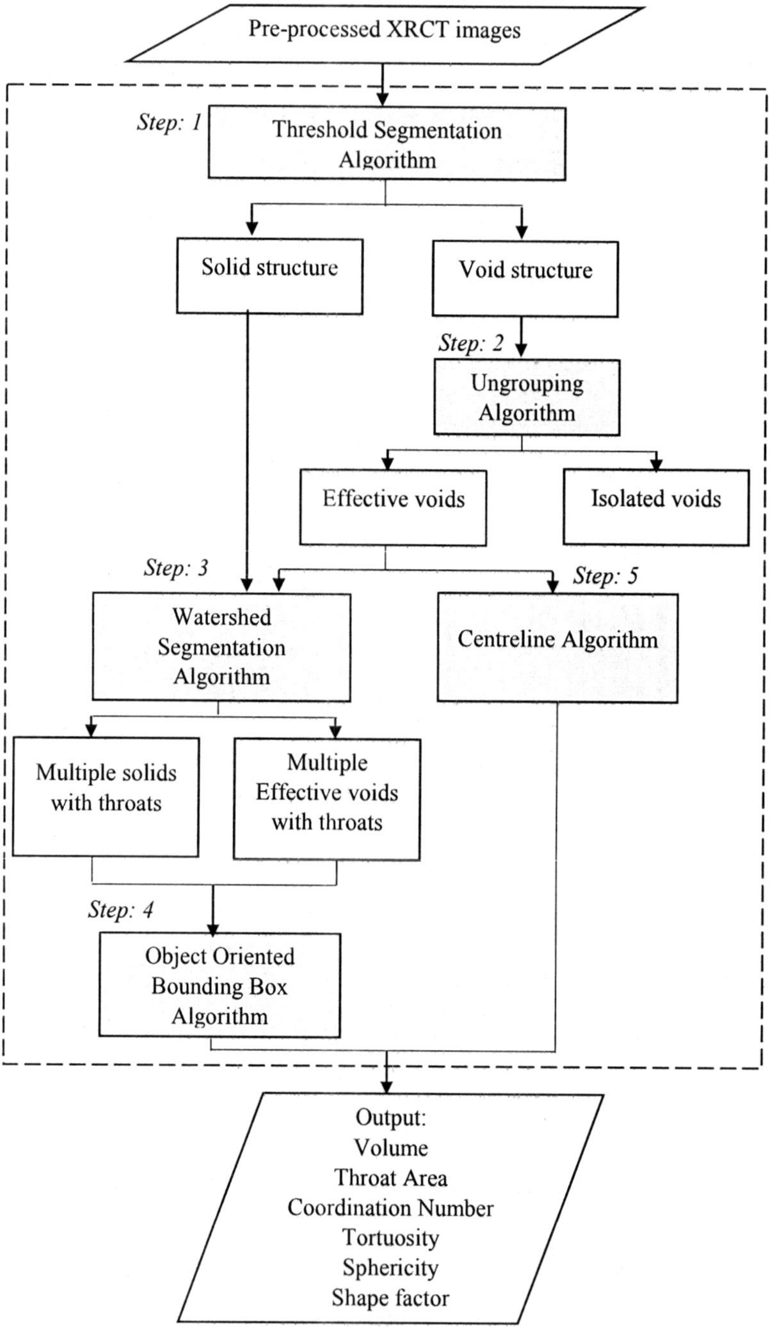

Fig. 1 Flowchart for pore network properties evaluation

and voids. The thresholding algorithms used in the current study includes histogram (intermodes, minimum), clustering (Otsu bilevel, trilevel, isodata), entropy (Kapur, Yen, Sahoo, Li) and laboratory (volumetric, discharge) algorithms [8]. The following sub-sections explain in detail the various other DIP algorithms shown in Fig. 1 such as ungrouping, watershed segmentation, object-oriented bounding box and centreline algorithms.

3.1 *Ungrouping Algorithm*

After the air voids are separated from the solids, we adopt an ungrouping algorithm to separate the effective or interconnected voids from the isolated voids. This algorithm is used in the permeability simulation models to remove the isolated voids since it does not contribute to the fluid flow simulations [15, 16]. The major pore structure connecting from the top to the bottom surface is considered as the effective or interconnected pore. When the pores do not have a connection zone with the interconnected pore, it can then be considered as isolated pores, regardless of their size. The lack of ungrouping algorithm results in the inaccurate measurement of average pore network properties due to the consideration of the isolated pores along with the effective voids as observed in the past studies [6, 17–19]. Figure 2 shows the three-dimensional segmented pore network structures of pervious concrete sample before and after ungrouping algorithm in Simpleware ScanIP N-2018.03.

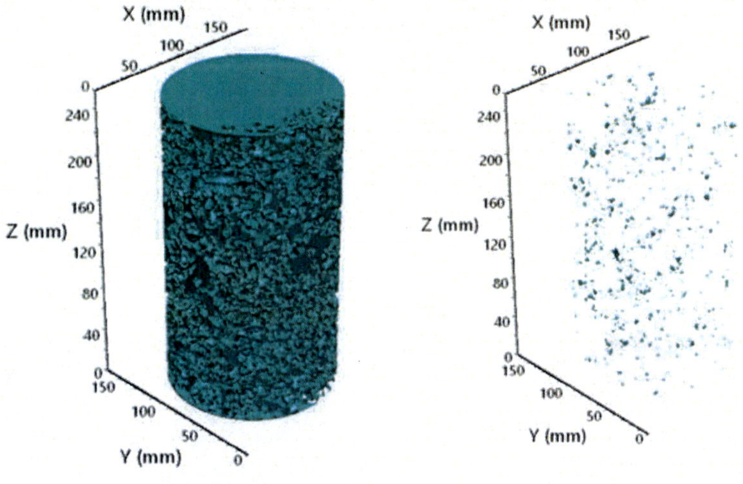

(a) Effective pores (b) Isolated pores

Fig. 2 Pervious concrete pore network structures after ungrouping algorithm

3.2 *Watershed Segmentation Algorithm*

The watershed segmentation algorithm is defined as a particle segmentation algorithm, in which the greyscale pixels of digital images are processed as topographic terrains having ridges and valleys and watersheds are identified as the boundaries separating different particles [20, 21]. The working procedure of the watershed segmentation algorithm is explained for a single XRCT image: (1) The input data includes the pre-processed XRCT image and the threshold segmented image, (2) The image gradient magnitude, defined as the first-order derivative of the greyscale values (using finite difference approach) with respect to the dimensions of the image, is first obtained from the pre-processed XRCT image, (3) The obtained image gradient magnitude is then overlapped to the threshold segmented image and considered as the topographic terrain (or height map of greyscale values) with ridges and valleys, (4) Finally, the valleys are filled with water (using a dilating algorithm) from the local minima point and the point or line of connection between two valleys are defined as the throats or watersheds. The above results are obtained from computations made within the software Synopsis' SimplewareTM ScanIP version N-2018.03.

Although the use of XRCT images to analyze pavement materials has increased over the past two decades, the current image processing techniques do not incorporate the watershed algorithm and this results in significant errors in the measurement of pervious pavement pore volume and shape properties [15, 16, 22, 23]. The watershed segmentation algorithm is necessary to avoid considering the entire volume of interconnected voids in the specimen as a single void structure, which may potentially result in significant errors in pore network properties. In this study, the evaluation of solid network properties was carried out for the combined cement and aggregate phases as both materials occupy similar greyscale values. Figure 3 shows the three-

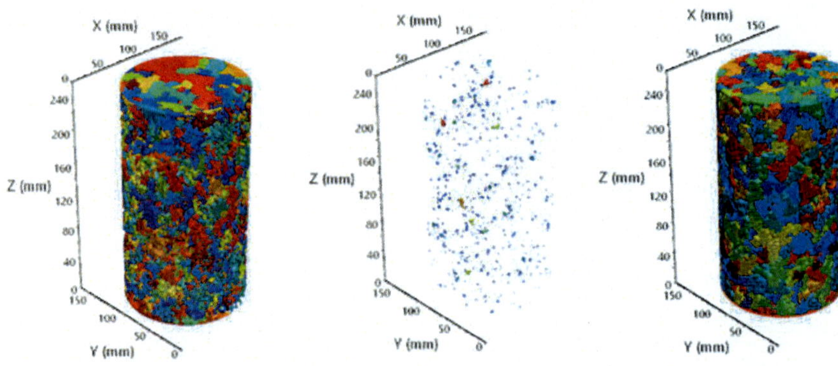

(a) Effective pores (b) Isolated pores (c) Solids

Fig. 3 Pervious concrete pore and solids network structures after watershed segmentation algorithm dimensional segmented pore and solid network structures of the pervious concrete sample after watershed segmentation in Simpleware ScanIP N-2018.03.

3.3 *Object Oriented Bounding Box Algorithm*

In addition to the above algorithms, the object-oriented minimum bounding box (OOBB) algorithm in Simpleware ScanIP N-2018.03 software was used in the current study to determine the shape factor characteristics. In this algorithm, the object (pore or solid) is placed in a virtual box (as shown in Fig. 4a) and rotated in three dimensions until the object occupies the smallest possible volume of the virtual box. Figure 4 illustrates the default axis-aligned and object-oriented bounding boxes. The use of an axis-aligned bounding box instead of an object-oriented bounding box will result in the significant overestimation of the virtual bounding box volume, thereby affecting the shape factor characteristics of the objects. The output of this OOBB algorithm includes the length, width and height of the bounding box, which is used to compute the shape factor for the pores and solids. This algorithm is similar to the digital sieving algorithm [24], in which the object is placed a cuboidal box with a square cross-section and rotated to get the least possible square cross-section.

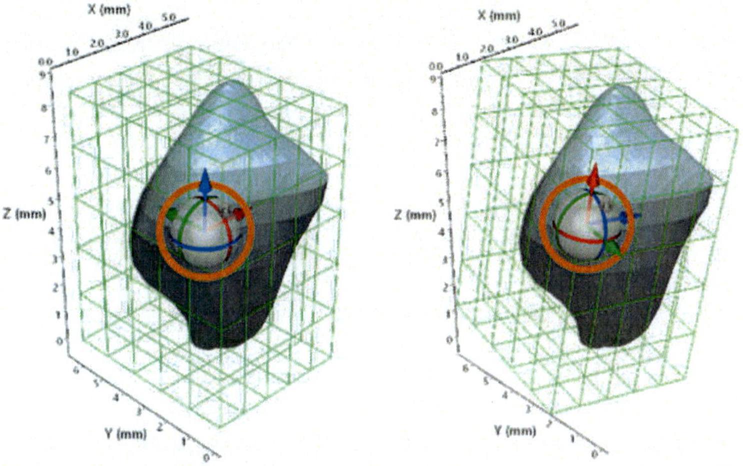

(a) Axis aligned bounding box (b) Object oriented bounding box

Fig. 4 Axis-aligned and object-oriented bounding boxes for shape factor calculation

3.4 Centreline Algorithm

The centreline algorithm in Simpleware ScanIP N-2018.03 software was used in the current study to determine the tortuosity characteristics. The effective pore structure obtained from the ungrouping algorithm is used to evaluate the tortuosity characteristics of the pervious concrete samples in the centreline algorithm. The working procedure of the centreline algorithm is explained as follows: (1) At first, the centreline of the effective pore structure from the top to the bottom of the sample will be created, (2) the calculation of the length for all the possible pore paths between the two ends of the sample is carried out, (3) The shortest distance between the two ends will then be obtained and used for the tortuosity determination (ratio of shortest fluid path to sample length).

4 Results and Discussion

4.1 Effect of Image Processing Algorithms on Pore Network Characteristics

In this section, the effect of various image processing algorithms such as thresholding, ungrouping and watershed algorithms on the determination of average pore volume characteristics are investigated. Table 1 compares the average pore volume properties of pervious concrete samples P1, P2 and P3 using various combinations of image processing algorithms. Table 2 shows the percentage errors of the average pore volume with respect to discharge-based thresholding, ungrouping and watershed segmentation algorithms. The details on the validation of the effective porosity and permeability with experimental results can be found in Jagadeesh [8].

The below observations can be made:

- An increase in the threshold value results in an increase of air void voxels thereby increasing the average pore volume properties. This is observed for all the combinations of the ungrouping and watershed algorithms and is in line with the findings on the effect of thresholding algorithms on porosity and permeability as reported in Jagadeesh et al. [8]. It is evident that several literatures do not mention the threshold algorithm used leading to difficulties in the comparison of results [15, 16, 22, 23, 25, 26].
- The ungrouping and watershed algorithms have a greater effect on the average pore volume properties and their percentage errors compared to the thresholding algorithms. For example, it can be seen for sample P1 in Table 2 that the percentage errors for various combinations of ungrouping and watershed algorithms using discharge thresholding algorithm vary from 0 to 197,895.546% (comparing row results). On the other hand, the error percentages for various

Table 1 Comparison of average pore volume using different DIP algorithms

Sample	Algorithms	Threshold	Average pore volume (mm^3)			
			Lack of watershed and ungrouping algorithm[a]	Use of ungrouping algorithm[b]	Use of watershed algorithm[c]	Use of watershed and ungrouping algorithm[d]
P1	Minimum	1204	568.800	785,500.000	192.068	405.527
	Li	1317	794.000	863,300.000	231.500	443.389
	Intermodes	1381	944.000	907,500.000	255.799	462.796
	Otsu trilevel	1471	1176.000	969,900.000	293.375	489.872
	Isodata	1574	1525.000	1,042,000.000	333.995	521.224
	Otsu bilevel	1583	1573.000	1,048,000.000	339.800	525.724
	Discharge	1632	1739.000	1,083,000.000	360.714	546.982
	Volumetric	1660	1870.000	1,103,000.000	372.776	561.324
	Kapur	1895	3098.000	1,278,000.000	484.254	656.104
	Sahoo	2056	3839.000	1,408,000.000	558.056	729.131
	Yen	2152	4045.000	1,491,000.000	595.310	771.539
P2	Discharge	1539	2736.529	927,200.000	582.970	845.988
	Volumetric	1776	3199.000	1,033,000.000	656.801	929.118
P3	Discharge	1620	1298.000	960,400.000	363.010	590.292
	Volumetric	1517	1151.000	905,700.000	324.620	546.257

[a] Single effective pore + multiple isolated pores [6, 17–19]
[b] Single effective pore [15, 16, 22, 23]
[c] Multiple effective pores + multiple isolated pores [25, 26]
[d] Multiple effective pores (current study)

thresholding algorithms using ungrouping and watershed algorithms varies from −25.861 to 41.054% (comparing column results).

The thresholding algorithm is selected based on the comparison with the laboratory experiments [8], whereas ungrouping and watershed algorithms are selected based on the importance of the effective pores and its understanding with the fluid flow characteristics. Overall, it has to be understood that the proper evaluation of the pore network properties has to be carried out using the discharge based thresholding, ungrouping and watershed segmentation algorithms to investigate its effect on non-Darcy permeability characteristics of pervious concrete specimens.

Table 2 Comparison of average pore volume percentage error using different DIP algorithms

Sample	Algorithms	Threshold	Percent error w.r.t. discharge, watershed and ungrouping algorithms			
			Lack of watershed and ungrouping algorithm[a]	Use of ungrouping algorithm[b]	Use of watershed algorithm[c]	Use of watershed and Ungrouping algorithm[d]
P1	Minimum	1204	3.989	143,506.188	−64.886	−25.861
	Li	1317	45.160	157,729.691	−57.677	−18.939
	Intermodes	1381	72.583	165,810.396	−53.234	−15.391
	Otsu trilevel	1471	114.998	177,218.449	−46.365	−10.441
	Isodata	1574	178.803	190,399.870	−38.939	−4.709
	Otsu bilevel	1583	187.578	191,496.798	−37.877	−3.886
	Discharge	1632	217.926	197,895.546	−34.054	0.000
	Volumetric	1660	241.876	201,551.974	−31.849	2.622
	Kapur	1895	466.381	233,545.714	−11.468	19.950
	Sahoo	2056	601.851	257,312.493	2.025	33.301
	Yen	2152	639.512	272,486.666	8.835	41.054
P2	Discharge	1539	223.471	109,499.663	−31.090	0.000
	Volumetric	1776	278.138	122,005.751	−22.363	9.826
P3	Discharge	1620	119.891	162,599.139	−38.503	0.000
	Volumetric	1517	94.988	153,332.538	−45.007	−7.460

[a] Single effective pore + multiple isolated pores [6, 17–19]
[b] Single effective pore [15, 16, 22, 23]
[c] Multiple effective pores + multiple isolated pores [25, 26]
[d] Multiple effective pores (current study)

4.2 *Effect of Mixture Parameters on Pore Network Characteristics*

Table 3 shows the obtained pore network properties of pervious concrete samples P1, P2 and P3 using the image processing algorithms illustrated in Fig. 1. Following observations can be made: (1) an increase in the nominal maximum aggregate size (P2 > P1) results in the increase of average pore volume, surface area, throat area, sphericity and shape factor and decrease in coordination number and tortuosity. The coordination number is defined as the number of contacts for a particular pore and the throats are defined as the geometric constrictions or the contacts between the voids; (2) The sample P3 exhibits the intermediate characteristics of P1 and P2 except for coordination number, shape factor and tortuosity. The major limitation in the calculation of the mean tortuosity is that the effect of the cross-sectional area of the

Table 3 Comparison of average pore properties for different pervious concrete samples

Average properties	Samples		
	P1	P2	P3
Volume (mm^3)	546.982	845.988	590.292
Surface area (mm^2)	1021.952	1227.445	984.176
Throat area (mm)	7.767	12.306	9.141
Coordination number	12.086	9.026	8.281
Sphericity	0.316	0.352	0.346
Shape factor	0.978	0.982	0.957
Tortuosity	1.373	1.363	1.138

interconnected pore paths has not been considered in the current study and will be used in the future using the weighted tortuosity parameter.

5 Conclusions

This study attempts to investigate the effect of digital image processing techniques on pore network properties of pervious concrete samples. Three different pervious concrete samples were produced in the laboratory and its internal pore network structure was obtained using medical XRCT and digital image processing. A detailed explanation of the advanced image processing algorithms such as watershed and ungrouping algorithms was carried out to evaluate the internal pore and solid network characteristics. The importance of the image processing algorithms on the pore network properties are investigated and the appropriate set of DIP algorithms were selected for further analysis to study the effect of pore network structure on computational fluid dynamics permeability simulation results. The developed image analysis methodology for the pervious concrete mixtures can be used to analyze the mechanisms of the fluid flow characteristics. Overall, the findings presented in this study can help in understanding the importance of pavement image processing techniques and their effects on pavement mixture quality control.

References

1. ACI Committee: ACI 522R-10. Repot on Pervious Concrete (2016)
2. Kayhanian M, Anderson D, Harvey JT, Jones D, Muhunthan B (2012) Permeability measurement and scan imaging to assess clogging of pervious concrete pavements in parking lots. J Environ Manage 95(1):114–123
3. Manahiloh KN, Muhunthan B, Kayhanian M, Gebremariam SY (2012) X-ray computed tomography and nondestructive evaluation of clogging in porous concrete field samples. J Mater Civ Eng 24(8):1103–1109

4. Ćosić K, Korat L, Ducman V, Netinger I (2015) Influence of aggregate type and size on properties of pervious concrete. Constr Build Mater 78:69–76
5. Kuang X, Ying G, Ranieri V, Sansalone J (2015) Examination of pervious pavement pore parameters with X-ray tomography. J Environ Eng 141(10):04015021
6. Chandrappa AK, Biligiri KP (2017) Relationships between structural, functional, and X-ray microcomputed tomography parameters of pervious concrete for pavement applications. Transp Res Rec 2629(1):51–62
7. Abera KA, Manahiloh KN, Nejad MM (2017) The effectiveness of global thresholding techniques in segmenting two-phase porous media. Constr Build Mater 142:256–267
8. Jagadeesh A, Ong GP, Su YM (2019) development of discharge-based thresholding algorithm for pervious concrete pavement mixtures. J Mater Civ Eng 31(9):04019179
9. Jagadeesh A, Ong GP, Su YM (2019) Evaluation of pervious concrete pore network properties using watershed segmentation approach. In: International airfield and highway pavements conference, Chicago, USA, July 21–24
10. Jagadeesh A, Ong GP, Su YM (2020) Effect of global thresholding algorithms on pervious concrete pore network properties using XRCT-based digital image processing. In: 9th international conference on maintenance and rehabilitation of pavements (MAIREPAV), Zurich, Switzerland, July 1–3
11. Jagadeesh A, Ong GP, Su YM (2020) Effect of thresholding algorithms on pervious pavement skid resistance. In: Advances in materials and pavement performance prediction (AM3P), San Antonio, USA, August 3–7
12. Ong GP, Jagadeesh A, Su YM (2020) Effect of pore network characteristics on non-Darcy permeability of pervious concrete mixture. Constr Build Mater 259:119859
13. ASTM: Standard test method for density and void content of hardened pervious concrete. ASTM C1754-12/C1754M. ASTM, West Conshohocken, PA (2012)
14. Fwa TF, Lim E, Tan KH (2015) Comparison of permeability and clogging characteristics of porous asphalt and pervious concrete pavement materials. Transp Res Rec 2511(1):72–80
15. Kutay ME, Aydilek AH, Masad E, Harman T (2007) Computational and experimental evaluation of hydraulic conductivity anisotropy in hot-mix asphalt. Int J Pavement Eng 8(1):29–43
16. Masad E, Al Omari A, Chen HC (2007) Computations of permeability tensor coefficients and anisotropy of asphalt concrete based on microstructure simulation of fluid flow. Comput Mater Sci 40(4):449–459
17. Gruber I, Zinovik I, Holzer L, Flisch A, Poulikakos LD (2012) A computational study of the effect of structural anisotropy of porous asphalt on hydraulic conductivity. Constr Build Mater 36:66–77
18. Chandrappa AK, Biligiri KP (2018) Pore structure characterization of pervious concrete using X-ray microcomputed tomography. J Mater Civ Eng 30(6):04018108
19. Chen S, You Z, Yang SL, Garcia A, Rose L (2021) Influence of air void structures on the coefficient of permeability of asphalt mixtures. Powder Technol 377:1–9
20. Kutay ME, Arambula E, Gibson N, Youtcheff J (2010) Three-dimensional image processing methods to identify and characterise aggregates in compacted asphalt mixtures. Int J Pavement Eng 11(6):511–528
21. Zelelew HM, Papagiannakis AT (2011) A volumetrics thresholding algorithm for processing asphalt concrete X-ray CT images. Int J Pavement Eng 12(6):543–551
22. Chen J, Yin X, Wang H, Ma X, Ding Y, Liao G (2018) Directional distribution of three-dimensional connected voids in porous asphalt mixture and flow simulation of permeability anisotropy. Int J Pavement Eng 1–13
23. Fedele V, Berloco N, Colonna P, Hertrich A, Intini P, Ranieri V, Sansalone JJ (2020) Computational fluid dynamics as a tool to estimate hydraulic conductivity of permeable asphalts. Transp Res Rec 2674(8):370–383
24. Hu C, Ma J, Kutay ME (2017) Three dimensional digital sieving of asphalt mixture based on X-ray computed tomography. Appl Sci 7(7):734

25. Zhang J, Ma G, Ming R, Cui X, Li L, Xu H (2018) Numerical study on seepage flow in pervious concrete based on 3D CT imaging. Constr Build Mater 161:468–478
26. Yu F, Sun D, Hu M, Wang J (2019) Study on the pores characteristics and permeability simulation of pervious concrete based on 2D/3D CT images. Constr Build Mater 200:687–702

Pavement Material Characterization and Modelling

A Comparative Investigation on the Effectiveness of a Wax and a Resin Based Curing Compound as an Alternate of Water Curing for Concrete Pavement Slab

Pankaj Goel and Rakesh Kumar

Abstract In this study, the effectiveness of two easily available concrete curing compounds in the Indian construction Industry namely; wax and synthetic resin-based, in comparison with water cured cubes and beams specimens of a paving concrete of 39 MPa compressive strength and 4.5 MPa flexural strength at 28-day has been reported. The concrete specimens were cast and kept for curing under different conditions such as in water, in the sunlight, and under the sunshade. Two coats of curing compounds as per the manufacturer's instruction were applied on the surfaces of the concrete specimens and put in the sunlight and under the sunshade. The compressive and the flexural strength in comparison with water cured concrete specimens were evaluated. The study has shown 12% lower compressive strength and about 7% lower flexural strength for concrete samples cured with resin-based curing compound than the water cured samples. The study also indicates a reduction in compressive and flexural strengths of about 33% and 11%, respectively for wax-based curing compound. Finally, the study shows an upper hand for the resin-based curing compound, however, neither of them has shown potential as an alternate for the conventional water curing.

Keywords Curing compound · Resin · Concrete · Compressive and flexural strength

1 Introduction

Concrete is a heterogeneous material, composed of cement, aggregates, water and air voids. The strength and the durability of a hardened concrete largely depend its matrix microstructure i.e., porosity, pore size distributions and its connectivity. The development of microstructure of concrete is mostly governed by the curing process. With an increase in voids, the strength and durability of concrete decreases. In order to reduce the voids in concrete two approaches are used, one is to use the water to cement ratio as low as possible. The second approach is to reduce the voids by

P. Goel (✉) · R. Kumar
Rigid Pavement Division, CSIR-Central Road Research Institute, New Delhi 110025, India

© The Author(s), under exclusive license to Springer Nature Switzerland AG 2022

H. R. Pasindu et al. (eds.), *Road and Airfield Pavement Technology*, Lecture Notes in Civil Engineering 193, https://doi.org/10.1007/978-3-030-87379-0_43

ensuring the hydration of cement as much as possible. The voids in concrete other than entrained and entrapped air, are created as the volume initially occupied by the water is vacated due to its consumption during the hydration process. The hydration process is a strength gaining process in which the voids especially capillary pores initially occupied by water are filled with the hydration products. The hydration of cement requires certain level of relative humidity (more than 90%) and temperature for a particular period of time (preferably 7 days or more). Majority of the cement hydration completes within first 7 days of concrete casting so during this period continuous water supply should be ensured as these hydration products primarily governs the quality of concrete. Curing of concrete maintains its temperature and moisture content, and ensures a successful development of concrete with desired properties. The strength and the durability of a hardened concrete largely depend on its matrix microstructure i.e., porosity, pore size distributions and its connectivity. The development of the microstructure of concrete is mostly governed by the curing process. With an increase in voids, the strength, and durability of concrete decrease. In order to reduce the voids in concrete two approaches are used, one is to use the water to cement ratio as low as possible. The second approach is to reduce the voids by ensuring the hydration of cement as much as possible. The voids in concrete other than entrained and entrapped air, are created as the volume initially occupied by the water is vacated due to its consumption during the hydration process. The hydration process is a strength gaining process in which the voids especially capillary pores initially occupied by water are filled with the hydration products. The hydration of cement requires a certain level of relative humidity (more than 90%) and a favorable temperature for a particular period of time (preferably 7 days or more). The majority of the cement hydration completes within the first 7 days of concrete casting so during this period continuous water supply should be ensured as these hydration products primarily govern the quality of concrete. Curing of concrete maintains its temperature and moisture content, ensures the successful development of desired properties of concrete because once curing stops strength the gaining process also stops [1, 2].

In the construction of a concrete pavement, a large surface area is exposed to the environment causing a rapid loss of water from the concrete surface in the plastic state resulting in the development of plastic shrinkage crack. Plastic shrinkage crack does occur when the rate of water loss due to evaporation from the concrete surface exceeds the rate at which bleed water is available. Improper curing of concrete causes insufficient hydration of the cementitious material and results in a reduction in mechanical properties as well as durability of concrete for its design life. On the other hand, proper curing not only controls the plastic shrinkage cracking of concrete but also makes the concrete more impermeable with enhanced abrasion-resistant. The effectiveness of curing depends on environmental conditions at the site (wind speed, temperature, relative humidity), materials used, method of construction, and availability of curing methods. Various curing methods are used such as ponding (conventional method), fogging, wet coverings, impervious paper, plastic sheets, curing compounds, etc. Among all the available curing methods, ponding is considered ideal for retaining heat and moisture. Nevertheless, the uses of curing compounds are also gaining popularity to control plastic shrinkage cracking particularly in a hot climatic conditions. The curing compounds usually consist of a synthetic resin, wax, chlorinated rubbers dissolved in a solvent or emulsified in water. They are sprayed to the pavement surface immediately after disappearance of the surface water, by evaporation of the solvent, an impermeable membrane forms that retards the loss of water from the concrete surface. Their effectiveness varies quite widely depending on the materials and strength of the emulsion [3]. Numerous researchers [4–10] have investigated the properties of concrete by using different curing compounds, and the factors affecting their performance. Curing compounds can efficiently maintain the moisture content and increase the degree of cement hydration [4]. Al-Gahtani [5] studied the effect of curing methods on the properties of plain and blended cement concretes and had reported that the strength development in concrete specimens covering by wet burlap was more than that in specimens cured by applying water or acrylic-based curing compounds. He further reported that concrete specimens cured by spraying curing compounds exhibited higher efficiency in decreasing plastic and drying shrinkage of a concrete strain than the specimens cured by covering with wet burlap only [5]. Wang et al. [6] reported that the effectiveness of curing compound mainly depends on the time of its application. They further concluded that among the curing compound studied chlorinated rubber performed the best, followed by solvent-based type, the least effective was the water-based type. Whitting and Synder [7] examined the effectiveness of different types of curing compounds (six) in retaining the water for hydration. They concluded that the application of curing compounds improved the concrete strength and reduced the permeability, relative to plastic sheeting and ponding. They further concluded that the performance of six curing compounds used in the study varied greatly, but none of them performed at par with the specimens cured with water or plastic sheeting. However, all the curing compounds performed better than the uncured specimens. Patil et al. [8], studied the effect of two different wax-based curing compounds on the strength and durability of concrete mixes of M20, M30, M40 and a self-compacting concrete. The compressive strength efficiency varies between 78–97% at 7 days and 80–95% at 28 days under different conditions in comparison with water-cured specimens [8, 9]. The results showed that both the curing compounds are less efficient in self-compacting concrete [8]. Princy and John [10], investigated the effectiveness of various curing compounds (acrylic resin-based and wax-based) on the properties of concrete. They achieved 96%–99% efficiency in compressive strength, 75–80% in flexural strength, and 84% in splitting tensile strength in comparison with the conventional curing.

enlargethispage-24ptIndia is a country that has the second largest road network in the world. Due to the overall economy of concrete pavement over flexible pavement, paradigms have been shifted towards the concrete pavement, consequently, most of the expressway and national highways are being constructed using concrete. This has opened a huge market for curing compound manufacturers. Several manufacturers of curing compounds claim a significant improvement in concrete properties without sufficient research data. Hence, in order to educate and disseminate the performance of curing compounds used in cement concrete road construction an extensive experimental study was conducted to evaluate the effectiveness of these curing compounds with reference to water curing, the salient findings have been reported in this paper.

2 Experimental Study

Ordinary Portland cement conforming to requirements of IS 8112 [11] was used in the study. A well-graded crushed quartzite coarse aggregate of nominal maximum size 20 mm was used and tested according to Indian standard IS-2386 [12]. The physical properties of aggregate such as specific gravity, bulk density and water absorption were 2.7, 1604 kg/m^3 and 0.46%, respectively. Mechanical properties of aggregate such as Impact, Crushing, and Abrasion loss (by Los-Angeles) values were 18%, 20% and 32%, respectively. Locally available land quarried sand was used as fine aggregate. Its specific gravity and water absorption was 2.71 and 0.42%, respectively. Potable water available in the CSIR-CRRI laboratory was used for mixing and curing of concrete specimens. A high range water-reducing admixture (HRWRA) Naphthalene based was used as a superplasticizer to achieve the desired workability. Two types of curing compounds: resin-based and wax-based was used in the study. The technical specifications of the curing compound (given by the manufacturer) are shown in Table 1.

2.1 Mix Proportions

The mix proportions used in this study are given in Table 2.

Table 1 Technical specification of curing compound

Details	Wax-based	Resin-based
Sp. gravity	1.02	1.10
Form	Wax-based liquid	Water-based liquid
Color	White	Pink
Coverage	4–6 m^2/l	10–15 m^2/kg
Drying time	Approx. 120 min. at 30 °C	Approx. 40–50 min./coat
Toxicity	Non-toxic	Non-toxic, Non-VOC

Table 2 Concrete mix proportions

Ingredients	Resin based mix proportion (kg/m^3)	Wax based, mix proportion (kg/m^3)
Cement	400	415
Fine aggregate	616	614
Coarse aggregate, 10 mm	596	494
Coarse aggregate, 20 mm	596	743
Water	160	166
W/C	0.38	0.40
HRWRA	–	0.38% of cement

2.2 Sample Preparation

150 mm cube, and 100 × 100 × 500 mm beam specimens were cast for the evaluation of compressive and flexural strengths of concrete respectively. The specimens were demolded after 24 h of casting. The compressive strength of concrete was determined as per Indian Standard IS-516 [13] at different ages 3, 7 and 28 days, while flexural strength of concrete was determined at age of 28 days as per Indian standards IS-516 [13]. Two concrete mixes shown in Table 2 were used in his study. From each mix 27 cubes of 150 mm and 9 beams of 100 × 100 × 500 mm were prepared. One set of 9 cubes and 3 beams were submerged in water for curing while the rest of the two sets i.e. 18 cubes and 6 beams were coated with curing compounds. After coating, 9 cubes and 3 beams were kept in sun and 9 cubes and 3 beams were kept in sunshade. Therefore, specimens from both mixes were water cured and cured with curing compounds.

2.3 Method of Spraying Curing Compound

The first coat of curing compound was sprayed over the top surface of the specimen just after a few hours of casting when surface water disappeared. The concrete specimens were demolded after 24 h of casting and curing compound was sprayed at the remaining five surfaces. The curing compound was sprayed using a low-pressure spray machine following the rate of application as prescribed by the manufacturer (5 m^2 per liter for wax-based and 13 m^2 per liter for resin-based) as shown in Fig. 1. The second and final coat of curing compound was again sprayed with the same application rate to all the surfaces just after 30–45 min after first application.

Fig. 1 Spraying of curing compound in progress

3 Results and Discussion

3.1 Compressive Strength

The average of three cubes crushing strength was recorded as the compressive strength of the mix. The compressive strength development with age, for samples cured with resin-based, wax-based curing compound exposed in different conditions along with water-cured samples are shown in Figs. 2 and 3, respectively.

It can be seen from Figs. 2 and 3 that the development in the compressive strength of concrete cured with ponding is higher up to 5–33% than the concrete specimens cured with curing compound and kept under sunshade and sunlight. It can be seen from Fig. 2 that, the efficiency of resin-based curing compound varies between 81 and 97% at different test ages under different exposure conditions. It can also be noticed that for wax-based curing compound the efficiency of compressive strength development is very low i.e., 67–82% in comparison with the specimens cured with water ponding. Since this reduction in efficiency of compressive strength of concrete was

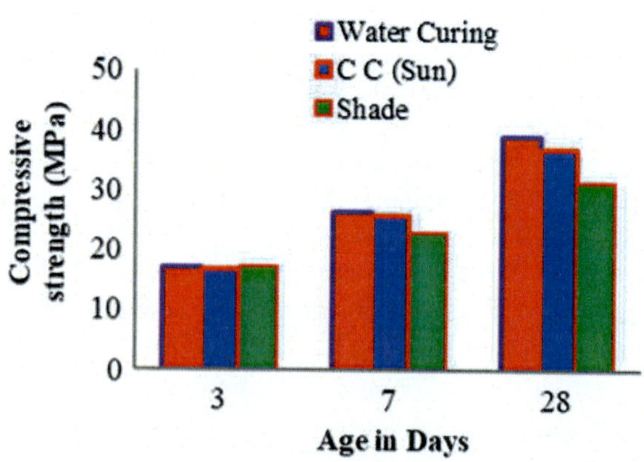

Fig. 2 Development in compressive strength with age (resin-based curing compound)

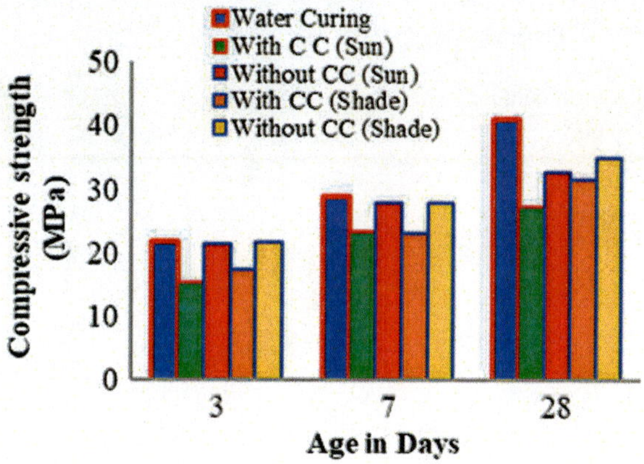

Fig. 3 Development in compressive strength with age (wax-based curing compound)

significant, to see the effect of wax-based curing compound mix was repeated with two additional sets of concrete specimens. One set of specimens was kept in sunlight and in sunshade without any curing while the other set was applied with a curing compound. Shocking results were noticed (also shown in Fig. 3); the compressive strength of concrete specimens kept without application of curing compound under the same exposure conditions (sunlight and in sunshade) were significantly higher (16–27%) than the concrete specimens applied with wax-based curing compound and kept in the same exposure conditions. It worth noting that the compressive strength of concrete specimens cured with resin-based curing compound was quite higher than the concrete specimens cured with wax-based curing compound under different exposure conditions. The reason for such behavior of the curing compounds may be attributed to the water retention efficiency of these curing compounds. The water retention efficiency of a resin-based curing compound is generally more than 90% while the same for a wax-based curing compound is about 60–70%. It was further observed that among all the cases, the compressive strength of concrete specimens cured with curing compound and kept under sunshade was slightly higher than those concrete samples kept under sunlight.

3.2 *Flexural Strength*

The flexural strength of the concrete beam specimens was determined as per IS-516. The third point loading system was used to determine the flexural strength of concrete. An average of three values was recorded as the flexural strength at 28 days and test results under the different curing conditions with resin-based and wax-based curing compounds are presented in Figs. 4 and 5, respectively.

It can be seen from Figs. 4 and 5 that flexural strength of concrete cured with ponding is higher (4.5–13%) than the concrete specimens cured with the curing compound and kept under sunlight and sunshade exposure conditions. The flexural

Fig. 4 Flexural strength of concrete under different curing conditions (with resin based)

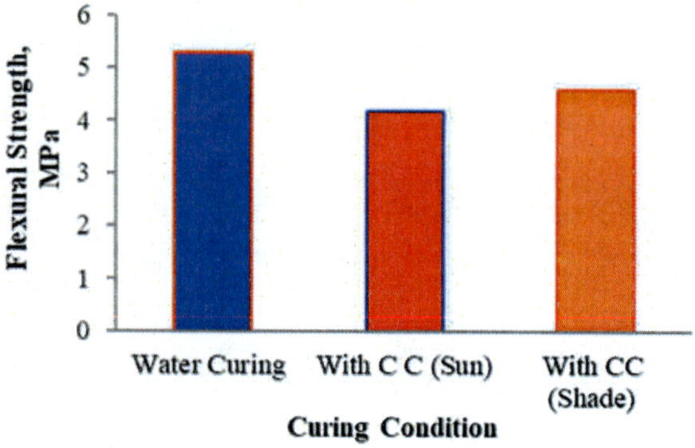

Fig. 5 Flexural strength of concrete under different curing conditions (with wax based)

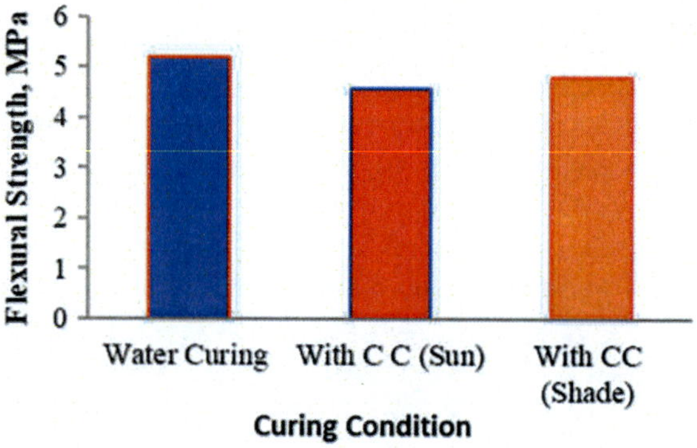

strength of concrete specimens cured with curing compound kept under the shade exposure condition is slightly higher than the concrete specimen cured with curing compound but kept in sunlight.

4 Conclusions/Recommendations

From the study following important conclusions may be drawn:

1. The compressive strength of concrete specimens cured with wax-based curing compound kept in sunlight and the sunshade is lesser than water cured concrete specimens at all the test ages. It is further observed that the compressive strength of concrete specimens not cured with curing compound but kept in the sunlight and sunshade are even higher than concrete specimens cured with curing compound.
2. The compressive strength of concrete specimens cured with the resin-based curing compound kept in sunlight and sunshade is 5–33% lesser than water cured samples at all ages.

3. From the compressive strength point of view resin-based curing compound performs much better than the wax-based curing compound.
4. The flexural strength of concrete specimens cured with curing compound and kept under the different exposure conditions is almost 4.5–13% lesser than water cured specimens.
5. Neither the wax-based nor the resin-based curing compound could be used as an alternative to the water curing of paving concrete.

Recommendation for Further Research

Further research on the use of curing compound related to the pavement concrete may be focused on shrinkage behavior in different environments, abrasion resistance of concrete besides microstructure study on cover zone concrete.

Acknowledgements The kind permission of the director CRRI, Professor Satish Chandra, to publish the work is highly acknowledged.

References

1. Neville AM (1997) Properties of concrete, 4th edn. Pitman Publishing Limited, London
2. Shetty MS, Concrete technology: theory and practice, 23rd revised edn. S Chand and Company, New Delhi, India
3. http://www.ccaa.com.au/imis_prod/documents/Library%20Documents/CCAA%20Datasheets/DS2006Curing.pdf
4. Ibrahim M, Shameem M, Al-Mehthel M, Maslehuddin M (2013) Effect of curing methods on strength and durability of concrete under the hot weather conditions. Cem Concr Compos 41:60–69
5. Al-Gahtani AS (2010) Effect of curing methods on properties of plain and blended cement concretes. Constr Build Mater 24(3):308–314
6. Wang J, Dhir RK, Levitt M (1994) Membrane curing of concrete: moisture loss. Cem Concr Res 24(8):1463–1474
7. Whitting MN, Synder MB (2003) Effectiveness of Portland cement concrete curing compounds. Transp Res Rec J TRB 1834:59–68
8. Patil S, Vaidya R, Math V (2016) Effect of curing compounds on strength and durability of concrete mixes. Int Res J Eng Technol (IRJET) 03(6):2874–2879
9. Gawatre WD, Sawant K, Mule R, Waydande N, Randeve D, Shirsath T (2017) Effectiveness of curing compound on concrete. IOSR J Mech Civ Eng 14(3):73–76. e-ISSN:2278-1684, p-ISSN: 2320-334X
10. Princy KP, John E (2015) Study on the effectiveness various curing methods on the properties of concrete. Int J Eng Res Technol (IJERT) 4(11):213–216, ISSN:2278-0181
11. IS: 8112-1989, Specification for 43 grade Ordinary Portland Cement. Bureau of Indian Standards, New Delhi
12. IS: 2386 (Part-3&4)-1963, Methods of test for aggregate for Concrete, determination of specific gravity, density, voids, water absorption and bulking & Mechanical properties. Bureau of Indian Standards, New Delhi
13. IS: 516-2000, Methods of tests for strength of concrete. Bureau of Indian Standards, New Delhi

An Assessment of the Polishing Behaviour of Road Surfacing Aggregates in Sri Lanka

H. V. Thenuwara, W. W. C. Jayasuriya, and H. R. Pasindu

Abstract Friction capacity between the tire-pavement interface, facilitates sufficient braking force ensuring road user safety. Inadequate skid resistance has been identified as the prominent pavement-related cause of highway crashes. Availability of friction at the wearing course is governed by pavement, vehicle and tire/driver condition parameters combined with environmental parameters, of which pavement-related characteristics concern a greater proportion. Crucial pavement-related characteristics include mineralogical and textural properties of road surfacing aggregates, where the polishing behaviour of coarse aggregates exhibits significant influence in the establishment of skid resistance of asphalt concrete pavings. Thus, many research have been conducted to determine the suitability of aggregates, in terms of resistance to polishing under trafficking. In the national context, broader attention is being developed towards road safety as one of the critical performance parameters due to the integration of an expressway system with the prevailing road network. However, the amount of such extensive study into pavement friction and properties of locally available aggregates, is scarce. No guidelines considering user safety have been specified in the Standard Specifications for Construction and Maintenance of Roads and Bridges by the Institute for Construction Training and Development, Sri Lanka. Therefore, initiatives to incorporate skid resistance in the national practice of aggregate selection for pavement construction, are indispensable. Hence, this study aims to establish the friction characteristics of Sri Lankan natural roadstones, through the Polish Stone Value Test involving the Accelerated Polishing Machine and the British Pendulum Tester, and to evaluate the observed polishing patterns through a petrographic analysis and microscopic investigation.

Keywords Pavement friction · Coarse aggregates · Polishing characteristics · Mineralogy · Aggregate properties

H. V. Thenuwara (✉) · W. W. C. Jayasuriya · H. R. Pasindu
Department of Civil Engineering, Faculty of Engineering, University of Moratuwa, Moratuwa, Sri Lanka
e-mail: thenuwarahv.20@uom.lk

H. R. Pasindu
e-mail: pasindu@uom.lk

© The Author(s), under exclusive license to Springer Nature Switzerland AG 2022

H. R. Pasindu et al. (eds.), *Road and Airfield Pavement Technology*, Lecture Notes in Civil Engineering 193, https://doi.org/10.1007/978-3-030-87379-0_44

1 Introduction

Pavement friction by definition, is the retarding force between the tire-pavement interface, that resists the relative motion of a vehicle tire in rotation [1]. Availability of satisfactory friction co-efficient at the pavement surface, ensures sufficient braking force on the vehicle against skidding during maneuvers. Inadequate friction capacity at the wearing course, has been recognized to be one of the prime causative factors of road crashes, apart from the vehicle, driver, other roadway infrastructure and environment related factors [2, 3]. Friction related pavement characteristics cause for about 28% of the total road crashes and 14% of all fatal crashes [4–6]. Hence, provision and maintenance of adequate road friction is indispensable to prevent roadway departure crashes, skidding, run-off-road and head-on collisions to facilitate safer riding [7–10].

The extent of friction co-efficient present at the tire-pavement interface is governed by many contributing constituents such as pavement, vehicle/tire condition properties and water film thickness, where friction related pavement characteristics concern a significant proportion [1]. Coarse aggregates comprise about 90% of the Hot Mix Asphalt (HMA) surfacings by weight, thus the crucial fiction-related pavement characteristics include mineralogical and textural properties of road surfacing aggregates [11, 12]. The induced polishing action of the exposed course aggregates of the wearing course over time due to repeated traffic application apart from adverse weather conditions and many other site-specific reasons, leads to friction deterioration of the pavement, resulting in the increase of highway crashes, specially under wet surface conditions [1, 3]. Therefore, the polishing resistance of coarse aggregates exhibits significant influence in the establishment of skid resistance of asphalt concrete pavings [13].

Pavement friction is a complex integration of two major frictional force components: adhesion resulted by the microtexture of individual aggregate particles, and hysteresis influenced by the macrotexture of the pavement surface. Although the hysteresis component is pre-dominant at greater vehicle speeds and wet weather conditions, adhesion contributes to the magnitude of the tire friction and skid resistance at all speeds for dry and wet surface conditions [1, 3, 14]. Microscopic asperities and textural properties of an individual aggregate define its polishing behavior and decay patterns when subjected to the traffic applications. In conclusion, polishing resistance of road surfacing aggregates is a key determining factor of the friction deterioration trends and long-term skid resistance of asphalt concrete pavements [13]. Hence, in addition to the physical and mechanical properties, polishing characteristics of road stones have to be evaluated in the aggregate selection process to ensure adequate and durable friction capacity at the pavement [15].

There are several laboratory methods available to assess the polishing characteristics of coarse aggregates, among which the Polished Stone Value (PSV) test is the most commonly used by many countries [14]. The polished stone value (PSV) test measures the resistance of roadstone to the polishing action of vehicle tyres under conditions similar to those occurring on the surface of a road, by means of accelerated polishing under controlled laboratory conditions [13]. The higher the PSV is, the more skid resistant is the aggregate. Many countries including the USA, the UK and China have defined specifications and provisions on the PSV of coarse aggregate, which is a key index in the aggregate selection process for the anti-sliding abrasion layer of an asphalt pavement [15–17]. As a step towards the incorporation of such user safety induced parameters in the design and construction stages of the pavement, countries such as Afghanistan, India, Bangladesh, Pakisthan, Thailand, Iran, Iraq and many other have established the frictional properties of their locally available road surfacing aggregates [12, 18–22].

Nonetheless, the amount of such extensive study into polishing characteristics of locally available aggregates, is scarce, whereas no relatable guidelines have been specified even in the latest issue of Standard Specifications for Construction and Maintenance of Roads and Bridges by the Institute for Construction Training and Development, Sri Lanka [23]. At present, around 12,380 km of national road network comprises of class "A" roads (trunk roads), class "B" roads (main roads) and class "E" roads (expressways) with maximum operating speeds ranging from70 km/h to 100 km/h [24–26]. An average of 38,000 crashes occur annually which result in around 3000 fatalities and 8000 serious injuries, of which about 600 crashes are recorded in the expressway system [27–29]. Ensuring satisfactory pavement friction is thus very important as speeding combined with inadequate braking force leads to a considerable amount of highway crashes. Therefore, counter-measures should be taken, from aggregate selection upto construction and periodic maintenance of especially major trunk roads, expressway network including the interchange ramps, critical horizontal curves and steep grades, intersections and other identified skid hazardous roadway sections [30].

However, in the national context, broader attention is being developed towards road user safety as one of the critical performance parameters due to the growing concern over the escalating number of speeding crashes. Therefore, initiatives to incorporate skid resistance in the national practice of aggregate selection for pavement construction, and to enable advanced planning of maintenance interventions in the Pavement Friction Management (PFM) of skid hazardous road sections, are demanding.

2 Research Objectives

As an initiative chapter to establish the polishing resistance of locally aviable road stones, encourage further experimental studies related to the topic, as well as to influence the incorporation of skid resistance of aggregates and user safety in the national practice of design, construction and maintenance of major trunk road network, this experimental study expects to,

1. Establish the friction characteristics of Sri Lankan natural roadstones, through the Polish Stone Value Test
2. Evaluate the observed polishing patterns through a petrographic analysis and microscopic investigation as PSV is meaningfully related to mineralogical properties of coarse aggregates.

3 Experimental Study

3.1 Specimen Preparation

The Industrial Mining License (IML) grade quarries of road surfacing aggregates located in Kaduwela (Colombo District), Anuradhapura (Anuradhapura District), Horana (Kalutara District), Hambantota (Hambantota District) and Kurunegala (Kurunegala District) account for the major contribution of the natural aggregate production for the road construction industry. Therefore, samples of aggregates were acquired from 10 major natural rock quarry and suppliers from the above mentioned locations, representing 5 rock origins, where aggregates were collected from 2 different major suppliers of the same rock source for each source.

In order to determine the polishing behaviour of the aggregates, 28 number (no.) of test specimens were prepared for 2 wheel runs in accordance with the ASTM-D3319-06, where 14 no. of specimens were subjected to polishing per each run (4 no. of specimens from each source of rock (2 no. of specimen from each supplier) + 4 no. of control stone specimen). The detailing of the material acquisition and specimen preparation is presented in Table 1.

3.2 Polish Stone Value Test

Polish Stone Value test was performed with the use of the Accelerated Polishing Machine and the British Pendulum Tester, in accordance with the ASTM-D3319-06 and ASTM-E303-93. The BPT was initially calibrated using the polished criggion stone specimens. The specimens of local aggregates were first conditioned for 2 h of accelerated polishing, prior to the test. The friction deterioration curves of the aggregates are shown in Fig. 1.

Before commencement of the process of polishing, Residual Polish Value (RPV) of the selected specimens was obtained. Polish Value (PV n) or British Pendulum Number (BPN) or Skid Resistance Number (SRV) of specimens after "n" number of hours of rotation under wet condition was taken using the BPT. 4 consecutive readings of BPN with no deviation greater than 3 points were recorded for each reading for the polishing intervals of 1.5 h during the 9 h of polishing.

Table 1 The location details of the material acquisition

Ganewalpola, Anuradhapura (Anuradhapura District)	Horana (Kalutara District)

(continued)

Table 1 (continued)

Kaduwela (Colombo District)	Hambantota (Hambantota District)

Kurunegala (Kurunegala District)	Control Stone

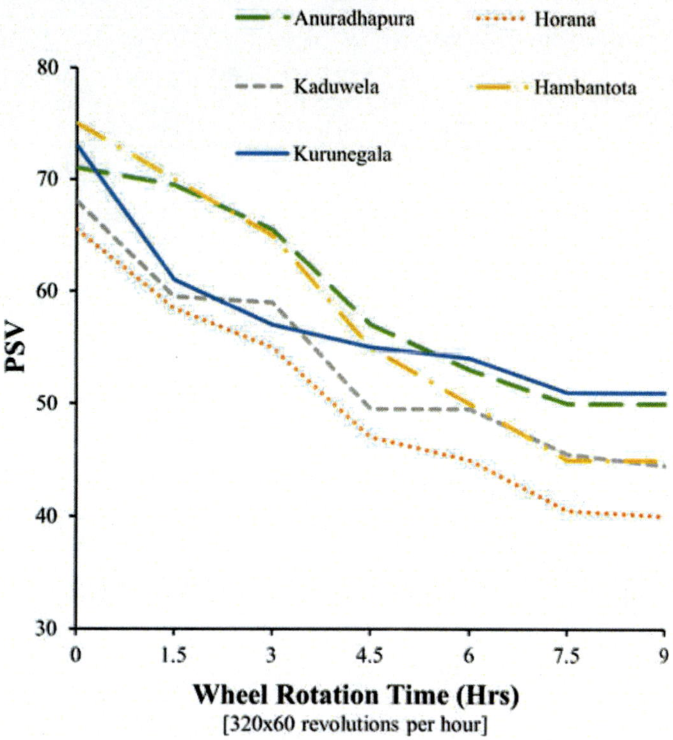

Fig. 1 Friction deterioration curves of the aggregate samples

3.3 Physical Properties

The test result summary of the physical properties of the aggregate samples determined by the basic laboratory testings (specific gravity and water absorption), is given in Table 2.

In order to test the compliance of the aggregates with the the Standard Specifications for Construction and Maintenance of Roads and Bridges by the Institute for Construction Training and Development (ICTAD)-Section 407 and 506, following mechanical properties of the aggregate samples, were evaluated as in Table 3.

Table 2 Physical properties of the aggregate samples

Rock origin	Specific gravity	Water absorption (%)
Anuradhapura	2.71	0.46
Horana	2.61	0.49
Kaduwela	2.61	0.42
Hambantota	2.67	0.45
Kurunegala	2.69	0.42

Table 3 Mechanical properties of the aggregate samples

Rock origin	LAAV (%)	ICTAD upper limit (%)	AIV (%)	ICTAD upper limit (%)	FI (%)	ICTAD upper limit (%)	SSS (%)	ICTAD upper limit (%)
Anuradhapura	38	40	23	30	20	35	1.8	12
Horana	35	40	16	30	15	35	1.8	12
Kaduwela	33	40	21	30	16	35	0.2	12
Hambantota	33	40	24	30	13	35	1.0	12
Kurunegala	37	40	24	30	17	35	2.2	12

1. Los Angeles Abrasion Value (LAAV)—ASTM C535
2. Aggregate Impact Value (AIV)—B.S.812:Part 112
3. Flakiness Index (FI)—B.S.812:Part 105.1
4. Sodium Sulphate Soundess (SSS)—ASTM C88.

3.4 Petrographic Analysis and Microscopic Investigation

Mineralogy of the aggregate samples were verified by the microscopic investigation of 30 μm thin sections prepared from 10 mm crushed rock chippings, using the Olympus BH polarising microscope attached with a microscope video camera. Table 4 elaborates a summary of the petrographic analysis of the aggregate samples.

4 Results and Discussion

In Sri Lanka, there are over 2500 Industrial Mining License (IML) grade quarries and most of the major aggregate suppliers of the road construction projects are from Hambanthota, Anurdhapuraya, Kurunegala, Kaduwela and Horana [31]. Therefore, natural aggregate samples representative of the above mentioned source points were acquired to be tested for the polishing resistance. All the aggregate samples from the five rock origins, is suitable for the wearing course application in terms of their physical and mechanical properties, as per the ICTAD specifications. Table 5 presents a summary of the experimental study outcomes.

As far as the friction deterioration curves are concerned, the PSVs of all the rock samples are scattered in the range of 40–75 during the 9 h polishng period, exhibiting intermediate levels of skid resistance capacities. In norm, coarse aggregates with PSV greater than or equal to 60, are regarded to fall into the high friction category.

Petrographic analysis of the rock samples concludes that all of them have a gneiss-based mineralogy and composition. Geologically, 90% of the natural rocks in Sri Lanka are made up of high grade metamorphic rocks of Precambrian age, which most consist of charnokites, migmatites and varieties of gneissic rocks [32]. Laboratory

Table 4 Mineralogical properties of the aggregate samples

Rock origin	Mineralogy (Petrography)	Thin section observation
Anuradhapura	Quartz, Feldspar, Biotite (Quatzo-feldspathic gneiss)	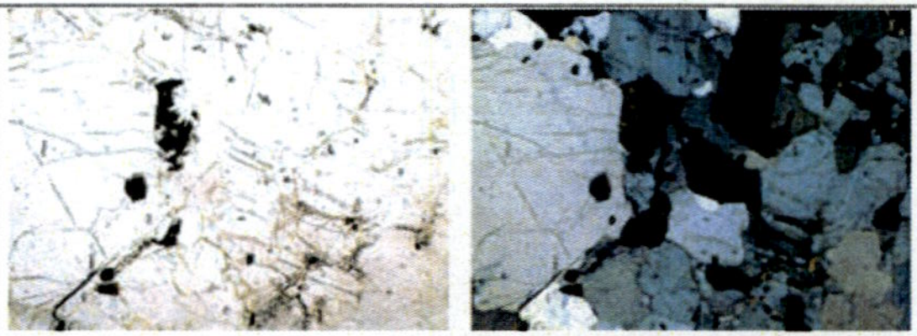
Horana	Plagioclase, Quartz, Biotite, Pyroxene (Biotite gneiss with pyroxene)	
Kaduwela	Plagioclase, Quartz, Biotite (Biotite gneiss)	

(continued)

Table 4 (continued)

Rock origin	Mineralogy (Petrography)	Thin section observation
Hambantota	Feldspar, Quartz, Hornblende, Biotite (Hornblende biotite gneiss)	
Kurunegala	Biotite, Hornblende, Plagioclase, Quartz (Biotite hornblende gneiss)	

Table 5 Laboratory test results summary of the aggregate samples

Rock origin	Petrography	Initial PSV before polishing	Final PSV after 9 h of polishing	Average PSV
Anuradhapura	Quatzo-feldspathic gneiss	71	50	60.5
Horana	Biotite gneiss with pyroxene	65.5	40	52.8
Kaduwela	Biotite gneiss	68	44.5	56.3
Hambantota	Hornblende biotite gneiss	75	45	60.0
Kurunegala	Biotite hornblende gneiss	73	51	62.0

investigations on gneiss-based coarse aggregates in Sri Lanka, have concluded that they are good quality, hard, tough and durable aggregates, as established in this study as well [33]. Gneiss-based aggregates are often termed as granite-based in the industry or literature due to their resemblance in many departments of physical and mechanical properties.

The texture of gneisses is of platy foliations, where coarser minerals are distributed in a softer mineral matrix. Hence, when subjected to polishing, the weaker minerals disintergrate exposing harder, unweathered minerals, which are able to retain their texture for a prolonged period and thus less susceptible to polishing [34, 35]. The presence of minerals such as Feldspar, Plagioclase and Pyroxene also contributes to the high skid resistance capacities due to their high siliceous content [36]. The thin section observations, justify the impressive degrees of polishing resistance of the rock samples collected.

5 Conclusion

There is no any form of specification nor national guidance over the selection of aggregates based on their polishing characteristics and there have been no studies carried out at the national level to assess the polishing characteristics of Sri Lankan roadstones. The major objective of this study was to evaluate the polishing patterns of local aggregates, where it has been established that the gneiss-based Sri Lankan roadstones showcase presence of wear-resistant particles and desirable frictional properties. The results and findings can be incorporated in the aggregate selection process specifically in the construction, maintenance and rehabilitation of major trunk roads and expressways and in the determination of intervention threshold levels in the Pavement Management System of the expressway network.

Since the study aims at determining the polishing behaviour of Sri Lankan roadstones under laboratory conditions, it is further recommended to evaluate the polishing patterns of natural aggregates under actual field conditions, to assist in determining the polishing patterns and decay curves of the pavement surface under site-specific conditions and in depth skid resistance oriented characterization of the roadstones.

References

1. Yager TJ, Rado Z, Hall JW, Smith KL, Titus-Glover L, Wambold JC (2009) Guide for pavement friction. National Cooperative Highway Research Program. https://doi.org/10.17226/23038
2. Najafi S (2015) Pavement friction management (pfm)-a step towards zero fatalities
3. Flintsch G, McGhee K, Izeppi EDL, Najafi S (2012) The little book of tire pavement friction. Apps.Vtti.Vt.Edu, no. September, pp 0–22
4. Smith KL, Larson RM, Flintsch GW, Sherwood J (2012) Improving highway safety through pavement friction management programs
5. Susan Herbel CM, Laing L (2010) Highway safety improvement program (HSIP) manual [Online]. Available http://safety.fhwa.dot.gov/hsip/resources/fhwasa09029/sec2.cfm.%0A; http://www.dot.state.mn.us/stateaid/trafficsafety/reference/2015-mndot-safety-handbook-large.pdf%5Cn; https://trid.trb.org/view/1364692
6. Larson RM (2004) Pavement functional pavement functional surface characteristics (PFSC)
7. Julian S, Moler F (2008) Public roads—gaining traction in roadway safety. FHWA-HRT-08-005, vol 72, no 1
8. Healow S (2011) Pavement friction management
9. Viner AHE, Sinhal R, Parry AR (2005) Linking road traffic accidents with skid resistance—recent UK developments, TRL paper reference PA/INF4520/05. Highways, no. August 2004, pp 1–13
10. Course AS, Maintenance P (2012) Pavement surface friction management g
11. Yang F et al (2019) An investigation of the polishing behavior of calcined bauxite aggregate. Coatings 9(11):1–16. https://doi.org/10.3390/coatings9110760
12. Shabani S, Ahmadinejad M, Ameri M (2013) Developing a model for estimation of polished stone value (PSV) of road surface aggregates based on petrographic parameters. Int J Pavement Eng 14(3):242–255. https://doi.org/10.1080/10298436.2012.693179
13. Hosking JR (1968) Factors affecting the results of polished-stone value tests. CroWthorne, Berkshire
14. Crisman B, Ossich G, Bevilacqua P, Roberti R (2020) Degradation prediction model for friction of road pavements with natural aggregates and steel slags. Appl Sci 10(1). https://doi.org/10.3390/app10010032
15. Xirouchakis D (2013) Correlations between mechanical and geometrical parameters in aggregates: a tool for quality assessment and control. Bull Geol Soc Greece 47(4):2069. https://doi.org/10.12681/bgsg.11091
16. de Izeppi EL, Flintsch GW, McGhee K (2010) Field performance of high friction surfaces. Assoc Princ Res Sci Virginia Transp Res Counc Transp Innov, no. June [Online]. Available http://www.virginiadot.org/vtrc/main/online_reports/pdf/10-cr6.pdf
17. Mahone DC, Sherwood C, The effect of aggregate type and mix design on the wet skid resistance of bituminous pavement: recommendations for virginia's wet accident reduction program
18. Qasim ZI (2016) Assessment the polishing of local aggregates by means of skid resistance for surface pavement layer. J Eng Sustain Dev 20(6)
19. Zahir H (2016) Performance of local aggregate in high friction surface treatment. Kansas State University Manhattan, Kansas
20. Liu J, Guan B, Chen H, Liu K, Xiong R, Xie C (2020) Dynamic model of polished stone value attenuation in coarse aggregate. Materials (Basel) 13(8). https://doi.org/10.3390/MA13081875

21. Khan KM, Ali F, Ahmed S, Pasha A, Mehmood I (2013) Skid resistance evaluation of aggregates from pakistan and Afghanistan. Int J Sci Eng Res 4(5):512–518 [Online]. Available http://www.ijser.org
22. Siriphun S (2016) Skid resistance of asphalt concrete based on thai aggregate. Suranaree University of Technology
23. Thenuwara HV, Pasindu H (2020) Methodology to incorporate frictional performance in the pavement management system of Sri Lankan expressways. https://doi.org/10.1109/MERCon50084.2020.9185214
24. National road master plan 2018–2027. Colombo (2018)
25. Corporation O (2015) EVJR1656 Ex-Post IV-10 Southern highway, no I, pp 1–43
26. ADB (2014) Sri Lanka: Southern transport development project summary report
27. Dharmaratne SD, Jayatilleke AU, Jayatilleke AC (2015) Road traffic crashes, injury and fatality trends in Sri Lanka: 1938–2013. Bull World Health Organ 93(9):640–647. https://doi.org/10.2471/blt.14.150193
28. Delivering road safety in Sri Lanka (2020). https://doi.org/10.1596/33341
29. Chinthanie RPD, Lanka S (2015) Accident analysis of Southern expressway
30. Geedipally SR, Pratt MP, Lord D (2019) Effects of geometry and pavement friction on horizontal curve crash frequency. J Transp Saf Secur 11(2):167–188. https://doi.org/10.1080/19439962.2017.1365317
31. Session V–Geology, soils and water resource management (36) rock quarrying and sustainable environment management dharmaratne P.G.R., no 36, p 2016
32. Wijayarathne GWLW, Thavaneeshan S, Sameera BHAT , Sandamali IGV, Chaminda SP, Jayawardena CL, Premasiri HMR and Samaradivakara GVI (2016) Identifications of most suitable locations for rock quarrying to supply aggregate requirements of the colombo port city development project
33. Jayawardena UDS, Dissanayake DMS (2008) Identification of the most suitable rocks types for manufacture of quarry dust in Sri Lanka. J Natl Sci Found Sri Lanka 36(3):215–218. https://doi.org/10.4038/jnsfsr.v36i3.157
34. Vaiana R, Capiluppi GF, Gallelli V, Iuele T, Minani V (2012) Pavement surface performances evolution: an experimental application. Procedia Soc Behav Sci 53:1149–1160. https://doi.org/10.1016/j.sbspro.2012.09.964
35. Dahir SHM, College CB, Factors influencing aggregate skid-resistance properties. In: Committee on Surface Properties-Vehicle Interaction-50th Annual Meeting, pp 136–148
36. Hartley A (1974) A review of the geological factors influencing the mechanical properties of road surface aggregates. Q J Eng Geol 7(1):69–100. https://doi.org/10.1144/GSL.QJEG.1974.007.01.05

Applicability of Gneiss Based Mineral Aggregates for the Use in High Friction Surface Treatment in Terms of Polishing Characteristics

H. V. Thenuwara, P. H. G. H. Buddhi, and H. R. Pasindu

Abstract Speed and safety management of highways plays a prominent role in ensuring road user safety, especially of the vulnerable categories. Pavement Friction Management (PFM) is being considered as the best possible option in developing the road infrastructure over the other prevalent strategies, with the focus of improving user safety. High Friction Surface Treatment (HFST) is widely being used as a remedial measure to improve the friction and texture retention capacity of the wearing course, as it provides economical and sustainable solutions in upgrading the pavement performance in both as an urgent response and in the long run as well. As an urgent remedial intervention, HFST can be incorporated in the national practice of safety management in Sri Lanka, to address the concerns over the escalating numbers of grievous highway crashes. In place of the most commonly used high skid resistant aggregates, gneiss-based mineral aggregate can be a viable alternative since it delivers similar functionality to that of granites. The study examines the polishing characteristics of widely used gneiss-based roadstones in the country through laboratory experiments involving the Polish Stone Value (PSV) test. Both qualitative and quantitative analyses of the specified physical and mechanical laboratory testing are conducted to evaluate the applicability of gneiss-based mineral aggregates in the use of HFST as an alternative skid-resistant roadstone in terms of their polishing behavior.

Keywords Coarse aggregates · Gneiss · High friction surface treatment · Highway safety · Polishing characteristics · Micro-texture · Mineralogy

H. V. Thenuwara (✉) · P. H. G. H. Buddhi · H. R. Pasindu
Department of Civil Engineering, Faculty of Engineering, University of Moratuwa, Moratuwa, Sri Lanka
e-mail: thenuwarahv.20@uom.lk

P. H. G. H. Buddhi
e-mail: 160074j@uom.lk

H. R. Pasindu
e-mail: pasindu@uom.lk

© The Author(s), under exclusive license to Springer Nature Switzerland AG 2022

H. R. Pasindu et al. (eds.), *Road and Airfield Pavement Technology*, Lecture Notes in Civil Engineering 193, https://doi.org/10.1007/978-3-030-87379-0_45

1 Introduction

Pavement friction is a crucial functional aspect of overall pavement performance, facilitating motorists the ability to control/maneuver their vehicles in a safe manner [1, 2]. In definition, the resistive force that obstructs the relative motion between a rotating vehicle tire and a pavement surface at the tire-pavement interface, is interpreted as pavement friction [1]. Therefore, friction co-efficient is incorporated in the geometric design of highways as a key input parameter, since a satisfactory degree of pavement friction is required to ensure user safety. Specifically, in the design, construction, and maintenance of accident-prone critical geometric elements of highway alignment (horizontal, spiral, and vertical curves, as well as intersections) provision and maintenance of adequate road friction is indispensable to prevent roadway departure crashes, skidding, run-off-road and head-on collisions for a safer riding [3–6].

Highway crashes, the ninth leading cause of death, which will escalate to the fifth by 2030, cause 1.3 million fatalities and 20–50 million grievous and non-grievous injuries worldwide annually [7, 8]. In addition to driver, vehicle, and highway infrastructure-related causative factors, inadequate pavement friction capacity catering to increased Stopping Sight Distance (SSD), has been identified as the major pavement-related cause of highway crashes. Friction-related pavement characteristics cause about 28% of the total road crashes and 14% of all fatal crashes [9–11]. Induced polishing action of the wearing course over time due to repeated traffic application apart from adverse weather conditions and many other site-specific reasons, leads to friction deterioration of the pavement, resulting in the increase of highway crashes, specially under wet surface conditions [1, 2, 12]. The coefficient of friction of a vehicle tire sliding over a wet pavement surface decreases exponentially as water film thickness increases and may lead to extreme conditions such as hydroplaning [1, 2].

Highway safety is a high priority user concern that is not being directly addressed, opposed in the case of mobility and user comfort. Three pillars of safety design are concerned with driver behavior, vehicle design, and highway infrastructure, where geometric design and Pavement Friction Management (PFM) of highways fall under infrastructure [7]. Pavement surface characteristics (network friction and texture) account for providing the minimum required skid resistance of the pavement to cater to the friction demand defined by the geometric design combined with the vehicle speed. Most current safety effort addresses the vehicle and driver issues whereas infrastructure effort is on geometric design issues, while focus on improvement and retention of pavement characteristics is minimal. Many comprehensive studies on crash trends, including the Federal Highway Administration (FHWA) have concluded that improvement of the pavement surface friction can result in an approximate 70% reduction of the number of crashes while contributing to an apparent decrease in the severity of crashes and their related consequences [3, 8, 13].

There are several restoration methods that can be used to reestablish the frictional properties of flexible pavements. The United States (US) Department of Transportation (DoT) of FHWA pooled fund study "Evaluation of Low-Cost Safety Improvements" recommends skid-resistive pavement surface treatments as one of the highly cost-effective pavement friction enhancements [14, 15]. According to the FHWA, High Friction Surface Treatment (HFST) is an appealing alternative since these High Friction Surface (HFS) systems can dramatically and immediately reduce the occurrence of crashes and fatalities at abrupt curves, variable superelevations, wet conditions, polished roadway surfaces, and excessive driving speeds [16].

2 Literature Review

2.1 *High Friction Surface Treatment*

High Friction Surface Treatment (HFST) is a specially designed process that can compensate driver error, poor geometric alignment of the roadway, and inadequate road surface skid resistance, hydroplaning, splash, and spray especially during wet weather, increasing driver awareness, and water drainage. The typical installation practice of HFSTs is the application of a single or double layer of hard, durable, and high Polished Stone Value (PSV) aggregates on hot-applied thermo-setting polymer resin binder or cold-applied epoxy at critical maneuver locations of the roadway [13].

HFST can resolve site-specific issues that require high friction demand by geometry, type of functional element, expected driver behavior or traffic volume, as in [8, 17].

- Intersections and roundabouts
- Exit-entrance ramps and toll plazas
- School crossings and crosswalks
- Sharp horizontal curves and steep slopes
- Bridge decks
- Identified crash-prone or skid hazardous areas.

In comparison, HFST excels micro surfacing technique as a countermeasure to improve pavement surface characteristics at the above-mentioned roadway locations, since it provides more friction than the latter would.

2.2 *Polishing Characteristics of Aggregates*

Friction related pavement characteristics are greatly defined and influenced by the textural properties of the exposed coarse aggregates of the wearing course. Microscopic asperities of the aggregate termed as microtexture (of the amplitude of planar deviations less than or equal to 0.5 mm within wavelength less than or equal to 0.5 mm), account for the adhesion component of the total friction force exerted on a rolling vehicle tire by the pavement, through intermolecular bonding between the tire tread and the aggregate surface [1, 2]. These irregularities facilitate the escape of thin water films in excess and skid resistance at all speeds for dry and wet conditions [1, 2, 18]. The degree of polishing of aggregates highly depends on these asperities at the micron scale, defined by the crystalline structure and the content of wear-resistant minerals of the aggregates. Microtexture is thus a function of aggregate particle mineralogy and petrology, which altogether contribute to the polishing characteristics and long-term skid resistance of mineral aggregates.

Therefore, specifications for the selection and use of aggregate for HFST are aimed at measuring the aggregate's ability to retain microtexture and resist polishing. The polished stone value (PSV) test measures the resistance of roadstone to the polishing action of vehicle tires under conditions similar to those occurring on the surface of a road, by means of accelerated polishing under controlled laboratory conditions. The higher the PSV is, the more skid-resistant is the aggregate. In general terms, high friction aggregates are such that those with PSV greater than or equal to 60 [19, 20].

2.3 *Commonly Used Aggregates for HFST*

The aggregates should resist abrasion and polishing due to trafficking action and disintegration caused by weathering, providing a durable, skid-resistant wearing course with an optimal level of pavement surface texture. As per the aggregate properties for HFST specified by the FHWA, the coarse aggregates should meet a minimum of PSV of 38, in accordance with the standard AASHTO T279 as similar to BS 812 part 114, BSEN1097-8, and ASTM D3319 [3, 8]. Commercially used common types of aggregates in HFST include [13, 21],

- Calcined bauxite
- Dolomite
- Granite
- Silica
- Steel slag
- Flint.

Calcined bauxite, a manufactured high wear and polishing resistant coarse aggregate with its PSV exceeding 70, has been predominantly used in the HFST application process [13]. However, many researchers have investigated the use of locally available aggregates in place of such artificial aggregate to optimize the cost of HFST process [8]. Granite is considered to have intermediate polishing resistance properties, with fairly good abrasion resistant qualities. Gneiss-based aggregates are often termed as granite-based in the industry or literature due to their resemblance in many departments of physical and mechanical properties.

Gneiss-based aggregates tend to behave similarly to igneous rocks of the same mineralogy and thus are widely used in the industry for road surfacing applications. Gneiss is a common and widely distributed type of rock, specifically in Asia, formed by high-grade regional metamorphic processes of either igneous or sedimentary rocks, whereas granite is of igneous rock origins. Common minerals found in a gneiss matrix are Biotite, Chlorite, Feldspar, Garnet, Graphite, Hornblade, Micas, Quartz, Quartzite, Silica, and Zircon, which can be discovered in granite-based aggregates as well [22]. It is a hard and durable type of aggregate with specific gravity ranging from 2.6 to 2.9. Medium to coarse-grained structure provides significantly less porous formation with Mohs' hardness value of 7, which reflects its impressive toughness and hardness properties [23, 24]. The texture of gneisses is of banded or platy foliations, with irregular fracture surfaces, where harder, coarser minerals are distributed in a softer mineral matrix. Hence, when subjected to polishing and weathering, the weaker minerals disintegrate, exposing harder, unweathered minerals, which can retain their texture for a prolonged period and thus restoring the skid resistance of the pavement [24, 25]. In conclusion, as far as textural and physical characteristics are concerned, gneiss-based aggregates are a vehement alternative to be used in the HFST applications.

2.4 *Experience of HFST in the United States*

Signature trial HFST application projects carried out in Pennsylvania, Wisconsin, Kentucky, and South Carolina DoTs have reported total crash reductions of 100%, 95%, 90%, and 57%, respectively, where Kentucky experienced a total crash reduction of 78%, with a wet weather crash reduction of 85%. In general, various state DOTs studies on HFST resulted in a 20–30% reduction in all crashes and a 50% reduction in wet weather crashes [21].

Proper installation of HFST can guarantee 7–12 years of service life, while 12–15 years on bridge decks and interstate highways with Average Daily Traffic (ADT) of 48,000–62,000 vehicles per day. Studies have shown that HFST delivers a high cost-benefit ratio of about 20–24 with a relatively excellent life cycle cost [26]. For large-scale projects and small bundled installations, unit cost of HFST can vary from 20 to 30 US Dollars per square meter [21]. Although the unit cost of HFST is higher than other skid-resistive pavement surface treatments, in the long run, it ensures maximum possible user safety and a durable pavement with minimum noise and rider comfort-related issues as well. Furthermore, the production and construction process related to HFST application concerns less adverse environmental impacts as well.

2.5 Status of PFM in the Context of Sri Lanka

According to the World Health Organization (WHO), 90% percent of the road accidents reported are taken place in low and middle-income countries, and highway crashes incur direct and indirect costs of about 3% of the Gross Domestic Product (GDP) in most countries [7]. Sri Lanka has a well-distributed road network of about 116,000 km, consisting of 12,380 km of National roads comprising of class "A" roads (trunk roads), class "B" roads (main roads), and Expressways (E) [27–29].

An average of 38,000 crashes occurs annually, which result in around 3000 fatalities and 8000 serious injuries, of which about 600 crashes are recorded in the expressway system [7, 30, 31]. Crash fatalities and injuries are expected to escalate steadily given the rapid growth in vehicle ownership unless urgently required countermeasures are implemented [32]. These numbers represent not only the loss of human lives but also the nation's working population apart from the infrastructure and property damage combined with other indirect economic losses. In a study conducted by the WHO, the estimated economic gains from achieving a 50% fatality reduction in the countries assessed, ranged between (7–22) % increase in GDP over the analysis time frame of 24 years, whereas the estimated population welfare gains were equivalent to (6–32) % of GDP [7]. The obvious need to implement strategies and measures to improve road user safety, can thus be well justified.

However, most of the safety management programs funded and guided by the government and the relevant authorities, emphasize on addressing the vehicle, driver or other vulnerable road user categories related to causative factors. The infrastructure-related issues are often left unaddressed specially those regarding inadequate geometric designs and crash-prone locations. Speed management by the erection of road signs, imposing speed limits only cannot effectively reduce the crash rate at such vulnerable roadway locations, where additional robust improvements of infrastructure are required. Sand sealing or any other slurry sealing methods and Double Bituminous Surface Treatment (DBST) carried out in the national maintenance routine of highways, are to optimize and retain the mobility, roughness and durability of the existing pavement surface, whilst there are no industry practices involved with the retention of pavement friction [12, 33]. Though Pavement Friction Management (PFM) under the Pavement Management System (PMS) has been incorporated into the national practice of maintenance of the expressway network of the country, focus and studies on possible restoration criteria are still scarce.

Hence, Pavement Friction Management, especially to reduce fatal highway crashes occurring at sharp horizontal curves, intersections, school crossings and identified crash-prone or skid hazardous areas, should be incorporated in the typical maintenance routine of the national road network ("A, B, E" class roads), in which case the application of High Friction Surface Treatment to improve pavement friction at such specifically identified accident-prone areas are highly recommended.

2.6 *Natural Coarse Aggregates in Sri Lanka*

Geologically, 90% of the natural rocks in Sri Lanka are made up of high-grade metamorphic rocks of Precambrian age, while the remaining are comprised of sedimentary rocks of predominantly Miocene age. Precambrian crystalline rocks, which most often occur as hills or mountains, consist of charnokites, migmatites, and varieties of gneissic rocks. There are 5 geological complexes prevalent in the country, and gneiss-based natural aggregates are the broadly available type of mineral aggregates [34].

The annual production of rock aggregates in the country, is around 8 million cubic meters and there are over 2500 Industrial Mining License (IML) grade quarries. Most of the major aggregate suppliers of the road construction projects in Sri Lanka are from Hambanthota, Anurdhapuraya, Kurunegala, Kaduwela, and Horana [34].

Investigations and studies on gneiss-based coarse aggregates in Sri Lanka have concluded that they are good quality, hard, tough, and durable aggregates complying with the Standard Specifications for Construction and Maintenance of Roads and Bridges by the Institute for Construction Training and Development (ICTAD) [35]. Yet, ICTAD has only defined specifications regarding the physical, and mechanical properties of roadstones, whereas there is no any form of specification nor national guidance over the selection of aggregates based on their polishing characteristics. Further, there have been no studies carried out at the national level to assess the polishing characteristics of Sri Lankan roadstones. Therefore, this study intends to evaluate the applicability of the most abundant type of mineral aggregates in Sri Lanka-Gneiss based coarse aggregates in the application of the HFST, in terms of their polishing behavior.

3 Methodology

The research objective of the study is to assess whether the locally available gneiss-based mineral aggregates comply with the requirements of polishing resistance specified for the aggregates in HFST by the FWHA. Following is a description in brief of the sequence, including the basic procedure, laboratory testing, and analysis methods adopted in the progression of the research.

- A thorough literature review, conducted to identify the best possible option of infrastructure development related approach-high friction surface treatment, to address the speed and safety management of high crash-prone, skid hazardous areas and vulnerable user category wise roadway elements of interest (school crossings).
- Isolation of the geological complexes in Sri Lanka with origins of gneiss-based mineral aggregates and material acquisition form the selected IML grade rock quarries/suppliers of highway construction industry.

Table 1 Details of the aggregate sources

Source of aggregate	District	Geological complex	Code
Ganewalpola, Anuradhapura	Anuradhapura	Wanni complex	A
Horana	Kalutara	Wanni complex	B
Kaduwela	Colombo	Wanni complex	C

- Determination of the physical properties of the collected aggregates through laboratory testing (specific gravity, water absorption, and thin section analysis tests).
- Establishment of the suitability of the aggregates for the wearing course application and surface treatment as per the ICTAD specifications by conducting to determine mechanical properties.
- Preparation of the PSV test specimens in accordance with the respective standards.
- Derivation of the friction deterioration curves of the aggregates through the Polish Stone Value test to determine the PSV values of gneiss-based aggregates.
- Determination of the texture retention capacities of the aggregates through evaluation of their micro-texture characteristics.
- Analysis of the laboratory experiment results in terms of the research problem.
- Recommendations to incorporate the research findings into national practice of safety management.

3.1 Material Acquisition

Aggregates were collected from 6 major suppliers of roadstones scattered in 3 Districts, whose sources of rock are located in the largest geological complex of the country. The rock sources of gneiss-based aggregates are located in Kaduwela (Colombo District), Anuradhapura (Anuradhapura District), and Horana (Kalutara District), representing the Wanni complex, the samples were acquired from 2 different major suppliers of the same rock source for each source. The detailing of the aggregate sources is indicated in Table 1.

4 Laboratory Testing

4.1 Physical Properties

Physical properties of the aggregates collected were first determined by basic laboratory testing (specific gravity, water absorption, and thin section observation). The test

Table 2 Physical properties of aggregates

Source of aggregate	Thin section observation	Mineralogy (Petrography)	SG	WA (%)
A	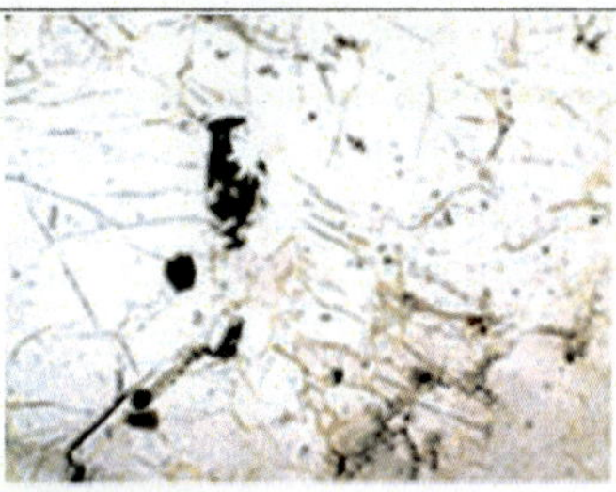	Quartz, Feldspar, Biotite (Quatzo-feldspathic gneiss)	2.71	0.46
B		Plagioclase, Quartz, Biotite, Pyroxene (Biotite gneiss with pyroxene)	2.67	0.49
C		Plagioclase, Quartz, Biotite (Biotite gneiss)	2.67	0.42

result summary is given in Table 2. The aggregates to be tested have the petrography of 3 sub-categories of gneiss.

- Biotite gneiss
- Quatzo-feldspathic gneiss
- Biotite gneiss with pyroxene.

Specific Gravity and Water Absorption (on Saturated Surface Dry (SSD) Basis) (ASTM C127)

Specific Gravity (SG) is defined as the ratio of the weight of aggregate to the weight of an equal volume of water, which is considered to be a measure of strength or quality of the material. Water Absorption (WA) is the change in the mass of an aggregate due to water absorbed in the pore spaces within the constituent particles, compared to the dry condition.

Thin Section Observation

Mineralogy of the aggregate samples was verified by the microscopic investigation of thin sections. The thin sections were produced from 10 mm crushed rock chippings bonded together using epoxy resin. The specimens were cut into slices, which were ground and polished to 30 μm thin sections. The sections were observed using the Olympus BH polarizing microscope attached to a microscope video camera, and images from the camera were sent to a video monitor and to a computer running the installed image processing and analysis software.

4.2 *Mechanical Properties*

Mechanical properties of the gneiss-based aggregates were then determined by the laboratory testing as per the ICTAD specifications (Section 407 and 506). Table 3 shows the test results of the aggregate samples.

Los Angeles Abrasion Value (ASTM C535)

The Los Angeles Abrasion Value (LAAV) is a measure of degradation of aggregates when subjected to a combination of actions involving abrasion, attrition, impact and grinding.

Aggregate Impact Value (B.S.812: Part 112)

Aggregate Impact Value (AIV) reflects the toughness of aggregates or the ability to resist fracture under pounding action or impact due to traffic loading.

Flakiness Index (B.S.812: Part 105.1)

Flakiness Index (FI) is a measure of the proportion of flat or flaky particles of the aggregate sample, that has a tendency to fracture along their weak dimension under trafficking, leading to gradation and stability-related issues.

Sodium Sulphate Soundness (ASTM C88)

Sodium Sulphate Soundness (SSS) delivers a measure of the resistance of aggregates to in-service weathering or disintegration.

Table 3 Mechanical properties of aggregates

Source of aggregate	LAAV (%)	AIV (%)	FI (%)	SSS (%)
A	38	22.5	20	1.8
B	35	15.8	15	1.8
C	33	20.8	16	0.2

4.3 Polishing Characteristics

Preparation of test specimens and laboratory testing with the use of Accelerated Polishing Machine and the British Pendulum Tester, was carried out to determine the polishing behavior of the gneiss-based aggregates.

Specimen Preparation

14 number (no.) of specimens (4 no. of specimens from each source of rock (2 no. of specimen from each supplier) + 2 no. of control stone specimen) were prepared in accordance with the ASTM—D3319-06 and the following were required.

- Coarse aggregate—ASTM D 75 (sieved through 12.7 mm and retained on 9.5 mm sieves)
- Fine sand—ASTM C 778 graded sand
- Metal molds and Sanding Block
- Mold release agent
- Bonding agent—polyester resin and catalyst.

One layer of coarse aggregates was placed in the mold in a tight packed manner, and the voids remaining were filled with sand. After being filled to the upper level by the bonding agent, the pressure was applied by the sanding block onto the mold. After completion of the hardening process, the mold was dismantled, and the specimen was taken off. Specimens were numbered from 1 to 14, and the direction of rotation was marked.

Polish Stone Value Test (ASTM—D3319-06)

By accelerating the rate of polish under controlled laboratory conditions using the Accelerated Polishing Machine, aggregates can be assessed before being used as a surfacing material. The degree of polish attained is measured using the British Pendulum Tester (BPT) and expressed as a PSV with values ranging from 30 to 80.

Accelerated Polishing Machine

Polishing of the 14 aggregate specimens was performed using the accelerated polishing machine with the use of the following supportive elements.

- Water
- Corn emery and flour emery (polishing agent)
- Specified rubber tire.

Fourteen specimens were placed using "O" rings onto the Road wheel in the numbered order in the clockwise direction. Water was applied at a rate of 4 l/h, and emery was applied at a rate of 0.48 kg/h. A load of 725 N was applied, and the polishing was performed for 9 h at a rate of 320 rpm. The specimens were first conditioned for 2 h through accelerated polishing, before the 9-h polishing was conducted.

British Pendulum Tester (ASTM—E303-93)

Initially, before the commencement of the process of polishing, the Residual Polish Value (RPV) of the selected specimens was obtained. Polish Value (PV n) or British Pendulum Number (BPN) or Skid Resistance Number (SRV) of specimens after "n" number of hours of rotation under wet condition was taken using the BPT. 4 consecutive readings of BPN with no deviation greater than 3 points were recorded for each reading for the polishing intervals of 1.5 h.

The PSV is calculated using the following equation. The PSVs of the 3 gneiss-based aggregate types are shown in the

$$\mathrm{PSV} = S + (52.5 - C) \tag{1}$$

where,

S mean of the four test specimens
C mean of the specimens of control stone.

The friction deterioration curves of gneiss-based aggregates are shown in Fig. 1. Table 4. discusses the prominent results from the polishing of the aggregates.

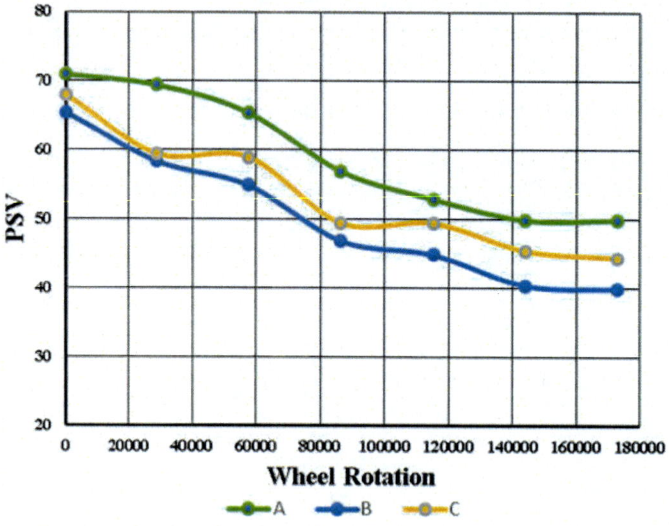

Fig. 1 Polishing characteristics (friction deterioration curve) of gneiss-based aggregates

Table 4 PSV results of the gneiss-based aggregates

Source of aggregate	Initial PSV before polishing	Final PSV after 9 h of polishing	Average PSV
A	71	50	60.5
B	65.5	40	52.8
C	68	44.5	56.3

4.4 Textural Characteristics

To further support the results and findings of the polish stone value test, textural properties of the gneiss-based aggregate samples were derived. Micro-texture of the aggregates was measured in terms of the root- mean square height (R_q) [36]. This determination was conducted on the specimens before and after each polishing interval using the Coordinate Measuring Machine (with a level of accuracy of 1 μm). Figure 2 illustrates the textural properties of the gneiss-based aggregates.

$$R_q = \sqrt{\frac{1}{N} \sum (Z_i - \overline{Z})^2} \tag{2}$$

where,

N Number of points
Z_i Generic ordinate (Z axis co-ordinate of the point)
$\overline{Z}$ Average of Z_i values.

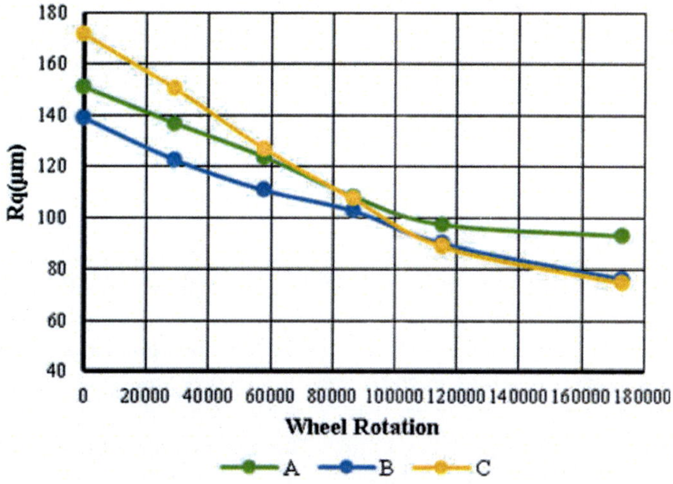

Fig. 2 Micro-texture deterioration curve of Gneiss-based aggregates

5 Results and Analysis

The mechanical properties of the gneiss-based aggregates all comply with the ICTAD specifications. The LAAVs are less than 40%, while the AIVs are less than 30%. The FI values are less than 35%, and the SSS values are less than 12%. The physical and mechanical properties indicate that all the sub-types of gneiss-based aggregates are suitable for pavement surfacing and surface treatment.

The results of the PSV test indicate intermediate levels of skid resistance, which is satisfactory to be applied in the HFST process, as per the FWHA specifications. The presence of minerals such as Feldspar, Plagioclase, and Pyroxene in the gneiss-based aggregate sources contributes to the high siliceous content resulting in high skid resistance capacities of the aggregates. The textural properties of the three types of gneiss-based aggregates also suggest the same in support of their reasonably good levels of friction retention. In general, aggregates with microtexture between 0.2 and 0.5 mm are considered to be high in skid resistance. The aggregate sources under the study showcase micro-texture levels between 0.17 and 0.7 mm, implicating the intermediate levels of friction and texture retention ability [13, 19].

6 Conclusion

The applicability of Gneiss-based mineral aggregate in the use of high friction surface treatment in terms of polishing characteristics, has been evaluated and established in this study. Gneiss-based mineral aggregate is one of the best possible alternatives to be used as the high polishing aggregates, in the application of HFST in Sri Lanka. The mineral aggregate exhibits a relatively high degree of polishing resistance with adequate friction and texture retention capacities. It will be a promising investment to reduce the number and the severity of highway crashes, if used for the infrastructure development and rehabilitation projects of the road sector.

7 Recommendations

HFST is a viable solution to improve the skid resistant properties of the pavement, specifically in terms of frictional and textural properties. Gneiss-based mineral aggregates can be used as the skid resistant aggregates in the application of high friction surface treatment, which can be successfully incorporated into national safety management practice and pavement friction management. Since the focus of this study was to evaluate the properties of gneiss-based mineral aggregates as required by the FHWA specifications of HFST application, in-depth experimental studies have to be carried out, especially on the mix design. Though there are ample amount of research and projects carried out on the application of HFST, further studies on the in-service performance of HFST with gneiss-based mineral aggregates at national level industry practice, are encouraged.

References

1. Yager T, Rado Z, Hall J, Smith KL, Titus-Glover L, Wambold J (2009) Guide for pavement friction. National Cooperative Highway Research Program. https://doi.org/10.17226/23038
2. Flintsch G, McGhee K, Izeppi EDL, Najafi S (2012) The little book of tire pavement friction. Apps Vtti Vt Edu, no. September, pp 0–22
3. Julian S, Moler F (2008) Public roads—gaining traction in roadway safety. FHWA-HRT-08-005, vol 72, no 1
4. Healow S (2011) Pavement friction management
5. Course AS, Maintenance P (2012) Pavement surface friction management g
6. Viner AHE, Sinhal R, Parry AR (2005) Linking road traffic accidents with skid resistance—recent UK developments. TRL Paper reference PA/INF4520/05. Highways, no. August 2004, pp 1–13
7. Delivering road safety in Sri Lanka (2020). https://doi.org/10.1596/33341
8. de Izeppi EL, Flintsch GW, McGhee K (2010) Field performance of high friction surfaces. Assoc Princ Res Sci Virginia Transp Res Counc Transp Innov, no. June 2010 [Online]. Available http://www.virginiadot.org/vtrc/main/online_reports/pdf/10-cr6.pdf
9. Smith KL, Larson RM, Flintsch GW, Sherwood J (2021) Improving highway safety through pavement friction management programs
10. Susan Herbel CM, Laing L (2010) Highway safety improvement program (HSIP) manual [Online]. Available http://safety.fhwa.dot.gov/hsip/resources/fhwasa09029/sec2.cfm.%0A; http://www.dot.state.mn.us/stateaid/trafficsafety/reference/2015-mndot-safety-handbook-large.pdf%5Cn; https://trid.trb.org/view/1364692
11. Larson RM (2004) Pavement functional pavement functional surface characteristics (PFSC)
12. Thenuwara HV, Pasindu H (2020) Methodology to incorporate frictional performance in the pavement management system of Sri Lankan expressways. https://doi.org/10.1109/MERCon50084.2020.9185214
13. Zahir H (2016) Performance of local aggregate in high friction surface treatment. Kansas State University Manhattan, Kansas
14. Merritt DK, Lyon C, Persaud B (2021) Evaluation of pavement safety performance. Washington, DC, 2015. Accessed 02 Apr 2021 [Online]. Available http://www.ntis.gov
15. McGee HW, Hanscom FR (2006) Low-cost treatments for horizontal curve safety. Washington, DC
16. Alhasan A, Nlenanya I, Smadi O, MacKenzie CA (2018) Impact of pavement surface condition on roadway departure crash risk in Iowa. Infrastructures 3(2). https://doi.org/10.3390/infrastructures3020014
17. Geedipally SR, Pratt MP, Lord D (2019) Effects of geometry and pavement friction on horizontal curve crash frequency. J Transp Saf Secur 11(2):167–188. https://doi.org/10.1080/19439962.2017.1365317
18. Crisman B, Ossich G, Bevilacqua P, Roberti R (2020) Degradation prediction model for friction of road pavements with natural aggregates and steel slags. Appl Sci 10(1). https://doi.org/10.3390/app10010032
19. Hosking JR (1968) Factors affecting the results of polished-stone value tests. CroWthorne, Berkshire
20. Xirouchakis D (2013) Correlations between mechanical and geometrical parameters in aggregates: a tool for quality assessment and control. Bull Geol Soc Greece 47(4):2069. https://doi.org/10.12681/bgsg.11091

21. Wilson B, Mukhopadhyay A (2016) Alternatice aggregates and materials for high fricion surface treatments. Texas A&M Transp Inst, no. May, 2016 [Online]. Available file:///C:/Users/dell/Downloads/dot_30989_DS1.pdf
22. Hartley A (1974) A review of the geological factors influencing the mechanical properties of road surface aggregates. Q J Eng Geol 7(1):69–100. https://doi.org/10.1144/GSL.QJEG.1974.007.01.05
23. Mahone DC, Sherwood C, The effect of aggregate type and mix design on the wet skid resistance of bituminous pavement: recommendations for virginia's wet accident reduction program
24. Dahir SHM, College CB, Factors influencing aggregate skid-resistance properties. In: Committee on Surface Properties-Vehicle Interaction-50th Annual Meeting, pp 136–148
25. Vaiana R, Capiluppi GF, Gallelli V, Iuele T, Minani V (2012) Pavement surface performances evolution: an experimental application. Procedia Soc Behav Sci 53:1149–1160. https://doi.org/10.1016/j.sbspro.2012.09.964
26. Yang G, Li QJ, Zhan Y, Yu W, Wang KCP, Peng Y (2019) Field performance evaluation of high friction surface treatments (HFST) in Oklahoma. Can J Civ Eng 46(12):1142–1150. https://doi.org/10.1139/cjce-2018-0521
27. National road master plan 2018–2027 (2018) Colombo
28. Corporation O (2015) EVJR1656 Ex-Post IV-10 Southern Highway, no I, pp 1–43
29. ADB (2014) Sri Lanka: Southern Transport Development Project Summary Report
30. Dharmaratne SD, Jayatilleke AU, Jayatilleke AC (2015) Road traffic crashes, injury and fatality trends in Sri Lanka: 1938–2013. Bull World Health Organ 93(9):640–647. https://doi.org/10.2471/blt.14.150193
31. Chinthanie RPD, Lanka S (2015) Accident analysis of southern expressway
32. Sandamal RMK, Ranawaka RKTK, Pasindu HR (2020) A framework to incorporate safety performance for low volume roads in pavement management systems. In: MERCon 2020—6 th International Multidisciplinary Moratuwa Engineering Research Conference, Proceedings, pp 407–412. https://doi.org/10.1109/MERCon50084.2020.9185235
33. Judith HLDMA (2003) Improvements in design and construction standards of surface dressings for national roads in Sri Lanka. University of Moratuwa
34. Wijayarathne GWLW, Thavaneeshan S, Sameera BHAT, Sandamali IGV, Chaminda SP, Jayawardena CL, Premasiri HMR and Samaradivakara GVI (2016) Identifications of most suitable locations for rock quarrying to supply aggregate requirements of the colombo port city development project
35. Jayawardena UDS, Dissanayake DMS (2008) Identification of the most suitable rocks types for manufacture of quarry dust in Sri Lanka. J Natl Sci Found Sri Lanka 36(3):215–218. https://doi.org/10.4038/jnsfsr.v36i3.157
36. Guan B, Wu J, Xie C, Fang J, Zheng H, Chen H (2018) Influence of macrotexture and micro-texture on the skid resistance of aggregates. Adv Mater Sci Eng 2018. https://doi.org/10.1155/2018/1437069

Chemical Properties of Modified Bitumen Incorporating Coconut Char

Rosmawati Mamat, Mohd Rosli Hainin, Norhidayah Abdul Hassan, Mohd Naqiuddin Mohd Warid, Mohd Khairul Idham Mohd Satar, and Noor Azah Abdul Rahman

Abstract Agriculture waste have many advantages such as high specific strength and modulus, low density, renewable nature, biodegradability and absence of health hazards. This advantage is good for construction industry because it can improve the properties of the composites. The coconut fruits contain 40% coconut husk, 30% fiber and 30% dust, consisted of flesh, shell and fiber and chemical composition such as cellulose, lignin, charcoal, acid and potassium. The advantages of coconut shell are strong, rigid and lightweight material, very economical as large amounts are available as agricultural waste material and environmentally friendly. This paper presents the performance of modified binder incorporating coconut char in terms of chemical properties. Fourier Transform Infrared Test (FTIR), Particle Size Analyzer Test (LPSA) and Elemental Analyzer Test (EA) were used to analyze the chemical properties on modified binder. Based on the findings, the increased content of char affects the performance of bitumen significantly, as the bitumen maintains stability and homogenous state even though placed in high temperature condition. The testing result value of FTIR and LPSA are increased with the increased content of coconut char. Based on the results, the modified mixture has the higher chemical components and porosity compared to original bitumen and coconut shell because of the modification.

Keywords Coconut char · Modified binder · Fourier transform infrared test · Particle size analyzer · Elemental analyzer

R. Mamat · M. R. Hainin (✉) · N. A. Hassan · M. N. M. Warid · M. K. I. M. Satar · N. A. A. Rahman
Faculty of Civil Engineering, Universiti Teknologi Malaysia, Skudai, Malaysia

M. R. Hainin
Universiti Malaysia Pahang, Pekan, Malaysia

© The Author(s), under exclusive license to Springer Nature Switzerland AG 2022

H. R. Pasindu et al. (eds.), *Road and Airfield Pavement Technology*, Lecture Notes in Civil Engineering 193, https://doi.org/10.1007/978-3-030-87379-0_46

1 Introduction

Asphalt is 100% reusable and while we are well on our way to achieving the goal of a truly circular economy in the sector, more remains to be done to improve the sustainability of road surfacing. Nonetheless, with roads increasingly being viewed as a service, comfort, safety and functionality are considered standard. When it comes to improve the sustainability of asphalt pavement, the waste hierarchy is a great tool to guide the decision-making process in selecting the most sustainable solutions. Many factors influence the behavior asphalt at low temperature such as environment, materials and also pavement structure. The consistency of the binder is an important factor for materials [7]. Temperature has a direct effect on bitumen, wherein at low temperature, bitumen acts like an elastic solid, while at high temperature, bitumen acts as viscous fluid. The behavior of bitumen must be measured through its physical, rheological and chemical properties; as bitumen is a viscoelastic material, the response depends largely on the temperature and loading time when stress is applied.

Nowadays, the environmental issue of biomass waste is of top concern [11]. Several researches have been conducted including the production of activated carbon from waste biomass. Coconut Shell had been discarded and grinded until it become a form of carbon. This study was conducted to determine the chemical and physical properties of coconut shell carbon and activated carbon to be used as an additive in asphalt bitumen.

In a study conducted by Licence [9], the pyrolysis oil obtained at these optimum process conditions were analyzed for physical and chemical properties to be used as an alternative fuel. As far as the proximate analysis is concerned, the contents of moisture, volatile matter, ash and fixed carbon are similar to the values reported for other types of waste [6]. The proximate and ultimate analysis and gross calorific values of the coconut shells are presented in Table 1. The results of this analysis showed that coconut shell has high potential to produce liquid fuel by pyrolysis conversion process based on investigation of the chemical properties [10].

Chemical changes during aging have been studied before and it is known that oxidation of bitumen produces carbonyls and sulfoxides. The production increases polarity and also causes increases in bitumen viscosity and softening point. Therefore, the result from aging test showed that chemical changes may differ largely between different bitumen, especially between unmodified and polymer modified binders.

Figure 1a shows the coconut shell collected for this study. The shell is first cleaned, then crushed to smaller sizes using an aggregate crusher machine. The crushed

Table 1 Proximate and ultimate analysis, and gross calorific values of solid coconut shell [6]

	Proximate analysis (wt%)			Ultimate analysis (wt%)					
Biomass samples	Combustible matter	Moisture	Ash	C	H	O	S	N	GCV (MJ/Kg)
Coconut shell	85.36	11.26	3.38	63.45	6.73	28.27	0.17	0.43	22.83

Fig. 1 **a** Coconut shell **b** Coconut char

coconut shell was burned at an optimum temperature of 450 °C for 5 min in a small furnace until it become in a form of carbon as shown in Fig. 1b. The objective in modifying the hot mix asphalt using coconut shell is to increase the specification of stability (permanent deformation resistance), durability (moisture damage and aging), fatigue cracking resistance, safety (adequate skid resistance), resistance to thermal cracking, permeability and flexibility.

Coconut Shell Carbon (CSC), commonly known as charcoal, is usually used as a fuel and as green material because it is safe to use [9]. It is one of the wood-based materials that can be used to produce charcoal. Charcoal derived from coconut shell has been applied in many things other than its use as fuel, such as gas purification, air filters, water filters and also in medicine [7]. This study evaluates the chemical properties of bitumen modified with CSC and coconut shell activated carbon (CSAC).

Activated carbon is obtained from other sources mainly due to small macrospores structure which renders it more effective for the adsorption of gas/vapor and for the removal of color and odor of compounds. It has high fixed carbon and low ash content. The adsorption behavior of activated carbon is determined not only by their porous structures but also by the chemical nature of its surface. Chemical properties describe the potential of some chemical change or reaction by virtue of its composition. Chemical change results in one or more substances from the original substances. The amount of chemical composition depends on the source of bitumen. The elements at the start of the reaction are rearranged into new product compounds or elements. Different elements or compounds are present at the end of the chemical change.

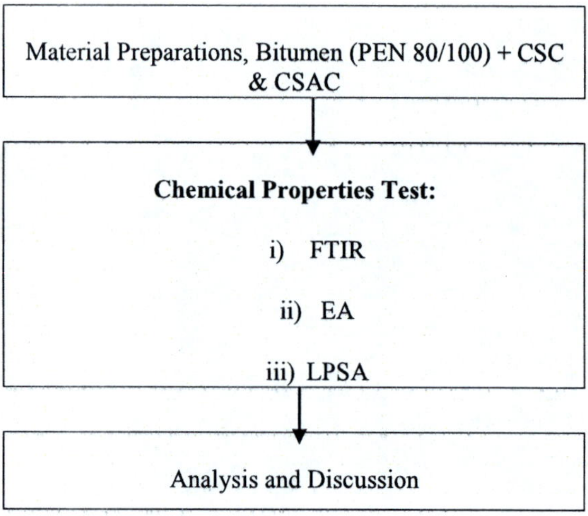

Fig. 2 Flow chart of methodology

2 Methodology

The chemical properties of the modified binder were determined through Fourier Transform Infrared Test (FTIR), Elemental Analyzer (EA) and Laser Particle Size Analyzer (LPSA). Figure 2 shows the flow chart of methodology for chemical properties test.

3 Materials and Method

3.1 Bitumen

The bitumen of penetration 80/100 was used in this study to produce a modified binder. The selection of bitumen contents is according to JKR Standard Specification for road [8], all the samples were ranged between 4 and 6% for AC 14 (Table 2).

Table 2 Characteristic of Bitumen PEN 80/100

Test	Standard test methods	80/100
Penetration (0.1 mm) at 25 °C	ASTM D5	85
Softening point (°C)	ASTM D36	46
Penetration index (PI)	–	−0.968
Ductility (mm) at 25 °C	ASTM D113	>100
Viscosity (mPa s) at 135 °C	ASTM D4402	306.5

3.2 *Production of CSC and Mixing Procedure*

3.2.1 Preparation of CSC

The coconut shell (waste) was collected from local area in Skudai, Johor and cut into small pieces and washed with water for removal of dust adhered to it. Then, it was dried in sunlight for 48 h or for 2 days. Dried materials were kept inside the furnace and it will be burnt in the furnace at 450 °C for removal of moisture and until it becomes charcoal. The charcoal is then crushed and sieved to 0.075 μm size. The CSAC (powder) were purchased from a supplier of activated carbon in Ipoh, Perak. After receipt of the material, the next process is to sieve the powder of CSAC to 0.075 μm size.

3.2.2 Mixing Process Between Bitumen and CSC and CSAC

Conventional bitumen 80/100 PEN was used in this study. The bitumen was supplied by Chevron Company in Selangor. The physical properties tests were performed on this bitumen prior to modification using CSC and CSAC. For the bitumen modification, the blending process of bitumen and binder was conducted by using a high shear mixer as shown in Fig. 3. The blending was carried out at the speed of 1000 rpm for 60 min at a temperature of 160 °C. A speed of 1000 rpm was used to ensure that the CSC and CSAC was well dispersed in the bitumen.

3.3 *Fourier Transform Infrared Test (FTIR)*

FTIR was used to determine the reaction of particles when exposed to infrared as shown in Fig. 4. The result obtained is used to determine the functional characteristic of the material. For bitumen sample, two polished salt plates were used to sandwich the bitumen sample. The plates were safely clamped and tested. Spectra or infrared spectrum was used for analysis by observing the absorbance infrared spectrum in order to obtain the qualitative and quantitative.

3.4 *Elemental Analyzer (EA) Test*

EA was conducted to analyze the chemical composition of particles. Information of chemical composition percentage can be obtained by using the analysis. Carbon and hydrogen are the main constituents found in the bitumen with a small amount of sulphur, nitrogen and oxygen. EA was used to test the chemical composition in

Fig. 3 Preparation of CSC

the bitumen as shown in Fig. 5. The change of composition from the bitumen tested was recorded.

3.5 Laser Particle Size Analyzer (LPSA) Test

Laser diffraction measures particle size distributions by measuring the angular variation in intensity of light scattered as a laser beam passes through a dispersed particulate sample. Large particles scatter light at small angles relative to the laser beam and small particles scatter light at large angles. Figure 6 shows the equipment of LPSA Test.

Fig. 4 Fourier transformation Infrared (FTIR)

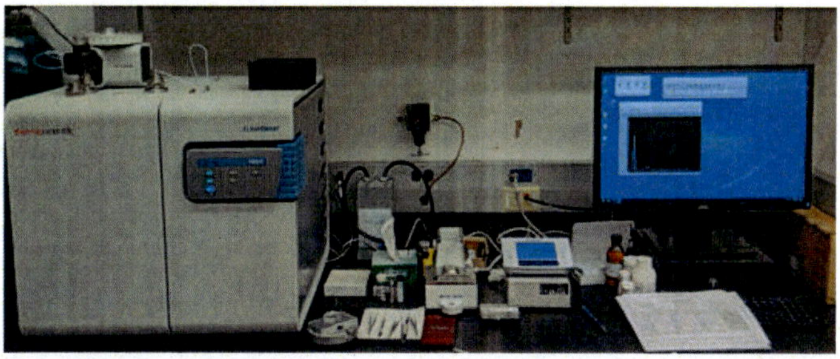

Fig. 5 Elemental analyzer (EA)

4 Results and Discussion

In this study, the chemical properties of the CSC and CSAC modified bitumen are evaluated through FTIR, EA and LPSA test. Results of the test are discussed in the following sections.

Fig. 6 Laser particle size analyzer (LPSA)

4.1 *Fourier Transform Infrared (FTIR)*

FTIR for asphalt binders can be used as an indicator of aging, because it allows the identification of the evolution of two chemical entities resulting from oxidation, the Carbonyl (C=O) and Sulfoxide (S=O) [12]. This test was conducted to investigate the change in chemical bond of the bitumen due to optimum bitumen content. CSC and CSAC are commonly used for modifying physical mechanical properties of bitumen. As shown in Figs. 7 and 8, FTIR test used in this study is capable to display wavenumber between 400 and 4000 cm^{-1}; only wavenumbers between 800 and 2800 cm^{-1} have been analyzed in this study. Figure 7 shows that smoothing CSAC recorded the wavenumber between 400 and 2400 cm^{-1}. The highest wavenumber is around 486 cm^{-1} and 0.55 Abs. In the range of 2700–400 cm^{-1}, several peaks were found. A strong functional group of the C=C bond of the aromatic index was identified near 2328 cm^{-1}. Meanwhile, at 2097 and 1991 cm^{-1}, C–H bond of the aliphatic index was observed. It was found significant in bitumen evolution and remains in a stable and homogenous state even though placed in high temperature condition. Figure 8 shows the spectrum for CSC, wavenumber that will recorded is between 600 and 2300 cm^{-1}. It shows the highest wavenumber is around 500 cm^{-1} and 1.7 Abs. In the range of 2050–400 cm^{-1}, several peaks were found. A strong functional group of the C=C bond of the aromatic index was identified near 2328 cm^{-1}. Meanwhile, at 2087 and 1992 cm^{-1}. C–H bond of the aliphatic index was observed. This is because the sample having different softening point temperature below 2 °C was categorized as stable and these samples were considered to be compatible.

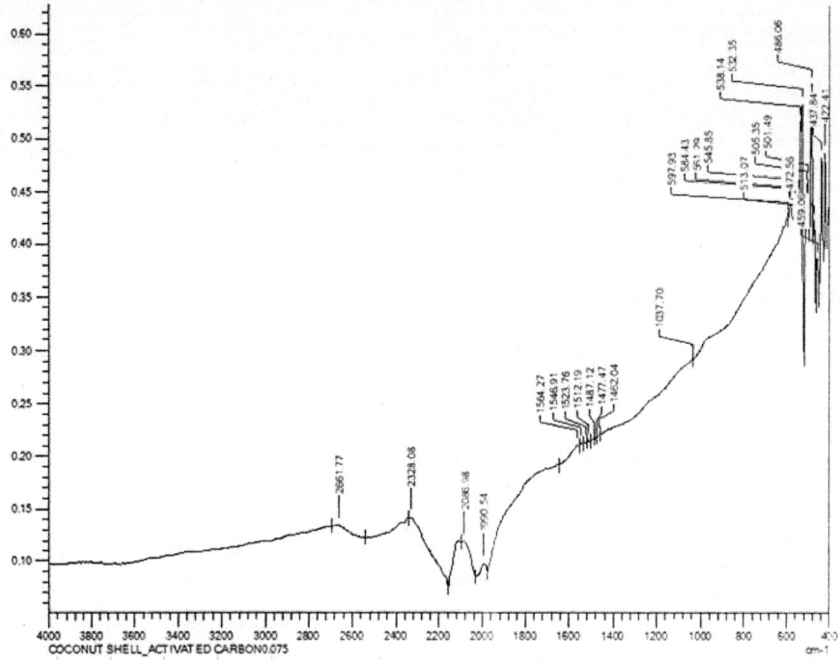

Fig. 7 FTIR spectrum for CSAC

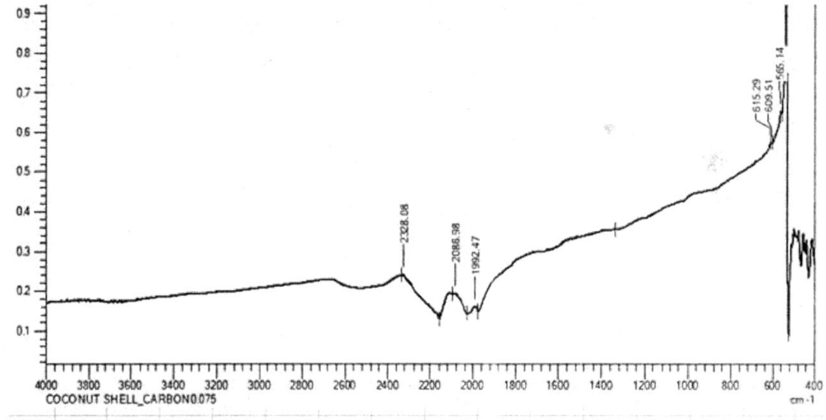

Fig. 8 FTIR spectrum of CSC

Table 3 Chemical components of CSC and CSAC

Elemental analysis (%)	CSC	CSAC
Carbon	54.3913	36.7021
Hydrogen	2.6811	4.6856
Nitrogen	0.3389	0.2578
Sulphur	0.241	0.1207

4.2 *Elemental Analyzer (EA)*

The EA of materials is mainly conducted as critical parameter in safety and product quality. The analysis was conducted to investigate the change in certain mixture based on chemical composition due to recycling process. This test is applied to bitumen content to determine the chemical composition in bitumen such as nitrogen, oxygen, carbon, hydrogen and sulphur. Table 3 presents the results of the chemical compositions between CSC and CSAC analyzed through the elemental analyzer. Based on the analysis, it was found that CSC and CSAC formed four main components which is carbon, hydrogen, nitrogen and sulphur. As a result, the CSAC mixtures have higher carbon, nitrogen and sulphur value than the CSC at 54%, 0.3% and 0.2%, respectively. It can be observed that CSC contains a very small amount of carbon, sulfur and nitrogen, 36%, 0.26% and 0.12%, respectively. It can be seen that the value of hydrogen for CSC was higher than the CSAC with value 4.7%. Based on the analysis, CSAC produced high carbon, nitrogen and sulphur with less hydrogen. This increase of carbon can be interpreted as carbon entering from the bitumen content and forming bonds, as well as bonding interactions with binder carbon molecules.

4.3 *Laser Particle Size Analyzer (LPSA)*

Figures 9 and 10 present the particle size and specific surface area results of CSC and CSAC. The results for all five percentages of modified bitumen of CSC and CSAC show the specific area of bitumen content consistently increase. The increase in specific surface area for modified bitumen of CSAC varied from 339.2 to 354.3 m^2/Kg with cumulative weight 15–20%. Meanwhile, the value of specific area for CSC varied from 35 to 77%. In general, modified bitumen prepared with CSC have the smaller particle size and greatest specific surface area, which indicated the highest storage stability. The result of particle size for CSC shows that the percentage of cumulative weight is higher when the specific area is lower. Meanwhile, the percentage of cumulative weight for CSAC is lower with bigger size of specific surface area. It can be seen that the specific area increases the contact between modifier (CSC and CSAC) with the bitumen, which affects the performance of the bitumen binder and also affects the performance of the mixtures.

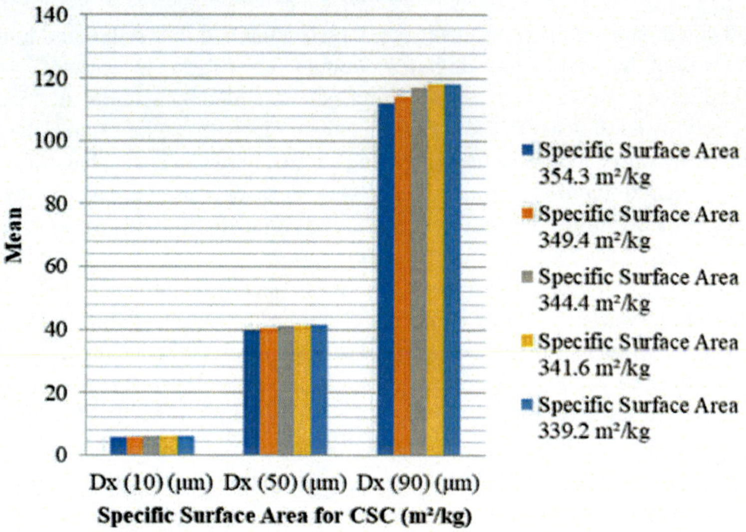

Fig. 9 Specific surface area for CSC

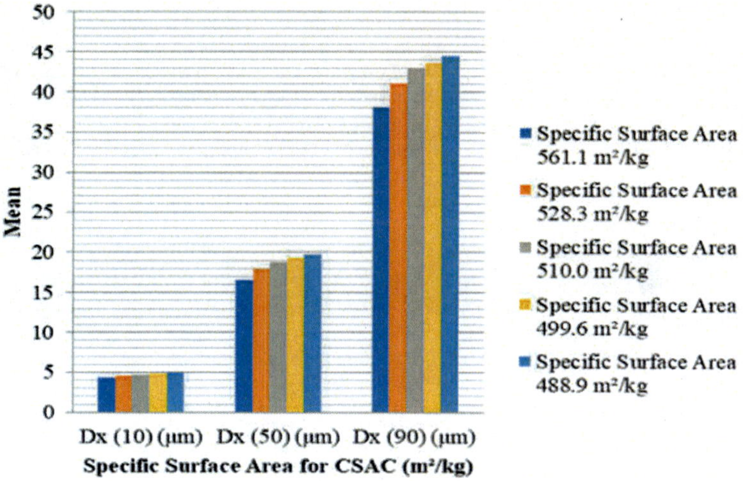

Fig. 10 Specific surface area for CSAC

5 Conclusions

Based on the results, it can be concluded that in terms of chemical point of view using elemental analyzer, the reduction in carbon, hydrogen, nitrogen and Sulphur compositions were observed. Moreover, result from Fourier transform infrared (FTIR) test showed significant bitumen performance, which remained in a stable and homogenous state even though placed in high temperature. The relative increase of C=O and S=O functional groups reflects asphalt aging, accompanied by a hardening (decrease in penetration and increase in the softening point). The LPSA result shows the larger specific surface area increases the contacting surface between the asphalt. Overall, it can be concluded that the coconut shell activated carbon was successfully prepared by chemical activation method. The results indicated that the adsorption capacity depends upon the pore size, surface area micropore volume of the activated carbon. Micro-mechanics of the bitumen surface are influenced by a synergistic effect of the chemical composition. In terms of chemical properties, value of carbon, hydrogen and nitrogen compositions were observed due to FTIR test. The results from Fourier transform infrared (FTIR) test showed that more carbonyl (C=O) and sulfoxide (S=O) were developed incorporating CSC and CSAC in bitumen mixtures. The results of chemical properties were supported by the chemical test which is EA and LPSA. From all the chemical tests, it was shown that the modified bitumen using CSC and CSAC can be selected to increase the performance of bitumen properties.

References

1. Ali AH, Mashaan NS, Karim MR (2013) Investigations of physical and rheological properties of aged rubberised bitumen. Adv Mater Sci Eng 2013:1–8. https://doi.org/10.1155/2013/239036
2. American Society of Testing and Materials (2012) ASTM D36. Standard Test Method for Softening Point of Bitumen (Ring-and-Ball Apparatus). ASTM International, West Conshohocken, PA, p 141
3. Asli H, Ahmadinia E, Zargar M, Karim MR (2012) Investigation on physical properties of waste cooking oil—Rejuvenated bitumen binder. Constr Build Mater 37:398–405. https://doi.org/10.1016/j.conbuildmat.2012.07.042
4. Da M, Lopes M, Zhao D, Chailleux E, Kane M, Gabet T (2013) Characterization of aging processes on the asphalt mixture surface. To cite this version: characterization of aging processes on the asphalt mixture
5. Das D, Samal DP, Meikap BC (2015) Preparation of activated carbon from green coconut shell and its characterization. J Chem Eng Process Technol 06(05). https://doi.org/10.4172/2157-7048.1000248
6. Demirbaş A (1999) Properties of charcoal derived from hazelnut shell and the production of briquettes using pyrolytic oil. Energy 24:141–150
7. Isacsson U, Zeng H (1997) Relationships between bitumen chemistry and low temperature behaviour of asphalt. Constr Build Mater 11(2):83–91. https://doi.org/10.1016/S0950-0618(97)00008-1
8. Jabatan Kerja Raya (2008) Standard specification for road works, Section-4: flexible pavement. Jabatan Kerja Raya, Malaysia, Kuala Lumpur
9. Licence CC (2011) Icme11-Abs-037 Pyrolysis of Coconut Shell for Bio-Oil 2011 (December):18–20
10. Lu X, Talon Y, Redelius P (2008) Aging of bituminous binders—laboratory tests and field data. The 4th Eurasphalt & Eurobitume Congress, pp 1–12. https://doi.org/10.13140/2.1.4101.2487

11. Mohd Iqbaldin MN, Khudzir I, Mohd Azlan MI, Zaidi AG, Surani B, Zubri Z (2013) Properties of coconut shell activated carbon. J Trop For Sci 25(4):497–503. https://doi.org/10.2307/23616990
12. Mohanty K, Naidu JT, Meikap BC, Biswas MN (2006) Removal of crystal violet from wastewater by activated carbons prepared from rice husk. Ind Eng Chem Res 45:5165–5171

Comparative Assessment of CMSDBC and HMSDBC Competency

Gautam Prakash and Sanjeev Kumar Suman

Abstract Over ninety percent of the total road network in India, including the airfield pavements, are flexible pavements with black-top surfaces. The construction of wearing and binder courses of a flexible pavement require high energy consumption during heating of bitumen and aggregate, with consequent giant emission of greenhouse gases during the preparation of hot-mixed bituminous mixtures. Earlier, in flexible pavements, the use of bitumen emulsion was restricted to only spray applications and as dust palliatives, but over the years, with the development of new types, grades, specifications, and availability of improved construction equipment and practices, emulsion-based cold mix technology offers a wide range of solutions. Judicial selection and appropriate use of these technologies can yield significant economic benefits, environmental benefits, and energy security as far as construction and maintenance of roads are concerned. Hence, this paper aimed to compare the strength, durability, and performance of a cold mix semi-dense bituminous concrete (CMSDBC) and hot mix semi-dense bituminous concrete (HMSDBC) pertaining to competency. Materials for the preparation of representative samples are coarse aggregate, fine aggregate, and cement as a filler, in suitable proportions mixed with bitumen or slow-setting cationic bitumen emulsion (where applicable). Tests like Marshall stability, Indirect tensile strength, moisture susceptibility, and rut depth using wheel tracker device were carried out according to standard guidelines. The results showed that CMSDBC yielded significant values, which are comparable with HMSDBC. It infers that CMSDBC is capable to be exercised in the practice for low traffic, low rainfall, moderate and cold climate regions.

Keywords Sustainability · Cold mix · Emulsion · Strength · Durability · Rut depth

G. Prakash (✉) · S. K. Suman
Civil Engineering Department, National Institute of Technology (NIT) Patna, Bihar 800005, India
e-mail: gautamp.phd18.ce@nitp.ac.in

S. K. Suman
e-mail: sksuman@nitp.ac.in

© The Author(s), under exclusive license to Springer Nature Switzerland AG 2022

H. R. Pasindu et al. (eds.), *Road and Airfield Pavement Technology*, Lecture Notes in Civil Engineering 193, https://doi.org/10.1007/978-3-030-87379-0_47

1 Introduction

Road construction is on the rise all over the country and is evident from the fact that there has been a rapid increase in India's rate of construction of roads per year since 2013–14 [1]. Though road construction is beneficial to the developmental growth of the nation, it has also resulted in a proportional increase in the production of greenhouse gases which eventually causes global warming. The primary reason of the emission of harmful gases is the adoption of hot mix asphalt (HMA) technology by the paving engineers for the construction of roads. In addition to the emission problem associated with the HMA technology, the requirement of intensive fuel energy demand for sustained heating of bitumen, aggregate, and the bituminous mixture during its preparation too is a deterrent. Considering all these factors, the use of cold mix asphalt (CMA) technology in flexible pavement has gained importance in many developed countries. CMA is produced by mixing bitumen emulsion, mineral filler, and un-heated aggregates at ambient temperatures (<40 °C). The non-heating of aggregates and bitumen during its preparation adds to the ease in laying of paving mix in wet and humid climatic conditions. It also makes CMA relatively pollution-free and economical [2]. The energy consumption and emission of harmful gases for construction of around 10 cm thick pavement are almost twice higher for HMA when compared with CMA [3]. Despite several advantages, CMA's use has mostly been limited to patch repairs and crack seals.

The engineering properties of CMA primarily depend on the aggregate source, emulsion type, curing condition, and time of curing [4]. MORTH [5] specification has almost followed the guidelines given by the Asphalt Institute for designing of CMA. Numerous researches have been conducted around the world to study the properties of CMA. Oruc et al. [6] suggested that the emulsified cold mixes with cement as filler may be used as a structural pavement layer. Thanaya et al. [7] reported that the addition of 1–2% of rapid-setting cement accelerated the early strength in addition to the enhanced mechanical performance of CMA. Pundhir et al. [8] prepared semi-dense bituminous concrete (SDBC) by cold mix method under varying test conditions and reported improvement in mix properties when curing time was increased. Abbas Al-Hdabi et al. [9] studied the effects of cement on gap-graded CMA and reported improved mechanical properties comparable with HMA within one-day curing. From field studies, it was reported that pavements constructed with CMA technology have excelled well for 10–15 years [10].

2 Research Objectives

The purpose of this study is to ascertain the competency of cold mix technology with respect to the conventional HMA technology by preparing semi-dense bituminous concrete (SDBC) mixtures. The research objectives are as follows:

- To evaluate and compare the marshall properties of cold mix SDBC (CMSDBC) mixtures and hot mix SDBC (HMSDBC) mixtures prepared at optimum binder contents.
- To evaluate and compare the performance of CMSDBC in comparison to HMSDBC based on mechanical properties such as tensile strength, moisture susceptibility, and rutting properties.

3 Materials and Methods

This section deals with the materials used and methods adopted in the construction of CMSDBC and HMSDBC mixtures. SDBC is a continuously graded mix, which can be used as a binder course and wearing course in flexible pavement. It consists of coarse aggregates, fine aggregate and filler, with suitable quantities of bitumen/emulsified bitumen binder.

3.1 Bitumen/Bitumen Emulsion

The bitumen used for the study was VG 30, which was utilized as the binder material for the preparation of HMA. For preparing CMA mixtures, cationic emulsification of bitumen was done with 65% residual bitumen. The percentage of residual bitumen was determined based on design procedures suggested by the Asphalt Institute [4]. The physical properties of bitumen and its emulsion is given in Table 1.

3.2 Aggregate

Three sets of natural stone aggregate, conforming to MORTH [5], have been used in this study, namely—20 mm nominal size, 10 mm nominal size, and stone dust with a specific gravity of 2.78, 2.72, and 2.70 respectively. These aggregates have been obtained from a quarry site at Gaya, Bihar, India. The physical properties of aggregates were tested and are presented in Table 2.

3.3 Filler

PPC 53 grade cement with a specific gravity of 3.04 was used as the filler material in this study. The gradation of the filler has met the required specifications as per IRC: 111 [22] and has been shown in Table 3.

Table 1 Physical properties of bitumen and emulsified bitumen

Property	Result	Testing standard
(a) *VG 30 bitumen*		
Specific gravity	1.02	IS 1202 [11]
Ductility at 25 °C (in cm)	87	IS 1208 [12]
Penetration at 25 °C (in mm)	68	IS 1203 [13]
Absolute viscosity at 60 °C (in poise)	2700	IS 1206 (part II) [14]
Kinematic viscosity at 135 °C (in cSt)	380	IS 1206 (part III) [15]
Softening point (in °C)	53	IS 1205 [16]
(b) *Emulsified bitumen*		
Specific gravity	1.01	IS 1202 [11]
Residual bitumen content	65%	IS 8887 (Annex J) [17]
Ductility at 27 °C (in cm)	98	IS 1208 [12]
Penetration at 25 °C (in mm)	82	IS 1203 [13]
Viscosity at 50 °C (in cSt)	125	ASTM D88 [18]

Table 2 Physical properties of aggregate

Property	Result	Testing standard
Aggregate impact value (%)	21.4	IS 2386 (part IV) [19]
Los Angeles abrasion value (%)	28.7	IS 2386 (part IV) [19]
Aggregate crushing value (%)	24.6	IS 2386 (part IV) [19]
Elongation index (%)	11	IS 2386 (part I) [20]
Flakiness index (%)	13	IS 2386 (part I) [20]
Water absorption (%)	0.4	IS 2386 (part III) [21]

Table 3 Gradation of the filler material

IS sieve (mm)	Cumulative % passing by weight of total aggregate	
	Obtained value	Required value as per specification
0.6	100	100
0.3	97.2	95–100
0.075	88.7	85–100

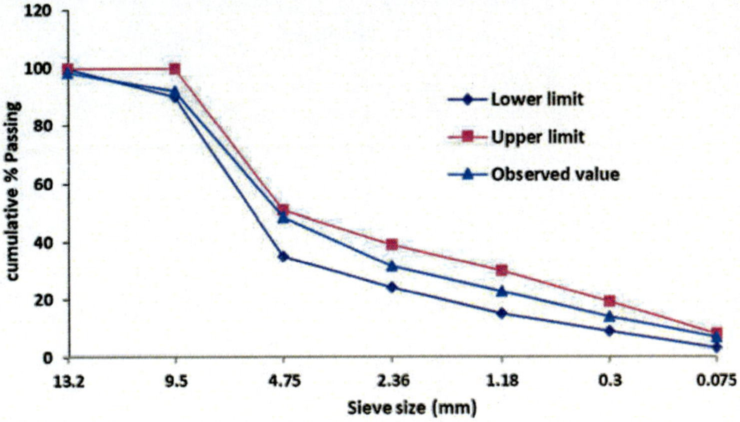

Fig. 1 Aggregate gradation used in CMSDBC and HMSDBC mixtures

3.4 Aggregate Gradation

Sieve analysis was conducted by the wet sieving method on the aggregates as per IS 2386 part I [20]. The combined gradation of aggregates and the filler must conform to the specified limits of CMSDBC and HMSDBC as per IRC:SP:100 [23] and IRC: 111 [22] respectively. Both CMSDBC and HMSDBC were prepared with a nominal maximum aggregate size of 9.5 mm and the same aggregate gradation. The final proportioning after combined aggregate gradation was selected at 5:47:45:3 for 20-mm nominal size aggregate, 10-mm nominal size aggregate, stone dust, and filler respectively. Figure 1 shows the plotted aggregate gradation chart with lower and higher limits.

3.5 Design Procedure of CMA

After the gradation of aggregates, the preparation of CMA includes the determination of various parameters to ascertain its suitability as compared to HMA. The designing procedure as per Asphalt Institute [4] and MORTH [5] has been discussed below by a step-by-step process.

- Calculation of initial residual bitumen content (IRBC) and initial emulsion content (IEC) by mass of the total mixture.

IRBC and IEC have been calculated by Eqs. 1 and 2 respectively.

$$IRBC = (0.05A + 0.1B + 0.5C) \times (0.7) \tag{1}$$

$$IEC = IRBC/X \tag{2}$$

where,

A percentage of aggregate retained on 2.36 mm sieve;
B percentage of aggregate passing 2.36 mm sieve and retained on 0.075 mm sieve;
C percentage of aggregate passing 0.075 mm sieve;
X percentage bitumen content in the emulsion.

- Determination of optimum pre-wetting water content (OPWC) by dry density test

In this test, pre-wetting of dry aggregates and filler material was done with suitable water content. After pre-wetting, emulsified bitumen of IEC value was added and mixed for few minutes to form a uniform coating. The process was repeated with varying amounts of water content at the same IEC value. The dry density of each sample was determined and the water content at which the maximum density was achieved was termed as OPWC.

- Determination of optimum total liquid content (OTLC) by dry stability test

Aggregate was mixed with emulsion (IEC value) and water (OPWC value) to prepare the mix. Then the mix was compacted at 50 marshall blows on both sides of the sample. More samples were prepared with varying PWC content. After compacting, the samples were kept in moulds for 24 h before extracting. Later, they were stored at 40 °C in an oven for 24 h. Then, they were kept at room temperature for 24 h. These samples were then tested for Marshall stability. The liquid content at which the highest stability value was achieved was reported as the OTLC.

- Determination of optimum residual bitumen content (ORBC) by soaked stability test

For the soaked stability test, the dry test samples were prepared as explained earlier in the dry stability test. After that, the dry samples were water-conditioned at room temperature for 48 h with a water level of half of the total thickness. The samples were inverted after the first 24 h and the other half was soaked in water. After conditioning, the samples were dried with a towel, and the Marshall stability test was performed. The residual bitumen content corresponding to the maximum soaked stability was reported as ORBC.

4 Results and Discussion

4.1 Determination of Optimum Compositions of HMSDBC and CMSDBC

Determination of optimum bitumen content (OBC) of HMSDBC mixture. The OBC for HMSDBC was determined by the Marshall method of mix design. The test procedure was followed as per ASTM D6927-15 [24]. The observed values of bitumen content at maximum stability, maximum bulk density, and at a mean air void percentage of 4% were selected for the calculation of OBC according to Eq. 3. The OBC was calculated to be 5.47%, sufficiently greater than the minimum prescribed value of 5% for HMSDBC as per IRC 111 [22]. The test results have been presented in Fig. 2.

$$OBC = \frac{BC_{\max\ density} + BC_{\max\ stability} + BC_{at\ 4\%\ air\ void}}{3} \tag{3}$$

Determination of optimum CMSDBC compositions. As per the CMA design procedure discussed in Sect. 3.5, the optimum compositions of CMSDBC were calculated based on IRBC and IEC values. From Eqs. 1 and 2, the IRBC value was calculated as 6.53% while, IEC was found to be 10.05%. Then, the OPWC was determined by the dry density method keeping a constant IEC and varying the pre-wetting water content as shown in Fig. 3. The OPWC was found to be 3.28%.

Further, OTLC was determined by dry stability test by varying the total liquid content while keeping the IEC value constant. It was found to be 6.48% (Fig. 4a). Also, the soaked stability test was conducted to determine the ORBC by varying the residual bitumen content while maintaining constant OTLC. It was found to be 4.81% (Fig. 4b). The obtained ORBC value was well above the minimum residual bitumen content of 4.5% for CMSDBC mixtures as per IRC: SP: 100 [23].

4.2 Marshall Properties

The Marshall properties such as Marshall stability, Marshall flow, percentage of air voids, voids in mineral aggregate (VMA), and voids filled with bitumen (VFB) of HMSDBC and CMSDBC mixtures were compared. The samples were prepared at their respective OBC and ORBC values of 5.47% and 4.81% for HMSDBC and CMSDBC respectively. The obtained values have been presented in Table 4.

The Marshall stability for HMSDBC was examined at 60 °C and the obtained value of 10.76 kN was above the minimum stability value of 9 kN as per IRC: 111 [22]. In the case of CMSDBC, IRC: SP: 100 [23] suggests testing the stability at 25 °C after 72 h of curing at room temperature. The obtained value of 5.73 kN was

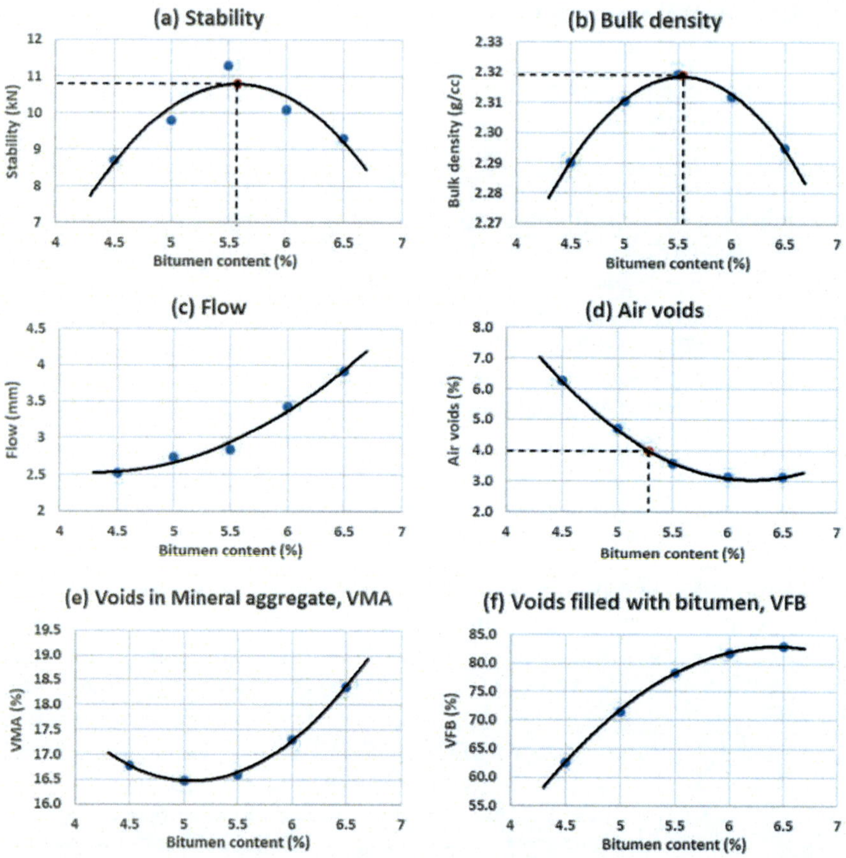

Fig. 2 Variation of (**a**) stability, (**b**) bulk density, (**c**) flow, (**d**) air voids %, (**e**) VMA, and (**f**) VFB in HMSDBC mixtures, with increase in bitumen content

found to be greater than the minimum required stability value of 5 kN. CMSDBC mixtures exhibited a higher flow value than HMSDBC due to the presence of water during its preparation. It requires a longer curing time to become stable. Air voids % in CMSDBC was found to be 8.32%, which was well above the range of 3–5% given by MORTH [5]. However, Gao et al. [25] have suggested that it is practically not possible to achieve an air void range as specified by MORTH [5] and for the design purpose generally an air void range of 7–10% must be adopted. The VMA and VFB values of both HMSDBC and CMSDBC were similar although the VMA value tends to be slightly higher in CMSDBC.

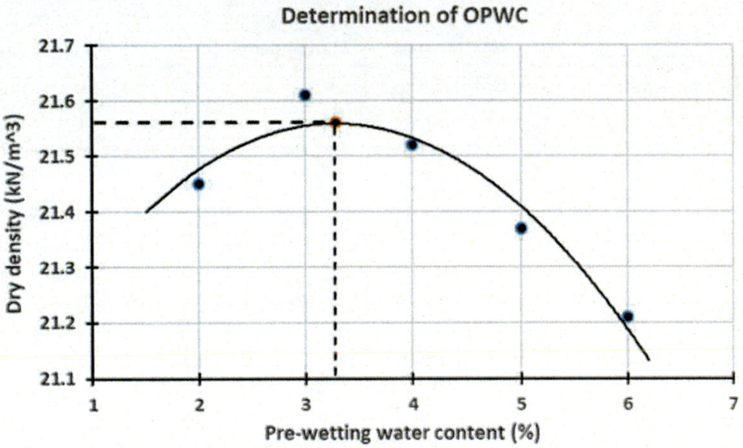

Fig. 3 Determination of OPWC in CMSDBC

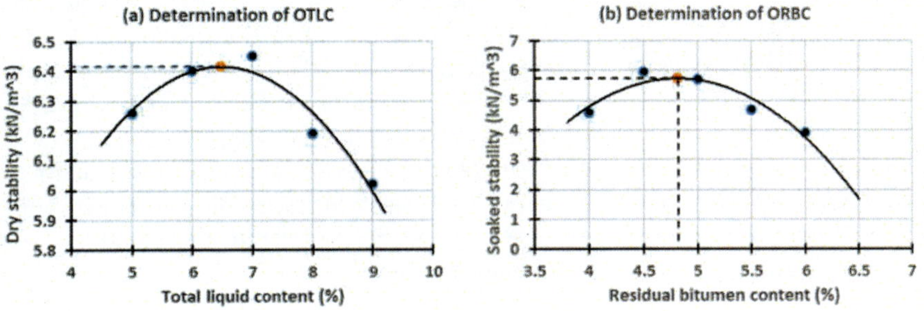

Fig. 4 Determination of (**a**) OTLC, and (**b**) ORBC in CMSDBC

Table 4 Marshall properties of HMSDBC and CMSDBC mixtures

Marshall properties	HMSDBC	CMSDBC
Marshall stability (kN)	10.76	5.73
Marshall flow (mm)	2.92	3.48
Air voids (%)	3.67	8.32
VMA (%)	16.61	16.92
VFB (%)	77.87	77.51

4.3 Tensile Strength Properties

The HMSDBC and CMSDBC mixtures were tested for tensile strength properties using the indirect tensile strength (ITS) test. For the ITS test, the mixtures were prepared at their respective OBC and ORBC values. The specimens were tested by adjusting the Marshall samples through two diametrically opposite loading strips.

Table 5 Tensile strength properties of HMSDBC and CMSDBC mixtures

Mixture type	Curing period (h)	ITS value (N/mm^2)
HMSDBC	24	0.92
	48	0.93
	72	0.95
CMSDBC	24	0.67
	48	0.72
	72	0.81

The compressive load induces a tensile load implicitly in the horizontal direction of the sample. A constant increasing compressive load is applied on the loading strips and the peak load was recorded. The specimen collapses along the diametric plane. The procedure adopted for the ITS test is outlined in ASTM D6931 [26]. To understand the effects of curing periods on tensile strength, all the samples were tested at three different curing periods of 24, 48 and 72 h in air. From the results, it was found that the CMSDBC mixtures exhibited lower indirect tensile strength than the HMSDBC mixtures (Table 5). In CMSDBC samples, a reduction of about 28%, 23% and 15% in the ITS values were observed with respect to HMSDBC samples at 24, 48 and 72 h of curing periods respectively. It was also observed that the tensile properties of CMSDBC can be enhanced with increased curing time (Table 5). The ITS value increased from 0.67 N/mm^2 (24 h.) to 0.72 N/mm^2 and 0.81 N/mm^2 at 48 h and 72 h curing periods respectively. This may be attributed to lower adhesion between the binder and aggregate due to the presence of moisture during its preparation.

Mathematically, the ITS is represented by Eq. 4.

$$ITS = \frac{2P}{\pi t D} \tag{4}$$

where,

P Maximum load (N);
t specimen height before test (mm);
D specimen diameter (mm).

4.4 Moisture Susceptibility

Moisture susceptibility is defined as the loss of serviceability of the bituminous mixture due to the presence of moisture. The retained Marshall stability (RMS) and tensile strength ratio (TSR) are responsible to control the moisture susceptibility problems. RMS is defined as the ratio of the stability of a water-conditioned Marshall specimen to the stability of an unconditioned Marshall specimen. Similarly, TSR can be defined as the ratio of the tensile strength of a water-conditioned Marshall

Table 6 Moisture susceptibility of HMSDBC and CMSDBC mixtures

Mixture type	RMS (%)	TSR
HMSDBC	92.94	0.91
CMSDBC	70.01	0.73

Specimen to the tensile strength of an unconditioned Marshall specimen. The conditioning period may vary according to the mixture type. According to IRC: 111 [22], the HMSDBC specimens were needed to be conditioned for 24 h at 60 °C in a water bath. While for CMSDBC, the specimens were needed to be water-conditioned at room temperature for 48 h with a water level of half of the total thickness as suggested in IRC: SP: 100 [23]. The specimen was inverted after the first 24 h and the other half was soaked in water. After conditioning, the samples were dried with a towel before carrying on with the susceptibility test. Table 6 presents the results of the RMS and TSR tests. The results showed that the CMSDBC mixture exhibited lesser moisture resistance than HMSDBC, although all the values were well within the range prescribed by IRC:111 [22] for and IRC: SP: 100 [23]. The RMS and TSR values of CMSDBC were found to be reduced by about 25% and 20% respectively in comparison to HMSDBC.

4.5 *Rutting Properties*

The bituminous mixtures were tested for rutting deformations at 60 °C using the wheel tracker device developed by Cooper Technology. The test specimen was prepared in a rut-testing mould with dimensions 305 mm × 305 mm × 50 mm. The required amount of bituminous mixture into the rut-testing mould is given by Eq. 5.

$$M = L \times l \times e \times \rho_m \times \left(\frac{100 - v}{100}\right) \times 10^{-6} \quad (5)$$

where,

- M mass of the bituminous mixture (kg);
- L Internal mould length (mm);
- l internal mould width (mm);
- e slab thickness (mm);
- ρ_m maximum density of bituminous mix (kg);
- v void content in slab (%).

The rut testing specimen was prepared for an air void content of 4%. After pouring the mixture into the mould, the compaction of the specimen was done, in accordance with EN 12697–33 [27], in a hydraulic steel roller compactor [28] developed by Cooper Technology. The compacted mixture was then placed into the wheel tracker

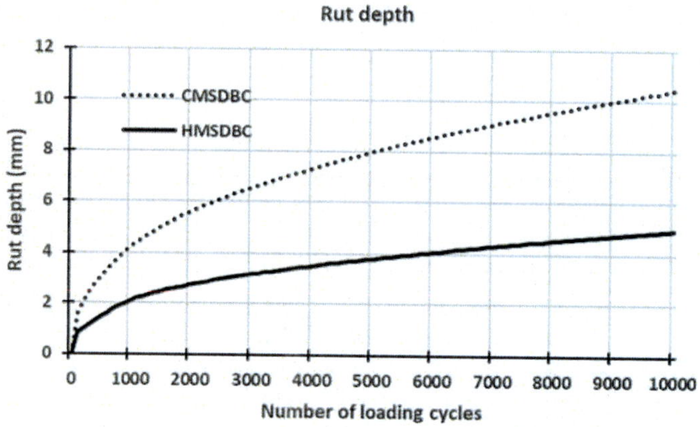

Fig. 5 Variation of rut depth with the number of loading cycles in HMSDBC and CMSDBC

device [29] developed by the Cooper technology, not before a minimum of 48 h has passed after compaction. The specimen was conditioned in the air at the testing temperature of 60 °C for a period of 4 h in the air-tight wheel tracker device. The conditioning was followed by the rolling of the testing wheel of diameter 200 mm and width 50 mm, on the compacted bituminous mixture up to 10,000 loading cycles. The temperature was maintained at 60 °C throughout the test. The whole process of rut testing was carried out in accordance with EN 12697–22 [30]. Figure 5 presents the variation in rut depth in HMSDBC and CMSDBC mixtures with increment in loading cycles. The result showed that CMSDBC has a higher rate of rut development than HMSDBC by about 53%.

Furthermore, other rut parameters like wheel tracking slope (WTS), dynamic stability index (DSI), and complex stability index (CSI) were also calculated from the results of the wheel tracker test to study the rutting characteristics of the mixtures (Table 7). WTS is defined as the amount of rut depth developed per 1000 load cycles. It is calculated by using Eq. 6.

$$WTS = \frac{d_{10000} - d_{5000}}{5} \tag{6}$$

where,

Table 7 Behaviour of rut parameters in HMSDBC and CMSDBC mixtures

Rutting parameters	HMSDBC	CMSDBC
Rut depth at 10,000 load cycles (mm)	4.95	10.44
WTS (mm/1000 load cycles)	0.228	0.492
DSI (load cycles/mm)	1608	730.9
CSI (load cycles/mm^2)	708.4	160.3

WTS wheel tracking slope (mm/1000 load cycles);
$d_{10,000}$ rut depth after 10,000 load cycles (mm);

d_{5000} rut depth after 5000 load cycles (mm).

DSI is defined as the number of load cycles to generate 1-mm rut depth during the last 15-min of one-hour testing. Whereas, CSI is defined as the number of load cycles to generate a deformation of an area of 1 mm^2 during the last 15-min of one-hour testing. Weaker the bituminous mixture, lower the values of DSI and CSI. Mathematically, DSI and CSI are calculated by Eqs. 7 and 8 respectively.

$$DSI = \frac{(t_2 - t_1)N}{d_2 - d_1} \tag{7}$$

$$CSI = \frac{(t_2 - t_1)N}{(d_2 - d_1)d_1} \tag{8}$$

where,

DSI dynamic stability index (load cycles/mm);
CSI complex stability index (load cycles/mm^2);
d_1 rut depth (mm) at time t_1;
d_2 rut depth (mm) at time t_2;
t_1 45 min.;
t_2 60 min.;
N running speed of wheel (26.8 cycles per minute).

The tests on rut parameters revealed that CMSDBC exhibit higher rut depth and WTS value than HMSDBC whereas higher DSI and CSI indices were reported in HMSDBC. The HMSDBC samples showed a rut depth of 4.95 mm after 10,000 cycles in comparison to 10.44 mm in CMSDBC. The WTS results showed that an average rut of 0.228 mm and 0.492 mm were developed per 1000 load cycles in HMSDBC and CMSDBC samples respectively. Regarding DSI, it took an average of 1608 load cycles to develop a rut of 1 mm in HMSDBC in comparison to 730.9 load cycles in CMSDBC. Similarly, the CSI was found to be 708.4 load cycles/mm^2 and 160.3 load cycles/mm^2 in HMSDBC and CMSDBC respectively. These results suggest that HMSDBC mixtures have better rutting characteristics in comparison to CMSDBC. The primary reason for such behavior is less aging of the binder during the cold mix preparation. The non-heating of the binder during its preparation result in insufficient aging, which ultimately leads to imparting reduced stiffness to CMSDBC mixtures.

5 Conclusions

The competency of cold mix with respect to hot mix technology has been assessed in this paper by laboratory evaluation of SDBC mixtures. The evaluated parameters in this assessment include Marshall properties, tensile strength properties, moisture susceptibility, and rutting characteristics. The bituminous mixtures were prepared at their respective OBC and ORBC values for HMSDBC and CMSDBC respectively. These optimum values have been determined before the assessment of the performance parameters of bituminous mixtures.

Both HMSDBC and CMSDBC mixtures have met the required specifications according to IRC 111 [22] and IRC SP 100 [23] respectively. The results showed that CMSDBC yielded significant values which are comparable, but inferior to HMSDBC on all the evaluated parameters stated above. Based on the test results, it was concluded that CMSDBC has higher air voids, lesser Marshall stability, lower tensile strength, increased moisture susceptibility, and reduced rutting properties than HMSDBC. These issues make the CMSDBC not favourable to be used for the surface course in heavy traffic, high rainfall and high temperature regions. However, they are capable to be exercised in the practice for low traffic, low rainfall, moderate and cold climate regions.

References

1. NHAI (2018) Highway construction touched 30 km/day in 2018–19. Times of India
2. Al-Busaltan S, Al Nageim H, Atherton W, Sharples G (2012) Mechanical properties of an upgrading cold-mix asphalt using waste materials. J Mater Civ Eng 24:1484–1491. https://doi.org/10.1061/(asce)mt.1943-5533.0000540
3. Goyer S, Dauvergne M, Wendling L et al (2012) Environmental data of cold mix using emulsified bitumen for a better selection of road materials. ISAP, Fr 1–12
4. Institute Manual Series No. 14 (AI MS 14) (1997) Asph cold Mix manual, Third Ed Part IV, Append F, USA, pp 106–122
5. Ministry of Road Transport and Highways (MORTH) (2013) Specifications for road and bridge works. Indian Roads Congr, pp 1–883
6. Oruc S, Celik F, Akpinar MV (2007) Effect of cement on emulsified asphalt mixtures. J Mater Eng Perform 16:578–583. https://doi.org/10.1007/s11665-007-9095-2
7. Thanaya INA, Zoorob SE, Forth JP (2009) A laboratory study on cold-mix, cold-lay emulsion mixtures. Proc Inst Civ Eng Transp 162:47–55. https://doi.org/10.1680/tran.2009.162.1.47
8. Pundhir NKS, Grover S, Veeraghavan A (2010) Cold Mix design of semi dense bituminous concrete. Indian Highw Indian Roads Congr 38:17–24
9. Al-Hdabi A, Al Nageim H, Seton L (2014) Performance of gap graded cold asphalt containing cement treated filler. Constr Build Mater 69:362–369. https://doi.org/10.1016/j.conbuildmat.2014.07.081
10. Redelius P, Östlund JA, Soenen H (2015) Field experience of cold mix asphalt during 15 years. Road Mater Pavement Des 17(1):223–242. https://doi.org/10.1080/14680629.2015.1068702
11. IS:1202 (1978) Indian standard methods for testing tar and bituminous materials: determination of specific gravity. Bur Indian Stand.
12. IS:1208 (1978) Indian standard methods for testing tar and bituminous materials: determination of ductility. Bur. Indian Stand.

13. IS:1203 (1978) Indian standard methods for testing tar and bituminous materials: determination of penetration. Bur. Indian Stand.
14. IS:1206 (Part II) (1978) Indian standard methods for testing tar and bituminous materials: determination of viscosity: Part II Absolute Viscosity. Bur. Indian Stand.
15. IS:1206 (Part III) (1978) Indian standard methods for testing tar and bituminous materials: determination of viscosity: Part III Kinematic Viscosity. Bur. Indian Stand.
16. IS:1205 (1978) Indian standard methods for testing tar and bituminous materials: determination of softening point. Bur. Indian Stand.
17. IS:8887 (2004) Bitumen Emulsion for Roads (cationic type)—specification. Bur Indian Stand, New Delhi
18. ASTM D88-07 (2013) Standard test method for Saybolt viscosity. In: ASTM International
19. IS 2386-4 (2002) Methods of test for aggregates for concrete, part 4: mechanical properties. New Delhi
20. IS 2386-1 (2002) Methods of test for aggregates for concrete, part I: particle size and shape. Bureau of Indian Standards, New Delhi, New Delhi
21. IS 2386-3 (2002) Methods of test for aggregate for concrete, part 3: specific gravity, density, voids, absorption and bulking
22. IRC: 111 (2009) Specifications for dense graded bituminous mixes. In: Indian Roads Congress. Indian Roads Congress, New Delhi, p 32
23. IRC:SP:100 (2014) Use of Cold Mix technology in construction and maintenance of roads using Bitumen Emulsion. Indian Roads Congr
24. ASTM D 6927-15 (2015) Standard test method for Marshall stability and flow of asphalt mixtures. ASTM Int Conshohocken, PA
25. Gao L, Ni F, Luo H, Charmot S (2015) Characterization of air voids in cold in-place recycling mixtures using X-ray computed tomography. Constr Build Mater 84:429–436. https://doi.org/10.1016/j.conbuildmat.2015.03.081
26. ASTM D 6931 (2018) 6931, Standard test method for indirect tensile (IDT) strength of asphalt mixtures. In: American society for testing and materials
27. EN 12697-33 (2003) Bituminous mixtures—test methods for hot mix asphalt—part 33: specimen prepared by roller compactor. Eur Stand, 1–9
28. CRT-RC-H2-1712-01 (2016) Hydraulic steel roller compactor—dual size
29. CRT-WTECO-A-1712-01 (2016) Wheel tracker ECO small device. Cooper Res Technol Ltd., Derbyshire, United Kingdom, pp 1–40
30. EN 12697-22 (2003) Bituminous mixtures—test methods for hot mix asphalt—part 22: wheel tracking. Eur Stand, 1–29

Effect of Hot Storage on Engineering Properties of Polymer Modified Binders and Asphalt Mixtures

Jian-Shiuh Chen, Shih-Hsiu Wei, and Wencheng Chang

Abstract Styrene butadiene styrene (SBS) is the polymer type most commonly used to manufacture polymer modified binders (PMBs). PMBs may need to be kept in a hot storage tank for an extended period of time because of adverse weather conditions or other uncertainties. Though PMBs show enhanced performance, a premature failure might occur in asphalt concrete during hot storage on exposure to a combination of heat and air. The aim of the current study is to examine the effect of hot storage on the engineering properties of PMBs and asphalt mixtures. SBS-based PMBs were stored in a storage tank at 180 °C up to 28 days in this study. It was estimated that more than four fifths the SBS polymer remained after long storage periods. There was a reduction in tenacity and elastic recovery test results, indicating that the degradation of the SBS polymer could occur after a long period of hot storage time. The other PMB test properties including toughness and storage stability did not vary significantly as the storage time was increased. Even though the SBS polymer in PMBs appeared to be degrading during storage, there was no change in the wheel tracking performance of the PMBs in asphalt mixtures. The cracking resistance obtained in semi-circular bend tests, by contrast, varied for the PMBs after hot storage. The results obtained in this study show that degradation of the polymer in a PMB during hot storage may not necessarily reduce the performance of asphalt mixtures.

Keywords Polymer modified binder · Hot storage · Storage tank · Polymer degradation

1 Introduction

Polymer modified binders (PMBs) are often used in asphalt pavements with the benefits of reducing temperature susceptibility and providing improved performance

---

J.-S. Chen (✉) · S.-H. Wei
Department of Civil Engineering, National Cheng Kung University, Tainan 701, Taiwan
e-mail: jishchen@mail.ncku.edu.tw

W. Chang
CECI Nova Technology Co., Ltd., Kaohsiung 806, Taiwan

© The Author(s), under exclusive license to Springer Nature Switzerland AG 2022

H. R. Pasindu et al. (eds.), *Road and Airfield Pavement Technology*, Lecture Notes in Civil Engineering 193, https://doi.org/10.1007/978-3-030-87379-0_48

compared with conventional asphalt binders. Improved performance benefits for the use of PMBs include reduced rutting and fatigue cracking in asphalt mixtures [1–3]. Styrene butadiene styrene (SBS) is the polymer type most commonly used to manufacture PMBs [4–6].

Because of uncertainties in construction projects caused by contractors, scheduling and environmental conditions, polymer modified binders may need to be stored at asphalt plant for an extended period of time. Even though SBS-based PMBs demonstrate enhanced service properties, they may show segregation, degradation or changes in engineering properties during hot storage on exposure to a combination of heat and air [6–11]. These problems are probably related to the complex nature of PMBs which ideally should perform properly in a short time during hot storage, but in some cases could be markedly affected by long storage periods [4, 11–15].

When stored at elevated temperatures, PMBs could segregate and lead to a layer of polymer floating on less modified bitumen. Studies of SBS-based PMBs subjected to aging tests such as thin film oven (TFO), rolling thin film oven (RTFO) or pressure aging vessel (PAV) tests indicated that test property changes can be attributed to the combined effect of bitumen oxidation and polymer degradation [15–16]. Gel permeation chromatography (GPC) experiments demonstrate that polymer degradation is due to a reduction in the molecular weight of the SBS when subjected to aging [17].

SBS-based PMBs may not be fully stable when kept at high temperatures during a long period of storage time, due to the differences in the solubility parameter and density between SBS and asphalt cement [5, 18]. There are numerous articles dedicated to the morphology and the stability behavior of PMBs [4, 19], the influence of aging [17, 21–22], the use of graft copolymers [18], the effect of the vulcanization reaction [2], the influence of polymer content [4, 23], molecular weight [16], and the chemical nature of the polymer [24]. Changes in binder test properties have been observed during hot storage; however, little information is available about whether the changes actually affect the performance of the PMBs on the road. This study is conducted to determine the effects of hot storage periods on the performance of PMBs and asphalt mixtures.

2 Experimental Design

Figure 1 presents the experimental design for testing PMBs and asphalt mixtures after different periods in hot storage. After PMB samples were prepared and stored at 180 °C for periods up to 28 days, their performance was monitored as a function of storage time. These samples were tested against the PMB specification to evaluate whether their performance could meet specified PMB requirements. The PMBs were used to evaluate the engineering properties of the binders including storage stability, morphology, elastic recovery, toughness, and multiple stress creep and recovery (MSCR). The performance of asphalt mixtures was assessed after the PMBs were incorporated into a 19-mm dense graded asphalt mix. Tests for asphalt mixtures

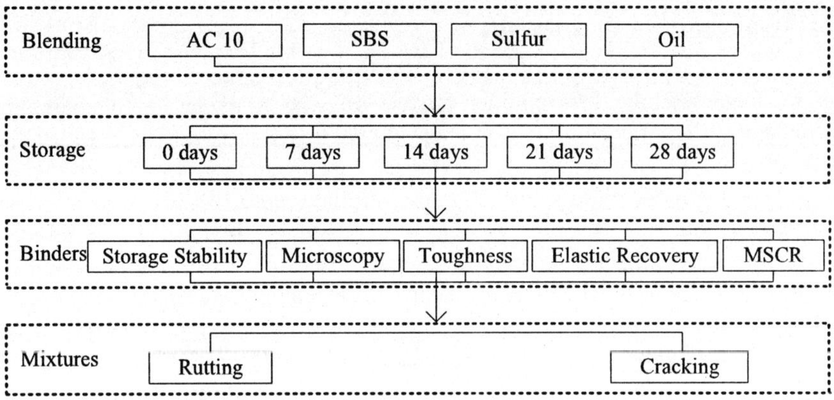

Fig. 1 Experimental testing plan

included rutting and cracking evaluation. Each test had at least two replicates, and the average value was used to generate the table and figures presented here.

2.1 Materials and PMB Sample Preparation

Asphalt cement AC-10 used in this study was obtained from the SIMOSA Oil Cooperation Ltd., Taiwan. The R&B softening point, the viscosity (at 60 °C) and the penetration (at 25 °C) values of the base binder were 36 °C, 986 poise, and 92 dmm, respectively. The content of saturates, aromatics, resins and asphaltenes in the AC-10 binder was 4.3%, 63.2%, 19.9%, and 12.6, respectively. Styrene–butadiene–styrene (SBS) polymer was obtained from the Chie-Mie Enterprise Ltd., Taiwan. Linear SBS with 160,000 g/mol average molecular weight was chosen as the modifier (31% by weight of styrene). A particular type of combining oil and sulfur were added as a crosslink agent to improve the miscibility of base asphalt and SBS polymer.

The PMB samples that were investigated in this study were prepared at 185 ± 5 °C using a Silverson high shear laboratory mixer (Model L5M) under an inert carbon dioxide atmosphere. PMB blending was performed under an inert atmosphere in order to ensure that the base asphalt used in the blends did not oxidise during the blending process. PMB samples were manufactured by initially adding appropriate masses of heated asphalt cement and polymer combining oil to a metal tin. The mix was then heated to 180 °C before a constant concentration of 5% polymer and 0.2% sulfur by weight of the PMB was added to the blend. This polymer level was chosen to be representative of that used to produce commercial SBS-based PMBs in Taiwan. It took one hour to homogenously incorporate the polymer into the blends.

2.2 Hot Storage Treatments

After manufacture, the SBS modified binders are exposed to elevated temperatures during storage and transportation in closed metal containers with vents for exit of built-up gases. The temperature and duration of exposure to elevated temperatures depends on the mode of transportation, distance between the PMB producing plant and customer site, and construction projects. In this study, polymer modified binders were stored in a storage tank free of direct contact with air to avoid over oxidation at 180 °C which is typical of field transport and storage temperature. To simulate storage conditions at an asphalt plant, the tank is equipped with a stirrer to homogenize the PMBs. Samples were taken for testing after they were stored for 0, 7, 14, 21 and 28 days. These five storage periods represented virgin, short term storage, medium term storage, long term storage, and very long term storage, respectively.

Immediately after the PMB blends were treated, they were split into different subsamples that were used during testing. Each of the PMBs was passed through a heated sample splitting funnel immediately after manufacture so that appropriate subsamples could be obtained for testing. The sample splitting funnel allowed equal amounts of each of the bulk samples to be transferred into two separate metal tins. The subsamples produced after the first sample splitting were then repeatedly passed through the sample splitting funnel and transferred into smaller metal tins until PMB subsamples were produced that were suitable for each type of test. Each PMB was stirred for one hour and sampled for testing.

2.3 Bitumen Testing

Storage Stability. The American Society for Testing and Materials (ASTM) D7131, Standard Practice for Determining the Separation Tendency of Polymer from Polymer Modified Asphalt, was used to determine the storage stability of polymer modified binders. An aluminium tube with 25 mm diameter and 140 mm height was filled with heated binder samples. The tube was vacuumed and sealed thoroughly to eliminate air oxidation and then placed vertically in a forced draft oven at 165 °C for 48 h. After that, the tube was kept in a refrigerator to be cooled at −10 °C for four hours. Each tube was cut horizontally into three equal parts. The top and bottom portions of the sample were then tested to determine the softening point per ASTM D36.

Microscopy. A Nikon YS100 optical microscope was used to observe the morphology of the PMB at 100× magnification. Samples were carefully poured inside a rubber mold that was immediately submerged in icy water at once to allow for rapid cooling and to preserve the morphology. Stripes of binders of approximately 20 mm × 40 mm × 1 mm were then seated on a metal plate and relocated to a freezer kept at −10 °C for 10 h. The sample was removed when frozen. After that, the stripes were separated in the central part to inspect the morphology. The morphology of the PMBs was to analyze the nature of the bitumen and polymer phase.

Elastic Recovery. The American Society for Testing and Materials (ASTM) D7131, Standard Practice for Determining the Separation Tendency of Polymer from Polymer Modified Asphalt, was used to determine the storage stability of polymer modified binders. An aluminium tube with 25 mm diameter and 140 mm height was filled with heated binder samples. The tube was vacuumed and sealed thoroughly to eliminate air oxidation and then placed vertically in a forced draft oven at 165 °C for 48 h. After that, the tube was kept in a refrigerator to be cooled at −10 °C for four hours. Each tube was cut horizontally into three equal parts. The top and bottom portions of the sample were then tested to determine the softening point per ASTM D36.

Toughness. This test was to measure the toughness and tenacity of polymer modified binders according to ASTM D 5801. A hemispherical head of specified size and shape was pulled from an asphalt sample at a rate of 50 cm/min. A continuous record of the force-versus-elongation curve was recorded and used to calculate the toughness and the tenacity of the sample. Toughness is defined in this procedure as the total work required to completely separate the tension head from the sample under the specified test conditions. Tenacity is a measure of the increasing force as the sample is stretched past the initial peak, and may indicate the type and amount of polymer used to modify the asphalt. It is defined as the work required to stretch the material after the initial resistance is overcome.

Multiple Stress Creep Recovery (MSCR). The MSCR test was conducted to evaluate the viscoelastic properties of PMBs according to ASTM D7405. This test was performed on a TA AR1500ex (TA Instruments) dynamic shear rheometer (DSR) using a constant stress creep of 1.0-s duration followed by a zero stress recovery of 9.0-s duration for 10 creep and recovery cycles at 64 °C. The non-recoverable compliance (Jnr) and percent recovery (%R) values were calculated at each loading cycle and then averaged at stress levels of 0.1 and 3.2 kPa.

2.4 *Asphalt Mixture Testing*

Wheel Tracking Test. The main aim of this study is to determine whether the segregation and degradation propensity of a PMB during hot storage affect the performance of the material on the road after the PMB was mixed prior to use. The PMBs for asphalt mix tests were representative of those that would contain the same proportions of each component used in the PMB formulation as the originally prepared blend. The limestone aggregate used for this study came from a local river. A 19-mm conventional dense graded asphalt mixture was prepared for both wheel tracking and semi-circular bend tests using the PMBs after hot storage treatments. The design binder content was 5.0% of the total asphalt mix weight. Wheel tracking test was performed by subjecting asphalt mixtures to 3000 passes of a loading wheel as per

Public Construction Commission Designation 02743 in Taiwan [25]. A slab specimen (300 × 300 × 50 cm) was prepared with an air voids content of 5.0% using a roller-press compactor. Lab-produced mixes were cured at 60 °C for 24 h, and tested at 60 °C for initial stability under the wheel tracking test. The wheel track consisted of a steel wheel (200 mm in diameter and 50 mm in width) applying a pressure of 689.4 kPa (100 psi) on top of a slab specimen. A wheel tracking test covered a loading distance of 230 ± 10 mm at the speed of 42 cycle/min. The dynamic stability (DS) used to characterize the rutting resistance of asphalt mixtures is calculated as expressed in Eq. 1. A higher dynamic stability value indicates better permanent deformation resistance.

$$DS = \frac{t_2 - t_1}{d_2 - d_1} \times N \tag{1}$$

where d_1 = deformation at t_1 minutes (mm), d_2 = deformation at t_2 minutes (mm), and N = speed of the wheel, 42 cycle/min in this study. Values of t_1 and t_2 were used in this study, which are at 45 and 60 min, respectively.

Semi-Circular Bend (SCB) Test. The SCB test was used to characterize the fracture properties of asphalt mixtures. The test utilized notched half-moon shaped specimens cut from asphalt mix cylinders with three notch depths of 25.4, 31.8, and 38.1 mm. Each SCB test was performed in triplicate using the method described in ASTM D 8044. The air voids of SCB specimens were controlled at 4%. During the test, a SCB specimen is supported by two bars on a flat surface and a monotonic load is applied to the curved surface above the notch. The ASTM standard specifies a vertical displacement rate of 0.5 mm/min. For data analysis, strain energy to failure was first calculated for each notch depth as the area under the load versus displacement data, and a linear regression was determined based on the strain energy versus notch depth results. Finally, the cracking parameter J-integral (Jc) was calculated by dividing the slope of the regression line by the specimen thickness. Asphalt mixtures with higher Jc values are expected to have better resistance to cracking than those with lower Jc values.

3 Results and Discussion

3.1 Storage Stability Results

Prior to the storage stability test, the softening point (SP) of each PMB sample was measured, and termed initial softening point and marked 'A' in Table 1. Following storage of the samples in a tube for two days at 165 °C, the softening points of the PMB in the top third of the tube (marked 'B') and the bottom third of the tube (marked 'C') were determined. A storage stability value was calculated as the softening point of the top third of the binder in the tube minus the softening point of the bottom

Table 1 Storage stability test results

Storage (days)	Initial SP (A)		Top SP (B)	
	Mean (°C)	COV (%)	Mean (°C)	COV (%)
(a)				
0	77.08	0.31	77.20	0.12
7	74.23	0.38	73.08	0.38
14	70.85	0.21	71.58	0.20
21	70.42	0.35	70.08	0.30
28	69.23	0.24	70.96	0.25
(b)				
0	76.52	0.35	0.68	0.38
7	72.20	0.25	0.88	0.28
14	70.85	0.18	0.73	0.19
21	69.03	0.31	1.05	0.33
28	69.85	0.28	1.11	0.26

third of the binder in the tube. Table 1 summarizes the storage stability test results of modified binders examined in the study.

The coefficient of variation (COV) is used as evaluation of the variability of measurements, and summarize in this table. Softening points of all test results show consistency since the COV values ranges from 0.12 to 0.38% in Table 1. One of the most adopted criteria to indicate the storage stability of polymer-modified binders is a difference of 3 °C or less in the softening point between the top and bottom samples [15–16, 20]. Examination of the data in Table 1 shows that hot storage did not much effect on storage stability over the range of storage periods. With the test results of between 0.68 and 1.11 °C, the PMBs are shown to have a low value in the storage stability test.

However, hot storage resulted in a reduction in softening point as listed in Table 1. Softening point continued to decrease with time but at a slower rate after 14 days. The reduction is likely to result from degradation in PMBs, thus leading to a loss in softening point during hot storage. It should be noted that a reduction in softening point in SBS-based PMBs is equivalent to a change in SBS concentration, the exact figure depending on the properties of the SBS and the asphalt fraction [2–5]. Furthermore, the softening point versus SBS concentration plots reported by other researchers [11, 19] indicate that a 19 °C softening point drop experienced by samples corresponded to an approximate 2% reduction in SBS concentration. Based on a 7.85 °C (77.08–69.23) decrease in softening point in this study, it is estimated that more than 80% the SBS polymer remained after 28 days of hot storage at 180 °C.

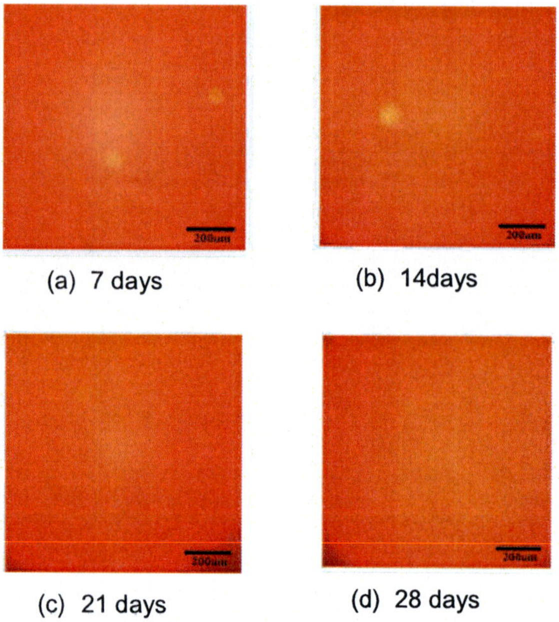

Fig. 2 Microscopic photographs after different periods of hot storage (100x)

3.2 *Morphology of Binders*

The microscopic photographs of binders observed at different storage periods are shown in Fig. 2. The PMBs exhibited single phase or close to homogeneous phase behavior when stored at 180 °C. The phase structure of the PMB was such that there would be little or no tendency to segregate as storage time increased. The PMBs were shown to be compatible, and could form a segregation resistant structure at the temperature at which they are likely to be stored and transported prior to final use. These microscopy observations agree with the storage stability results in Table 1. The morphology photographs indicate that the storage stability value of the PMBs barely increased after very long term storage. Thus, the SBS polymer network remained and kept its modification effect after hot storage for 28 days.

3.3 *Toughness Results*

Figure 3 shows the effect of storage time on the toughness and tenacity values of the PMBs. Each bar represents the average of three replicates, and the error bars indicate one standard deviation from the average value. The toughness of the PMBs is the total area under the force-versus-elongation curve and represents its ability to absorb energy before failure. After 28 days of hot storage, the toughness value was relatively unchanged, and exceeded the minimum toughness value of 20 N-m.

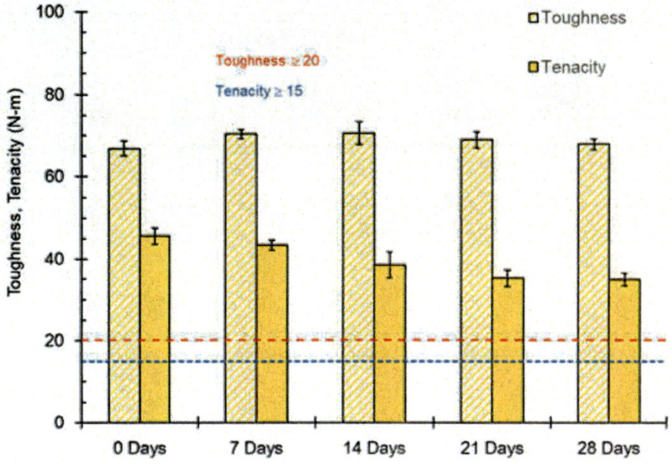

Fig. 3 Toughness and tenacity of PMBs after different periods of hot storage

The tenacity value of the PMBs decreased after hot storage. The reduction in the tenacity value was due to the fact that the aging of the PMBs consisted of two parts. On the one hand, the oxidation of the base asphalt could lead to an increase in stiffness, which might contribute to the toughness. On the other hand, the degradation of SBS polymer might have a softening effect on PMBs, which would cause a decrease in tenacity. As hot storage proceeded, the tenacity value of the PMBs gradually decreased. The minimum tenacity should be no less than 15 N-m. After stored for 28 days, the PMBs still met this minimum specification.

3.4 Elastic Recovery Results

The elastic recovery test is used to measure the percentage of recovery strain of the PMBs after elongation. Figure 4 shows that the elastic recovery of the PMBs decreased about 15% after 28 days of hot storage. It is likely that the degradation of the SBS polymers during hot storage resulted in a decrease of the elastic recovery. For PMBs, with the increase in storage time, the elastic recovery decreased gradually from 93 to 78% after 28 days of hot storage. The reason the elastic recovery slowly reduced was that a single phase of the SBS polymer modifier was beneficial to the elasticity of the PMBs. These test results are in good agreement with the morphologies of the PMBs presented in Fig. 2. Although a homogeneous phase in PMB could enhance the elastic recovery of the PMBs, an increase in storage time might lead to a negative effect on the adhesion characteristics of the PMBs, thus influencing the flexibility.

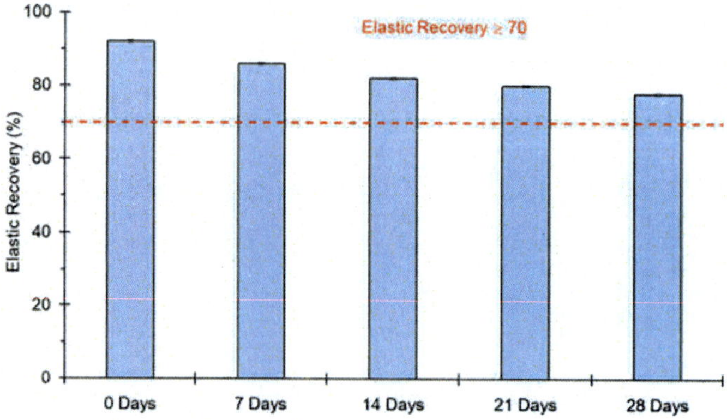

Fig. 4 Elastic recovery of PMBs after different periods of hot storage

3.5 Multiple Stress Creep and Recovery (MSCR) Results

To evaluate the high-temperature properties of the PMBs after different storage periods, the MSCR tests were conducted at 64 °C. Figure 5 shows that the PMBs had the lowest non-recovery compliance (Jnr) values at 3.2 kPa when stored at 0 days, which coincided with the evolution of the elastic recovery. A Jnr index is used to characterize the changes in PMBs because of hot storage. The Jnr index is defined as the ratio of the Jnr at a stress level for any storage time to the Jnr at this stress level of the PMB stored at 0 days. This index calculated from data generated at 3.2 kPa shear stress in this study. The Jnr index increased unexpectedly as storage time increased. An increase in Jnr of 15–26 times the Jnr of the binder stored at 0 days is observed for the PMBs. As mentioned previously, the effect of storage time on the PMBs includes

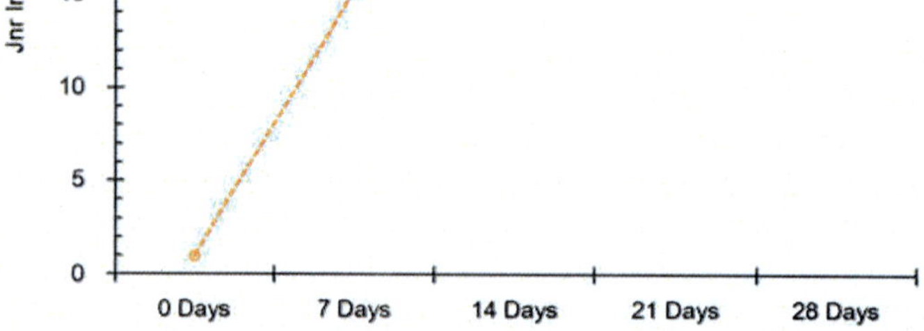

Fig. 5 Jnr index of PMBs after different periods of hot storage

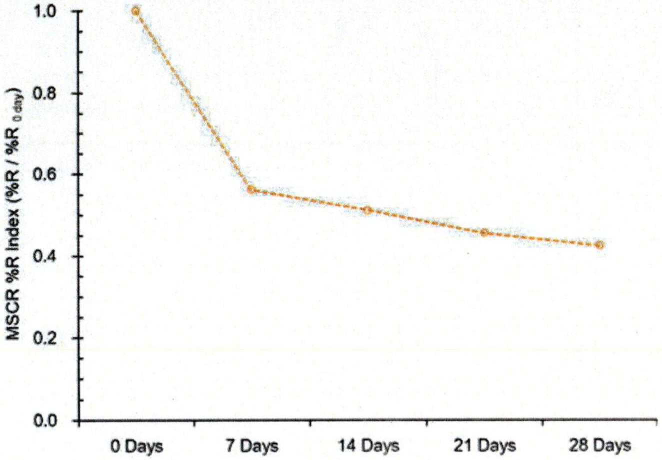

Fig. 6 %R index of PMBs after different periods of hot storage

oxidation of the base binder and degradation of the SBS polymer. The degradation of the SBS polymer seems to play the primary role, which led to an increase in Jnr.

Test results in Fig. 5 indicate that the PMBs had a high increase in Jnr index and, as time of hot storage increased, the rate of increase in Jnr decreased. The degradation of SBS happened likely in the beginning, but slowed after 14 days of hot storage. This implies that the SBS polymer still had a considerable modification effect after stored for 28 days. The oxidation of base binder appears to play a minor role because the PMBs were not exposed to fresh air in the tank during hot storage.

Figure 6 shows the MSCR percent recovery (%R) decreasing at a decreasing rate with increasing storage time for two stress levels. The trends are plotted using a %R index that is calculated similarly to the Jnr index above but using %R. The %R index decreases as the storage period is extended. The degradation of the polymer network could be the cause for the decreased %R. The use of the MSCR %R might lead to a false negative detection of an elastomeric polymer concentration with PMBs that have been stored for a very long period of time because the dynamic shear rheometer uses %R as an indicator of polymer content. It is suggested that PMBs samples be immediately taken for MSCR testing after pumped into a storage tank at asphalt plant.

Figure 7 shows the same data as Figs. 5 and 6 but plotted on a Jnr versus %R plot. A control line (defined by the equation $\%R = 29.371 \times Jnr^{\wedge}-0.2633$) superimposed for the combined data suggests that the %R of the PMBs is related to the storage period of the binder. PMB binders initially met the specification requirement of the E grade, but the PMB was below the control limit after 14-day storage. This observation suggests that PMBs be stored less than 14 days to meet the minimum %R value.

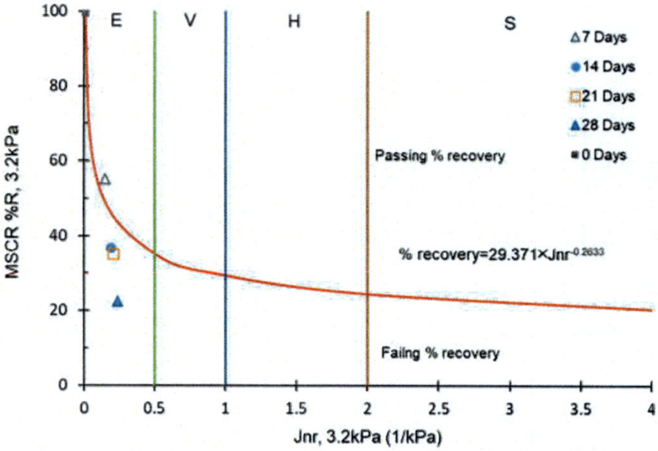

Fig. 7 MSCR Jnr versus %R at 64 °C after different periods of hot storage

3.6 *Wheel Tracking Test Results*

Wheel tracking tests were performed on asphalt mixtures containing PMBs stored by different periods. Plots of average wheel tracking depth versus number of loading passes are shown in Fig. 8. The asphalt mixtures prepared using the PMB stored at 0 days was slightly less resistant to rutting than those which contained PMBs stored more than 7 days. The results shown in Fig. 8 indicate that each of the PMBs stored beyond 7 days showed almost the same resistance to permanent deformation in asphalt mixtures as their average wheel tracking depths after 3000 loading wheel

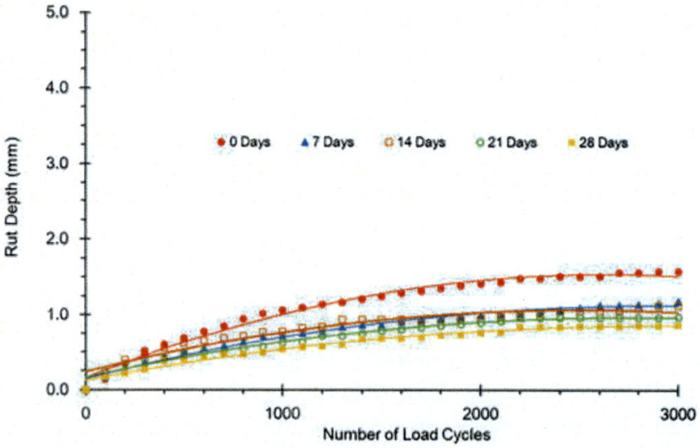

Fig. 8 Effect of hot storage periods on rut depth

passes were all within 1 mm. As expect, the wheel tracking test results showed less rutting with the binder being aged for long periods of time.

Wheel tracking performance is evaluated by the dynamic stability (DS) which is defined as the resistance to plastic deformation due to traffic loading. The DS value was 5311, 5721, 5712, 5682 and 5689 cycle/mm for asphalt mixtures produced by the PMBs stored after 0, 7, 14, 21 and 28 days, respectively. The dynamic stability greater than 1500 cycle/mm is recommended for a dense-graded asphalt mix. The PMB was highly modified and had low Jnr-values per Fig. 7; therefore, little rutting could be seen at the test temperature.

The use of PMBs shows satisfactory rutting performance. The rutting data from different storage periods are basically similar partly due to the fact that these mixtures were prepared with the same crushed limestone aggregate. Based on the results of wheel tracking tests, the wheel tracking performance of the PMBs is not significantly affected by their propensity to degrade during hot storage.

3.7 *Semi-circular Bend (SCB) Test Results*

Semi-circular bend tests were performed using the same PMBs that were subjected to wheel tracking tests to investigate whether the propensity of a PMB to degrade during storage affected its resistance to cracking. Figure 9 presents the cracking parameter J-integral results with a loading rate of 0.5 mm/min. The cracking resistance of the asphalt mix with the PMB stored at 0 days was slightly higher as compared with that of the asphalt mix with the PMB stored for 28 days. However, there was not a consistent trend. Specimens with the PMB stored for 21 days had higher J-integral results than those with the PMBs with shorter storage periods. Asphalt mixtures containing the PMB after 7 days of hot storage had the lowest J-integral results.

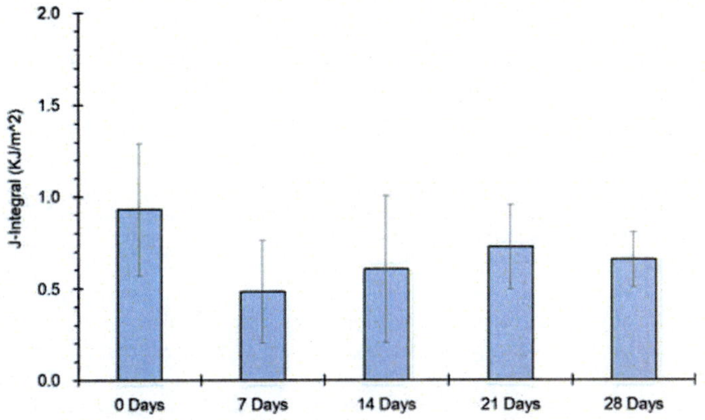

Fig. 9 Microscopic photographs after different periods of hot storage (100x)

Based on the SCB results obtained during this study, it does not appear that hot storage will necessarily result in reduced resistance to cracking performance on the road.

4 Conclusions and Recommendations

This study was to evaluate the effect of storage periods on the performance of SBS-based PMBs and asphalt mixtures. Little segregation occurred in PMBs following hot storage for 28 days. Low results in the storage stability test indicated that a considerable amount of SBS polymer remained after very long term storage. Evidence of polymer degradation during hot storage was mainly derived from observing that the tenacity, elastic recovery and percent recovery values of the PMBs decreased over time. The degradation of SBS happened likely at the beginning and slowed after hot storage for 14 days. The results obtained in wheel tracking tests indicated that the resistance of the PMBs to permanent deformation in an asphalt mix was not affected by their propensity to degrade during hot storage. The cracking performance of the PMBs was not significantly influenced by the storage periods. Based on the results obtained in this study, it appears that the propensity of a PMB to degrade during hot storage would not necessarily affect its performance on the road in terms of its resistance to rutting and cracking in asphalt mixtures. It is suggested that PMBs samples be immediately taken for MSCR testing after pumped into a storage tank at asphalt plant, and PMBs be stored less than 14 days to meet the requirement for percent recovery. Future work is recommended to evaluate compatibilizers to mitigate the polymer degradation of SBS-modified binders during hot storage.

References

1. Liu G, Nielsen E, Komacka J, Greet L, van de Ven M (2014) Rheological and chemical evaluation on the ageing properties of SBS polymer modified bitumen: from the laboratory to the field. Constr Build Mater 51:244–248
2. Kilger A, Swiertz D, Bahia HU (2019) Long-term aging performance analysis of oil modified asphalt binders. Transp Res Rec 2673(12):404–412
3. Chen JS, Wang TJ, Lee CT (2018) Evaluation of a highly-modified asphalt binder for field performance. Constr Build Mater 171:539–545
4. Zhang Z, Han S, Han X, Cheng X, Yao T (2020) Comparison of SBS-modified asphalt rheological properties during simple-aging test. J Mater Civ Eng 32(9):04020241
5. Larsen DO, Alessandrini JL, Bosch A, Cortizo MS (2009) Micro-structural and rheological characteristics of SBS–asphalt blends during their manufacturing. Constr Build Mater 235:2769–2764
6. Airey GD (2003) Rheological properties of styrene butadiene styrene polymer modified road Bitumens. Fuel 82(10):1709–1719
7. Brule B, Brion Y, Tangu A (1988) Paving asphalt polymer blends: relationship between composition, structure and properties. J Assoc Asphalt Paving Technol 57:41–64

8. Masson JF, Collins P, Woods JR, Bundalo-Perc S, Al-Qadi I (2006) Degradation of bituminous sealants due to extended heating before installation: a case study, report NRCC-48663. Institute for Research in Construction, Ottawa, Canada
9. Cong P, Wang J, Zhou Z (2020) Effects of thermal degradation on polymer modified asphalt binders during storage and transportation. Constr Build Mater 248:118694
10. Chang CM, Chang YJ, Chen JS (2009) Effect of mixture characteristics on cooling rate of asphalt pavements. J Transp Eng 135(5):297–304
11. Geng J, Li H, Sheng Y (2014) Changing regularity of SBS in the aging process of polymer modified asphalt binder based on GPC analysis. Int J Pavement Res Technol 7(1):77–82
12. Chen JS, Lin CH, Stein E, Hothan J (2004) Development of a mechanistic-empirical model to characterize rutting in flexible pavements. J Transp Eng 130(4):519–525
13. Cortizo MS, Larsen DO, Bianchetto H, Alessandrini JL (2004) Effect of thermal degradation of SBS copolymers during the ageing of modified asphalts. Polym Degrad Stab 86:275–282
14. Singh SK, Kumar Y, Ravindranath SS (2018) Thermal degradation of SBS in bitumen during storage: influence of temperature, SBS concentration, polymer type and base bitumen. Polym Degrad Stab 147(1):64–75
15. Chen JS, Liao MC, Lin CH (2003) Determination of polymer content in modified bitumen. Mater Struct 36(263):594–598
16. Yan C, Huang W, Ma J, Xu J, Lv Q, Lin P (2020) Characterizing the SBS polymer degradation within high content polymer modified asphalt using ATR-FTIR. Constr Build Mater 233:117708
17. Lin P, Huang W, Liu X, Apostolids P, Wang H, Yan C (2020) Laboratory evaluation of the effects of long-term aging on high-content polymer-modified asphalt binder. J Mater Civ Eng 32(7):04020157
18. Fu H, Xie L, Dou D, Li L, Yu M, Yao S (2007) Storage stability and compatibility of asphalt binder modified by SBS graft copolymer. Constr Build Mater 21:528–533
19. Carcer IA, Masegosa RM, Vinas MT, Sanchez-Cabezudo M, Salom C, Prolongo MG, Contreras V, Barcelo F, Paez A (2014) Storage stability of SBS/sulfur modified bitumens at high temperature: influence of bitumen composition and structure. Constr Build Mater 52:245–252
20. Chen JS, Tsai CJ (1999) How good are linear viscoelastic properties of asphalt binder to predict rutting and fatigue cracking? J Mater Eng Perform 8(4):443–449
21. Mouillet V, Lamontagne J, Durrieu F, Planche JP, Lapalu L (2008) Infrared microscopy investigation of oxidation and phase evolution in bitumen modified with polymers. Fuel 87(10):1270–1280
22. Hossain R, Wasiuddin NM (2019) Evaluation of degradation of SBS modified asphalt binder because of RTFO, PAV, and UV aging using a novel extensional deformation test. Transp Res Rec 2675:447–457
23. Valkering CP, Vonk W (1990) Thermoplastic rubbers for the modification of bitumens: improved elastic recovery for high deformation resistance of asphalt mixes. In: 15th Australian road research board conference, 1990, Darwin, Northern Territory, vol 15(2), pp 1–20
24. Zani L, Giustozzi F, Harvey J (2017) Effect of storage stability on chemical and rheological properties of polymer-modified asphalt binders for road pavement construction. Constr Build Mater 145:326–335
25. Public Construction Commission (2019) Wheel Tracking Test. PCC Designation: 02743, Executive Yuan, Taipei, Taiwan (in Chinese)

Effect of Segregation Tendency of Aggregates on Macroscopic Properties of Asphalt Mixture Based on Composite Geometric Characteristics

Jinfei Su and Peilong Li

Abstract The composite geometric characteristics of coarse aggregates can effectively evaluate the combined effect of geometric morphology and size distribution in particle system. The composite angularity indices can be used to evaluate the segregation tendency of aggregates. To analyze the effect of aggregate segregation on the macroscopic properties of asphalt mixtures on the microscopic level, five segregated asphalt mixtures were designed, and the macroscopic properties were measured by a set series of tests encompassing Superpave gyratory compactor test, immersion Hamburg wheel-tracking test and indirect tensile test. The results show that the heavier the coarse aggregate segregation, the higher composite geometric indices, which are closely related to the macroscopic properties of segregated asphalt mixture. The composite angularity and textures play a greater role in the contact properties of the particle system and the macroscopic properties of the asphalt mixture than the composite shape. The large composite geometric indices imply strong contact among coarse aggregate, impeding the process of compaction. The low segregated asphalt mixture has most stable skeleton-bonding structure formed by the interaction of the geometric morphology and asphalt binder, and its resistance to rutting deformation, water stability and low-temperature crack resistance are best.

Keywords Road engineering · Macroscopical properties · Aggregate segregation · Particle contact · Composite geometric characteristic

---

J. Su · P. Li (✉)
Highway School, Chang'an University, Xi'an 710064, Shaanxi, China
e-mail: lipeilong@chd.edu.cn

J. Su
e-mail: jinfeisu@chd.edu.cn

P. Li
Key Laboratory of Road Structure and Material Ministry of Transport, Chang'an University, Xi'an 710064, China

© The Author(s), under exclusive license to Springer Nature Switzerland AG 2022

H. R. Pasindu et al. (eds.), *Road and Airfield Pavement Technology*, Lecture Notes in Civil Engineering 193, https://doi.org/10.1007/978-3-030-87379-0_49

1 Introduction

The uneven distribution of aggregate determines the compaction effect and mechanical properties of asphalt mixture to a great extent, which is related to the pavement distress and durability [1–3]. Zeng et al. [4] used X-Ray CT to observe the internal structure of asphalt mixture, segmented the image based on the EM algorithm of Gaussian mixture model (GMM), and analyzed the aggregate distribution. The distribution coefficient of coarse particle can represent the uniformity of the asphalt mixture. Senthilmurugan et al. [5] concluded that the uneven distribution of aggregate in the transverse direction is significantly higher than that in the longitudinal direction. Jia et al. [6] reported that the aggregate segregation affected the density distribution. Zhang et al. [7, 8] analyzed the influence of the uniformity of aggregate distribution on the void and density distribution, and the void uniformity coefficient was proposed to evaluate the compaction uniformity of asphalt mixture. Li et al. [9] demonstrated that the aggregate segregation exercised great impact on the water stability, high temperature stability, low-temperature anti-cracking and tensile strength of asphalt mixture, but the mesoscopic essence of the aggregate affecting the macro performance of asphalt mixture was not elucidated. Jimmy [10] proposed a new statistical analysis method to observe intelligent compaction data, and he found that the compaction uniformity index could be used to evaluate the compaction non-uniformity of asphalt mixture caused by aggregate segregation. Jaime et al. [11] used finite element method to reconstruct the three-dimensional (3D) model of asphalt mixture. The results showed that the uneven aggregate distribution was an important factor affecting the fracture properties of asphalt mixture.

The interaction among particles is closely related to the geometric morphologies, size, content of each of the file aggregates, leading to aggregate segregation [12–14]. The angularities and surface textures of coarse aggregates in particle system contact and rub with each other, resulting in the change in the size distribution of aggregates, and consequently, the aggregate segregation occurs. Asphalt binder has a significant bonding/lubrication effect on the contact interface among aggregates, reducing the uneven distribution of aggregates to some extent. However, interaction among aggregates is the main reason for uneven distribution in asphalt mixture [15]. Wu et al. [16] implemented discrete element method to reconstruct the segregated asphalt mixture. The coarse aggregate segregation changed in the contact force and the number of contact points inside the asphalt mixture, which in turn deteriorated the mechanical properties of the mixture. Hao [17] reported that the coarse aggregate with a size of 4.75–13.2 mm has great influence on the aggregate segregation of asphalt mixture [18]. During the process of mixing and transporting the asphalt mixture, the interaction among aggregates is the mesoscopic manifestation of the contact and friction of all particles with different sizes under the bonding and lubrication effect of asphalt, and therefore, only analyzing the effect of a part of aggregates on segregation tendency has limitations.

The composite geometric characteristics of coarse aggregates reflect the contribution of the geometric morphology on the contact strength of particle system, and

Table 1 The bulk specific gravity of coarse aggregates

Sieve size (mm)	19	16	13.2	9.5	4.75
Bulk specific gravity (g/cm^3)	2.684	2.688	0.695	0.690	0.702

Table 2 The apparent relative gravity of fine aggregates

Sieve size (mm)	2.36	1.18	0.6	0.3	0.15	0.075
Apparent relative gravity (g/cm^3)	2.73	2.77	2.773	2.765	2.717	2.715

the composite angularity index can be used to effectively evaluate the aggregate segregation tendency of aggregate-asphalt system [19, 20]. Therefore, this investigation designed five segregated asphalt mixtures to carry out gyration compaction test, immersion Hamburg wheel-tracking test and indirect tensile test. Based on the particle characteristics, the macroscopic properties of asphalt mixtures were analyzed combined the interface interaction between aggregates and asphalt binder. The research works are helpful to improve the level of gradation design and performance prediction of asphalt mixture.

2 Materials and Mixture Design

2.1 *Materials*

The aggregate was limestone from Shaanxi and the test was conducted according to Test Method of Aggregate for Highway Engineering (JTG E42-2005), with the specific gravities of the aggregates presented in Tables 1 and 2. The mineral filler was the finely ground limestone with the apparent relative gravity of 2.817 g/cm^3.

Esso 90# base asphalt was selected, and the properties were presented in Table 3 in accordance with the relevant requirements of the Test Procedure for Highway Engineering Asphalt and Asphalt Mixtures (JTJ 052-20). The mixing and compaction temperatures of asphalt mixture were determined as 165 °C and 145 °C respectively according to the viscosity-temperature curve of asphalt.

2.2 *Gradation Design*

To simulate the segregation of asphalt mixture for practical engineering, five AC-20 asphalt mixtures with different segregation degrees was designed according to the previous studies [21]. Based on the segregation tendency of coarse aggregates, the involved asphalt mixtures were divided into five class of heavy, middle, low, no coarse aggregate segregation and fine aggregate segregation, which are respectively

Table 3 The properties of asphalt

Indicators		Technical requirements	The test results
Penetration/0.1 mm	15 °C, 5 s, 100 g	80 ~ 100	33
	25 °C, 5 s, 100 g,		9
	0 °C, 5 s, 100 g,		132
Ductility (50 mm/min,15 °C)/cm		≥100	≥150
Softening point (Ring and Ball Method)/°C		≥45	46.5
Density (15 °C) g/cm^3		≥1.01	1.038
Relative density (25 °C)g/cm^3			1.034
Apparent viscosity/Pa.s	135 °C		0.360
	175 °C		0.073

named as H, M, L, N, and F asphalt mixtures. N asphalt mixture was designed by Superpave method and the optimum asphalt content is determined. The gradation curve is shown in Fig. 1, and optimum asphalt content (OAC) is 3.9%.

N asphalt mixture was divided into three parts, A, B and C with 4.75 mm and 9.5 mm as the key sieve pores. When the aggregate was within the range of 0–4.75 mm, the asphalt mixture was denoted as A, with B when the aggregate was within the range of 4.75–9.5 mm, and C within 9.5–26.5 mm. The corresponding asphalt content of A, B and C was measured by mixture combustion method, which was respectively 5.6%, 4.79% and 2.43%.

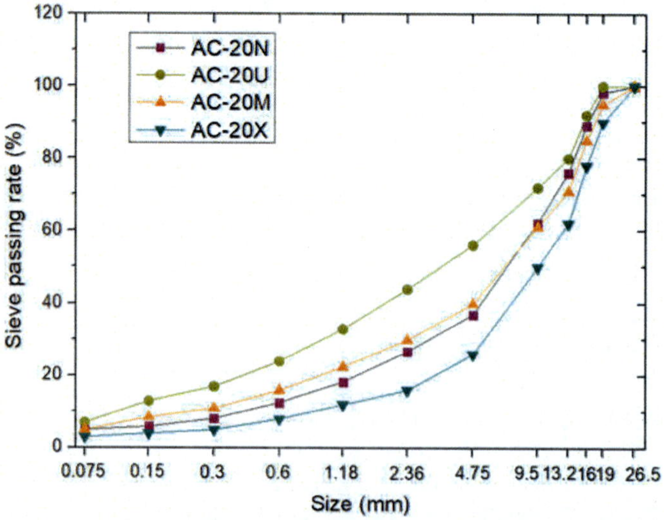

Fig. 1 The gradation curve of asphalt mixture without segregation

Table 4 The proportion and asphalt dosage of asphalt mixture segregation

Segregated asphalt mixtures	Proportion	Asphalt content (%)		
	A: B: C	Sample 1	Sample 2	Average value
F	24: 43: 33	4.14	4.28	4.21
N	19.7: 37.3: 43	3.92	3.88	3.90
L	14: 32: 54	3.58	3.54	3.56
M	8: 26: 66	3.23	3.13	3.18
H	3: 19: 78	2.86	2.98	2.92

According to the asphalt mixture segregation criteria presented by literature [21], combine the three parts of asphalt mixture A, B, and C in different proportions to obtain mixtures with different segregation degrees. The proportion of each part and asphalt content are present in Table 4.

3 Testing Methods

3.1 AIMS Test

The composite geometric indices *CI* of coarse aggregates in particle system can be used to predict the aggregate segregation tendency of asphalt mixture [20]. To elucidate the macroscopic properties of segregated asphalt mixtures on the on the micro-meso-scopic level, the model was established to calculate the *CI* encompassing composite shape index CI_{SP}, composite texture index CI_{TX} and composite angularity index CI_{GA}, as shown in Eq. (1), Eq. (2) and Eq. (3).

$$CI_{SP} = \frac{\sum_{i=1}^{m}[(\frac{a_i}{g_i \times V_{Wi}} \times C_{wi} \times d_i) \times SP_i]}{\sum_{i=1}^{n}(\frac{a_i}{g_i \times V_{Wi}} \times C_{wi})} = \frac{\sum_{i=1}^{m}[(\frac{a_i}{g_i \times V_{Wi}} \times C_{wi} \times \frac{P_{i+1}}{1.11}) \times SP_i]}{\sum_{i=1}^{n}(\frac{a_i}{g_i \times V_{Wi}} \times C_{wi})} \quad (1)$$

$$CI_{TX} = \frac{\sum_{i=1}^{m}[(\frac{a_i}{g_i \times V_{Wi}} \times SA_{wi} \times d_i) \times TX_i]}{\sum_{i=1}^{n}(\frac{a_i}{g_i \times V_{Wi}} \times SA_{wi})} = \frac{\sum_{i=1}^{m}[(\frac{a_i}{g_i \times V_{Wi}} \times SA_{wi} \times \frac{P_{i+1}}{1.11}) \times GA_i]}{\sum_{i=1}^{n}(\frac{a_i}{g_i \times V_{Wi}} \times SA_{wi})} \quad (2)$$

$$CI_{GA} = \frac{\sum_{i=1}^{m} [(\frac{Ma_i}{g_i \times V_{Wi}} \times V_{Wi} \times d_i) \times GA_i]}{\sum_{i=1}^{n} (\frac{Ma_i}{g_i \times V_{Wi}} \times V_{Wi})} = \frac{\sum_{i=1}^{m} [(\frac{a_i}{g_i} \times \frac{P_{i+1}}{1.11}) \times GA_i]}{\sum_{i=1}^{n} (\frac{a_i}{g_i})} \quad (3)$$

where, M is the total mass of mineral aggregates; g_i is the specific gravity of i-th file aggregates; C_{wi} is the weighted perimeter of aggregate shape; SA_{wi} is the weighted surface area of aggregate shape; V_{wi} is the weighted volume of aggregate angularity; d_i is the average size of i-th aggregates. SP_i, TX_i and GA_i are sphericity, texture index and angular gradient of i-th aggregates respectively, which were measured by AIMS II system.

3.2 Superpave Gyratory Compactor Test

The compaction effect of asphalt mixture is the mesoscopic manifestation of particle rolling and slip, which is closely related to particle characteristics [22]. The superpave gyratory compactor tester (SGC) can actually simulate the compaction process of asphalt mixture. The size of the specimen was ϕ 150 mm × 100 mm.

Compaction energy index (*CEI*) reflecting the energy required in the process of compaction, which can be used to evaluate the compaction properties of asphalt mixture. It was calculated using Eq. (4).

$$\text{CEI} = \int_0^{N_{des}} aN^b dN = \left[\frac{a}{b+1} N^{b+1}\right]_0^{N_{des}} \quad (4)$$

3.3 Immersion Hamburg Wheel-Tracking Test

Immersion Hamburg wheel-tracking test (HWTD) can simulate the effect of wheel load on asphalt pavement under high temperature and water environments.

The specimens formed by SGC were cut into the samples with a diameter of 150 mm and a height of 62 ± 2 mm. A HWTD test required two samples, and the water-bath temperature was 50 °C. When the rut depth reached to 20 mm, the tests automatically end. And then the load cycles at failure and creep rate were obtained to evaluate the high temperature stability of asphalt mixture. In addition, Fig. 2 shows the typical curve of HWTD test, and the intersection of creep curve and strip curve is the stripping point. The cycle number, rutting depth and spalling rate at spalling failure can be used to evaluate the water stability of asphalt mixture.

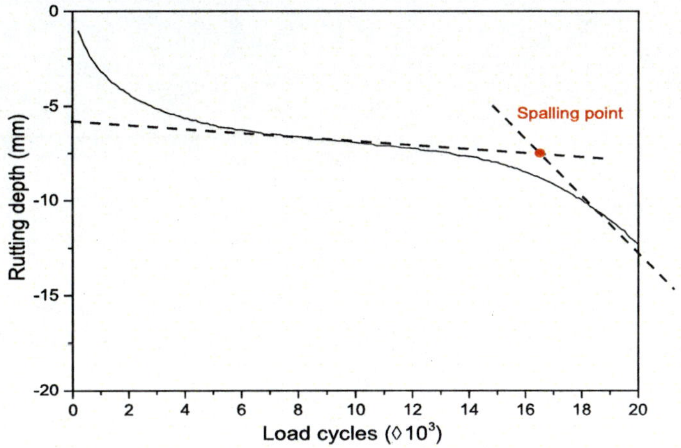

Fig. 2 Typical curve of immersion HWTD test

3.4 Indirect Tensile Test

The indirect tensile test (ITT) was conducted to simulate the cracking process of asphalt mixture at low temperature, and the tensile strength R_T, destructive tensile strain ε_T and modulus of failure stiffness S_T were calculated to evaluate the low-temperature crack resistance. The specimen formed by SGC was cut into a cylinder sample with the size of ϕ 150 mm × 40 mm. The test parameters were test temperature of −10 °C and the loading rate of 1 mm/min.

4 Results and Discussion

4.1 The Relationship Between the Parameters of Composite Geometric Characteristic and the Compaction Properties

As illustrated in Fig. 3, *CEI* presents a downward trend with the increase of composite shape index CI_{SP}. The higher the *CEI* value is, the more energy is needed to densify asphalt mixture; and lower *CEI* value indicates that the mixture has good workability. Aggregate particles roll and slip under the action of head of SGC, especially, the large-sized particles likely tend to rotate in space [23]. For particles with the same shape, the larger the particle size is, the greater the contribution to contact among particles is. CI_{SP} mainly affects the rotational inertia of particles, hindering the compaction of the asphalt mixture. In the process of compaction, the state of asphalt mixture changes from loose flow to forming stable with rotation of particles. The higher the CI_{SP} of particle system, the greater the probability of the particle rotation, which

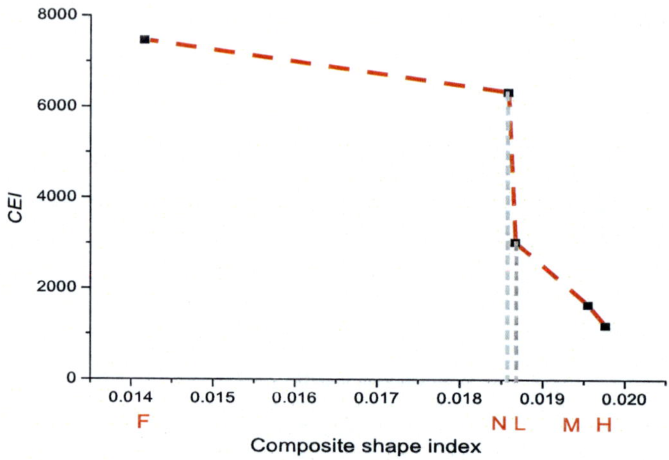

Fig. 3 The relationship between of CI_{SP} and *CEI*

makes the asphalt mixture easy to be compacted to reach design height or design volume of air voids.

Figure 4 delineates that CI_{TX} has linear negative correlation with *CEI*. The F asphalt mixture has lowest CI_{TX} but the largest *CEI*, whereas conversely order of CI_{TX} and *CEI* for H asphalt mixture. The number of coarse aggregates in F asphalt mixture is less and the particle size is mainly concentrated in 4.75–9.5 mm, producing moderate or weak force chain with the contact and friction among aggregates. The compaction energy was dissipated by those force chains during compaction process of asphalt mixture, and consequently, more compaction energy is needed to input

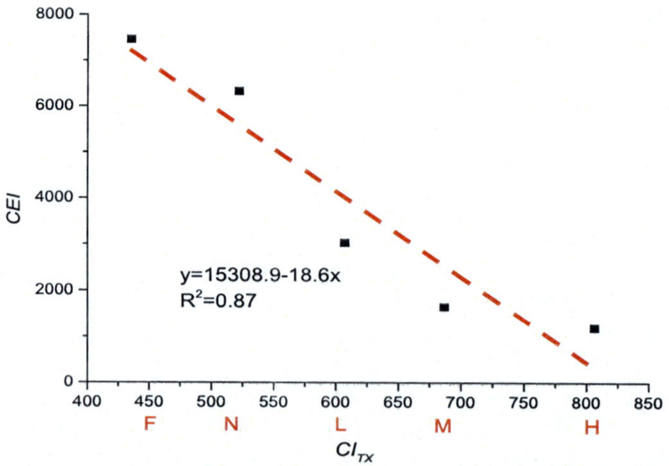

Fig. 4 The relationship between CI_{TX} and *CEI*

into F asphalt mixture. With the increase of segregation degree of coarse aggregates, a part of fine aggregates in aggregate-asphalt system is gradually replaced by coarse aggregates, leading to the decrease of specific surface area of all aggregates and the content of asphalt mortar. CI_{TX} is the unity of texture characteristics and the size distribution of coarse aggregates. The larger number of coarse aggregates in the particle system, the higher the CI_{TX}. The interstitial content of asphalt mortar decreases with the rise of CI_{TX}. The migration space of coarse aggregate becomes larger, and the dissipated energy by weak and moderate force chains is less, and therefore, the coarse aggregate is prone to rolling and sliding under the action of compaction effort, and the *CEI* declines with increasing the CI_{TX}.

Based on Fig. 5, the influences of three parameters on *CEI* are in the same manner, particularly, CI_{GA} has a good linear correlation with *CEI*, and the correlation coefficient is as high as 0.91. CI_{GA} is the unity of the angularity characteristics and the size distribution of large-sized coarse aggregates in particle system. The lubrication effect of asphalt binder on coarse aggregate contact is obviously greater than bonding effect at compaction temperature of 155 °C [24]. Therefore, the larger the particle size, the greater the contribution of particle angularity, and the more significant the lubrication effect of asphalt. The CI_{GA} is a key factor affecting the contact properties of particle system. The contact among particles is the strongest in H particle system, meanwhile, the lubrication effect of asphalt binder on the contact is the most significant, leading to the lowest *CEI*. As a result, the compaction energy of asphalt mixture is determined by the combined action of contact effect among particles and bonding/lubrication effect of asphalt, and the lubrication effect on workability is greater than contact effect for H asphalt mixture. The CI_{TX} and CI_{GA} can reflect the coarse aggregate distribution and contact properties, while the influence of asphalt on compaction properties is significant. Thus, CI_{GA} and CI_{TX} are helpful to analyze the micro-meso-scopic compaction process, but they cannot be used to predict the compaction properties of asphalt mixture.

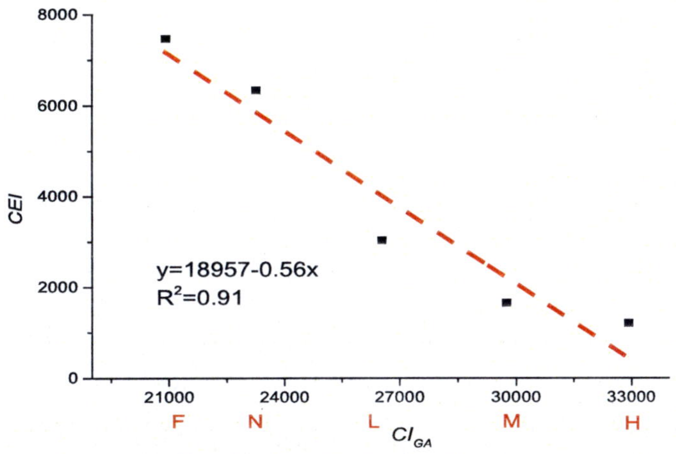

Fig. 5 The relationship between CI_{GA} and *CEI*

4.2 *The Relationship Between the Parameters of Composite Geometric Characteristic and High-Temperature Creep Properties*

As can be observed from Figs. 6, 7 and 8 that the load cycles at failure initially increased and then decreased with the increase of the degree of coarse aggregate segregation, achieving the maximum value for the L asphalt mixture. The creep rate bucks the trends with load cycles. As a result, the shear performance of F,

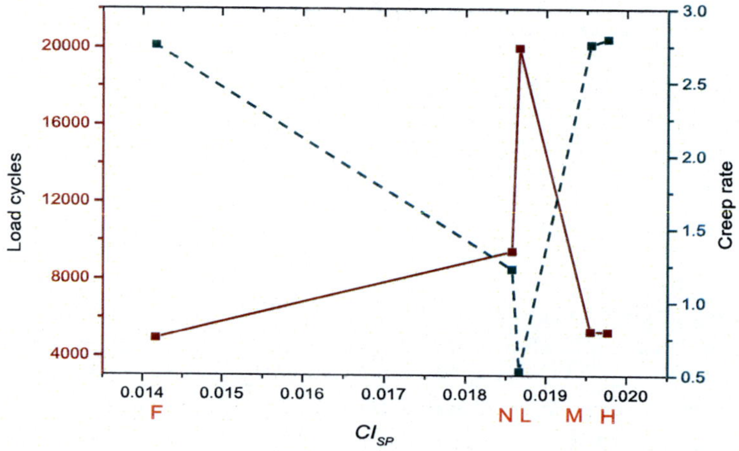

Fig. 6 The influence of CI_{SP} on the parameters of resistance to shear deformation

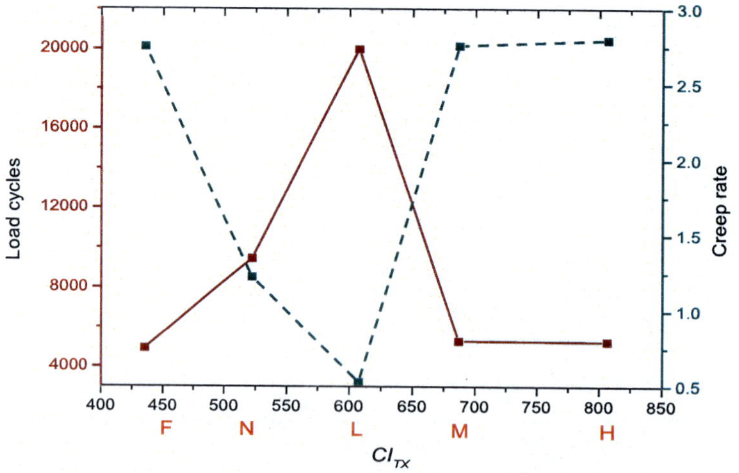

Fig. 7 The influence of CI_{TX} on the parameters of resistance to shear deformation

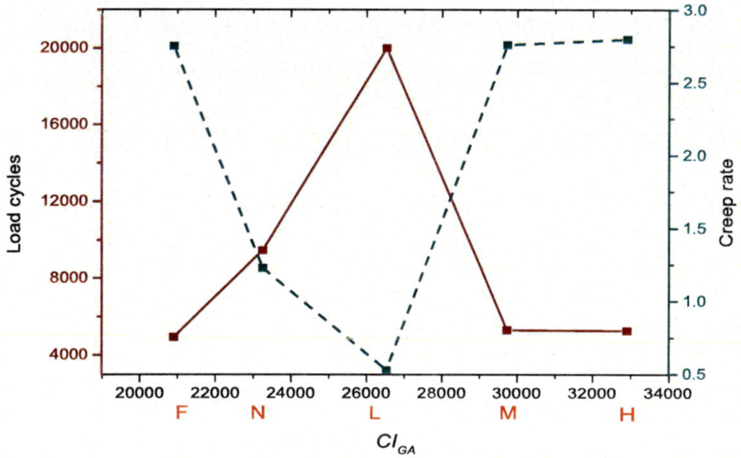

Fig. 8 The influence of CI_{GA} on the parameters of resistance to shear deformation

M and H asphalt mixtures is poor. The number of fine aggregates is largest in F asphalt mixture, leading to the worst skeleton, thus, the shear deformation under the repeated action of wheel load occur easily. For M and H asphalt mixtures has larger coarse aggregates and less fine aggregates, and the asphalt displayed poor bonding and stabilizing effects on the skeleton. In the transition process of asphalt mixture from loose flow state to stable forming state, when the asphalt mixture changes from a loose flow state to a stable forming state, the interaction between asphalt and aggregate evolves continuously. The interference effect of fine aggregates on coarse aggregate skeleton structure gradually transforms into filling effect of asphalt mortar and bonding effect of asphalt mastic. The asphalt mixture is regarded as an aggregate-asphalt system composed of coarse aggregate, asphalt mastic and asphalt mortar. The skeleton structure of coarse aggregates and the bonding effect of mastic are the principal factors affecting the high-temperature stability. Asphalt mortar extends the skeleton structure, leading to little effect of skeleton on resisting the external load, which is named filling interference effect. The bonding effect of asphalt mastic is insufficient to stabilize the skeleton structure, which is named the skeleton failure effect. The heavier the segregation tendency of coarse aggregates is, the larger the composite geometric parameters are, so the more significant the contact-friction effect among particles is. The M and H asphalt mixture have similarly poor resistance to shear deformation, indicating that the bonding effect on particle contact is getting worse and worse with the increase of segregation tendency of coarse aggregates.

L asphalt mixture has larger CI_{TX} and CI_{GA} than those of N asphalt mixture, but almost equal value of CI_{SP}. The results indicate that the total number of particles in L and N asphalt mixture is nearly equal, but the L asphalt mixture has higher number of large-sized coarse aggregates. The L asphalt mixture with best rutting resistance has strongest skeleton bonding strength resulting from the coupling interaction between the particle contact effect and bonding/lubrication effect of asphalt binder.

4.3 The Relationship Between the Parameters of Composite Geometric Characteristic and Water Stability

Figure 9 delineates the water stability parameters of involved five asphalt mixtures at spalling point. The order of cycle number of asphalt mixture is L > N > M > H > F, and the order of rutting depth of rutting is N < F < L < M < H, with L < F < H < N < M for spalling rate. Conducting comprehensive analysis of three parameters indicates that the water stability of L asphalt mixture is the best, followed by that with N asphalt mixture, and the worst is that with M and H asphalt mixture.

When the moisture enters the internal structure of asphalt mixture through the voids, dynamic pore water pressure is produced under the vehicle load, causing microcracks in the asphalt coated on the aggregate surface or filled in the voids. And then, moisture in turn invades the contact interface between aggregates and asphalt, reducing the adhesion, as a result, the water stability of asphalt mixture becomes worse. With the increase of segregation tendency of coarse aggregates, the volume of air voids (*VV*) of asphalt mixture becomes larger and larger. There is an optimal range of *VV* with the best water stability. The contact effect of coarse aggregate and the filling effect of asphalt mortar are important sources of voids. In this investigation, the water stability of the segregated asphalt mixture will be elucidated based on the particle characteristics.

As exhibited in Fig. 10, the larger the composite geometric parameters of coarse aggregate, the higher the *VV* of asphalt mixture, and the *VV* evinces the robust correlations with CI_{TX} and CI_{GA} with the correlation coefficients as high as 0.95.

There are more coarse aggregates in particle system with higher CI_{TX} and CI_{GA}, but less fine aggregates, leading to strong contact, and therefore, the large *VV* of asphalt mixture is measured. Asphalt invades the surface texture of coarse aggregate

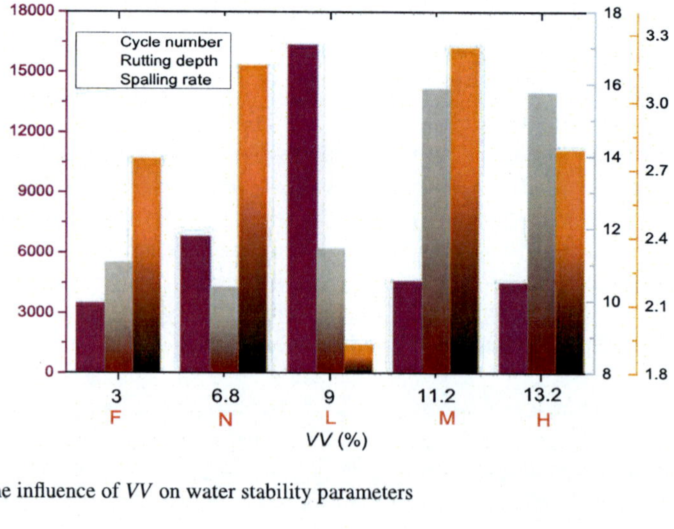

Fig. 9 The influence of *VV* on water stability parameters

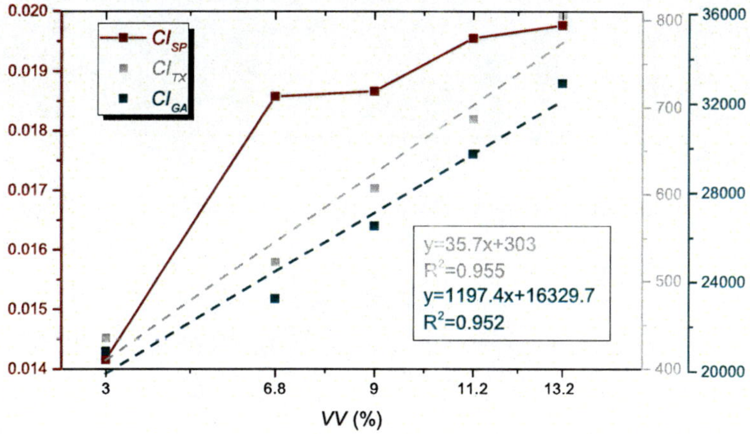

Fig. 10 The relationship between *VV* and composite geometric parameters

and lubricates the contact and friction between particles during addition of asphalt, increasing the contact area and improving the densification of skeleton structure to a certain extent. In addition, asphalt is combined with fine aggregates with large specific surface area to form asphalt mortar to fill skeleton void. The CI_{TX} and CI_{GA} of M and H asphalt mixture are relatively larger, and the contact and friction among coarse particles is significant, while the filling effect of asphalt mortar is poor, yielding large *VV*. As a result, CI_{TX} and CI_{GA} of particle system are closely related to the formation of void in asphalt mixture and can be used to predict the size of void. The F asphalt mixture has the smallest *VV*, but the largest number of fine aggregates causing the thin asphalt film. Under the action of dynamic pore water pressure, asphalt film is washed and corroded by moisture, and consequently, asphalt coated on the surface of the coarse particles easily spalls. The *VV* of L asphalt mixture is at a medium level, and the moisture penetrating into the internal structure is less than M and H asphalt mixture. On the other hand, L asphalt mixture has a certain number of coarse aggregates to form a stable skeleton structure, exercising a slowing effect on asphalt spalling under the coupling effect of water and load. Therefore, L asphalt mixture shows the best water stability.

4.4 The Relationship Between the Parameters of Composite Geometric Characteristic and Low-Temperature Crack Resistance

As depicted in Fig. 11, the R_T of involved asphalt mixture except for F asphalt mixture decreases with the increasing of the segregation tendency of coarse aggregates, with larger ε_T of L and F mixture, and the largest S_T of N asphalt mixture. The

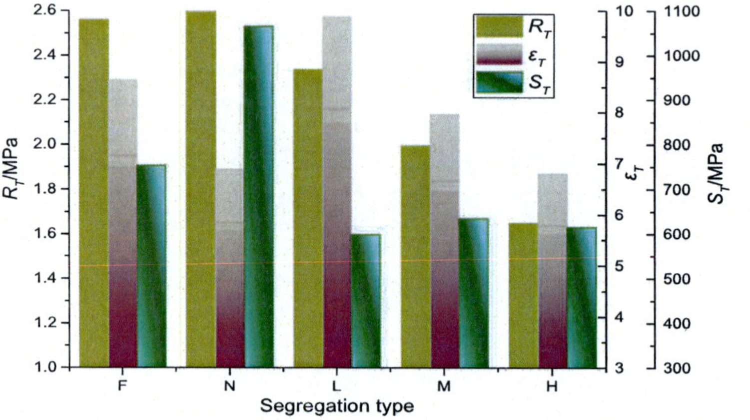

Fig. 11 The parameters of crack resistance of the segregated asphalt

comprehensive analysis of those parameters reflects that the low-temperature crack resistance of the dense structure is better than that of the skeleton structure, and best for the L asphalt mixture. The micro-cracks initially appear on the upper and lower surfaces of the mixture specimen along with axial loading, and then the local stress concentration occurs at the tip of the micro-cracks. When the stress transfers through voids, the micro-cracks will expand rapidly. Conversely, for dense structure with less voids, the stress transfer is impeded, slowing down the crack propagation speed.

Figure 11 illuminates that the *VV*, CI_{TX} and CI_{GA} of M and H mixture are larger. Compared with the N particle system, the larger CI_{TX} and CI_{GA} are, the higher the number of fine aggregates replaced by large-sized coarse aggregates, producing large-sized voids inside the asphalt mixture, and consequently, the crack propagation speed of M and H type mixture is relatively faster. F asphalt mixture has the smallest CI_{TX}, implying the least coarse aggregates and most fine aggregates. And therefore, large number of contact points occur and increases the direction of stress transfer, dissipating the stress.

There are adequate contact points in L asphalt mixture to effectively dissipate stress, and the skeleton bond strength is the largest, yielding the best low-temperature crack resistance.

4.5 *The Division Method of Asphalt Mixture Aggregate Segregation Tendency*

This investigation takes mesoscopic contact as a bridge built between the microscopic particle characteristics and the macroscopic properties of segregated asphalt mixture. The contact among the angularity and friction of the surface texture determine the contact strength of particle system, which in turn affects the macroscopic properties of the asphalt mixture.

The model built in Sect. 3.1 can accurately calculate the *CI* of all particle combination by inputting content of variously-sized aggregates. Aggregate segregation changes the size distribution of particles, resulting in different *CI*. This investigation delimits degree of segregation for asphalt mixtures adopting *CI*, so as to predict the compaction properties and road performance of asphalt mixture. The maximum value calculation method is used to normalize *CI* and the calculation method is as follows:

$$CI' = \frac{CI}{CI_{\max}} \tag{4}$$

where, CI' is non-dimensional composite geometric index, $CI_{\max}$ is the maximum value of *CI*. $CI_{\max}$ is the *CI* of SMA-20L particle system in accordance with the common gradations of the top and middle layers of asphalt pavement recommended in the Technical Specification for Construction of Highway Asphalt Pavement (JTG F40-2004). The normalization results of *CI* are presented in Table 5.

The shape affects the rotational inertia of particles and plays a support role for angularity and surface texture, indicating that it displays slight effect on the macroscopic properties. Therefore, CI_{TX} and CI_{GA} are utilized to divide the degree of segregation for asphalt mixtures, with the results presented in Table 6.

Table 5 The normalization results of *CI*

Indexes	Degree of segregation				
	F	N	L	M	H
CI_{SP}'	0.092353	0.12115	0.121731	0.127507	0.128909
CI_{TX}'	0.197395	0.236718	0.275325	0.311424	0.365658
CI_{GA}'	0.580243	0.645348	0.736289	0.825303	0.913184

Table 6 The division results of segregation tendency

Indexes	Delimiting results				
	F	N	L	M	H
CI_{TX}'	0–0.2	0.2–0.25	0.25–0.3	0.3–0.35	0.35–1
CI_{GA}'	0–0.6	0.6–0.7	0.7–0.8	0.8–0.9	0.9–1

5 Conclusion

This investigation focusing on the AC-20 gradation designed five asphalt with different degree of segregation. and the composite geometric index is calculated. The influence of aggregate segregation on the macroscopic properties of asphalt mixture is analyzed based on the particle characteristic. The main conclusions are as follows:

(1) Particle shape affects the compaction process of asphalt mixture by determining the rotation inertia of particles, and angularity and texture are the principal factors affecting contact strength among particles. The heavier the coarse aggregate segregation is, the more significant the lubricating effect of asphalt on the texture and angularity of particles is.

(2) The skeleton-bonding system is the external manifestations of coupling interaction of particle contact effect and bonding/lubrication effect of asphalt. The filling interference effect of asphalt mortar in L asphalt mixture is smaller than that in F and N asphalt mixture, and the skeleton failure effect is smaller than that in M and H asphalt mixture, leading to the best resistance to shear deformation.

(3) The voids are the foremost factor affecting the water stability of asphalt mixture, resulting from the large-sized coarse aggregates. The heavier segregation of coarse aggregate produces higher CI_{GA} and VV, indicating poor water stability of asphalt mixture.

(4) The smaller the CI_{TX} is, the larger the number of contact points inside the asphalt mixture is, and the easier the stress will be dissipated. The larger the CI_{GA} is, the faster the micro-cracks expand. The low-temperature crack resistance of the dense structure is better than that of the skeleton structure.

(5) The maximum value calculation method is adopted to normalize the composite geometric index. $CI_{TX}^{'}$ and $CI_{GA}^{'}$ can be used to divide the degree of segregation for asphalt mixtures.

References

1. Chang C, Baladi GY, Wolff TF (2001) Using pavement distress data to assess the impact of construction on pavement performance. Transportation Research Record 1761. Transportation Research Board, National Research Council, Washington, DC, pp 15–25
2. Tashman L et al (2002) X-ray tomography to characterize air void distribution in Superpave gyratory compacted specimens. Int J Pavement Eng 3(1):19–28
3. McCuen RH, Azari H, Shashidhar N (2001) Statistical characterization of aggregate in-homogeneity in asphalt concrete using computer simulation. Transportation Research Record 1757. Transportation Research Board, National Research Council, Washington, DC, pp 119–126
4. Zeng LW, Zhang SX, Zhang XN (2014) The research on aggregate microstructure uniformity image processing of asphalt mixture based on computer scanning technology. Adv Mater Res 831:393–400

5. Thyagarajan S, Tashman L, Masad E et al (2010) The heterogeneity and mechanical response of hot mix asphalt laboratory specimens. Int J Pavement Eng 11(2):107–121
6. Jia J, Wan YP, Liu HH et al (2019) Evaluation of compaction uniformity of the paving layer based on transverse and longitudinal measurements. Int J Pavement Eng. Published online
7. Zhang ZQ, Huang SL, Zhang K (2017) Accurate detection method for compaction uniformity of asphalt pavement. Constr Build Mater 145:88–97
8. Zhang K, Zhang ZQ, Luo YF et al (2017) Accurate detection and evaluation method for aggregate distribution uniformity of asphalt pavement. Constr Build Mater 152:715–730
9. Li XL, Chen SY, Xiong KY et al (2018) Gradation segregation analysis of warm mix asphalt mixture. J Mater in Civ Eng 30(4):04018027
10. Si JZ (2018) Intelligent compaction for quality control and acceptance for soil and base compaction through statistical analysis. Transp Res Rec 1–8
11. Jaime W, Silvia C, Andrew B (2017) Influence of material heterogeneity in the fracture of asphalt mixtures. Int J Pavement Eng 20(7):747–760
12. Feng C, Denis J, Manfred NP (2020) Vibration-induced aggregate segregation in asphalt mixtures. Mater Struct 53:27
13. Yang XH, Bi YQ, Zhang JP et al (2016) Impacts of aggregate morphological characteristics on asphalt mixture performance based on experimental tests. In: 4th Geo-China international conference on sustainable civil infrastructures—innovative technologies for severe weathers and climate changes, Shandong, Peoples R China, vol 266, pp 34–41
14. Gao JF, Wang HN, Bu Y et al (2020) Muhammad Irfan. Influence of coarse-aggregate angularity on asphalt mixture macroperformance skid resistance, high-temperature, and compaction performance. J Mater Civ Eng 32(5):04020095
15. Cong L, Shi JC, Wang TJ et al (2019) A method to evaluate the segregation of compacted asphalt pavement by processing the images of paved asphalt mixture. Constr Build Mater 224:622–629
16. Wu WL, Tu ZX, Zhu ZH et al (2019) Effect of gradation segregation on mechanical properties of an asphalt mixture. Appl Sci 9:308
17. Feng XJ, Song Ye B, Hao PW (2013) A new laboratory method to characterize gradation segregation of large stone asphalt mixtures. Constr Build Mater 38:1199–1203
18. Peng YH, Hu JY, Hu SF (2014) Influence of AC-25 particle size distribution on gradation segregation. J Traff Transp Eng 14(05):1–7+18
19. Ding Z, Li PL, Wu X et al (2020) Evaluation of the contact characteristics of graded aggregate using coarse aggregate composite geometric indexes. Constr Build Mater 247
20. Su JF, Li PL, Sun SF et al (2021) Effects of composite geometric characteristics of coarse particles on interface interactions of aggregate-asphalt system. Constr Build Mater Accept
21. Zhang K, Cui WS, Zhang ZQ et al (2014) Study on asphalt mixture segregation criteria based on laboratory tests. J Wuhan Univ Technol 36(05):55–61
22. Su JF, Li PL, Sun SF et al (2020) Effect of contact slip behaviour of particle on compaction characteristics of the asphalt mixture. Int J Pavement Eng. Published online
23. Li PL, Su JF, Gao P et al (2019) Analysis of aggregate particle migration properties during compaction process of asphalt mixture. Constr Build Mater 197:42–49
24. Su JF, Li PL, Wei XF et al (2020) Interface transformation behavior of bonding/lubrication of aggregate-asphalt system. J Mater Civ Eng 32(12):04020380

Effects of Moisture Damage Sensitivity of Asphalt Mixtures Incorporating Treated Plastic as Additive

Siti Nur Naqibah Kamarudin, Mohd Rosli Hainin, Muhammad Naqiuddin Mohd Warid, Mohd Khairul Idham Mohd Satar, Noor Azah Abdul Raman, and Kabiru Usman Rogo

Abstract The issue of moisture damage due to water intrusion in conventional road pavement is crucial due to increasing traffic load and volume, especially in the dry process. Poor workability occurs between the aggregate and binder interaction, increasing air voids in the asphalt mixture. Consequently, reduce the strength of the asphalt mixture. Hence, the additive material is one possible approach to reduce the issue. The use of treated plastic (Low-density Polyethylene) (TP) has a high potential to improve the interaction of aggregate and binder during the mixing process to against moisture damage. Further investigation on the effect of different ratios of TP content (0, 2.5, 5.0, 0.75, and 1.0% by weight of total aggregate) in asphalt mixture was studied. Mechanical performances on moisture damage of the optimum TP content in asphalt mixture were focused and evaluated with retained stability and tensile strength ratio tests. The finding revealed that 0.75% of TP content increased the resistance of moisture damage in asphalt mixtures compared to the conventional mixture.

Keywords Moisture damage · Additive · Waste polymer · Hot mix asphalt

1 Introduction

Moisture damage in asphalt mixtures occurs due to excessive or improper runoff of surface waters resulting in the deterioration of the entire pavement structure [1, 2] besides the aging issue [3, 4]. Hence led to permanent deformation issues such as rutting, fatigue cracking, and stripping. Moisture damage is mainly caused by a loss of adhesion or cohesive forces as two primary modes within the binder and aggregate, which are supposed to hold the latter together [5]. These issues had forced many road agencies and researchers to modify the conventional mixture besides applying new

S. N. N. Kamarudin · M. N. M. Warid · M. K. I. M. Satar · N. A. A. Raman · K. U. Rogo
Department of Geotechnics and Transportation, School of Civil Engineering, Faculty of Engineering, Universiti Teknologi Malaysia (UTM), Skudai, Johor, Malaysia

M. R. Hainin (✉)
College of Engineering, Universiti Malaysia Pahang (UMP), 26600 Pekan, Pahang, Malaysia

© The Author(s), under exclusive license to Springer Nature Switzerland AG 2022

H. R. Pasindu et al. (eds.), *Road and Airfield Pavement Technology*, Lecture Notes in Civil Engineering 193, https://doi.org/10.1007/978-3-030-87379-0_50

technology such as warm mix asphalt [6]. The use of polymer as an additive becomes one alternative to make the conventional mixture more superior and perform longer. However, polymer-modified mixtures increase initial expenses by about 20% higher than conventional mixtures [7], but it lowers maintenance costs since it provides more durable pavement [8]. Hence, many researchers started to concern about to use of waste materials such as waste cooking oil [9], steel slag [10, 11], palm oil fuel ash [12], crumb rubber [13]. These waste materials showed high potential to be converted as an additive in road construction and increased the performance of conventional mixture besides reducing the environmental problem.

There are two methods to add additives in asphalt mixture. It is either by wet or dry method. Basically, in the wet method, the additive blended with bitumen at high temperature before adding to warm aggregate. Meanwhile, in the dry method, the additive was added into warm aggregate before mixed with hot bitumen. Due to the polymer properties, many researchers had preferred to use the wet method to obtain a homogenous modified asphalt [14]. This is because the addition of polymer as additive by dry method had been reported to perform inconsistent performance related to the short mixing time, cause a short polymer-bitumen interaction time, and lead to inadequate bitumen modification [15]. Nonetheless, less complicated, easy to apply in plants, and increased the amount of additive used in asphalt mixture compared to wet method become the main factors to many researchers to interest in dry method application [16].

Waste plastic polyethylene is a non-biodegradable product and such a worrisome problem due to its improper disposal at landfills due to the high demand of plastic users in daily life [17]. This worldwide issue caused the use of waste plastic as an additive in asphalt mixture received well attentions in many kinds of research. Based on previous studies, the use of waste plastic polyethylene type High-Density Polyethylene [18] and Low-Density Polyethylene [19] increased resistance to permanent deformation and better moisture damage under high traffic loads and in hot climate countries [20, 21] due to its high rigidity [22]. More than that, the application of the dry method increased by about 0.43% the use of waste plastic in asphalt mixture compared to the wet method. Treated waste plastic was more rough and porous surface than non-treated plastic, which gave high potential in increasing the interaction between the additive and asphalt materials during the dry method. It also increased the amount of waste plastic by about 0.36% by weight of the total mixture compared to non-treated plastic [23].

In this study, the effects of moisture damage sensitivity of the asphalt mixtures incorporating treated plastic additives were evaluated. The significance of this study was to improve the performance of asphalt mixture and enhance the use of waste plastic in asphalt mixture as an alternative to reduce the environmental problem regarding excessive and improper disposal of waste plastic at landfills. It is not only contributed to the global warming issue but also a health issue.

2 Experimental Work

2.1 Materials

Aggregate and asphalt binder. Aggregate with gradation nominal maximum aggregate size 14 mm (AC 14) as shown in Fig. 1 from Hanson Quarry Sdn Bhd was used. For asphalt binder, 60–70 penetration bitumen was used from Kemaman Bitumen Company. The properties of bitumen were listed in Table 1.

Additive. The additive was supplied by Ahn Vertex Sdn Bhd. Waste plastic polyethylene additive was made by the waste plastic bag. The additive was physically treated (grafting method) and specially designed in powder form as shown in Fig. 2, size passes through sieve size 0.425 mm. The melting temperature was 104.3 °C.

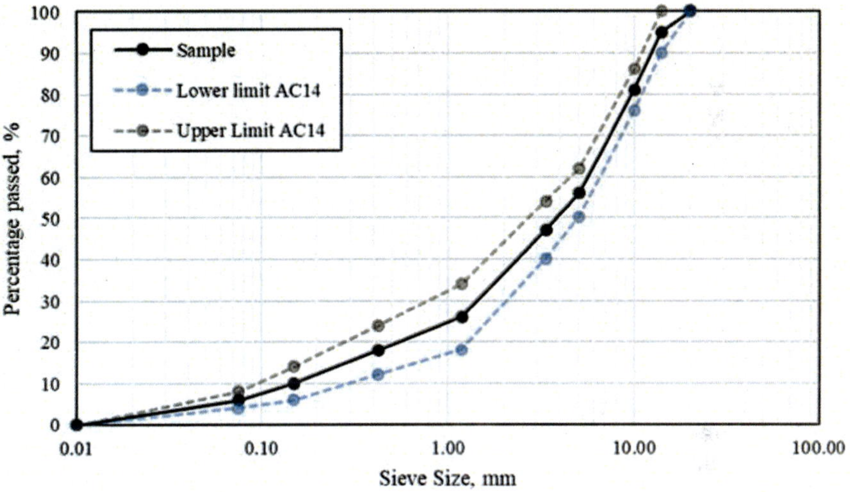

Fig. 1 Aggregate at AC14 gradation

Table 1 The bitumen properties

Properties	Results
Density (g/cm^3)	1.03
Penetration (PEN) at 25 °C	60.9
Softening point (°C)	50
Viscosity (Pa.s)	0.7 at 135 °C
Dynamic shear rheometer, G*/sin δ (kPa)	1.685 at 64 °C

Fig. 2 Treated Plastic (TP) additive

2.2 *Sample Preparations*

All samples were prepared according to ASTM D 6926-10 [24] with an average height of 65 mm and a diameter of 100 mm. The heated aggregate was mixed with treated plastic (TP) content (0, 0.25, 0.50, 0.75, and 1.0% by weight of total aggregate) before hot bitumen with optimum bitumen content 4.9% was added under mixing temperature ±180 °C. Each sample was compacted with 75 blows per face using a Marshall compactor at temperature ±145 °C. Compacted samples had cooled to room temperature for a day before mechanical tests were performed.

2.3 *Marshall Stability and Flow Test*

According to ASTM D6927 [25], the Marshall test was conducted by a universal compressive machine, as shown in Fig. 3. For prior, all samples were immersed in the water bath for 45 min at a temperature of 60 °C. The conditioned samples were tested with a maximum load at a constant loading rate of 50.8 mm/minute to measure the stability. The load applied was increased until it reached a maximum of stability and flow then the load began to decrease. The properties were compared with Jabatan Kerja Raya's specification [26].

Fig. 3 Universal compression machine with Marshall loading fixture

2.4 *Moisture Damage on Compacted Mixtures*

Immersion Marshall Test. According to ASTM D1075 [27], six samples were prepared similar to the Marshall sample and divided into two subsets with three samples each. One subset was soaked in the water bath at 60 ± 1 °C for 30 min. Another subset was soaked in the water bath at 60 ± 1 °C for 24 h and designated as conditioned samples. All samples were then loaded to failure with applied load to the sample at a constant compression rate of 50 ± 5 mm/min using a universal compression machine, designated as unconditioned samples. Retained Marshall stability (RMS) was calculated based on Eq. (1). RMS value represented the possibility of moisture damage occurred in the asphalt mixture.

$$RMS = \frac{MS_2}{MS_1} \times 100\% \quad (1)$$

where,

MS_2 The average Marshall stability for conditioned samples, kN.
MS_1 The average Marshall stability for unconditioned samples, kN.

Tensile Strength Ratio Test. According to ASTM D4867/D4867M [28], six samples were prepared for each mixture design with a similar procedure in preparation of Marshall mix design and compacted to void content to 7 ± 1% and also was divided into two subsets with three each. One subset was conditioned at a constant temperature of 25 ± 1 °C in the water bath for 20 min and labelled as dry samples. For another subset was conditioned under 35 kPa pressure by the vacuum pump to achieve the saturation level between 55 and 80%. After that, the samples were immersed in the water bath at 60 ± 1 °C. After 24 h, the samples were then conditioned with a soaked water bath for 1 h at 25 ± 1 °C.

Then, indirect tensile strength (ITS) was determined as referred to ASTM D6931 [29]. The dry and wet conditioned samples were placed at two parallel blades, and uniform tensile stress with constant diametral load at 50 mm/min was applied, as shown in Fig. 4, until the maximum load was reached. The maximum load was recorded.

Indirect Tensile Strength (ITS) was determined. Each sample was placed under pressure load by two parallel blades of the machine and underwent uniform tensile stress with constant diametral load at 50 mm/min until the maximum load was reached

Fig. 4 Universal compression machine with ITS fixture

and the maximum load was recorded. The maximum load was recorded, and the tensile strength ratio (TSR) was calculated by Eq. (2). TSR value indicated the effect of water on the tensile strength of the asphalt mixture.

$$TSR = \frac{ITS_{wet\ conditioned}}{ITS_{dry\ conditioned}} \quad (2)$$

3 Results and Discussions

3.1 Marshall Stability and Flow Test

Table 2 summarizes the result of Marshall properties for all mixtures. All mixtures met all the JKR specifications except 1.0% of TP content, which failed in flow value. The data shows that the stability, flow, and stiffness reduced when 0.25% of TP content was added to the asphalt mixture, yet increasing as the TP content increased. However, the stability and stiffness decreased when the TP content was added more than 0.75%. On the other hand, the VTM value slightly increased, while the VFB value decreased as the TP content increased except 1.0% of TP content. These results depicted that the mixture became more viscous and harder as TP content increased, which can be related to its higher stability and stiffness value [30]. Overall, 0.75% TP content was selected as the optimum content, although 0.25 and 0.50% TP content also met all the requirements. It is because 0.75% showed a greater improvement with 12 and 5% in stability and stiffness performance than other mixtures. Moreover, based on these findings, it can be suggested that treated plastic by dry process showed consistency in Marshall performance due to its finer size and porous surface led to well dispersion into binder matrix during the mixing process. In addition, its low melting temperature compared to mixing temperature made the TP additive easily melt with less effort in a short period.

Table 2 Validation of Marshall mixture specification

JKR specification [26]	Stability, kN	Flow, mm	Stiffness, kN/mm	Void in total mix (VTM), %	Void filled with bitumen (VFB), %
	> 8.0	2.0–4.0	>2.0	3.0–5.0	70–80
0 % (Control)	15.89 ✓	3.5 ✓	4.54 ✓	4.0 ✓	74 ✓
0.25 %	12.53 ✓	3.0 ✓	4.25 ✓	4.3 ✓	72 ✓
0.50 %	13.21 ✓	3.2 ✓	4.18 ✓	4.6 ✓	71 ✓
0.75 %	17.72 ✓	3.7 ✓	4.75 ✓	4.8 ✓	70 ✓
1.0 %	16.96 ✓	4.1 x	4.14 ✓	4.9 ✓	74 ✓

Table 3 Summary of control and TP mixture for immersion Marshall test

Type of mixture	Bulk specific gravity	Air voids, %	Stability 30 min at 60 °C MS_1 (kN)	Flow F_1 (mm)	Stability 24 h at 60 °C MS_2 (kN)	Flow F_2 (mm)	RMSF, % (MS_2/F_2) (MS_1/F_1)	RMS, %
Control	2.344	4.0	15.90	3.5	14.90	3.5	93.7	93.7
0.75%TP	2.342	4.8	17.72	3.7	15.23	3.4	93.5	86.0

3.2 *Moisture Damage on Compacted Mixtures*

Immersion Marshall Test. Table 3 summarizes retained Marshall stability (RMS) and retained Marshall stability and flow (RMSF) for the conditioned and unconditioned mixture. Meanwhile, Fig. 5 illustrates the graph of Marshall stability at both conditioned, while Fig. 6 plots the RMS and RMSF value for 0.75% TP mixture compared to the control mixture.

Based on Fig. 5, as expected, a 0.75% TP mixture showed insignificant higher of MS_1 and MS_2 for both conditioned and unconditioned mixture than the control mixture. However, by referring to Fig. 6, the RMS value of 0.75% TP mixture was lower than the control mixture. Simultaneously, the RMSF value of both mixtures showed an insignificant difference. Higher RMS value indicated high resistance to moisture damage when intrusion by water [31]. Although 0.75% TP mixture obtained lower RMS value than control mixture, higher RMS value with more than 75% indicated high resistance to moisture damage. These findings can be suggested that the addition of 0.75% of TP content in the mixture increased the internal strength due to its higher in MS_1 and MS_2 even though the performance to resist moisture damage

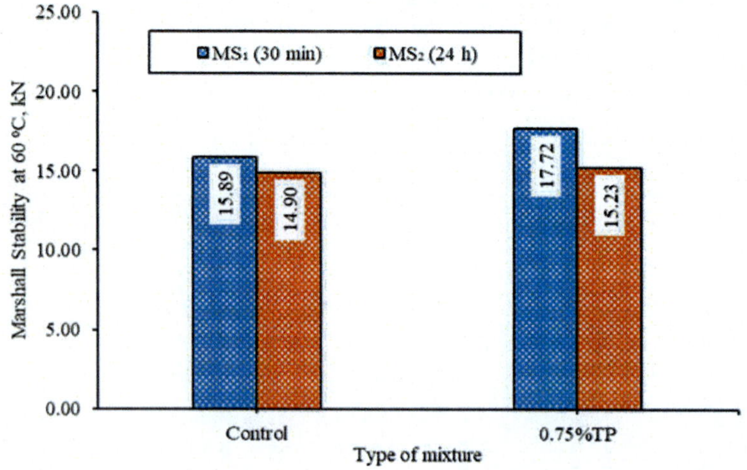

Fig. 5 Marshall stability at 60 °C for 30 min and 24 h of control and 0.75% TP mixture

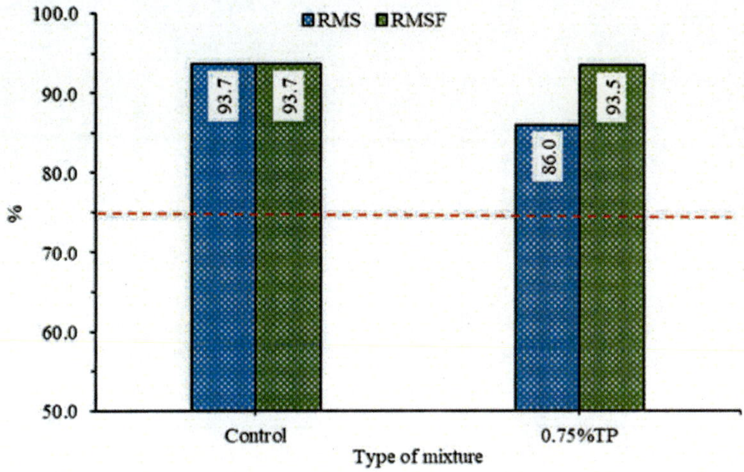

Fig. 6 RMS and RMSF value of control and 0.75% TP mixture

was slightly lower than the control mixture. It can be related to the properties of polyethylene where the molten polyethylene caused stiffen the binder when polymer-binder interaction occurs during the mixing process [32].

Tensile Strength Ratio Test. Figures 7 and 8 present the comparison of ITS and TSR value for wet and dry conditioned control and 0.75% TP mixture, respectively. Tensile strength of dry and wet conditioned generally increased when 0.75% of TP content was used. The increase in wet conditioned strength was less than the increase in dry conditioned strength for both mixtures. Concurrently, the TSR value of 0.75% TP mixture was slightly higher than the control mixture, represented its moisture

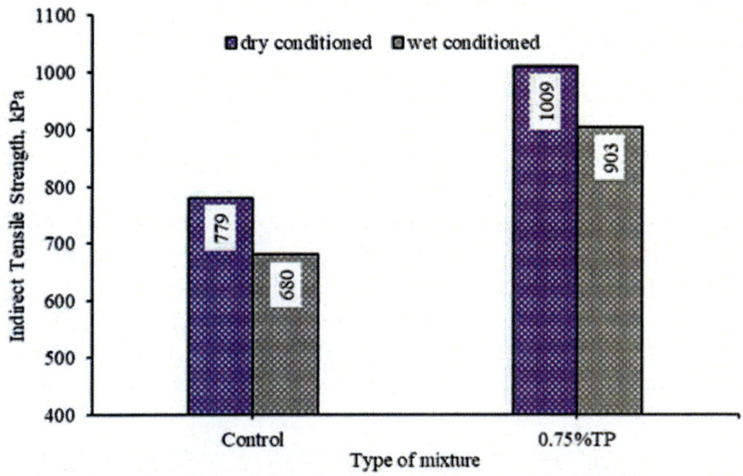

Fig. 7 ITS value of control and 0.75% TP mixture

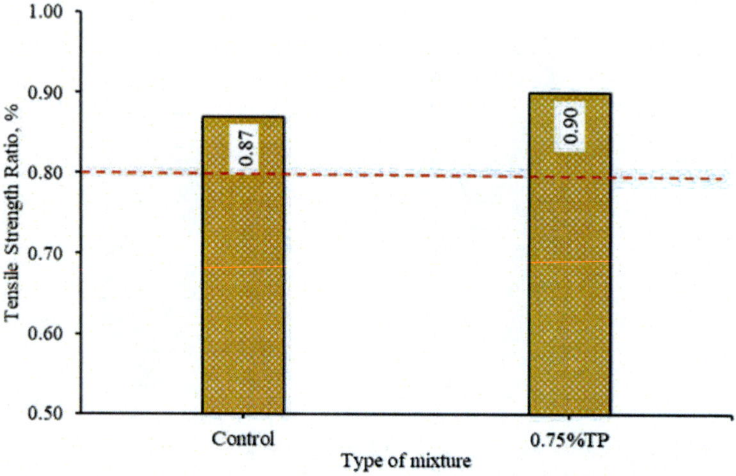

Fig. 8 TSR value of control and 0.75% TP mixture

resistance of mixture increased. Notably, the TSR value of more than 0.80 depicted that the mixture was good in resisting moisture damage. The serviceability of flexible pavements improved when the resistance to moisture damage higher [33]. Despite that, the higher average ITS and TSR value of 0.75% TP mixture can be related to its higher stability and stiffness performance. It can be reviewed, the addition of polymer additive increased the strength of the mixture due to its higher adhesion of binder and additive to the aggregates, thus good bonding between the materials [9, 34].

4 Conclusions

Based on the findings, as compared to the control mixture, it can be concluded that:

(a) Marshall properties showed 0.25, 0.50 and 0.75% met all the specifications. 0.75% TP mixture was selected as the optimum content due to an increase in stability and stiffness with 12 and 5%.
(b) The addition of 0.75% of TP content in the mixture showed lower RMS and no improvement in RMSF value indicated the mixture slightly more susceptibility to moisture damage than the control mixture. However, the RMS and RMSF values of more than 75% represented the mixture still good in resistance to moisture damage, which can be related to its higher strength of mixture due to its higher stability in both conditioned.
(c) 0.75% TP mixture had higher ITS and TSR values, indicating that the mixture's moisture resistance increased due to its higher tensile strength against water intrusion in the mixture than the control mixture.

In a nutshell, the addition of TP improved the moisture resistance in the asphalt mixture. Hence, the road service life can be increased. The finding enhanced the use of waste plastic in asphalt mixture to produce sustainable roads, thus reducing the environmental problem. However, it is recommended to study further on stripping resistance and the mixture's performance using nano-size of TP additive in detail.

Acknowledgements This work was supported/funded by Ministry of Higher Education under Fundamental Research Grant Scheme (FRGS/1/2019/TK01/UTM/02/1).

References

1. Lantieri C, Lamperti R, Simone A, Vignali V, Sangiorgi C, Dondi G (2015) Mobile laser scanning system for assessment of the rainwater runoff and drainage conditions on road pavements. Int J Pavement Res Technol 8(1):1
2. Hainin MR, Idham MK, Yaro NSA, Hussein SOAE, Warid MNM, Mohamed A, Naqibah SN, Ramadhansyah PJ (2018) Performance of hot mix asphalt mixture incorporating Kenaf fibre. IOP Conf Ser Earth Environ Sci 140(1):012–092
3. Idham MK, Mohd Rosli H, Yaacob H, Warid MNM, Abdullah ME (2013) Effect of aging on resilient modulus of hot mix asphalt mixtures. Adv Mater Res 723:291–297
4. Abdullah ME, Zamhari KA, Buhari R, Nayan MN, Hainin MR (2014) Short term and long term aging effects of asphalt binder modified with montmorillonite. Key Eng Mater 594:996–1002
5. Little DN, Allen DH, Bhasin A (2018) Chemical and mechanical processes influencing adhesion and moisture damage in hot mix asphalt pavements. In: Modeling and design of flexible pavements and materials. Springer, Cham, pp 123–186
6. Abdullah ME, Zamhari KA, Shamshudin MK, Mohd Rosli H, Idham MK (2013) Rheological properties of asphalt binder modified with chemical warm asphalt additive. Adv Mater Res 671:1692–1699
7. Behnood A, Gharehveran MM (2019) Morphology, rheology, and physical properties of polymer-modified asphalt binders. Eur Polym J 112:766–791
8. Brasileiro L, Moreno-Navarro F, Tauste-Martínez R, Matos J, Rubio-Gámez MDC (2019) Reclaimed polymers as asphalt binder modifiers for more sustainable roads: a review. Sustainability 11(3):646
9. Azahar WNAW, Jaya RP, Hainin MR, Bujang M, Ngadi N (2017) Mechanical performance of asphaltic concrete incorporating untreated and treated waste cooking oil. Constr Build Mater 150:653–663
10. Oluwasola EA, Hainin MR, Aziz MMA (2016) Comparative evaluation of dense-graded and gap-graded asphalt mix incorporating electric arc furnace steel slag and copper mine tailings. J Clean Prod 122:315–325
11. Aziz MMA, Hainin MR, Yaacob H, Ali Z, Chang FL, Adnan AM (2014) Characterisation and utilisation of steel slag for the construction of roads and highways. Mater Res Innovations 18(sup6):S6-255
12. Hainin MR, Jaya RP, Ali Akbar NA, Jayanti DS, Yusoff NIM (2014) Influence of palm oil fuel ash as a modifier on bitumen to improve aging resistance. J Eng Res 2(1):34–46
13. Ling TC, Nor HM, Hainin MR, Lim SK (2010) Long-term strength of rubberised concrete paving blocks. Proc Inst Civ Eng Constr Mater 163(1):19–26
14. Khurshid MB, Qureshi NA, Hussain A, Iqbal MJ (2019) Enhancement of Hot Mix Asphalt (HMA) Properties using waste polymers. Arab J Sci Eng 44(10):8239–8248
15. Martin-Alfonso JE, Cuadri AA, Torres J, Hidalgo ME, Partal P (2019) Use of plastic wastes from greenhouse in asphalt mixes manufactured by dry process. Road Mater Pavement Des 20(1):S265–S281

16. Movilla-Quesada D, Raposeiras AC, Silva-Klein LT, Lastra-González P, Castro-Fresno D (2019) Use of plastic scrap in asphalt mixtures added by dry method as a partial substitute for bitumen. Waste Manage 87:751–760
17. Awoyera PO, Adesina A (2020) Plastic wastes to construction products: status, limitations and future perspective. Case Stud Constr Mater 12:e00330
18. Köfteci S (2016) Effect of HDPE based wastes on the performance of modified asphalt mixtures. Procedia Eng 161:1268–1274
19. Merbouh M, Glaoui B, Mazouz A, Belhachemi M (2014) Effect of addition of plastic wastes on the creep performance of asphalt. In: Eurasia Waste Management Symposium, Istanbul, Turkey
20. Nejad FM, Azarhoosh A, Hamedi GH (2014) Effect of high density polyethylene on the fatigue and rutting performance of hot mix asphalt—a laboratory study. Road Mater Pavement Des 15(3):746–756
21. Ziari H, Nasiri E, Amini A, Ferdosian O (2019) The effect of EAF dust and waste PVC on moisture sensitivity, rutting resistance, and fatigue performance of asphalt binders and mixtures. Constr Build Mater 203:188–200
22. Hasan MRM, You Z, Satar MKIM, Warid MNM, Kamaruddin NHM, Ge D, Zhang R (2018) Effects of titanate coupling agent on engineering properties of asphalt binders and mixtures incorporating lldpe-caco3 pellet. Appl Sci 8(7):1029
23. Kamarudin SNN, Hainin MR, Warid MNM, Satar MKIM, Jaya RP (2021) The usage of treated plastic as additive to improve the asphalt mixture's performance by using dry mix method. Key Eng Mater 879:126–135
24. ASTM D6926 (2011) American society for testing and materials: standard practice for preparation of bituminous specimens using Marshall apparatus. ASTM International, West Conshohocken
25. ASTM D6927 (2015) American society for testing and materials: standard test method for Marshall stability and flow of asphalt mixtures. ASTM International, West Conshohocken
26. Jabatan Kerja Raya Malaysia (JKR) (2008) Standard Specification for Road Works, Section 4: Flexible pavement. No. JKR/SPJ/2008-S4. Malaysia
27. ASTM D1075-11 (2011) American society for testing and materials: standard test method for effect of water on compressive strength of compacted bituminous mixtures. ASTM International, West Conshohocken
28. ASTM D4867/D4867M-09 (2014) American society for testing and materials: standard test method for effect of moisture on asphalt concrete paving mixtures. ASTM International, West Conshohocken
29. ASTM D6931-12 (2012) American society for testing and materials; standard test method for indirect tensile (IDT) strength of bituminous mixtures. ASTM International, West Conshohocken
30. Romastarika R, Jaya RP, Yaacob H, Nazri FM, Agussabti, Ichwana, Jayanti DS (2017) Effect of black rice husk ash on the physical and rheological properties of bitumen. In: International conference on applied physics and engineering, vol 1875, pp 1–8
31. Zhang H, Gao Y, Guo G, Zhao B, Yu J (2018) Effects of ZnO particle size on properties of asphalt and asphalt mixture. Constr Build Mater 159:578–586
32. Madzlan N, Habib NZ, Kamaruddin I (2014) Creep behavior of polyethylene modified bituminous mixture. APCBEE Proc 9:202–206
33. Hamzah MO, Kakar MR, Hainin MR (2015) An overview of moisture damage in asphalt mixtures. J Teknol 73(4)
34. Anurag K, Xiao F, Amirkhanian SN (2009) Laboratory investigation of indirect tensile strength using roofing polyester waste fibers in hot mix asphalt. Constr Build Mater 23(5):2035–2040

Efficient Numerical Methodology for the Determination of Thermal Conductivity of Asphalt Mixtures

Huichen Wang, L. Chu, Lei He, and T. F. Fwa

Abstract Despite the importance of the thermal conductivity (k) to the study of pavement thermal behaviors, it is still a challenging task for engineers to come up with an efficient method for its determination without laborious laboratory measurements. Recently some computer-aided approaches were proposed to overcome the mentioned difficulty but still suffered from time-consuming and inefficient computation. This study presents a hybrid methodology for the determination of k value with a totally automatic program, employing genetic algorithm (GA) to achieving efficient model calibrations, coupled with the finite element simulation to make estimations of the thermal conductivity of the asphalt mixtures. The improved performance of the proposed method is demonstrated by a 2-D row-column structure model example to achieve a higher accuracy and efficiency existing approaches under the same conditions.

Keywords Asphalt concrete · Thermal conductivity · Finite element method · Row-column model

1 Introduction

As one of the key properties of asphalt concrete, thermal conductivity has a major influence on the temperature field in an asphalt pavement structure [1, 2]. The structural response of an asphalt pavement is affected by daily and seasonal temperature variations because of the viscoelastic characteristics of asphalt binder [3, 4]. Hence, to better address the issues of cracking and rutting of asphalt pavements, the thermal conductivity of asphalt mixtures has been a key material property of interest for in-service pavement performance research [5, 6]. Besides the analysis of load-induced

H. Wang · L. Chu (✉) · L. He · T. F. Fwa
School of Highway, Chang'an University, Xi'an, China
e-mail: longjiachu@chd.edu.cn

T. F. Fwa
National University of Singapore, Singapore, Singapore

© The Author(s), under exclusive license to Springer Nature Switzerland AG 2022
H. R. Pasindu et al. (eds.), *Road and Airfield Pavement Technology*, Lecture Notes in Civil Engineering 193, https://doi.org/10.1007/978-3-030-87379-0_51

pavement response under different temperature conditions, it permits one to determine not only the time window for effective compaction [7, 8], but also the required cooling period before opening to traffic [9–11] in actual construction.

The thermal conductivity k of a heterogeneous material is determined by its composition. Instead of experimental methods, researchers have made numerous attempts to predict k by means of mathematical modelling [12]. There are five basic models developed: specifically, the Series, the Parallel, the Maxwell-Eucken and the Effective Medium Theory (EMT) [13, 14]. As for the domain of asphalt paving mixtures, numerical models have recently emerged as a promising tool for accurate determination of the temperature field in a pavement structure.

In recent years, finite element methods (FEMs) have been successfully applied by researchers to perform thermal analysis of engineering materials. Islam et al. [15] made use of cylindrical specimens of asphalt mixtures with 150 mm in diameter and 170 mm in height for laboratory testing. Compared to the results of experimental data, they validated the feasibility of FEM for thermal conductivity prediction with a promising agreement between these two results. This approach, however, is not practically suitable for pragmatic application because of the excessive reliance on testing, which suffers from the requirement of special devices that are usually not commonly available in standard pavement engineering laboratories. In addition, since the changes of air void content of asphalt pavement during its service phase, the corresponding thermal conductivity would vary with time, which calls for a large amount of experimental testing that has to be done for monitoring the varying thermal conductivity.

In view of the predicament of laboratory testing, Chen et al. [16, 17] conducted their studies by only using finite element models. They achieved a satisfying agreement between predictions and actual measurement values with errors generally within 10% under the condition of a steady-state one-dimensional heat transfer. The very similar study by Mirzanamadi et al. [18] also reported satisfactory estimates of k for asphalt mixtures by using 2-D model of asphalt mixtures with aggregates of circular cross sections, different from the polygonal aggregates in Chen's method [16]. These two approaches share the same time and effort consuming step of generating microstructures of an asphalt mixture. The effectiveness of the FEM thermal conductivity determination approach has been exemplified by Chu et al. [19, 20]. They indicated that the microstructure of asphalt structure made no difference theoretically to the overall effective thermal conductivity. Using the commercial software ABAQUS [21], they adopted a straightforward row-column FEM heat transfer model, coupled with genetic algorithm (GA) to realize the k prediction of an asphalt mixture with any air void ratio. Finally, they reported excellent results with lower error rates and higher efficiency compared with Chen's method [16] under the same condition, which throws light on the application of thermal conductivity determination at practice. Despite the success of the application of row-column FE model, there is still room for improvement in their study as follows:

1. The software Matlab [22] and ABAQUS [21] are not compatible, leading to a low efficiency because of the required manual participation in the GA cycles.

2. The procedure is not friendly to untrained operators.

To overcome the mentioned difficulties, this study proposed a hybrid methodology for FEM determination of k value with an integrated framework to make better estimations of the thermal conductivity of the asphalt mixtures.

2 Objective

The proposed analysis was based on the conceptual framework proposed by Chu et al. [19, 20]. The objective of this paper is to improve the computational efficiency of their 2-D model and lower the training requirements for users in determining the thermal conductivity of an asphalt mixture. This approach planted a novel GA scripts (in Python computer language) into the software ABAQUS [21], achieving a totally automatic computation process with only limited necessary parameters, which significantly reduced the difficulty in practical applications, as well as the processing time and manual participation required.

3 Framework of Hybrid GA-FE Methodology

The proposed hybrid GA-FE framework (depicted in Fig. 1) begins with setting the k ranges of asphalt and aggregate, while the k of air is considered as a constant at a given temperature. The percentages by volume of asphalt binder, aggregates and air voids are also required as inputs to the models generating. With the required parameters input, the program would generate a regular row-column finite element mesh to represent the body of asphalt mixture modeled using the finite element software ABAQUS [21]. The row-column model is generated by making random volume-based assignments of the cells to asphalt, aggregate and air based on their percent shares by volume in the asphalt mixture. Each of the cell is assigned the appropriate k value of the constituent assigned to it. Although the asphalt mixture model is represented by three constituents in the examples in this paper, the proposed approach can be easily extended to asphalt mixtures with more constituents.

GA is a computational method based on an analogy to natural selection and biological evolution, aiming to finally identify the 'fittest' solution (i.e. the best solution). The fittest solution is evaluated by means of a fitness function. The fitness function of the GA formulation is defined to minimize the errors of the predicted k value of the asphalt mixture. The optimization process of GA is essentially a search process by generating new trial solutions (known as offspring solutions) through GA operations such as crossover and mutation of trial solutions. GA trial-and-error search process continues until a predefined termination criterion, i.e. a pre-determined value of fitness function, is reached. The final solution with the 'fittest' fitness function gives the output k value of the asphalt mixture.

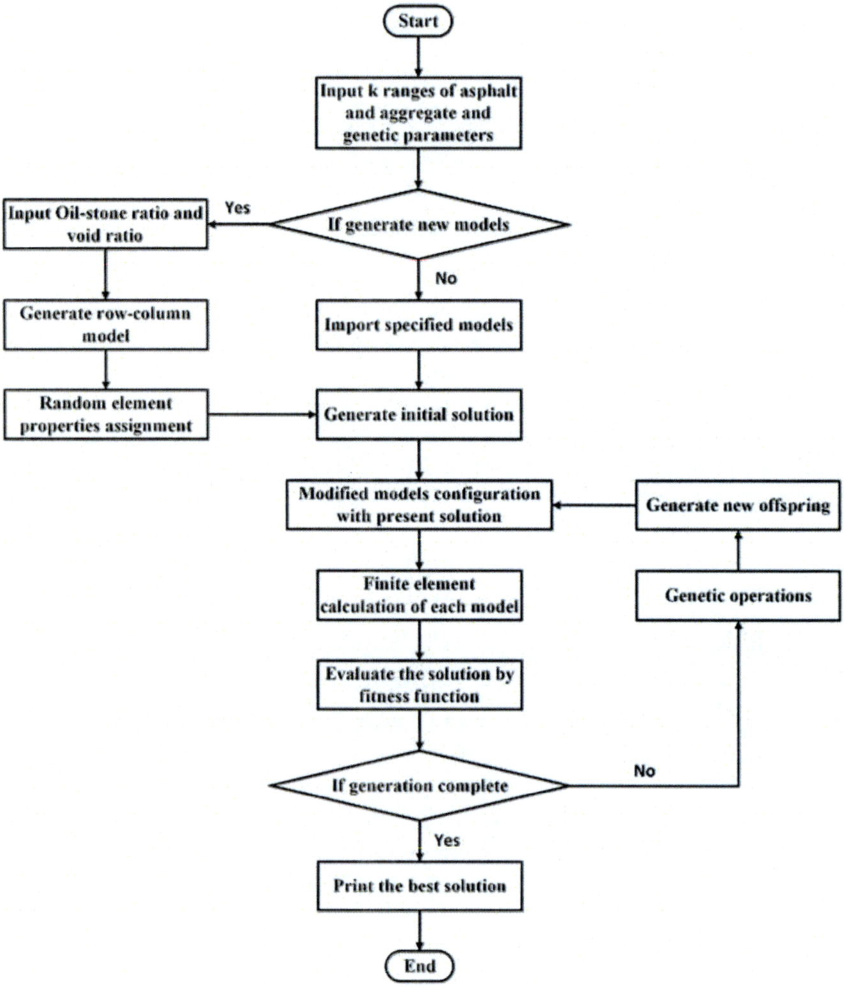

Fig. 1 The framework of hybrid GA-FE methodology

4 Determination of Asphalt Mixture Thermal Conductivity k

4.1 Random Assignment of Element Properties

For illustration, regular row-column models (as depicted in Fig. 2a) with 100 × 100 mm in dimensions with uniform regular mesh elements each measuring 1 × 1 mm are adopted in this paper to illustrate the calculation process of the proposed approach. Two steps are involved. The first step is to determine the number of cells

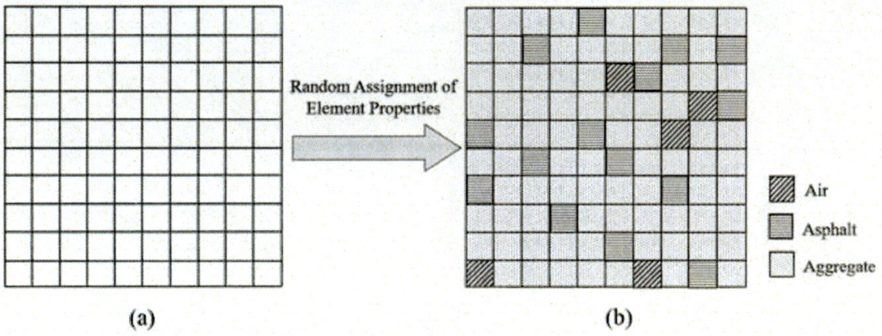

Fig. 2 Random assignment of element properties

that will be occupied by volume proportion of each constituent of the asphalt mixture considered. For example, in a 10,000 cells model with volume shares of 15% asphalt binder, 80% aggregates, and 5% air voids.

The cell numbers of the three constituents are 1500, 8000 and 500, respectively. The second step is to randomly assign the three constituents to the cells according to their volume shares, and each cell to its corresponding constituent property, as shown in Fig. 2b. After these two steps, the model is ready for thermal analysis.

4.2 *Finite Element Thermal Analysis*

The commercial software ABAQUS [21] was used to analyze the finite element model. For the purpose of determining k, the problem of 2-D steady-state heat conduction is considered. As shown in Fig. 3, plate A and plate B, which firmly located on the upper and lower sides of the model, are held at constant temperatures of TA and TB respectively, while the other two sides are completely insulated from heat transfer. Due the temperature differential between the two plates, heat conduction will take place under the defined conditions, which can be appropriately solved by Fourier's law [23]. In a given cell i with its thermal conductivity k_i, heat flux can take place in x- and y-directions, respectively. The heat flux of each cell i at a given direction can be calculated as follows:

$$q_{yi} = -k_j \frac{\partial T}{\partial y} \quad q_{xi} = -k_j \frac{\partial T}{\partial x} \tag{1}$$

where q_{xi} and q_{yi} represent the heat flux (W/m^2) in x direction and y direction respectively, k_j is the thermal conductivity (W/m·K) of the cell, T is the temperature in degree K. In this example, the one-dimensional heat transfer is in the y direction, the corresponding equations are as follows:

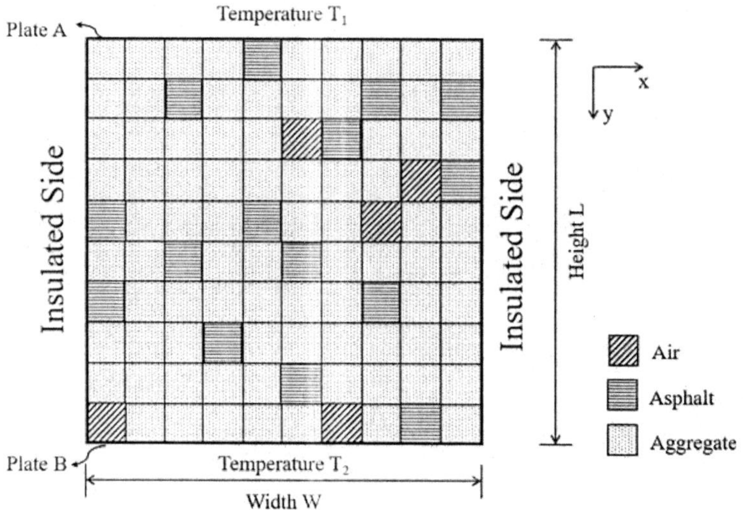

Fig. 3 Schematic diagram of 2-D heat conduction model

$$Q = \frac{\sum_{j=1}^{m} q_{yj} w_j}{W} \tag{2}$$

$$k = \frac{QL}{T_1 - T_2} = \frac{L \cdot \sum_{j=1}^{m} q_{yj} \cdot w_j}{W(T_1 - T_2)} \tag{3}$$

where Q (W/m^2) is the total heat flux accepted by plate B, the heat flux of each cell j in the y direction is represented as q_{yj} with its width w_j (m), W (m) is the width of model, L represents the distance between plate A and plate B in degree m, and T1 and T2 represent the temperatures of plate A and plate B respectively.

5 Genetic Algorithm Formulation in Model Calibration

5.1 Genetic Representation of the Variables

The GA program presented in this example is for solving a problem in the most general case that the k values of asphalt and aggregate are unknown but the k values of asphalt mixtures with different asphalt-aggregate volume shares are known. The most distinctive feature of GA is that the variables of interest are represented by a string structure of numeric code known as genes, which are usually in either binary form or integer form. Different genes are concatenated to from a set of strings known as chromosome, which can be simply regarded as a set of solutions to the problem

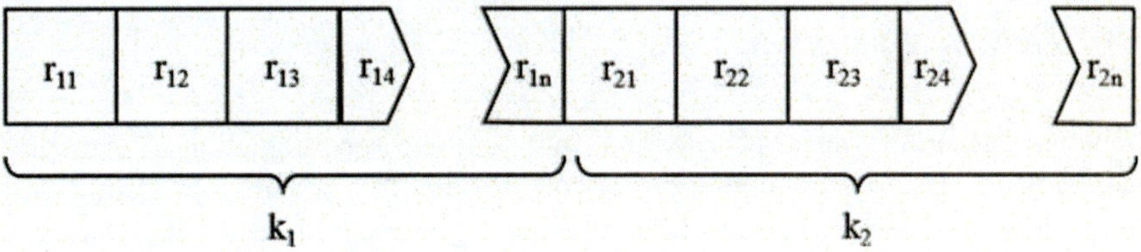

Fig. 4 Chromosome structure with k_1 of asphalt and k_2 of aggregate (r_{ij} represents the jth numeric code of the ith value k)

studied. This problem has two main variables to be determined: k1 of asphalt and k2 of aggregate as shown in Fig. 4. The values of k would vary with the genetic operations conducted by their chromosome like mutation and recombination during the generation of population. Some constraints must be considered to obtain the expected results. Specifically, the allowable ranges of k values of asphalt and aggregate respectively must be specified. This study takes the k of air as 0.026 W/m·K at 25 °C, the k of PG58-34 asphalt as 0.25 W/m·K, and the k_a of crushed gravel aggregates within the range from 1.73 to 3.98 W/m·K [19].

5.2 *Generation of Initial Population*

The GA optimization process requires to be driven by an initial pool of population, which is generated by random selection of k_a values from the preset interval of variables. In this example, the program will keep randomly picking k_a values from the allowable range until the number of initial population is reached.

5.3 *Fitness Function*

The fitness function is to evaluate the difference between the measured and predicted values. Considering that individuals with higher fitness will have a greater chance of survival, the fitness function is defined as follows:

$$f = \sum\nolimits_{n=1}^{N} \frac{1}{|PredictedKn - MeasuredKn|} \tag{4}$$

where f is the total fitness of overall models to current solution of k values.

5.4 Genetic Operations

Genetic operations consist of selection, mutation and recombination as shown in Figs. 5 and 6. Each of the operations is conducted with a specified probability for the purpose of selecting the best individual during every population of generation, leading to the convergence of variables to the problem. Genetic operations result in changes in the string sequence of genes, which in turn influence the values represented by them. Evaluation is performed to measure the adaption of each individual that will be ranked by its fitness value for selection, after which offspring is generated.

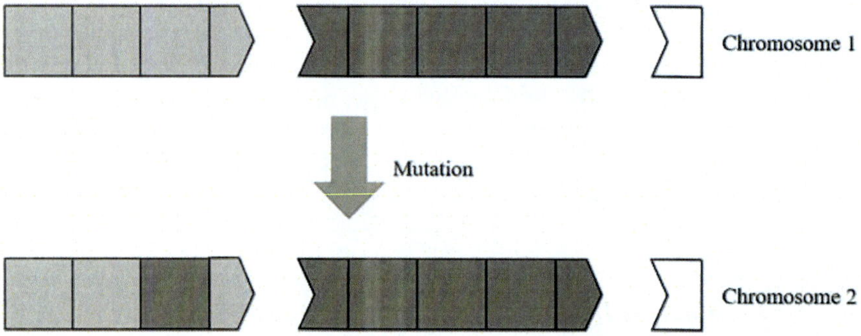

Fig. 5 Mutation of gene

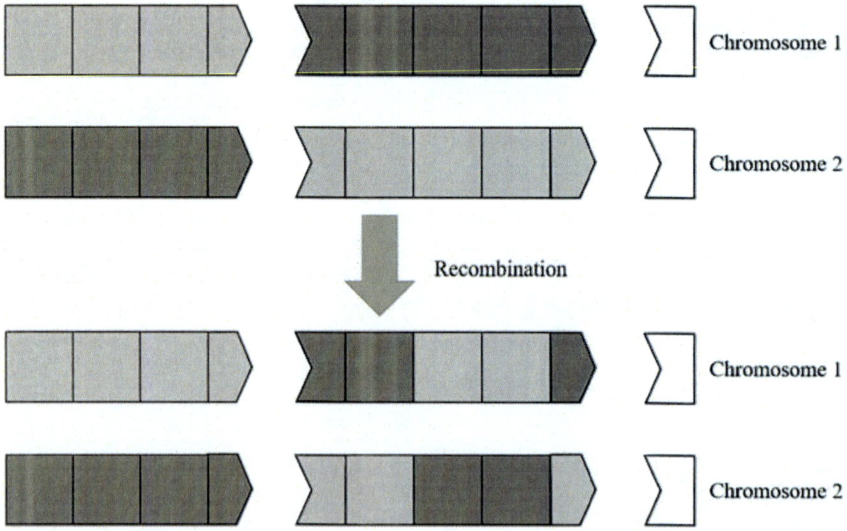

Fig. 6 Recombination of chromosome

5.5 *Parameters of GA*

There are totally four process parameters of GA in this study: recombination probability, mutation probability, number of generations and size of population.

The probabilities of recombination and mutation ensure the searching efficiency and the finding of global instead of local optima. While mutation is usually a small probability event in the process, it is of essence in the case of a small population, in which recombination alone cannot produce high quality offspring. The size of population represents the number of attempting solutions in each cycle of the GA process. Besides, a stopping criterion is required for terminating the GA program. It is usually either the cycle when the objective function is achieved, or the maximum number of iteration generations defined by the user.

6 Results and Analysis

For illustration of the influence of GA parameters on the optimization calculation, this example considered a recombination probability of 0.8 and a mutation probability of 0.5 with different generation numbers and population sizes. As shown in Table 1, the final adopted number of generations is 5 and the size of population is 20 to achieve a balance between computing time and accuracy of calculation. Group 1 shows the adjustment of parameters for better calculation performance, while group 2 contains the repeated prediction under the same conditions to validate the robustness of the GA formulation.

The detailed comparison of the results with those of Chu et al. [19] are summarized in Table 2. The measured values were from the experimental work from Mrawira and Luca [24]. For the problem in this study, all the simulation runs were carried out on a 3GHZ CPU workstation with 64 GB memory. It can be seen from Fig. 7 that these two lines of results practically coincide with each other, both indicated

Table 1 Simulation results

Group	GA parameters				Results	
	CBPB	MTPB	NGEN	POPS	Time (min)	Average error rate (%)
1	0.8	0.5	10	30	2038	4.28
	0.8	0.5	10	20	1444	4.25
	0.8	0.5	5	20	703	4.31
2	0.8	0.5	5	20	711	5.27
	0.8	0.5	5	20	705	4.22
	0.8	0.5	5	20	698	4.46

Note CBPB—Recombination probability, MTPB—Mutation probability; NGEN—Number of generations; POPS—Size of population

Table 2 Results comparison with Chu et al. [19]

Property	Asphalt mixture										
	CG1a	CG1b	CG1c	CG1d	CG2a	CG2b	CG2c	CG3a	CG3b	CG3c	CG3d
% V_{air}	7.8	5.5	3.8	3.4	7.2	5.0	3.1	8.2	5.5	4.2	3.6
% $V_{asphalt}$	13.9	14.2	14.5	14.5	14.0	14.3	14.6	13.8	14.2	14.4	14.5
% V_{agg}	78.3	80.3	81.7	82.1	78.8	80.7	82.3	78.0	80.3	81.4	81.9
Measured k (W/m·°C)	1.70	1.79	2.01	2.07	1.69	1.76	2.11	1.84	1.87	1.96	1.98
Results in Chu (W/m·°C)	1.80	1.90	1.96	1.98	1.84	1.91	2.00	1.79	1.90	1.96	1.98
Results in this paper (W/m·°C)	1.81	1.90	1.97	1.99	1.84	1.93	1.99	1.81	1.90	1.95	1.98

Note V_{air}, $V_{asphalt}$, V_{agg} refer to volume-based proportion of air, asphalt and aggregate respectively. Data from asphalt mixture with crushed gravel aggregate by Mrawira and Luca et al. [24]

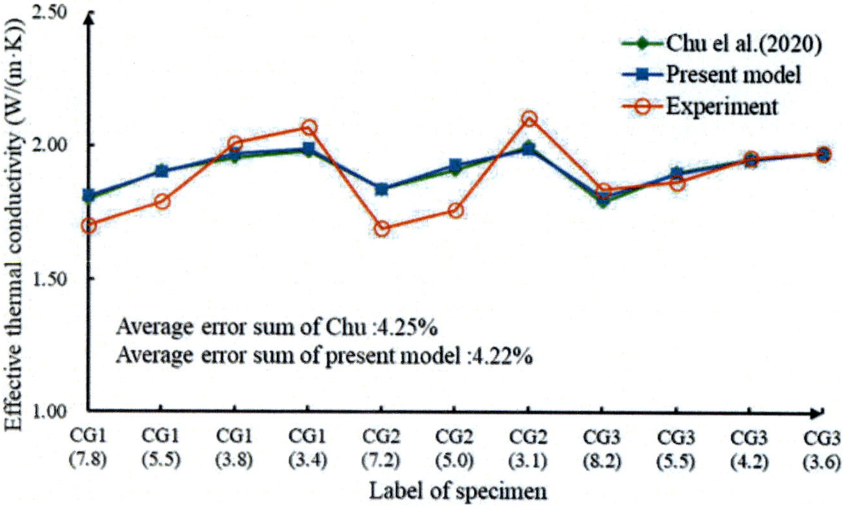

Fig. 7 Thermal conductivity values of CG mixtures

acceptable accuracy in predictions. The average error rate of this proposed method is slightly lower than that of Chu et al. [19] with 4.22% compared to 4.25%. It is noted that the method in this study only took about 11 h and 45 min on average, which is substantially more efficient in comparison with Chu et al.'s 48 h or longer.

7 Conclusion

This article has presented a proposed framework of hybrid GA-FE methodology for the numerical determination of thermal conductivity of asphalt mixtures, with the aim to improve the efficiency of calculation and reduce the manual participant in the procedure as far as possible. A totally automatic program has been developed to plant the GA optimization procedure into ABAQUS software calculation process to realize the iterative loop of GA and FE for the optimization search of the 'fittest' solution for k of the asphalt mixture studied. Examples has been presented to demonstrate to the much efficient performance of this method compared to the original solution process of Chu et al. [19]. The results suggest that the proposed method is much more user friendly, and could achieve higher efficiency with no loss in accuracy prediction, both of which are of high significance for practical applications.

Acknowledgements The authors gratefully acknowledge financial support by the Fundamental Research Funds for the Central Universities, China. Project No. 300102219307 and Shannxi Science and Technology Project 2020JQ-390.

References

1. Gui J, Phelan PE, Kaloush KE, Golden JS (2007) Impact of pavement thermophysical properties on surface temperatures. J Mater Civ Eng 19(8):683–690
2. Hassn A, Aboufoul M, Wu Y, Dawson A, Garcia A (2016) Effect of air voids content on thermal properties of asphalt mixtures. Constr Build Mater 115:327–335
3. Hall MR, Dehdezi PK, Dawson AR, Grenfell J, Isola R (2011) Influence of the thermophysical properties of pavement materials on the evolution of temperature depth profiles in different climatic regions. J Mater Civ Eng 24(1):32–47
4. Minhoto M, Pais J, Pereira P, Picado-Santos L (2005) Predicting asphalt pavement temperature with a three-dimensional finite element method. Transp Res Rec J Transp Res Board 1919:96–110
5. Chadbourn BA, Luoma JA, Newcomb DE, Voller VR (1996) Consideration of hot mix asphalt thermal properties during compaction. In: Quality management of hot mix asphalt. ASTM International
6. Herb W, Velasquez R, Stefan H, Marasteanu MO, Clyne T (2009) Simulation and characterization of asphalt pavement temperatures. Road Mater Pavement Des 10(1):233–247
7. Masad E, Koneru S, Scarpas T, Kassem E, Rajagopal KR (2010) Modeling of hot-mix asphalt compaction: a thermodynamics-based compressible viscoelastic model (No. FHWA-HRT-10-065). Turner-Fairbank Highway Research Center
8. Tavman IH (1996) Effective thermal conductivity of granular porous materials. Int Commun Heat Mass Transf 23(2):169–176
9. Jendia SM, Jarada A (2006) Traffic opening time and time available for compaction for fresh asphalt layer using slab specimens model. IUG J Nat Stud 14(1)
10. Wang J, Carson JK, North MF, Cleland DJ (2006) A new approach to modelling the effective thermal conductivity of heterogeneous materials. Int J Heat Mass Transf 49(17–18):3075–3083
11. Zhu B, Chu L, Fwa TF (2019) Cooling characteristics of warm mix asphalt for multi-lift pavement repair and resurfacing. Int J Pavement Eng 2:1–12

12. Progelhof RC, Throne JL, Ruetsch RR (1976) Methods for predicting the thermal conductivity of composite systems: a review. Polym Eng Sci 16(9):615–625
13. Maxwell J (1873) A treatise on electricity and magnetism. Nature 7(182):478–480
14. Wang J, Carson JK, North MF, Cleland DJ (2006) A new approach to modelling the effective thermal conductivity of heterogeneous materials. Int J Heat Mass Transf 49(17/18):3075–3083
15. Islam R, Tarefder RA (2014) Determining thermal properties of asphalt concrete using field data and laboratory testing. Constr Build Mater 67(pt.b):297–306
16. Chen J, Wang H, Li M, Li L (2016) Evaluation of pavement responses and performance with thermal modified asphalt mixture. Mater Des 111:88–97
17. Chen J, Zhang M, Wang H, Li L (2015) Evaluation of thermal conductivity of asphalt concrete with heterogeneous microstructure. Appl Therm Eng 84:368–374
18. Mirzanamadi R, Johansson P, Grammatikos SA (2018) Thermal properties of asphalt concrete: a numerical and experimental study. Constr Build Mater 158:774–785
19. Chu L, He L, Fwa TF (2020) Determination of thermal conductivity of asphalt paving mixtures using finite element method. Constr Build Mater 243:118250
20. Chu L, He L, Fwa TF (2020) 3-dimensional finite element modelling of asphalt mixtures for thermal conductivity determination. Int J Pavement Eng. https://doi.org/10.1080/10298436.2020.1834109
21. ABAQUS (2016) ABAQUS Standard Theory Manual, Pawatucket, RI, USA
22. Mathworks Inc. (2018) MATLAB User Manual Version R2016. Natick, MA
23. Ozisik MN (1979) Basic heat transfer. McGraw-Hill Book Company, New York
24. Mrawira DM, Luca J (2006) Effect of aggregate type, gradation, and compaction level on thermal properties of hot-mix asphalts. Can J Civ Eng 33(11):1410–1417

Influence of Surface Treated Montmorillonite Nano Clay on Oxidative Aging Properties of Bitumen

Sidharth Kumar Patra, Mahabir Panda, and Aditya Kumar Das

Abstract The present study was undertaken to address the influence of surface-treated Montmorillonite Nano Clay (MNC) on aging and temperature susceptibility of base bitumen. This study used a control VG-30 grade bitumen as base binder with different percentages of MNC by weight of bitumen (3–6% at an increment of 0.5%). Surface modification of MNC was done with Octadecyl amine and Aminopropyl-triethoxysilane. The binder's aging susceptibility was characterized using different aging indices determined based on conventional and performance parameters. The outcome of the aging indices indicates a significant improvement in temperature and aging resistance potential of base bitumen with the incorporation of NC. It can also be stated that an increase in aging resistance can be effective in delaying the intermediate temperature cracking of bitumen. The overall study inferred that the addition of MNC can be considered to sustain bitumen's oxidative aging.

Keywords Bitumen · Montmorillonite Nano clay · Temperature susceptibility · Aging resistivity

1 Introduction

Oxidation of bitumen is considered one of the common reasons for early attainment of cracking in the bituminous pavement at intermediate temperature conditions. In general, a conventional bitumen may not sustain the cracking by reducing the aging impact subjected to a wide range of traffic loading and extreme climatic circumstances. Adverse field conditions profoundly influence the aging of bitumen, which makes the bitumen a stiffer binder [4, 9]. Thus, the stiffer binder may sustain the high-temperature rutting/permanent deformation; whereas, a binder with high stiffness may susceptible to intermediate temperature cracking.

S. K. Patra · M. Panda
National Institute of Technology Rourkela, Odisha 769008, India

A. K. Das (✉)
Siksha 'O' Anusandhan University, Bhubaneswar, Odisha 751030, India

© The Author(s), under exclusive license to Springer Nature Switzerland AG 2022

H. R. Pasindu et al. (eds.), *Road and Airfield Pavement Technology*, Lecture Notes in Civil Engineering 193, https://doi.org/10.1007/978-3-030-87379-0_52

Many additives/modifiers (e.g., polymers, crumb rubber, hydrated lime, polypropylene acids, rejuvenators, etc.) have been utilized to overcome such a problem through modification of bitumen [3, 8, 13]. However, phase separation between modifiers and bitumen has been reported as a serious concern. On the other hand, recent developments in research works helped to identify nanomaterial's' utilization, and are considered as answer for preparing a good-performing bitumen binder. Nanomaterials are known to be very popular in the civil engineering applications as they have the high surface area and high surface energy that help in developing a higher number of interaction points in the bitumen matrix and hence, enhance the long-term performance of bitumen and bituminous mixes. Among different nanomaterials, Nano clay, Nano hydrated lime, carbon nanotubes, and Nano silica have gained significant interest as they have the beneficial impact in enhancing the performance of bitumen [1, 7, 10, 12].

Moreover, previous studies have reported nanomaterials' potential impact in addressing the rheological issues as they possess high functional density, high strain resistance, and higher catalytic properties [6, 11, 14]. Similarly, Montmorillonite Nano clay (MNC) has received considerable attention in pavement engineering as it is readily available and environment friendly. However, the effects of MNC are limited to preliminary studies which merely explain the influence on aging and temperature susceptibility of bitumen. These issues have motivated the authors to carry out the present study. This study presents a laboratory investigation related to surface-treated Montmorillonite Nano clay on aging susceptibility of bitumen using conventional and rheological studies. The aging behavior of MNC modified bitumen was evaluated using different aging indices based on various physical and rheological properties. Overall, extensive laboratory studies were taken up in order to develop a sound ground on the effectiveness of MNC on the aging susceptibility of bitumen at high and intermediate temperature conditions.

2 Materials

A control viscosity grade bitumen (VG-30) and surface treated Montmorillonite Nano clay (NC) were selected. These two materials were collected from local sources. The basic properties of VG-30 and NC are presented in Tables 1 and 2 respectively. The surface modification of MNC was done with 15–35% (by wt.) octadecyl amine

Table 1 Basic properties of VG-30 bitumen

Properties	Penetration (dmm) at 25 °C	Softening point, °C	Ductility, cm at 25 °C (RTFO sample)	Absolute viscosity, P	Kinematic viscosity, cSt
Result	60	47.5	≥100	2689.3	367.2
Criteria (IS 73, 2013)	≥45	≥47	≥75	2400–3600	≥ 250

Table 2 Basic properties of surface modified Montmorillonite Nano clay

Properties	Appearance (colour)	Size	Density
Specification	White to off white powder	≤20 μ	200–500 kg/m^3

and 0.5–5% (by wt.) aminopropyltriethoxysilane. The surface-treated MNC shows a highly intercalated and exfoliated structure, and can help in uniform dispersion in the bitumen matrix. Additionally, the exfoliated layers may act as an oxygen barrier that can increase the aging resistance of the bitumen [2].

The required quantity of preheated MNC was added to preheated VG-30 bitumen at 155 ± 5 °C, and stirred using a shear mixer at a speed of 4000 rpm for about 1 h. The mixing temperature and time were framed based on the literature on nanomaterials [1, 5]. Dosages of MNC that were added to bitumen were varied from 3 to 6% at an increment rate of 0.5%. The initial trials on dosages of MNC indicated that the addition of MNC up to 2.5% by weight of bitumen did not result in a significant change in the performance parameters. A total of eight different bitumen binders were prepared and referred as: VG-30 (control bitumen), NC 3 (VG-30 + 3% NC), NC 3.5 (VG-30 + 3.5% NC), NC 4 (VG-30 + 4% NC), NC 4.5 (VG-30 + 4.5% NC), NC 5 (VG-30 + 5% NC), NC 5.5 (VG-30 + 5.5% NC), and NC 6 (VG-30 + 6% NC). The conditioning exercise of bitumen specimens was done to simulate the service conditions of bitumen in the field and to understand the impact of MNC on bitumen aging resistance. The short-term aging (STA) was done using a Rolling Thin Film Oven (RTFO) as per ASTM D 2872, and long-term aging (LTA) of bitumen was done using a Pressure Aging Vessel (PAV) as per ASTM D 6521.

3 Experimental Program

3.1 Aging Behavior of Bitumen Binder Using Aging Indices

The aging susceptibility properties of modified and unmodified bitumen were evaluated using aging index (AI) values. It has been reported that bitumen exhibiting relatively less value of AI is known to be less susceptible to aging and vice versa [3]. The determination of aging indices based on different parameters of bitumen samples are discussed hereafter.

Increase in Softening Point This study considered the increase in softening point (tested as per ASTM D 36) of unaged and short-term aged bitumen specimen to understand the influence of MNC on aging susceptibility of bitumen and was given in Eq. (1).

$$\Delta T = T_1 - T_0 \tag{1}$$

where T_0 and T_1 are softening points of unaged and short-term aged bitumen specimens, respectively.

Ductility Retention Rate This study determined the ductile aging index as a ratio of ductility values of bitumen before and after short-term aging condition, and is expressed as in Eq. (2).

$$DRR = D_1/D_0 \times 100 \quad (2)$$

where D_0 and D_1 are ductility values of unaged and short-term aged bitumen specimens respectively.

Activation Energy Aging Index Activation energy parameter is an indication of temperature susceptibility of bitumen based on viscosity measured at elevated temperature conditions. This study determined the change in activation energy using Eq. 3 to evaluate the aging index values.

$$EAI = (Ea_1 - Ea_0)/Ea_0 \times 100 \quad (3)$$

where Ea_0 and Ea_1 are activation energies of unaged and short-term aged bitumen specimens respectively.

Rutting Parameter Aging Index The rutting parameter (G*/Sinδ) infers the rutting resistance property with respect to bitumen's stiffness at high-temperature conditions. This study evaluated aging index as a ratio of the G*/Sinδ of unaged and short-term aged bitumen samples, tested as per ASTM D7175.

$$AI = \frac{\left|(G^*/\mathrm{Sin}\delta)_{aged}\right|}{\left|(G^*/\mathrm{Sin}\delta)_{unaged}\right|} \quad (4)$$

where $(G^*/\mathrm{Sin}\delta)_{unaged}$ and $(G^*/\mathrm{Sin}\delta)_{aged}$ are rutting parameters of bitumen specimens before and after short-term aging respectively.

Non-recoverable Creep Compliance Aging Index Non-recoverable creep compliance (J_{nr}) implies the permanent deformation resistance of bitumen at high-temperature conditions. This study evaluated aging index as a ratio of J_{nr} of unaged and short-term aged bitumen samples, tested as per ASTM D7405.

$$AI = \frac{\left|(J_{nr})_{aged}\right|}{\left|(J_{nr})_{unaged}\right|} \quad (5)$$

where $(J_{nr})_{unaged}$ and $(J_{nr})_{aged}$ are rutting parameters of bitumen specimens before and after short-term aging respectively.

Fatigue Parameter Index The fatigue parameter (G*Sinδ) implies the fatigue damage resistance of bitumen at intermediate temperature conditions. This study evaluated aging index as a ratio of the G*Sinδ of unaged and long-term aged bitumen samples, tested as per ASTM D7175.

$$AI = \frac{|(G^* \, Sin\delta)_{aged}|}{|(G^* \, Sin\,\delta)_{unaged}|} \tag{6}$$

where $(G^*Sin\delta)_{unaged}$ and $(G^*Sin\delta)_{aged}$ are rutting parameters of bitumen specimens before and after long-term aging respectively.

Carboxyl Index Carbonyl index is determined for the base binder to understand the impact of short-term and Long-term aging on base bitumen. An increase in carbonyl index encourages a decrease in ductile behavior and an increase in brittleness in the bitumen property. This study evaluates the carbonyl index using absorbance vs wavenumber plot and using Eq. (7).

$$I_{C=0} = \frac{\text{area of the carbonyl band centred around wavenuber } 1690-1740\ \text{cm}^{-1}}{\text{area of he spectral bands between } 2000 \text{ and } 600\ \text{cm}^{-1}} \tag{7}$$

where $I_{c=0}$ is the carbonyl index of the bitumen specimens.

4 Results and Discussions

4.1 *Increase in SP (ISP)*

Bitumen exhibiting a comparatively lower ISP value indicates a lower degree of aging susceptibility and vice versa. Figure 1a shows the change in ISP values over increase in MNC dosages. The addition of MNC to base bitumen resulted in a decrease in ISP value. For example, the ISP value for the base bitumen (VG-30) is 4.57 °C, which

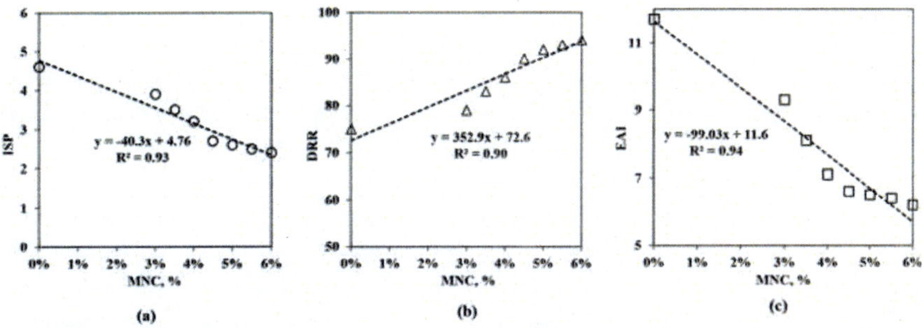

Fig. 1 **a** Variations of ISP over MNC dosages, **b** variations of DRR over MNC dosages, **c** variations of EAI over different MNC dosages

decreased to 2.4 °C for bitumen with 6% MNC. Such a resulting trend in ISP implies that MNC can increase the aging resistance of base bitumen.

4.2 Ductility Retention Rate (DRR)

Bitumen having a relatively higher DRR value indicates a lower degree of aging susceptibility and vice versa. Figure 1b shows the change in DRR values over increase in MNC dosages. It was noticed that the addition of MNC to base bitumen resulted in an increase in DRR value. For instance, the DRR value for the base bitumen (VG-30) is 75%, which increased to 94% for bitumen with 6% MNC. Such a resulting trend in DRR values implies that MNC can increase the aging resistance of base bitumen. It can also be stated that an increase in aging resistance may help the bitumen to improve the ductile behavior and thus, the early attainment of cracking can be delayed.

4.3 Activation Energy Aging Index (EAI)

The activation energy has a good correlation with the temperature susceptibility of bitumen. Bitumen with a relatively higher value of activation energy possesses a lower degree of temperature dependency. It is observed from this study that the addition of MNC decreased the temperature susceptibility of base bitumen. Besides, lower the activation energy index (EAI), the higher the bitumen's aging resistance is. Figure 1c illustrates that increased addition of MNC decreased the EAI value. Consequently, the temperature susceptibility and aging susceptibility of base bitumen reduced after the inclusion of MNC.

4.4 Rutting Parameter Aging Index

The aging index (AI) of base and MNC modified bitumen binders was calculated using Eq. (4) at 46, 52, 58, and 64 °C. Results showed that the AI decreases with an increase in MNC (Fig. 2), which indicates a lower degree of aging susceptibility of base bitumen at high-temperature conditions. At 58 °C, the addition of MNC up to 6%, resulted in a decrease in AI value from 2.02 to 1.61. It is inferred from such variations in AI value that MNC has the anti-aging ability to improve base bitumen's aging resistance. It has been reported that silicate platelet structure of NC helps in slowing down the effect of oxidation [2]. Similarly, the present study indicated that addition of MNC led to decrease in rate of increase in $G^*/\mathrm{Sin}\delta$ value of aged bitumen, and thus there is a significant improvement in aging resistance with decreasing value of AI.

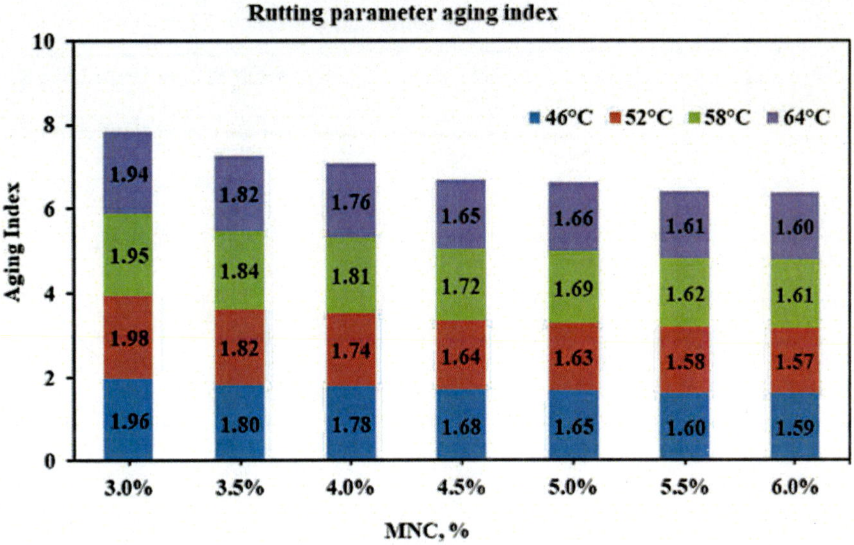

Fig. 2 Variations of rutting parameter aging index over different MNC dosages

4.5 *Non-recoverable Creep Compliance Aging Index*

The AI of base and MNC modified bitumen binders was determined at 64 °C and presented in Fig. 3 using Eq. (5). The resulting trend indicated that the AI value decreases with an increase in MNC dosages, implying a lower degree of aging susceptibility of base bitumen. Such a reduced trend in the aging resistance of base bitumen may be due to the intermolecular structure of exfoliated layers and mineralogical interaction of MNC with the bitumen matrix. Besides, such a strong interaction phase may be a favorable condition in decreasing the aging effect, which is evident from reduction of AI values over MNC dosages. MNC has silicate platelets structure that might reduce the effect of oxidation and thus, the stiffness of aged bitumen did not have significant improvement. This resulted in a notable improvement in aging resistance with decreasing value of AI [2].

4.6 *Fatigue Parameter Aging Index*

The aging index (AI) based on fatigue parameters was calculated using Eq. (6) at 20 and 25 °C. AI values were found to decrease with an increase in MNC dosages (Fig. 4), which indicates a lower degree of aging susceptibility of base bitumen at intermediate temperature conditions. For instance, At 20 °C, the addition of MNC up to 6% resulted in a decrease in AI value from 2.20 to 1.67. It can be stated from such variations in AI value that MNC has the anti-aging ability to improve

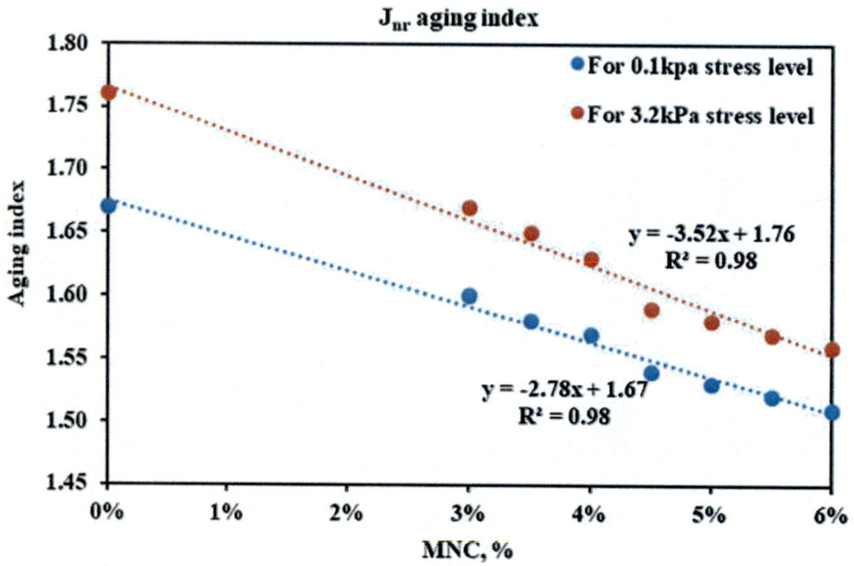

Fig. 3 Variations of J_{nr} aging index over different MNC dosages for 0.1 and 3.2 kPa stress levels

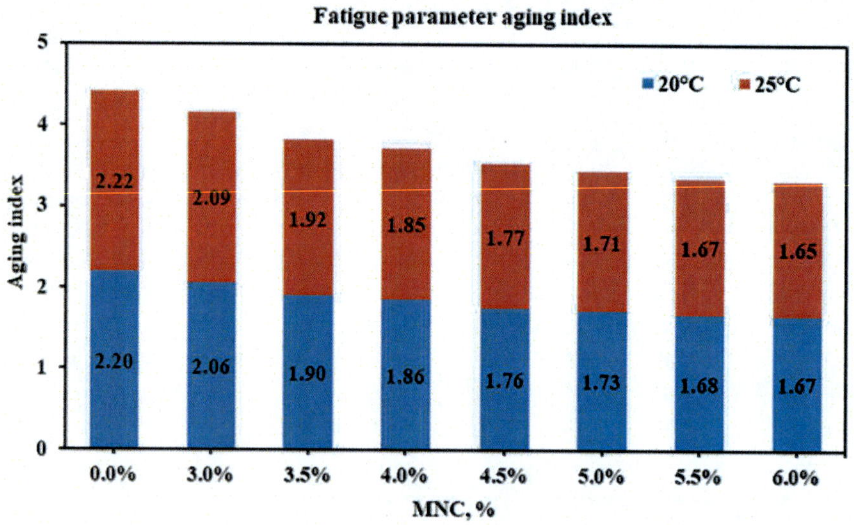

Fig. 4 Variations of fatigue parameter aging index over different MNC dosages

aging resistance of base bitumen. Moreover, such encouraging trends signify that increase in aging resistance may help the bitumen to enhance the elastic response at intermediate temperature conditions and thus, the early attainment of cracking can be delayed. Further, the present study utilizes surface treated Montmorillonite

Nano clay that possesses highly exfoliated and intercalated structure. Such structure might help in better dispersion of MNC in the bituminous binder matrix, and as the exfoliated layers can act as an oxygen barrier it may considerably improve the aging resistance too [2].

4.7 Carbonyl Index

The carbonyl index of base and MNC modified bitumen binders was calculated using Eq. (7), considering both short-term and long-term aging conditions. The Carbonyl index parameter is considered one of the most influential parameters to evaluate the oxidative aging impact on bitumen. Lower the carbonyl index is, higher is the aging resistance of bitumen. Figure 5 illustrates the carbonyl index values of bitumen binder over different dosages of MNC. The increase in MNC dosages reduced the carbonyl index of base bitumen, indicating an increase in aging resistance irrespective of short-term and long-term conditions (Fig. 5).

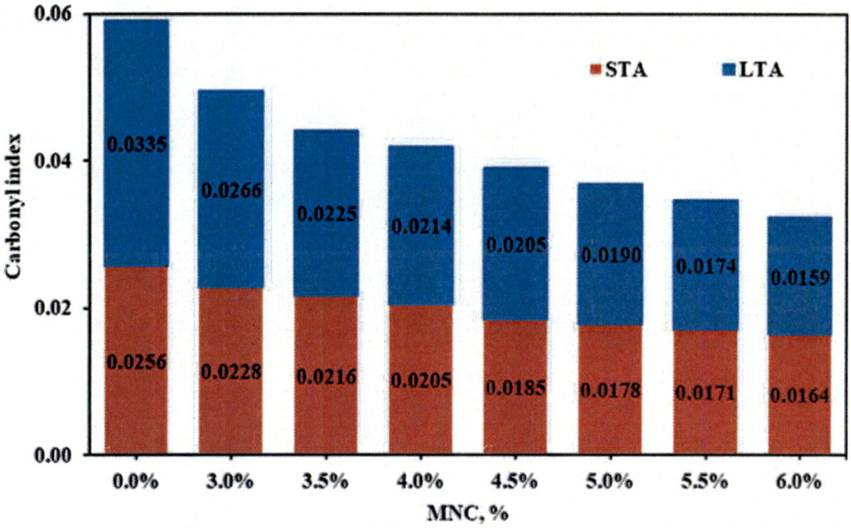

Fig. 5 Variations of carbonyl index over different MNC dosages for short-term and long-term aged bitumen binders

5 Conclusions

In the present study, an attempt has been made to evaluate the influence of surface-treated Montmorillonite Nano clay on aging susceptibility of bitumen binder associated with high and intermediate temperature conditions. Results of aging indices evaluated based on conventional parameters indicated that the NC modified binder can effectively be less susceptible to oxidative aging. Aging index based on G*/Sinδ and J_{nr} showed improved resistance to oxidative aging for NC modified bitumen. Consequently, it can be stated that the NC modified bitumen binder can withstand aging at high-temperature conditions. The resulting values of aging based on fatigue parameter and carbonyl index of NC modified bitumen binder inferred that the addition of NC can have an effective contribution in increasing the long-term aging resistance.

References

1. Ashish PK, Singh D (2018) Study on understanding functional characteristics of multi-wall CNT modified asphalt binder. Int J Pavement Eng 1–14
2. Bujdák J, Slosiariková H (1992) The reaction of montmorillonite with octadecylamine in solid and melted state. Appl Clay Sci 7(4):263–269
3. Das AK, Panda M (2017) Investigation on rheological performance of sulphur modified bitumen (SMB) binders. Constr Build Mater 149:724–732
4. Das PK, Balieu R, Kringos N, Birgisson B (2015) On the oxidative ageing mechanism and its effect on asphalt mixtures morphology. Mater Struct 48(10):3113–3127
5. Kavussi A, Barghabany P (2016) Investigating fatigue behavior of nanoclay and nano hydrated lime modified bitumen using LAS test. J Mater Civ Eng 28(3):1–7
6. Liu G, Wu S, Van de Ven M, Molenaar A, Besamusca J (2010) Characterization of organic surfactant on montmorillonite nanoclay to be used in bitumen. J Mater Civil Eng 22(8):794–799
7. Moeini AR, Badiei A, Rashidi AM (2019) Effect of nanosilica morphology on modification of asphalt binder. Road Mater Pavement Des 1–17
8. Parvez MA, Wahhab HIAA, Shawabkeh RA, Hussein IA (2014) Asphalt modification using acid treated waste oil fly ash. Constr Build Mater 70:201–209
9. Read J, Whiteoak D, Hunter RN (2003) The shell bitumen handbook. Thomas Telford
10. Saltan M, Terzi S, Karahancer S (2017) Examination of hot mix asphalt and binder performance modified with nano silica. Constr Build Mater 156:976–984
11. Yang Q, Liu Q, Zhong J, Hong B, Wang D, Oeser M (2019) Rheological and micro-structural characterization of bitumen modified with carbon nanomaterials. Constr Build Mater 201:580–589
12. Zhambolova A, Vocaturo AL, Tileuberdi Y, Ongarbayev Y, Caputo P, Aiello I, Rossi C O, Godbert N (2020) Functionalization and modification of bitumen by silica nanoparticles. Appl Sci 10(17):6065, 1–11
13. Zhang D, Zhang H, Zhu C (2017) Effect of different rejuvenators on the properties of aged SBS modified asphalt. Pet Sci Technol 35(1):72–78
14. Zhu C, Zhang H, Shi C, Li S (2017) Effect of nano-zinc oxide and organic expanded vermiculite on rheological properties of different bitumens before and after aging. Constr Build Mater 146:30–37

Physical and Rheological Properties of Cup Lump Rubber Modified Bitumen

Anwaar Hazoor Ansari, Fauzan Mohd Jakarni, Ratnasamy Muniandy, Salihudin Hassim, Zafreen Elahi, and Mohamed Meftah Ben Zair

Abstract Pavement with long service life and low maintenance and rehabilitation costs are favored; thus, the need for high-quality bitumen is growing. This situation has led to the use of polymers as modifiers to enhance the properties of pavement to be more resistive to high temperature and stress during service life. Recently natural rubber in coagulated form has been identified as a biopolymer due to its high resilience and elastomeric properties. Therefore, this research was conducted to investigate the physical and rheological performance of bitumen with cup lump natural rubber with four percentages (3, 6, 9 and 12%) by weight of bitumen. The raw rubber was processed and treated before blending with bitumen. The modified bitumen was tested for physical and rheological properties. The incorporation of cup lump rubber decreased the penetration and increased the softening point. Furthermore, it also enhances the rheological properties in term of complex modulus (G*) and phase angle (δ).

Keywords Bitumen · Biopolymers · Natural rubber · Cup lump · Complex modulus

1 Introduction

The modification of the bitumen for road construction is essential to prolong the service of roads. The base bitumen cannot accommodate the high stress due to high traffic volume and axle load [1]. Unfortunately, the distresses starting to appear on the pavement before reaching the maximum designed service life of the pavement. Different polymer modifications have been reported overtime for the road industry

A. H. Ansari (✉) · F. M. Jakarni (✉) · R. Muniandy · S. Hassim · Z. Elahi · M. M. Ben Zair
Department of Civil Engineering, Faculty of Engineering, Universiti Putra Malaysia, 43400 UPM Serdang, Selangor, Malaysia
e-mail: gs52876@student.upm.edu.my

F. M. Jakarni
e-mail: fauzan.mj@upm.edu.my

© The Author(s), under exclusive license to Springer Nature Switzerland AG 2022

H. R. Pasindu et al. (eds.), *Road and Airfield Pavement Technology*, Lecture Notes in Civil Engineering 193, https://doi.org/10.1007/978-3-030-87379-0_54

to improve performance [2]. The polymer modified binders can enhance the properties of the asphalt mixture. The modified binders substantially improve the mixture properties in term of elasticity, adhesion, cohesion, and stiffness. Among the polymers, bio-renewable polymers are of keen interest of area [3]. One of the renewable polymers is natural rubber which led to the growing interest in bitumen modification particularly for the pavement industry [4, 5].

Worldwide more than 50,000 multipurpose products are produced by using natural rubber (NR) [6]. After the discovery of vulcanization, the NR became a technically necessary reserve and an essential raw material [7, 8]. NR is produced from a runny, milky white liquid called natural rubber latex (NRL) that drips from the bark of tropical and subtropical plants through tapping. Almost 99% of global NR comes from the tree species Hevea brasiliensis. Mainly, there are two major techniques of collecting NR; the first is the preservation of latex as a liquid by adding some quantity of ammonia and proceed to further usage as slabs. The second is collecting latex in hard form by letting it coagulate in the collection cups by itself or by adding formic acid to convert to faster to cup lump natural rubber (CLNR) [9, 10].

Most of the research on the incorporation of the NR into the bitumen modification was performed in the United Kingdom in the 1950–1960s [11]. NR production has increased from 6.9 to 13 million metric tons from 2001 to 2017. However, the price is decreasing since 2011 and is probable to go on to decrease (Board, 2018). The Variability in the production and price of NR has generated a substantial need for the alternate consumption of NR [12]. Therefore, the utilization of NR as a modifier for bitumen can benefit the farmers of the rubber tree and it can enhance the road performance in a cost-effective way. Conversely, Hevea brasiliensis is essentially pure poly-cis-1,4 isoprene (comprising 99.9% of cis 1,4 structural units) and its particles vary in size from 0.15 to 3 mm; it has the chemical formula C_5H_8 with a molecular weight of 105–106 Da [13]. Figure 1 shows the chemical structure of cis-1,4- polyisoprene in NRL extracted from Hevea brasiliensis.

NR is a noteworthy polymer that can be inexpensively incorporate into bitumen [12]. NR enhance the stability of flexible pavements due to their good tear strength, and fatigue resistance [14, 15]. Thus, some research-supported NR as a polymer

a b c

Fig. 1 Processing of coagulated NR: **a** fresh cup lump rubber, **b** water extraction, **c** drying of small pieces of rubber modifier due to its good low-temperature properties and insignificant loss of hysteresis on dynamic loading [15]. Consequently, NR improved the flexible pavement structures and increase the service life span with minimal annual maintenance [16]. It helps in disintegrating low temperature developed stresses although, at higher temperatures, it controls the flow of the asphalt mixture by functioning as a membrane and enhancing the shear strength. Fresh rubber in the form of a cup lump has been found to modify bitumen as a biopolymer. Raw rubber in the form of a cup lump has been identified to modify bitumen as a biopolymer. The chemical properties of CLNR are similar to those of NRL. In addition, CLNR is appropriate for mixing with bitumen because of its low water content [5].

Additionally, CLR is suitable for mixing with bitumen as of its low water content. On the other hand, the latex form of natural rubber has 50–60% water. The processing method of the CLR is quite simple in comparison with the other forms of NR. Due to the initial phase of implementation and evaluation very less literature is available on CLR modified binder. However, the implementation of the natural rubber with different types on the field has already been tried to evaluate the possible techniques and methodologies [17]. The proper stability and homogeneity of the bitumen and rubber is the keen concern of the researchers before any field implication of a larger scale. Conversely, Clearer insight and practices for the cup lump rubber modified bitumen is required to gain more broad data for the future perspective. This study objective is to investigate the physical and rheological performance of the CLR modified bitumen. The bitumen is modified with the incorporation of CLR at 3, 6, 9 and12% by weight of the binder and compared with the base bitumen at intermediate and high-temperature properties.

2 Experimental Works

2.1 Materials

The 60/70 penetration grade bitumen is used in this study and was supplied by Kajang rocks groups, Selangor, Malaysia. The coagulated form of natural rubber (cup lump rubber) was collected from the agricultural farm, Universiti Putra Malaysia. The cup lump rubber was processed for water extraction and then cut into pieces of size <2.5 cm. CLR freshly collected has some moisture which was also reduced by air drying for 48 h and then oven drying at 60 °C for 6 h before using for modification as shown in Fig. 1. The basic properties of bitumen and rubber used are in Table 1 together with the Jabatan Kerja Raya Malaysia (JKR) specification.

Table 1 Basic physical properties of 60/70 grade bitumen

Material	Properties	Testing protocol	Specifications	Results
60/70 grade bitumen	Penetration (dmm)	ASTM D5	60–70	62.0
	Softening point test (°C)	ASTM D36	49–52	46.4
	Rotational viscosity @ 135 °C Pa s	ASTM D4402	3.0 max	0.377
	Specific gravity	ASTM D70	-	1.030

2.2 *Preparation of the Modified Bitumen*

The cup lump rubber (CLR) was added at five different percentages (0, 3, 6, 9, and 12%) by weight of the base bitumen. The cup lump rubber modified bitumen (CLRMB) were blended at a temperature of 160 °C and rpm of 1900 using IKA RW20 DZM stirrer for 90 min. The CLR was added incrementally to the temperature stabilized bitumen at 160 °C with low rpm of 500 and gradually speed was increased to blend for 90 min.

2.3 *Conventional Physical Binder Tests*

Penetration

A penetration test was conducted for base and cup lump rubber modified bitumen. According to ASTM D5, enough melted binder was poured into a cylindrical container (55 mm diameter and 35 mm in depth). The penetrometer needle was allowed to penetrate for about 5 s together with a load of 100 g at a temperature of 25 °C to measure the hardness consistency of the binder. The results were measured at tenths of a millimetre that a standard penetrometer needle penetrates the bitumen sample.

Softening

ASTM D36 standard was used to perform the softening point test of the base and the cup lump rubber modified bitumen. The physical test measures the temperature phase change of the binder. The binder softening was measured by placing the steel balls on steel rings brass filled with bitumen. The sample temperature was stabilized at 5 °C. This test was performed to determine the degree of softening the asphalt binder attains by placing the steel ball on the sample in steel ring brass. The two rings data was collected when it reached to flat plate of 2.36 mm below the ring.

Rotational Viscosity

The Brookfield viscometer was used to measure the rotational viscosity of the base and the modified binder at temperature 135 and 165 °C and ASTM D4402 was followed for the standard procedure. The test helps to ensure the mixing and compaction temperature for the binders and the enhancement in the binder properties upon modification. The binder was heated enough to fill the preheated sample chamber which requires about 8–11 g of binder and then was placed in the controlled chamber for conditioning. After the sample was stabilized at desired temperature the spindle was inserted into the chamber and conditioned for 15 min to achieve the equilibrium. The spindle rotation was continued for 5 min before the values were measured.

Temperature Susceptibility

The temperature susceptibility of CLRMB was determined by calculating the penetration index (PI). The lower susceptibility with the temperature corresponds with the higher values of the PI of the binder. The softening point and penetration results were incorporated to calculate the PI of the base and modified binders. The equation for the PI is shown in Eq. (1).

$$PI = \frac{1952 - 500\log Pen - 20S.P}{50\log Pen - S.P - 120} \quad (1)$$

Penetration at 25 °C is represented as Pen and the softening point is as S.P

Dynamic Shear Rheometer

The rheological properties of the modified and base binder could be measured by a dynamic shear rheometer (DSR). The DSR used in this study is Haake Rheo stress1. The unaged binders were tested in temperature sweep protocol at 40–82 °C to characterize the complex modulus G* and phase angle (δ). The test was performed in accordance with the AASHTO T315.

3 Results and Discussion

Based on the conventional and rheological testing the results and analysis of the bitumen are as follow.

3.1 Penetration and Softening

The measured results of the penetration at a test temperature of 25 °C and the experimental softening point results of base and modified bitumen are illustrated in Fig. 2. It was revealed that the penetration values for the base binder are substantially high in comparison with the cup lump rubber modified bitumen. The decrement in the penetration is attributed to the hardness of the binder which is mainly due to the incorporation of the cup lump rubber. The penetration of the base binder was 63 dmm which is decreased to 53 dmm on 3% addition of cup lump rubber which is about

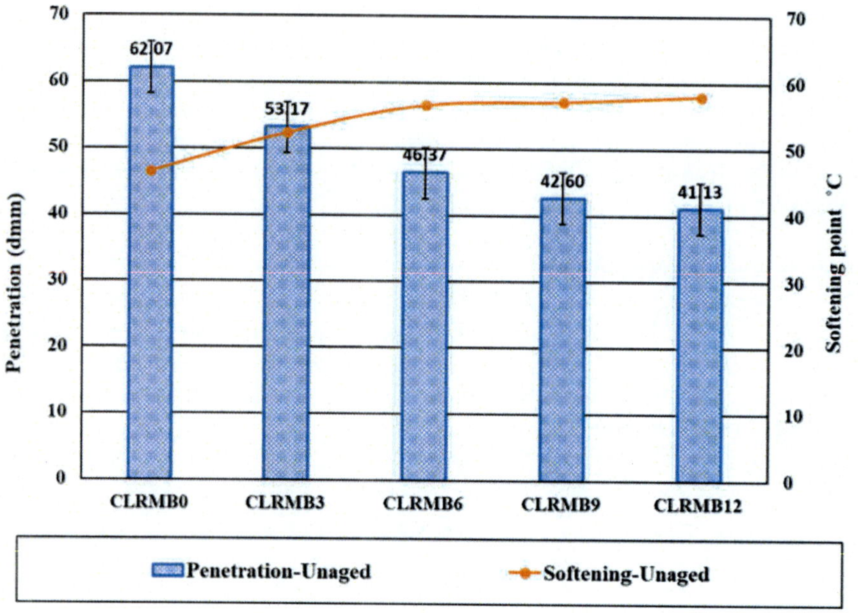

Fig. 2 Penetration and softening point results for CLR modified bitumen

15%. However, the enhancement has been observed on increasing the percentage of CLR regardless of the CLRMB12 which shows almost the same as CLRMB9. The improvement in the resistance towards the penetration is due to the CLR macromolecules dispersion in the bitumen matrix which ultimately holds the bitumen molecules and shared the resistance. On the other hand, the incorporation of the CLR enhances the softening temperature of the modified binders as shown in Fig. 2. This indicates the strong effect of CLR on the properties of the binder which is also consistent with the values of penetration after modification. It has been observed from the results that CLRMB3 has softening temperature of 52.3 °C which increased from 46.6 °C of base bitumen. Similar results were observed for a higher percentage of CLR regardless of CLRMB12. The poor compatibility between binder and rubber on higher dosage cause a similar and decrement in results for the penetration and softening point, respectively.

3.2 *Temperature Susceptibility*

The penetration index (PI) is the most frequently used method to measure the temperature susceptibility of the binders. The normal range for the base binders for high-temperature sensitive binders is approximately −3 and for the low temperature-sensitive binders the PI value is near +7. The penetration index of the base bitumen

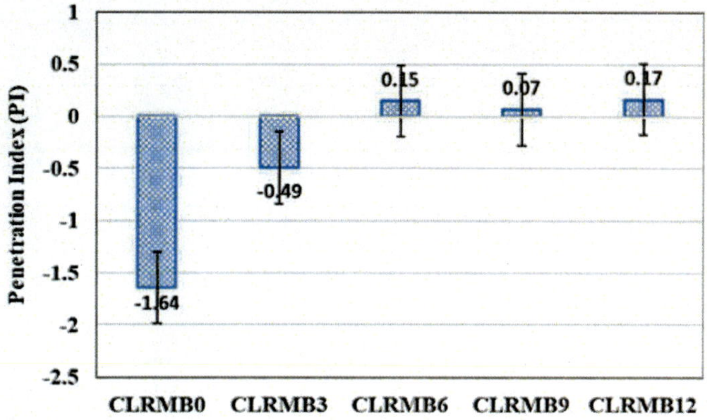

Fig. 3 Penetration index results

is −1.64 which is reduced to 0.07 for the higher content of CLR. The sharp increment in the PI shows that cup lump rubber is less susceptible to temperature variations (Fig. 3).

3.3 Rotational Viscosity

As presented in Fig. 4 the measure values for the viscosities for the base and modified binder and can be stated that all the CLRMB have higher values in comparison with the base binder. The viscosity values for the modified binder are 0.65, 1.18, 1.23, and 2.05 Pa s at 3–12% of CLR at 135 °C. The modified values are higher than the base bitumen viscosity at cited temperature. However, the same trend has also been observed for the viscosities at higher temperature or mixing temperature. The expected results satisfied the hypothesis of rubber and bitumen reactivity which and to form the cross-linking structure between to increase the binder viscosity. However, the sudden increase in the viscosity value for the CLRMB12 is obvious due to the non-compatibility between bitumen and cup lump rubber. The higher quantity makes bitumen incompetent to absorb the bitumen and the remaining rubber particles are in a binder to form a non-homogenous binder. The same results have also been encountered by some other studies related to rubber or natural rubber incorporation. It has also been observed that modification of binder up to CLRMB9 seems perfect for industrial usage as the values are not crossing the limitation of 3 Pa s.

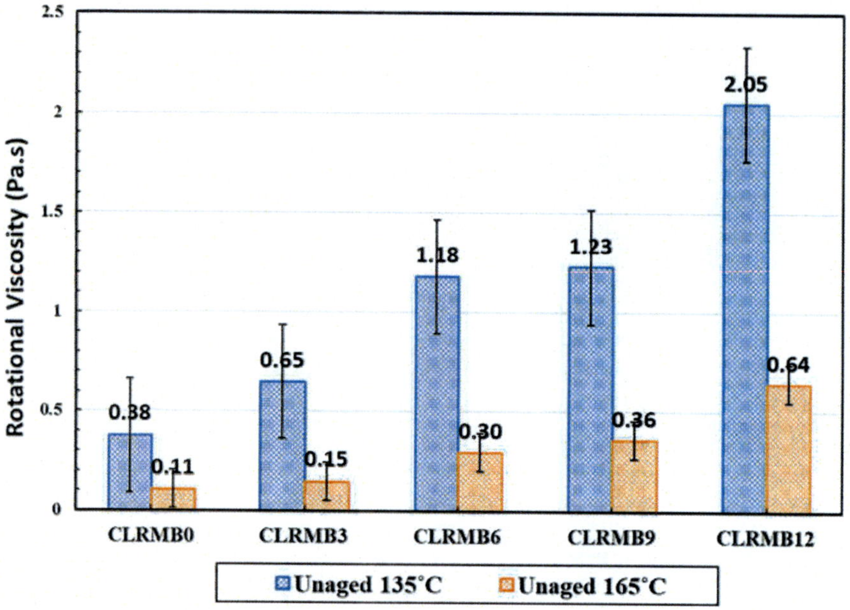

Fig. 4 Rotational viscosity results at 135 and 165 °C

3.4 Rheological Properties of CLR Modified Bitumen.

The influence of the cup lump rubber on the binder properties in term of rheological behaviour has evaluated with a dynamic shear rheometer. The temperature sweep protocol was followed at a temperature range of 40–82 °C to determine the temperature-related behaviour of rubberised bitumen at a frequency of 1.5 Hz. The complex modulus G^* represents the total deformation of the binder under the dynamic loading conditions. Figure 5 shows the plots for the complex modulus versus the range of temperatures. The higher G^* represents the higher rigidity of the binder and which is favoured in the pavement industry in accordance with the super pave limitations. It is shown in Fig. 5 that the complex modulus is lower at low temperature for CLRMB0 in comparison with the other modified bitumen.

However, when the temperature has increased the performance of the base bitumen is lower than the other CLRMB binders at 3–12%. The performance of CLRMB is not in accordance with the limitation as its complex modulus is very high which seems not practical for the pavement industry as it requires high temperature to produce a good compatible mixture. On the other hand, the complex modulus of the modified binder at temperature 40–46 °C is almost similar but they behave differently at high temperature due to the elastomeric properties of cup lump rubber. The CLRMB3 and CLRMB6 have shown better performance on the higher temperature and their grading as well regardless of CLRMB6 on temperature 76 °C which shows a slightly

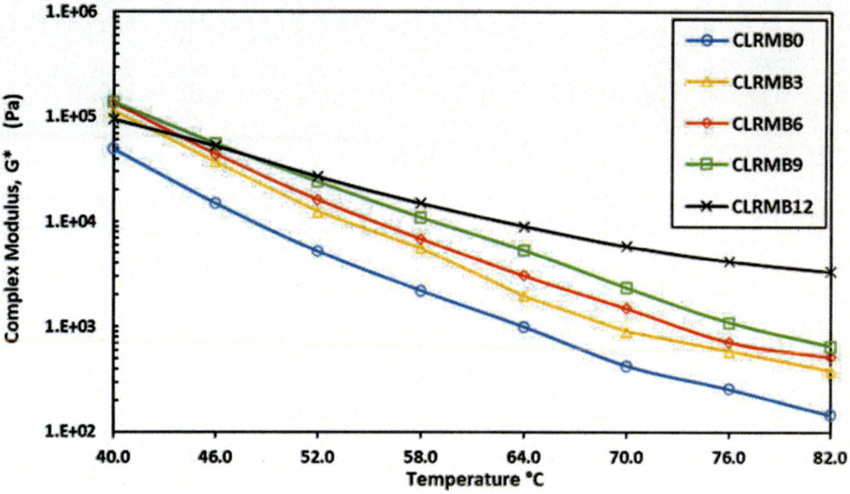

Fig. 5 Complex modulus isochronal plots at 1.5 Hz CLR modified bitumen

decrement. It has been observed clearly for the CLRMB12 that due to the incompatibility of the CLR and bitumen on higher dosage the values of G^* is very high which complies with the stiff behaviour of bitumen. The same behaviours on a higher dosage of CLR were encountered for the penetration and softening points results. Figure 6

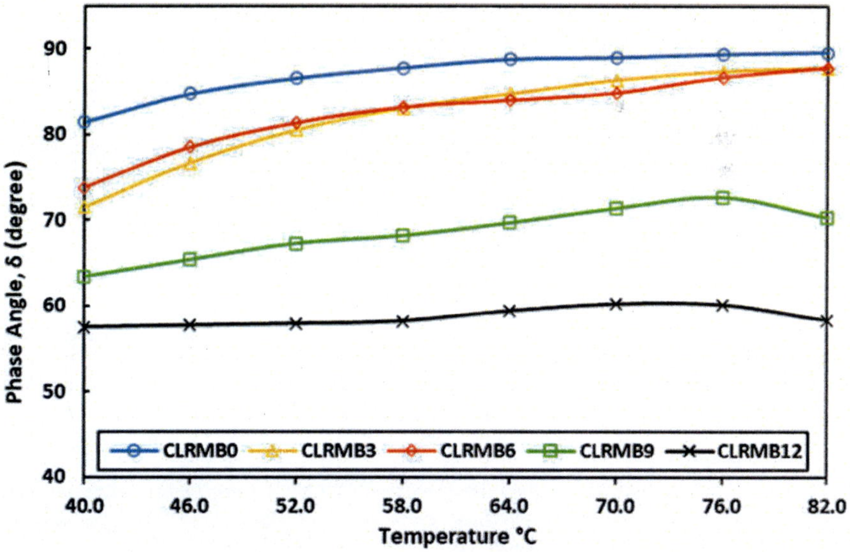

Fig. 6 Phase angle (δ) isochronal plots at 1.5 Hz CLR modified bitumen

illustrates the phase angle (δ) of the base and CLR modified binders. The temperature variation changes the phase angle of the binder. The higher the values for the phase angle (δ) more viscous are the binder in performance and the lower values represents the elastic behaviour of the bitumen. It is shown in Fig. 6 that the addition of CLR reduces the phase angle (δ) of the binder and a similar outcome has also been observed at high temperatures. It is obvious and according to the hypothesis that CLR enhance the elastic behaviour of the bitumen due to the cross-linking structure produce while blending at a higher temperature. However, improved elasticity is also essential to encounter the fatigue and rutting issue of the pavement.

4 Conclusion

Based on the bitumen modification analysis it can be concluded that the binder properties have been improved on the incorporation of the cup lump rubber. The physical and rheological properties shown an enhanced performance at all temperatures. The higher viscosity of the binder on the addition of CLR shows the structure compatibility of both materials. However, the increased content of the CLR produces the non-compatible modified binder which some unexpected results. The rheological analysis in term of complex modulus G^* and phase angle (δ) further strengthens the physical analysis of the binder. Thus, the incorporation of the CLR shows a promising avenue for further research work.

References

1. Ben Zair MM, Jakarni FM, Muniandy R, Hassim S (2021) A brief review: application of recycled polyethylene terephthalate in asphalt pavement reinforcement. Sustainability 13:1303. https://doi.org/10.3390/su13031303
2. Saowapark W, Jubsilp C, Rimdusit S (2019) Natural rubber latex-modified asphalts for pavement application: effects of phosphoric acid and sulphur addition. Road Mater Pavement Des 20:211–224. https://doi.org/10.1080/14680629.2017.1378117
3. Ansari AH, Jakarni FM, Muniandy R, Hassim S (2021) A review on the application of natural rubber as asphalt modifier. IOP Conf Ser Mater Sci Eng 1075:012031. https://doi.org/10.1088/1757-899X/1075/1/012031
4. Ali Omar H, Nur NI, Ceylan H, Sajuri Z, Jakarni FM, Ismail A (2016) Investigation of the relationship between fluidity and adhesion strength of unmodified and modified bitumens using the pull-off test method. Constr Build Mater 122:140–148. https://doi.org/10.1016/j.conbuildmat.2016.06.053
5. Ansari AH, Jakarni FM, Muniandy R, Hassim S, Elahi Z (2021) Natural rubber as a renewable and sustainable bio-modifier for pavement applications: a review. J Clean Prod 289:125727. https://doi.org/10.1016/j.jclepro.2020.125727
6. Nair KPP (2010) Rubber (Hevea brasiliensis). In: The agronomy and economy of important tree crops of the developing world. Elsevier, pp 237–273. https://doi.org/10.1016/B978-0-12-384677-8.00008-4
7. Mathew MRSN (1992) Natural rubber, biology, cultivation and technology. Elsevier

8. Nakao T, Kohjiya S (2014) Computer simulation of network formation in natural rubber (NR). In: Chemistry, manufacture and applications of natural rubber. Elsevier, pp 216–246. https://doi.org/10.1533/9780857096913.1.216
9. Bhowmick AK, Stephens HL (2000) Handbook of elastomers. CRC Press, New York. https://doi.org/10.1201/9781482270365
10. Franta I (2012) Elastomers and rubber compounding materials. Elesvier, New York
11. Wen Y, Wang Y, Zhao K, Sumalee A (2017) The use of natural rubber latex as a renewable and sustainable modifier of asphalt binder. Int J Pavement Eng 18:547–559. https://doi.org/10.1080/10298436.2015.1095913
12. Zborowski A, Kaloush KE (2007) Predictive equations to evaluate thermal fracture of asphalt rubber mixtures. Road Mater Pavement Des 8:819–833. https://doi.org/10.1080/14680629.2007.9690101
13. Kohjiya S, Ikeda Y (2014) Natural rubber: production, properties and applications. Elsevier
14. Al-Humairi SFI, Alias AH, Haron NA, Hassim S, Mohd Jakarni F (2021) Sustainable pavement: a review on the usage of pavement as a mitigation strategy for UHI. IOP Conf Ser Mater Sci Eng 1075:12010. https://doi.org/10.1088/1757-899x/1075/1/012010
15. Ruggles C (2005) The use of natural rubber latex in modified asphalt road binders in the UK
16. Polacco G, Stastna J, Biondi D, Zanzotto L (2006) Relation between polymer architecture and nonlinear viscoelastic behavior of modified asphalts. Curr Opin Colloid Interface Sci 11:230–245. https://doi.org/10.1016/j.cocis.2006.09.001
17. Othman Z, Hainin MR, Warid MNM, Idham MK, Kamarudin SNN (2018) Cup lump modified asphalt mixture along jalan Kuala Lumpur-Kuantan, daerah Temerloh, Pahang. In: MATEC web Conference, vol 250, p 02007. https://doi.org/10.1051/matecconf/201825002007

Rheological Properties of Modified Binder Incorporating Coconut Char

Rosmawati Mamat, Mohd Rosli Hainin, Norhidayah Abdul Hassan, Mohd Khairul Idham Mohd Satar, Mohd Naqiuddin Mohd Warid, and Noor Azah Abdul Rahman

Abstract Coconut char derive from coconut shell is a waste product that is anticipated to be potential alternative for modified binder. Coconut char has great potential to be used as the base material for modifier because it is safe to use. In this study the coconut char was used as additive in asphalt modified binders. This paper presents the rheological properties of modified bitumen incorporating coconut char. Rheology can be defined as the fundamental measurements associated with the flow and deformation characteristics of bitumen, evaluation of the flow and permanent deformation of time-temperature dependent materials, such as bitumen, that are stressed through the application of force. The modified bitumen was tested under different aging conditions using Rotational Viscometer (RV), Rolling Thin Film Oven (RTFO) and Dynamic Shear Rheometer (DSR). Coconut char was blended with 5, 10, 15, 20 and 25% by weight of the penetration grade 80/100. DSR was employed to measure the rutting resistance, the G*/sin at temperature 46–76 °C. Based on the findings, it is shown that the increase in coconut char percentage can effectively give highest G*/sin and affect the rutting performance under ageing condition. It also found that modified bitumen samples enhanced the rheological properties of the bitumen in terms of fatigue cracking and rutting.

Keywords Coconut char · Modified binder · Rheological properties

1 Introduction

Sustainability in construction is one of the efforts that have been practiced by developed countries to create more healthier environment and to reduce the environmental impact of a construction. Using materials that are categorized as agricultural waste

R. Mamat (✉) · M. R. Hainin · N. A. Hassan · M. K. I. M. Satar · M. N. M. Warid · N. A. A. Rahman
Faculty of Civil Engineering, Universiti Teknologi Malaysia, Johor Bahru, Malaysia

M. R. Hainin
Universiti Malaysia Pahang, Pekan, Malaysia

© The Author(s), under exclusive license to Springer Nature Switzerland AG 2022
H. R. Pasindu et al. (eds.), *Road and Airfield Pavement Technology*, Lecture Notes in Civil Engineering 193, https://doi.org/10.1007/978-3-030-87379-0_55

material in construction industry holds a high potential of reducing global environmental pollution. Nowadays the expectations of roads and their surface are changing. New technologies related to pavement technologies include reclaimed asphalt pavement (RAP), recycled asphalt shingles (RAS), mixtures containing waste from industrial processes, and bio-oil asphalt modifiers [13]. Today, researchers are looking for binder tests that not only could demonstrate the mixture's performance but also study the ease and time taken to conduct the test. Also, researchers have identified how to improve the durability, safety and efficiency of asphalt pavement through asphalt modification due to increasing traffic load and traffic volume. Improving the properties of asphalt binders is an effective method to enhance pavement strength. Many researchers have conducted studies to evaluate the physical and rheological properties of modified asphalt binder.

Future pavement surfaces will have to meet increased demands in terms of quality, safety and sustainability. The asphalt industry has been encouraged to change the function of road economy to a new transition i.e. make, use, reuse, reduce, remake and recycle [6]. Challenges are also faced by the contractors involved in road construction industry due to recent price increases in the cost of crude oil due to low availability and high demand. Today it is becoming more difficult to find natural resources. The usage of waste materials in cement, concrete and other construction materials has numerous indirect benefits such as reduction in landfill cost, saving in energy, and protecting the environment from possible pollution effect [14]. Increasing energy costs and the strong worldwide demand for petroleum has encouraged the development of alternative binders to modify or replace asphalt binders using waste materials. Coconut shell is mostly used as filler in construction materials [1, 10, 12, 15, 16].

1.1 Coconut Char

Figure 1a shows the coconut shell collected for this study. As shown, the shells have been cleaned, then crushed to smaller sizes using an aggregate crusher machine. The crushed coconut shell was burned at an optimum temperature of 450 °C for 5 min in a small furnace until it becomes a form of carbon as shown in Fig. 1b. The objective of modifying the hot mix asphalt using coconut shell is to increase the specification of stability (permanent deformation resistance), durability (moisture damage and aging), fatigue cracking resistance, safety (adequate skid resistance), resistance to thermal cracking, permeability and flexibility.

Charcoal coconut shell has been used in various research fields because it is efficient, environmentally friendly, economical and has potential to enhance the properties of construction materials. According to [11], most bitumen consists of 82–88% carbon content. Thus, charcoal coconut shell has the potential to be used as replacement or additive in the bitumen due to their chemical element compatibility [17].

Coconut shell carbon, commonly known as charcoal, is usually used as a fuel and green materials because it is safe to use [11]. It is one of the wood-based materials

Fig. 1 a Coconut shell. b. Coconut char

that can be used to produce charcoal. Charcoal derived from coconut shell has been applied in many things other than its use as fuel, such as gas purification, air filters, water filters and also in medicine. This study evaluates the performance of coconut shell carbon (CSC) and coconut shell activated carbon (CSAC) from coconut shell as a modified binder.

The rheological properties of bitumen are measured using conventional tests including viscosity, storage stability, flash point and tests after thin film oven aging and pressure aging vessel. Rheology can be defined as the fundamental measurements associated with the flow and deformation characteristics of bitumen, evaluation of the flow and permanent deformation of time-temperature dependent materials, such as bitumen, that are stressed through the application of force [1]. Asphalt mixtures that deform and flow too readily may be susceptible to rutting and bleeding, while those that are too stiff may be exposed to fatigue and cracking. The study of bitumen rheology is not a new field and has been extensively studied all over the world. Several road pavement distresses are related to rheological bitumen properties (Table 1).

Table 1 Physical and rheological properties of 80/100 PG bitumen [2]

Test properties	Results
Viscosity at 135 °C (Pa s)	0.65
$G^*/\sin\delta$ at 64 °C (kPa)	1.35
Ductility at 25 °C (cm)	>100
Softening point (°C)	47
Penetration at 25 °C (d-mm)	88
Ductility after RTFOT at 25 °C (cm)	>100
RTFOT aged $G^*/\sin\delta$ at 64 °C (kPa)	6.022
PAV aged $G^*/\sin\delta$ at 25 °C (kPa)	3122.5

The rheological properties of the modified binder were determined through the rheological performance test which is Dynamic Shear Rheometer (DSR), Rotational Viscometer (RV). The physical properties and chemical structures of asphalts will be changed when exposed to heat, oxygen, and ultraviolet (UV) light, which is called aging and method to accelerated aging of bitumen binder using Rolling Thin Film Oven (RTFO) and Pressure Aging Vessel (PAV).

2 Materials and Method

2.1 Bitumen

The bitumen of penetration 80/100 were used in this study to produce a modified binder. The selection of bitumen contents is according to JKR Standard Specification for road [9], all the samples were ranged between 4 and 6% for AC 14.

2.2 Production of CSC and CSAC

The coconut shell (waste) was collected from local area in Skudai and cut into small pieces and washed with water for removal of dust adhered to it. Then, it was dried in sunlight for 48 h or for 2 days. Dried materials were kept inside the furnace and it will be burnt in the furnace at 450 °C for removal of moisture and until it becomes charcoal. After that, the next process is to crush the charcoal and sieve to 0.075 μm size. The CSAC (powder) were purchased from a supplier of activated carbon in Ipoh, Perak. After receiving the material, the next process is to sieve to a powder of CSAC to 0.075 μm size (Fig. 2).

Carbonization process of coconut shell activated carbon, in which the sampled materials were further oven-dried for one hour at 105 °C, has been reported elsewhere. The crushed samples were carbonized for one hour in muffle furnace at different optimum devolatilization temperature of between 250‒750 °C depending on the type and nature of the materials used. Chemical activation was carried out on each of the carbonized samples. A weighed quantity of 30 g of potassium trioxocarbonate [K_2CO_3] and 30 g of sodium hydrogen trioxocarbonate [$NaHCO_3$] were dissolved in 30 ml of distilled water. The carbonized sample was weighed and mixed thoroughly with the prepared solution of K_2CO_3 and $NaHCO_3$. The activated carbons thus obtained were finally smoothened, stored in well-fitted airtight corked bottles and properly labeled before being characterized.

Fig. 2 The method of preparation of coconut char

2.3 Storage Stability Test

The storage stability test is used in order to evaluate the stability and compatibility of the modifier in asphalt binder at certain temperature. An aluminium tube of 25 mm diameter × 14 mm High filled with 50 g of modified asphalt was placed on vertical position with controlled temperature of 163 ± 5 °C for 48 ± 1 h. The sample tube was cooled in the freezer until it reached the temperature of −6.7 ± 5 °C for a minimum of 4 h. The tube was cut into 3 sections with the same measurement. The ring ball softening test was conducted to determine the possible differences of the sample characteristics. The test was conducted using the top and bottom cut tube. The physical properties of asphalts, including softening point, penetration and viscosity were measured in accordance with ASTM D36 [3] and D4402 [4].

2.4 Dynamic Shear Rheometer (DSR)

The dynamic shear rheometer (DSR) was used to characterize the behavior of asphalt viscous and elasticity when exposed to medium and high temperature. The test was conducted in accordance with ASTM D1715 [5]. Complex shear modulus G* and phase angle ծ value obtained the viscoelastic when exposed to temperature and frequency of loading. The test was conducted at temperature increment 6 °C ranged from 46 to 82 °C and decrement of 3 °C ranged from 40 °C to 22 °C for the fatigue of cracking performance. Samples of 1 mm thick plate and 25 mm diameter top plate was used to obtain RTFO unaged results. For the PAV samples, 2 mm thick plate and 8 mm diameter top plates were used. Sample was sandwiched with two circular plates and stress applied on top of oscillating plate. Frequency of 1.59 Hz was used and one cycle of oscillation applied. The stain was computed and the value of G* and ծ recorded.

2.5 Rotational Viscometer (RV)

Viscosity is the inverse of fluidity and used to determine required torque of the cylindrical spindle that is submerged in the bitumen with specific weight and temperature. The test was measured when the rotation of the spindle was constant. The test was conducted in accordance to ASTM D4402 [4]. The test was conducted using cylindrical Spindle number 27 with rotational speed of 20 RPM on specimen temperature 135 and 165 °C to obtain the material viscosities value when rotational speed was held constant.

2.6 Rolling Thin Film Oven (RTFO)

Rolling Thin Film Oven (RTFO) Test measures the effect of heat and air on a moving film of hot mix asphalt binder, simulating short-term aging that occurs during production and paving operations. This is important for investigating and predicting early age HMA pavement behaviour and distress.

2.7 Pressure Aging Vessel (PAV)

The Pressure Aging Vessel (PAV) is an aging procedure proposed by the Strategic Highway Research Project (SHRP) to simulate long-term field oxidative aging of asphalt binders. The development of the PAV included evaluation of the factors affecting the PAV procedure as well as validation of the PAV as a procedure to

Table 2 Storage stability test of modified bitumen

Content (%)	Difference top and bottom		Specification <2.2
	CSC	CSAC	
5	1	1	Pass
10	−1	−1	Pass
15	0	1	Pass
20	0	−1	Pass
25	−1	0	Pass

mimic field aging of binders. The evaluation of the rheological changes indicates that aging changes the rheological type of asphalts by changing the shape of the master rheological curve while the temperature dependency, as measured by the time-temperature shift function, does not appear to change significantly as a result of aging.

3 Results and Discussion

3.1 Storage Stability

The result of storage stability test was shown in Table 2. The difference between the top and bottom part of the modified bitumen pass specification was less than 2.2 °C [2]. This shows that the difference in softening point of temperature of top and below 2.2 °C was categorized as stable and it is considered to have compatibility between bitumen. However, the differences between the top and bottom part of modified bitumen for different percentage of CSC and CSAC contents seem to be inconsistent.

3.2 Dynamic Shear Rheometer (DSR)

The DSR is used to characterize the visco-elastic properties of the binder and measures the G* and δ by subjecting a small sample of binder. According to SHRP, G* and δ values are used to determine the rutting and fatigue of the bitumen. The G*/sin δ is often used to characterize the rutting resistance of an asphalt pavement at a high-performance temperature. A greater G*/sin δ value indicates a pavement having a better permanent deformation resistance. Figure 3 shows the G*/sin of the asphalt binder modified with CSC and CSAC at temperature from 46 to 76 °C. It can be observed that both modified binders show higher rutting resistance compared to control. The binder with CSAC has the highest G*/sin δ at amount of 25% compared

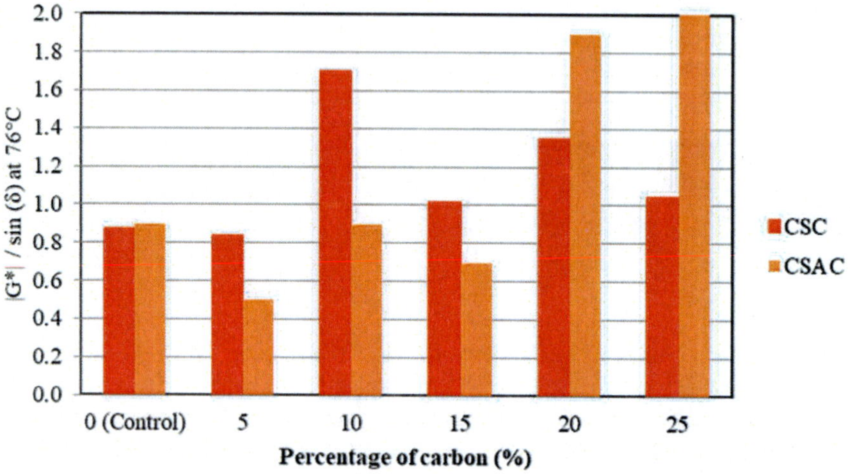

Fig. 3 G*/sin δ values at reference temperature 76 °C

to CSC. The results were simplified in Fig. 3 at reference temperature 76 °C. The G*/sin δ increased with the increasing of CSC and CSAC. Based on the graph, the CSC shows higher rutting resistance compared to CSAC at amount 5% to 15%. Meanwhile, the opposite trend was observed at amount 20 and 25%. The CSAC has higher rutting resistance compared to CSC.

3.3 *Rotational Viscometer (RV)*

Figures 4 and 5 present the rotational viscometer of control bitumen mixture and modified bitumen mixtures CSC and CSAC with results of viscosity at 135 and

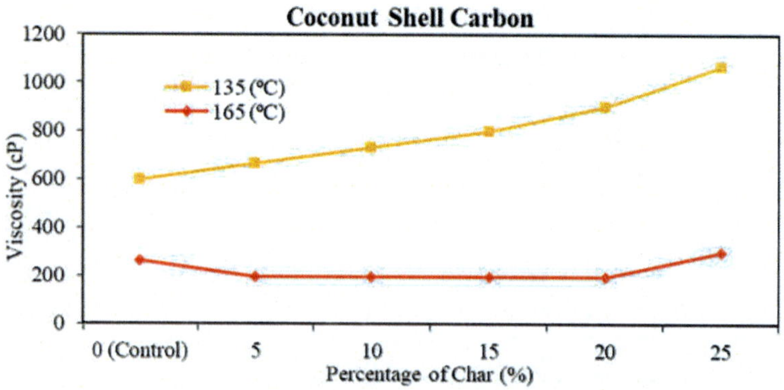

Fig. 4 Viscosity of bitumen containing CSC at different temperatures

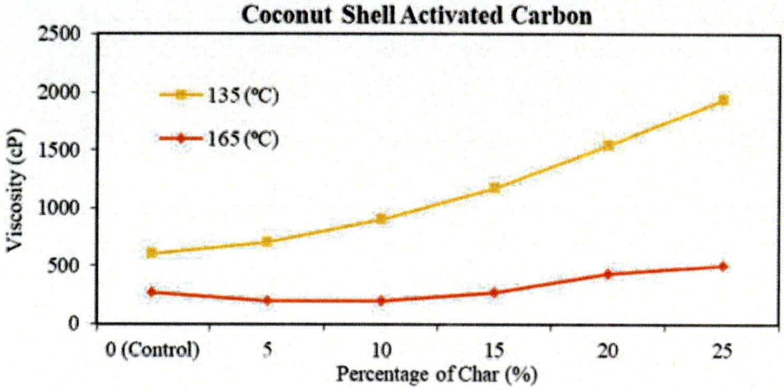

Fig. 5 Viscosity of bitumen containing CSAC at different temperatures

165 °C. The graph shows the modified bitumen of CSAC have a higher viscosity than the CSC for both temperatures. At temperature 135 °C, the viscosity increased with increasing of CSC and CSAC content from 5 to 25%. Meanwhile, for CSC at temperature 165 °C, the value of viscosity is similar from 5 to 15% and viscosity for CSAC content have same result for 5 and 10%, but the value of viscosity increases at 15–25% content of CSAC. The effect of the additive can be observed in modified bitumen. The additive generally decreased the viscosity of temperature 165 °C. The rotational viscosities were slightly higher for modified binder samples. The test result also indicates that influence of CSC content on modified asphalt binders increase the viscosity. Therefore, it can be postulated that the effect of the additive on the bitumen's viscosity may depend on the type and source of the specific bitumen and its effect is independent of the bitumen´s initial penetration grade.

3.4 Rolling Thin Film Oven (RTFO)

Figures 6 and 7 show the results of rutting index G* / sin δ for control and short term-ageing of modified bitumen. For short term ageing samples, the failure temperature is defined when the G*/sin δ is less than 2.2 kPa. CSC samples show that only 25% of CSC fail at 82 °C, while the control sample and 5–20% modified bitumen of CSC fail at 70 °C. As can be seen, the modified bitumen of CSAC have maintained their performance grade by resisting the rutting until 76–82 °C although after short term ageing process. It was found that activated carbon can demonstrate similar capability of modification on un-aged binder but lower as compared after short term aging. It may be concluded that modification of bitumen using activated carbon proves to be a more effective modifier in terms of increasing rutting resistance of the binder.

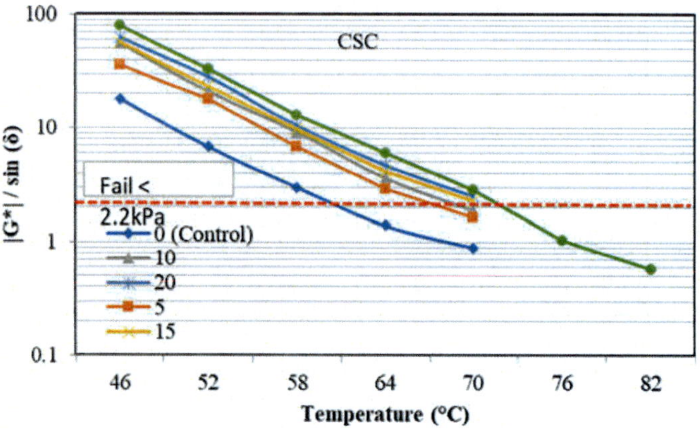

Fig. 6 Rutting parameter of short-term ageing (RTFO) samples for CSC

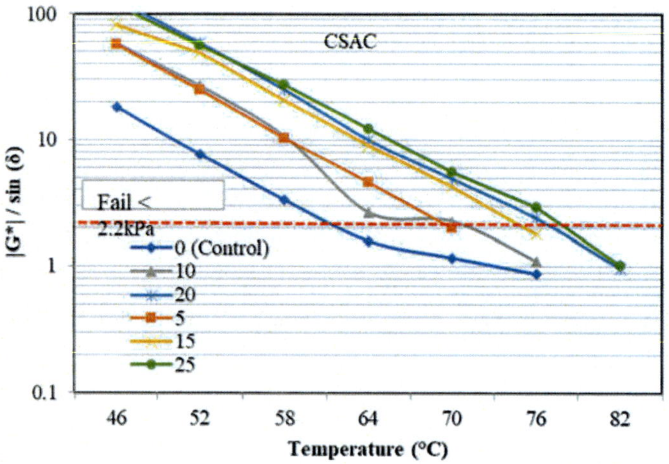

Fig. 7 Rutting parameter of short-term ageing (RTFO) samples for CSAC

3.5 *Pressure Aging Vessel (PAV)*

Figures 8 and 9 show the results of rutting index G*/sin δ for control and short term-ageing of modified bitumen. It can be seen that the 25% of CSC and CSAC have higher performance than the control and other samples, while 5% of CSC and CSAC were lower than the control samples. It is indicated that at 25% of CSC and CSAC, the bitumen is stiffer which caused failure at earlier temperature compared to the control samples and also other samples. Fatigue cracking resistance of the bitumen binder remained unchanged with up to 5% addition of the modifier (CSC and CSAC) whereas the fatigue resistance decreased with the addition of CSC and CSAC.

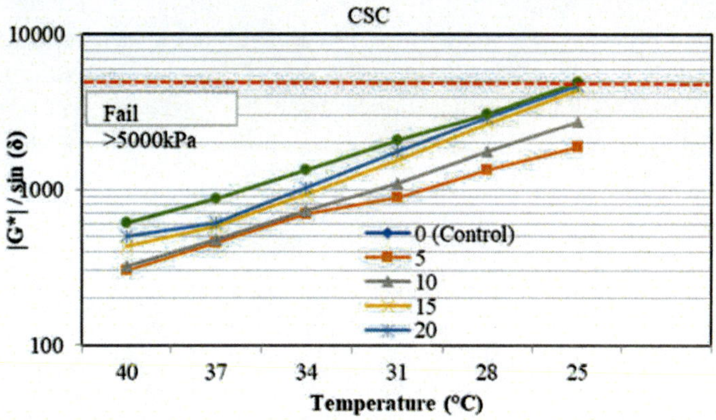

Fig. 8 Rutting parameter of long-term ageing (PAV) samples for CSC

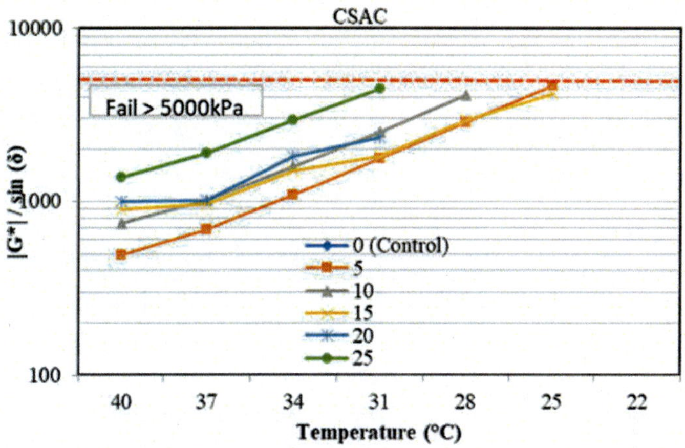

Fig. 9 Rutting parameter of long-term ageing (PAV) samples for CSAC

4 Conclusions

The consistency of asphalt binder containing CSC and CSAC has been compared to the conventional asphalt binder. The content of CSC and CSAC affect the rheological properties. The result shows that both modified asphalt binders have high rutting resistance compared to the control bitumen. The modification seems to be more significant at high CSC and CSAC contents. These high values indicate that the modified binder is stiffer than the pure binder. All the modified binders, those incorporating the CSC and CSAC, achieve the best values and improved rheological properties. This suggests that interaction between this coconut carbon is efficient enough to affect the thermal and mechanical properties of the asphalt. These changes were a result of the ability of the modifier to increase the friction between the layers of the asphalt binder. The viscosity of the modified binders increased, while their penetration decreased; this behavior indicates an enhancement of the stiffening effect and the high temperature properties of the asphalt.

References

1. Agunsoye JO, Talabi SI, Bello SA, Awe IO (2014) The Effects of Cocos Nucifera (Coconut shell) on the mechanical and tribological properties of recycled waste aluminium can composites. Tribol Ind 36(2):155–162
2. Ali AH, Mashaan NS, Karim MR (2013) Investigations of physical and rheological properties of aged rubberised bitumen. Adv Mater Sci Eng 2013:1–8. https://doi.org/10.1155/2013/239036
3. American Society of Testing and Materials (2012a) ASTM D36. Standard test method for softening point of bitumen (ring-and-ball apparatus). ASTM International, West Conshohocken, PA, p 141
4. American Society of Testing and Materials (2012b). ASTM D4402. Standard test method for viscosity determination of asphalt at elevated temperatures using a rotational viscometer. ASTM International, West Conshohocken, PA
5. American Society of Testing and Materials (2008) ASTM D7175. Standard test method for determining the rheological properties of asphalt binder using a dynamic shear rheometer. ASTM International, West Conshohocken, PA. American Society of Testing and Materials
6. He GP, Wong WG (2008) Effects of moisture on strength and permanent deformation of foamed asphalt mix incorporating RAP materials. Constr Build Mater 22(1):30–40. https://doi.org/10.1016/j.conbuildmat.2006.06.033
7. Hunter RN, Self A, Read J (2015) The shell bitumen handbook, 6th edn. ICE Publishing, London, UK
8. Husseinsyah S, Mostapha M (2011) The effect of filler content on properties of coconut shell filled polyester composites. Malaysian Polymer J 6(1):87–97
9. Jabatan KR (2008) Standard specification for road works, Section-4: flexible pavement. Jabatan Kerja Raya Malaysia, Kuala Lumpur
10. Kumar S, Kumar B (2012) Study of mechanical properties of coconut shell particle and coir fibre reinforced epoxy composite. Int J Adv Eng Res 4(2):39–62
11. Licence CC (2011). Icme11-Abs-037 pyrolysis of coconut shell for bio-oil, pp 18–20
12. Madakson PB, Yawas DS, Apasi A (2012) Characterization of coconut shell ash for potential utilization in metal matrix composites for automotive applications. Int J Eng Sci Technol 4(3):1190–1198
13. Mohammad LN, Louise I, Stewart R, Materials E, Transportation L, Rouge B, Elseifi M. A, Guillory L, Associate D, Cooper SB, Challa H, Naidoo P, Technologies GA, Christian P (2013) Laboratory evaluation of asphalt mixtures containing bio-binder technologies
14. Oyejobi DO, Abdulkadir TS, Yusuf IT, Badiru MJ (2012). Effects of palm kernel shells sizes and mix ratios on lightweight concrete. J Res Inf Civil Eng 9(2)
15. Sapuan SM, Harimi M, Maleque MA (2003) Mechanical properties of epoxy/coconut shell filler particle composites. Arab J Sci Eng 28(2):171–181
16. Sarki J, Hassan SB, Aigbodion VS, Oghenevweta JE (2011) Potential of using coconut shell particle fillers in eco-composite materials. J Alloys Compounds. 509(5):2381–2385 (Elsevier B.V.)
17. Zhao S, Huang B, Shu X, Ye P (2014) Laboratory investigation of biochar-modified asphalt mixture. Transport Res Rec: J Transport Res Board 2445:56–63

Sustainable Pavement Materials

A Brief Review: Application of Recycled Polyethylene Terephthalate as a Modifier for Asphalt Binder

Mohamed Meftah Ben Zair, Fauzan Mohd Jakarni, Ratnasamy Muniandy, Salihudin Hassim, and Anwaar Hazoor Ansari

Abstract The management of the waste materials produced on the earth is one of the highly substantial environmental issues. Hence, the increasing usage of plastic every year is one of the concerns across the entirety of the world. The limiting landfill conditions and higher production of the waste plastic materials stress the researcher to manage it alternatively. Conversely, the growing number of heavy vehicles on roads cause road distress to asphalt pavements that contain pure asphalt binders that are insufficiently resistant. Civil engineers and scientists have attempted to address such issues by enhancing the performance of asphalt pavement via modification of asphalt binders' properties. This paper summarized the possible utilization of waste polyethene terephthalate (PET) for binder modification. It has been evaluated that the PET can significantly improve the binder properties. The chemical modified PET showed better performance in term of physical and rheological properties. In addition, the compatibility of PET with binder has been highlighted with modified chemical structures and via morphological assessment of modified binder. In addition to solving landfilling issues of waste PET plastic, such recycling approaches generated in these studies can also be used by asphalt researchers to explore future avenues of improvement for the characterization of asphalt binder as a paving material.

Keywords Modified asphalt binder · Recycling · Polyethylene terephthalate (PET) · Viscosity · Rheological characteristic · FTIR

1 Introduction

At present, plastic is an indispensable utility in our lives. It is one of the most significant inventions during the age of industrialization. Since the year 1856, it has become one of the most widely used man-made materials [1]. With the increase in the global

M. M. Ben Zair (✉) · F. M. Jakarni (✉) · R. Muniandy · S. Hassim · A. H. Ansari
Department of Civil Engineering, Faculty of Engineering, Universiti Putra Malaysia, UPM Serdang, 43400 Selangor, Malaysia

M. M. Ben Zair
Department of Civil Engineering, Faculty of Engineering, Misurata University, Misurata, Libya

© The Author(s), under exclusive license to Springer Nature Switzerland AG 2022
H. R. Pasindu et al. (eds.), *Road and Airfield Pavement Technology*, Lecture Notes in Civil Engineering 193, https://doi.org/10.1007/978-3-030-87379-0_56

population, innovation in packaging and utility sectors, and changes in the lifestyles of people, there is a tremendous increase in the consumption of plastics over the last couple of decades [2]. In addition to consumer products, plastic polymers are also utilized to generate synthetic fibers, coatings, sealants, foams and adhesives [4]. The usage of plastic in Europe is mainly for packaging 38%, construction and building 21%, automotive 7%, electronic and electrical 6% and others including leisure and medical 28% [3].

Considering its negative effect on the environment, especially marine life, there is global and political focus on the worsening of plastic waste and exploration of sustainable methods of managing such waste. The concerns regarding the use of plastics and disposal of the associated waste includes: animals getting entangled in plastics or ingesting plastics, non-sustainable manner of present usage; and waste left in landfills and encroaching natural habitats [4]. The staggering increase in global production of plastic from 1.5 million tonnes back in the 1950s to 335 million tonnes in 2016 reported an equal increase in plastics being released into all facets of the environment [5]. Plastic waste has become a global pollutant [6]. In 2015, approximately 6300 metric tonnes of plastic waste had been generated. Around 9% was recycled, 12% had been incinerated, and 79% ended up in landfills or the natural environment [7]. In 2018, world plastics production was 359 million tonnes. Distributed on following, 18% North America, 4% South America, 17% Europe, 7% middle east 51% Asia. China is considered the largest manufacture with 30% of plastics produced around the world [8].

The increase of waste production harms the sustainability of the environment [9]. Some of the major global concerns include plastic debris that is brought by inland wastewaters or waterways, and that are blown into the ocean by wind [10]. The migration of potentially toxic waste and plastic additives in soil, water, and air take place in every phase of the plastic lifecycle [11, 12]. Such continuous discharge of plastic waste into the landfills and environment is a direct consequence of high production of single-use plastic, in addition to the improper regulation of management cycle, that includes collection, treatment, recycling as well as disposal [13]. In recent times, additives were used to enhance the mechanical properties and rheological properties of asphalt mixture. For example, introducing polymers to asphalt binder enhances the rheological properties and reduce its susceptibility in the variation of temperature. As such, under low temperature, its mechanical strength increases. This is apparent from the increase in the lifetime of pavement in the subtropical and tropical areas [14, 15]. The rheological and physical properties of asphalt binder are important. As such, this literature review aims to evaluate the application of waste polyethylene terephthalate (PET) as a Thermoplastic modifier. PET at various contents used as low-cost bitumen modifiers to improve the processing and performance properties of asphalt binder. It also promotes significantly interaction between thermoplastic polymer and bitumen matrix.

2 Polyethylene Terephthalate (PET)

PET is essentially a semi-crystalline, thermoplastic polyester that has good transparency, safety as well as strength [16]. The intermediates, pure ethylene glycol (EG) and terephthalic acid (TPA), are obtained from crude oil for the generation of PET. Upon heating together, the initial product formed is a monomer bis (2-hydroxyethyl) terephthalate (BHET) that is combined together with polymers of low molecular weight (oligomers). There is further reaction of the mixture, which distils out any extra ethylene glycol and generates the PET [17].

Polyethylene Terephthalate (PET) is arguably a significant modern plastic of the polyester family that is commonly used as packaging and container material (i.e., bottles, electrical and electronic devices, automobiles products, housewares, sporting goods, and lighting products etc.) [18, 19]. In the year 2015, PET made up almost 7% of the global plastic demand at approximately 18.8 million tons [20]. In addition, almost 62% of all bottles that are collected for the purpose of recycling is composed of PET bottles [21]. However, only 50% of secondary PET is collected and processed globally [22]. Compared to virgin PET, there are important environmental advantages of recycled PET [23]. At present, the world focuses on recycling in order to lower PET waste and promote rational use of resources [24]. The recycling steps for the PET is described in Table 1. The accumulated amount of generated plastic waste and disposed between 1950 and 2015, as well as the projected amount by the year 2050 is illustrated in Fig. 1. Close to 16% of this amount had been put through recycling efforts up to 2015. Projections estimate that approximately thirty-three of the plastic waste will have been recycled by the year 2050 [7].

Table 1 Details of recycling steps [25]

Form of recycling steps	Description
Primary recycling	a. Most conventional technique of recycling PET. Also termed as re-extrusion b. Inexpensive and requites uncontaminated scraps. Only handles single type waste c. The recycled waste or scrap is either utilized as second grade material or mixed with virgin material
Secondary recycling	a. Include elimination of contaminates and sorting of arid separation of waste. Also called mechanical recycling process b. Reduce size via extrusion by heat, reforming, crushing and grinding
Tertiary recycling	a. Chemical recycling is carried out via solvolytic chain cleavage b. Tins process can either be a partial depolymerization on to its oligomers or total depolymerization to its initial monomers

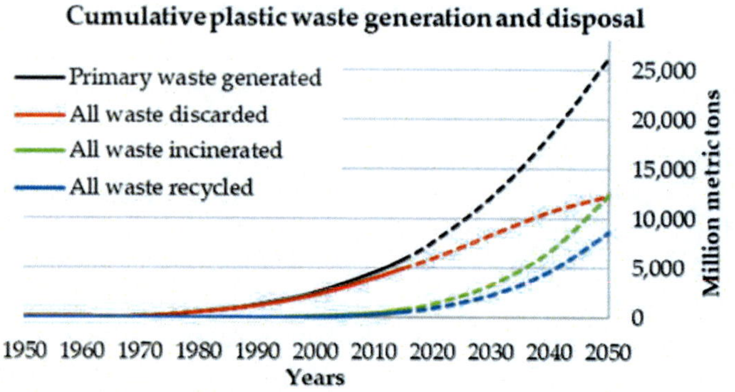

Fig. 1 The accumulated amount of the plastic waste generated and disposed between 1950–2015 and the projected amount by 2050 [7]

3 PET Waste for Asphalt Binder Modification

One of the most widely utilized plastics is PET, especially in beverage bottles owing to its excellent barrier properties to trap moisture and gas. Despite the recyclable nature of PET bottles, a PET bottle of single use possesses a short-life span, as in the case when the recycling industry fails to match waste and consumption [26]. Moreover, PET plastic waste may be utilized in asphalt binder [27, 28]. Such PET plastic waste has always been the focus of research within the field of civil engineering. Nonetheless, the focus of past investigations has only covered the incorporation of PET plastic waste via the dry method as a substance of filler and fine aggregate. As such, usage of waste PET was explored by utilizing it as additive within SMA (stone mastic asphalt) and as fine particles within asphalt mixture [29, 30].

PET that is recycled may be utilized as enhanced modifier for asphalt binders. Such usage may be a regarded as a technique of reusing PET waste. This presents the opportunity to address potential environmental risks, and also lowers cost of construction that otherwise would have occurred should polymers be utilized within the mixture of asphalt [31]. The asphalt interaction (adhesion) is the major factor for deterioration of pavement in most mixtures of asphalt [32]. In addition, it has been reported that PET had a positive impact such as the provision of high temperature properties for the asphalt mixture [33]. Nonetheless, the PET is not able to consistently raise the elasticity of asphalt binder, especially during unexpected and dramatic lowering of temperature. The low and intermediate temperature properties of the asphalt binder deteriorated as a result of this condition [34].

PET was integrated with asphalt binder in order to assess the effect of PET on the properties of asphalt binder. There has been an observation that the incorporation of PET is able to improve the consistency binder and increase its elastic behaviour. Nonetheless, increasing the viscosity of modified binder produced increased the mixing and compaction temperature. Despite the fact that this is not economical in the short run, it is still able to lead to the building of roads with fewer requirements for maintenance. The incorporation of PET has reduced modified the oxidation of asphalt binder and produce some anti-ageing effect [35].

3.1 *Physical Properties of PET-Modified Asphalt Binder*

Asphalt binder is essentially an engineering material that is utilized as a binder for distinct purposes that include airfield runaways and highways. It is composed of heavy hydrocarbons that board into four groups namely aromatics, asphaltenes, saturates and resins [36]. In addition, binders that are used for construction of pavement are mostly obtained from fossil fuels [37]. Such complex chemical form provides binder with favourable properties such as thermoplasticity and viscoelasticity that lead to distinctive rheological behaviour. Within pavement engineering, there were efforts to utilize physically recycled waste PET within asphalt pavement. Under such applications, PET waste is broken down into smaller parts and then incorporated directly to the hot asphalt binder or mixture at the time of production. Observations found that the integration of scrap waste PET enhance the fatigue resistance, running resistance and Marshall stability, as well as simultaneously lowers the moisture susceptibility of asphalt mixture and binder [38].

Different investigations have concentrated on the impact of PET on the physical properties of the binder. The impact of utilizing Polyethylene Terephthalate (PET) bottle waste as an modifier of asphalt binder was assessed by Guru et al. [39]. Thin Liquid Polyol PET (TLPP) and Viscous Polyol PET (VPP), which are two novel additive materials, were chemically sourced from PET waste bottles. They were utilized in separated enhancement of the base asphalt. Outcomes show that except for the 1% (w/w) VPP, the VPP and TLPP experienced increase in penetration and reduced the softening point of base binder. Similarly, Mahrez and Karim [40] noted similar outcomes when incorporating 2, 4, 6, and 8% of PET by weight of binder in powder form to the asphalt binder produced a lowering of penetration values by 35, 30, 21 and 14%. In addition, as shown in Fig. 2, the incorporation of 2–8% PET level increases the softening point in ranges from 5 to 13%. Another study

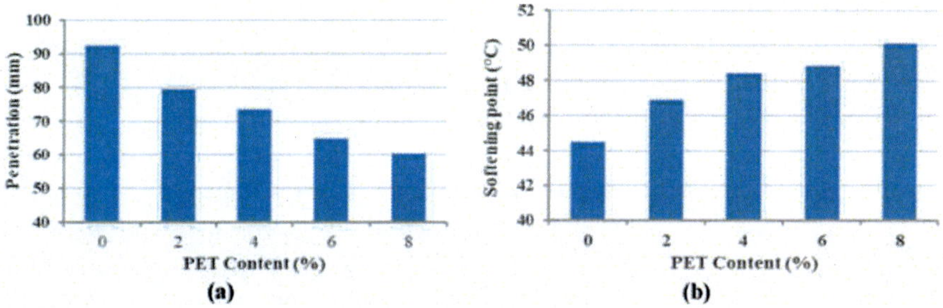

Fig. 2 **a** Penetration outcomes assessments of the base and modified asphalts; **b** softening point assessment outcomes of the base and modified asphalts [40]

by Padhan and Gupta [41] sought to run a novel method of in situ polymerization of monomers bis(2-hydroethyl) terephthalamide (BHETA). Outcomes highlight the lowering of values of penetration as well as increase in point of softening, similar to past studies [39–41]. Table 2 summarizes the physical properties of PET modified binder at various concentrations.

Furthermore, it was found by Guru et al. [39] that under all test temperatures, there was a reduction in viscosity as TLPP increased. The effect of utilizing TLP in reducing the working temperature of asphalt was affirmed by the 32.5% reduction of viscosity at 160 °C for 10% PET in modified mixture. The viscosity is also increased via the integration of PET into the mixture as illustrated in Fig. 3. The rutting process which takes place in hot places that are prone to heavy traffic, significantly reduces the performance of the pavement. As such, the asphalt mixtures of hot places should use high viscosity binder. The rutting issues are alleviated by incorporating PET wastes to binder. In another similar investigation, Leng et al. [42] conducted viscosity tests under three temperatures: 180, 165 and 150 °C. The incorporation of the BHETA additive increases the viscosity of Crumb Rubber Modified Asphalt Binder (CRMA) at higher temperatures. The increase in viscosity is relatively small upon incorporation of BHETA additive at small percentages (3–5%). Nonetheless, there is a staggering increase in viscosity following the addition of 7% of BHETA.

In another analogous research by Abuaddous et al. [48], measurements of viscosity were carried out by running rotational viscosity assessment at two testing temperatures, which are 165 and 135 °C. From the test outcomes, the viscosity of asphalt binder increased with the incorporation of recycled polyethylene terephthalate (RPET) at both testing temperatures. The increase in viscosity of asphalt binder is due to the increase in consistency upon incorporation of RPET. As such, in comparison with control asphalt binders, asphalt binder with additional RPET would have superior field performance in terms of rutting.

3.2 Rheological Properties of PET-Modified Asphalt Binder

Complex shear modulus (G) and phase angle (δ) and can be used to assess the viscoelastic behaviours of the binder at high to moderate temperatures. Softer asphalt corresponds to lower complex modulus while the high modulus binder is relatively more viscous and stiffer. While integrated into the mixture, there may be a reduction of rutting problems [40]. Phase angle of lower values provided greater elastic mixtures, thus improving the rutting and aging resistance of the mixture. The rutting nature of asphalt binder is usually evaluated by the phase angle (δ) and complex modulus (G) [39]. As concluded by Mahrez and Karim [40], the PET modified mixture possessed greater G* than the control binder, thus improving the resistance of binder. This may be due to the increasing in rutting resistance. In another study, Karahrodi et al. [49] reported that by utilizing the Waste polyethylene terephthalate (wPET) and ground tire rubber (GTR) blend as cheaper binder

Table 2 Summary of the reviewed literature on the physical properties of PET modified binder at various concentrations

References	Type of material	Binder grade	Replacement (%)	Penetration (mm)	Softening point (c)	Viscosity
Guru et al. [39]	• TLPP • VPP	40–50	0, 1, 2, 3, 5, and 10 0, 1, 2, 3, 5, and 10	(45–61) (35.56%) ↑ (45–60) (33.33%) ↑	(51.5–49.8) (3.30%) ↓ (51.5–49.8) (3.30%) ↓	–
Mahrez and Karim [40]	PET powder	80–100	0, 2, 4, 6 and 8	92.4–60.4 (34.63%) ↓	(44.4–50.1) (12.84%) ↑	Viscosity (mPas) at 135 °C (325–450.59) (38.64%)↑
Silva et al. [35]	Micronized PET	50–70	0, 4, 5, and 6	(52.9–24.4) (53.88%) ↓	(39.6–61) (54.04%)↑	–
Padhan and Gupta [41]	BHETA-PU	80–100	0, and 2	(84, 66) (21.43%) ↓	(46, 52) (13.04%) ↑	Viscosity (Poise) at 150 °C (1.57, 2.95) (87.90%) ↑
Leng et al. [42]	CR-BHETA	60–70	0, 3, 5, 7 of BHETA mixed with 10 of CR	(42–35) (16.67%) ↓	(53–62) (16.98%) ↑	• Viscosity (cP) at 150 °C (1183–2220) (87.66%) ↑ • Viscosity at 165 °C (700–1200) (71.43%) ↑ • Viscosity at 180 °C (500–800) (60%) ↑
Padhan et al. [43]	• BAET • BHETA	80–100	0, 0.5, 1, and 2	(84–71) (15.48%) ↓ (77–84) (9.09%) ↑	(46–48) (4.35%) ↑ (47–45) (4.26%) ↓	–

(continued)

Table 2 (continued)

References	Type of material	Binder grade	Replacement (%)	Penetration (mm)	Softening point (c)	Viscosity
Leng et al. [44]	PET	60–70	Virgin binder 0%, virgin binder with 2% PET, and PET modified binder mixed with 15, 25 and 40% RAP binders	(69–39) (43.48%) ↓	(48–57.5) (19.79%) ↑	• Viscosity (cP) at 135 °C (483–1180) (144.30%) ↑ • Viscosity at 150 °C (250–604) (141.6%) ↑ • Viscosity at 165 °C (188–356) (89.36%) ↑
Sreeram et al. [45]	Synthesise PET	60–70	0, 2, and 4	(70–55) (21.43%) ↓	(48–53) (10.42%) ↑	• Viscosity (cP) at 135 °C (415–501) (20.72%) ↑ • Viscosity at 150 °C (202–251) (24.26%) ↑ • Viscosity at 165 °C (150–202) (34.67%) ↑

(continued)

Table 2 (continued)

References	Type of material	Binder grade	Replacement (%)	Penetration (mm)	Softening point (c)	Viscosity
Padhan et al. [46]	PET additives and SBS modified binder (PS- MB)	60–70	0, 3 and 5 mixed with 4% of SBS	(47–64) (36.17%) ↑	(62–50) (19.35%) ↓	• Viscosity (Cp) at 135 °C (1145–456) (60.17%) ↓ • Viscosity at 150 °C (610–234) (61.64%) ↓ • Viscosity at 165 °C (400–457) (14.25%) ↑
Ameli et al. [47]	Anti-Stripping Agents (ASA), Ground Tire Rubber (GTR), and PET	85–100	• Virgin binder, PC1, PC2, PC3 • Virgin binder, PC1A, PC2A, PC3A • Virgin binder, PC1B, PC2B, PC3B • Virgin binder, PC1C, PC2C, PC3C	(67.5–45) (33.33%) ↓ (67.5–45) (33.33%) ↓ (67.5–41) (39.26%) ↓ (67.5–37) (45.19%) ↓	(45–59) (31.11%) ↑ (45–65) (44.44%) ↑ (45–58) (28.89%) ↑ (45–61) (35.56%) ↑	(1.22–2) (63.93%) ↑ (1.22–2.17) (77.87%) ↑ (1.22–2.37) (94.26%) ↑ (1.22–2.11) (72.95%) ↑
Abuaddous et al. [48]	PET particles	60–70	0, 5, 10, 15, 20	(69–58) (15.94%) ↓	(47.25–51) (0–7.94%) ↑	• Viscosity at 135 °C (420–760) (80.95%) ↑ • Viscosity at 165 °C (55–207) (276.36%) ↑

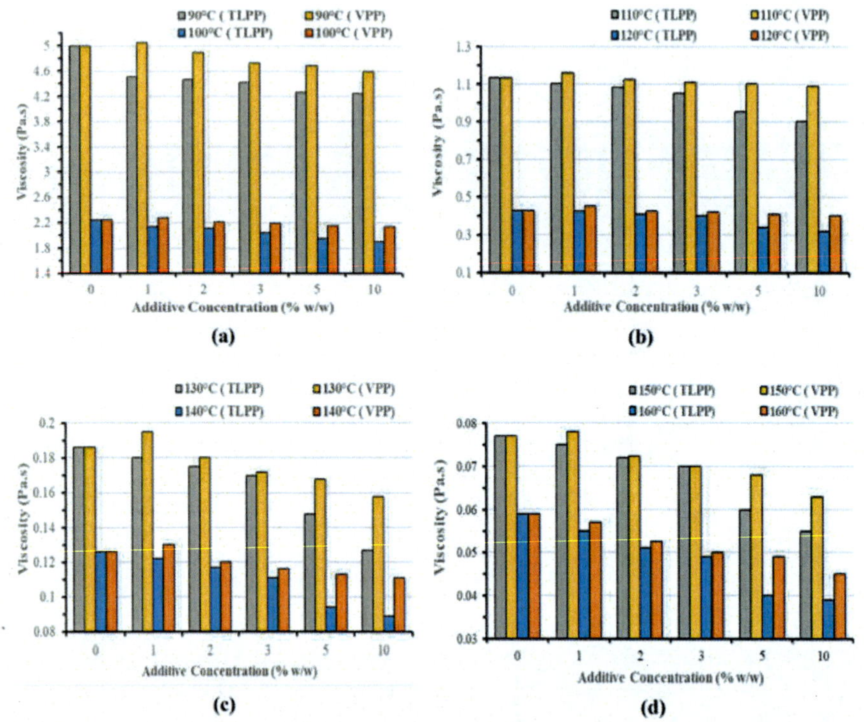

Fig. 3 Outcomes of viscosity assessment for the base and modified asphalts; **a** between 90 and 100 °C; **b** between 110 and 120 °C; **c** between 130 and 140 °C; **d** between 150 and 160 °C [39]

modifiers. The (G*/sinδ) parameter was raised by the use of wPET/GTR blend. This denotes a greater resistance against permanent degeneration or rutting of binder enhanced with wPET/GTR blends, that is pronounceable for binder integrated with 7% of wPET/GTR75/25 and wPET/GTR90/10 as shown in Fig. 4. This phenomenon is related to the thermoplastic nature of wPET material, which permits superior dispersion of wPET-rich blends within the bitumen matrix.

Leng et al. [42] was assessed the impact of utilizing bis (2-hydroxyl ethylene) terephthalamide (BHETA) obtained from PET material as a performance-enhancer to enhance the rheological properties of asphalt binder. Outcomes for both RTFO aged and unaged binders show that the 5 wt% of BHETA was mixed with 10% of CR (10CR-5BH), 3 wt% of BHETA was mixed with 10% of CR (10CR-3BH), and 7 wt% of BHETA was mixed with 10% of CR (10CR-7BH) blends have (G*/sinδ) values are marginally greater. In another related study Padhan and Gupta [41] obtained the same results for polyurethane-based Polymer modified bitumen (BHETA-PU) and styrene-butadiene-styrene (SBS) modified asphalt binder under RTFO unaged binders. Similar outcomes were found by Abuaddous et al. [48] who assessed the usage of RPET as a modifier asphalt binder. The outcomes show that following incorporation of RPET, the asphalt binder's high-temperature rutting improved, as

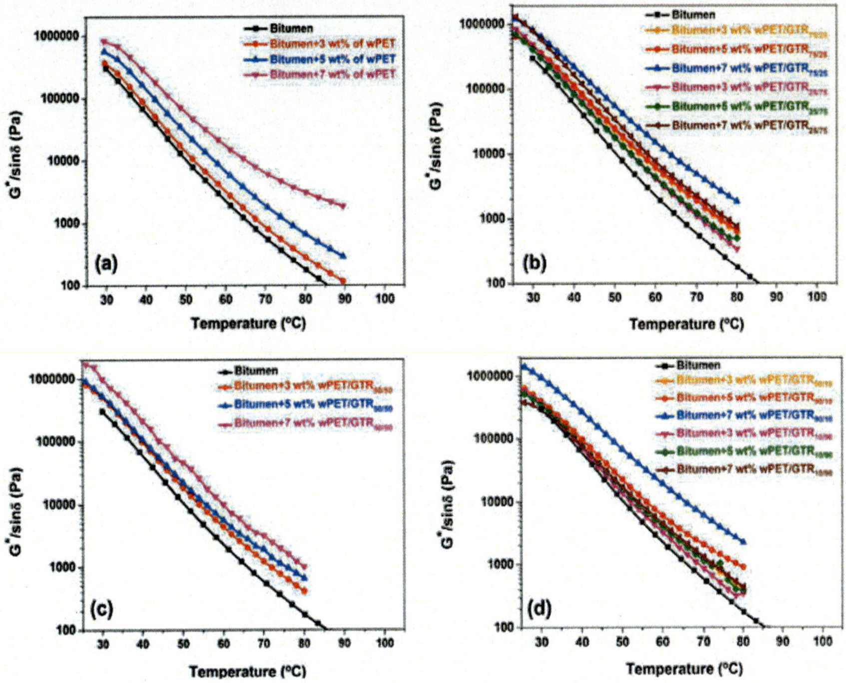

Fig. 4 Temperature dependence of (G*/sinδ) for pristine and polymer modified binder: **a** binder + wPET; **b** binder + wPET/GTR75/25 and binder + wPET/GTR25/75, **c** binder + wPET/GTR50/50, **d** binder + wPET/GTR90/10 and binder + wPET/GTR10/90 [49]

detected through DSR rutting parameter (G*/sinδ). The high-temperature performance grade of asphalt binder improved by one grade from 64 to 70 °C (for control asphalt binder) at 20 and 15% RPET. Padhan et al. [46] used chemically recycled PET to generate additives as well as utilized with SBS in order to enhance asphalt binder. The results were in the same vein as previously discussed.

PET modified asphalt binder was also investigated for the intermediate temperature to evaluate the fatigue parameter. Padhan and Gupta [41] obtained the satisfactory outcomes produced by the (G*Sinδ) at 19 °C for all three binders (SBS based PMB, base binder and BHETA-PU). As such, outcomes indicate that all the blends attain the similar grading of performance. In a separate investigation carried out by Padhan et al. [46], PET was chemically recycled to generate additives as well as utilized with SBS in order to enhance asphalt binder. In order to characterize the properties modified of asphalt binder, superpave binder tests fatigue factor (G*sinδ) was conducted and showed improved outcomes for fatigues results. In addition, compared to SBS modified binder (S-MB), the 3% of PET additives, and 4% of SBS modified binder (PS-MB3) led to better fatigue cracking resistance. As such, as shown in Fig. 5, the level of additive utilized influences the degree to which PET additives enhance the fatigue resistance of SBS modified binder.

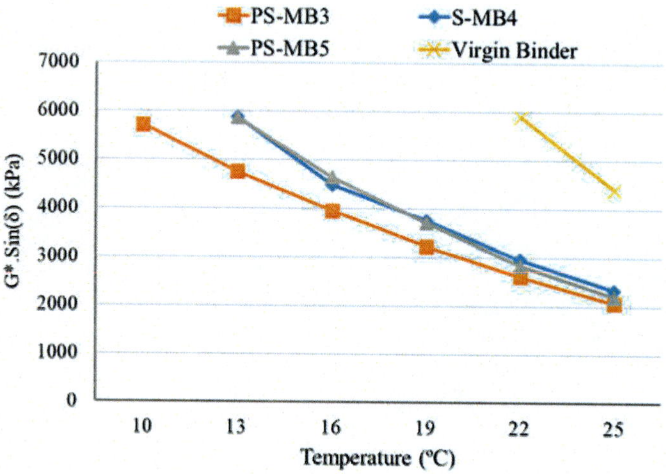

Fig. 5 Impact of utilizing SBS and PET on fatigue performance of asphalt binder [46]

Leng et al. [42] who assessed the BHETA and CRMA modification with asphalt binder concluded the fatigue resistance parameter. As such, the base CRMA, the BHETA modified CRMA provides better fatigue resistance. This suggests that in comparison with virgin binder. In another study Abuaddous et al. [48] reported that all RPET percentages, the fatigue parameters (G*sinδ) was lower than 5000 kPa-Superpave limit. The RPET modifier negatively influenced the asphalt binder performance under low temperature. The asphalt grade of asphalt binder at low temperature dropped from −22 to −16 °C (for control asphalt binder) at 20 and 15% RPET. It is therefore favorable to utilize RPET modifier during hot weathers.

3.3 *Morphological and Chemical Analysis of PET Modified Asphalt Binder*

Scanning Electron Microscope (SEM) fitted with Fourier Transform Infrared (FTIR) and Energy Dispersive Spectrometer (EDS) is able describe chemical structure. Based on an investigation carried out by Guru et al. [39], upon examination of the SEM images, it is observable that removal of volatile components led to samples' porosity. Such uniformity can be found in all asphalt samples SEM images in Fig. 6. The EDS outcomes highlight that sulphur was set as 0.513% for the base asphalt, 2.765% for VPP modified asphalt as well as 3.706% for TLPP modified asphalt. SEM images of S-MB and PS-MB samples, as well as of the PET additives were showed by Padhan et al. [46]. The SEM image of PS-MB highlights the fact that the PET additives were well dispersed in the asphalt binder matrix. Possible novel structures were found in the modified binder upon integrating S-MB with PET additives. The

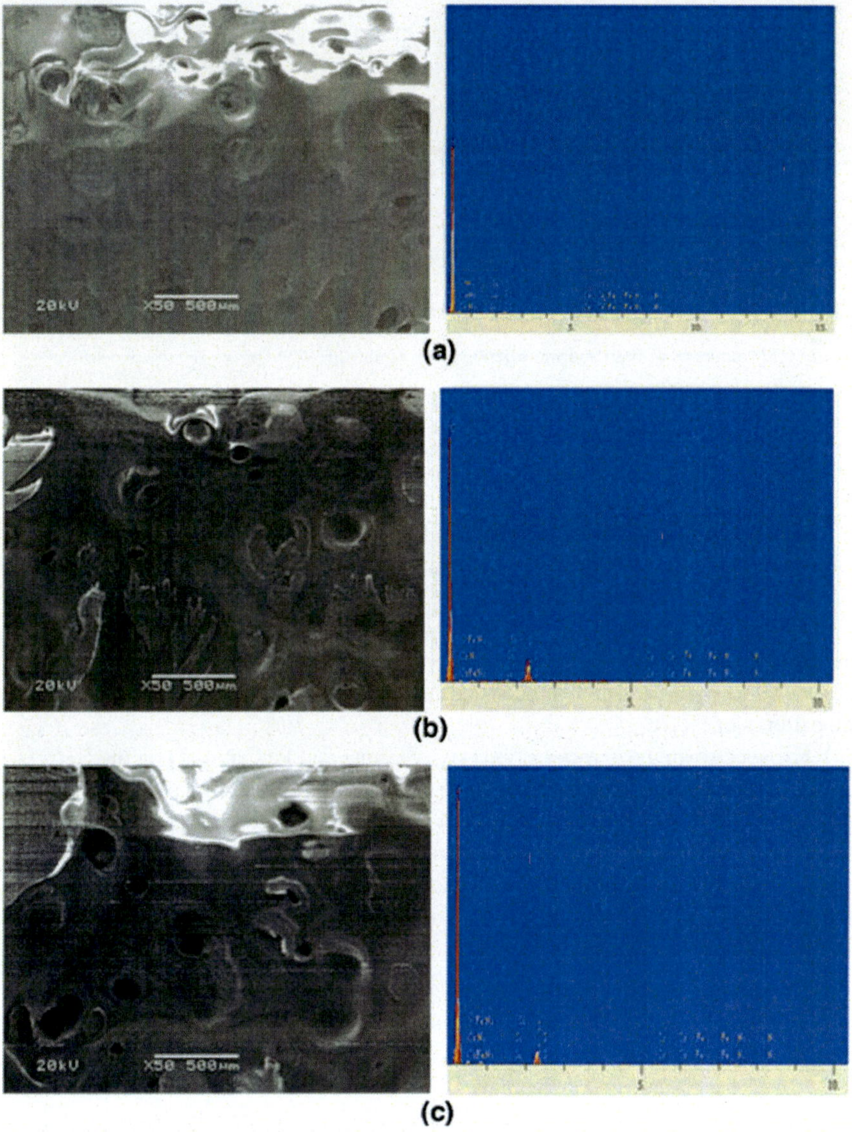

Fig. 6 EDS analyses and SEM images of asphalt samples; **a** base asphalt; **b** TLPP modified asphalt; **c** VPP modified asphalt [39]

white particles within the asphalt matrix which are the PET additives were glued to the S-MB binder in a uniform manner.

According to an investigation carried out by Guru et al. [39], Fig. 7 provides the modified asphalt samples and FTIR patterns of base binder. The patterns show that the modified samples had distinct PET additives bond structures (VPP, TLPP).

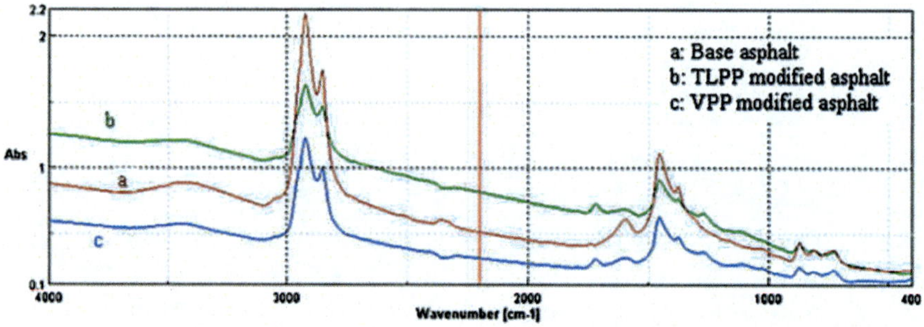

Fig. 7 FTIR patterns of base and modified asphalt samples [39]

Recordings of C=O bonds generated by the additives were apparent at 1800 cm^{-1}. As per subsection preparation of PET additive, this outcome supports the fact in which esterification occurs with glycolysis reaction existing C=O bonds. The FTIR pattern of base asphalt had the C=C double bond at 1600 cm^{-1}. Following modifications stage, bond cleavage occurred at 1600 cm^{-1} for the base asphalts while the alkenes was transformed into alkanes. As such, C=C double bond at 1600 cm^{-1} appeared unfit for the modified asphalt samples. Additive properties to the modified asphalts are reflected by the SO_2 bond at 1300 cm^{-1}, that were unavailable within the base asphalt FTIR pattern. Other bond structures appeared to be similar with respect to the patterns of FTIR.

FTIR spectra analysis was carried out by Gupta and Padhan [41] on the BHETA, BHETA-PU and base binder samples. The following properties of the modified binder are illustrated by the IR spectra obtained of BHETA. The FTIR spectra of the base bitumen, BHETA and BHETA-PU samples are shown in Figs. 8, 9 and 10 respectively. Characteristic absorption band between 3300 to 3446 cm^{-1} is shown by N–H

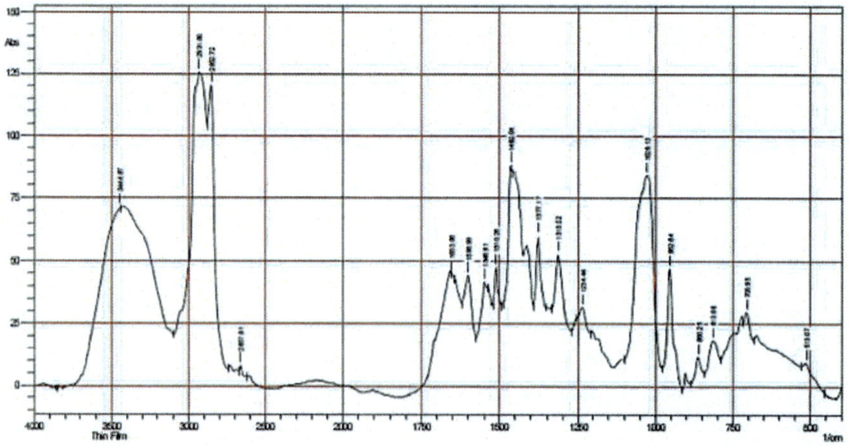

Fig. 8 IR spectra of BHETA [41]

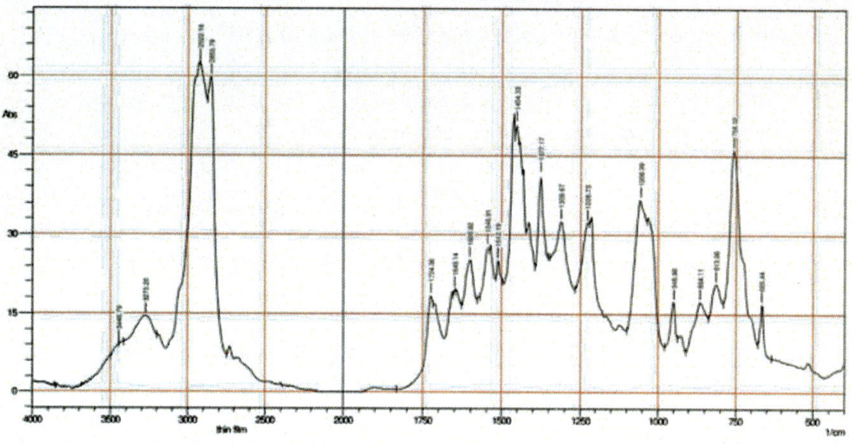

Fig. 9 IR spectra of BHETA-PU [41]

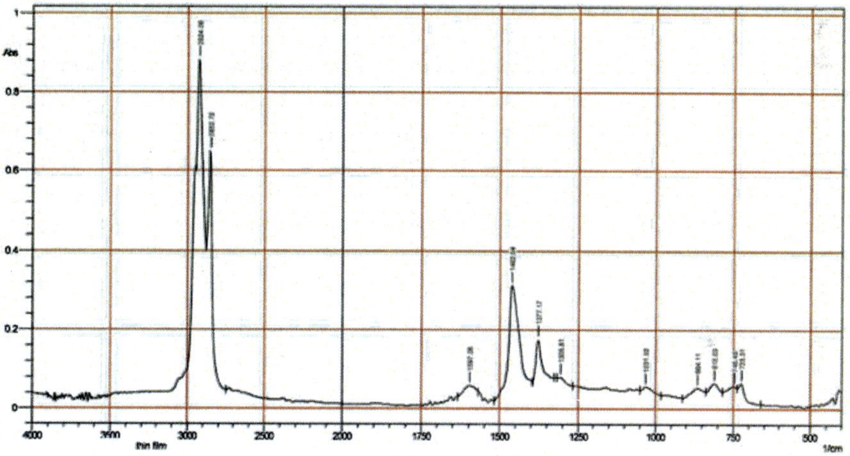

Fig. 10 IR spectra of base bitumen [41]

group involved in the hydrogen bond. The absorption bands of non-bonded N–H group are apparent at approximately 3446 cm^{-1}. Two carbonyl bands are projected by the usual FTIR bands of polyurethane linkage: one at 1724 cm^{-1} linked to free BHETA's CAO groups and another at 1711 cm^{-1} linked to bonded groups of CAO. The formation of polyurethane linkage is confirmed by the building of novel absorbance bands at 3346, 3300 and 3446 cm^{-1} for N–H group absorption bands and carbonyl bands at 1724 cm^{-1} as well as 1711 cm^{-1} seen within FTIR analysis of BHETA-PU, that is non-existent in both base binders and BHETA additives. The generated novel polymer is polyurethane, as confirmed by the FTIR study.

An investigation by Leng et al. [42] on the impact of BHETA and crumb rubber on the chemical nature of asphalt via FTIR studies. Outcomes prove that the bands seen at 1012 cm^{-1} and 1600 cm^{-1} are a result of the C-O_{str} as well as C-C_{str} stretches, respectively. At the same time, those at approximately 2850 cm^{-1} and 2919 cm^{-1} are attributed to Aliphatic C-H_{str} as well as Alkyl C-H_{str} stretches, respectively. CRMA binders possessing BHETA additives had distinct peaks at 1645, 3300 and 1550 cm^{-1}, in comparison with base CRMA binder. Clear shifting of such peaks may be attributed to chemical interaction between crumb rubber and BHETA.

4 Conclusion

The aim of the review is to summarize all literature and to promote the usage in the binder as modifier. It uncovers the possible utilization of PET in the binder to reduce the recyclability issue of PET and the constructing the sustainable pavements by utilization minimum crude oil for a higher performance property. Investigations in the past have shown that it is one of the many waste products that may be used in the pavement industry. The physical and rheological properties of PET modified binder have been enhanced and demonstrate greater properties to a certain degree. Along with the ductile properties of the modified binder, the penetration and resistance to temperature has been enhanced. As for the rheological properties, there is also a higher viscosity behaviour for the high addition of PET, which improved the performance grade of binder. The high and low temperature properties have improved due the addition of PET together with the other polymers. In a nutshell the consumption of the PET in the binder can results in many benefits to the construction of the highway pavements.

References

1. Geetha S, Narayanamoorthy S, Kureethara JV, Baleanu D, Kang D (2020) The hesitant Pythagorean fuzzy ELECTRE III: an adaptable recycling method for plastic materials. J Clean Prod 125281. https://doi.org/10.1016/j.jclepro.2020.125281
2. Vanapalli KR, Bhattacharya J, Samal B, Chandra S, Medha I, Dubey BK (2021) Optimized production of single-use plastic-Eucalyptus wood char composite for application in soil. J Clean Prod 278:123968. https://doi.org/10.1016/j.jclepro.2020.123968
3. Hahladakis JN, Velis CA, Weber R, Iacovidou E, Purnell P (2018) An overview of chemical additives present in plastics: migration, release, fate and environmental impact during their use, disposal and recycling. J Hazard Mater 344:179–199. https://doi.org/10.1016/j.jhazmat.2017.10.014
4. Yawo R, Mishra N, Ramudhin A (2020) Resources, conservation & recycling the role of technological innovation in plastic production within a circular economy framework. Resour Conserv Recycl 163:105094. https://doi.org/10.1016/j.resconrec.2020.105094
5. Gideon C, Faggio C (2019) Microplastics in the marine environment: current trends in environmental pollution and mechanisms of toxicological profile. 68:61–74

6. Hohn S, Acevedo-Trejos E, Abrams JF, Fulgencio de Moura J, Spranz R, Merico A (2020) The long-term legacy of plastic mass production. Sci Total Environ 746:141115. https://doi.org/10.1016/j.scitotenv.2020.141115.
7. Geyer R, Jambeck JR, Law KL (2017) Production, use, and fate of all plastics ever made. Sci Adv 3:25–29. https://doi.org/10.1126/sciadv.1700782
8. PlasticsEurope (2019) Plastics—The facts 2019. Anal Eur Plast Prod. Demand waste data
9. Muniandy R, Ismail DH, Hassim S (2018) Performance of recycled ceramic waste as aggregates in hot mix asphalt (HMA). J Mater Cycles Waste Manag 20:844–849. https://doi.org/10.1007/s10163-017-0645-x
10. Jambeck JR, Geyer R, Wilcox C, Siegler TR, Perryman M, Andrady A, Narayan R, Law KL (2015) Plastic waste inputs from land into the ocean. 347
11. Chae Y, An Y (2018) Current research trends on plastic pollution and ecological impacts on the soil ecosystem: a review*. Environ Pollut 240:387–395. https://doi.org/10.1016/j.envpol.2018.05.008
12. Xanthos D, Walker TR (2017) International policies to reduce plastic marine pollution from single-use plastics (plastic bags and microbeads): a review. Mar Pollut Bull 118:17–26. https://doi.org/10.1016/j.marpolbul.2017.02.048
13. Wagner TP (2017) Reducing single-use plastic shopping bags in the USA. Waste Manag 70:3–12. https://doi.org/10.1016/j.wasman.2017.09.003
14. Ansari AH, Jakarni FM, Muniandy R, Hassim S, Elahi Z (2021) Natural rubber as a renewable and sustainable bio-modifier for pavement applications: a review. J Clean Prod 289:125727. https://doi.org/10.1016/j.jclepro.2020.125727
15. Ali Omar H, Md Yusoff NI, Sajuri Z, Ceylan H, Jakarni FM, Ismail A (2015) Evaluating the effect of mixing process on nano-clay modified binders using the pull-off test method. Appl Mech Mater 802:357–362. https://doi.org/10.4028/www.scientific.net/amm.802.357
16. Park S, Thanakkasaranee S, Shin H, Ahn K, Sadeghi K, Lee Y, Tak G, Seo J (2020) Preparation and characterization of heat-resistant PET/bio-based polyester blends for hot-filled bottles. Polym Test 91:106823. https://doi.org/10.1016/j.polymertesting.2020.106823
17. Al-Sabagh AM, Yehia FZ, Eshaq G, Rabie AM, ElMetwally AE (2016) Greener routes for recycling of polyethylene terephthalate. Egypt J Pet 25:53–64. https://doi.org/10.1016/j.ejpe.2015.03.001
18. Zair MM, Ben, Jakarni FM, Muniandy R, Hassim S (2021) A brief review: application of recycled polyethylene terephthalate in asphalt pavement reinforcement. Sustain 13:1–17. https://doi.org/10.3390/su13031303
19. Awaja F, Pavel D (2005) Recycling of PET. Eur Polym J 41:1453–1477
20. Taniguchi I, Yoshida S, Hiraga K, Miyamoto K, Kimura Y, Oda K (2019) Biodegradation of PET: current status and application aspects. https://doi.org/10.1021/acscatal.8b05171
21. ACC American Chemistry Council, 2017 (2018): 2018 United States national postconsumer plastic bottle recycling report. ACC American Chemistry Council
22. Polymers RT (2017) Methods of recycling, properties and applications of recycled thermoplastic polymers. https://doi.org/10.3390/recycling2040024
23. Shen L, Nieuwlaar E, Worrell E, Patel MK (2011) Life cycle energy and GHG emissions of PET recycling: change-oriented effects. Int J Life Cycle Assess 16:522–536. https://doi.org/10.1007/s11367-011-0296-4
24. Shen L, Nieuwlaar E, Worrell E (2011) Life cycle energy and GHG emissions of PET recycling : change-oriented effects. 522–536. https://doi.org/10.1007/s11367-011-0296-4.
25. Lettieri P, Baeyens J (2009) Recycling and recovery routes of plastic solid waste (PSW): A review. 29:2625–2643. https://doi.org/10.1016/j.wasman.2009.06.004
26. Shukla SR, Harad AM, Jawale LS (2008) Recycling of waste PET into useful textile auxiliaries. 28:51–56. https://doi.org/10.1016/j.wasman.2006.11.002
27. Navarro R (2007) The influence of polyethylene in the mechanical recycling of polyethylene terephtalate. 5:110–116. https://doi.org/10.1016/j.jmatprotec.2007.04.126
28. Ameri M, Nasr D (2017) Performance properties of devulcanized waste PET modified asphalt mixtures. Pet Sci Technol 35:99–104

29. Moghaddam TB, Soltani M, Karim MR (2014) Stiffness modulus of polyethylene terephthalate modified asphalt mixture: a statistical analysis of the laboratory testing results. https://doi.org/10.1016/j.matdes.2014.11.044
30. Soltani M, Baghaee T, Rehan M, Baaj H (2015) Analysis of fatigue properties of unmodified and polyethylene terephthalate modified asphalt mixtures using response surface methodology. Eng Fail Anal 58:238–248. https://doi.org/10.1016/j.engfailanal.2015.09.005
31. Leng Z, Padhan RK, Sreeram A (2018) Production of a sustainable paving material through chemical recycling of waste PET into crumb rubber modified asphalt. J Clean Prod 180:682–688. https://doi.org/10.1016/j.jclepro.2018.01.171
32. Aziz MMBA, Muniandy R, Abdullah K, Mahmud AR, Khalid K, Ismail A (2010) Preliminary determination of asphalt properties using microwave techniques. J Eng Appl Sci 5:70–81
33. Moghaddam TB, Karim MR, Syammaun T (2012) Dynamic properties of stone mastic asphalt mixtures containing waste plastic bottles. Constr Build Mater 34:236–242. https://doi.org/10.1016/j.conbuildmat.2012.02.054
34. Ahmadinia E, Zargar M, Karim MR, Abdelaziz M, Shafigh P (2011) Using waste plastic bottles as additive for stone mastic asphalt. Mater Des 32:4844–4849
35. Rodrigues G, Carvalho MW, Costa DB (2015) Use of micronized polyethylene terephthalate (PET) waste in asphalt binder use of micronized polyethylene terephthalate (PET) waste in asphalt binder. 6466. https://doi.org/10.1080/10916466.2015.1079538
36. Read J, Whiteoak D (2003) The shell bitumen handbook. Thomas Telford
37. Tayh SA, Muniandy R, Hassim S, Jakarni F (2017) Aging and consistency characterization of bio-binders from domestic wastes. Int J Appl Eng Res 12:2613–2622
38. Garcia-Morales M, Partal P, Navarro FJ, Gallegos C, Garcı M (2006) Effect of waste polymer addition on the rheology of modified bitumen. 85:936–943. https://doi.org/10.1016/j.fuel.2005.09.015.
39. Gürü M, Çubuk MK, Arslan D, Farzanian SA, Bilici I (2014) An approach to the usage of polyethylene terephthalate (PET) waste as roadway pavement material. 279:302–310.
40. Mahrez A, Karim MR (2010) Rheological evaluation of bituminous binder modified with waste plastic material
41. Padhan RK, Gupta AA (2018) Preparation and evaluation of waste PET derived polyurethane polymer modified bitumen through in situ polymerization reaction. Constr Build Mater 158:337–345. https://doi.org/10.1016/j.conbuildmat.2017.09.147
42. Leng Z, Padhan RK, Sreeram A (2018) Production of a sustainable paving material through chemical recycling of waste PET into crumb rubber modified asphalt. J Clean Prod 180:682–688. https://doi.org/10.1016/j.jclepro.2018.01.171
43. Padhan RK, Mohanta C, Sreeram A, Gupta A (2018) Rheological evaluation of bitumen modified using antistripping additives synthesised from waste polyethylene terephthalate (PET). Int J Pavement Eng 1–9. https://doi.org/10.1080/10298436.2018.1519192
44. Leng Z, Sreeram A, Padhan RK, Tan Z (2018) Value-added application of waste PET based additives in bituminous mixtures containing high percentage of reclaimed asphalt pavement (RAP). J Clean Prod 196:615–625. https://doi.org/10.1016/j.jclepro.2018.06.119
45. Sreeram A, Leng Z, Padhan RK, Qu X (2019) Eco-friendly paving materials using waste PET and reclaimed asphalt pavement. 3733. https://doi.org/10.1080/1023697X.2018.1534617
46. Padhan RK, Leng Z, Sreeram A, Xu X (2020) Compound modification of asphalt with styrene-butadiene-styrene and waste polyethylene terephthalate functionalized additives. J Clean Prod 277:124286. https://doi.org/10.1016/j.jclepro.2020.124286
47. Ameli A, Maher J, Mosavi A, Nabipour N, Babagoli R, Norouzi N (2020) Performance evaluation of binders and Stone Matrix Asphalt (SMA) mixtures modified by Ground Tire Rubber (GTR), waste Polyethylene Terephthalate (PET) and Anti Stripping Agents (ASAs). Constr Build Mater 251:118932. https://doi.org/10.1016/j.conbuildmat.2020.118932
48. Abuaddous M, Taamneh MM, Rabab SR (2020) The potential use of recycled polyethylene terephthalate (RPET) plastic waste in asphalt binder
49. Karahrodi MH, Jazani OM, Paran SMR, Formela K, Saeb MR (2017) Modification of thermal and rheological characteristics of bitumen by waste PET/GTR blends. Constr Build Mater 134:157–166. https://doi.org/10.1016/j.conbuildmat.2016.12.134

A Review on Utilization of C & D Waste for Road and Pavement Construction in the International Context and Applicability of the Findings to Sri Lanka

K. P. Arandara, S. T. Priyadarshana, G. N. Paranavithana, and R. Dissanayake

Abstract In this paper, the utilization of C & D (Construction and Demolition) waste for roads and pavements construction is reviewed in the international context. Further, the results are assessed to see the relevance to Sri Lanka. The relevant references gathered from the databases are referred to this study. The different countries and regions including Asia, Australia, Middle-East, Europe, and the USA, are covered with 67 papers. Those references are critically reviewed by applying PRISMA (Preferred Reporting Items for Systematic Reviews and Meta-Analysis) methodology and key C & D waste types utilized for roads and pavements construction are found out. The paper discusses Recycled Concrete Aggregates (RCA), Crushed Clay Bricks (CCB), and Recycled Asphalt Pavement (RAP) as the C & D materials that commonly are used for roads and pavements construction. Also, policies, laws, regulations, and procedures adopted in multiple countries, applicable to the C & D waste sector are discussed in the paper. The findings are evaluated in the latter part of the paper and proposed applicability/rationales to Sri Lanka are examined simultaneously. The authors conclude that the specified three types of CDW based recycled materials could be utilized across the cross-section of roads and pavements through effective methods and applications in several counties and the same approach is applicable in Sri Lanka also. This study confirms that very significant policy support in terms utilization of CDW is evident in most countries and the particular learnings could be applied in Sri Lanka to improve the current state appropriately.

K. P. Arandara (✉) · S. T. Priyadarshana · G. N. Paranavithana
Department of Civil Engineering, Open University of Sri Lanka, Colombo, Sri Lanka
e-mail: kokila@leanandgreensolutions.com

S. T. Priyadarshana
e-mail: mspri@ou.ac.lk

G. N. Paranavithana
e-mail: gnpar@ou.ac.lk

R. Dissanayake
Department of Civil Engineering, University of Peradeniya, Peradeniya, Sri Lanka
e-mail: ranjith@eng.pdn.ac.lk

© The Author(s), under exclusive license to Springer Nature Switzerland AG 2022
H. R. Pasindu et al. (eds.), *Road and Airfield Pavement Technology*, Lecture Notes in Civil Engineering 193, https://doi.org/10.1007/978-3-030-87379-0_57

Keywords C & D waste · Recycled concrete aggregate · Crushed clay aggregates · Recycled asphalt pavement · C & D waste policies · C & D waste applications

1 Introduction

As per the Oxford Advanced Learner's Dictionary, one of the different definitions given for the word "waste" is "materials that are no longer needed and are thrown away" [17]. Even though there is no single definition for "Construction and Demolition Waste" (C & D Waste or CDW), there are several definitions suggested by several researchers and organizations. The CDW materials consisting of the debris generated from the construction, renovation, and demolition of buildings, roads, and bridges [55]. Also, CDW is one of the heaviest and most voluminous of wastes generated by the construction sector [21]. Further, it has been recorded that CDW represents one of the most significant components of the solid waste streams in the United States [57]. More importantly, around 35% of all global waste is accounted by CDW. Also, the percentages of CDW generation vary according to the country and the region. For example, the portions generated in Spain, United Kingdom, Australia, Japan and Italy are 70%, 50%, 44%, 36%, and 30% of the total waste respectively [6]. Similarly, CDW places massive pressure and negatively impacts human living environments [66]. Because of land constraints, the urban and local governments face challenges due to CDW [53]. Also, very recently, ICSBE, Sri Lanka reported that the estimated total volume of CDW within the Western Province of Sri Lanka in 2019 is 124, 517 m^3 [9].

The above statistics indicate that CDW is a vital topic and therefore, different solutions are tested and practiced by several countries to control the impact of CDW. It has been recorded that reusing, recycling, and revalorizing CDW can conserve natural resources and reduce the volume of waste disposal in landfills [49]. The recycled CDW products are more diversified, mainly counting recycled aggregate, recycled concrete, recycled bricks for several applications including recycled road base materials [1]. Likewise, there are evidences for the processing of CDW to produce them as raw materials for road construction.

In the Sri Lankan context, the supply of fine and coarse aggregates becomes a most significant problem due to the natural resource limitations. Also, the embodied energy generated due to distance and transportation matters are highly substantial [7]. Also, in general, the rural roads and pavements development projects could be received less attention with respect to highway constructions where consume a high volume of building materials including sand and natural aggregates. Therefore, this literature review intends to realize the utilization of CDW for roads and pavements construction in the international context and to discuss the applicability of the findings to Sri Lanka.

2 Aim

This review aims to identify the utilization of C & D waste for roads and pavements construction in the international context and to discuss the applicability of the findings to Sri Lanka.

3 Objectives

There are three objectives of this review and those objectives are;

1. Characterization in the utilization of the C & D waste materials for road and pavement constructions in the identified countries
2. To understand the rationality for C & D waste-related policies that have been implemented in specified countries and states
3. To discuss the applicability of the findings to Sri Lanka.

4 Methodology

The Preferred Reporting Items for Systematic Reviews and Meta-Analysis (PRISMA) method was adopted in this research. The research question considered for PRISMA is "How the C & D wastes are utilized for road and pavement construction in the international context and how the findings can be applied to Sri Lanka?" Figure 1 shows the PRISMA flow chart which describes the criteria that are considered to choose the most relevant research papers/publications for this literature review.

The relevant references published in publications with Science Citation Index (SCI) are identified and gathered from the databases including WOS (Web of Science), PUBMED, Sciencedirect, and from other open-sources (Google Scholar and Researchgate). Those gathered publications are filtered to exclude publications related to duplications, book chapters, and editorial letters. As the third step, as shown in Fig. 1, eligibility is addressed on meta-data and key words while proceeding with abstract reading and body skim reading. Finally, those eligible publications are included in this review paper.

5 Findings in International Context

Based on the above methodology, finally authors referred to 67 publications and the respective findings are given below. In par with the review's first objective, the identified types and methods of applications of C & D Waste Materials for road and pavement constructions in different countries are summarized in this section.

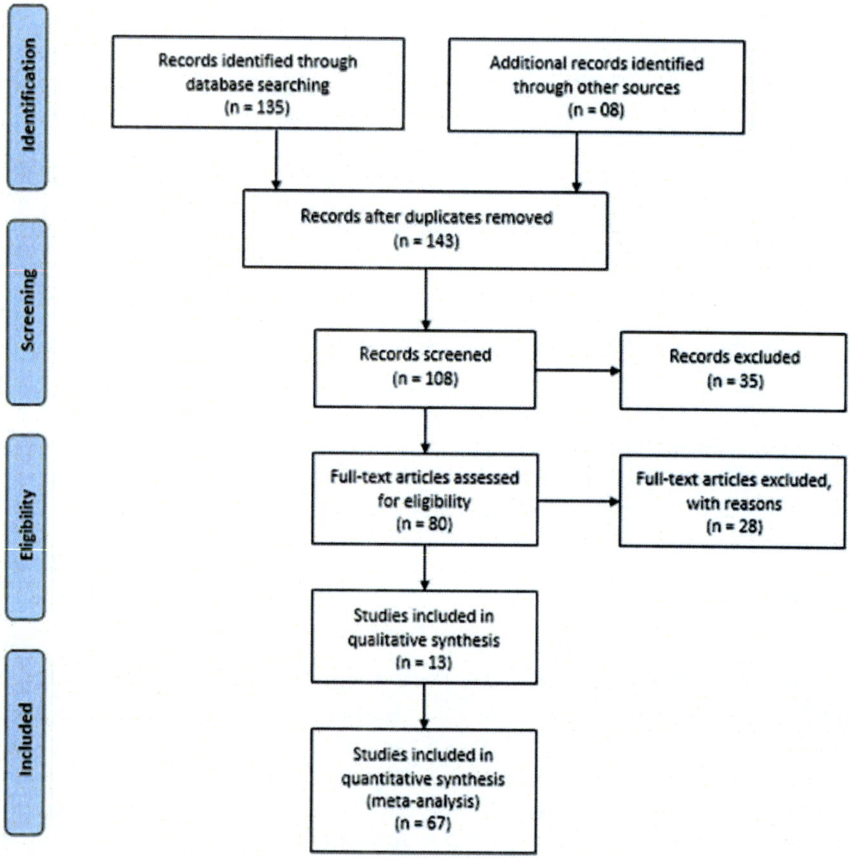

Fig. 1 The PRISMA flow chart. *Source* Modified from Pati and Lesa [46]

5.1 Different Types of Materials

The materials are generated in varying amounts based on factors [57] and [5] say that the classification of C & D waste varies with territorial contexts. More importantly, the recycling of demolition waste used for roads could be a sustainable solution [48]. The use of C & D waste in pavement structure may range from subgrade to road surface. These wastes may also be used in unpaved rural roads. Comparatively, concrete wastes, crushed clay bricks, and recycled asphalt pavement are the three critical types of C & D waste [22]. In the CDW recycling point of view also, the above three CDW types are highlighted [1] and the 60–70% (by weight) of CDW is composed of concrete and masonry [15].

Table 1 recaps the details of the said three significant types of C & D waste materials and it indicates the primary CDW materials, the secondary materials produced from primary materials, standard abbreviations used for them, and the evidence for the facts.

Table 1 CDW material types utilized to roads and pavements

No.	C & D waste type (Primary)	Name of produced material (Secondary)	Evidence (References)
1	C & D concrete waste	Recycled aggregates (RA)	(1) Naik and Moriconi [41] (2) Wagih et al. [61] (3) Kox et al. [27] (4) Wijayasundara [62] (5) Tuladhar et al. [58] (6) Courard et al. [8]
		Recycled concrete aggregate (RCA)	
		Recycled concrete fine aggregates (RCFA)	
		Recycled concrete coarse aggregates (RCCA)	
2	C & D Bricks	Crushed clay bricks (CCB)	(1) Apebo et al. [2] (2) Otoko [45] (3) Arisha et al. [3] (4) Liang et al. [29]
		Waste clay brick (WCB)	
		Recycled clay masonry (RCM)	
		Recycled brick aggregates (RBA)	
		Crushed clay bricks fine aggregate (CCBFA)	
		Crushed clay bricks coarse aggregate (RCBCA)	
3	C & D Asphalt	Recycled asphalt (RA)	(1) US Department of Transportation [60] (2) Nordic Road Forum [43] (3) Zaumanisa and Mallick [67] (4) Sharma et al. [52] (5) Yadava and Ahamad [64] (6) Mrugacz [39] (7) Turkvilma [59] (8) Limitless Paving and Concrete [32] (9) Kemper [26] (10) Edge Environment Pty Ltd. [31]
		Recycled asphalt pavement (RAP)	
		Reclaimed Asphalt pavement	

5.2 *Methods of Applications of the Materials*

The methods and applications pertaining to the appropriate secondary materials and the references are presented in Table 2.

Table 2 Methods and applications of the CDW materials pertain to roads and pavements

No.	Name of produced material (Secondary)	Methods and application	Evidence (Reference)
1	Recycled aggregates (RA)	To produce a subbase with a soaked CBR value of at least 30%	Poon and Chan [47]
2	Recycled concrete aggregate (RCA)	1. As a base-course material for road construction 2. For sub-bases of the roads under specified condition 3. As an adhesive to asphalt in order to enhance the properties of asphalt mixture	(a) Richardson and Jordon [50] (b) Jayakody et al. [20] (c) Hou et al. [18]
3	Recycled concrete fine aggregates (RCFA)	To increase the strength of subgrade soil	Ho et al. [23]
4	Crushed clay bricks (CCB)	Cement stabilized sub-base for blending with soil for unbound systems such like embankments and fillings	(a) Liang et al. [29] (b) Mueller and Stark [40]
5	Crushed clay bricks fine aggregate (CCBFA)	As fine aggregate for concrete and especially, for low weight concrete and under specified conditions for concrete with higher compressive strength	Apebo et al. [2]
6	Recycled asphalt (RA)	To utilize differently through different processes	(a) Kandhal and Mallick [24] (b) Karlsson and Isacsson [25] (c) Zaumanis and Mallick [67]
7	Recycled asphalt pavement (RAP)	To use in granular sub base course, granular or stabilized base course, bituminous base course or binder course and also in Wearing or surface course	Yadava and Ahmad [64]
8	Reclaimed asphalt pavement	To produce a recycled hot mix	US Department of Transportation [60]

6 Policy Support in the International Context

In general, a waste management policy provides a national framework for waste and resource recovery in a country or territory. It outlines roles and responsibilities for collective action by businesses, governments, communities, and individuals. It is noticed that C & D waste management policies follow rationalities and alterations in different countries and regions including Asia, Australia, Middle-East, Europe and the USA.

6.1 Rationale for Policies

Based on the "Polluter Pays Principle (PPP)", some governments have implemented a waste charging scheme as a significant economic policy to impose a levy to the direct polluters who should be responsible for C & D waste disposal [63]. There are laws that cover solid waste management-related provisions in China [1]. AECOM further states that in National and Local governmental levels in China, respective Authorities and Ministries work hand-in-hand to align with C & D policies. Hong Kong has been committed to formulate a construction waste management policy by applying the latest principles (e.g., 3R principles and polluter pays principle) [33].

The Environmental Code of the Republic of Kazakhstan is the primary directive that regulates waste management-including construction and demolition waste management in Kazakhstan [59]. The National Waste Policy that was endorsed by Australian environment ministers in November 2009 is a collaborative approach that aims to avoid the generation of waste, reduce the amount of waste for disposal, manage waste as a resource and ensure that waste treatment, disposal, recovery and re-use is undertaken in a safe, scientific and environmentally-sound manner. In addition, waste management and resource recovery in Australia is dependent on the regulatory framework of the states and territories [10].

Also, Abu Dhabi has taken several policy initiatives and for examples, Abu Dhabi Environment Policy Agenda (ADEPA) and Abu Dhabi Waste Management Master Plan 2040 (ADWMMP—2040) [16]. Also, UAE has developed advanced systems to meet the UAE goal of sustainability in the UAE building sector [35].

Europe has developed specific policies or regulations regarding the management of C & D waste [51]. Also, West Framework Directive of the EC follows the member states by 2020 to re-use and/or recycle C & D waste to the degree of 70% [13]. Further, a detailed policy and framework conditions about C & D waste are described in the ECU C & D Waste Protocol [11]. The State and county regulatory agencies have passed legislation that puts strict controls on C & D waste disposal practice, illegal dumping and ground water protection [34]. The Disposal Bans and Mandatory Recycling was declared in the United States [42].

6.2 Alterations of Policies

The policies to be reviewed, updated, and altered time to time to gain the effectiveness. The first CDW management-related national policy document was issued in May 1995 in China, and this is called the "city appearance and environmental sanitation management regulations". It defines the management actions for urban construction site materials and equipment. Following that, several related policies were issued by the Ministry of Housing and Urban–Rural Development and other governing bodies. Accordingly, significant alterations are evident in several years, including the year 2003, 2005, 2008, 2009, 2011 and 2014 [19]. Bao et al. [5] emphasize

Table 3 C & D concrete and brick waste generation in 2019 Western province, Sri Lanka

Source [9]			Figures calculated by the authors
No.	C & D waste type	Waste generation in 2019 (m^3)	Percentage (%) out of total C & D waste generation in Western province (13,089.29 m^3)
1	Concrete	3029.63	22.16%
2	Bricks	1021.43	7.47%
3	Asphalt	Numbers not found	NA

some improvements and they indicate that providing more government support is an indispensable facilitator for promoting on-site recycling in Hong Kong. A team of researchers suggests some changes and alterations to the existing C & D waste policies and regulations in Australia, to ensure that any relevant policies will achieve maximum consensus among stakeholders [54].

In UAE, the government had to produce smart policies [35]. The construction of new facilities for recycling or disposal is enormously contentious, fuelling ongoing battles between waste exporting and waste importing states. This particular issue for New York City, which no longer has any disposal facilities and must export all the waste that does not recycle [48].

7 Applicability of the Findings to Sri Lanka

The findings' applicability and availability are discussed and summarized in the Tables 3, 4 and 5.

8 Conclusions

It is identified that the C & D wastes are reprocessed and utilized in different ways for road and pavement constructions throughout the world. It is concluded that there are three specified types of CDW that are used for roads and pavements and those types are concrete wastes, crushed clay bricks, and recycled asphalt pavement. Further, it is found that there are effective methods and applications associated with recycled CDW materials under the above three significant types of CDW for road and pavement construction. Accordingly, the specified CDW based reprocessed materials could be utilized across the cross-section of roads and pavements. The majority of those methods and applications could be considered and can be applied in Sri Lanka. Further, it is confirmed that the policy support is very much significant in most countries in terms of utilization of CDW. Those policies and rules are reviewed and altered to suit to cater present demands. In Sri Lankan point of view, the experience of

Table 4 C & D waste types and proposed applicability to Sri Lanka

No.	C & D waste type	Name of secondary material/product	Application of secondary material/product for roads and pavement	Proposed applicability to Sri Lanka (rationalities)
1	C & D Concrete Waste	Recycled concrete fine aggregates (RCFA)	To replace natural fine aggregates (sand) in mass-concrete production for concrete roads and pavements	(a) Significant C & D concrete waste generation is evident (b) The RCA Centre (COWAM Centre in Galle) are readily available (c) Sri Lanka is in a crisis of Sand (d) This would be a sustainable alternative
			To improve subgrade	The above (a), (b) & (e) Natural soil would be remained
			To blend with crushed clay bricks to produce subbase	The above (a), (b) & (e)
		Recycled concrete coarse aggregates (RCCA)	To replace natural coarse aggregates (granite) in mass-concrete production for concrete roads and pavements	The above (a), (b), (d) & (f) Natural granite aggregate sources are limited to Sri Lanka (g) The limited high quality natural granite to be conserved to future major projects
			To use as an adhesive to asphalt to enhance the properties of asphalt mixture	The above (a), (b), (d) & (h) Significant C & D asphalt waste could be evident (numbers not found) (i) This could be a leading alternative to reduce consumption of virgin asphalt
2	C & D bricks	Crushed clay bricks fine aggregate	To replace river sand in mass-concrete for roads	The above (a), (b), (c) & (d)

(continued)

Table 4 (continued)

No.	C & D waste type	Name of secondary material/product	Application of secondary material/product for roads and pavement	Proposed applicability to Sri Lanka (rationalities)
		Crushed clay bricks coarse aggregate	To apply in cement stabilized sub-base to construct roads and pavements	The above (a), (b) & (e)
			To use in unbound system like embankments and fills related to roads and pavements	The above (a), (b) & (e)
3	C & D asphalt	Recycled asphalt pavement (reclaimed asphalt pavement) (RAP)	To blend with virgin aggregates to use in any sub-layer from subgrade to Surface course of roads	The above (a), (b), (e) & (h)

such policy support is minimal and the remaining policies could be enhanced further by referring to the findings in the international context. Finally, this paper concludes that there is a room for improvements on the results of findings to carry out further research on the same topic but with different key words like C & D waste-related society, economics, and the environment.

Table 5 C & D waste-related policy support in the international context and availability of them in Sri Lanka

No.	Referred Country or region	Policy/law/regulation/strategy	Availability in Sri Lanka (rationalities)
1	China	Environmental protection law	Available
2		Cleaner production promotion law	**Evidence Not Found**
3		Solid waste pollution prevention law	**Evidence Not Found**
4		Circular economy promotion law	**Evidence Not Found**
5		Building law	Available
6		Ministry of housing and urban–rural development	Available
7		National development and reform commission (NDRC)	**Evidence Not Found**
8		Ministry of industry and information technology	Available

(continued)

Table 5 (continued)

No.	Referred Country or region	Policy/law/regulation/strategy	Availability in Sri Lanka (rationalities)
9		Ministry of science and technology	Available
10		Ministry of finance	Available
11		Ministry of state administration of taxation	Available
12	Hong Kong	waste management policy by applying the latest principles (e.g., 3R principles and polluter pays principle)	**Evidence Not Found**
13	Australia	Country-wide construction waste management practices	**Evidence Not Found**
14		National waste policy	**Evidence Not Found**
15	UAE	TADWEER for municipal solid, commercial, medical, agricultural, industrial, construction and demolition waste, ADNOC for oil and gas waste and FANR for radioactive waste	Not very much applicable
16	Netherlands	Forbidden the disposal in landfills of the waste that can be incinerated	**Evidence Not found**
17	Germany	Forbidden the disposal in landfills of the waste that can be recycled	**Evidence Not Found**
18	Denmark	Voluntary agreement with good practices and environmental management systems for demolition activities	**Evidence Not Found**
19	Europe	The construction industry to be obligated by law to select and recover pre-defined types of C & D waste if the quantity generated exceeds a certain amount of waste	**Evidence Not Found**
20		West framework directive of the EC which requires that the member states by 2020 should re-use and/or recycle C & D waste to the degree of 70%	**Evidence Not Found**
22	The USA	Disposal bans & mandatory recycling	**Evidence Not Found**

Acknowledgements The authors would like to thank the previous authors/researchers who researched on construction and demolition waste. Also, researchers would show gratitude to the ICPT 2021 Conference and its organizing committee for their corporation to take this publication to the proceedings. Finally, the authors would acknowledge everyone those who extended their helping hands in various ways to make this publication a success.

References

1. AECOM (2018) People's Republic of China: construction and demolition waste management and recycling. Ministry of Housing and Urban-Rural Developments, ADB, Shatin, HK, China
2. Apebo NS, Agunwamba JC, Ezeokonkwo JC (2014) The suitability of crushed over burnt bricks as coarse aggregates for concrete. Int J Eng Sci Innov Technol (IJESIT) 3(1):315–321
3. Arisha A, Gabr A, Badawy SE, Shwally S (2016) Using blends of construction & demolition waste materials and recycled clay masonry brick in pavement. Procedia Eng 143:1317–1324
4. Banerjee C (2015) C & D waste processing in India—Delhi shows the way. Infra Construction and Equipment Magazine, India, Delhi
5. Bao Z, Lee WMW, Lu W (2020) Implementing on-site construction waste recycling in Hong Kong: barriers and facilitators. Science of the Total Environment
6. Chen J, Su Y, Si H, Chen J (2018) Managerial areas of construction and demolition waste: a scientometric review. Int J Environ Res Public Health 15(11)
7. Chintha J (2011) Embodied energy of alternative building materials and their impact on life cycle cost parameters, s.l.: Researchgate
8. Courard L, Rondeux M, Zhao Z, Michel F (2020) Use of recycled fine aggregates from C & DW for unbound road sub-base. Materials-MDPI 13
9. Dissanayake R, Paranavithana G, Gajanayake P, Indula P (2020) Construction demolitions waste generation survey in Western Province, Sri Lanka, Colombo. International Center for Sustainable Built Environment, Sri Lanka
10. Edge Environment Pty Ltd. (2012) Construction and demolition waste guide—recycling & re-use across the supply chain. Commonwealth of Australia, Australia
11. European Commission (2016) EU construction & demolition waste management protocol. Entrepreneurship and SMEs, European Commission, Belgium
12. FCM (2005) Reuse and recycling of road construction and maintenance materials. Federation of Canadian Municipalities and National Research Council, Ontario
13. Ganguly P (2012) Construction and demolition waste handling in the EU. Littera Scripta 5(2):205–217
14. Gary Liss & Associates (2002) C & D recycling plans and policies: a model for local government recycling and waste reduction. California 916:341–6300: The California Integrated Waste Management Board
15. Giorgi S, Lavagna M, Campioli A (2018) Guidelines for effective and sustainable recycling of construction and demolition waste. ABC Department, Politecnico Di Milano, Milan, Italy
16. Hittini BY, Shineika AI (2019) Constrution waste management in UAE: an exploratory study. WIT Trans Ecol Environ 238:679–686
17. Hornby AS (2000) Oxford advanced learner's dictionary of current English. Oxford University Press, Chennai
18. Hou Y, Ji X, Li J, Li X (2018) Adhesion between Asphalt and recycled concrete aggregate and its impact on the properties of asphalt mixture. J Mater MDPI 11
19. Huanga B et al (2018) Construction and demolition waste management in China through the 3R principle. J Resourc Conserv Recycl 129:36–44
20. Jayakody S, Gallage C, Kumar A (2012) Assessment of recycled concrete aggregates for road base and sub-base. The GEOMATE International Society, Kuala Lumpur

21. Jayakody S, Zimar AMZ, Ranweera RALM (2018) Potential use of recycled construction and demolition waste aggregates for non-structural concrete applications. Natl Sci Found Sri Lanka 46(2):205–216
22. Jia X, Ye F, Huang B (2015) Utilization of construction and demolition wastes in low-volume roads for rural areas in China. J Transp Res Board, pp 39–47
23. Ho HJ, Iizuka A, Shiba E (2020) Chemical recycling and use of various types of concrete waste: a review. J Cleaner Prod
24. Kandhal PM, Mallick RB (1997) Pavement recycling guidelines for state and local governments-participant's reference book. National Center for Asphalt Technology, Auburn. Report Number FHWA-SA-98-042
25. Karlsson R, Isacsson R (2003) Investigations on bitumen rejuvenator diffusion and structural stability. J Assoc Asphalt Paving Technol 72:463–501
26. Kemper Equipment (2020) Kemper equipment (Online). Available at: https://www.kemperequipment.com/blog/benefits-of-recycled-asphalt-product/. Accessed 18 Nov 2020
27. Kox S et al (2019) Experimental evaluation of the high-grade properties of recycled concrete aggregates and their application in concrete road payment construction. Case Stud Construct Mater 11
28. Kumara W, Ekanayake L (2009) https://core.ac.uk/. (Online). Available at: https://core.ac.uk/download/pdf/141731567.pdf/. Accessed 17 Sep 2020
29. Liang C et al (2019) Potential activity of recycled clay brick in cement stabilized subbase. Appl Sci 9 (5208)
30. Liang C et al (2019) Potential activity of recycled clay brick in cement stabilized subbase. J Appl Sci MDPI, 9
31. Lim AJ, Cao Y, Costa DD-d (2020) Recycled materials in roads and pavements, a guide for local councils. Local Government NSW and Waste Transformation Research Hub and School of Chemical and Biomolecular Engineering, NSW
32. Limitless Paving & Concrete (2019) Limitless paving & concrete (Online). Available at: https://limitlesspavingandconcrete.com/how-to-recycle-and-reuse-asphalt/. Accessed 20 Nov 2020
33. Lu W, Tam VWY (2013) Construction waste management policies and their effectiveness in Hong Kong: a longitudinal review. Renew Sustain Energy Rev 23:214–223
34. Macozoma DS (2001) Review of the US secondary construction materials market: any lessons for South Africa and other developing countries? International Council for Research and Innovation in, Florida
35. Mawed M, Nuaimi MSA, Kashawni G (2020) Construction and demolition waste management in the UAE: application and obstacles. Int J GEOMATE 18(70):235–245
36. Mishra B (2015) A study on use of reclaimed asphalt pavement (RAP) materials in flexible pavement. Int J Innov Res Sci Eng Technol 4(12):12170–12177
37. Misra S (n.d) Business opportunities in C & D waste handling. Department of Civil Engineering, Indian Institute of Technology Kanpur, Kanpur
38. Mithushan S, Nishanthan JA, Niwunhella AGMA (2017) Effect of partial replacement of conventional ingredients in concrete with alternative materials. Matara, Sri Lanka, s.n.
39. Mrugacz J (2019) Wolf paving (Online). Available at: https://www.wolfpaving.com/blog/recycled-asphalt-learn-the-benefits-and-process-of-recycling-pavement/. Accessed 18 Nov 2020
40. Mueller HA, Stark U (2015) Recycling of clay Brick Debris. University Weimar, Faculty of Civil Engineering, Weimar
41. Naik TR, Moriconi G (2005) Environmental-friendly durable concrete made with recycled materials for sustainable concrete construction. J Struct Eng 61(01):237–244
42. NERC (2012) Summary of U.S. Sate and Municipal C & D regulations and requirements. VT 05301: Northeast Recycling Council, Inc. (NERC)
43. NVF (2012) Sustainable asphalt. Nordic Road Forum, Finland
44. NYC DCC (2003) Construction and demolition waste manual. NYC Department of Design & Construction, NYC
45. Otoko GR (2014) Use of crushed clay bricks aggregate in concrete. Int J Eng Technol Res (Online) 2(4):1–9

46. Pati D, Lesa N (2018) How to write a systematic review of the literature. Health Environ Res Des J 11(1):15–30
47. Poon CS, Chan D (2006) Feasible use of recycled concrete aggregates and crushed clay brick as unbound road sub-base. Constr Build Mater 20:578–585
48. Rahman MT, Mohajerani A, Giustozzi F (2020) Recycling of waste materials for asphalt concrete and bitumen: a review. Mater Mol Divers Preserv Int 13
49. Rey ID et al (2016) Feasibility of using unbound mixed recycled aggregates from CDW over expansive clay subgrade in unpaved rural roads. Materials MDPI 9(931)
50. Richardson BJE, Jordan DO (1994) Use of recycled concrete as a road pavement material within Australia. Queensland, Australian Road Research Board
51. Sáez PV, Merino MDR, Amores CP, Antonio ADS (2011) European legislation and implementation measures in the management of construction and demolition waste. Open Constr Build Technol J 5:156–161
52. Sharma U, Giri HK, Kathri A (2018) Use of recycled asphalt material for sustainable road construction. Indian Highways, pp 29–35
53. Sheth JT, Patel K, Devkar GA (2016) Analysis of construction and demolition waste management policies from sustainability perspectives. Ahmedabad, s.n.
54. Shooshtarian S et al (2020) Landfill levy imposition on construction and demolition waste: Australian stakeholders' perceptions. J Sustain MDPI 12
55. Shrivastava S, Chini A (2011) Construction materials and C & D waste in India. School of Building Construction, University of Florida, USA
56. Storey J (2008) Construction materials stewardship, the status quo in selected countries. Centre for Building Performance Research, Victoria University of Wellington, Wellington
57. Townsend T (2017) Benefits of construction and demolition debris recycling in the United State, Chicago, USA: www.cdrecycling.org
58. Tuladhar R, Marshall A, Sivakugan N (2020) Use of recycled concrete aggregate for pavement construction. Advances in construction and demolition waste recycling. Woodhead Publishing, Queensland, pp 181–197
59. Turkyilmaz A et al (2019) A comprehensive construction and demolition waste management model using PESTEL and 3R for construction companies operating in Central Asia. Sustain Mol Divers Preserv Int 11
60. US Department of Transportation (2011) Statistical analysis of performance of recycled hot mix asphalt overlays in flexible pavement rehabilitation. VA 22101-2296: US Department of Transportation
61. Wagih AM, Karmoty HZE, Ebid M, Okba SH (2012) Recycled construction and demolition concrete waste as aggregate for structural concrete. HBRC J Hous Build Natl Res Center 9:193–200
62. Wijayasundara M (2019) Recycled concrete. Elsevier Inc., Melbourne
63. Wing V, Yan Tam Y, Lu W (2016) Construction waste management profiles, practices, and performance: a cross-jurisdictional analysis in four countries. Sustainability 8:190
64. Yadava AK, Ahmad SA (2019) Characterization of Indian reclaimed asphalt pavement (rap) on the basis of sources of rap and gradation characteristics of rap aggregates. Int J Innov Technol Explor Eng (IJITEE) 8(7):2402–2408
65. Yang Y, Zhao Y (2015) Referring sustainable waste management strategy drives life-cycle sustainable building benefits of China. Atlantis Press, Kunming
66. Yuan H (2017) Barriers and countermeasures for managing construction and demolition waste: a case of Shenzhen. J Clean Prod 157:84–93
67. Zaumanis M, Mallick RB (2015) Review of very high-content reclaimed asphalt use in plant-produced pavements: state of the art. Int J Pavement Eng 16(1):39–55

Carbon Footprint Assessment of Steel Slag Asphalt Pavement in Singapore

See Soo Loi, Swarna Kamala Subramaniam, and Hao Xusheng

Abstract Electric Arc Furnace (EAF) Steel Slag is a by-product generated from the steel-making process which is being used as aggregates for the road construction wearing course. Replacement of naturally occurring aggregates with locally available EAF steel slag as a road construction material would promote sustainability of road construction, reduce carbon footprint and minimise the environmental impact of the steel making industry. Since 1994, 100% of steel slag generated in Singapore has been fully recycled into aggregates and used in the asphalt wearing course on roads by LTA and PUB and within the ports by PSA. Climate change is one of the greatest challenges of our time and a major threat to the livelihoods of living creatures on earth. Greenhouse gasses generated from human activities is no doubt the biggest contributor of climate change observed since the mid-twentieth century. Carbon footprint is one of the most widely used tools to quantify greenhouse gases for assessing the environmental impacts of a production activity. The amount of carbon dioxide and other greenhouse gases associated with a given product can be calculated and expressed as CO_2 equivalent. This paper briefs the applications of steel slag aggregate for asphalt pavement in Singapore and compares the carbon footprint of locally produced steel slag aggregates with imported granite aggregates. The environmental benefits of using steel slag have been assessed through the comparison of embodied CO_2 emission values for steel slag aggregate and granite aggregate. Data used in the calculations include electricity usage, diesel consumption, transportation and use of explosives. The analysis reveals that carbon dioxide equivalent emissions decreases as the use of EAF steel slag aggregates increases. The transportation of granite aggregates into Singapore represents a significant contribution to its overall carbon footprint. Thus, using steel slag aggregates generated locally in Singapore would contribute a smaller carbon footprint to the asphalt wearing course making process and reduces the disposal of non-incinerable waste to the landfill site.

S. S. Loi (✉) · S. K. Subramaniam · H. Xusheng
NSL Chemicals Ltd., 26 Tanjong Kling, Singapore 628051, Singapore
e-mail: ssl@nslchemicals.com.sg

H. Xusheng
e-mail: hx@nslchemicals.com.sg

© The Author(s), under exclusive license to Springer Nature Switzerland AG 2022

H. R. Pasindu et al. (eds.), *Road and Airfield Pavement Technology*, Lecture Notes in Civil Engineering 193, https://doi.org/10.1007/978-3-030-87379-0_58

Keywords Steel slag · Carbon footprint · Greenhouse gasses · Road construction

1 Introduction

Electric Arc Furnace (EAF) Steel Slag is a by-product generated from the steel-making process. Steel Slag is formed from the reaction of flux such as calcium oxide with inorganic non-metallic components present in the steel scrap. It subsequently undergoes a physical process of crushing and separation to produce the required gradation for further road construction use. The amount of steel slag available in Singapore for reuse is estimated to be 50–60k tons per year [1].

Steel slag can be beneficially used as road surfacing aggregates when it has been correctly treated and processed. The formulation of road mixes using steel slag as aggregates have shown to give better rut resistance, mechanical stability, and higher affinity to asphalt, thereby making the wearing course of the road more durable. Since 1994, 100% of steel slag generated in Singapore has been fully recycled into aggregates and used in the asphalt mix for the wearing course of roads [1]. The rough surface texture contributes to higher skid resistance, as reflected from the polished stone value (PSV) in Table 1.

The conventional disposal method of EAF steel slag is landfilling. Over the years, landfilling has become less favourable as it occupies a wide area of land that will eventually boost-up the disposal costs. The use of EAF steel slag in road construction applications significantly reduce the demand for natural aggregates and would divert waste materials from landfills and stockpiles [3].

Singapore is a small island country where scarcity of land and resources is of critical concern. The country relies heavily on imported natural resources such as granite and sand from neighbouring countries for constructional use. Besides being not a sustainable resource, like steel slag, the transportations of such raw materials over long distances would inevitably result in higher carbon footprint [4].

Table 1 Aggregate properties of granite and steel slag [2]

Properties	Granite	Slag
pH	7	11–12
Polished stone value	49–54	60–61
LA abrasion value (%)	>35	17–25
Aggregate impact value (%)	21–31	10–14
Aggregate crushing value (%)	27	15–21
Flakiness index	22	3–16
Elongation index	28	7

Replacement of naturally occurring aggregates with locally available EAF steel slag as a road construction material would promote sustainability of road construction and minimise the environmental impact. Significant amounts of virgin naturally occurring material are being used annually in roadwork construction and development. On top of that, traditional quarry materials are becoming scarce due to high worldwide demand from civil and construction sector and is also an environmentally sensitive products to export.

Climate change is one of the greatest challenges of our time and a major threat to the livelihoods of living creatures on earth. Natural resource consumption to sustain that industry's growth increases steadily, contributing to environmental deterioration, for example, the rise in the atmosphere's temperature as well as that of the oceans, which has led to the well-known climate crisis of global warming [5].

Greenhouse gasses generated from human activities is no doubt the biggest contributor of climate change observed since the mid-twentieth century. Carbon footprint is one of the most widely used tools to quantify greenhouse gases for assessing the environmental impacts of a production activity. The amount of carbon dioxide and other greenhouse gases associated with a given product can be calculated and expressed as CO_2 equivalents [6].

The advantages of using a renewable resource such as steel slag, in the reduction of GHG emissions can be quantified using embodied CO_2 emission calculations. This paper briefs the applications of steel slag aggregate for asphalt pavement in Singapore and compares the carbon footprint of locally produced steel slag aggregates with imported granite aggregates.

2 Objective

The objective of this investigation is to determine the carbon footprint of EAF steel slag aggregates and compare the carbon footprint of locally produced steel slag aggregates with imported granite aggregates.

3 Materials and Methods

In the present study, the carbon footprint involved the quantification of the carbon dioxide (CO_2) released from the raw material to distribution of aggregates, including explosives, transportation, electricity, and diesel consumption. Figure 1 shows the methodology overview and key steps for carbon footprint analysis.

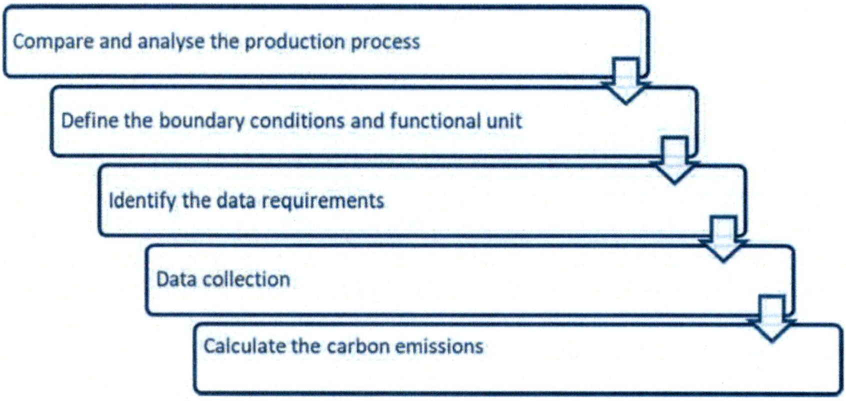

Fig. 1 Overview and key steps for carbon footprint analysis

3.1 System Boundaries

This investigation only covers raw material generation, processing and distribution of steel slag and granite aggregates. The production of asphalt concrete and the entire road making process are out of the scope of this investigation, Fig. 2 shows the system boundaries of the present study.

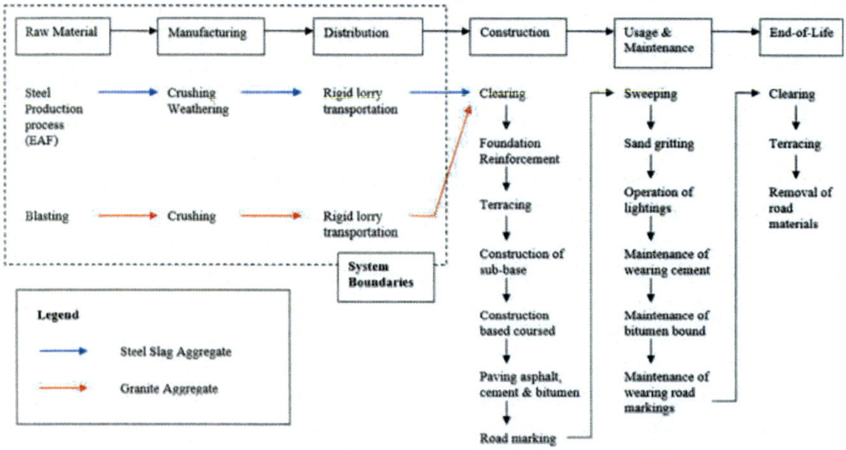

Fig. 2 Schematic diagram of asphalt making process, and the system boundaries considered

3.2 Functional Unit

The functional unit used in this study is 1 kg of aggregates leaving the gates of production facilities.

3.3 Allocation

Since steel slag is a by-product of the steel making process, it has undergone a primary process, which is the steel making process, before undergoing a secondary process, which is crushing. When a production system produces several products or by-products, material and energy flows and the associated environmental burdens must be partitioned between them to accurately reflect their individual contribution to the environmental impacts [7]. However, to do an apple-to-apple comparison, there will be no allocation for steel slag aggregates from the steel making process, since the embodied energy in naturally-occurring materials such as granite cannot be accounted for.

3.4 Data Collection

Data for steel slag aggregate processing were obtained from NSL Chemicals Ltd steel slag production plant in Singapore, while data for granite aggregate production were obtained from the quarrying and production facility of Eastern Pretech in Malaysia.

4 Life Cycle Inventory (LCI)

This phase involved data collection and calculation procedures to quantify the system's inputs and outputs. Carbon footprint assessment for steel slag aggregate and granite aggregate, expressed in kg CO_2-*e*/kg, in Tables 2 and 3, which includes diesel for crushing process, transportation for the raw material transport, electricity usage and explosive agents for quarry (granite).

4.1 Steel Slag Aggregate

The data were obtained from NSL Chemical's Roadstone production located at Tanjong Kling, Singapore. Embodied carbon for steel slag aggregate is **0.01101** kgCO_2e/kg. The calculations are shown in Table 2.

Table 2 Calculations for LCI of steel slag aggregates

No.		Usage		Emission factor		Emissions ($kgCO_2e/kg$)
1	Diesel	12,258[a]	Litre/month	2.48	$kgCO_2$/litre	0.00507
2	Electricity	15,000[b]	kWh/month	0.7341	$kgCO_2e$/kWh	0.00184
3	Transportation	17.7[c]	km	0.000232	$kgCO_2$/kg km	0.00411
					Total	**0.01101**

[a] Average monthly diesel consumption
[b] Average monthly electricity consumption
[c] Distance from steel slag aggregate manufacturing plant in Tanjong Kling to asphalt manufacturing plant in Sg Kadut

Table 3 Calculations for LCI of granite aggregates

No.		Usage		Emission factor		Emissions ($kgCO_2e/kg$)
1	Explosives	0.000187	kg/kg	0.132	$kgCO_2$/kg	0.00002
2	Diesel	0.001065[a]	litre/kg	2.48	$kgCO_2$/litre	0.00264
3	Electricity	0.002[b]	kWh/kg	0.5295	$kgCO_2e$/kWh	0.00106
4	Transportation	50.8[c]	km	0.000232	$kgCO_2$/kg km	0.01179
					Total	**0.01551**

[a] Average monthly diesel consumption
[b] Average monthly electricity consumption
[c] Distance from granite aggregate manufacturing plant in Johor to asphalt manufacturing plant in Sg Kadut

4.2 Granite Aggregate

Embodied carbon for granite aggregate is **0.01551** $kgCO_2e/kg$. The calculations are shown in Table 3.

5 Impact Assessment

5.1 Steel Slag Aggregate

Since steel slag is a waste material, the use of steel slag as aggregates in asphalt mix reduces the need for land filling, thus conserves space. It also reduces the use of raw materials such as granite, thus avoiding resource depletion.

5.2 Granite Aggregate

The use of natural granite aggregate contributes to the depletion of raw materials, which is granite rock. Since granite is mined from quarries, it destroys the natural habitats of wildlife in the area when a virgin quarry is open. Explosives used also contribute to noise pollution to the surrounding population.

6 Results and Discussions

The environmental benefits of using steel slag have been assessed through the comparison of embodied CO_2 emission values for steel slag aggregate and granite aggregate. Based on (Fig. 3) the comparison graph of carbon footprint of steel slag aggregate and granite aggregate, usage of granite aggregates contributes higher carbon emissions. For steel slag, the main carbon emission comes from the transportation, diesel for handling and followed by the electricity for crushing. Granite may consume lower energy to crush due to its lower hardness, but it creates additional emission due to the use of explosives at the quarry.

The transportation cost on average is much higher for granite because of the longer travelling distance. Steel slag is generally used locally as high cost of material production makes long-distance transportation of the material economically not feasible. Using locally generated steel slag can significantly reduce the carbon footprint of the product. More intangible benefits include the conservation of natural resources and biodiversity by avoiding over-exploitation.

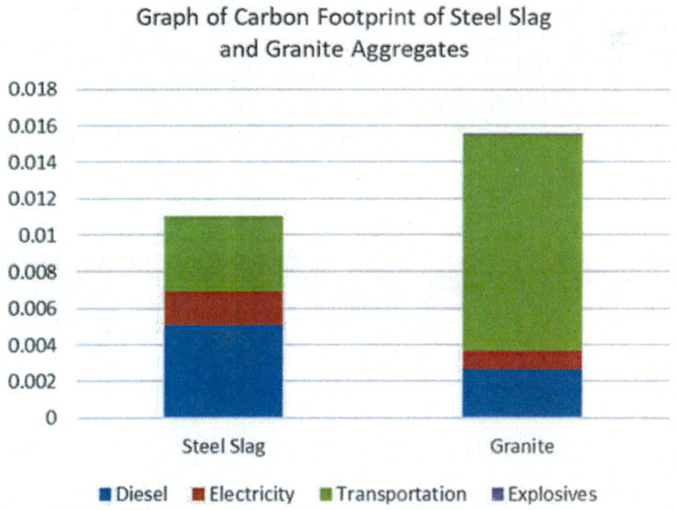

Fig. 3 Comparison graph of carbon footprint of steel slag and granite aggregates

7 Steel Slag Aggregates Application in Singapore

NSL Chemicals started research and utilization of steel slag as roadstone in collaboration with Nanyang Technological University ("NTU") and Land Transport Authority ("LTA") in 1991. The engineered product was on field trials in 1993 and commercially used on Singapore roads in 1994. Low noise steel slag premix developed in 2013 in collaboration with National University of Singapore and currently being used on express ways in Singapore. The research project received a distinguish awards for being part of the Green & Sustainable Road Pavement initiatives.

Around 1.6million tons of steel slag aggregates were supplied by NSL Chemicals to LTA, PSA and Changi Airport to resurface Singapore roads since 1994. This has saved the use natural aggregates with steel slag aggregates and prevented approximately 700 Km^3 of supposed waste material from being disposed at the landfill site.

Our steel slag business has contributed in creating an environmentally friendly products for the road construction market, particularly the resurfacing of roads, in Singapore (Fig. 4). NSL Chemical's steel slag business helps to conserve landfill sites and promotes sustainable development through its lowered carbon footprint when compared to imported aggregates. The primary focus of this business segment is to mass recycle steel slag, the waste generated from steel mills, to save landfill sites for dumping.

Fig. 4 Steel slag road pavement in Singapore

8 Limitations

This investigation only covers raw material generation, processing and distribution of steel slag and granite aggregates. The asphalt mix production is not included in the carbon footprint calculation due to lack of operational data.

9 Conclusions

The analysis reveals that carbon dioxide equivalent emissions decreases as the use of EAF steel slag aggregates increases. The transportation of granite aggregates into Singapore represents a significant contribution to its overall carbon footprint. Thus, using steel slag aggregates generated locally in Singapore would contribute a smaller carbon footprint to the asphalt wearing course making process and reduces the disposal of non-incinerable waste to the landfill site.

References

1. Sustainable construction, a guide on the use of recycled materials, building & construction authority (BCA), Singapore (2008)
2. Yiying H, Keng Boon S, Soo Loi S, Tien Fang F, Chin Chong Y (2015) Steel slag aggregate for asphalt pavement. In: 9th international conference on road and airfield pavement technology, China
3. Teo P, Zakaria S, Salleh S, Taib M, Mohd Sharif N, Abu Seman A, Mohamed J, Yusoff M, Yusoff A, Mohamad M, Masri M, Mamat S (2020) Assessment of electric arc furnace (EAF) steel slag waste's recycling options into value added green products: a review. Metals 10(10):1347–1368
4. Shan X, Zhou J, Chang V, Yang E (2017) Life cycle assessment of adoption of local recycled aggregates and green concrete in Singapore perspective. J Clean Prod 164:918–926
5. Reham R, Nehdi M (2005) Carbon dioxide emissions and climate change: policy implications for cement industry. Environ Sci Policy 8(2):105–114
6. Loyarte-López E, Barral M, Morla J (2020) Methodology for carbon footprint calculation towards sustainable innovation in intangible assets. Sustainability 12(4):1629
7. Chen C et al (2010) LCA allocation procedure used as an initiative method for waste recycling: an application to mineral additions in concrete. Resour Conserv Recycl 54:1231–1240

Effects of Recycled Plastics on the Performance of Polymer Modified Bitumen (PMB)

Aye Nandar Win Maung, Yang Pin Kelvin Lee, Nyok Yong Ho, Guo Wei Tiah, and Kai Yi

Abstract Polymer modified bitumen (PMB) has been widely used in asphalt mixtures as it provides better performance and durability as compared to unmodified binders. Styrene Butadiene Styrene (SBS) is one of the most common additives used in PMB. With the advent of bitumen technology, PMB can be further enhanced to improve its performance in terms of properties such as viscosity, ageing resistance, stiffness etc. This study aims to evaluate the effects of recycled plastics on the performance of SBS-modified binder. The recycled plastics involved in this study as partial substitute of SBS are low density polyethylene (LDPE) and high density polyethylene (HDPE). The analysis was conducted using a series of tests which include elastic recovery, dynamic shear test (before and after short term ageing), viscosity test, multiple-stress creep and recovery test and storage stability.

Keywords Polymer modified bitumen · Styrene Butadiene styrene · Recycled plastic · Elastic recovery · Dynamic shear · Multiple-stress creep and recovery · Viscosity · Storage stability

1 Introduction

Bitumen has been modified by using various modifiers as it can enhance binder performance and durability, thereby reduces pavement maintenance costs and lengthen the expected design life of pavement. The properties of modified bitumen are dependent on the type of modifier used and compatibility with base bitumen. There is a wide variety of modifiers such as styrene–butadiene–styrene (SBS), styrene-butadiene-rubber (SBR), styrene–isoprene-styrene (SIS), crumb rubber, latex rubber, polyethylene (PE), polypropylene (PP), ethylene–vinyl-acetate (EVA), sulphur, polyphosphoric acid (PPA) and wax. To date, SBS is the most common modifier for bitumen modification. Although all modifiers tend to improve the bitumen properties to some extent, there are still some limitations, such as high cost, low ageing resistance, poor

A. N. Win Maung (✉) · Y. P. K. Lee · N. Y. Ho · G. W. Tiah · K. Yi
Samwoh Innovation Centre Pte Ltd., 25E Sungei Kadut Street 1, Singapore 729333, Singapore
e-mail: aye.maung@samwoh.com.sg

© The Author(s), under exclusive license to Springer Nature Switzerland AG 2022
H. R. Pasindu et al. (eds.), *Road and Airfield Pavement Technology*, Lecture Notes in Civil Engineering 193, https://doi.org/10.1007/978-3-030-87379-0_59

workability and storage stability in the development of modified bitumen [1]. The modifier must be effective, practicable, ecological and economic. Moreover, effect of modifiers in the process (mixing temperature, mixing time, etc.) of bitumen modification is also a concern. This study aims to evaluate the effects of recycled plastics as partial substitute for SBS on the performance of SBS-modified binder. The analysis was conducted according to a series of test which include penetration, softening point, rotational viscosity, elastic recovery, dynamic shear, multiple stress creep and recovery (MSCR) and storage stability.

1.1 Styrene–Butadiene–Styrene (SBS)

SBS is one of the successfully proven polymers for modified bitumen production. It enhances the stiffness and flexibility of the binder. SBS can be soften on heating, become harden after cooling and are able to elastically recover from permanent deformation once the load is removed [2]. SBS is incorporated into bitumen, through mixing and shearing at high temperature to uniformly disperse, and the amount of SBS needed is highly dependent on the specification and performance grade required. Most studies showed that SBS modified bitumen improves rutting resistance at high temperature and crack resistance at low temperature, and also improves the ageing resistance [3]. However, differences in properties such as molecular weight, density, polarity and solubility in between SBS and bitumen lead to the segregation during hot storage (160 °C) and result in poor storage stability [1].

1.2 Recycled Plastics

Waste plastics has been widely recycled in various ways due to environmental issues and interest in the incorporation of recycled plastics into construction materials has increased. Recycled plastics can be used in flexible pavement construction by mixing together with hot bitumen and aggregates in asphalt production (dry mix) and as a modifier by mixing with hot bitumen (wet mix). Studies have shown that the use of recycled plastics can improve the properties of asphalt mixtures with similar results as compared to modified bitumen with virgin polymers, and has added value of environmental advantages [4, 5]. Some of the recycled plastics include low density polyethylene (LDPE), high density polyethylene (HDPE), polypropylene (PP), ethylene–vinyl acetate (EVA), acrylonitrile–butadiene–styrene (ABS), polyethylene terephthalate (PET) and polyvinyl chloride (PVC) [4].

2 Methodology/Materials and Testing Programs

2.1 Materials Used and Samples Preparation

The base bitumen used in this study was penetration grade 60/70 with the penetration value of 62 dec.mm, softening temperature of 48 °C and dynamic viscosity at 135 °C of 0.41 Pa S. SBS linear block copolymer with bound styrene of 31% mass was selected as a modifier. The HPDE and LDPE are recycled plastics resins produced from a local plastic waste processing company. The various modifiers are shown in Fig. 1.

The modifiers were added with SBS to the base bitumen in various proportions as shown in Table 1. The SBS modified bitumen was prepared using 6% SBS (by mass of binder) which is used as a control sample. LDPE and HDPE were mixed with SBS in various proportions as shown in Table 1. For the purpose of comparison among the various modifiers, the total polymer/modifier content was fixed at 6% by mass of binder. The samples were prepared in the laboratory by using the Silverson

Fig. 1 Types of modifiers

Table 1 Formulation of modified binder samples

Base bitumen	Notation	SBS (%)	LDPE (%)	HDPE (%)	Remark
Penetration grade 60/70	SBS (control)	6	–	–	SBS modified bitumen
	SBS-LDPE (3-3)	3	3	–	SBS-LDPE modified bitumen
	SBS-LDPE (4-2)	4	2	–	
	SBS-LDPE (5-1)	5	1	–	
	SBS-HDPE (3-3)	3	–	3	SBS-HDPE modified bitumen
	SBS-HDPE (4-2)	4	–	2	
	SBS-HDPE (5-1)	5	–	1	

high shear mixer at temperature of 170–190 °C for 1 h followed by the blending for 4 h. Binder tests were performed after 24 h curing at room temperature.

2.2 *Testing Programs*

The modified binder samples were tested according to various parameters which include penetration, softening point, rotational viscosity, elastic recovery, dynamic shear, multiple stress creep and recovery (MSCR) and storage stability. A summary of the tests is shown in Table 2. The MSCR test was conducted at specified temperature to determine the presence of elastic response and non-recoverable compliance in modified bitumen under shear creep and recovery at two stress levels. Studies have shown that the non-recoverable compliance (J_{nr}) determined from MSCR provides a better correlation to rutting as compared to the rutting resistance factor (G*/sin δ) determined from the dynamic shear test [6, 7].

The storage stability of the modified binder samples was tested as polymers are likely to show some level of instability when stored in a static mode with time [8]. Differences in softening temperature and penetration value between top and bottom part of samples indicate the stability of modified bitumen after stored at high temperature for 3 consecutive days.

Table 2 Testing plan for bitumen properties

Description	Test temperature	Test method
Standard penetration	25 °C	ASTM D5
Softening point temperature	-	ASTM D36
Rotational viscosity	135, 165 °C	ASTM D4402
Elastic recovery	25 °C	ASTM D6084
Dynamic shear	Performance grade temperature	ASTM D7175
Rolling thin film oven (Short term ageing)	163 °C	ASTM D2872
Multiple-stress creep and recovery	Performance grade temperature	ASTM D7405
Storage stability	180 °C	BS EN 13399

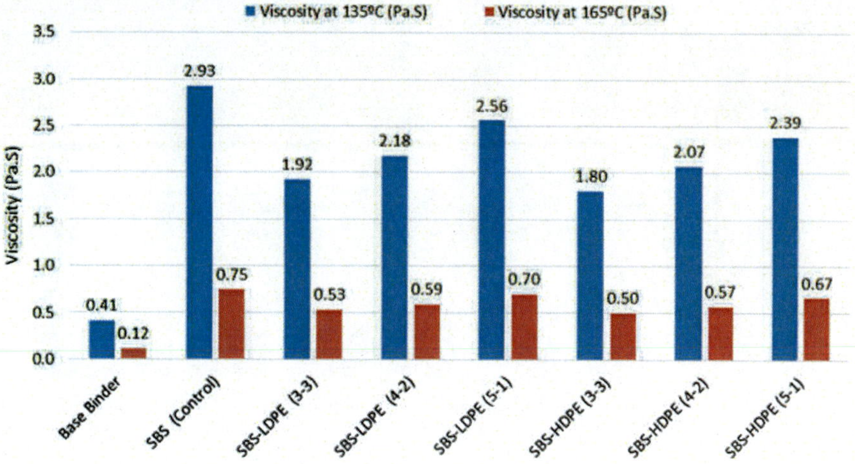

Fig. 2 Dynamic viscosity of modified binders

3 Results and Discussion

3.1 *Dynamic Viscosity*

Dynamic viscosity of modified binder is related to the mixing and compaction temperatures of asphalt mixture. The viscosity of the modified binder samples was measured by rotational viscometer at 135 and 165 °C. Figure 2 shows the viscosity values of the modified binders and base binder. Based on the results, both SBS-LDPE and SBS-HDPE binders showed lower viscosities compared to the control SBS binder.

3.2 *Penetration*

Penetration test is a common test to measure the consistency of modified binder at 25 °C and higher penetration value indicate softer consistency. All the modified binders exhibited significant lower penetration values as compared to the base binder. All the modified binders showed comparable penetration values of between 35 and 45 dec.mm (Fig. 3).

3.3 *Softening Point*

The softening point test measures the temperature at which bitumen starts to flow. Bitumen with higher softening point temperature ensures that they will not flow

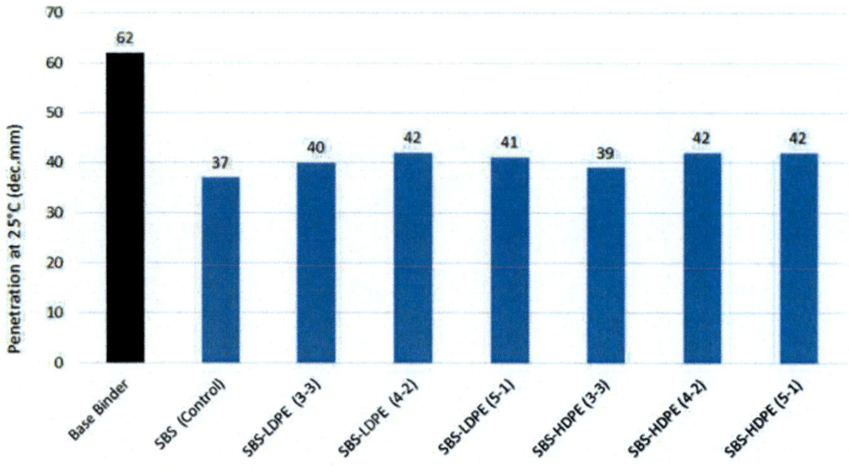

Fig. 3 Penetration value of modified binders

during service. The higher softening point temperature, the lesser temperature susceptibility which influence on the resistance to permanent deformation of the mixtures. Thus, the softening point of the modified bitumen can be used to analyse improvements in the performance at high in service temperatures [4]. All the modified binders showed higher softening point temperatures as compared to the based binder. Binders with higher SBS contents generally showed higher softening point temperatures. In comparison, SBS-LDPE and SBS-HDPE modified binders showed similar performance (Fig. 4).

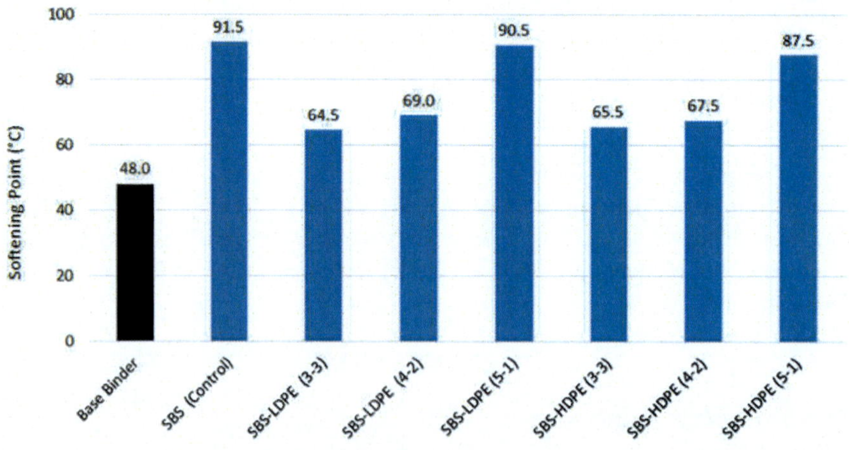

Fig. 4 Softening point temperature of modified binders

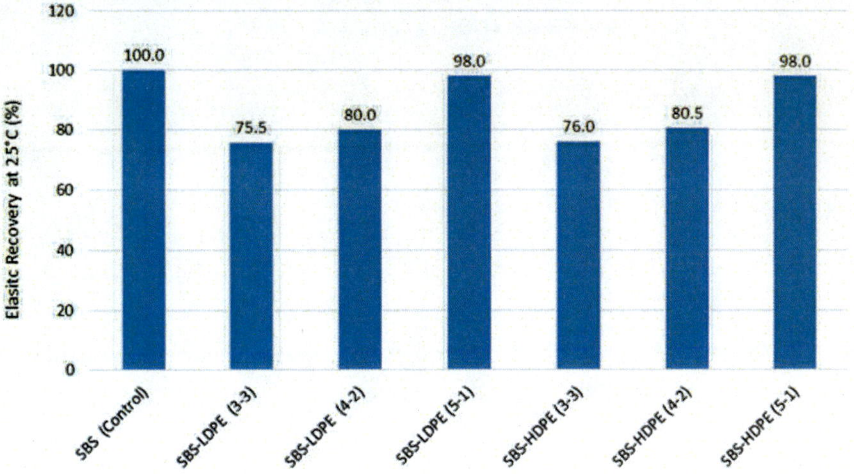

Fig. 5 Elastic recovery of modified binders

3.4 *Elastic Recovery*

The elastic recovery of a bituminous binder was measured by the recoverable strain determined after severing an elongated briquet specimen of the binder at specified temperature of 25 °C, in confirming that a material has been added to the binder to provide a significant elastomeric characteristic. The SBS binder showed 100% elastic recovery while the modified binders with LDPE and HDPE showed about 75–98% recovery. Binders with higher SBS contents showed higher elastic recovery. Both SBS-LDPE and SBS-HDPE samples showed similar elastomeric behaviour. The replacement of SBS with LDPE and HDPE resulted in reduction of elastic recovery by up to 24.5% for binders with 3% replacement of SBS (Fig. 5).

3.5 *Dynamic Shear Rheometer*

The rutting resistance factor (G*/sin δ) of modified binders before and after a short-term ageing using the RTFOT was measured by the dynamic shear rheometer (DSR) at 76 °C. The results are shown in Fig. 6.

As compared to the control SBS binder, the modified binders with LDPE and HDPE showed a reduction in the G*/sin δ values for all results measured before and after RTFOT. Nonetheless, all the binders meet the PG76 requirement of at least 1.0 and 2.2 kPa before and after ageing respectively. Binders with higher SBS contents generally showed higher G*/sin δ values.

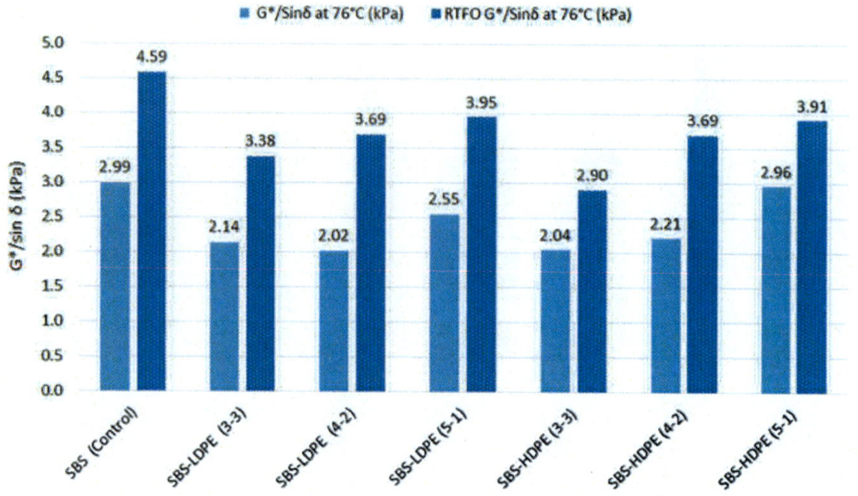

Fig. 6 Effect of modifiers on G*/sin δ values

Comparing the performance between two modifiers (LDPE and HDPE), SBS-HDPE (5-1) with 1% HDPE showed the highest G*/sin δ values before and after RTFOT respectively.

3.6 *Multiple-Stress Creep and Recovery (MSCR)*

The MSCR test involved repeated creep and recovery testing at different stress level to evaluate the potential permanent deformation of the modified binders. The tests were conducted on RTFOT samples at 70 and 76 °C. The J_{nr}@3.2 kPa (which evaluates the rutting resistance) and % recovery (which evaluates the elastic response) were determined.

The results for J_{nr}@3.2 kPa are shown in Fig. 7. Binders modified with SBS alone showed the lowest Jnr (0.08 and 0.26 kPa^{-1} at 70 and 76 °C respectively). The other modified binders showed less than 1.0 kPa^{-1} at 70 °C (i.e. Very Heavy traffic rating) and less than 2.0 kPa^{-1} (i.e. Heavy traffic rating) at 76 °C.

Comparing the performance between two recycled plastics modifiers (LDPE and HDPE), SBS-HDPE (4-2) with 2% HDPE showed the lowest J_{nr}@3.2 kPa values at both 70 and 76 °C indicating least temperature sensitivity for a 6 °C increase in test temperature.

To analyse the stress sensitivity of the modified binders, the $J_{nr\,diff}$ was determined from the MSCR test which is defined as the difference between J_{nr}@3.2 kPa and J_{nr}@0.1 kPa. Generally, the $J_{nr\,diff}$ should not be greater than 75% of the J_{nr}@0.1 kPa according to AASHTO M332 specification [9]. The reason for this requirement is

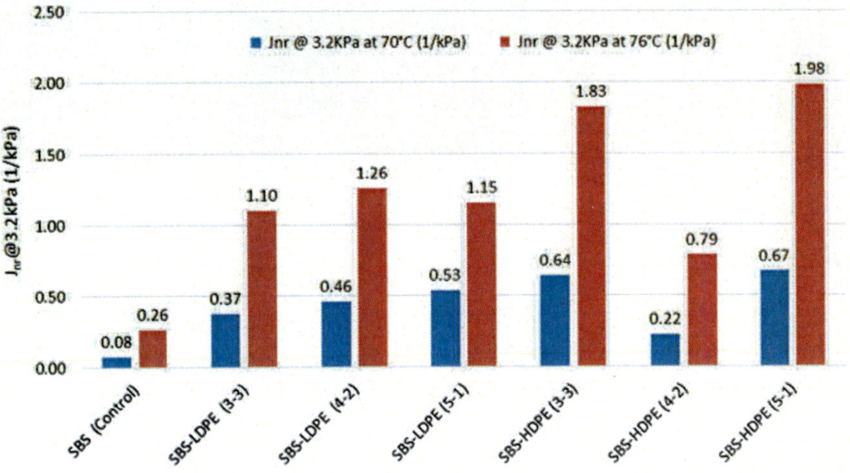

Fig. 7 Non recoverable creep compliance J_{nr} @3.2 kPa of modified binders

that some binders are very sensitive to stress level and exhibit yielding at higher stress level [10].

The results for $J_{nr\ diff}$ are shown in Fig. 8. The results generally increase as test temperature increases. In this case, all the results exceeded the 75% requirement. These findings were also reported by other studies which indicated that polymer modified bitumen with high polymer content could result in high stress sensitivity [11]. Moreover, there have been reported challenges with meeting $J_{nr\ diff}$ specification especially for modified binder with low J_{nr} value of less than 0.5 kPa^{-1} [7]. And AASHTO M332 (2020) remove the $J_{nr\ diff}$ specification requirement for Jnr value

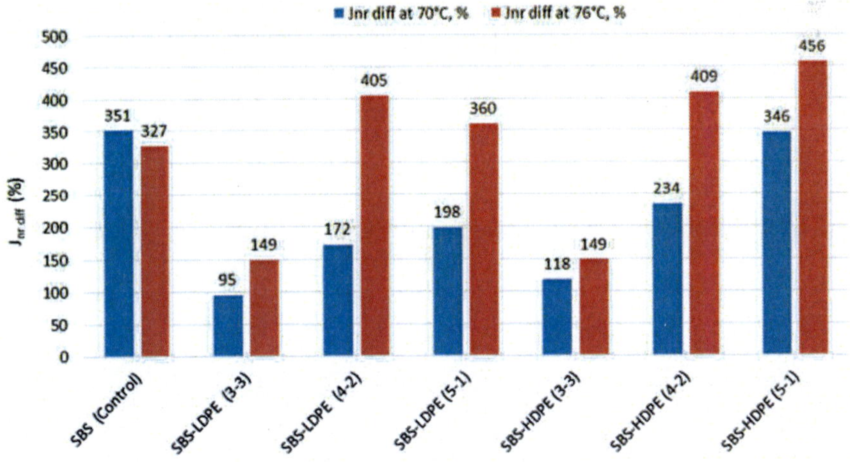

Fig. 8 $J_{nr\ diff}$ of modified binders

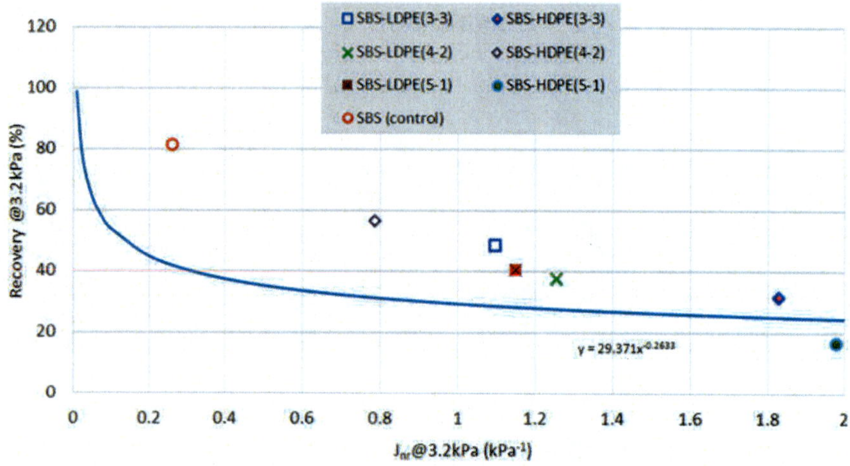

Fig. 9 Non recoverable creep compliance versus percent recovery

less than 0.5 kPa^{-1} [12]. Nonetheless, the current findings can still provide a relative comparison of the stress sensitivity among the various modifiers.

As compared to the SBS binder, recycled plastics generally showed lesser $J_{nr\,diff}$ at 70 °C which indicates that these modifiers can help to reduce the stress sensitivity of the modified binder. At 76 °C, recycled plastics seemed to reduce stress sensitivity as the plastic content increases.

The J_{nr}@3.2 kPa (at 76 °C) versus % recovery curve were plotted according to AASHTO R92-18 [13] as shown in Fig. 9. The results above the curve indicated that these modified binders have significant elastic response. Among the binders, SBS-HDPE (5-1) did not meet the requirements.

Figure 10 shows the Quadrant plot [14] which analyzes the elastic recovery and MSCR recovery at the same time. Minimum MSCR % recovery ($R_{3.2}$) requirement varies based on the binder grade and minimum elastic recovery requirement is based on the agency's specifications [14]. It can be seen that the SBS binder and SBS-HDPE (4-2) both meet the recovery requirement but the rest of the modified binders failed to meet the minimum MSCR % recovery even though their elastic recovery values are high. This highlighted that the % recovery can provide a better evaluation of the elastic response of binder as compared to ER test as reported by previous studies [6, 7].

3.7 *Storage Stability*

Modified bitumen which consist of mainly bitumen and at least one additional modifier, are known to display phase separation under certain conditions. Storage

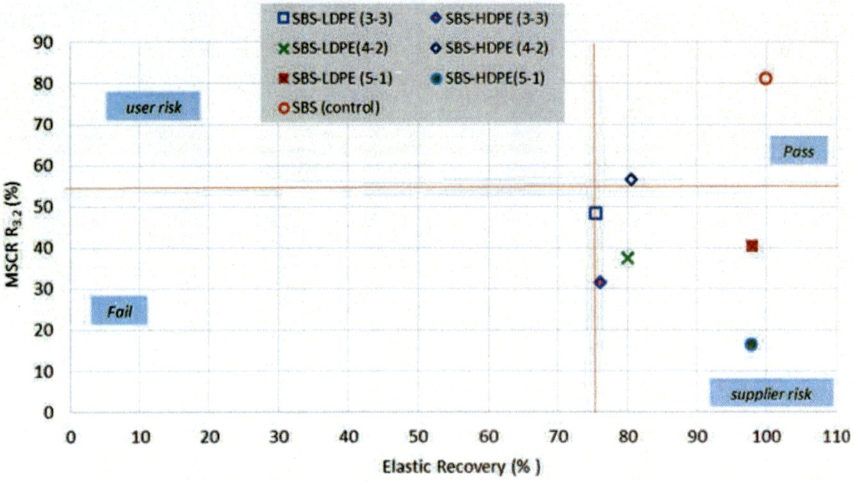

Fig. 10 Elastic recovery versus MSCR percent recovery

stability test is to determine the tendency of polymer to separate from polymer modified asphalt under static heated storage conditions. Large differences in test results between top and bottom specimens indicate that there is a degree of incompatibility between the polymer and the base asphalt [8].

The effects of storage were evaluated in this study by difference in properties of softening point temperature and penetration between top and bottom of the samples. The results are shown in Fig. 11.

As compared to the control SBS, SBS-HDPE binders and SBS-LDPE binders showed significantly higher tendency of separation with up to 18.5 °C difference in softening points and up to 4 dec.mm difference in penetration values at LDPE content of 3%.

4 Conclusions and Recommendations

The effects of recycled plastics (LDPE and HDPE) on SBS modified binder were evaluated by various properties tests and the key observations are as follow:

- All polymer modified binders meet the performance grade PG76 specification by AASHTO M332.
- Both recycled plastics reduce the viscosity of SBS modified binder.
- Based on MSCR test, all modified binders met the J_{nr} criteria for heavy traffic (Jnr@3.2 less than 2 kPa^{-1}). SBS-HPDE (4-2) with 2% HDPE showed the best performance among the modifiers (LDPE and HDPE) with Jnr@3.2 kPa less than 1.0 kPa^{-1} and % recovery more than 55%.

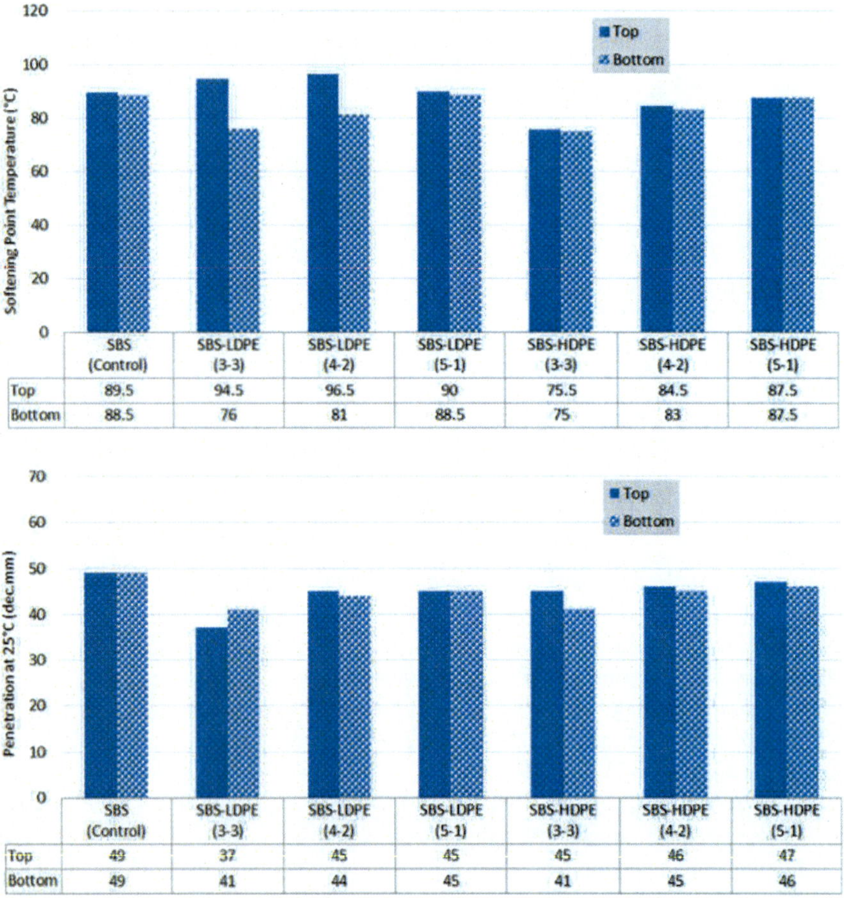

Fig. 11 Storage stability of modified binders

- Modified binder with SBS—recycled plastics (LDPE/HDPE) are less stress sensitive as compared to SBS binder.
- In terms of storage stability, SBS-LDPE binders are less stable compared to SBS and SBS-HDPE binders.
- Further studies on the effect of long-term binder ageing, low temperature performance as well as asphalt concrete performance tests should be carried out on recycled plastics with SBS modified binder.

References

1. Zhu J, Birgisson B, Kringos N (2014) Polymer modification of bitumen: advances and

challenges. Eur Polym J 54:18–38
2. Stroup-Gardiner M, Newcomb DE (1995) Polymer literature review. Minnesota Department of Transportation, St Paul, Minnesota. Report No.: MN/RC-95/27
3. Maarten MJ, Jacobs Marc JA, Stet Andre AA (2002) Molenaar. Decision model for the use of polymer modified binders in asphalt concrete for airfields. Presentation for the 2002 federal aviation administration airport technology transfer conference
4. Costa LMB, Hugo Silva MRD, Oliveira JRM, Fernandes SRM (2016) Incorporation of waste plastic in asphalt binders to improve their performance in the pavement. Int J Pavement Res Technol 6(4)
5. Kelvin LYP, Lwin MA, Xuechun W, Yong HN (2019) Evaluation of hot mix asphalt incorporating plastic waste. In: 11th international conference on road and airfield pavement technology, Kuala Lumpur 10–12 July
6. Al-Abdul Wahhab HI, Dalhat MA, Habib MA (2016) Storage stability and high-temperature performance of asphalt binder modified with recycled plastic. Road materials and pavement design
7. Stempihar J, Gundla A, Shane Underwood B (2017) Interpreting stress sensitivity in the multiple stress creep and recovery test. American Society of Civil Engineers
8. Federal Highway Association (2011) US Department of transportation. The multiple stress creep recovery (MSCR) procedure. FHWA-HIF-11-038
9. AASHTO (2014) Standard specification for performance graded asphalt binder using multiple stress creep recovery (MSCR) test. AASHTO M332, Washington
10. https://asphalt.mydigitalpublication.com/publication/?i=534997&article_id=3215983&view=articleBrowser&ver=html5
11. White G (2017) Grading highly modified binders by multiple stress creep recovery. Road Mater Pavement Des 18(6):1322–1327
12. AASHTO (2020) Standard specification for performance graded asphalt binder using multiple stress creep recovery (MSCR) test. AASHTO M332, Washington
13. AASHTO (2018) Standard practice for evaluating the elastic behavior of asphalt binder using the multiple stress creep recovery (MSCR) test. AASHTO R92, Washington
14. Hossain Z, Ghosh D, Zaman M, Hobson K (2016) Use of the multiple stress creep and recovery (MSCR) test method to characterize polymer modified asphalt binders. J Testing Eval 44(1)

Evaluation of Rutting Properties of Asphalt Binders and Mixtures with Tire Pyrolytic Char

Abhinay Kumar, Rajan Choudhary, and Ankush Kumar

Abstract Waste or end-of-life tires are generated in huge quantities all over the world and pose a serious challenge for the environment. Pyrolysis is a quite effective chemical route for recycling of waste tires, and produces three-phase products: liquid pyrolytic oil, gases, and solid tire pyrolytic char (TPC). The TPC is generally considered a by-product of waste tire pyrolysis process and has recently gained interest as an additive/modifier to road asphalt binders. Rutting in thick asphalt layers is a commonly observed distress on heavy-duty flexible pavements, which are quite preferred in India as well as throughout the globe. Rutting in the asphalt-bound pavement layers refers to the accumulation of permanent deformation under the influence of high pavement temperatures and heavy traffic. Asphalt binders play a significant role in imparting rutting resistance to an asphalt concrete pavement layer. Evaluation of rutting resistance of TPC modified binders and mixtures is thus quite important to support its wide-spread use in construction of asphalt pavements. This study first evaluated the rutting performance of TPC modified binders (prepared at five TPC contents) at multiple temperatures (40, 50, 60, and 70 °C) and stress levels through five rheological parameters: (1) Superpave rutting parameter; (2) Shenoy rutting parameter; (3) zero shear viscosity; (4) non-recoverable creep compliance (J_{nr}) from multiple stress creep and recovery (MSCR) test; and (5) non-recoverable strain rate ($\Delta\varepsilon_{nr}$) from the MSCR test. A dynamic shear rheometer was used to measure the TPC modified asphalt binder rutting characteristics. Asphalt mixtures were then fabricated using the TPC modified binders and evaluated for rutting performance using the Hamburg wheel-tracking device. Correlation analysis was performed for binder rutting parameters and mixture rut depth. Overall, the findings indicate that the addition of TPC enhances the rutting performance of both asphalt binders and the asphalt mixtures.

Keywords Tire pyrolytic char · Waste tire · Modified asphalt · Rutting · MSCR · HWTD

A. Kumar · R. Choudhary (✉) · A. Kumar
Indian Institute of Technology Guwahati, Guwahati, Assam, India
e-mail: rajandce@iitg.ac.in

© The Author(s), under exclusive license to Springer Nature Switzerland AG 2022

H. R. Pasindu et al. (eds.), *Road and Airfield Pavement Technology*, Lecture Notes in Civil Engineering 193, https://doi.org/10.1007/978-3-030-87379-0_60

1 Introduction

There is an increasing emphasis across the globe, and in India, related to the use and development of alternative technologies for recycling scrap tires toward a circular tire economy [1, 2]. India generates about 6–7% of the world's total waste tires estimated at 1.5 billion annually [3, 4]. Pyrolysis is receiving a wide popularity as an alternative technology for effective end-of-life treatment and extraction of useful products from waste tires. The thermochemical decomposition of waste tires through pyrolysis produces gases, liquid pyrolytic oil, and solid carbonaceous char [referred in this study as the tire pyrolytic char (TPC)]. While the process provides an opportunity to recover alternative energy products in the form of liquid pyrolytic oil and gases, the TPC is generally considered a by-product and finds limited applications [5, 6]. The TPC originates from carbon black and inorganic filler materials added during tire manufacturing in addition to the coke material generated during pyrolysis reactions [2, 6]. Being a carbonaceous product, its use as an asphalt binder additive/modifier will find an economical and large-scale application route. Asphalt binders are regularly modified in an effort to enhance the performance and service life of pavements against increasing axle loads, tire pressures, and temperature variations. Even though the earliest works on the use of TPC in asphalt date back to 1995–1997 [7, 8], the material has received a renewed attention for asphalt modification since the past 4–5 years [9–11].

Rutting in thick asphalt pavement layers is a commonly observed distress on heavy duty flexible pavements and refers to the accumulation of permanent deformation in the asphalt pavement layers under the influence of high pavement temperatures (encountered during summer) and heavy traffic. India being a tropical country with high air temperatures recorded up to 50 °C, it is not surprising that rutting is one of the two most common distresses (the other being fatigue cracking) observed on the country's highways [12, 13]. An improvement in the rheological properties of asphalt binders through asphalt binder modification with the use of one or more additives to a neat binder is an effective approach to mitigate asphalt mixture rutting. Several research efforts have been made in the last three decades in the direction of recognizing binder rheological properties (or parameters) that can accurately represent the rutting resistance of asphalt mixtures [14, 15]. Some of the rutting parameters include: Superpave rutting parameter; Shenoy rutting parameter; zero shear viscosity (ZSV); and parameters (such as compliance) from repeated creep recovery (RCR) and multiple stress creep and recovery (MSCR) tests.

Although rutting is a high-service temperature phenomenon, it should be appreciated that the phenomenon does not occur at a single temperature and may mobilize over a range of 'high' temperatures that should reflect the high summer temperatures observed on a project location. Evaluation of rutting resistance of TPC modified binders under various temperatures and stress levels is thus important to support its use in road construction. Some previous studies have attempted the evaluation of rutting resistance of TPC modified asphalt binders [5, 9, 10]. However, the evaluation was either done at a single temperature or when performed at multiple temperatures, only the Superpave rutting parameter was used.

Therefore, the rutting performance of the TPC modified binder and that of the corresponding asphalt mixture needs to be studied and compared. To fulfil this research need, the main objective of the present study is the characterization of rutting performance of TPC modified binders and mixtures. In this study, four temperatures (40, 50, 60, and 70 °C) were used for the evaluation of binder rutting parameters of TPC modified binders at various dosages (0% (control), 5%, 10%, 15%, and 20% by weight of binder). The rutting performance of TPC modified binders was first evaluated at multiple temperatures using five binder rheological parameters: (1) Superpave rutting parameter; (2) Shenoy rutting parameter; (3) ZSV; (4) non-recoverable creep compliance (J_{nr}) from MSCR test; and (5) non-recoverable strain rate ($\Delta\varepsilon_{nr}$) from the MSCR test. A dynamic shear rheometer (DSR) was used to evaluate the TPC modified asphalt binders. Asphalt mixtures were then fabricated using the TPC modified binders and evaluated for rutting performance using the Hamburg wheel-tracking device (HWTD) test. Correlation analysis was performed for binder rutting parameters and mixture rut depth.

1.1 Background of Binder Rutting Parameters

The Superpave rutting parameter (G*/sin δ) is based on the complex shear modulus (G*) and phase angle (δ) measured on DSR at a particular frequency and strain. The expression for G*/sin δ assumes rutting as a stress-controlled, cyclic loading phenomenon and is based on minimizing the dissipated energy in each cycle of loading (W_c) given by Eq. (1):

$$W_C = \pi\,.\sigma_0^2\cdot\frac{1}{G^*/\sin\delta} \tag{1}$$

where, σ_0 is the shear stress. Equation (1) indicates that to reduce the work dissipated per loading cycle, the parameter G*/sin δ should be increased, and therefore a higher value of this parameter indicates a better rutting resistance. Despite being included in the Superpave PG specifications for asphalt binders [16], many researchers have highlighted the limitations of the parameter and have put forth some refinements.

Shenoy [17] proposed one such refinement termed as the Shenoy rutting parameter presented in Eq. (2). The parameter is based on unrecovered strain and emphasizes the effect of elasticity in terms of better sensitivity to δ. A lower unrecovered strain in the binder is associated with an increased value of this parameter, and is desirable for an improved rutting resistance.

$$Shenoy\,rutting\,parameter = \frac{G^*}{1-\frac{1}{\tan\delta\sin\delta}} \tag{2}$$

ZSV is another parameter recommended for rutting characterization and it is an intrinsic characteristic of binder viscosity measured when the shear rate approaches zero and the binder reaches a steady state of flow. Some of the frequently used models to estimate ZSV use complex viscosity (η^*) measurement from DSR with varying frequency. The Cross model, presented in Eq. (3), was used determine ZSV for asphalt binders from a frequency sweep test:

$$\eta^* = \frac{\eta_0 - \eta_\infty}{1 + (K\omega)^m} + \eta_\infty \tag{3}$$

where, η^* = complex viscosity at frequency ω (rad/s), η_0 = zero shear viscosity, η_∞ = infinite shear viscosity at infinite frequency, K and m = model parameters. A higher numerical value of ZSV is desirable for better resistance against rutting.

The most recent advancement in the rutting characterization of binder is the MSCR test that acknowledges the significant role of non-recovered deformation of bitumen on the rutting performance. The test methodology applies loading pulses with rest periods at multiple stress levels to simulate the actual traffic variables. The standard MSCR test [18] comprises 10 cycles at two different stress levels (0.1 and 3.2 kPa) where one cycle includes a 1 s creep and a 9 s recovery time without any time lag between the cycles. Non-recoverable creep compliance (J_{nr}) is the MSCR parameter used for rutting characterization and is the ratio of the unrecovered strain at the end of an MSCR cycle to the applied stress (Eq. 4). Percent recovery in an MSCR cycle is calculated from peak/creep strain and the unrecovered strain as shown in Eq. (5).

$$J_{nr} = \frac{\varepsilon_u}{\sigma} \tag{4}$$

$$Percent\ recovery = \frac{\varepsilon_p - \varepsilon_u}{\varepsilon_p} \times 100 \tag{5}$$

where, ε_p denotes the peak/maximum strain, ε_u denotes the unrecovered/residual strain and σ is the applied stress level in each cycle. Recent advances in the MSCR methodology for characterization of modified binders have indicated that a higher stress level than the currently used 0.1 and 3.2 kPa is needed to achieve a state of stress representative of that encountered by modified binders in pavements. A third stress level of 10 kPa was used in this study, as also used in previous study [19]. Further, thirty MSCR cycles were used, as recommended in some recent studies [20, 21], and the J_{nr} and recovery data were obtained from the last five creep-recovery cycles.

Moreno-Navarro et al. [20] pointed that the J_{nr} is calculated as the average value obtained for a specific number of creep-recovery cycles; however, it does not provide information on how the rutting behavior evolves with the increasing number of MSCR cycles. They proposed a new parameter called the non-recoverable strain rate ($\Delta\varepsilon_{nr}$) given by Eq. (6) and measured in the units of %/cycle:

$$\Delta\varepsilon_{nr} = \frac{\varepsilon_{nr,30} - \varepsilon_{nr,15}}{30 - 15} \tag{6}$$

where $\varepsilon_{nr,30}$ = cumulative non-recoverable strain after 30 cycles and $\varepsilon_{nr,30}$ = cumulative non-recoverable strain after 15 cycles. $\Delta\varepsilon_{nr}$ is the fifth binder rutting parameter used in this study. A higher $\Delta\varepsilon_{nr}$ would indicate a higher rate of the development of non-recoverable strain in the binder with the progression of MSCR cycles, and hence a lower rutting performance. In this manner, both J_{nr} and $\Delta\varepsilon_{nr}$ have the same interpretation in terms of binder rutting resistance: lower values indicate better resistance.

It can be noted that the first three rutting parameters (Superpave, Shenoy, and ZSV) are measured within the linear viscoelastic domain of the binders (at low strains), whereas the MSCR based parameters (J_{nr} and $\Delta\varepsilon_{nr}$) at higher stress levels are measured in the nonlinear domain.

2 Materials and Methodology

A straight-run (neat) Viscosity Graded (VG30) binder meeting the requirements of Indian Standard IS 73 [22] was used as the base binder in this study. Table 1 shows the basic properties of the base binder. The high-temperature PG grade of the binder was found as PG 64. The TPC was supplied by Innova Engineering & Fabrication (Mumbai, India) in powdered form, which was again sieved in the laboratory on a 75 μm sieve and only the particles passing the sieve were used for binder modification. The industrial-scale pyrolysis process of waste tires is described elsewhere [11]. TPC particles had a specific gravity of 1.71. Figure 1a shows the physical

Table 1 Properties of base asphalt binder (VG30)

Property	Requirements[a]	Results
Penetration at 25 °C, 100 g, 5 s, 0.1 mm	min 45	51.6
Softening point (R&B) (°C)	min 47	52.7
Flash point (COC) (°C)	min 220	280
Solubility in trichloroethylene (%)	min 99	>99
Absolute viscosity at 60 °C (poise)	2400–3600	3410
Kinematic viscosity at 135 °C (cSt)	min 350	525
Properties of RTFO aged residue		
Ductility, 25 °C (cm)	min 40	>100
Viscosity ratio 60 °C	max 4	1.15

[a] Requirements according to IS: 73 [22]

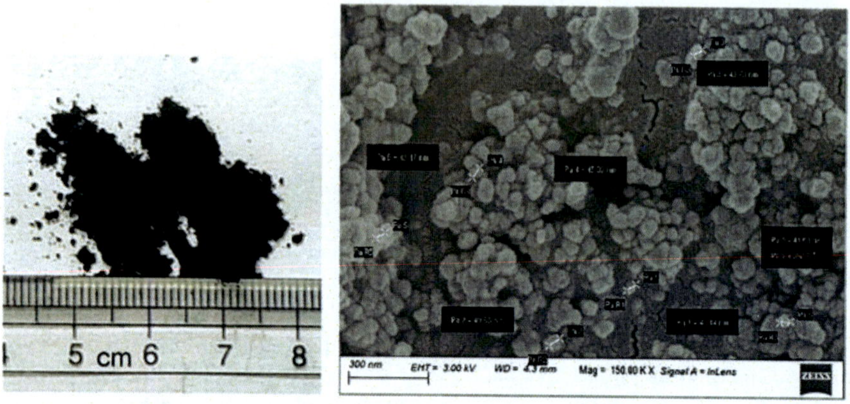

(a) (b)

Fig. 1 **a** Physical appearance of TPC, **b** SEM image of TPC

appearance of TPC as black powdered material, whereas Fig. 1b shows the scanning electron microscopy (SEM) image of the TPC. TPC exhibits hierarchical morphological features, that is there are primary particles (sizes smaller than 100 nm) as well as aggregates and agglomerates of the particles in irregular shapes. The primary particles tend to form aggregates due to covalent bonds and van der Waal forces [23].

A high-shear mixing device (make: IKA T-25 Digital Ulta-Turrax) with a rotor–stator assembly was deployed for binder modification at four TPC dosages (5, 10, 15, and 20 by binder weight). TPC was added when the base binder attained a temperature of 160 °C and the mixing was performed for 30 min at a 12,000 rpm shear rate. Sulfur was then added to all TPC-asphalt blends for desired storage stability at 0.3% by weight of the modified binder. The addition of sulfur was followed by an additional 15 min high-shear mixing. The control and TPC modified binders were short-term aged using a rolling thin film oven (RTFO) at 163 °C for 85 min following ASTM D2872 [24]. The short-term aged binders so obtained were then used for rheological rutting characterization.

A measuring spindle of 25 mm diameter with a gap of 1 mm between the spindle and bottom plate was used during all DSR testing. Figure 2a shows the DSR setup. Superpave and Shenoy rutting parameters were measured in accordance with ASTM D6373 [16] (10 rad/s frequency and 10% strain). To measure ZSV, frequency sweep tests were conducted from 0.1 to 100 rad/s. A low strain of 0.1% was used to ensure the linearity of response of the binders at all frequencies and the four test temperatures (40, 50, 60, and 70 °C). The Cross model was used to fit the complex viscosity (η^*) versus frequency data and estimate the ZSV. As described earlier, the MSCR test consisted of 30 creep-recovery cycles with 1 s creep and 9 s recovery. The test was performed at three stress levels: 0.1, 3.2, and 10.0 kPa. MSCR J_{nr}, recovery, and non-recoverable strain rate ($\Delta\varepsilon_{nr}$) were measured. Each test was performed in duplicate and the average results were reported.

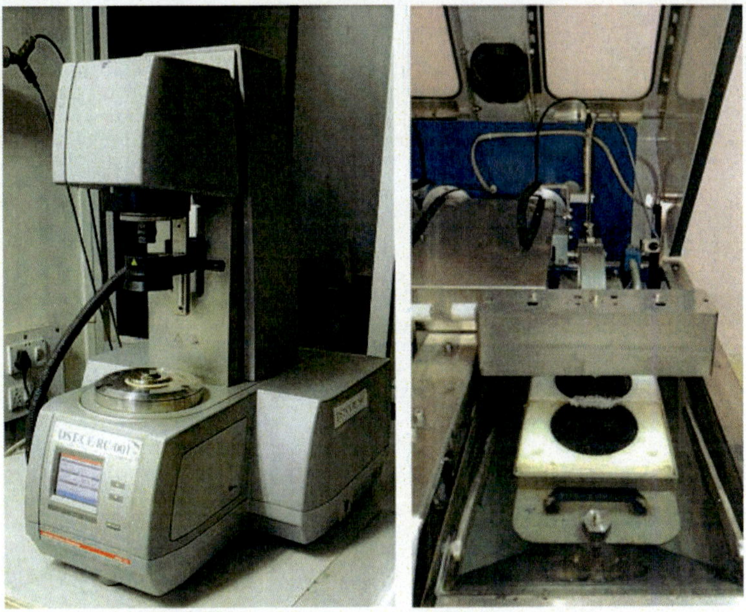

(a) (b)

Fig. 2 **a** DSR setup, **b** HWTD test device

The asphalt mixture specimens were prepared using a dense-graded Bituminous Concrete (BC) gradation of 13.2 mm nominal maximum aggregate size (NMAS), specified by MoRTH [25]. To allow proper comparison with respect to the TPC dosages, the binder content (5.5% by mix weight) and mix volumetrics were kept constant across the BC mixtures. The HWTD test was conducted following the AASHTO T324 [26] specifications on a pair of identical 150 mm diameter and 60 mm height gyratory compactor specimens fabricated at controlled air voids of 7 ± 0.5% (Fig. 2b). The specimens were submerged in water at 50 °C and conditioned for 30 min before the application of 10,000 passes of steel wheel carrying a 705 N load. The rut depth at the end of 10,000 passes was then used as the main mixture rutting performance parameter in this study.

3 Results and Discussion

3.1 Binder Rutting Parameters

Figures 3 and 4 show the plots of Superpave and Shenoy rutting parameters at the four test temperatures. The Superpave and Shenoy rutting parameters, determined

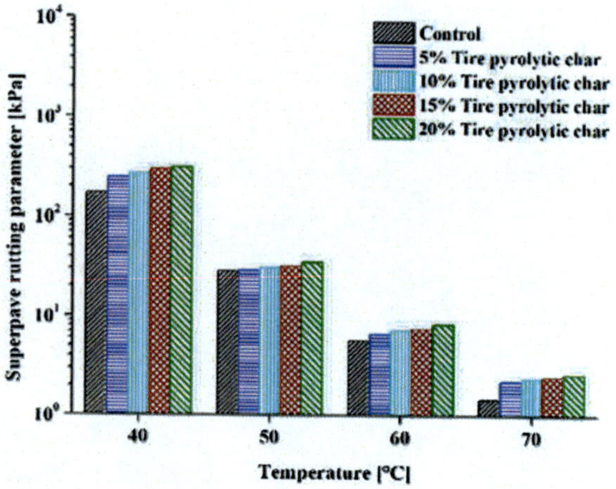

Fig. 3 Superpave rutting parameter

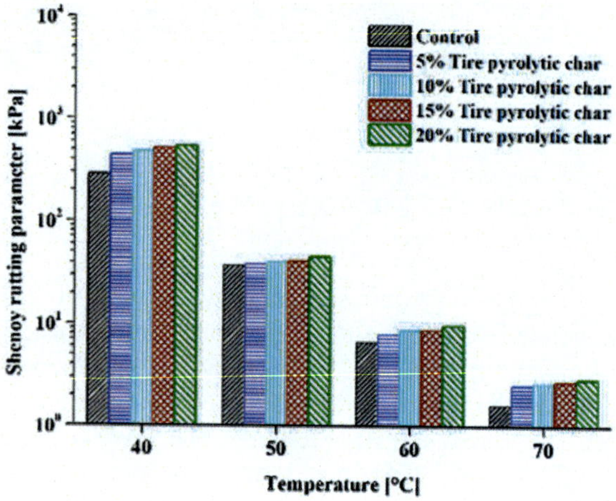

Fig. 4 Shenoy rutting parameter

through the $G*$ and δ values, showed a similar trend with respect to TPC dosage at all four test temperatures: the parameters increase with an increase in TPC content and decrease with the increase in temperature. The results indicate an improved rutting resistance of the binders with the addition of TPC. Wang et al. [10] also found similar results and reported the increased rutting resistance of binder with TPC due to the higher stiffness facilitated by the absorption of light fractions of the binder because of the high surface area of TPC. Similar improved rutting results with TPC were also found by Feng et al. [9] and Li et al. [5]. Averaged over all TPC dosages, the percent increase in Shenoy parameter over the Superpave parameter is 75%, 31%, 17%, and 10% at 40, 50, 60, and 70 °C temperatures, respectively. The Shenoy parameter takes into consideration the emphasized influence of phase angle (or elasticity) on the potential of binder against permanent deformation, leading to a higher value of the Shenoy rutting parameter than the Superpave.

The ZSV was estimated using the Cross model fit to the complex viscosity versus frequency data. The model parameters of the Cross model given in Eq. (3) were estimated using nonlinear curve fitting analysis. Figure 5 shows the Cross model fit to the data for all binders at 60 °C. It is seen that the model fits the data well with a high coefficient of determination ($R^2 \approx 0.99$). Good fits were also obtained at other temperatures but are not presented here for brevity. The ZSVs of all TPC modified binders at the four temperatures were determined and are presented in Fig. 6. As per the trends observed in Fig. 6, the TPC modification increased the ZSV values at all the four test temperatures, which are also consistent with Superpave and Shenoy rutting parameters.

The results of MSCR J_{nr} at the three stress levels are shown in Fig. 7 for the control and modified binders with different TPC dosages at the four test temperatures. In general, the J_{nr} values decreased with the addition of TPC indicating that the TPC

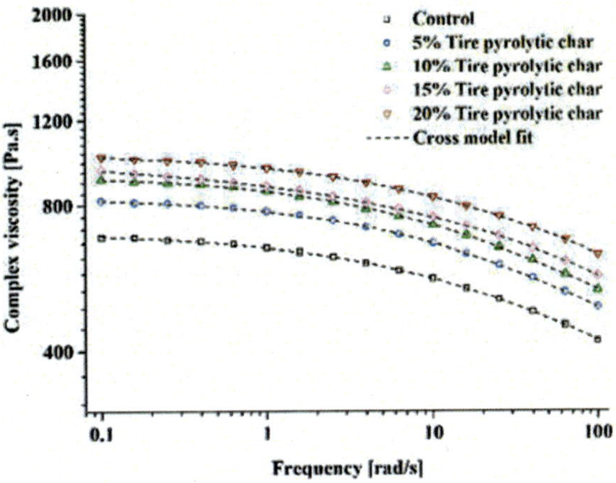

Fig. 5 Cross model fit at 60 °C

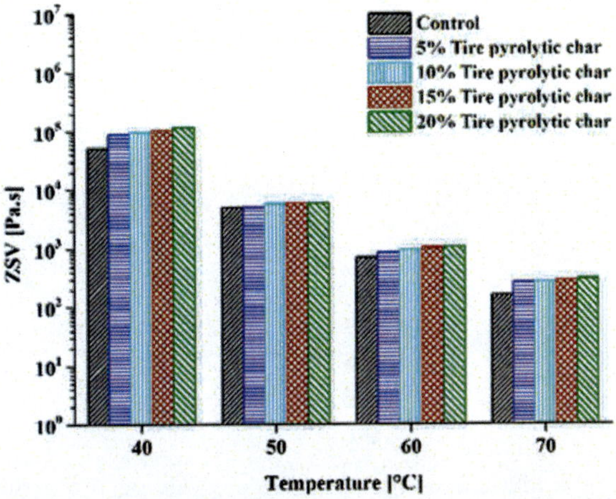

Fig. 6 ZSV results

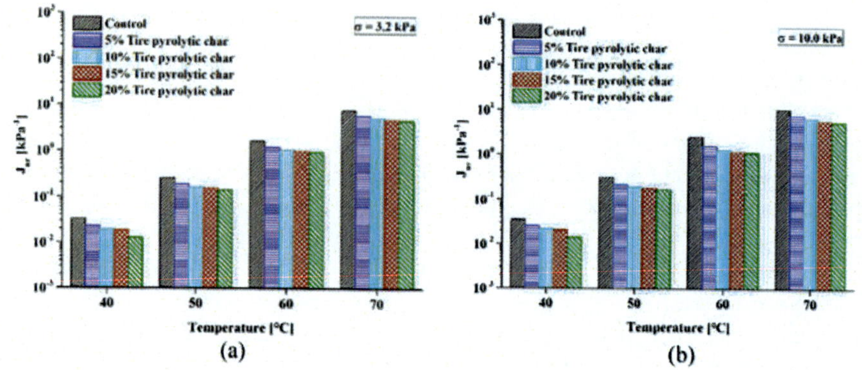

Fig. 7 MSCR J_{nr} results at **a** σ = 3.2 kPa and **b** σ = 10.0 kPa

modification lead to a lower plastic (unrecovered strain) at each stress level and each test temperature, and hence a better permanent deformation resistance. For example, the J_{nr} value of the control binder was found to be 1.58 kPa^{-1} at 3.2 kPa stress level at 60 °C. J_{nr} values with 5, 10, 15, and 20% TPC contents were 1.17, 1.02, 0.97, and 0.91 kPa^{-1}.

Figure 8 shows the results of MSCR percent recovery for all binders. A slight improvement in recovery values is observed on the addition of TPC. As expected the recovery decreases with an increase in test temperature and stress level. The recovery at 70 °C temperature is omitted because negative recoveries were observed. Possible reasons for negative recovery include instrument inertia and tertiary creep response of the binder under the combined influence of high stress levels and high temperatures [27]. The current MSCR standard ASTM D7405 [18] recommends negative recovery to be treated as zero recovery. Similar findings have been reported in a previous study on TPC modified asphalt [10]. Figure 9 presents the results of the second MSCR rutting parameter $\Delta\varepsilon_{nr}$. The addition of TPC particles decreases the

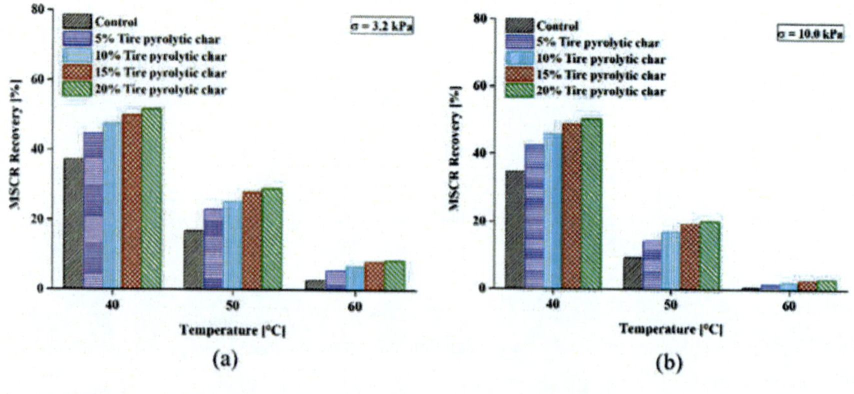

Fig. 8 MSCR recovery results at **a** σ = 3.2 kPa and **b** σ = 10.0 kPa

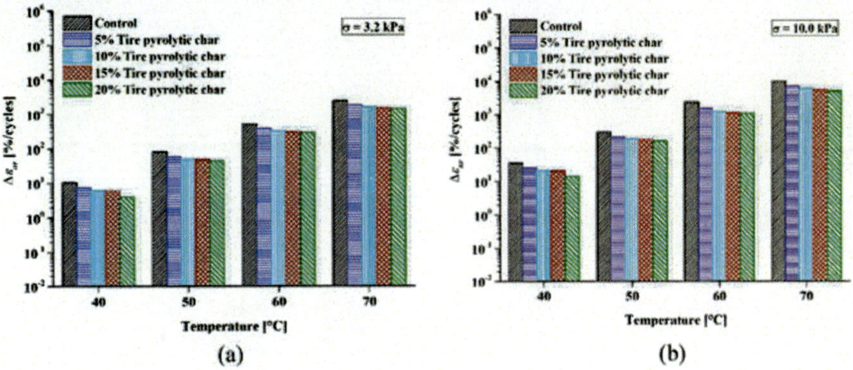

Fig. 9 MSCR $\Delta\varepsilon_{nr}$ results at **a** σ = 3.2 kPa and **b** σ = 10.0 kPa

$\Delta\varepsilon_{nr}$ values, which is beneficial for rutting resistance as it indicates a lower rate of progression of the non-recovered strain with the evolution of creep-recovery cycles. Similar to J_{nr}, the $\Delta\varepsilon_{nr}$ parameter also increases with an increase in temperature and stress level. Averaged over the four temperatures and three stress levels, the TPC dosages of 5%, 10%, 15%, and 20% decrease the $\Delta\varepsilon_{nr}$ values by 27%, 38%, 41%, and 49%, respectively. The numerical values of $\Delta\varepsilon_{nr}$ are much higher than J_{nr} as the non-recovered strain values [constituting the numerator in Eq. (6)] increase rapidly with an increase in the stress level and temperature.

3.2 *Mixture Rut Depth from HWTD Tests*

The HWTD test was performed at 50 °C on the asphalt mixture specimens fabricated with the control and TPC modified binders. The rut depth after 10,000 passes was the main parameter used to characterize the mixture rutting performance. Figure 10 presents the results of rut depth after 10,000 passes. The TPC dosages of 5%, 10%, 15%, and 20% decreased the rut depth by 12%, 25%, 46%, and 65%, respectively, when compared to the control mixture. The mixture rutting results are consistent with the trend shown by the TPC modified asphalt binder rheological parameters.

3.3 *Rutting Resistance Improvement (RRI) with TPC Modified Binders*

To estimate and compare the improvements in the five binder rutting parameters at different TPC dosages, a rutting resistance improvement (RRI) percentage was calculated. RRI was calculated as the percent increase (for Superpave, Shenoy, and ZSV) or percent decrease (for MSCR J_{nr} and $\Delta\varepsilon_{nr}$) in the binder rutting parameter

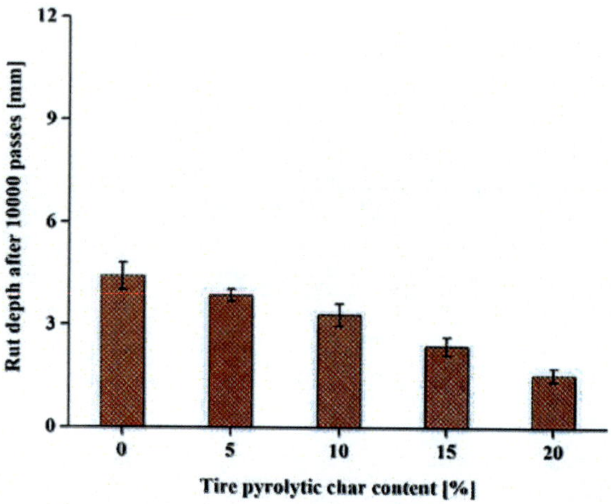

Fig. 10 Rut depth from HWTD

with respect to that of the control binder. The trends of RRI for the TPC modified binders at the four test temperatures are shown in Fig. 11. The trend of RRI increases with the increase in TPC dosages for all five methods, indicating that a higher TPC content imparts a better RRI. At each temperature the RRI values corresponding to MSCR J_{nr} and $\Delta\varepsilon_{nr}$ are very close to one another, implying that both J_{nr} and $\Delta\varepsilon_{nr}$ are equally sensitive to the TPC modification at the four test temperatures. Another observation is that the TPC dosages of 10 and 15% yield closer RRI values as seen for temperatures 50, 60, and 70 °C. The RRI for Superpave, Shenoy, and ZSV are close to each other at 50, 60, and 70 °C temperatures, while ZSV RRI is considerably higher at 40 °C than RRI for other parameters. Further, Fig. 11b also includes the RRI for HWTD rut depth calculated as a percent decrease in rut depth of mixtures with TPC modified binders compared to the control mixture. It can be interpreted from Fig. 11b that the TPC modification has a more profound effect on the mixtures' rutting resistance than the binders' at TPC dosages of 15 and 20%.

3.4 Correlation Between Binder and Mixture Rutting Parameters

Table 2 shows the correlation matrix between all binder rutting parameters and the mixture rut depth. Each cell in Table 2 is the Pearson correlation coefficient (R). Since the HWTD test was performed at 50 °C, the correlations were made with binder rutting parameters measured at 50 °C only. It is seen that both MSCR based parameters J_{nr} and $\Delta\varepsilon_{nr}$ have a very good correlation with HWTD rut depth (R > 0.96), whereas Superpave, Shenoy, and ZSV parameters have almost the same and good correlation (R $\approx$ 0.90) with mix rut depth. Further, the correlation coefficients between J_{nr} and $\Delta\varepsilon_{nr}$, and that between and Superpave and Shenoy parameters

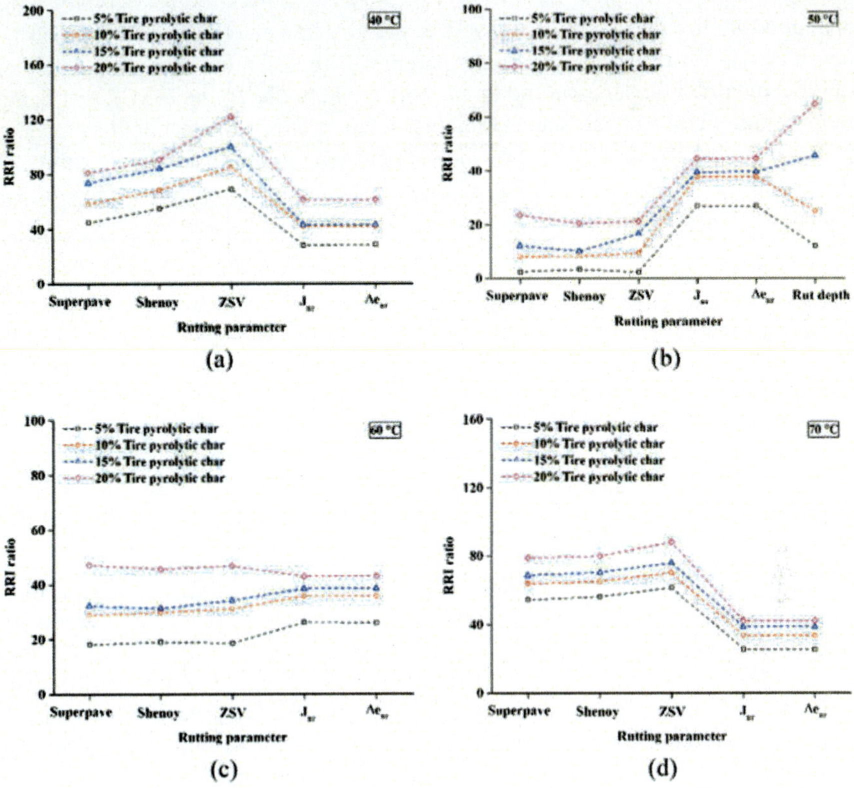

Fig. 11 RRI results at **a** 40 °C, **b** 50 °C, **c** 60 °C, and **d** 70 °C

Table 2 Correlation result between rutting parameters and the mixture rut depth

	Superpave	Shenoy	ZSV	Jnr_3.2	Jnr_10	enr_3.2	enr_10	HWTD
Superpave	1							
Shenoy	0.996	1						
ZSV	0.851	0.850	1					
Jnr_3.2	−0.796	−0.828	−0.868	1				
Jnr_10	−0.781	−0.814	−0.846	0.999	1			
enr_3.2	−0.796	−0.828	−0.869	0.999	0.999	1		
enr_10	−0.782	−0.815	−0.847	0.999	0.999	0.999	1	
HWTD	−0.901	−0.912	−0.888	0.964	0.961	0.965	0.961	1

Superpave: G*/sin δ; Shenoy: Shenoy rutting parameter; *Jnr_x*: MSCR J_{nr} at ‘*x*’ kPa stress level; *enr_x*: MSCR $\Delta\varepsilon_{nr}$ at ‘*x*’ kPa stress level; HWTD: rut depth from the HWTD test are found higher than 0.99. Both MSCR based parameters J_{nr} and $\Delta\varepsilon_{nr}$ can be considered as the recommended binder parameters representing the rutting potential of TPC modified binders and mixtures. This is likely due to the MSCR test being more representative of the actual vehicular loading (since the test applies loading pulses followed by rest/recovery periods in between), and also due to the ability of the test to measure permanent deformation in the nonlinear response range of the binders.

4 Conclusions

Based on the results and analysis, the following conclusions can be drawn:

- The modification of asphalt binder with TPC particles improved the rutting resistance of the neat binder, as evidenced by all five binder rutting parameters (G*/sin δ, Shenoy parameter, ZSV, MSCR J_{nr}, and MSCR $\Delta\varepsilon_{nr}$) at the four test temperatures.
- The rutting resistance of asphalt mixtures fabricated with TPC modified was also higher than the control mix as found from the Hamburg wheel tracking test on the asphalt mixtures.
- Both MSCR parameters J_{nr} and non-recoverable strain rate ($\Delta\varepsilon_{nr}$) had a very good correlation with the HWTD rut depth (coefficient of correlation >0.96).
- At each test temperature, the RRI values corresponding to MSCR J_{nr} and $\Delta\varepsilon_{nr}$ were very close to one another, implying that both J_{nr} and $\Delta\varepsilon_{nr}$ were equally sensitive to TPC modification at the four test temperatures.

Pyrolysis is a clean way to dispose of tire wastes and also a sustainable method to produce energy-rich products. The tire pyrolytic char generated as a by-product of the process shows a good potential for use as a sustainable and economical asphalt binder modifier and will also give way for its large-scale utilization. More in-depth studies are needed to characterize and understand other performance properties of the resulting mixtures such as fatigue life, resistance to moisture-induced damages, and aggregate-binder bonding.

Acknowledgements The authors gratefully acknowledge Innova Engineering & Fabrication (Mumbai, India) for providing the tire pyrolytic char.

References

1. Murugan S, Ramaswamy MC, Nagarajan G (2009) Assessment of pyrolysis oil as an energy source for diesel engines. Fuel Process Technol 90(1):67–74
2. Williams PT (2013) Pyrolysis of waste tyres: a review. Waste Manage 33(8):1714–1728

3. Martínez JD, Puy N, Murillo R, García T, Navarro MV, Mastral AM (2013) Waste tyre pyrolysis–a review. Renew Sustain Energy Rev 23:179–213
4. Mishra L (2016) Turning waste tyre into 'green steel'. The Hindu. https://www.thehindu.com/business/Turning-waste-tyre-into-%E2%80%98green-steel%E2%80%99/article14518524.ece. Last accessed 2021/01/28
5. Li C, Fan Z, Wu S, Li Y, Gan Y, Zhang A (2018) Effect of carbon black nanoparticles from the pyrolysis of discarded tires on the performance of asphalt and its mixtures. Appl Sci 8(4):624
6. Xu J, Jiaxue Y, Jianglin X, Chenliang S, Wenzhi H, Juwen H, Guangming L (2020) High-value utilization of waste tires: a review with focus on modified carbon black from pyrolysis. Sci Total Environ 742:140235
7. Lesueur DIDIER, Dekker DL, Planche JP (1995) Comparison of carbon black from pyrolized tires to other fillers as asphalt rheology modifiers. Transp Res Rec 1515:47–55
8. Park T, Lee K, Salgado R, Lovell CW, Coree BJ (1997) Use of pyrolyzed carbon black as additive in hot mix asphalt. J Transp Eng 123(6):489–494
9. Feng ZG, Rao WY, Chen C, Tian B, Li XJ, Li PL, Guo QL (2016) Performance evaluation of bitumen modified with pyrolysis carbon black made from waste tyres. Constr Build Mater 111:495–501
10. Wang H, Lu G, Feng S, Wen X, Yang J (2019) Characterization of bitumen modified with pyrolytic carbon black from scrap tires. Sustainability 11(6):1631
11. Kumar A, Choudhary R (2020) Use of waste tyre pyrolytic products for asphalt binder modification. Int J Pavement Eng Asphalt Technol 21:35–51
12. Ray M, Of the world's 15 hottest places, 10 are in India. https://www.hindustantimes.com/india-news/of-the-world-s-15-hottest-places-10-are-in-india/story-i7z7pGDp8J6Tf9aN6LLg3H.html. Last accessed 2021/02/06
13. Nagabhushana MN, Tiwari D, Jain PK (2013) Rutting in flexible pavement: an approach of evaluation with accelerated pavement testing facility. Procedia Soc Behav Sci 104:149–157
14. Domingos MDI, Faxina AL (2016) Susceptibility of asphalt binders to rutting: literature review. J Mater Civ Eng 28(2):04015134
15. Hajikarimi P, Rahi M, Moghadas Nejad F (2015) Comparing different rutting specification parameters using high temperature characteristics of rubber-modified asphalt binders. Road Mater Pavement Des 16(4):751–766
16. ASTM D6373 (2021) Standard specification for performance-graded asphalt binder. ASTM international, West Conshohocken
17. Shenoy A (2001) Refinement of the superpave specification parameter for performance grading of asphalt. J Transp Eng 127(5):357–362
18. ASTM D7405 (2020) Standard test method for multiple stress creep and recovery (MSCR) of asphalt binder using a dynamic shear rheometer. ASTM international, West Conshohocken
19. Golalipour A, Bahia HU, Tabatabaee HA (2017) Critical considerations toward better implementation of the multiple stress creep and recovery test. J Mater Civ Eng 29(5):04016295
20. Moreno-Navarro F, Tauste R, Sol-Sánchez M, Rubio-Gámez MC (2019) New approach for characterising the performance of asphalt binders through the multiple stress creep and recovery test. Road Mater Pavement Des 20(sup1):500–520
21. Kumar A, Choudhary R, Kumar A (2020) Characterisation of asphalt binder modified with ethylene–propylene–diene–monomer (EPDM) rubber waste from automobile industry. Road Mater Pavement Des 1–25
22. IS 73 (2013) Paving bitumen—specification. Bureau of Indian Standards, New Delhi
23. Cardona N, Campuzano F, Betancur M, Jaramillo L, Martínez JD (2018) Possibilities of carbon black recovery from waste tyre pyrolysis to be used as additive in rubber goods-a review. In: IOP conference series: materials science and engineering, vol 437. IOP Publishing, pp 012012
24. ASTM D2872 (2012) Standard test method for effect of heat and air on a moving film of asphalt (rolling thin-film oven test). ASTM international, West Conshohocken
25. MoRTH (2013) Specifications for road and bridge works (Fifth Revision). Indian Roads Congress, Govt. of India, Ministry of Road Transport and Highways, New Delhi

26. AASHTO T324 (2019) Standard method of test for Hamburg wheel-track testing of compacted asphalt mixtures (AASHTO)
27. Liu H, Zeiada W, Al-Khateeb GG, Shanableh A, Samarai M (2021) Use of the multiple stress creep recovery (MSCR) test to characterize the rutting potential of asphalt binders: a literature review. Constr Build Mater 269:121320

Evaluation of Warm Mix Asphalt with Reclaimed Asphalt Pavement in Rhode Island

Neha Shrestha, Kang-Won Wayne Lee, and George Veyera

Abstract The potential for using Warm Mix Asphalt (WMA) in reclaimed asphalt pavement (RAP) is becoming more interesting topic nowadays due to economic and environmental benefits. WMA technology allows reductions in production and compaction temperatures guaranteeing relevant environmental cost saving benefits. The objectives of the present study were to investigate and evaluate the performance of a typical additive in WMA pavement with RAP on rutting, fatigue cracking and thermal cracking resistance, which was used on Rhode Island (RI) Route 102. The asphalt binder was tested at different dosages of WMA additive using Dynamic Shear Rheometer (DSR), Rolling Thin Film Oven (RTFO), Pressure Aging Vessel (PAV), Multiple Stress Creep Recovery (MSCR), and Bending Beam Rheometer (BBR). It was found that 0.70% additive would lessen pavement damage in case of rutting, fatigue cracking and thermal cracking. After that, two HMA specimens, with and without RAP and two WMA specimen with and without RAP were prepared using Superpave Gyratory Compactor (SGC). These specimens were tested with the Asphalt Mixture Performance Tester (AMPT) and developed master curves for each specimen. It was observed that WMA mixtures with RAP, could perform better in fatigue resistance but expected to have poor rutting performance than HMA and HMA-RAP. This study indicates that addition of WMA additives performs better in fatigue resistance.

Keywords Warm mix asphalt · Reclaimed asphalt pavement · Dynamic modulus · Superpave gyratory compactor · Master curve · Fatigue

1 Introduction

Increasing truck traffic loads and traffic volumes have led to high demand for more durable and sustainable pavements. It has been estimated that truck traffic on US highways will continue to increase, surpassing all other modes of freight shipment

N. Shrestha · K.-W. W. Lee (✉) · G. Veyera
University of Rhode Island, Kingston, RI 02881, USA
e-mail: leekw@uri.edu

© The Author(s), under exclusive license to Springer Nature Switzerland AG 2022

H. R. Pasindu et al. (eds.), *Road and Airfield Pavement Technology*, Lecture Notes in Civil Engineering 193, https://doi.org/10.1007/978-3-030-87379-0_61

soon. Tractor trailers and heavy vehicles account for much of the damage done to highways [1]. Thus, a University of Rhode Island (URI) research team has been developing a durable pavement structure based on performance analyses as shown in Fig. 1. In addition, an asphalt mixture that will have high thermal cracking (TC) resistance and top-down fatigue cracking (TDFC) resistance was researched for wearing of surface course. Another mixture was evaluated for intermediate course that will have high fatigue cracking resistance in the asphalt layer in the present study. The final step is to develop an asphalt mixture that will have high bottom-up fatigue cracking resistance (BUFC) for asphalt base layer.

Since there is routinely a need for considerable pavement rehabilitation, substantial volumes of removed asphalt concrete, also called reclaimed asphalt pavement (RAP), are typically accumulated. The disposal of this material consumes a considerable volume of land and creates the potential for environmental pollution. These excess materials, in turn, create the opportunity for greater amounts of old asphalt pavements to be re-used and reclaimed, reducing the need for virgin materials, and providing an environmentally friendly alternative. Additionally, accumulated RAP materials and growing traffic volumes have resulted in the desire to find more sustainable pavement strategies and techniques.

RAP consists of granular pavement materials containing a mixture of bitumen and aggregates produced after the removal of the asphalt pavement by partial or full depth milling. Asphalt pavement rehabilitation involves milling and resurfacing of the existing asphalt pavement to mitigate rutting, cracking, potholes, and other distresses. Natural resources are conserved by using RAP, which also reduces costs, and is environment friendly. Millions of tons of RAP are stockpiled in the northeastern United States. Several studies indicating that RAP materials could be effectively used to resist rutting, thermal cracking, and temperature susceptibility using different recycling techniques [2–5]. One of the advantages of RAP utilization is that it does not have to be heated to a high temperature to prevent further aging (of its already aged binder), and the ability for rapid pavement restoration. In fact, to prevent RAP from getting heated to a very high temperature (very high referring to conventional virgin mix production temperature, e.g., 150 °C; recommended ranges of generally

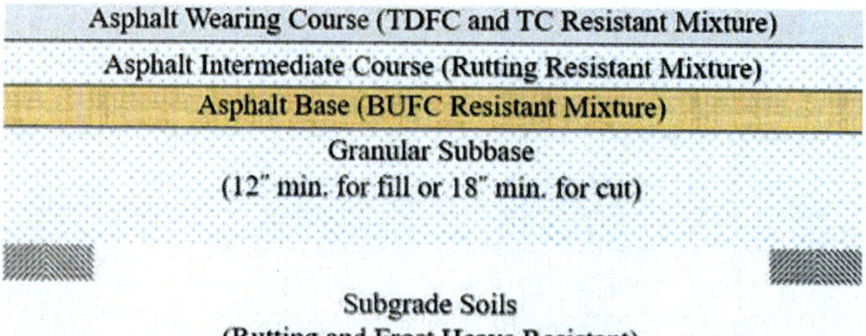

Fig. 1 Proposed durable asphalt pavement section for New England

accepted RAP temperatures are 110–135 °C), the material will be introduced in a separate spot in the recycling drum plant, so that it does not encounter the heater flame, but instead is heated by contact with the superheated (190 °C) virgin aggregates [6].

Warm Mix Asphalt (WMA) is a recent broad category of technology in the asphalt industry that is aimed at reducing energy consumption in the production and construction of asphalt pavements [7, 8]. WMA is produced at temperatures of about 15–40 °C, lower than those required to produce hot mixed asphalt (HMA). This technology provides a method of attaining low viscosity in the asphalt at relatively low temperatures [7]. Energy savings provide two major benefits: lower construction costs, and reduced emissions, which makes asphalt pavement construction more affordable and more environmentally conscious [8]. Incorporating WMA additives in RAP mixtures can increase the sustainability benefits, and also enhance asphalt pavement performance against rutting [5]. In addition, WMA allows high proportions of RAP to be used in asphalt mixtures [5]. WMA uses additives (organics/chemicals) or a water-based foaming process to reduce the temperature needed to produce asphalt materials [9, 10].

The Rhode Island Department of Transportation (RIDOT) has done several trial projects using different types of WMA additives to build stronger, more durable asphalt roads, with reduced physical maintenance requirements, providing reduced costs. Route 102 in Coventry, RI is one of the trial sections conducted by RIDOT using warm mix chemical additives. Typically, RI regular HMA asphalt pavements are designed for a 20-year life, and generally consist of four layers (subgrade soils, granular subbase, asphalt base, and asphalt surface). The routine maintenance and rehabilitation of pavement requires a effective strategy to meet design life requirement and beyond. Thus, a new pavement structure has been proposed as shown in Fig. 1. However, at this time, there are no specific layer materials to use based on performance. The present study was an initial exploration of suitable asphalt mixtures for each layer based on performance with and without RAP.

2 Objective of the Study

The primary objective of this study was to develop WMA mixtures with RAP in RI durable pavement structures for improved performance, and to investigate the feasibility of using a typical WMA additive in the asphalt layer. Specific research objectives were as follows:

1. Compare rheological properties of asphalt binder contained WMA additive with the asphalt binder,
2. Explore longer lasting asphalt mixtures including WMA additive with and without RAP,
3. Determine the dynamic moduli of asphalt specimens to develop Master Curves, and

Table 1 Unmodified and modified asphalt binders

Specimen name	Description
0-0	PG 58-28 without WMA additive
0-5	PG 58-28 with 0.5% WMA additive
0-7	PG 58-28 with 0.7% WMA additive
0-9	PG 58-28 with 0.9% WMA additive

4. Predict the rutting and cracking performance of pavement structures with WMA additive containing RAP.

3 Materials and Specimen Preparation

3.1 Asphalt Binder

Rhode Island typically uses PG 58-28 as the base asphalt binder when mixtures contain more than 15–25% of RAP as per RIDOT standard specifications [11]. For this study, a PG 58-28 asphalt binder without any polymer modification was acquired from a local distributor in Providence.

3.2 WMA Additive

A new generation chemical WMA additive, that has been commonly used by the asphalt paving industry in New England States, provides a temperature reduction in the range of 50–75 °C lower than typical HMA. Inclusion of the WMA additive helps to integrate recycled asphalt materials more easily in asphalt mixtures [12]. A typical chemical WMA additive has been used for this study, and specimens were prepared based on the manufacturer's recommendations. Table 1 shows both the unmodified and modified asphalt binder.

3.3 Mineral Aggregates

It is rare to obtain a desired aggregate gradation from a single aggregate stockpile. Therefore, Superpave mix designs usually draw upon several different aggregate stockpiles and blend them together in a ratio that will produce an acceptable final blended gradation. Superpave mix designs typically use 3 or 4 different aggregate stockpiles. The present study used four types of aggregates sources and one RAP source. Within the Superpave mix design requirements, gradations are identified based upon NMAS (Nominal Maximum Aggregate Size), which is defined as one

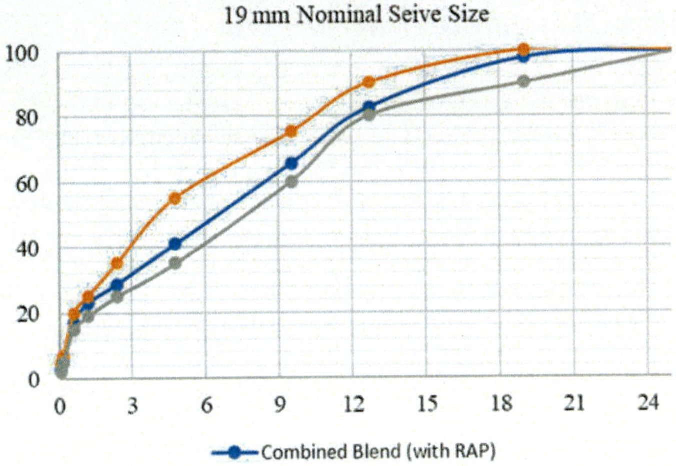

Fig. 2 Blend of aggregates passed through RIDOT specification

sieve size larger than the first sieve size that retains more than 10%. In accordance with RIDOT specifications, mixes containing RAP should have 19 mm NMAS. Figure 2 shows the blended aggregates with percentages that meet the RIDOT specification.

Specimen Preparation
WMA binders were prepared using a high shear mixer (i.e., mechanical stirrer with attached heater). Asphalt was heated to 138 °C (280°F) in the oven until the asphalt binder became liquid. Then, asphalt binder was removed from the oven, and the typical WMA additive was slowly introduced at a low mixing speed. The mixture was then sheared at a rotation speed of 150 rpm for about 30 min to ensure a homogenous blended mix.

4 Laboratory Testing Methods

4.1 Dynamic Shear Rheometer (DSR) Test

The DSR test was used to evaluate the viscous and elastic behaviors of asphalt binders at temperatures starting from medium to high. The DSR measures a specimen's complex shear modulus (G^*) and phase angle (δ). The complex shear modulus (G^*) can be the specimen's total resistance to deformation when repeatedly sheared, while the phase angle (δ), is the lag between the applied shear stress and the resulting shear strain. In Superpave, high temperature grade has been defined as the temperature where $G^*/\sin\delta$ of bitumen before and after aging is higher than 1.0 and 2.2 kPa, respectively.

4.2 Rolling Thin-Film Oven (RTFO) Test

The RTFO test was used to produce short-term aged asphalt binder for physical property testing following AASHTO T 240 [13]. Asphalt binder is exposed to oven temperatures at 163 °C (325°F) for 85 min to simulate manufacturing and placement aging. This aging is expected to represent typical conditions where the asphalt is considerably aged from the first exposure to the plant burner and contact with hot aggregates, throughout hauling and paving, until the final compaction takes place. Research leading to the development of the RTFO test indicated that an aging time of 85 min produces aging effects comparable to average field conditions.

4.3 Multiple Stress Creep Recovery (MSCR) Test

The MSCR test is the most recently method for binders to evaluate their non-recoverable creep compliance (J_{nr}), which is the ratio of non-recoverable strain to applied stress. Field loading conditions were simulated following AASTHO M332 procedures, which is why J_{nr} can be considered as a rutting parameter [13]. This approach provides a high temperature binder specification that more accurately represent the asphalt binder rutting performance. A major benefit of the MSCR test is that it eliminates the need to run additional tests such as elastic recovery, toughness and tenacity, and force ductility. This test is specifically designed to indicate the effects of polymer modification additives on the asphalt binder behavior. The MSCR test uses the well-established creep and recovery test concept to evaluate the binder's potential for permanent deformation.

4.4 Bending Beam Rheometer (BBR) Test

The BBR test provides a measure of the low temperature stiffness and relaxation properties of asphalt binders. These parameters give an indication of an asphalt binder's ability to resist low temperature cracking. PAV residue was tested at low temperature using the BBR to measure creep stiffness (S) and logarithmic creep rate (m) at 60 s as per AASTHO T313 [13]. The specification requires that the stiffness of the testing specimen (S) should be less than 300 MPa, and the value of m should be greater than 0.300 to resist low temperature cracking.

4.5 Indirect Tensile Test

The tensile properties of bituminous mixtures are evaluated in the Indirect Tensile test where a cylindrical specimen is loaded by a compressive load along a diametric plane at a constant rate acting parallel to and along the vertical diametrical plane of the specimen through two opposing loading strips. This loading configuration develops a relatively uniform tensile stress perpendicular to the direction of the applied load, and along the vertical diametrical plane, ultimately causing the specimen to be tested in tension by splitting along the vertical diameter. The Superpave mix design method was used to prepare the specimen for this test. HMA and WMA specimens with 20% RAP, were prepared at the optimum binder content (OBC) at a temperature of 25 °C for a period of 2 h in air. These specimens were then mounted on a conventional Marshall testing apparatus, loaded at a deformation rate of 51 mm/min with the load at failure recorded for each case. A 13 mm (0.5 in) wide loading strip was used for the 101 mm (4 in) diameter specimens to provide a uniform loading that produces a nearly uniform stress distribution. The static indirect tensile strength of specimens was determined using the procedure outlined in ASTM D 6931 [14].

4.6 Dynamic Modulus

For linear viscoelastic materials such as HMA, the stress–strain relationship under a continuous sinusoidal loading is defined by its complex dynamic modulus ($|E^*|$). This is a complex number that relates stress and strain for linear viscoelastic materials subjected to continuously applied sinusoidal loading in the frequency domain. The complex modulus is defined as the ratio of the amplitude of the sinusoidal stress (at any given time, t, and angular load frequency, ω), $\sigma = \sigma_o \sin(\omega t)$, and the amplitude of the sinusoidal strain $\varepsilon = \varepsilon_o \sin(\omega t - \delta)$, at the same time and frequency, that results in a steady state condition.

$$|E*| = \sigma/\varepsilon = \sigma_o \sin \omega t / \varepsilon_o \sin(\omega t - \delta) \tag{1}$$

where,

σ_o = peak (maximum) stress,
ε_o = peak (maximum) strain,
δ = phase angle (deg),
ω = angular velocity,
t = time (s), and
i = imaginary component of the complex modulus.

Stiffness (dynamic modulus) is a key material property that determines strains and displacements in pavement structures. Stiffness data for a HMA mix as obtained from the $|E^*|$ test provide very important information about the linear viscoelastic behavior of that particular mix over a wide range of temperatures and loading frequencies.

4.7 Master Curves

The mechanical behavior of viscoelastic materials such as asphalt composites is dependent on the temperature and the loading frequency at which the material is tested [15]. To compare test results of various mixes, it is useful to normalize one of these variables. Data collected at different temperatures can be "shifted" relative to the time of loading, so that the various curves can be aligned to form a single master curve. Master curves are constructed using the principle of time–temperature superposition. The data at various temperatures are shifted with respect to time until the curves merge into a single smooth function. The master curve for dynamic modulus as a function of time developed in this manner describes the time dependency of the material. The amount of shifting at each temperature required to form the master curve describes the temperature dependency of the material. The greater the shift factor, the greater the temperature dependency (temperature susceptibility) of the mixture.

Time–Temperature Superposition of $|E^*|$—In the Mechanistic-Empirical Pavement Design Guide (MEPDG) the stiffness of HMA at all temperature levels and time rate of loading, is determined from a master curve constructed at a reference temperature (generally taken as 70°F) [16]. Master curves are constructed using the principle of time–temperature superposition [17]. The data at various temperatures are shifted with respect to time until the curves merge into single smooth function. The master curve of the modulus as a function of time formed in this manner describes the time dependency of the material. The amount of shifting at each temperature required to form the master curve describes the temperature dependency of the material. In general, the master modulus curve can be mathematically modeled by a sigmoidal function described by:

$$\mathrm{Log}\left|E^*\right| = \delta + \frac{\alpha}{1 + e^{\beta+\gamma(\log t^r)}} \tag{2}$$

where,

t_r = reduced time of loading at reference temperature,
δ = minimum value of $|E^*|$,
$\delta + \alpha$ = maximum value of $|E^*|$, and
β, γ = parameters describing the shape of the sigmoidal function.
The shift factor can be shown in the following form:

$$a(T) = \frac{t}{t_r} \text{ or } \log(f_r) = \log(f) - \log[a(T)] \tag{3}$$

where,

$a(T)$ = shift factor as a function of temperature,
t = time of loading at desired temperature,
t_r = reduced time of loading at reference temperature, and
T = temperature of interest.

For improved precision, a 2nd order polynomial relationship between the logarithm of the shift factor, i.e., Log $a(T_i)$and the temperature in °F is used. The relationship can be expressed as follows:

$$\text{Log a}(T_i) = aT_i^2 + bT_i + c \quad (4)$$

where,

$a(T_i)$ = shift factor as a function of temperature T_i,
T_i = temperature of interest, °F, and
a, b and *c* = coefficients of the second order polynomial.

|E*|**$_{Measurements}$ ***from Master Curves—Following the new MEPDG's input Level-1 approach, |*E**| master curves of all mixtures were constructed for a reference temperature of 70°F using the principle of time–temperature superposition [17]. The data at various temperatures were shifted with respect to time until the curves merged into a single sigmoidal function representing the master curve using a 2nd order polynomial relationship between the logarithm of the shift factors, log $a(T_i)$ and the temperature. The time–temperature superposition was done by simultaneously solving for the four coefficients of the sigmoidal function (δ, α, β, and γ) as described in Eq. 2, and the three coefficients of the 2nd order polynomial (*a, b,* and *c*) as described in Eq. 4. The Microsoft Excel™ "Solver" function was used to conduct the nonlinear optimization for simultaneously solving these seven parameters.

5 Testing Results and Analysis

5.1 Rutting Resistance

Rutting is a wheel path surface depression resulting from plastic or permanent deformation in each pavement layer. AASHTOWare Pavement ME Design (PMED) computes the rut depths within the HMA, unbound aggregate layers, and subgrade soils. Since the New England States have strong subgrade soils, and use good quality granular subbase materials, the present study concentrated on investigating (intermediate course) HMA permanent deformations only. To resist rutting, an asphalt binder should be stiff, and it should be elastic. Therefore, the complex shear modulus elastic portion, $G^*/\sin\delta$ should be high. When rutting is of greatest concern, a minimum value for the elastic component of the complex shear modulus is specified. Intuitively, the higher the G^* value the stiffer the asphalt binder is, and the lower the δ value, the greater the elastic portion of G^* is. Therefore, the higher the rutting parameter, the better the performance against rutting.

From DSR results of unaged binder as shown in Fig. 3, asphalt binder containing 0.7 and 0.9% additive has the highest rutting parameter values compared to other binders. This suggests that that (0.7–0.9) % WMA binder has better performance against rutting.

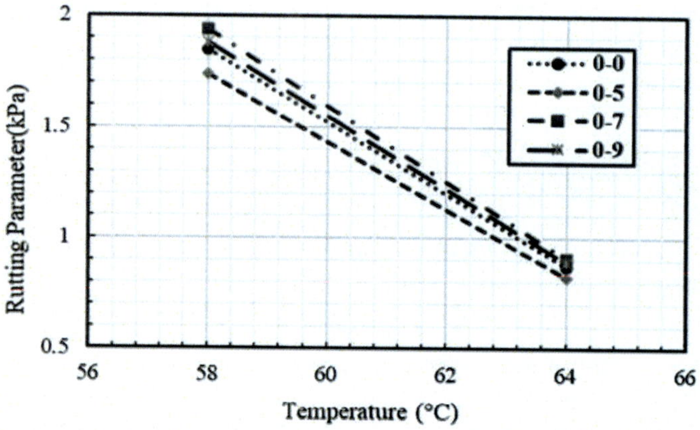

Fig. 3 Rutting parameter of asphalt binder and asphalt binder contained WMA additive

In the MSCR test, two separate parameters can be determined during each loading cycle: non-recoverable creep compliance (J_{nr}), and percentage of recovery (R). In addition to determining J_{nr}, the MSCR test can be used to determine the amount of recovery in an asphalt binder during the creep-recovery testing. MSCR Recovery provides an indication of the delayed elastic response of the asphalt binder. A high delayed elastic response indicates the asphalt binder has a significant elastic component at the test temperature. If the asphalt binder meets the appropriate J_{nr} specification, then it would be expected that the binder will minimize its contribution to rutting.

For standard traffic loading, J_{nr} (at 3.2 kPa) is required to have a maximum value of 4.0 kPa^{-1}. Also, AASHTO MP19 maintains a requirement that the difference in J_{nr} values between 0.1 and 3.2 kPa shear stress should not exceed a ratio of 0.75, and Table 2 shows that the results agree with these conditions. Also, from the MSCR test results (Table 2), asphalt binder containing 0.9% additives showed the higher recovery percentage, and lower non-recoverable creep compliance as compared with the regular asphalt binder.

Table 2 Results of MSCR tests for unmodified and modified binders

Asphalt binder	$R_{0.1}$(%)	$R_{3.2}$(%)	R_{DIFF}(%)	J_{NR}0.1(KPA^{-1})	J_{NR}3.2(KPA^{-1})	J_{NR}diff (%)
0-0	20.79	11.53	44.53	0.680	0.787	15.80
0-5	16.36	6.65	59.37	1.090	1.290	18.42
0-7	18.09	8.53	52.87	0.883	1.040	17.54
0-9	25.34	13.41	47.06	0.599	0.725	21.06

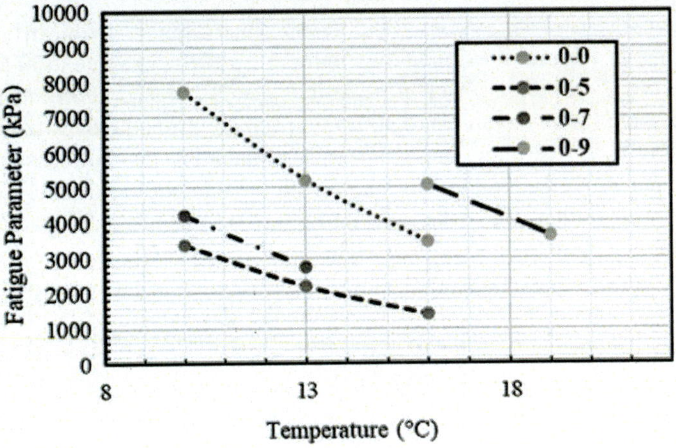

Fig. 4 Fatigue resistance of asphalt binder and asphalt binder with WMA additive

5.2 *Fatigue Resistance*

Fatigue damage in asphalt pavements is a complex phenomenon occurring from repeated bending resulting in asphalt pavement micro-damage. This micro-damage is a competitive process between micro-cracking and healing, manifested as a reduction in asphalt pavement stiffness, degrading the load-carrying capacity and the ability to resist further damage. Eventually, micro-cracks coalesce into macro-cracks that appear in the wheel path. To resist fatigue cracking, an asphalt binder should be elastic (able to dissipate energy by rebounding and not cracking), but not too stiff (excessively stiff substances will crack rather than deform-then-rebound). Therefore, the complex shear modulus viscous portion, $G^*\sin\delta$ should be a minimum. When fatigue cracking is of greatest concern (late in a HMA pavement's life), a maximum value for the viscous component of the complex shear modulus is specified. Figure 4 shows that 0.5 and 0.7% have the lowest $G^*\sin\delta$ values compared with the unmodified asphalt binder, which indicates that (0.5–0.7) % WMA binders showed better performance against fatigue cracking.

5.3 *Resistance Against Low Temperature Cracking*

The BBR test is used to determine an asphalt binder's creep stiffness as a function of time. Because low temperature cracking is a phenomenon found mostly in older pavements, the test is run on the long-term aged residue from the PAV. As temperature becomes low, the asphalt binders start to become stiffer. Stiffness is the extent to which an object resists deformation in response to an applied force. Also, creep stiffness describes the strain in response to stress at low temperature. And, if this creep stiffness is too high, cracking will occur. Since a higher creep stiffness value indicates higher thermal stresses, a maximum creep stiffness value (300 MPa) was specified (Fig. 5). Since a lower *m*-value indicates a reduced ability to for stress relaxation, a minimum *m*-value of 0.300 was specified. From the BBR test results as shown in Fig. 6, it was found that 0.5% WMA binder has less stiffness and higher relaxation parameters (*m*-value) compared with the unmodified asphalt binder, and 0.7% WMA binder has nearly the same stiffness and same relaxation parameter (*m*-value) compared with the unmodified asphalt binder. Therefore, it can be concluded that the (0.5–0.7) % WMA additives containing binders indicated better performance against lower temperature cracking. Also, from the above results in terms of rutting, fatigue cracking and thermal cracking, it can be concluded that 0.7% WMA additives was the optimal dosage rate for further study.

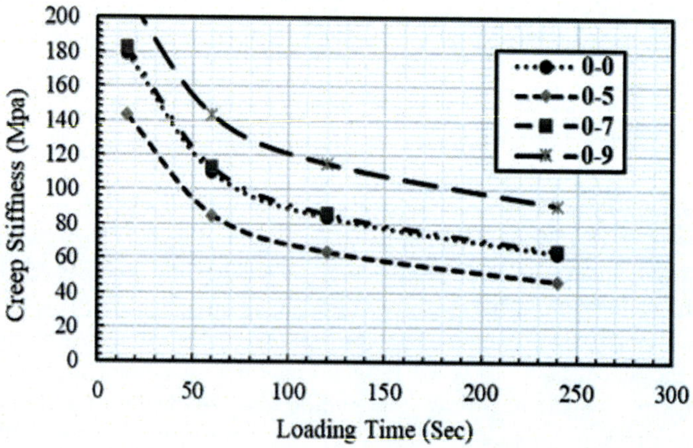

Fig. 5 Creep stiffness of asphalt binder and asphalt binder contained WMA additive

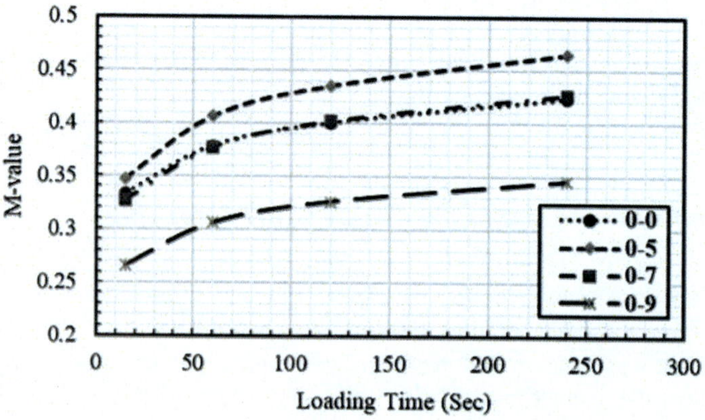

Fig. 6 m-value of asphalt binder and asphalt binder contained WMA additive

5.4 Preparation of Asphalt Mixtures

A Superpave gyratory compactor (SGC) is used as compaction devices and the compaction effort in mix design is based on expected traffic. It is used to prepare the SuperPave mix design and to find out the volumetric properties that relates to the performance of asphalt pavement. Once the binder and aggregates were selected for the asphalt mixture, then they were combined to produce the optimum mixture properties. The SGC is intended to simulate the rolling effect of field paving on the mixture, and it is more consistent with the field stress state of the mixture, being closer to the actual field conditions. In accordance with guidelines to make the specimens for specified traffic loading, less than 0.3 million ESALs were used. Several trial blends were evaluated to determine the optimum binder content (OBC).

5.5 Density and Voids Calculations

The following two measures of densities were determined:

Bulk specific gravity (*Gmb*), and Theoretical maximum specific gravity (*TMD, Gmm*).

5.6 OBC Selection

The OBC was determined at 4% air voids. From Fig. 7, the OBC for HMA with RAP was at 5.3, and 5.6% for WMA (0.7% additive) with RAP.

5.7 Indirect Tensile Test Results

Eight specimens were prepared and tested for indirect tensile strength. Results for the HMA and WMA mixtures at unconditioned OBC specimens are given in Table 3. The test results showed that the average tensile strength of HMA mixtures with 20% RAP is 677.5 kPa with a standard deviation of 299.1 kPa, whereas WMA is 588.6 kPa with a standard deviation 110.9 kPa. The test results showed a variation in strength, which could be attributed to specimen preparation, material handling, instrument handling, and other potential sources of error. However, it was found that HMA mixtures have high average tensile strengths compared with WMA mixtures.

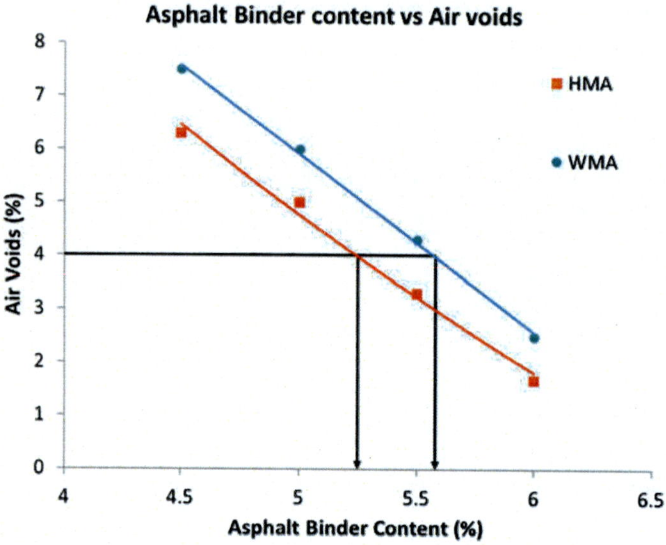

Fig. 7 Air void effect on different asphalt binder content

Table 3 Results of indirect tensile strength tests of WMA and HMA specimens

Specimen #	Specimen thickness (mm)	Specimen diameter (mm)	Maximum load (N)	IDT strength (kPa)
HMA #1	55.9	150	11,676.8	873.3
HMA #2	48.3	150	10,123.8	876.7
HMA #3	48.1	150	5224.3	453.6
HMA #4	50.8	150	6154.4	506.3
WMA #1	42.0	150	5255.5	523.6
WMA #2	50.8	150	6434.7	529.4
WMA #3	49.5	150	6483.7	547.1
WMA #4	48.3	150	8708.7	754.3

5.8 Dynamic Modulus Test Results

Four specimens were prepared and tested using the Asphalt Mixtures Performance Tester (AMPT). Testing details are shown in Table 4, and AMPT results are given in. Figure 8 Results indicate that the dynamic modulus of all WMA and HMA mixtures decreases significantly with the increase of temperature and exhibits frequency dependency under the same temperature conditions. Higher loading frequency means the higher dynamic modulus for asphalt mixtures, which illustrates the viscoelastic properties of that asphalt mixtures [18].

Table 4 Dynamic modulus of HMA and WMA specimens

Specimen A	Asphalt binder PG 58-28	HMA mixtures for Base with 20% RAP
Specimen B	Asphalt binder PG 58-28 with 0.7% Evotherm	WMA mixtures for base with 20% RAP
Specimen C	Asphalt binder PG 64-28	HMA mixtures for base without RAP
Specimen D	Asphalt binder PG 64-28 with 0.7% Evotherm	WMA mixtures for base without RAP

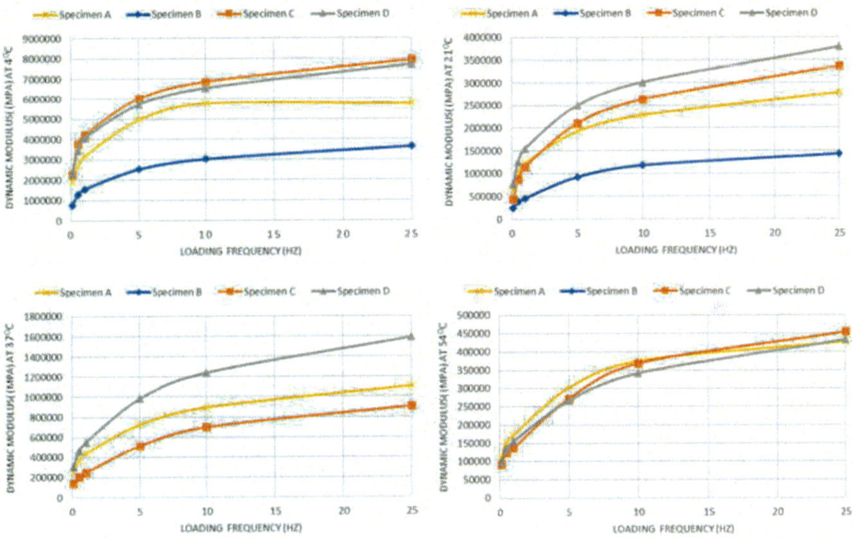

Fig. 8 Dynamic modulus of HMA and WMA specimens

Four specimens were prepared according to the AASTHO T 342-1 specification [13]. Specimens were tested in the AMPT machine to determine the dynamic modulus of the respective mixtures. Dynamic modulus testing requires a 150 mm high by 100 mm diameter specimen at a target air void content, cored from a 175 mm high by 150 mm diameter specimen.

In this study, 21 °C (70°F) is taken as the reference temperature and $|E^*|$ master curves of all mixes were developed at the reference temperature following the principle of time–temperature superposition. The data at various temperatures were shifted in line with frequency until the curve merges into a single sigmoidal function, representing the master curve (Fig. 9). The least-squares method is used for determining the coefficients of the sigmoidal functions (α, β, γ and δ) (Table 5). The shift factor is a function of temperature and is independent of strain level so it can be used within the linear viscoelastic range to be predict material behavior at any strain level.

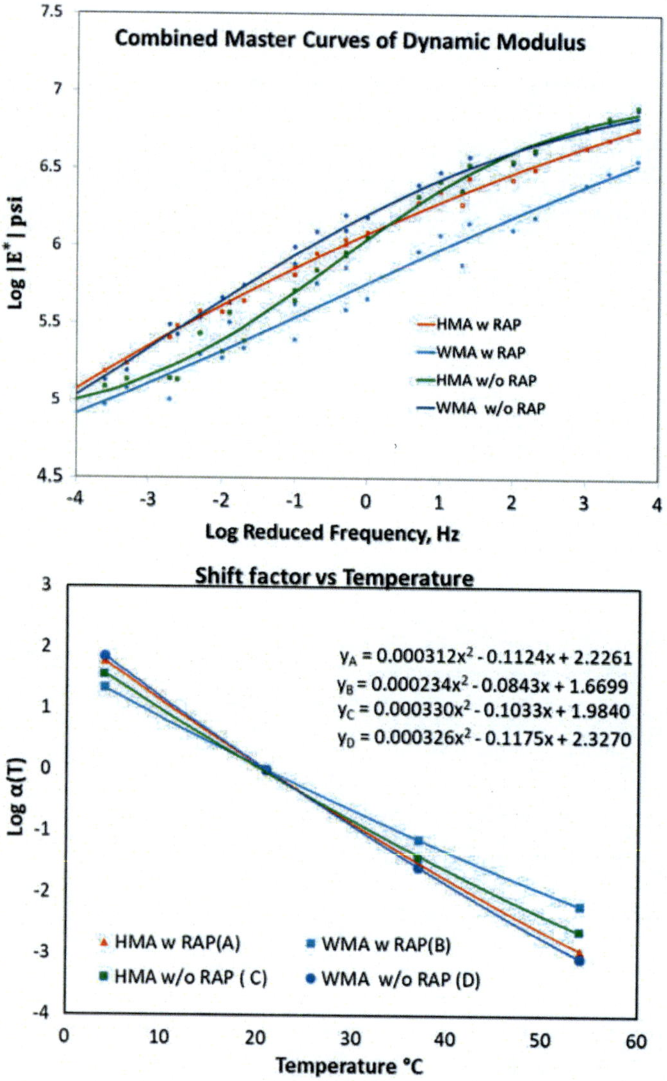

Fig. 9 Master curves and shift factor plots of HMA and WMA

Table 5 Master curve parameters

Specimen	δ	α	β	γ
HMA with RAP	4.568099	2.490947	−0.445790	−0.468820
WMA with RAP	4.629264	2.267807	0.0602610	−0.645040
HMA without RAP	4.84628	2.190182	−0.155180	−0.820180
WMA without RAP	4.088297	3.098425	−0.768200	−0.417560

Figure 9 shows combined master curve for four different mixes plotted together. This figure showed that at high temperature and low frequency, HMA-RAP indicates the highest reduced modulus, and WMA without RAP has second highest modulus compared to other mixes. These results suggest that WMA without RAP can be expected to improve performance against rutting compared with HMA without RAP and WMA-RAP [19]. Again, at high frequency and low temperature, WMA-RAP shows the lowest reduced modulus compared with all other mixes. Results for WMA-RAP mixtures can also be expected to improve the performance against fatigue cracking compared with other mixtures but was found to have poor rutting resistance compared with HMA and HMA-RAP.

6 Conclusions and Recommendations

Based on the laboratory test results and analyses from this study, the following conclusions can be made:

1. The addition of WMA additive enhanced the resistance properties of the asphalt binder against rutting, fatigue, and temperature cracking.
2. Combined Master Curves showed that WMA-RAP performed better in fatigue cracking resistance than HMA and HMA-RAP.
3. It was found that there was an average 20% drop oi tensile strength in WMA-RAP as compared to HMA-RAP mixtures.
4. There is need for further investigation and improvement regarding rutting performance using WMA additive.

Acknowledgements The present study was sponsored by USDOT Region 1 University Transportation Center (UTC)—Transportation Infrastructure Durability Center (TIDC) Grant. The RIDOT also provided partial support. Special thanks to Mr. Jose Lima, Mr. Michael Byrne, and Dr. Kate Wilson of the Material Section of the RIDOT. Also, the authors would like to acknowledge the technical support provided by the University of Rhode Island (URI), and the University of Connecticut (UConn), especially Mr. James Mahoney.

References

1. Lee KW, Peckham W (1990) Assessment of damage caused to pavements by heavy trucks in New England. Transp Res Record 1286
2. Lee KW, Todd B, Mueller M, Singh A (2016) A rational mix-design procedure for cold in-place recycling asphalt mixtures and performance prediction. J Mater Civil Eng. American Society of Civil Engineers (ASCE) 28
3. Lee KW, Soupharath N, Shukla A, Franco C, Manning F (1999) Rheological and mechanical properties of blended asphalt with recycled asphalt pavement binder. J Assoc Asphalt Paving Technol 68:89–128

4. Lee KW, Mueller M, Singh A (2014) Cold in-place recycling as a sustainable pavement practice. J Civil Eng Architect 8(6):680–692
5. Saleh M (2016) Laboratory evaluation of warm mix asphalt incorporating high RAP proportion by using evotherm and sylvaroad additives. Constr Build Mater 114:580–587
6. Mallick RB, Kandhal PS, Bradbury RL (2008) Using warm-mix asphalt technology to incorporate high percentage of reclaimed asphalt pavement material in asphalt mixtures. Transp Res Rec 2051(1):71–79
7. Zhang K, Wen H, Hobbs A (2015) Laboratory Tests and numerical simulations of mixing superheated virgin aggregate with reclaimed asphalt pavement materials. Transp Res Rec J Transp Res Board 2506:62–71
8. Hernandez F, Lee KW (2012) Sustainable warm-mix asphalt. In: Proceedings of US-Korea conference, Los Angeles, CA
9. Sebaaly P, Hajj EY, Piratheepan M (2015) Evaluation of selected warm mix asphalt technologies. Road Mater Pavement Des 16(sup1):475–486
10. Rubio MC, Martínez G, Baena L, Moreno F (2012) Warm mix asphalt: an overview. J Clean Prod 24:76–84
11. RIDOT (2004) Standard specifications for road and bridge construction. Rhode Island Department of Transportation
12. Kuang Y (2012) Evaluation of Evotherm as a WMA technology compaction and anti-strip additive. Citation is incomplete
13. American Association of State Highway and Transportation Officials (AASHTO) Specifications and testing procedures (2018)
14. American Societies of Testing Materials (ASTM) Specifications and testing procedures (2018)
15. Cross SA, Jakatimath Y, KC S (2007) Determination of dynamic modulus master curves for Oklahoma HMA mixtures (FHWA/OK 07 (05)). Oklahoma State University
16. AASHTO (2015) Mechanistic-empirical pavement design guide—a manual of practice, 2nd edn. Publication Code: MEPDG-2, ISBN: 978-1-56051-597-5
17. Pellinen TK, Witczak MW (2002) Stress dependent master curve construction for dynamic (complex) modulus (with discussion). J Assoc Asphalt Paving Technol 71:281–309
18. Zhu H, Sun L, Yang J, Chen Z, Gu W (2011) Developing master curves and predicting dynamic modulus of polymer-modified asphalt mixtures. J Mater Civ Eng 23(2):131–137
19. Shrestha N (2019) Performance prediction of warm mix asphalt pavement containing reclaimed asphalt pavement in Rhode Island. Open Access Master's Theses. Paper 1874

Feasibility of Usage of Waste Plastic as a Bitumen Modifier: A Review

Shankani Gunarathna and W. K. Mampearachchi

Abstract Bitumen is prone to deformations and cracking under severe temperatures and under fatigue. To overcome these drawbacks bitumen is modified by adding various additives. SBS and crumb rubber are some of the conventional bitumen modification materials. Recently, waste plastic has also been identified as a viable material to be used in bitumen modification. This would decrease the cost of bitumen modification as well as would provide a sustainable solution for waste plastic disposal. It has been found that incorporation of waste plastic in bitumen modification could increase the fracture resistance, rutting resistance and thermal stability of asphalt pavements. However, the ductility, elasticity and creep resistance were found to be decreased in waste plastic modified bitumen. Further, increased viscosity due to incorporation of waste plastic in bitumen modification had induced poor workability as well as higher mixing and paving temperatures. Increased temperatures not only cause ageing of bitumen but also cause emissions which could affect adversely on the environment. This paper critically reviews the properties of waste plastic modified bitumen with respect to several factors: types of plastic, modification methods, optimum plastic percentage, mixing temperature etc. Further, the paper discusses about the remedial measures that are being explored to improve the aforementioned drawbacks of waste plastic modified bitumen.

Keywords Bitumen · Modification · Recycle · Sustainable paving · Waste plastic

1 Introduction

Bitumen as a binding agent in flexible pavements is used widely due to its effective adhesive and viscoelastic characteristics [1–4]. However, bitumen is prone to viscoelastic deformations, cracking under severe temperatures and vulnerable for

S. Gunarathna (✉) · W. K. Mampearachchi
Department of Civil Engineering, University of Moratuwa, Moratuwa, Sri Lanka
e-mail: gunarathnaaasu.19@uom.lk

© The Author(s), under exclusive license to Springer Nature Switzerland AG 2022

H. R. Pasindu et al. (eds.), *Road and Airfield Pavement Technology*, Lecture Notes in Civil Engineering 193, https://doi.org/10.1007/978-3-030-87379-0_62

cracking under fatigue [5, 6]. Hence, bitumen modification has been studied extensively to enhance the performance of the bitumen. Styrene butadiene styrene (SBS) and crumb rubber are some widely and commonly used bitumen modifiers [7, 8].

Recently, studies have been done to investigate the incorporation of recycled materials in the modification of bitumen [5, 9]. Among the materials studied, usage of waste plastic in modification process was found to increase the fatigue resistance and rheological properties of bitumen successfully [10, 11]. The incorporation of polyethylene in the bitumen modification indicated that the flow and stability of asphalt were increased and the rutting potential was reduced [12–16]. However, it was observed that the ductility and elasticity of the bitumen were lowered with the incorporation of High density polyethylene (HDPE) and Low density polyethylene (LDPE) in the bitumen modification [14, 16]. Further, the creep recovery is decreased with the incorporation of HDPE in bitumen modification [13].

Waste plastic is considered as a significant environmental issue [17]. Therefore, incorporation of waste plastic in bitumen modification will reduce the waste plastic bags in landfills and could provide a sustainable solution for the disposal of plastic.

2 Bitumen Modification

Bitumen acts as the binder material for asphalt concrete. Asphalt concrete is a flexible pavement type and consisted with bitumen (binder) and aggregates (coarse and fine). Even though the asphalt concrete is the mostly used paving material it is susceptible for viscoelastic deformations, cracking under severe temperatures and vulnerable for cracking under fatigue. Therefore, modification of bitumen to improve the service life and improve the performance by eradicating the above negative effects is a well-researched area [18].

Polymers and fibers are being used as modifiers of bitumen. Elastomer, crumb rubber and plastomer are the three main types of polymer modifiers [19–21]. When selecting a polymer modifier, it should blend with bitumen and enhance the rutting, cracking and creeping resistance in high temperatures and brittle in low temperatures. Further, the viscosity of the modified bitumen should not be very high at mixing temperature [22]. The conventional bitumen modification materials are Styrene butadiene styrene (SBS) and crumb rubber [5, 7, 8, 23].

2.1 *Elastomer Bitumen Modifiers*

The elastomer polymers have the capability to return to the initial state after been subjected to increasing tensile strength which will increase the elongation up to 1300% of its original length [19]. They are amorphous and cross-linked polymer above its glass transition [24]. 75% of the bitumen modification is executed by use of elastomer polymers [20]. When elastomer polymer is mixed with bitumen two phases will be formed at the service temperature; Phase one is a polymer rich phase which is elastic and soft while Phase two is asphalt rich phase which is brittle and not elastic [19]. When polymer is added 3–6% by weight, polymer may swell and become more dominant and governs the bled properties [19].

The most common elastomer bitumen modifier is Styrene butadiene styrene (SBS) and identified as most suitable bitumen because SBS increase the elasticity properties and resistance to oxidation and thermal cracking [19]. However, in some research it is discovered that the workability of asphalt mixture and cost effectiveness had been reduced due to incorporation of SBS [8]. Due to the increasing cost of polymer modifiers many researchers are leaning towards harvesting waste materials as bitumen modifiers.

2.2 *Crumb Rubber Bitumen Modifiers*

Incorporation of crumb rubber as a bitumen modifier has been researched from about early 1960s [25]. It was found that in high temperature rubber particles will absorb the bitumen and swell allowing liquid concentration to increase [5]. Through the research conducted throughout the years it has discovered that the incorporation of rubber will decrease the potential for rutting and cracking as well as the maintenance cost and the traffic noise [26–28]. However, it is important to use the optimum rubber content, or the increment of rubber could result an undesirable higher level of flexibility and the tensile strength could be lowered by introducing rubber as itself to the bitumen [29]. Further, storage stability of crumb rubber modified bitumen in higher temperature is low and more research need to be conducted in this avenue [30]. Nevertheless, through research it is evident that the crumb rubber modifier is more effective in bitumen modification than SBS [8].

2.3 *Plastomer Bitumen Modifiers*

Plastomers are preferably termed as thermoplastic elastomers. When compared to elastomers, plastomers have high early strength and less flexibility resulting to fracture under higher strains [19]. Considering the bitumen modifiers, 15% of the bitumen modifiers are plastomer bitumen modifiers [20]. Polyethylene, Polypropylene (PP) and Polyvinyl Chloride (PVC) are some of the thermoplastics that have been in use. Incorporation of polymers as bitumen modifiers could increase the stiffness of the bitumen and temperature susceptibility leading to increase resistance to rutting [31, 32]. Plastomers can be used in either wet method or dry method where plastomers could dissolve in bitumen or use to coat aggregates respectively [12].

3 Use of Plastics as Bitumen Modifiers

3.1 *Polyethylene*

It is a semi crystalline material and it consisted with a long chain of carbon atoms with two hydrogen atoms attached to each carbon atom [12]. Mainly two types of polyethylene can be found: High density polyethylene (HDPE) and Low-density polyethylene (LDPE). Polyethylene is the highest ranked plastic production; hence it is the highest ranked plastic found in the waste with no proper disposal method [33]. Many studies have been conducted to investigate the potential of both HDPE and LDPE as bitumen modifiers.

Polyethylene is shredded and mixed into the base bitumen sample using a high-speed stirrer which is rotating at 3500 rpm [16]. The mixing temperature is 160–170 °C for 20–60 min [16, 20]. The modified bitumen sample is passed through an ASTM 100 sieve at 165 °C to verify the uniformity of the mixture [16]. The optimum percentage of polyethylene to be incorporated in the bitumen is identified as 5% by weight [16].

HDPE has an average density of 0.95 g/cm^3 and compared to LDPE it has higher impact resistance and higher tensile strength. It was discovered that when HDPE is incorporated in bitumen modification, it decreases the penetration and increase the viscosity and softening point (see Fig. 1) [20]. This could be due to the formation of dispersed polymer network as shown in Fig. 2 [20, 34]. Further HDPE demonstrate shear thickening behavior due to agglomeration of HDPE leading to a higher rutting resistance [20]. Further, it was found that HDPE modified bitumen has a higher elastic modulus and complex viscosity which will result in enhanced the dynamic viscoelastic properties [35]. Moreover, HDPE modified bitumen exhibited enhanced wear resistance and enhanced impact value [15]. Therefore, it is evident that the incorporation of HDPE enhances the properties of bitumen. However, HDPE incorporation in bitumen modification has found to decrease creep recovery and elastic recovery resulting increased potential for fatigue cracking [13].

LDPE has an average density of 0.92 g/cm^3 and high corrosion resistance and low moisture permeability [12]. However, stiffness, strength and temperature resistance of LDPE is very low [12]. When LDPE is incorporated in bitumen it was found that the flow and stability of bitumen has been increased [14]. Further, less plastic deformation, higher shear resistance, higher resilient modulus and higher resistance to cracking potential at lower temperature were discovered in LDPE modified bitumen [16]. LDPE modified bitumen exhibit similar decrease in penetration, increase in viscosity and softening point as the HDPE modified bitumen (see Fig. 1) [20]. However, the ductility and elasticity were observed to be decreased with the content of LDPE added [14, 16].

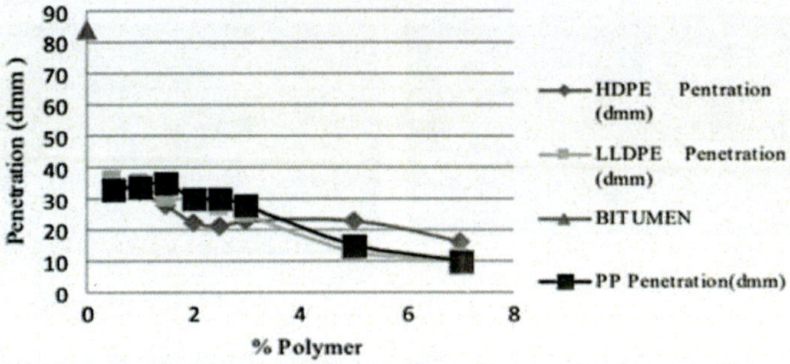

(a)

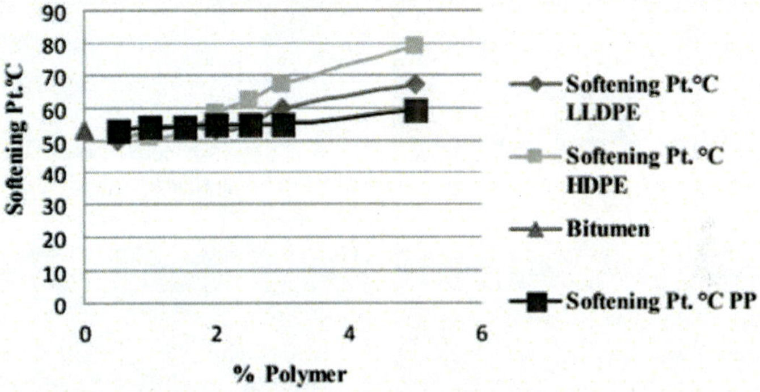

(b)

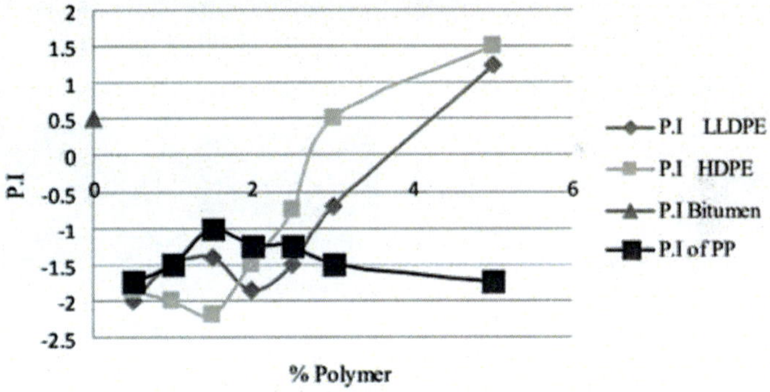

(c)

Fig. 1 Properties of Plastic modified bitumen; **a** penetration results for HDPE and LDPE, **b** softening point results for HDPE and LLDPE, **c** penetration index results for HDPE, LDPE, PP and Bitumen [20]

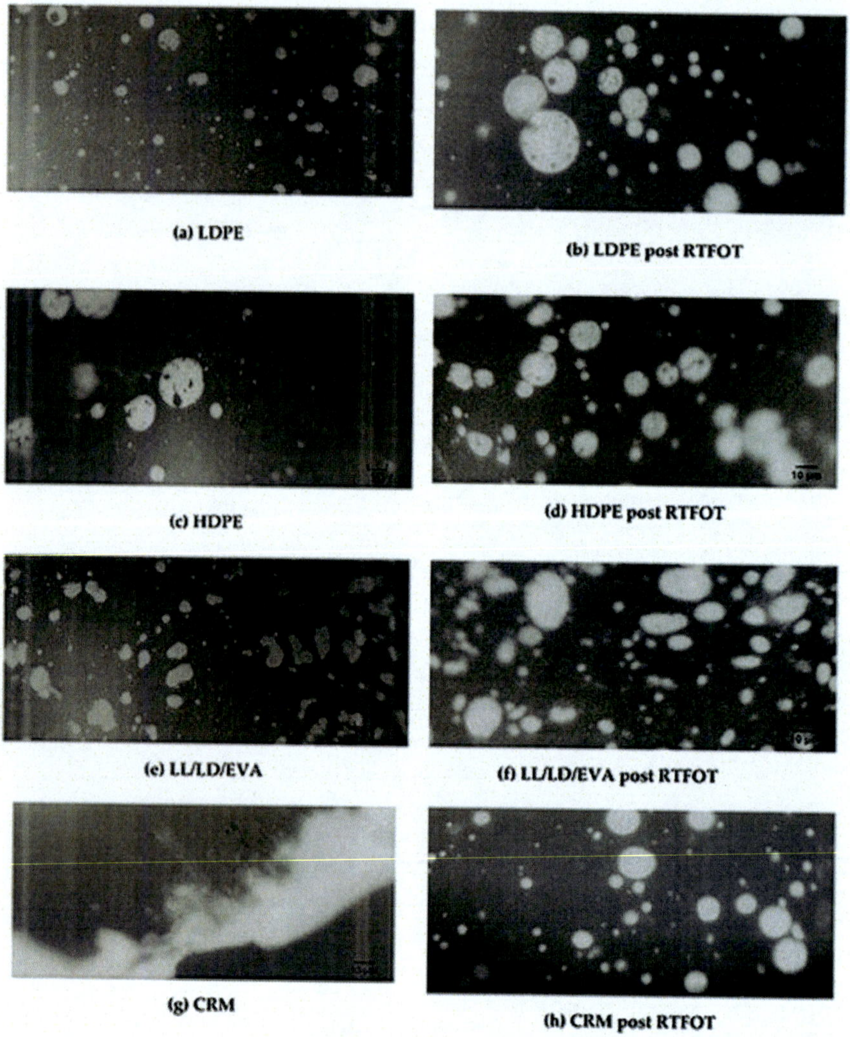

Fig. 2 Micrographs of the pslastic modified bitumen produced, before and after short-term aging: **a** bitumen modified with LDPE before aging; **b** bitumen modified with LDPE after aging at the RTFOT; **c** bitumen modified with HDPE before aging; **d** bitumen modified with HDPE after aging at the RTFOT; **e** bitumen modified with LDPE-LLDPE-EVA before aging; **f** bitumen modified with LD/LL/EVA after aging at the RTFOT; **g** bitumen modified with CRM before aging; **h** bitumen modified with CRM after aging at the RTFOT [34]

3.2 *Polypropylene*

Polypropylene is a thermoplastic and can be incorporated in the bitumen modification as a fibrous material [5, 36]. Further, polypropylene is the second ranked plastic in the world contributing to second ranked waste plastics [33]. Polypropylene is mixed with the base bitumen sample in the 163 °C temperature at a speed of 500 rpm for 2 h in constant mixer to obtain bitumen modification [20, 37]. The optimum percentage of polypropylene by weight was found as 0.125% for bitumen modification which demonstrated higher performance than SBS bitumen modification for same weight by ratio [38].

Incorporation of polypropylene as a bitumen modifier had increased the Marshal Stability values as well as the flow. Further, polypropylene bitumen modification had favourable impact on enhancement in rutting resistance, fatigue resistance and resistance to reflective cracking [38, 39].

However, melting point of polypropylene is 160 °C which is lower than the modification temperature. Hence, process of bitumen modification might not be completed properly to achieve the modification of bitumen [40].

3.3 Polyethylene Terephthalate (PET)

Polyethylene terephthalate (PET) is the third rank in the plastic production and about 18% of the plastic production is PET [33]. Hence, PET had been identified as a major contributor for plastics waste.

Mainly in PET bitumen modification, the aggregates were pre heated up to 200 °C for 2 h and bitumen was pre heated up to 150 °C for 1 h and mixed for about 5 min, then PET is blended to the mixture for 2 min in the 165 °C temperature [41]. The optimum weight by percentage of PET is found as 6% for bitumen modification [41].

Incorporation of PET bitumen modification has resulted in increased Marshal stability, resistance to rutting and fatigue as shown in Fig. 3 [42]. Further, it was found that PET has a low glass transition temperature and semicrystalinity which will allow to use it as an excellent polymer additive in bitumen modification [43].

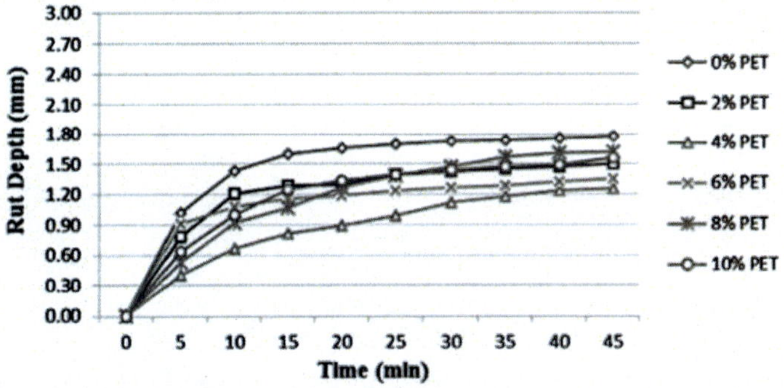

Fig. 3 Wheel track results for PET modified bitumen [42]

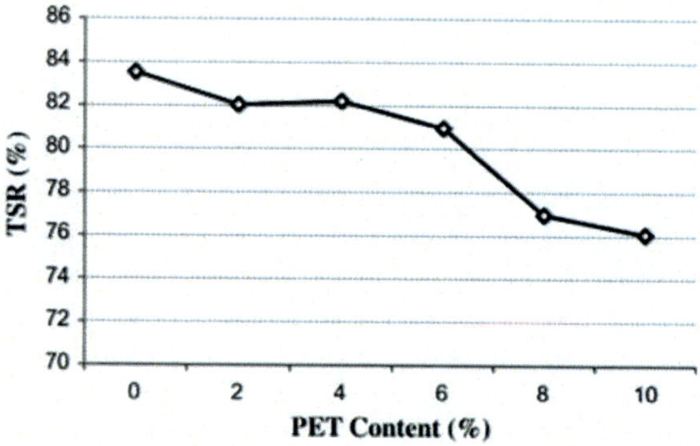

Fig. 4 TSR for asphalt mixtures for PET modified bitumen [41]

However, incorporation of PET in the bitumen modification did not increase the elasticity of the bitumen [41]. Further, the surface texture of the PET modified bitumen degraded due to phase separation [44]. Also, as shown in Fig. 4., the moisture susceptibility of the mixture is not improved due to the crystal form of PET after mixing that holds the sticky binder on its surface and decreases the asphalt film thickness around the aggregate [41].

3.4 Polyester Fiber Modified Bitumen

Polyester is a plastic fibrous material which are been extensively used in textile industry [45]. Hence, the polyester waste has been accumulated in the world. Therefore, research has been conducted to study the feasibility of incorporation of polyester in bitumen modification.

The polyester fibers are pre-heated to 165 °C and added slowly to the base bitumen sample and mixed by a constant mixer at a temperature of 165 °C and a speed of 500 rpm for 2 h [6, 46].

When polyester is incorporated in the bitumen modification, it was found that the viscosity of the bitumen was increased, and the fatigue potential had been reduced [6]. Further, the polyester reinforced bitumen has a higher stiffness, higher strength and it is susceptible to higher temperatures [46]. Polyester fibers create a spatial network and through this network the reinforce effect is functioned (see Fig. 5) [47]. Further, the reinforcement effect may depend on the fiber's size, orientation, shape and the tensile strength [47]. Furthermore, polyester fibers are chosen over polypropylene fibers due to their higher melting point and when strong and durable reinforcement is needed in higher temperatures [48].

Moreover, reinforcing bitumen by polyester may increase the resistance to fatigue damage [6], enhance the toughness and resistance to creep [49, 50] and incorporation of fibers in bitumen increase the elasticity of bitumen as shown in Table 1 [51].

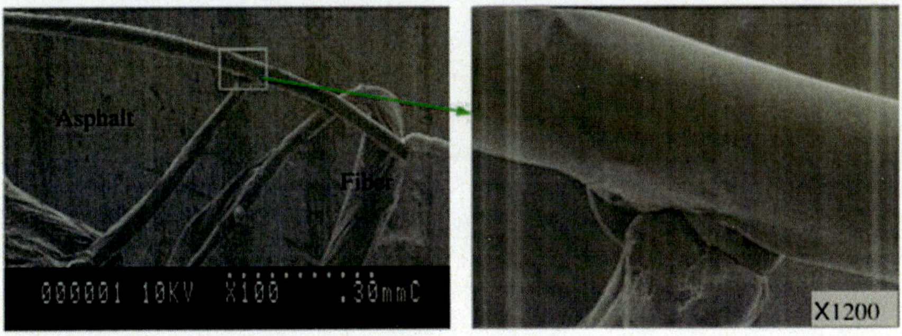

Fig. 5 SEM microstructure of fiber modified bitumen [47]

Table 1 DSR test results of fiber modified asphalts [51]

Fiber modified asphalt	Temperature (°C)	G^*(kPa)	Phase angle δ (°)	G^*/sin δ (kPa)
No fiber	82	0.412	89.6	0.412
Polyester fiber I	82	0.945	81.9	0.955
Polyester fiber II	82	1.380	80.6	1.399
Polyacrylonitrile fiber	82	0.952	82.2	0.961
Lignin fiber	82	2.460	69.5	2.626
Asbestos fiber	82	0.591	87.2	0.592

4 Bitumen Modification by Waste Plastics and Crumb Rubber

In the latest studies, it was found that the incorporation of waste plastic in the crumb rubber bitumen modification could increase the storage stability of the modified bitumen [10, 11].

Incorporation of Polyethylene in crumb rubber bitumen modification increase the bitumen performance in high temperature while crumb rubber increase the resistance to cracking in low temperature [52, 53]. Hence, incorporation of polyethylene and crumb rubber in bitumen modification has enhanced bitumen performance in both low and high temperatures.

Further, incorporation of PET in crumb rubber bitumen modification had been researched. Introduction of PET in crumb rubber bitumen modification increased the storage stability of the modified bitumen. Further, it improves rutting resistance, creep resistance and elastic and recovery properties [10]. However, effect on thermal stability was not profound [10].

5 Conclusion

In conclusion, it was found that use of waste plastic in bitumen modification increase the fatigue resistance, rutting resistance, creep resistance and resistance to reflective cracking. However, use of plastic in bitumen modification increase the viscosity which results in poor workability. Further, incorporation of polyethylene and polypropylene in bitumen modification in higher temperatures are incompatible due to their lower melting point.

However, incorporation of different materials in the bitumen modification would impact favourably in bitumen modification by complementing each other and eradicating the issues raised by the other material as in use of plastic in crumb rubber bitumen modification.

Therefore, it is recommended to research on incorporation waste plastic with blend of other bitumen modification materials to overcome the drawbacks identified.

References

1. Jacob H (1989) Classification, structure, genesis and practical importance of natural solid oil bitumen ("migrabitumen"). Int J Coal Geol 11(1):65–79
2. Mehrotra AK, Svrcek WY (1986) Viscosity of compressed Athabasca bitumen. Can J Chem Eng 64(5):844–847
3. Rubio MC et al (2012) Warm mix asphalt: an overview. J Clean Prod 24:76–84
4. Svrcek WY, Mehrotra AK (1982) Gas solubility, viscosity and density measurements for Athabasca bitumen. J Can Petrol Technol 21(04)
5. Sulyman M, Sienkiewicz M, Haponiuk J (2014) Asphalt pavement material improvement: a review. Int J Environ Sci Dev 5(5):444
6. Wu S, Ye Q, Li N (2008) Investigation of rheological and fatigue properties of asphalt mixtures containing polyester fibers. Constr Build Mater 22(10):2111–2115
7. Airey G, Singleton T, Collop A (2002) Properties of polymer modified bitumen after rubber-bitumen interaction. J Mater Civ Eng 14(4):344–354
8. Kök BV, Çolak H (2011) Laboratory comparison of the crumb-rubber and SBS modified bitumen and hot mix asphalt. Constr Build Mater 25(8):3204–3212
9. Jamshidi A et al (2017) Performance of pavements incorporating industrial byproducts: a state-of-the-art study. J Clean Prod 164:367–388
10. Leng Z, Padhan RK, Sreeram A (2018) Production of a sustainable paving material through chemical recycling of waste PET into crumb rubber modified asphalt. J Clean Prod 180:682–688
11. Ling T et al (2019) Value-added application of waste rubber and waste plastic in asphalt binder as a multifunctional additive. Materials 12(8):1280
12. Awwad MT, Shbeeb L (2007) The use of polyethylene in hot asphalt mixtures. Am J Appl Sci 4(6):390–396
13. Costa LM et al (2019) Using waste polymers as a reliable alternative for asphalt binder modification–performance and morphological assessment. Constr Build Mater 198:237–244
14. Kajugaran S, Weragoda V (2016) Development of polymer modified asphalt using filler. In: 2016 Moratuwa engineering research conference (MERCon). IEEE
15. Khurshid MB et al (2019) Enhancement of hot mix asphalt (HMA) properties using waste polymers. Arab J Sci Eng 1–10
16. Punith V, Veeraragavan A (2007) Behavior of asphalt concrete mixtures with reclaimed polyethylene as additive. J Mater Civ Eng 19(6):500–507

17. Mortimer G (2017) Getting rid of plastic bags: a windfall for supermarkets but it won't do much for the environment. Conversation 17
18. Mathew TV, Rao KK (2007) Chapter 23 pavement materials: bitumen, in pavement materials: bitumen, introduction to transportation engineering, NPTEL
19. Elseifi MA, Flintsch GW, Al-Qadi IL (2003) Quantitative effect of elastomeric modification on binder performance at intermediate and high temperatures. J Mater Civ Eng 15(1):32–40
20. Habib NZ et al (2011) Rheological properties of polyethylene and polypropylene modified bitumen. Int J Civil Environ Eng 3(2):96–100
21. Sitinamaluwa HS, Mampearachchi WK (2014) Development of a polymer-modified bitumen specification based on empirical tests–case study for Sri Lanka. Road Mater Pavement Des 15(3):712–720
22. Özen H et al (2008) Laboratory performance comparison of the elastomer-modified asphalt mixtures. Build Environ 43(7):1270–1277
23. Airey GD (2003) Rheological properties of styrene butadiene styrene polymer modified road bitumens☆. Fuel 82(14):1709–1719
24. Painter P, Coleman M (1997) Crystallization, melting and glass transition. Fundamentals of polymer science. In: Painter PC, Coleman MM (eds) 2nd ed. Lancaster, Pennsylvania: Technomic Publishing Company, Inc., pp 259–303
25. Scofield L (1989) The history, development, and performance of asphalt rubber at ADOT: special report. 1989: Arizona Department of Transportation
26. Huang B et al (2002) Louisiana experience with crumb rubber-modified hot-mix asphalt pavement. Transp Res Rec 1789(1):1–13
27. Lee S-J et al (2007) Laboratory study of the effects of compaction on the volumetric and rutting properties of CRM asphalt mixtures. J Mater Civ Eng 19(12):1079–1089
28. Liang RY, Lee S (1996) Short-term and long-term aging behavior of rubber modified asphalt paving mixture. Transp Res Rec 1530(1):11–17
29. Xiao F, Amirkhanian SN (2008) Resilient modulus behavior of rubberized asphalt concrete mixtures containing reclaimed asphalt pavement. Road Mater Pavement Des 9(4):633–649
30. Shu X, Huang B (2014) Recycling of waste tire rubber in asphalt and portland cement concrete: an overview. Constr Build Mater 67:217–224
31. Newman K (2004) Polymer-modified asphalt mixtures for heavy-duty pavements: fatigue characteristics as measured by flexural beam testing. In: FAA worldwide airport technology transfer conference, Atlantic city, NJ, USA
32. Ota Vacin C (2002) Investigation of polymer modified asphalt by shear and tensile compliances. In: Material characterization for inputs into AASHTO 2002 guide session of the 2004 annual conference transportation association, Canada, Québec City, Québec
33. Zia KM, Bhatti HN, Bhatti IA (2007) Methods for polyurethane and polyurethane composites, recycling and recovery: a review. React Funct Polym 67(8):675–692
34. Celauro C, Saroufim E, Mistretta MC, La Mantia FP (2020) Influence of short-term aging on mechanical properties and morphology of polymer-modified bitumen with recycled plastics from waste materials. Polymers 12(9):1985
35. González O et al (2002) Rheological techniques as a tool to analyze polymer– bitumen interactions: bitumen modified with polyethylene and polyethylene-based blends. Energy Fuels 16(5):1256–1263
36. Abtahi SM, Sheikhzadeh M, Hejazi SM (2010) Fiber-reinforced asphalt-concrete–a review. Constr Build Mater 24(6):871–877
37. Tapkın S, Çevik A, Uşar Ü (2009) Accumulated strain prediction of polypropylene modified Marshall specimens in repeated creep test using artificial neural networks. Expert Syst Appl 36(8):11186–11197
38. Abtahi S et al (209) A comparative study on the use of SBS polymers and polypropylene fibers modifying asphalt concrete structures. In: International conference sustainable aggregates, asphalt technology and pavement
39. Tapkın S et al (2009) Repeated creep behavior of polypropylene fiber-reinforced bituminous mixtures. J Transp Eng 135(4):240–249

40. Huang H, White TD (1996) Dynamic properties of fiber-modified overlay mixture. Transp Res Rec 1545(1):98–104
41. Ahmadinia E et al (2012) Performance evaluation of utilization of waste polyethylene terephthalate (PET) in stone mastic asphalt. Constr Build Mater 36:984–989
42. Garcia-Morales M et al (2006) Effect of waste polymer addition on the rheology of modified bitumen. Fuel 85(7–8):936–943
43. Kulo g ˇ lu N (1999) Effect of astragalus on characteristics of asphalt concrete. J Mater Civil Eng 11(4): 283–286
44. Padhan RK et al (2013) Poly (ethylene terephthalate) waste derived chemicals as an antistripping additive for bitumen–an environment friendly approach for disposal of environmentally hazardous material. Polym Degrad Stab 98(12):2592–2601
45. Anurag K, Xiao F, Amirkhanian SN (2009) Laboratory investigation of indirect tensile strength using roofing polyester waste fibers in hot mix asphalt. Constr Build Mater 23(5):2035–2040
46. Chen J-S, Lin K-Y (2005) Mechanism and behavior of bitumen strength reinforcement using fibers. J Mater Sci 40(1):87–95
47. Chen H, Xu Q (2010) Experimental study of fibers in stabilizing and reinforcing asphalt binder. Fuel 89(7):1616–1622
48. Maurer DA, Malasheskie GJ (1989) Field performance of fabrics and fibers to retard reflective cracking. Geotext Geomembr 8(3):239–267
49. Nai-sheng G, Ying-hua Z, Lue-lun S (2006) Effect of fiber contents on toughness of polyester fiber asphalt concrete. J Traffic Transp Eng 4
50. Ye Q, Wu S (2009) Rheological characteristics of polyester fiber modified asphalt mastic. J Highw Transp Res Dev 9:007
51. Fitzgerald RL (2000) Novel applications of caron fiber hot mix asphalt reinforcement and carbon-carbon pre-forms. Michigan Technological University Michigan
52. Yao Z et al (2018) Integrated utilization of recycled crumb rubber and polyethylene for enhancing the performance of modified bitumen. Constr Build Mater 170:217–224
53. Zhang J et al (2019) Experimental evaluation of crumb rubber and polyethylene integrated modified asphalt mixture upon related properties. Road Mater Pavement Des 20(6):1413–1428

Influence of Using Waste Materials as Partial Replacement of Coarse Aggregates in Concrete Paver Blocks

Ann George, I. C. Athira, Nithya Susan Joseph, Sneha Joseph, Vineetha Unnikrishnan, Maneesha Sara Mathew, Sophia Roy, and Maria Rebecca Thomas

Abstract Interlocking concrete block pavement has been widely used in place of flexible and rigid pavements owing to its easiness in assembling, replacement and transportation. It also has excellent drainage capacity. Its application includes sidewalks, car parks, cycle tracks and other lightly trafficked areas. The production of Interlocking concrete block pavement leads to extensive exploitation of natural resources, for the procurement of aggregates. Utilization of waste materials as partial replacement of aggregates conserves the scarce aggregates and reduces pollution. The objective of this study is to check the feasibility of using waste materials as a sustainable and economically viable option for producing concrete paver blocks. This study investigates the effect of using two easily available waste materials in Kerala viz. coconut shell and Ethylene Vinyl Acetate rubber as a partial replacement alternative of coarse aggregates in the production of concrete paver blocks. This paper investigates the effect of replacement of various percentages viz. 0, 5 10, 15 and 20% of coconut shell and rubber on the properties of the paver blocks. The effect on 7th day, 14th day and 28th day compressive strength as well as water absorption was studied. The optimum percentage of replacement was determined. Also, between coconut shells and scrap rubber the better alternative for replacement of coarse aggregate was evaluated. The compressive strength on replacement with coconut shells keeps on increasing till the optimum percentage of 15% is reached and then decreases gradually while for rubber there is an increase in trend till the optimum percentage of 5% is reached and then decreases. Replacement with coconut shell is a better alternative than rubber since more aggregates can be replaced.

Keywords Interlocking Block Pavement · Coconut shell · EVA rubber · Compressive strength

A. George (✉) · I. C. Athira
Department of Civil Engineering, Mar Baselios College of Engineering and Technology, Nalanchira, Thiruvananthapuram 695015, India
e-mail: ann.george@mbcet.ac.in

N. S. Joseph · S. Joseph · V. Unnikrishnan · M. S. Mathew · S. Roy · M. R. Thomas
Department of Civil Engineering, Mar Baselios College of Engineering and Technology, Nalanchira, Thiruvananthapuram 695015, India

© The Author(s), under exclusive license to Springer Nature Switzerland AG 2022

H. R. Pasindu et al. (eds.), *Road and Airfield Pavement Technology*, Lecture Notes in Civil Engineering 193, https://doi.org/10.1007/978-3-030-87379-0_63

1 Introduction

Interlocking Concrete Block Pavement have been extensively used in many countries for quite a long time owing to its easiness in assembling, replacement and transportation. It also has excellent drainage capacity. Paver is a multi-shaped piece of brick, concrete or tile. Some of the applications include pedestrian side-walks, footpaths, cycle tracks, car parks and malls and other lightly trafficked areas. For light traffic the pavement can consist of blocks 60 mm thick laid over a sand bedding 20–30 mm and a base course 200 mm thick.

The production of Interlocking concrete block pavement, leads to extensive exploitation of natural resources for procuring aggregates. Utilization of waste materials as partial replacement of aggregates conserves the scarce aggregates and reduces pollution.

Coconut shell is a form of agricultural solid waste with possible use as coarse aggregate in the production of concrete. India occupies the premier position in the world in coconut production with an annual production of 13 billion nuts, followed by Indonesia and Philippines. Although research has been carried out to understand the mechanical properties of concrete paver blocks yet those involving coconut shells as replacement alternative for aggregate is limited in extent. Hence, further research is needed for better understanding of the behavior of coconut shells as aggregate in concrete.

Ethylene Vinyl Acetate (EVA) Rubber popularly known as expanded rubber or foam rubber is used as padding for various sports accessories such as ski boots, bicycle saddles, hockey pads, boxing gloves and helmets, water ski boots, fishing rods etc. It is typically used as a shock absorber in sports shoes. EVA is lightweight, easy to form, odourless, glossy and cheaper compared to natural rubber. EVA is much softer than rubber and shows rubber-like properties owing to its flexibility. Hence it can be used as partial replacement alternative for coarse aggregate in interlocking concrete block pavement.

This study investigates the effect of using two easily available waste materials in Kerala viz. Coconut shell and Ethylene Vinyl Acetate rubber as a partial replacement alternative of coarse aggregates in the production of concrete paver blocks.

2 Review of Earlier Works

Jain et al. studied the effect of replacement of coarse aggregate in concrete by coconut shells by trying 10%, 20%, 30% and 40% replacement using M20 mix. The effect of 7 day and 28 day compressive strength was evaluated and it was found that the strength decreases with increase in percentage of coconut shells. Annadurai et al. found that properties of coconut shell aggregate concrete was almost comparable to conventional concrete. The study was conducted on reinforced concrete pipes and it was found that it behaved well under hydrostatic pressure. Studies conducted by

Annadurai on concrete mixes with varying percentage of coconut shells indicate that the density and compressive strength decreases whereas the deflection increases but it gives a warning before failure. The dry density, deflection characteristics and 7 day, 14 day and 28 day compressive strength were evaluated. It was also found that coconut shell aggregate concrete requires pretreatment for water absorption before its use as an aggregate in concrete making. The compressive strength was found to increase with age and it was found that coconut shell aggregate concrete doesn't deteriorate once coconut shell aggregates are encapsulated in concrete. Studies conducted by Ramachandrudu et al. indicates that the density, workability and compressive strength decreased with increased percentage of coconut shell, however the voids and the water absorption increased. Discarded rubber tyres were chipped to 10–20 mm size by Bompa et al. to replace the coarse aggregates in concrete. The increase in use of rubber decreases the density and compressive strength, however an increase in toughness was observed. Torgal et al. studied the effect of using rubber tyres as partial replacement alternative for sand. Metakaolin and fly ash were used as partial replacement alternative for cement which was able to minimize the strength loss due to addition of rubber. The durability was assessed in terms of capillary water absorption and resistance to sulphuric acid attack. The results indicate that upto 15% of replacement with rubber is possible. Tung-Chai Ling et al. varied the percentage of rubber from 5 to 50% as a replacement alternative for sand. The increase in percentage of rubber decreased both the density as well as compressive strength. The investigations by Khorami et al. shows that addition of rubber decreases the stiffness and load bearing capacity of concrete. There was a decrease in compressive strength and modulus of elasticity and increase in water absorption with the addition of rubber. Chaikaew et al. studied the effect of using crumb rubber to replace coarse and fine aggregates in concrete pedestrian block. It was found that blocks were lighter had more flexibility with better energy absorption. However, they were less strong compared to conventional concrete blocks. It was noted that both compressive and flexural strength are found to decrease with rubber content, while the toughness increases.

3 Methodology

The coconut shells are obtained from oil mills and houses and the scrap rubber are procured from the local market. The coconut shells are crushed, and scrap rubber are chipped to a size range of 6–12 mm. According to IRC SP: 63-2004, for areas of light traffic, paver blocks made with M35 grade of concrete is used. The proportion of the ingredients for M35 mix is found by preparing various trial mixes, keeping water cement ratio constant and by varying the cement content.

3.1 Material Testing

The tests conducted includes aggregate impact test, aggregate crushing test, abrasion test and shape test. Specific gravity of all the materials is found out to determine the required weight for each mix proportion. The other tests include Sieve analysis, Standard consistency and setting time. Compressive strength test and water absorption test were also carried out.

3.2 Mix Design

Mix design is the process of selecting suitable ingredients of concrete and determining their relative proportions, with the object of producing concrete of certain minimum strength and durability as economically as possible. Trial mixes with 350, 370 and 400 kg of cement were prepared. Cubes of 15 cm x 15 cm x 15 cm were then cast and tested. The compressive strength of the resulting mixes were observed to be less than the required value. Therefore, trial mix with 420 kg of cement was adopted as it conformed to the requirements of M35 mix.

3.3 Casting of Paver Blocks

The paver blocks of size 200 × 100 × 60 mm were cast. The effect of replacement of various percentages viz. 0, 5 10, 15 and 20% of coconut shell and rubber on the properties of the paver blocks were studied by determining the 7th day, 14th day and 28th day compressive strength as well as water absorption characteristics. The optimum percentage of replacement was found out. Also, between coconut shells and scrap rubber the better alternative for replacement of coarse aggregate was evaluated.

Figure 1 shows the casting of paver blocks.

The test specimen are then stored in moist air for 24 h and after this period the specimens were removed from the moulds and kept under water for 28 days.

4 Results

The various ingredients used for the production of concrete paver blocks are cement, coarse aggregate, fine aggregate and water. Waste coconut shell and rubber chips were used to partially replace the coarse aggregates in concrete. The properties of the constituents are summarized below.

Fig. 1 Casting of paver block

Table 1 Characteristics of cement

Name of the test	Test result	Standard value
(1) Standard consistency test	28%	26–33%
2) Setting time	Initial setting time = 45 min Final setting time = 430 min	Initial setting time not less than 30 min and final setting time not more than 600 min
(3)Specific gravity	3	3.15

4.1 Cement

Portland Pozzolana cement was used in the investigation. The physical properties of the cement are tabulated in Table 1. The results conform to IS specifications.

4.2 Fine Aggregate

Due to the scarcity of river sand, M sand was used for the entire work. M sand (less than 4.75 mm) having specific gravity 2.59 was used as fine aggregate. The grading curve of M sand was plotted as shown in Fig. 2. From the grading curve, it can be seen that the fine aggregate used conforms to the specifications of Zone 2.

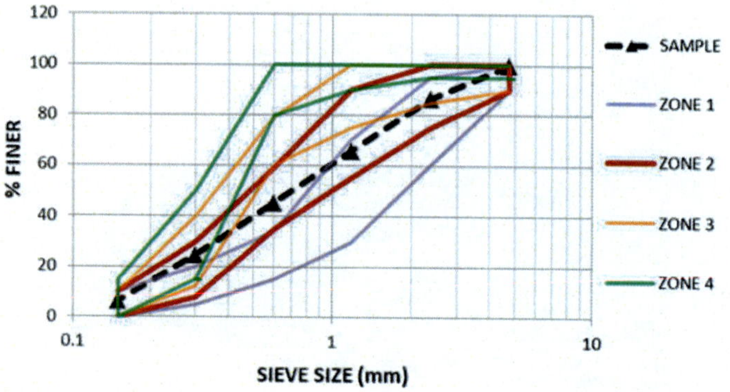

Fig. 2 Grading curve of fine aggregate

Table 2 Characteristics of coarse aggregate

Name of the test	Test result	Standard value
(1) Crushing strength	23.64%	30%
(2) Abrasion test	33.84%	35%
(3) Impact test	14%	10–20% (very tough)
(4) Shape test	Flakiness index:13.26% Elongation index:26.53%	Flakiness index 35%
(5) Specific gravity	2.77	2.5–3.2
(6) Water absorption	0.4%	4.1–2%

4.3 *Coarse Aggregate*

Crushed aggregate of maximum 12 mm size was used as coarse aggregate in the present work. The size of coarse aggregates ranges between 6 and 12 mm. The properties of the coarse aggregate obtained are tabulated in Table 2. The result conforms to IS recommendations.

4.4 *Coconut Shell*

The coconut shell used in this work ranges in size from 6 to 12 mm. Figure 3 shows the crushed shells of coconut. The properties of the coconut shells obtained are tabulated in Table 3.

Fig. 3 Crushed shells of coconut

Table 3 Characteristics of coconut shell

Name of the test	Test result
(1) Specific gravity	1.3
(2) Water absorption	1.3%

4.5 Rubber

Rubber used in this work is of size ranging from 6 to 12 mm. Figure 4. shows the chipped pieces of rubber. The properties of the rubber are given in Table 4.

The particle size distribution curve for Coarse aggregate, coconut shell and rubber is shown in Fig. 5.

4.6 Water Absorption

The water absorption characteristics for paver blocks were studied by varying the percentage of coconut shell and rubber by 0, 5 10, 15 and 20% respectively. The results of water absorption test are shown in Fig. 6 and Fig. 7 for coconut shell and rubber respectively.

Coconut shells are capable of absorbing water owing to their porous nature. The replacement of coarse aggregate with coconut shells shows a decrease in percentage of water absorption initially. The paver blocks corresponding to 15% replacement of coconut shells gives the least water absorption value. This decrease in water absorption is considered to be due to a part of dry aggregates being replaced by pretreated coconut shells, which absorb less moisture than dry aggregates. The water absorption trend shows an increase after 15%. This can be mainly due to the formation

Fig. 4 Chipped pieces of rubber

Table 4 Characteristics of rubber

Name of the test	Test result
(1) Specific gravity	0.9
(2) Water absorption	1.03%

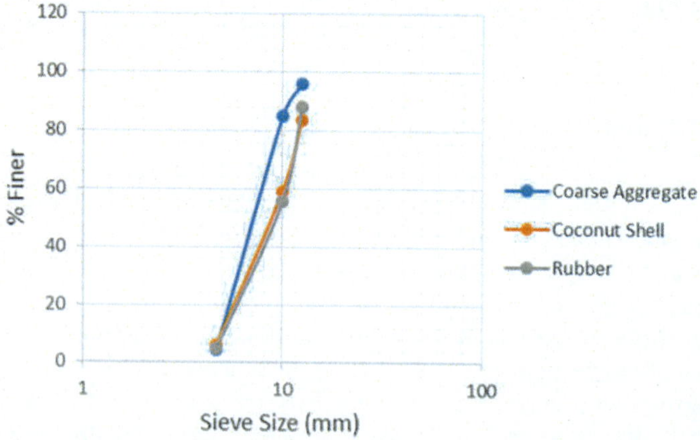

Fig. 5 Particle size distribution curve of coarse aggregate, coconut shell and rubber

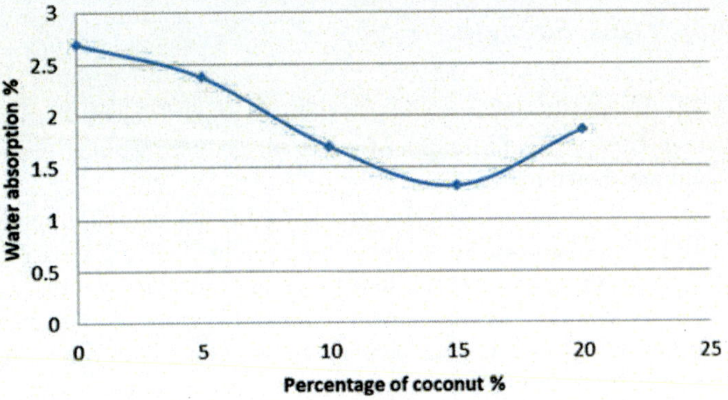

Fig. 6 Water absorption of paver blocks with varying percentage of coconut shell

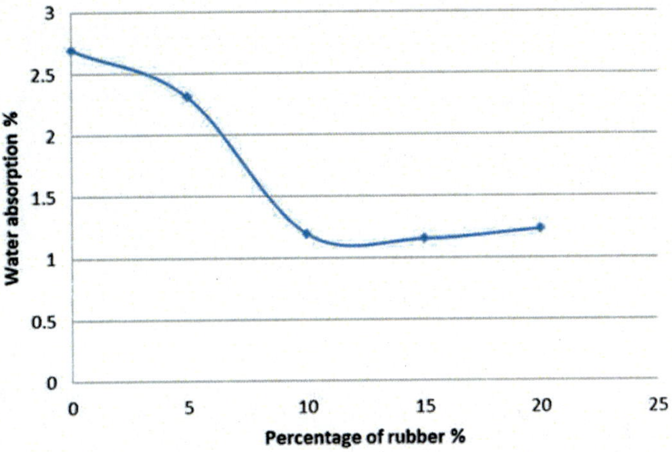

Fig. 7 Water absorption of paver blocks with varying percentage of rubber

of minor cracks in the cement around the coconut shells which leads to increased water permeability.

EVA rubber does not absorb much water. Hence there is a decrease in the water absorption by the paver block with higher percentage of rubber. But beyond a certain point further increase in the percentage of rubber reduces the bond with the cement and there is a chance of formation of minute cracks around the rubber pieces. This can lead to the penetration of water into the block and causes slight increase in the water absorption by the paver block.

4.7 Compressive Strength

The 7-day, 14- day and 28-day compressive strength of paver block were found out.

The average Compressive Strength of paver blocks with varying percentage of Coconut Shell are shown in Fig. 8.

The compressive strength keeps on increasing till the optimum percentage of coconut shells (15%) is reached and then decreases gradually. The decrease is because of the flaky nature of coconut shells that increase voids between the aggregates which decreases the bond between cement paste and aggregates. The compressive strength of paver blocks with various percentages of rubber is shown in Fig. 9. The compressive strength shows an increase in trend till optimum percentage of rubber

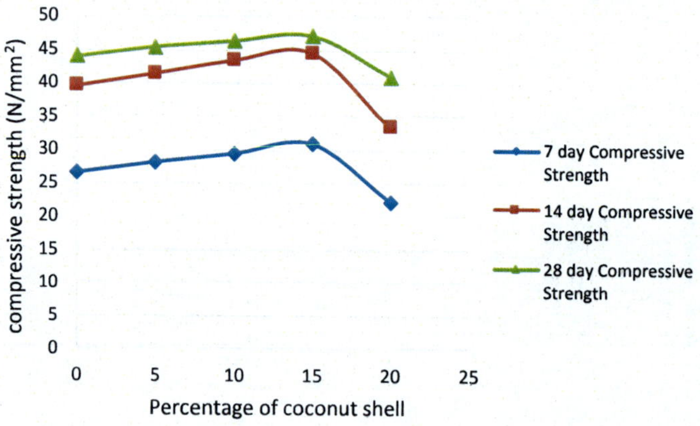

Fig. 8 Compressive strength of paver blocks with varying percentage of coconut shells

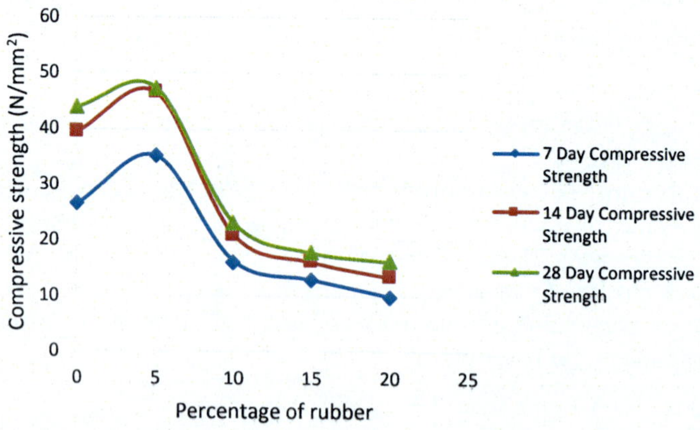

Fig. 9 Compressive strength of paver blocks with varying percentage of rubber

Table 5 Cost analysis

Materials	No of blocks	Unit rate (Rs./piece)	Amount (Rs.)
Conventional paver block	1,87,500	11.79	22,09,743
5% rubber replaced paver block	1,87,500	11.59	21,73,350
15% coconut replaced paver block	1,87,500	11.20	21,00,563

(5%) is reached and then decreases. The bond between the EVA rubber and cement paste is low and therefore the strength gradually decreases.

From the above results, the optimum percentage of replacement of coarse aggregates for coconut shell is 15% and that for rubber is 5%. Compressive strength of coconut shells is observed to be more when compared to rubber. The optimum percentage of coconut shell obtained is 15% and therefore it is the best alternative than rubber since more aggregates can be replaced.

4.8 Cost Comparison

The cost of paver block based on the market value of materials is as shown in Table 5.

For paving one kilometer single lane road, there will be a cost reduction of 5% and 2% by considering optimum percentage replacement of coconut shell and rubber respectively. The cost of paver block having 15% replacement by coconut shell is found to be the least. Thus, the production of these paver blocks on large scale is economical.

5 Conclusion

An experimental study on utilization of waste materials such as Coconut Shells and EVA Rubber for replacing coarse aggregates in paver blocks was carried out. The following conclusions can be drawn from the study.

- In case of replacement with coconut shell, the compressive strength keeps on increasing till the optimum percentage of coconut shells, of 15% is reached and then decreases gradually.
- For optimum percentage replacement of coarse aggregates by coconut shell, the 7 day, 14 day and 28 day compressive strength was found to increase by 16.6%, 12.6%, and 7.2% compared to control mix. Although there is increase in compressive strength with age, the rate of gain in strength declines with age, when compared to the control mix.

- In case of replacement with coconut shell, the percentage of water absorption decreases gradually. The paver blocks corresponding to 15% replacement of coconut shells gives the least water absorption value.
- For optimum percentage replacement of coarse aggregates by coconut shell, there was 51% decrease in water absorption value compared to the control mix.
- In case of replacement with rubber, the compressive strength shows an increase in trend till the optimum percentage of rubber, of 5% is reached and then decreases.
- For optimum percentage replacement of coarse aggregates by rubber, the 7 day 14 day and 28 day compressive strength was found to increase by 33%, 17.6% and 6.4% compared to control mix. Although there is increase in compressive strength with age, the rate of gain in strength declines with age, when compared to the control mix.
- The 28 day compressive strength for optimum percentage replacement with coconut shell was found to be more than that for rubber.
- In case of replacement with rubber, there is a decrease in the water absorption by the paver block with higher percentage of rubber.
- For optimum percentage replacement of coarse aggregates by rubber, there was 14% decrease in water absorption value compared to the control mix.
- For paving one kilometer single lane road, there will be a cost reduction of 5% and 2% by considering optimum percentage replacement of coconut shell and rubber respectively. The cost of paver block having 15% replacement by coconut shell is found to be the least.
- Hence coconut shell is a potential material which can be used as partial replacement alternative for coarse aggregate in paver block production. Compared to rubber, more replacement of coarse aggregate is possible. Also replacement with coconut shell is more economical.

6 Scope for Future Work

The viability of using coconut shell and rubber has been studied by testing its compressive strength (7 days, 14 days and 28days) and water absorption. However more aspects like durability, and, hardness can also be studied in future. Also the strength of the paver blocks after 60 days, 90 days & 120 days can also be studied.

References

1. Kanojia A, Jain SK (2017) Performance of coconut shell as coarse aggregate in concrete. Constr Build Mater 140(5):150–156
2. Gunasekaran K, Annadurai R, Chandar SP, Anandh S (2016) Study for the relevance of coconut shell aggregate concrete non-pressure pipe. Ain Shams Eng J 8(4):523–530
3. Gunasekaran K, Annadurai R, Kumar PS (2013) Plastic shrinkage and deflection characteristics of coconut shell concrete slab. Constr Build Mater 43(1):203–207

4. Yerramala A, Ramachandrudu C (2012) Properties of concrete with coconut shells as aggregate replacement. Int J Eng Inventions 1(6):21–31
5. Bompa BDV, Elghazouii AV, Teran AM, Stafford PJ (2017) Behaviour of rubberised concrete members in asymmetric shear tests. Constr Build Mater 159(4):361–375
6. Azevedo F, Torgal P, Aguiar CI, Camoer AF (2012) Properties and durability of HPC with tyre rubber water. Constr Build Mater 28(8):186–191
7. Ling TC (2011) Prediction of density and compressive strength for rubberized concrete blocks. Constr Build Mater 25(2):4303–4306
8. Ganjian E, Khorami M, Maghsondi AA (2009) Scrap-tyre-rubber replacement for aggregate filler in concrete. Constr Build Mater 23(4):1828–1836
9. Sukontasukkul P, Chaikaew C (2006) Properties of concrete pedestrian block mixed with crumb rubber. Constr Build Mater 20(2):450–457
10. IS 10262: 2009; Guidelines for concrete mix design proportioning, Bureau of Indian Standards, New Delhi
11. IS 15658:2006; Precast concrete blocks for paving, Bureau of Indian Standards, New Delhi
12. IRC SP: 63-2004; Guideline for the use of interlocking concrete block pavement, The Indian Roads Congress, New Delhi
13. IS 2386(Part 1)-1963; Methods of test for aggregates for concrete, Bureau of Indian Standards, New Delhi
14. IS 2386(Part 4)-1963; Methods of test for aggregates for concrete, Bureau of Indian Standards, New Delhi
15. IS 1489(Part 1)-1991; Specifications for Portland Pozzolana cement, Bureau of Indian Standards, New Delhi

Rutting and Cracking Performance of Asphalt Concrete Incorporating Plastic Waste and Crumb Rubber

Aung Lwin Moe, Yang Pin Kelvin Lee, Nyok Yong Ho, and Xuechun Wang

Abstract Typically, asphalt pavements encounter two major types of distresses, rutting and cracking. Rutting occurs mostly in early stage or immature age while cracking occurs eventually after aging when the pavement becomes stiffer. Generally, stiffer asphalt pavements have good rutting resistance and softer pavements have good cracking resistance. It is important to achieve a balanced asphalt mixture having both good rutting and cracking resistances. This study aims to evaluate the use of two types of recycled materials, plastic waste and granulated crumb rubber, on the rutting and cracking properties of asphalt. Penetration Grade 60/70 binder was used as base binder, and for the control-mix. Different types of samples were prepared with Marshall and gyratory compactors and the asphalt properties were determined with respect to resilient modulus, cracking resistance and Hamburg wheel tracking test, before and after simulated aging in the laboratory.

Keywords Rutting · Cracking · Recycled materials · Plastic waste · Crumb rubber · Resilient modulus · Cracking resistance · Hamburg wheel tracking test

1 Introduction

Majority of roads in Singapore are flexible pavements, where rutting and cracking have long been a major concern for flexible asphalt pavements.

Rutting is one of the most common permanent pavement deformations due to repetitive traffic loads, which gradually accumulate small pavement deformations that appear as longitudinal depressions in the wheel paths of roadways [1]. Not only is rutting a concern for driving safety, but also it reduces pavement strength and the service life, increasing the difficulty of pavement maintenance as well as its cost. To improve rutting performance of asphalt pavement, asphalt must possess enough

A. L. Moe (✉) · Y. P. K. Lee · N. Y. Ho · X. Wang
Samwoh Innovation Centre Pte Ltd, 25E Sungei Kadut Street 1, Singapore 729333, Singapore
e-mail: moeaung.lwin@samwoh.com.sg

Y. P. K. Lee
e-mail: kelvin.lee@samwoh.com.sg

© The Author(s), under exclusive license to Springer Nature Switzerland AG 2022

H. R. Pasindu et al. (eds.), *Road and Airfield Pavement Technology*, Lecture Notes in Civil Engineering 193, https://doi.org/10.1007/978-3-030-87379-0_64

structural integrity to overcome the shear forces exerted on it by the tremendous weight of vehicles. In other words, asphalt need to have a much stiffer performance profile. However, experience has shown that stiffer asphalt mix can have a negative effect on the fracture characteristics of the pavement. Hence, it is important to achieve a balanced asphalt mixture having both good rutting and cracking resistances.

In Singapore, plastic is the largest category of waste disposed. In 2019, only 4 percent of the 930,000 tonnes of plastic waste generated was recycled [2]. Plastic is a unique material with many benefits; it is cheap, versatile, light weight and resistant. However, plastic pollution is having a negative impact on our oceans and wildlife health [3]. Waste plastics have a vital role in the current asphalt applications. Trial sections of roads made with waste plastic modified asphalt have been widely adopted. Various studies were conducted to evaluate the performance of the waste plastic modified asphalt pavement. The results showed that the ideal content of waste plastic is 6–8% to improve the rutting and aging resistances [4].

Another type of waste that generated in Singaporeis scrap tires; about 33,000 tonnes was generated in Singaporein 2019. The scrap tires are usually processed into crumbs for making surfaces (e.g. for playgrounds) and into chips for use as a solid fuel. Some of them are disposed in landfills, which caused environmental problems. Therefore, there is a need to recycle scrap tires for more beneficial applications. One of the most beneficial applications of scrap tires is to process it into crumb rubber and use in the production of asphalt concrete. The use of crumb rubber not only alleviate waste disposal problem; it offers the benefit of resource recovery. On the other hand, crumb rubber has been known over decades as one of the best asphalt binder modifiers. Various studies show that the use of crumb rubber has a significant effect on increasing the fracture energy of asphalt mixtures and its use improves the performance of asphalt by increasing the resistance to cracking growth [5, 5].

Previous studies were conducted to evaluate the use of plastic waste and crumb rubber in asphalt concrete [7–9]. This paper presents a study that was carried out to evaluate the use of two types of recycled materials, plastic waste and granulated crumb rubber, on the rutting and cracking properties of hot mix asphalt. Test samples were prepared using straight bitumen, penetration grade (Pen) 60/70 for both control samples and samples incorporating recycled materials. Dry mixing method was used for both plastic waste and granulated crumb rubber, where recycled material were added to the mixing chamber together with hot aggregate and bitumen. Performance test samples were prepared with Marshall and gyratory compactors. The performance properties were evaluated by means of resilient modulus, cracking resistance and Hamburg wheel tracking tests, with and without simulated aging.

2 Materials

In this study, recycled materials used were processed low-density polyethylene (LDPE) plastic waste and granulated crumb rubber (CR) as shown in Fig. 1. The gradation of the crushed granite used is shown in Table 1.

Fig. 1 Processed LDPE (left) and granulated crumb rubber (right)

Table 1 Aggregate gradation

BS Sieve (mm)	Percent passing
19	100
12.5	79–99
9.5	68–88
4.75	48–68
2.36	33–53
1.18	20–40
0.6	14–30
0.3	9–21
0.15	6–16
0.075	3–6

Straight bitumen Pen 60/70 was used for both control mix and mixes incorporating recycled material. The gradation of the granulated crumb rubber is shown in Table 2.

Table 2 Sieve result of ground crumb rubber

BS Sieve (mm)	Percent passing
2.36	100.0
1.18	78.1
0.6	32.6
0.3	14.1
0.15	11.9
0.075	11.7
Pan	0

3 Experimental Design

Based on a previous study, processed low-density polyethylene LDPE plastic waste with dosage of 8% of bitumen weight was used for asphalt mix incorporating plastic waste (L) [9]. Crumb rubber (C) with dosage of 12% of bitumen weight was used for asphalt mix incorporating crumb rubber. For the mixture incorporating both LDPE and crumb rubber (M), half of the respective dosages which are 4% LDPE and 6% CR of bitumen weight were used. A control mix (V) was also prepared without the waste plastic and crumb rubber. The abbreviation of the asphalt mixtures and components are shown in Table 3.

All samples were prepared with 5.5% bitumen content. Dry mixing method was used for asphalt samples incorporating recycled materials where the recycled materials were added to the laboratory mixer directly. After the test specimens were prepared, performance tests were conducted on two set of the samples, without aging set and after aging set.

Simple stimulated aging process was conducted for the samples to be tested after aging, where the samples were conditioned in a forced draft oven at 85 °C for 5 days to reflect a critical duration of field exposure about 7–10 years [10].

3.1 Tests

Marshall samples were prepared with Marshall compactor and laboratory test were carried out to access the performances. Gyratory compactor was used to prepare 150 mm diameter samples which were used for Hamburg double wheel tracking test. Samples were divided into two groups, the first set was tested without aging process and second set was tested after aging process. Types of test performed are shown in Table 4.

Table 3 Abbreviations and components

Mix	Recycled material (% of bitumen by weight)		Bitumen
	LDPE	CR	
V	–	–	Pen 60/70
C	–	12%	
M	4%	6%	
L	8%	–	

Table 4 Type and purpose of tests

Type of test (test temperature)	Test method	Samples per test	Test purpose
Resilient modulus (25 °C)	ASTM D4123	3	To evaluate the stiffness of asphaltic concrete
Indirect tensile strength (25 °C)	ASTM D6931	3	To evaluate the relative quality of asphaltic concrete in conjunction with laboratory mix design testing and for estimating the potential for rutting or cracking
Cracking resistance (25 °C)	IDEAL-CT	3	To evaluate the cracking resistance of asphaltic concrete
Hamburgwheel tracking test (60 °C)	AASHTO T324	1	To evaluate the susceptibility of asphaltic concrete to deform under load by measuring the rut depth formed by repeated passes of a loaded wheel at a fixed temperature

4 Results

4.1 Resilient Modulus

The resilient modulus test was conducted according to ASTM D4123 at 25 °C at a loading frequency of 0.33 Hz by using a Universal Testing Machine (UTM-25). Test results are shown in Tables 5.

The results showed that the resilient moduli values after aging were higher than the resilient moduli without aging. Resilient modulus is a measure of stiffness level of the compacted asphalt mix. After aging, stiffness level increased due to oxidation of the bitumen. Based on the test results, asphalt incorporating crumb rubber (CR) has lower resilient moduli while asphalt incorporating waste plastic (L) has higher resilient moduli compared to the Control (V) mix, for both without ageing and after aging.

4.2 Indirect Tensile Strength

Indirect tensile strength (ITS) was carried out by loading the specimen across its vertical diametrical plane between a pair of steel loading strips. The indirect tensile strength was calculated by the maximum (failure) load. ITS results are shown in Table 6.

Table 5 Results of resilient modulus test

Sample ID	Resilient modulus (MPa)	
	Without aging	After aging
V-1	4123.5	
V-2	3474.5	
V-3	3242.0	
V-4		5199.5
V-5		4921.0
V-6		4938.5
Average	3613.3	5019.7
C-1	2632.5	
C-2	4095.5	
C-3	3058.0	
C-4		3122.5
C-5		5644.0
C-6		3632.5
Average	3262.0	4133.0
M-1	4478.8	
M-2	4463.5	
M-3	4386.5	
M-4		5154.0
M-5		5240.5
M-6		5442.0
Average	4442.9	5278.8
L-1	5752.5	
L-2	5253.5	
L-3	4888.0	
L-4		6528.5
L-5		4957.0
L-6		5818.5
Average	5298.0	5768.0

The ITS results showed similar trend as the resilient modulus test presented in the previous section. All the ITS results increased after aging. Compared to the control mix (V), asphalt incorporating crumb rubber (CR) showed lower ITS while asphalt incorporating waste plastic (L) has higher ITS. Asphalt mixes incorporating both crumb rubber and waste plastic (M) showed comparable ITS to the Control (V) mix.

Table 6 Results of indirect tensile strength test

Sample ID	Indirect tensile strength (kPa)	
	Without aging	After aging
V-1	1342.7	
V-2	1418.3	
V-3	1321.3	
V-4		1480.3
V-5		1493.8
V-6		1565.1
Average	1360.7	1513.1
C-1	1362.5	
C-2	1051.0	
C-3	1099.9	
C-4		1212.2
C-5		1528.6
C-6		1251.5
Average	1171.1	1330.7
M-1	1371.5	
M-2	1473.5	
M-3	1293.0	
M-4		1452.6
M-5		1560.1
M-6		1605.7
Average	1379.3	1539.5
L-1	1571.2	
L-2	1536.0	
L-3	1547.6	
L-4		1662.3
L-5		1533.9
L-6		1683.6
Average	1551.6	1626.6

4.3 *Cracking Resistance (IDEAL-CT Index)*

The IDEAL-Cracking Test (CT) is a simple test to determine the cracking behavior of asphalt mixtures. There are various types of tests available to measure the cracking performance, but the IDEAL-CT index is the most straight forward method [11]. The CT index can be calculated after the indirect tensile strength test, by using the failure load and the loading curve.

It was found that CT index or cracking resistance reduced after the aging process. Aging caused the asphalt mixture stiffen by oxidation of bitumen and possibly other components except the aggregate. Even stiffened samples after aging have higher indirect tensile strength value compare to without aging samples, the narrow area under curve and steeper down-slope of the curve contribute the lower CT index which are used for CT index calculation. Therefore, once the samples have undergone aging process, cracking resistance reduced.

The indirect tensile strength curves are shown in Fig. 2, and the average CT index results are shown in Table 7. Results showed that asphalt incorporating plastic waste LDPE (L) has the lowest crack resistance while asphalt mix incorporating crumb rubber (C) showed highest crack resistance. The Control (V) and asphalt mixes (M) showed comparable performance in terms of crack resistance.

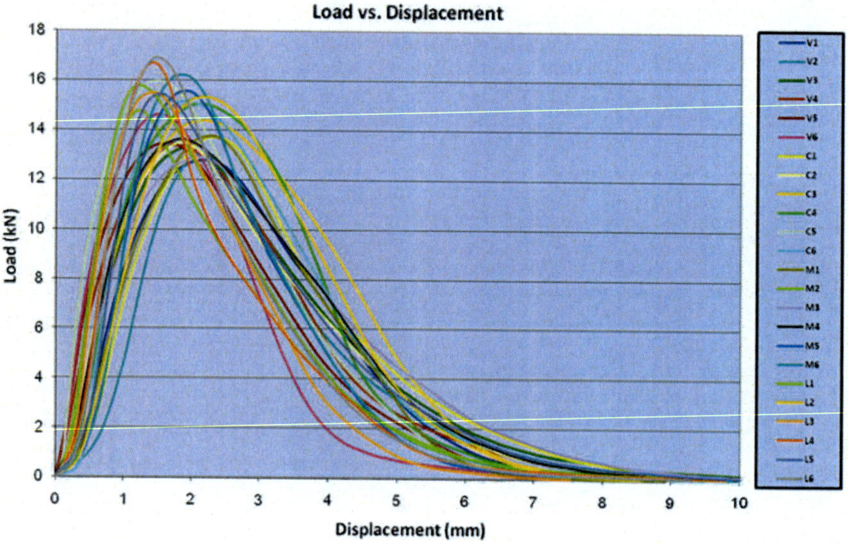

Fig. 2 Indirect tensile strength curves

Table 7 Cracking resistance (IDEAL-CT index) values

Sample type	Without aging	After aging
V	54.8	38.6
C	60.9	47.8
M	48.5	37.9
L	38.6	24.5

4.4 Hamburg Wheel Tracking Test

The Hamburg wheel tracking test is used to evaluate rutting and moisture damage of compacted asphalt mixture samples. Gyratory compactor was used to compact 150 mm diameter and 60 mm thickness test samples. After the samples were cooled to room temperature, arc of the two circular samples were cut and fitted in the mould and submerged in 60 °C water test-chamber as shown in Fig. 3. The steel wheel with 47 mm width and 705 N load tracked across the submerged sample for 10,000 cycles (20,000 passes), or until 20 mm of rut depth occurs. Speed of the steel wheel was 26 cycles or 52 passes per minutes. Rut depth was measured and recorded continuously with LVDT (Linear Variable Differential Transformer) while the temperature of the water was maintained at 60 °C across the test. Image of test samples after the test is shown in Fig. 4. The test results are shown in Table 8 and the plot of the results are shown in Fig. 5.

Fig. 3 Hamburg wheel track test samples

Fig. 4 Hamburg wheel track test samples after the test

Table 8 Hamburg Rut depth results

Sample type	Without aging		After aging	
	Rut depth (mm)	No. of passes	Rut depth (mm)	No. of passes
V	20.0	5,260	20.0	10,624
C	20.0	11,478	20.0	18,928
M	18.9	20,000	7.0	20,000
L	9.3	20,000	4.9	20,000

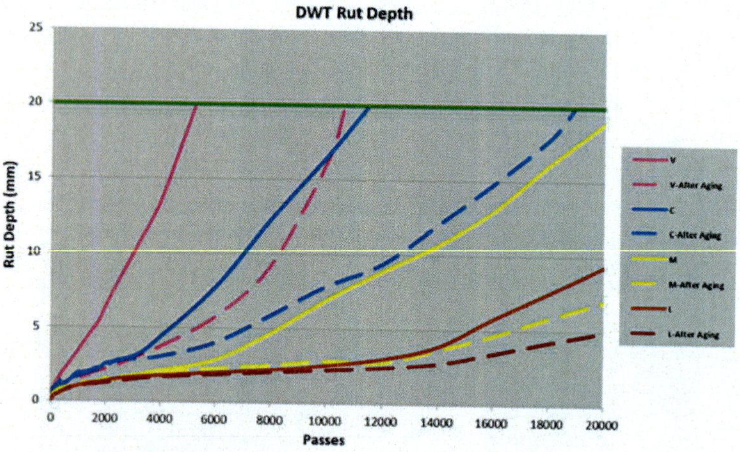

Fig. 5 Hamburg wheel track test results

Results showed that asphalt mix incorporating plastic waste (L) has the highest rutting resistance (i.e. lowest rut depth) while the control mix (V) showed the lowest rutting resistance. Asphalt mix (M) incorporating both plastic waste and crumb rubber showed rutting resistance between that of crumb rubber mix (C) and waste plastic mix (L). The trend is similar for both without aging and after aging samples.

5 Conclusions

This study evaluates the use of plastic waste and crumb rubber in asphalt concrete. Plastic waste increases the resilient modulus, indirect tensile strength and rutting resistance of asphalt concrete but reduces the crack resistance. On the other hand, crumb rubber increases the crack and rutting resistance of asphalt concrete but reduces the resilient modulus and indirect tensile strength.

Asphalt with a combination of both plastic waste and crumb rubber showed comparable or better performance compared to the control mix, for both without ageing and after ageing. As such, this combination provides a good potential to enhance both the rutting and cracking performance of asphalt concrete.

References

1. Ahmad J, Abdul Rahman MY, Hainin MR (2011) Rutting evaluation of dense graded hot mix asphalt mixture. Int J Eng Technol IJET-IJENS 11(5)
2. National Environment Agency (2019) NEA Waste Statistics and Overall Recycling, Singapore. https://www.nea.gov.sg/our-services/waste-management/waste-statistics-and-overall-recycling
3. Ritchie H, Roser M (2018) Plastic pollution. Published online at OurWorldInData.org, https://ourworldindata.org/plastic-pollution
4. Mashaan NS, Chegenizadeh A, Nikraz H, Rezagholilou A (2021) Investigating the engineering properties of asphalt binder modified with waste plastic polymer. Ain Shams Eng J 12:2
5. Ma F, Dai JS, Fu Z, Liu J, Dong WH, Huang Z (2020) A new type of Crumb Rubber asphalt mixture: a dry process design and performance evaluation. Multidisciplinary Digital Publishing Institute
6. Sayed NH, Ahadi S, Aslan h, Kuyumcu ZC (2020) Investigation of the effect of Crumb Rubber additive on the fracture of characteristics of asphalt mixtures in control and asphalt rubber mixtures. In: 8th international symposium on innovative technologies in engineering and science, Turkey
7. Ho NY, Lee YPK, Moe AL (2016) Recycling of scrap tires for use in asphalt concrete in Singapore. In: 8th international conference on maintenance and rehabilitation of pavements (MAIREPAV). Singapore
8. Moe AL, Lee YPK, Ho NY, Wang XC (2018) Development of environmental friendly rubberized asphalt. In: 11th Asia pacific transportation and environment conference, Malang, Indonesia
9. Lee YPK, Moe AL, Wang XC, Ho NY (2019) Evaluation of hot mix asphalt incorporating plastic waste. In: 11th international conference on airfield and pavement tehnology, Kuala Lumpur, Malaysia
10. Kim YR, Underwood S, Farrar MJ, Glaser RR (2015) Long-term aging of asphalt mixtures for performance testing and prediction. Interim Report to the National Cooperative Highway Research Program (NCHRP) Project NCHRP 09-54
11. Zhou FJ, Im S, Sun LJ, Scullion T (2017) Development of an IDEAL cracking test for asphalt mix design and QC/QA. Road Mater Pavement Design 18:sup4

The Volumetric Challenge of Crumbed Rubber Modified Asphalt Mixtures

Greg White and Andy Kidd

Abstract Significant research has demonstrated the potential benefits associated with crumb rubber modification of bituminous binders for asphalt road surface production. Most of this previous research has focused on higher (20–30%) dosages of crumbed rubber and there is now an increased interest in lower (5–15%) dosages for improved binder and asphalt ageing to extend the time between periodic local road surfacing. For Marshall designed asphalt mixtures, the volumetric composition of the mixture is important and isolating the effect of the crumb rubber modification from the effect of an increased modified binder content is critical to understanding the economic viability of low dosage crumb rubber as an anti-ageing additive. Testing of a typical 10 mm dense graded asphalt mixture showed that the rubber is partially dissolved into the bitumen and both complete dissolution (100%) or no dissolution (0%) based calculations have a significant effect on Marshall volumetric properties such as binder film thickness, air voids and voids in the aggregate. Further work is required to determine a reliable method for determining the degree of crumb rubber dissolution on a case-by-case basis.

Keywords Crumbed rubber · Asphalt · Mixture · Volumetric · Binder content

1 Introduction

Governments are continually seeking practical measures to address the increasing burden of both domestic and industrial waste streams. Australia is no exception, with a Government imposed progressive waste export ban commencing in 2021 [1]. As a result, there has been an increased interest in recycling or otherwise using various waste materials in the construction of infrastructure, including road pavements [2].

G. White (✉) · A. Kidd
University of the Sunshine Coast, Sippy Downs, QLD, Australia
e-mail: gwhite2@usc.edu.au

A. Kidd
Brisbane City Council, Brisbane, QLD, Australia

© The Author(s), under exclusive license to Springer Nature Switzerland AG 2022

H. R. Pasindu et al. (eds.), *Road and Airfield Pavement Technology*, Lecture Notes in Civil Engineering 193, https://doi.org/10.1007/978-3-030-87379-0_65

One of the keys to successful recycling and reuse of waste materials in road pavements is to replace expensive raw materials with economically available wastes or byproducts that do not impact pavement performance [3]. Another key to success is replacing relatively expensive raw materials with relatively economically processed waste from clean waste streams [4]. For these reasons, partial and full replacement of Portland cement in concrete, as well as the bituminous binder in asphalt mixtures, is of great interest.

In asphalt mixtures, the most common historical approach to material recycling has been the re-use of processed reclaimed asphalt (RA). RA is commonly stockpiled, crushed, tested and recycled back into new asphalt at the production plant [5]. Typically, 10–20% RAP is incorporated, with higher RAP percentages also considered when the RA is available in greater quantities [6]. However, in recent times, other recycled materials have been incorporated into asphalt mixtures. Waste printer toner [7], crushed glass [8], processed plastic [9] and crumbed tyre rubber [10] have all been reported. Some of these approaches are viewed as using road infrastructure as a horizontal landfill. However, crumb rubber is one technology that has a long history and has been demonstrated to improve the engineering properties of the asphalt mixture, compared to otherwise identical control mixtures [10].

The most common use of crumb rubber in asphalt mixtures has been to improve the elasticity and cracking resistance [11]. That is, crumb rubber has been used to replace conventional elastomeric polymers for improving the engineering properties of the asphalt mixture. It has also been similarly used to modify sprayed seal bitumen for both crack resistance and temperature sensitivity [11]. However, in recent years, crumb rubber has been considered for use at lower dosages, as a binder ageing retarder. In this application, the crumb rubber is not replacing an expensive conventional polymer, but is added to the unmodified binder, for use in local road asphalt surfacing, to extend the time between resurfacing due to binder ageing and mixture weathering.

It is commonly accepted that the crumb rubber does not fully digest into the bituminous binder the way conventional modifiers do, resulting in a multi-phase binder. When conducting volumetric mixture designs, such as the Marshall method [12], it is important to understand whether the rubber should be included in the binder content or not. When using low contents of crumb rubber in otherwise unmodified binder, this has a significant impact on the economic viability of this technology.

This research explores the challenge of volumetric mixture design using low dosage crumb rubber modified binder. A typical dense graded local road surfacing mixture was produced in the laboratory with a range of crumb rubber dosages. The effect on the mixture volumetrics were calculated for Marshall compacted samples. The effect of performing the volumetric calculations based on different assumptions about the degree of rubber dissolution in the binder was also considered.

2 Background

2.1 *Local Road Asphalt Mixtures*

Dense graded asphalt is arguably the most commonly used type of asphalt. Mixtures with 14 mm and 10 mm nominal maximum aggregate size are typically used for local road surfacing and these mixtures generally do not require polymer or other modification. Rather, a conventional bitumen is usually used as the binder. Furthermore, these local road mixtures do not usually fail due to structural and load related issues. That is because the traffic is mainly light vehicles, with only occasional rubbish collection trucks, emergency vehicles and commuter bus traffic. As a result, most of these roads are resurfaced because of age related weathering of the asphalt surface.

2.2 *Marshall Mixture Design*

Outside of countries that prefer Superpave and other performance-property related asphalt mixture design, the Marshall method of design remains commonplace and well known. The Marshall mixture design method was developed in 1939 by Bruce Marshall for the Mississippi Highways Department [12]. The primary aims of the Marshall design method include:

- A densely graded aggregate skeleton with approximately 15% voids in the mineral aggregate.
- Filling the voids in the mineral aggregate with bituminous binder, to retain 3–5% air voids.
- The Marshall Stability, Marshall Flow and volumetric properties are determined over a range of binder contents and the optimal binder content selected (Fig. 1).

Grading, bitumen content, compacted density, Marshall Flow and Marshall Stability are the primary design criteria and quality assurance parameters. Marshall samples are compacted using a standard Marshall hammer. This is a controlled energy approach because the standard hammer is dropped over a standard distance for a standard number of times. Consequently, a more workable mixture will have higher density and lower air voids content after compaction. This is different to gyratory and other compaction methods that can compact samples to a pre-determined air void content.

2.3 *Crumbed Rubber Modified Binder*

As stated above, crumbed tyre rubber is a well-established technology for the productive use of a waste material that can also improve the performance of asphalt mixtures

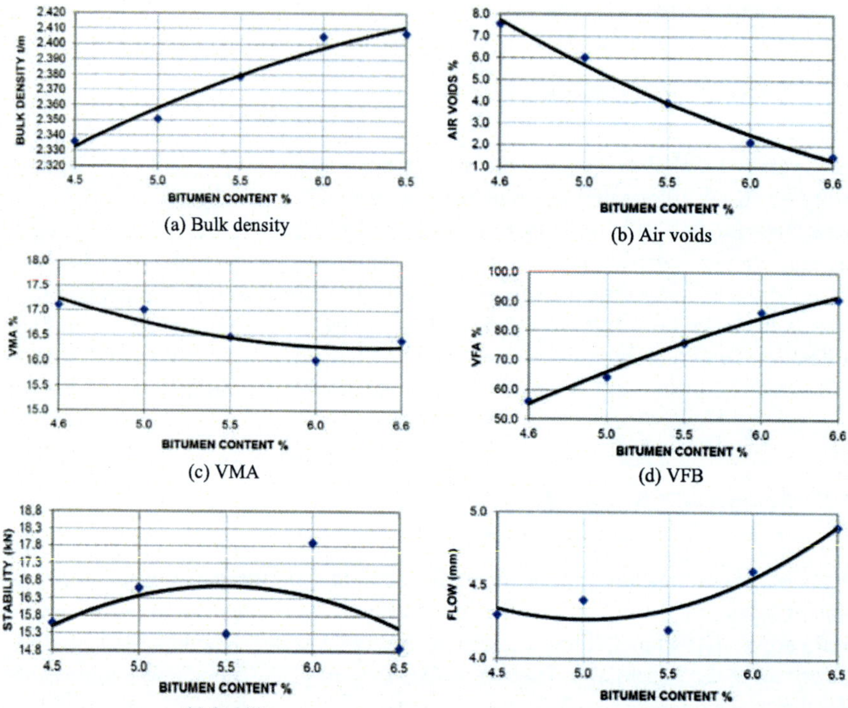

(a) Bulk density

(b) Air voids

(c) VMA

(d) VFB

(e) Stability

(f) Flow

Fig. 1 Marshall design parameters as a function of binder content

[11]. For example, crumbed rubber modified PG 64–22 unmodified binder was found to produce asphalt mixtures with deformation resistance, cracking resistance and elastic modulus values that exceeded otherwise identical asphalt mixture produced with PG 76–22 polymer modified binder [10]. However, despite these positive laboratory results, field trials have produced variable results and highlight the complexity of crumbed rubber modified binders.

First there is the source and size of the rubber crumbs. Smaller crumbs produce a more homogenous binder, but the cost of processing is higher. Second, there is the method of blending, which can be wet or dry. Dry mixing is where the rubber is added directly to the asphalt production plant with the aggregate, while wet mixing sees the crumb rubber pre-blended into the bitumen prior to asphalt production [13]. Wet blending produces a more homogenous asphalt mixture and increases dissolution into the bitumen, but the crumbed rubber can segregate out during storage and transportation [11]. Finally, for wet mixed crumbed rubber modification, the mixing can be performed in a high shear mill or a low shear mixer, which also affects the degree of dissolution and subsequent storage and transportation stability [13].

The practical issues identified above primarily relate to the degree of dissolution of the crumbed rubber into the bituminous binder and the associated post-blending stability. Unlike conventional polymers, crumbed rubber does not completely dissolve when mixed into bitumen. Rather, a multi-phase product consisting of solid rubber particles and bitumen containing dissolved rubber is produced [11]. At the interface between the solid rubber and the solid bitumen is a gel phase consisting of semi-liquid rubber (Fig. 2). The relative volume of each phase of crumbed rubber modified bitumen depends on the rubber type, particle size, bitumen-rubber compatibility, and the time and temperature of mixing.

Furthermore, how the different phases of crumb rubber should be considered in a volumetric-based asphalt mixture design method is not clear. When crumb rubber is used to improve the engineering properties of asphalt mixtures, this issue is of limited practical importance. That is because the crumb rubber is replacing a conventional polymer, which is very expensive compared to the bitumen. The bitumen volume is retained and the crumb rubber is considered an additional material, meaning that the total binder content is increased by the volume of the crumb rubber. However, the associated additional cost remains low compared to the cost of the conventional polymer being replaced.

One of the secondary benefits reported for crumb rubber modified asphalt mixtures is a reduced rate of environmental ageing [14]. Although this is attributed to the presence of the crumb rubber, it may simply reflect the increased binder film thickness in the mixture. Even if the crumb rubber only partially dissolves, this increases the volume of modified binder, which in turn increases the binder film thickness, which reduced the exposure of the binder to the environment and improves the age-related durability of the mixture.

However, now that crumb rubber is being considered for use in asphalt mixtures for local road surfacing, primarily to improve surface durability, the economic proposal is less clear. As explained above, local roads do not usually justify improved asphalt mixture engineering properties and therefore a conventional bitumen is commonly used as the binder. Consequently, any crumb rubber that is added to the mixture,

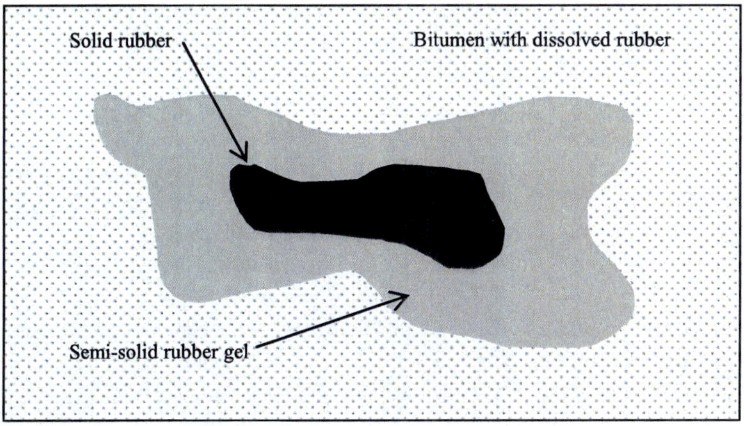

Fig. 2 Multi-phase partial dissolution of crumbed rubber in bitumen

without a commensurate reduction in the combined binder content, means the cost of the mixture increases, because crumb rubber modified binder is usually more expensive than conventional bitumen, even though it is less expensive than polymer modified binder.

It follows that there are only two ways for crumb rubber modified binder to be economically viable in local road asphalt mixtures. The first is a reduction in the bitumen content, to maintain a consistent total binder content, without reducing the performance of the mixture. Practice has shown this is not usually the case because an increase in the total binder content is usually necessary to make a workable asphalt mixture. The second is if the additional cost associated with maintaining the bitumen content consistent, is offset by longer lasting asphalt surface mixtures due to reduced age-related weathering. In this case, the total binder content increases, noting that the modified binder is also more expensive, by mass, than the conventional binder. Either way, understanding how the multi-phase crumb rubber modified binder effects volumetric mixture design is important.

3 Materials and Methods

The local government authority in Brisbane (Queensland, Australia) was interested in understanding the potential benefit of low (5–15%) crumb rubber modification of the bitumen in their asphalt mixtures for local road surfacing [15]. The associated research includes laboratory and field trials to evaluate the engineering property improvement and surface life extension associated with different levels of crumb rubber dosage. This preliminary laboratory trial was designed to inform the volumetric consideration of crumb rubber modification in order to isolate, as far as possible, the effect of the crumb rubber modification from the volumetric change in asphalt mixtures when the bitumen content is fixed.

3.1 Materials

Crumb rubber was used to modify the Class C170 bitumen used to produce otherwise identical BCC Type 2 asphalt. C170 is an Australian viscosity graded unmodified bitumen meeting the requirements of AS 2008. BCC Type 2 asphalt is a local 10 mm sized dense graded (Fig. 3) mixture used for surfacing light trafficked roads within the Brisbane region. It contains 10% crushed and screeded RA. The asphalt is designed according to the Marshall method and generally has 5.7% binder content (by mass) and a Marshall air voids content of 3%.

A commercial source of fine crumb rubber (0.2–0.4 mm) was added to the C170 bitumen at 190 °C and blended with a Silverson mixer for 2 h at 300 rpm. Crumb rubber was added to the bitumen at dosages of 5%, 10% and 15% by weight of

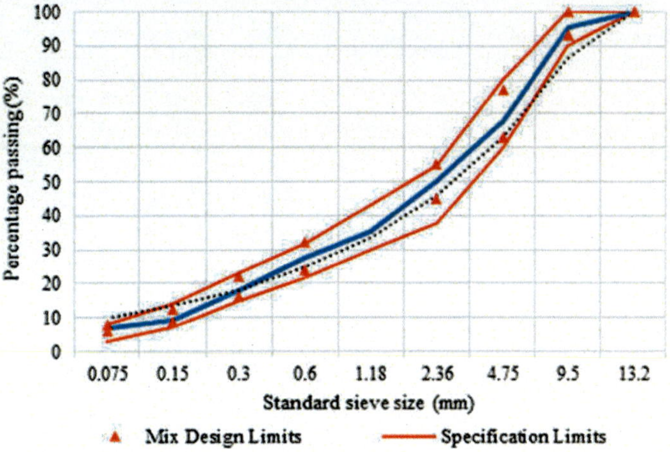

Fig. 3 BCC Type 2 10 mm nominal dense grade asphalt mix design

the unmodified bitumen. Some C170 was also retained in unmodified form for the production of control asphalt samples.

The aggregate fractions were heated to 160 °C and then combined according to the target gradation (Fig. 3) and various binder contents were combined with the aggregate, with mixing performed in a standard laboratory mixer. Bitumen contents ranged from 4.8% to 6.4%. Depending on the crumbed rubber dosage, this required variable amounts of binder to be added to the mixtures, with the greatest quantity of binder required when the bitumen content was 6.4% and the crumbed rubber dosage was 15% (Table 1).

3.2 Methods

The bulk asphalt samples were sub-sampled and Marshall specimens were compacted by 50 blows (per face) of a standard Marshall hammer. Compaction was performed at 142 °C. Specimens were tested for Marshall properties according to AS 2891.5, including compacted density (t/m^3), air voids (%), voids in the mineral aggregate (VMA) (%), voids filler with binder (VFB) (%), Stability (kN) and Flow (mm). Extracted binder content (% by mass), effective binder volume (BEF) (%) and the binder film thickness (BFT) (μm) were also calculated.

Initially, all calculations were performed assuming the crumb rubber was completely dissolved in the bitumen (100% dissolution). That is, the solid rubber and gel phases were assumed to be part of the modified binder volume. Second, the calculations were reperformed assuming the rubber was not dissolved at all

Table 1 Target bitumen and binder content for various mixtures

Bitumen content (%)	Crumb rubber dose (%)	Binder content (%)
4.8	15	5.5
5.0	0	5.0
5.0	10	5.5
5.0	15	5.8
5.2	5	5.5
5.5	0	5.5
5.5	5	5.8
5.5	10	6.1
5.5	15	6.4
5.8	0	5.8
5.8	5	6.1
5.8	10	6.4
5.8	15	6.7
6.1	0	6.1
6.1	5	6.4
6.1	10	6.7
6.1	15	7.0
6.4	15	7.3

Note Bitumen content refers to C170 bitumen, while binder content refers to the modified binder, which included both the C170 bitumen and the wet mixed crumb rubber. These binder contents were added and exclude any binder from RA

(0% dissolution). That is, all crumb rubber was assumed to remain solid and function effectively as fine aggregate. Finally, the calculations were performed assuming different degrees of dissolution, including 20%, 40%, 60% and 80%.

4 Results

The volumetric results for the 100% dissolution of crumb rubber are summarised in Fig. 4. The equivalent volumetric results for 0% dissolution are summarised in Fig. 5. The Marshall Stability and Flow results are in Fig. 6.

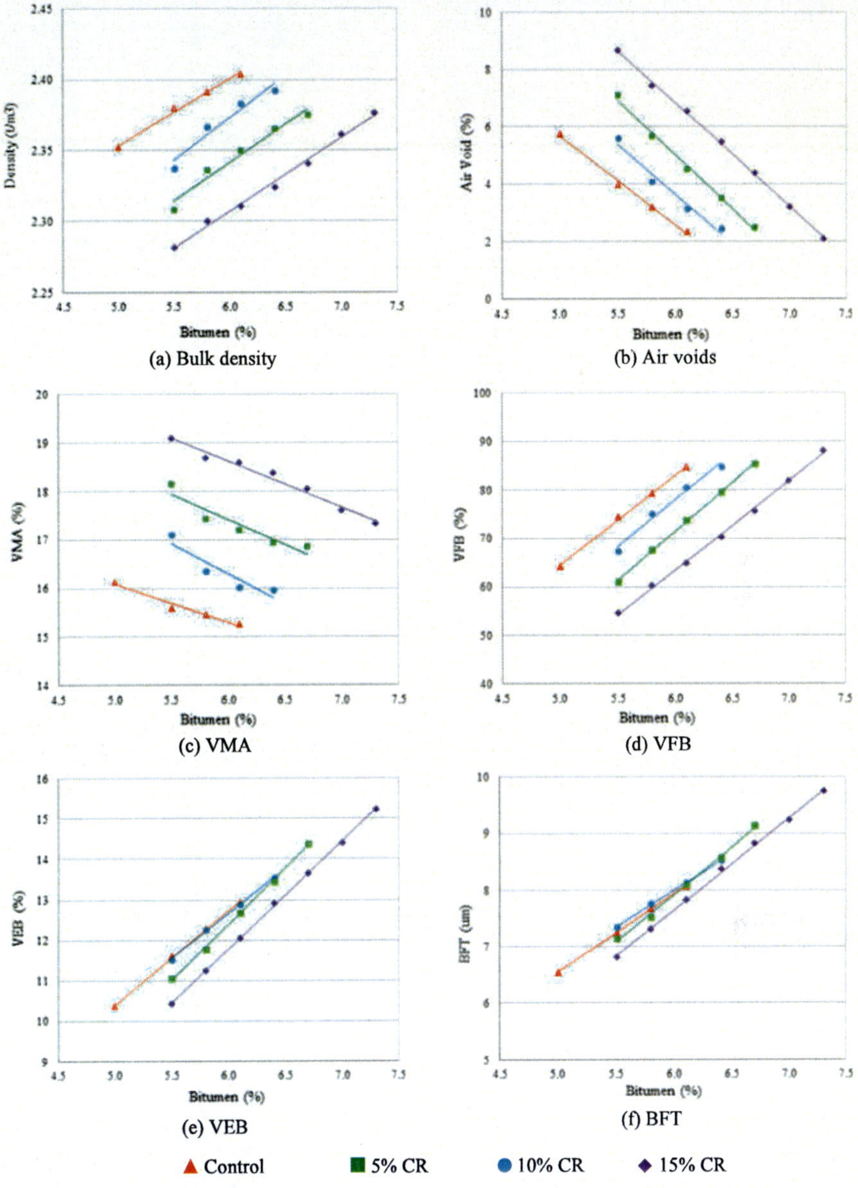

Fig. 4 Volumetric properties for 100% crumb rubber dissolution calculation

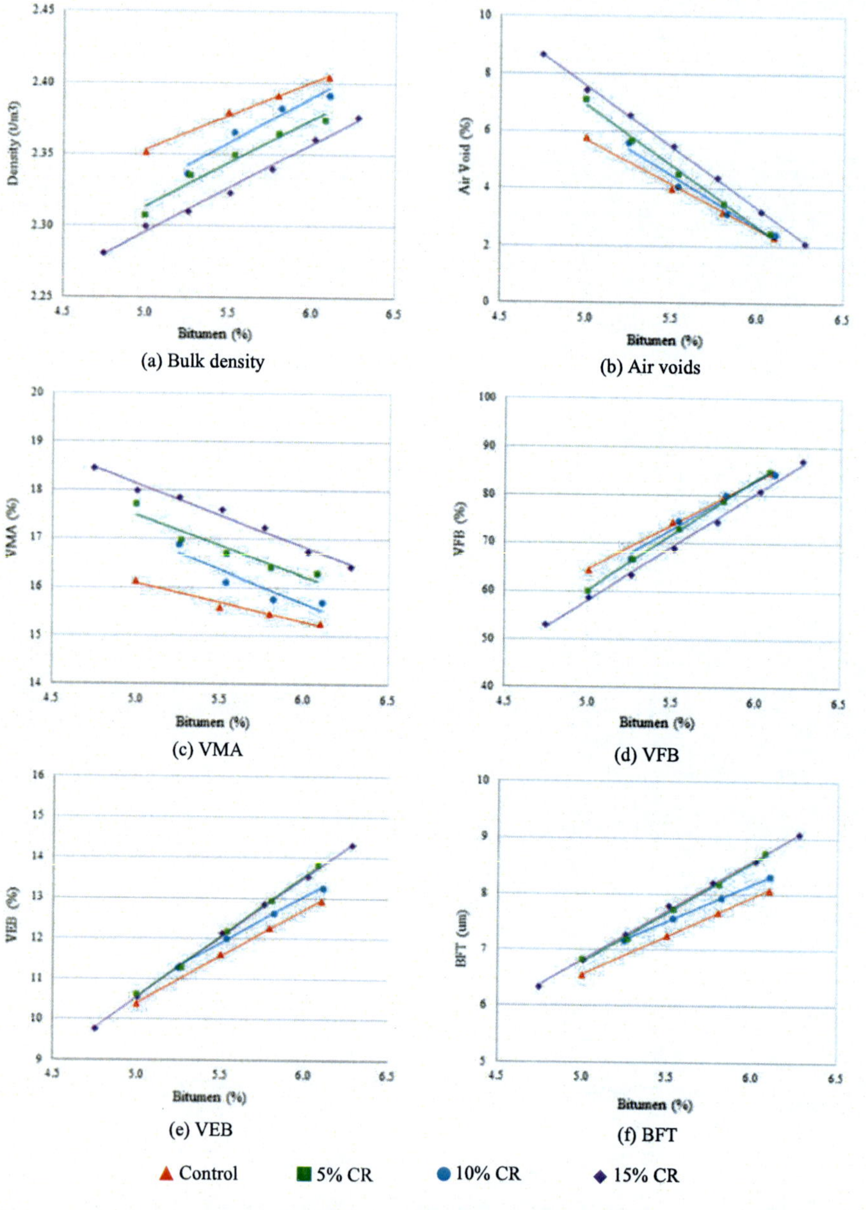

Fig. 5 Volumetric properties for 0% crumb rubber dissolution calculation

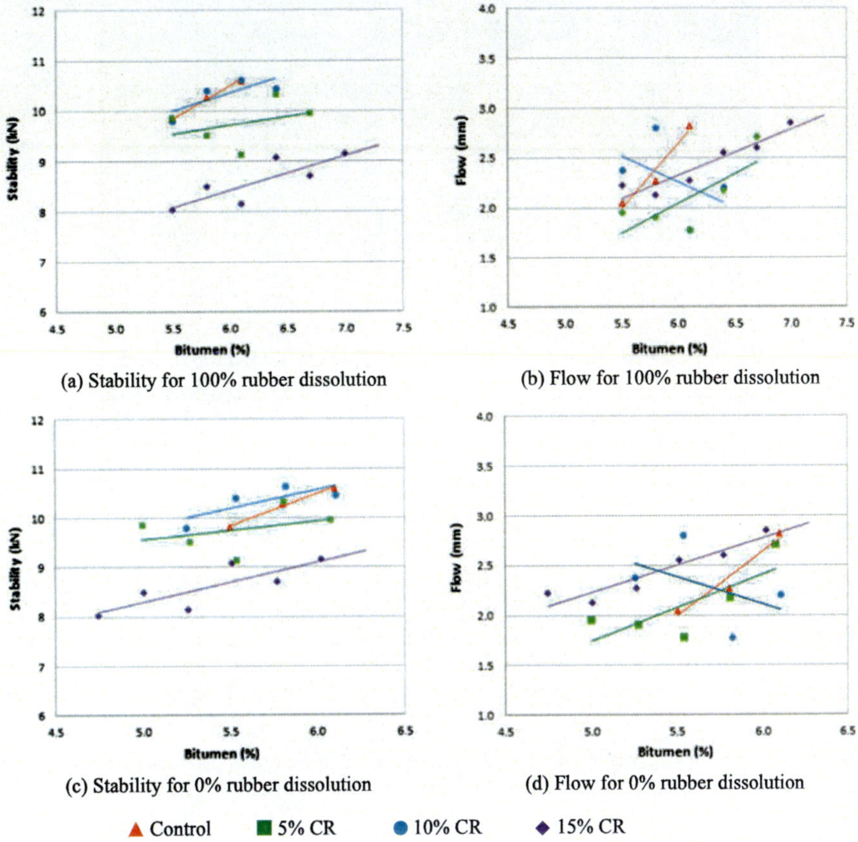

(a) Stability for 100% rubber dissolution

(b) Flow for 100% rubber dissolution

(c) Stability for 0% rubber dissolution

(d) Flow for 0% rubber dissolution

Fig. 6 Marshall stability and flow results

5 Discussion

5.1 Partial Dissolution of Crumb Rubber

During the laboratory testing, the partial dissolution of the crumb rubber was verified. Figure 7 shows the C170 bitumen and the crumb rubber modified binder. The semi-dissolved crumb rubber particles are clear. Furthermore, Fig. 8 shows solid rubber particles removed after toluene extraction of the modified binder. This emphasizes the importance of underlying the impact of the partial dissolution of crumb rubber on asphalt volumetrics. The quantity of solid rubber recovered from the extracted binder in this instance was equivalent to approximately 20% effective dissolution. That is, approximately 80% of the rubber particles remained undissolved and were recovered on a 75 μm sieve.

Fig. 7 Crumb rubber modified (left) and unmodified bitumen (right) binders

Fig. 8 Solid rubber particles recovered from extracted binder

5.2 *Effect of Crumb Rubber on Volumetric Properties*

Higher crumb rubber dosage contents in a fixed bitumen content significantly changed the volumetric composition of the asphalt mixtures. The density reduced as the rubber dosage increased, as shown in Fig. 4a. The maximum density reduced because the rubber particle density is less than the bitumen particle density and much less than the aggregate particle density. Furthermore, the air voids increased for 5–15% crumb rubber (Fig. 4b) because the binder viscosity increased and therefore the fixed compaction energy associated with the Marshall specimen preparation method did not densify the modified mixtures to the same degree as the unmodified mixtures. That is, the asphalt was less workable. Because the volumetrics are mathematically linked, the VMA increased (Fig. 4c) and the VFB decreased (Fig. 4d) as the

(a) (b)

Fig. 9 Bulk asphalt samples with **a** 5.5% and **b** 7.0% binder content, both with 15% crumb rubber dosage

crumb rubber dosage increased. In contrast, the binder film thickness was relatively unchanged (Fig. 4f) because the increased binder volume was spread further around the less dense mixture, resulting in a comparable average thickness of binder coating the aggregate particles. These effects were calculated assuming the full digestion of the crumb rubber into the bitumen. That also implies that the volume of bituminous binder was held constant regardless of the crumb rubber dosage, requiring the C170 bitumen content to be reduced by the amount of crumb rubber added. However, in practice, the samples with high rubber dosage and lower bitumen content were fragile and were visually assessed as being deficient of bituminous binder compared to typical dense graded asphalt mixtures (Fig. 9). Additionally, approximately double the mixing time was required to adequately coat all the aggregate particles when binder content was held constant, which is indicative of a 'dry' or lean mixture. Consequently, it was concluded that the crumb rubber modified binder content must be increased significantly to produce viable asphalt mixtures, despite the additional cost associated with a greater volume of the more expensive modified binder.

5.3 *Effect of Crumb Rubber on Marshall Stability and Flow*

The Marshall Stability and Flow were more variable than the volumetric properties (Fig. 6). As the crumb rubber dosage increased, the Stability was generally unaffected until the crumbed rubber dosage was 15%, for which the Stability was significantly lower (Fig. 6a). The Flow results were even more variable with little trend observed regardless of the crumb rubber dosage (Fig. 6b). However, this probably says more about the repeatability of the Marshall Flow test on elastomeric modified binders, rather than specifically about the effect of crumb rubber.

5.4 *Effect of Dissolution on Volumetric Properties*

When the same volumetric composition results were recalculated assuming the rubber remained solid (ie. 0% dissolution) the reported binder contents all reduced because the rubber no longer counted towards the binder content. The change from full dissolution (100%) to no dissolution (0%) was also calculated for 20%, 40%, 60% and 80% partial dissolution (Fig. 10). The effect was most evident for the VMA (Fig. 10a) and BFT (Fig. 10b). The BFT is critically important to the ability to isolate the effect of crumb rubber modification from the effect of increasing the binder content. This in turn affects the economic viability of crumb rubber for asphalt durability related to binder ageing and mixture weather.

The bulk density and air voids results were unaffected because the value is the same whether the crumb rubber is considered as solid (aggregate) or liquid (binder). Similarly, the measured Marshall properties were unaffected by the basis of the volumetric calculations, except the binder contents were all reduced when the rubber

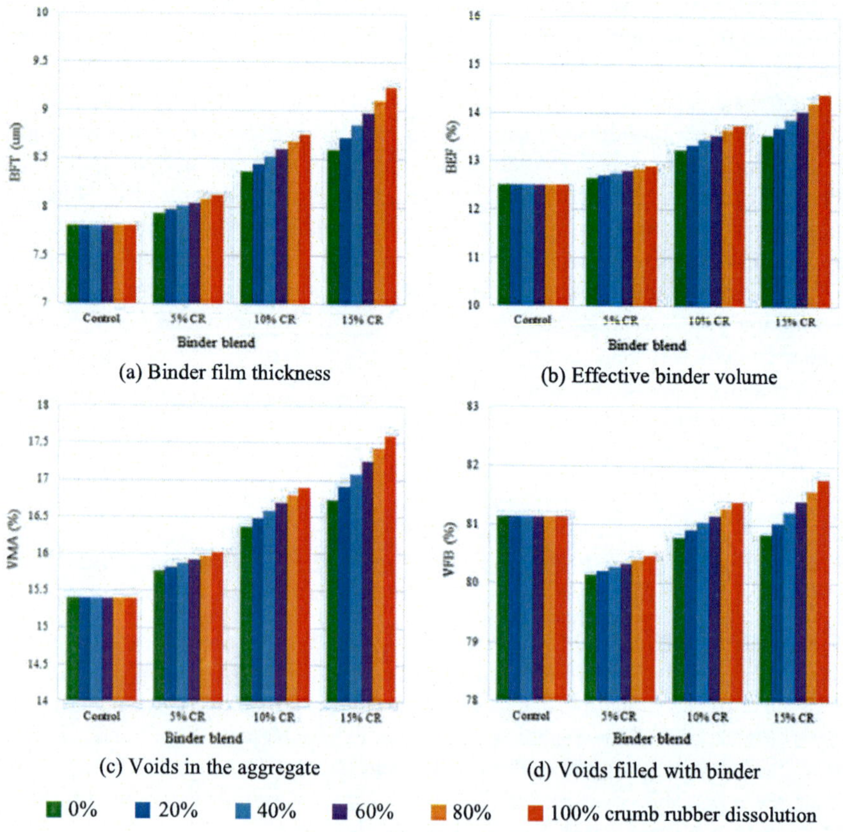

Fig. 10 Effect of crumb rubber dissolution and dosage on volumetric properties

was assumed to be only partially dissolved. Overall, as the crumb rubber dosage increased, selecting the appropriate degree of dissolution became more important for calculating realistic volumetric properties.

5.5 *Isolating the Crumb Rubber Effect*

It is well established that the binder film thickness has a significant effect on the rate of binder oxidation. Therefore, for local roads that are generally resurfaced due to environmental erosion, isolating the effect of the binder content and film thickness, from that associated with the crumb rubber modification is important.

It is clear that complete dissolution of the crumb rubber is an unrealistic assumption and results in unworkable mixtures. However, assuming no dissolution results in significantly more expensive mixtures and is also unrealistic. Consequently, the only way to isolate the effect of crumb rubber modified binder on the age-related erosion and durability of asphalt mixtures for local road surfacing, is to determine the actual degree of dissolution. This is expected to be dependent on the crumb rubber particle size, the method of blending and the binder-rubber compatibility. These parameters will need to be determined on a case-by-case basis.

6 Conclusion

Crumb rubber has been associated with reduced ageing of bituminous binders and asphalt mixtures used for road surfacing. However, this has generally been for higher (20–30%) crumb rubber dosages, where the crumb rubber replaces expensive polymers for engineering property improvement. For lower (5–15%) dosages proposed for improved local road asphalt surface weathering and ageing, isolating the effect of the binder film thickness from that of the crumb rubber is critically important to understanding the economic viability of crumb rubber modification. However, further work is required to develop a reliable test to measure the degree of rubber dissolution, which is expected to be required on a case-by-case basis. Until that issue is resolved, there is no way to isolate the effect of the increased binder content from the modification of the binder itself. This issue is unique to crumb rubber modified binders because other polymer dosages are lower (2–5%) and the polymers fully dissolve into the bitumen.

References

1. Skatsoon J (2019) Timetable set down for waster export ban. Government News. 11 Nov 2019. https://www.governmentnews.com.au/timetable-set-down-for-waste-export-ban/
2. Foley M (2021) Waste export ban drivers rubbish research into local manufacturing. The Sydney Morning Herald. 2 Jan 2021. https://www.smh.com.au/politics/federal/waste-export-ban-drives-rubbish-research-into-local-manufacturing-20210101-p56r7h.html
3. Jamshidi A, White G (2020) Evaluation of performance of challenges of use of waste materials in pavement construction: a critical review. Appl Sci 10(226):1–13
4. White G, Magee C (2019) Laboratory evaluation of asphalt containing recycled plastic as a bitumen extender and modifier. J Traffic Transp Eng 7(5):218–235
5. Austroads: Maximising the reuse of reclaimed asphalt pavement—outcomes of year two: RAP mix design, AP-T-286-15, Sydney, Australia, 2 Feb 2015
6. Pires GM, Del Barco Carrion AJ, Airey GD, Presti DL (2017) Maximising asphalt recycling in road surface courses: the importance of a preliminary binder design. Tenth International conference on the bearing capacity of roads, railways and airfields. Athens, Greece
7. Yildirim Y, Korkmaz A, Prozzi J (2003) The toner-modified asphalt demonstrative projects. Research report FHWA/TX-05/5-3933-01-2, center for transportation research, The University of Texas in Austin, Texas, USA
8. Jamshidi A, Kurumisawa K, Nawa T, Jize M, White G (2007) Performance of Pavements incorporating industrial byproducts. J Clean Prod 164:367–388
9. White G (2020) A synthesis of the effects of two commercial recycled plastics on the properties of bitumen and asphalt'. Sustainability 12(8594):1–20
10. Bennert T. Maher, A. Smith J (2004, July) Evaluation of crumb rubber in hot mix asphalt. Research Report BAY RU9247. https://cait.rutgers.edu/wp-content/uploads/2018/05/bay-ru9247.pdf
11. Marais HIJ, Botha B, Hosfink W, van Heeerden J (2017) Latest developments in crumb rubber modified bitumen for use in asphalt and seals—the South African experience. International Flexible Pavement Conference. Melbourne, Victoria, Australia
12. White TD (1985) Marshall procedures for design and quality control of asphalt mixture. Asphalt Pavement Technol 54:265–285
13. Bahia HU, Davies R (1994) Effect of Crumb Rubber modifiers (CRM) on performance-related properties of asphalt binders. Asphalt Paving Technology. Saint Louis, Missouri, USA. 21–23 Ma 1994
14. Grobler J, Harrison J, Chen S (2021) Facilitating the use of Crumb Rubber modified asphalt on local government roads in South East Queensland. Report prepared for Tyre Stewardship Australia, ARRB Transport Research, Brisbane, Australia
15. Kidd A, Stephenson G, White G (2021) Towards the use of crumb rubber modified asphalt for local government roads. In: International flexible Pavement symposium, Australian flexible Pavement Association, 3–5 August 2021 (Article in Press)

Utilization of Waste Cooking Oil as a Sustainable Product to Improve the Physical and Rheological Properties of Asphalt Binder: A Review

Mahmoud Ibrahim Eldeek, Fauzan Mohd Jakarni, Ratnasamy Muniandy, and Salihudin Hassim

Abstract Yearly, vast numbers of new roads are constituted and existing ones are restored on a regular basis. The usage of recycled asphalt pavement (RAP) aids to minimise dependency on natural resources. Hence, organic regenerator such as waste cooking oil (WCO), have recently been proposed as a sustainable commodity to enhance the recycling of aged asphalt. This paper summarised the potential of WCO being utilized as a renewable commodity in order to boost asphalt binder recycling. Physical properties tests for instance ductility test, softening point test, penetration test, flash & fire point test and specific gravity tests; are commonly adopted alongside test for rheological properties for example dynamic shear rheometer (DSR), Multiple Stress Creep Recovery (MSCR) tests linear amplitude sweep (LAS) and have all been adopted to enhance the binder properties. Generally, this paper offers an abbreviation of the favourable and unfavourable effects of WCO utilisations and indicates that it could be used as an asphalt rejuvenator that is both environmentally friendly and able to create high quality pavements.

Keywords Asphalt binder · Reclaimed asphalt pavement (RAP) · Recycling · Rejuvenator · Pavement · Waste cooking oil (WCO)

1 Introduction

The populace grows in tandem with the ever increasing advancement in healthcare leading to inevitable escalation for food supply and a relative growth of kitchen waste. Edible cooking oil is essentially a widely sought after commodity which can be sourced from plants and animals alike. Regardless of the source, WCO is generated

M. I. Eldeek (✉) · F. M. Jakarni · R. Muniandy · S. Hassim
Department of Civil Engineering, Faculty of Engineering, Universiti Putra Malaysia, 43400 UPM Serdang, Selangor, Malaysia

F. M. Jakarni
e-mail: fauzan.mj@upm.edu.my

M. I. Eldeek
Department of Civil Engineering, Faculty of Engineering, Misurata University, Misurata, Libya

© The Author(s), under exclusive license to Springer Nature Switzerland AG 2022
H. R. Pasindu et al. (eds.), *Road and Airfield Pavement Technology*, Lecture Notes in Civil Engineering 193, https://doi.org/10.1007/978-3-030-87379-0_66

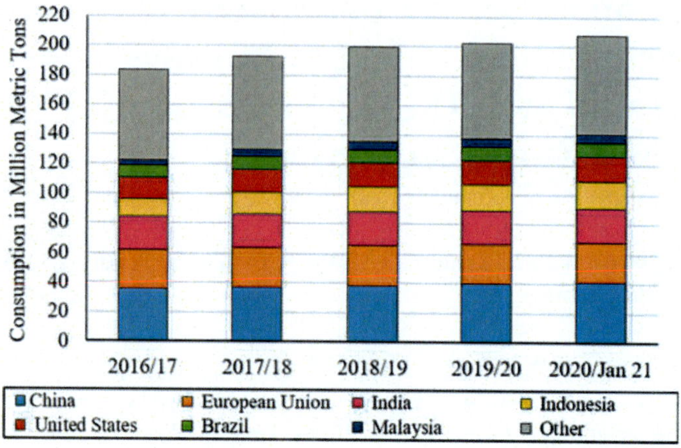

Fig. 1 5 Year Worldwide Data of Vegetable Oil Consumption from 2016 to 2020

after utilization of such oils i.e., after cooking and frying [1]. As of 2020, vegetable oil consumption on a global scale has reached a whopping 207.94 million metric tons as shown in Fig. 1; and this figure is on a steady increase at a rate of 2% annually [2]. With a yearly production of 41.09 million metric tons, China stands as the largest WCO producer followed by the European Union at 27.46 million tons. By virtue of the above data, there is a significant rise in WCO levels and waste management becomes a reciprocal obligation of nations on a global scale.

Efforts to recycle or reuse WCO is substantially low; at only 20–30% of the total produced [3]. Ignorant users who discard WCO directly into sewerage systems, waste bins, kitchen sinks or flushing them out to land and water bodies are essentially causing harm to others and the environment as WCO must be treated and properly managed [4, 5]. Such insensitive acts cause adverse effects to the environment whereby contaminants would be released which disrupts oxygen supply for aquatic life and therefore threatens the entire ecosystem. Contaminants would snowball causing a blockage of the sewerage system resulting in and overflow and ultimately triggers a hike in the cost of operations and maintenance of waste water treatment plants. With reference to previous literature focused on the United States [6], it was cited that a 25% hike in the cost of water treatment operations [7] when there was a 40% blockage in sewerage system resulting from WCO being disposed down the kitchen sink.

Bailey et al. [8] saw to the patent of using WCO as an asphalt rejuvenator which very much supported the postulate of WCO utilization in rejuvenation of aged binders with their asphalt mixture. Following its success, more research [9–12] were conducted to study the utilization of WCO in rejuvenation of aged asphalt binders and amelioration of its characteristics to levels closest to the virgin binder. Asphalt binder whether aged or virgin can be mixed directly with MCO or post-filtration to remove any impurities; it is also known to have shown improved outputs when subjected to chemical treatment [1]. Unsaturated fatty acids make up a major part of

WCO which are homologous to maltenes; the lower molecular weight constituents of asphalt binders.

In the course of its duration of use, asphalt binders will gradually lose the lighter oil constituents i.e. maltenes and accumulates heavier oil constituents i.e. asphaltenes; WCO is capable of equalising the loss of maltenes in aged binders [13, 14]. As the ideal choice for integration in asphalt rejuvenation, WCO possess a few desirable aspects, namely it is found in excesses—in 2020 alone, there is a recorded 209.14 million metric tons of consumption of its various forms and the fact that it is partially indistinguishable from bitumen further supports the notion [2].

2 WCO Physical and Chemical Properties

The temperature of cooking oil is oftentimes brought up to the range of 160–200 °C during the process of frying or cooking and generally with the presence of air and moisture [15]. One would expect changes such as the increase in density, total polar material (TPM), specific heat, saturated fatty acid components and viscosity levels due to heating which causes breakdown responses i.e. hydrolysis, polymerization, and oxidation [5]. Reactions observed as these degeneration reactions prolong would be the diminishing presence of double bonds or quantity of monounsaturated and poly-unsaturated constituents, peroxide value, antioxidants (Vitamin E), smoke point, iodine value, saponification value, and volatile components. Should the oil be subjected to further use, it would undergo further degradation to the point it becomes unsuitable for frying or cooking and should ideally be disposed [5, 15–18].

The significant components of WCO are identified from the probable constituents of WCO shown in Table 1 which are oleic acid, linoleic acid and palmitic acid at

Table 1 Probable chemical constituents of WCO [1, 9, 21]

Formulation of fatty acids	Type of free fatty acid	Possible % in waste cooking oil (WCO)	Type of Saturation
C18:1 (Cis 9)	Oleic acid	43.67	Unsaturated
C16:0	Palmitic acid	38.35	Saturated
C18:2 (Cis)	Linoleic acid	11.39	Unsaturated
C18:0	Stearic acid	4.33	Saturated
C14:0	Myristic acid	1.03	Saturated
C18:3 alpha	γ- Linolenic acid	0.37	Unsaturated
C12:0	Lauric acid	0.34	Saturated
C18:2 t	Linolenic acid	0.29	Unsaturated
C20:1	Cis-11-Eicosenoic acid	0.16	Unsaturated
C21:0	Heneicosanoic acid	0.08	Saturated
	Total	100	

Table 2 Physio-chemical composition of WCO and virgin cooking oil [20]

Properties	Values of virgin Cooking oil (palm cooking oil)	Values of waste cooking oil (WCO)
Acid value (mg KOH/gm)	0.3	4.03
Calorific value (J/gm)	–	39,658
Saponification value (mg.KOH/gm)	194	177.97
Peroxide value (meq/kg)	< 10	10
Density (gm/cm^3)	0.898	0.9013
Kinematic viscosity (mm^2/s)	39.994	44.956
Dynamic viscosity (MPa s)	35.920	40.519
Flash point (°C)	161–164	222–224
Moisture content (wt. %)	0.101	0.140

percentages of 43.67%, 11.39% and 38.35%, respectively [1, 9]. Nonetheless, deviations should be expected whereby the configurations of fatty acids in WCO would be inconsistent following chemical transformations that take place during frying and cooking since the fatty acid configuration of different fresh and commercial oils would be distinguishable from one another [19]. Results in Table 2 depicts the difference in physio-chemical composition of WCO and virgin cooking oil described by Ullah et al. [20]. The likelihood of thermal cracking or catalytic cracking is increased by the availability of long-chain acids e.g. monounsaturated acid such as oleic acid and saturated acid such as palmitic acid, resulting in the formation of hydrocarbon chains [21].

Therefore, it is deduced that the physical and chemical constituents would be inconsistent as it is highly dependent on various factors such as the quality and kind of cooking oil, sort of food cooked, the food constitution, temperature of cooking, frequency of oil usage and period of cooking.

3 Basic Properties Tests of WCO Modified Asphalt Binder

Table 3 depicts the conventional physical properties tests which are conducted to

Table 3 Summary of literature on the consequence of waste oil addition on the physical characteristics

Ref	Rej type and content %	Other additives					Binder used/tested	Control binder
		Penetration test	Softening Point	Ductility	Specific Gravity	Flash &Fire Point		
[22]	WVO 1–3	↑	↓	↑	↓	↓	Binder penetration grade obtained from 6 years old RAP	40/50 Penetration grade
[23]	WVCO 2,4,6,8,10	↑	↓	–	–	–	Artificially aged Virgin Binder using PAV, RTFO, and oxi-aging method	70/100 Penetration grade
[24]	WCO 1–5	↑	↓	↑	–	–	Artificially aged Virgin binder in RTFO at different aging times	PEN 70
[25]	WCO 2,3,3.5,4	↑	↓	–	–	–	Extracted RAP binder	60/70 Penetration grade
[26]	WVO 5,10,15,20	↑	↓	↑	–	↑	Artificially aged Virgin binder in PAV and RTFO	PEN 70
[27]	WVO 1,2,4,6,8	↑	↓	↑	↓	↑	Virgin binder 60/70 PG	60/70 Penetration grade
[28]	WCO 6	↓	↑	↓	–	–	Virgin binder artificially aged in RTFO and PAV	80/100 Penetration grade
[29]	WVO 1,3,5	↑	↓	–	–	–	Virgin binder	50/70 Penetration grade

(continued)

Table 3 (continued)

Ref	Rej type and content %	Other additives					Binder used/tested	Control binder
		Penetration test	Softening Point	Ductility	Specific Gravity	Flash &Fire Point		
[30]	WVO 2–6	↑	↓	–	–	–	Binder was extracted from 12 years old RAP	50/70 Penetration grade
[31]	WCVO 2,4,6,8, 10	↑	↑	↑	–	–	Binder recovered from RAP	PG 64–22
[1, 32]	WCO 3,4,5	↑	↓	–	–	–	Virgin binder 60/70 PG	60/70 Penetration grade
[33]	WCO 3,4,6	↑	↓	–	–	–	Virgin binder 60/70 PG	60/70 Penetration grade
[10, 34]	WCO 1–5	↑	↓	–	–	–	Aged 50/60, 40/50 Penetration grade	80/100 Penetration
[35]	WCO 2, 3, 4, 6	↑	↓	–	–	–	Artificially aged Virgin binder with RTFO aging mode and propeller mixer aging mode	Original Bitumen

evaluate the physical characteristics of binders such as softening point test (ASTM D36), penetration test (ASTM D5), flash & fire point test (ASTM D92), specific gravity test (ASTM D70) and ductility test (ASTM D113).

Consistency and deformation resistance of bitumen at 25 °C can be determined during penetration testing while thermal susceptibility at a greater temperature which is usually 45 to 85 °C can be identified from the softening point test. The specific gravity test allows volume correction that shifts with a turn in temperature and the flash & fire points test is aims at promoting risk-free operation at the asphalt plant. Adhesion properties of asphalt binders are addressed via the ductility test.

Asphalt pavement undergoes various changes as it ages such as turning hard, brittle and experiencing an increased tendency in cracking. The rejuvenation process promotes the reduction of the aged asphalt's brittleness at various concentrations.

Past studies which have been conducted to analyse the consequence of the inclusion of WCO on the physical characteristics have been summarized in Table 3. From examination of Table 3, it can be inferred that after addition of WCO, the softening point declines as penetration and ductility rises. The ductility point, softening point and penetration value of the intended bitumen grade was successfully made possible by addition of WCO.

From the studies conducted, it was deduced that to reconstitute the penetration value of aged bitumen groups (50/60, 40/50) to levels similar to that of virgin binders (at 80/100 penetration grade), it would require 1% and 3% of WCO respectively and to attain the intended binder softening point for the aforementioned groups it would require 4% and 2% of WCO respectively [10, 34]. In tests conducted to enhance the characteristics of 60/70 penetration grade and to reconstitute the characteristics of extracted RAP binders to levels of targeted binder grade, 2% of WCO was required to successfully attain the said results and 3.5% of WCO was required for the restitution of the intended softening point [1, 25, 27, 32, 33]. The consistency of an asphalt binder can be evaluated from penetration testing. WCO was included as a rejuvenator at mixed concentrations in efforts to soften the aged binder.

Studies have been conducted by Asli, Zargar and Azahar [9, 12, 32] on different penetration graded binders binders (30/40, 40/50, 50/60, 60/70) and different percentages (1% to 5%) of WCO by weight of the binders; aged and unaged binders. Figure 2. and 3. show that the penetration value rises in proportion to the percentage of WCO added while the softening point is inversely proportional to the percentage of WCO added. In order to attain the original penetration grade of 80/100, the aged binder of bitumen groups Pen50/60, Pen40/50 and Pen 30/40 required 1%, 3% and 4% of WCO respectively. Figure 3 shows that about 4%, 2% and 1% of WCO was needed by age bitumen groups Pen30/40, Pen40/50 and Pen50/60 respectively in order to achieve the 80/100 grade bitumen.

As for untreated samples, a greater penetration rate was observed and for treated WCO, a greater softening point temperature was attained. Contrastingly, the increase of WCO concentrations caused the softening point value for both, treated and untreated WCO to decrease; as anticipated with the maximum temperature reaching

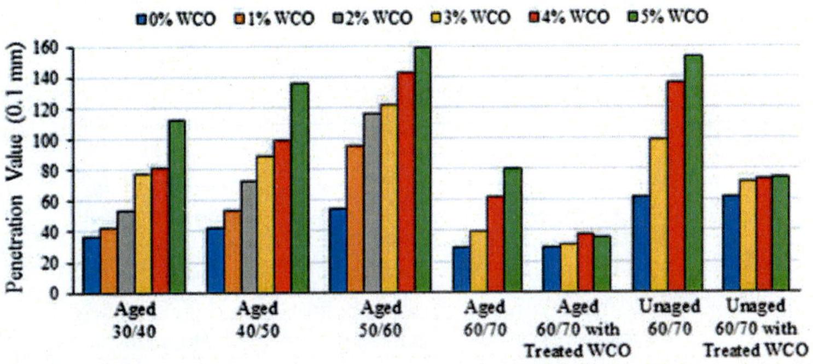

Fig. 2 Relationship of Penetration Value with different percentages of WCO [9, 12, 32]

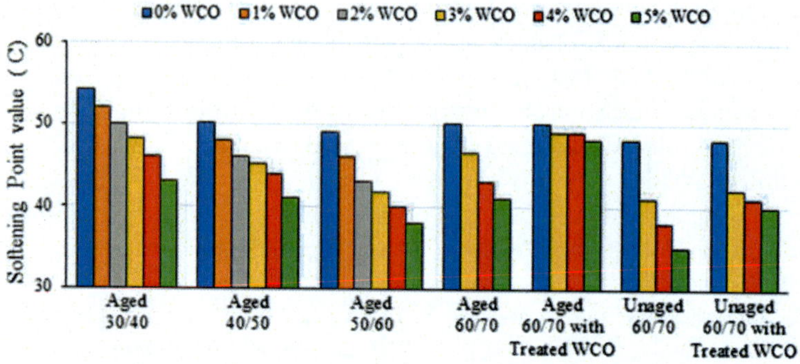

Fig. 3 Relationship of Softening Point Value with different percentages of WCO [9, 12, 32]

40 and 45 °C for 3% untreated and 3% treated WCO respectively. Alteration of WCO and improved chemical bonding with asphalt could be the contributing factors. Therefore, it can be inferred those older bituminous binders require a greater dosage of rejuvenator to reconstitute the intended softening point and penetration value.

Ductility test would allow for the evaluation of the flexibility and tensile deformation of asphalt. Studies by Chen et al. [36] as depicted in Fig. 4 revealed a sharp increase in ductility with the addition of WCO but this trend was followed by a sudden reduction with the oil percentage is 6% or more. This could be due to the non-uniformity in the mixing of WCO with control asphalt. As the grade of asphalt changes, the ductility changes significantly as observed with (Pen60/80) and (Pen40/60) grade.

Several studies [24, 26, 27, 31, 37, 38] had showed that ductility improves as WCO dosage is increased in the aged binder. Two studies [25, 32] had reviewed the consequence of WCO dosage on the specific gravity whereby the inclusion of WCO in the binder brought about a reduction in overall specific gravity. The flashpoint of asphalt binders can change with the level of WCO dosage. Furthermore, greater ductility shows the characteristic of the binders' appealing deformability and its superior effectiveness to withstand thermal cracking of asphalt binder [39].

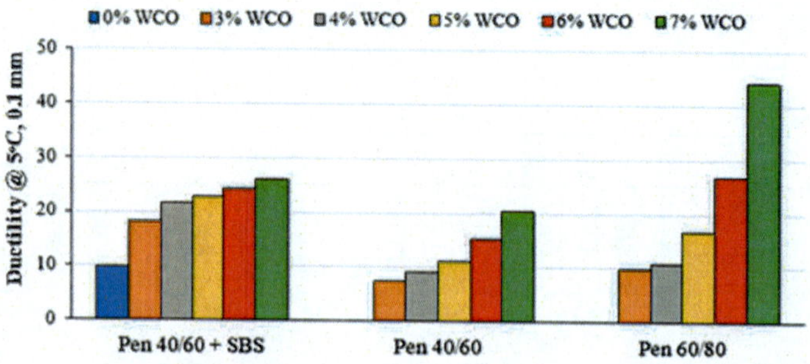

Fig. 4 Relationship of Softening Point Value with different percentages of WCO [36]

Based on three studies conducted on the fire and flash point variations, contrasting results were exhibited in terms of effective change in a flashpoint with the use of WCO as two of the studies [27, 38] showed a hike in flash point temperature of rejuvenated binder and one other proclaimed a conceivable decline in the fire and flash point of the rejuvenated binder [26].

4 Rheological Testing of WCO Modified Binder

4.1 Effect of WCO on the Viscosity and Mixing and Compaction Temperature

In the event of high binder viscosity, the risk of thermal cracking increases while in events of very low viscosity, the risk of rutting and fatigue increases [40]. As a general rule, Brookfield rotational viscometer (ASTM D4402) would be adopted to gauge the viscosity levels of asphalt binder.

It was Mahrez et al. [35] who pioneered the investigation of WCO inclusion on the viscosity levels at mixed temperatures. It was established that the rotational viscosity of the binder gradually reduces with the hike in temperature and concentration of WCO. It was observed that the viscosity levels of RTFO aged binder had declined by half in the presence of 6% WCO. In past investigations [9, 10, 12, 34], 1–5% of WCO was used in the aged bitumen groups (penetration grade 50/60, 40/50, 30/40) and the viscosity level was compared to that of the control binder (penetration grade 80/100) at mixed temperatures ranging from 90 °C to 180 °C.

The studies show that in order to achieve a viscosity level similar to that of the control binder at 110 °C, 2, 4, and 5% WCO was required. With an increase in temperature; such as at 135 and 150 °C, the volume of WCO required to achieve the viscosity of the control binder increases as well. There were similar studies conducted [1, 25, 27, 32, 41] where RAP binder was used and the control binder (penetration grade 60/70) had been artificially aged and rejuvenated with WCO at 1–8 wt.% and the viscosity levels of the rejuvenated binder was established at several temperatures i.e. ranging from 135 to 165 °C and compared with the control binder. The results revealed that at WCO concentration between 1 and 4%, the rejuvenated binder was able to exhibit viscosity levels close to the control binder at 165 °C.

Figure 5 illustrates the rotational viscosity test results of WCO based binder at several temperature points. Based on data collected it was determined that addition of WCO with increasing temperature resulted in a constant decrease in viscosity; whereby the maximum reading was recorded at 1.074 Pa s at the 120 °C mark. At 150 °C, a sudden change in viscosity was recorded; at a 21.5% drop from its former temperature and subsequently the levels steadily decreased. At 180 °C, a downward trend similar to that observed at 150 °C was recorded, albeit viscosity levels were different from that recorded at the 120 °C mark.

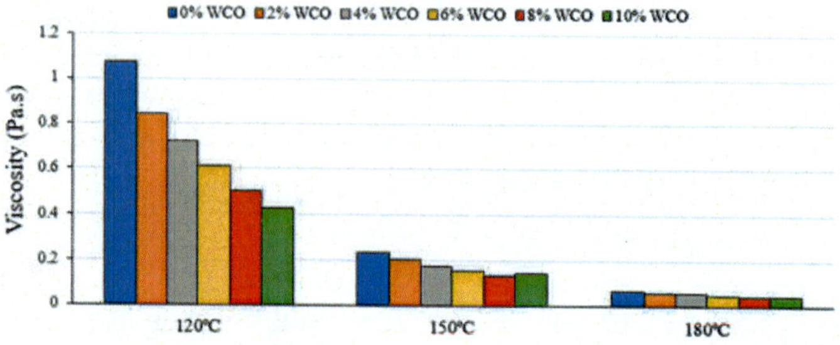

Fig. 5 Relationship of Viscosity with different percentages of WCO at different temperature [8]

In a few other studies conducted [1, 3, 22–30, 35–37, 40–42] where multiple control binders were studied, it was established that at various temperatures and WCO concentrations, viscosity is inversely proportional to the level of WCO present. Even though this is not unexpected, the results further solidifies the findings that WCO which is flexible in nature can produce constant results in aspects of a decline in viscosity.

Undoubtedly at lower viscosities, the mixing and compaction temperature of the mix can be reduced. Thus from an economic perspective, such reduction in mixing and compaction temperature would be much preferred; especially in the construction industry for its cost saving aspects. Nevertheless, though an excessive amount of rejuvenator can reduce viscosity levels, the aspect of poor serviceability would manifest. With lower serviceability, the product would be significantly softer and prone to rutting. Taking everything into consideration i.e. environmental, energy and economic aspects, low viscosity binders are more advantageous since they promote manipulability (mixing and compaction temperatures), demands lower energy thus increasing savings and necessitate less fuel which is environmental friendly.

4.2 *Rutting and Fatigue Resistance*

Rutting and fatigue resistance can be established by examination of the rheological characteristics of a binder:

Dynamic Shear Rheometer (DSR)

The dynamic shear rheometer (DSR) (ASTM D7175) test is able to show the rheological characteristics whereby any alteration in the characteristics of binders or the shear resistance is noted. The technique involves the measurement of the phase angle (δ) and complex shear modulus (G^*) of unaged binders. Complex G^* represents the total resistance of the binder to the level of deformation as a result of shear loading and δ is the delay period between shear strain and shear stress [41]. Rutting resistance at extreme temperature is demonstrated in the formula, $G^*/\sin\delta$, and fatigue resistance at a moderate temperature is demonstrated by G*sinδ as portrayed in the Superpave program [44].

A list of references can be found in Table 4; whereby WCO is used as a rejuvenator and the consequences on δ, G*, G*sin δ and G*/sin δ were examined at various WCO dosages and frequency-dependency (isothermal) in different control systems with aged or extracted binders. Based on the table, in order to reconstitute the G* of the aged binders (penetration grade 60/70, 50/60, 40/50 and 30/40), WCO inclusion of 1%, 1% and 3% were required respectively. As for the δ, in order to correlate the phase angle of virgin binder (penetration grade 80/100), the dosage of WCO required is 1%, 3%, 4% and 5% respectively [9, 10, 34, 45, 46].

In studies by Chen et al. [42], it was highlighted that rutting resistance of RAP extracted binder and artificially aged binders with resemblance to binders with penetration grade 40–60 and PG 64–22 would require 5% of WCO. These findings coincided with Ji et al. [31] whereby when RAP binders with added 6% WCO exhibited rutting and fatigue characteristics matching the virgin PG 64–22 bitumen. In another study, Azahar et al. [41] presented results of reduced G* and increased δ with the use of WCO of higher acidity and water content, consequently decreasing the general resistance to rutting and raises the fatigue resistance of WCO-rejuvenated binder.

Succinctly, conducted studies and established that fatigue cracking resistance of asphalt binder can be enhanced by increasing the concentration of WCO. With the increase in WCO concentration in aged asphalt binder, the deformation resistivity and recovery efficiency of the control and aged binders decreases because of the increased phase angle and reduced complex shear modulus. Consequently, there is a reduction of the G*/sinδ value leading to a decline of the rutting failure temperature whereas the general fatigue resistivity of the binder increases [3, 10, 12, 23, 32, 35, 37, 47–49].

Multiple Stress Creep Recovery (MSCR)

Contemporarily, Multiple Stress Creep Recovery (MSCR) test has been adopted to test the rutting potential and fatigue characteristics of the binder [11, 54, 55]. The test is accepted in determination of permanent deformation potential of the binder at extreme temperatures. Two crucial parameters i.e. the percentage recovery (%) and non-recoverable creep compliance (Jnr) were addressed by the MSCR test in the outcome of modified asphalt binder under tension. Rutting resistance was clearly represented by the non-recoverable creep compliance (Jnr); a high Jnr value is equivalent to decreased rutting resistance and enhanced thermal resistance of the binder [54]. With an increase in the WCO dosage with the binder, it is expected that the Jnr value would increase, consequently decreasing the rutting resistance and improving the thermal resistance [11].

MSCR test results during the study of cumulative strain curve of time strain responses at various tensions (0.1 kPa and 3.2 kPa), exhibited higher cumulative deformation by virgin binders compared to aged binders, however the cumulative deformation of aged asphalt binders can be increased with the inclusion of WCO. As more WCO is added, a greater cumulative deformation is experienced [26]. Analogous to a study by Wang et al. [56] on the Jnr value of the binders under two creep

Table 4 Literature study of complex shear modulus, phase angle, and the optimum dosage of WCO made identical parameters to the control binder

Ref	Rej Type and Content%	the optimum dosage of WCO made identical parameters to the control binder								Binder Tested	Control Binder
		Effect of increase in WCO%				Isochronal behavior (constant frequency)					
		G*	δ	G*/sin δ	G*sin δ	G*	δ	G*/sin δ	G*sin δ		
[50]	WCO 2–6							6%	6%	Artificially aged and extracted RAP binder	VG 30
[46]	WCO 3–5							5%		Virgin binder	60/70 Pen. grade
[24]	WCO 1–5					1–3%		1–3%		Artificially aged Virgin binder in TFOT at different aging times	PEN 70
[23]	WVCO 2–10							7%	7%	Artificially aged Virgin binder using RTFO, PAV and oxiaging method	70/100 Pen. grade
[25]	WCO 2, 3, 3.5, 4							3.5%	3.5%	Extracted RAP binder	60/70 Pen. grade
[34]	WCO 1–3							2%		Virgin binder	80/100 Pen. grade
[31]	WVCO 2–10							6%	6%	Binder recovered from RAP	PG 64–22
[32]	WCO 3–5							4%		Virgin binder 60/70 PG	60/70 Pen. grade
[51]	WCO 2–10	4%	4%	4%	4%	4%	4%	4%	4%	Virgin binder	50: 50 TLA: TPB

(continued)

Table 4 (continued)

Ref	Rej Type and Content%	the optimum dosage of WCO made identical parameters to the control binder								Binder Tested	Control Binder
		Effect of increase in WCO%				Isochronal behavior (constant frequency)					
		G*	δ	G*/sin δ	G*sin δ	G*	δ	G*/sin δ	G*sin δ		
[52]	WVO 12								12%	Extracted RAP binder (PG 94e12)	PG 64–22
[53]	WCO 5, 10	5%	5%	5%						Binder extracted from RAP	PG 64–22
		5%	5%	5%						Mass ratio recovered aged	
[42]	WEVO 3–7	6%	6%	6%						Artificially aged Virgin binder by RTFO, PAV	60/80 Pen. grade
		5%	5%	5%							40/60 Pen. grade
		4%	4%	4%							40/60 Pen. + SBS
[12]	WCO 1–4					3%	4%			Aged binder 40/50 Penetration grade	80/100 Pen. grade
[10]	WCO 1–5					1%	1%			Aged 60/70	80/100 Pen. grade
						1%	3%			Aged 50/60	
						3%	4%			Aged 40/50	
						3%	5%			Aged 30/40	

tension levels, the results proved that the Jnr value increases by two-folds with the addition of 5 wt.% WCO. Additionally, examination of the time-strain curves depict the detrimental consequences of bio-oil inclusion on the permanent deformation resistance.

Briefly, adding WCO to the binder causes an increase in the softness of the binder and a hike in the non-recoverable creep compliance but a decline in the permanent deformation resistance. In consequence, as a matter of precaution, rutting resistance should be considered based on climatic conditions of the site before addition of WCO into the binder.

Linear Amplitude Sweep (LAS)

Presently novel tests such as the Linear Amplitude Sweep (LAS) has been favoured to study the rutting potential and fatigue characteristics of the binder [11, 26, 54, 55] The LAS test is carried out at moderate temperatures to evaluate the level of binder resistance to the damage sustained as a consequence of cyclic load with a linear intensification of strain at a constant frequency [57]. The test comprised of two stages which are the (1) frequency sweep test intended to determine the alpha "α" parameter for damage analysis and the (2) amplitude sweep test for examination of the damage accumulated in samples against the loading cycles.

Literature on LAS tests on binders containing WCO are somewhat scarce. The fatigue life extrapolation as logarithmic LAS cycles to failure vs. strain (%) was studied in Zaumanis et al. [55]. The results revealed that the virgin binder exhibited poor response and low fatigue life at 2.5% strain. However, as the percentage was increased i.e. at about 10%, the response improved and exhibited better fatigue life than RAP and rejuvenated binders. That said, at meagre strain %, RAP binders have high fatigue life which declines to the minimum level at 10% strain. Contrastingly, waste vegetable oil-rejuvenated binder possess excellent fatigue life at both lower and higher strain %, which is on par with virgin binders. Studies by Wang et al. [56] explored the ramifications of WCO inclusion on the fatigue life and discovered that addition of WCO to the binder promotes its aversion to fatigue against the loading cycles. Inevitably, it can be inferred that WCO inclusion to the binder grants fatigue improvements on the basis of rheological test results.

4.3 *Thermal Cracking Resistance*

Some parameters affiliated with the cracking resistance of asphalt pavement are the m-value and creep stiffness, which often describe the binder's performance at low temperature. These parameters are usually measured by a bending beam rheometer (BBR) (ASTM D6648) and carried out on a PAV-aged binder. Creep stiffness defines the estimated levels of stress as a consequence of thermal contraction and the m-value describes the creep rate or simply the capability to relax stresses. With comparison to virgin binders, creep stiffness improves at low temperature in contrast to the m-value which declines; the latter also increases the susceptibility of thermal cracking.

The results of BBR of the aged, rejuvenated and virgin binders at −18 °C were examined by Yu et al. [58]. In comparison with the virgin binder, the outcome of mixing rejuvenated asphalt with waste vegetable oil (WVO) produced a tenfold reduction in creep stiffness and more than double the increment of the m-value. Further to that, Ji et al. [31] examined the impact on creep stiffness in relation to the dosage of WCVO rejuvenator to the asphalt binder. The outcome indicated that at low temperatures, the stiffness of the rejuvenated binder constantly declined with the increase in WCVO concentrations; implying that the thermal cracking aversion levels had been improved and the risk of thermal cracking had been lowered.

The relationship between the quality of WCO and thermal cracking was examined by Zhang et al. [59]. It was deduced that as the WCO keeps getting cyclically reheated, the value of stiffness rose while the m-value declined. Other comparable studies [25, 26, 60] proved that inclusion of WCO was advantageous in conditions of low-temperature characteristics. In brief, it can be observed that WCO dosing gives better thermal cracking aversion to the asphalt binder; albeit both creep stiffness and m-values are dependent on the quality and kind of WCO used as asphalt rejuvenators. Zhang et al. [59] examined different aged, virgin and WCO rejuvenated asphalts to draw an analogy between the stiffness and the m-value. The results exemplify that from − 12 to − 18 °C, the stiffness increases while the m-value declines regardless of the type of binders which increases the susceptibility of thermal cracking. Contrastingly, other studies [61, 62] have revealed an enhancement in thermal cracking aversion with the use of WCO rejuvenators.

Since low temperatures result in stiffer binders, inclusion of WCO rejuvenators could soften the binder and promote a decline of the G* value. With the decline of the G* factor, the binder would be able to attain improved cracking resistance. Succinctly, it can be inferred that the effect on thermal cracking aversion at critically low temperatures of rejuvenated asphalt would be reliant on the type and properties of the WCO used for asphalt rejuvenation.

5 Conclusions

This review scrutinizes the numerous aspects of rejuvenating aged asphalt with WCO to revive its initial characteristics. Of late, various experiments have been performed to assess the relevancy of adopting WCO as an asphalt rejuvenator and the feasibility of using it with asphalt mixes. With reference to the present available literature. The physical characteristics of WCO rejuvenated bitumen are tremendously influenced by the composition of the oil. (a) A reduction in overall viscosity of the bitumen brings about higher ductility, penetration value and lower specific gravity and softening point. (b) Greater thermal resistance as a result of improved flexibility at lower temperatures. (c) Binders are more prone to permanent deformation. Rheological assessment of WCO-based rejuvenated bitumen features. (a) a decline of stiffness and rutting resistance during addition of WCO in the binder. (b) enhanced resistance to thermal cracking and greater bitumen fatigue.

This review paper underlines the efficacy improvements on asphalt binders using WCO as a rejuvenator. Several literatures had featured reassuring outcomes with enhancements in terms of rheological and physical characteristics of asphalt binders. Nevertheless, there exists numerous constrains in adoption of WCO in asphalt pavements; summarised hereunder:

- Establishment of optimal application rates of WCO rejuvenators are still inconclusive.
- Data on rutting and cracking resistance is yet to be comprehensive especially since literature for cold climate is still lacking.
- Another aspect that deserves more scrutiny is the consequence of mixing WCO with other modifiers to achieve enhanced pavement performance.

References

1. Azahar WNAW, Jaya RP, Hainin MR et al (2017) Mechanical performance of asphaltic concrete incorporating untreated and treated waste cooking oil. Constr Build Mater 150:653–663. https://doi.org/10.1016/j.conbuildmat.2017.06.048
2. United States Department of Agriculture Foreign Agricultural Service (2021) Oilseeds: world markets and trade. World Production, Markets, and Trade Report
3. Chen M, Xiao F, Putman B et al (2014) High temperature properties of rejuvenating recovered binder with rejuvenator, waste cooking and cotton seed oils. Constr Build Mater 59:10–16. https://doi.org/10.1016/j.conbuildmat.2014.02.032
4. Kabir I, Yacob M, Radam A (2014) Households' awareness, attitudes and practices regarding waste cooking oil recycling in Petaling, Malaysia. IOSR-JESTFT 8:45–51
5. Sanli H, Canakci M, Alptekin E (2011) Characterization of waste frying oils obtained from different facilities. In: Proceedings of World Renewable Energy Congress—Sweden, 8–13 May 2011, Linköping, Sweden vol 57, pp 479–485. https://doi.org/10.3384/ecp11057479
6. Knothe G, Krahl J, Van Gerpen JBT-TBH (Second E (2010) 8—current status of the biodiesel industry. AOCS Press, pp 299–399
7. Tsoutsos T, Tournaki S, Gkouskos Z et al (2019) Quality characteristics of biodiesel produced from used cooking oil in southern Europe. ChemEngineering 3:19
8. Bailey HK, Philips P (2010) asphalt rejuvenation, rejuvenation of reclaimed asphalt field of the invention the invention relates to reclaimed asphalt compositions and rejuvenation thereof with ester-functional compositions having cyclic content. US Pat 34586:A12010
9. Asli H, Ahmadinia E, Zargar M, Karim MR (2012) Investigation on physical properties of waste cooking oil—rejuvenated bitumen binder. Constr Build Mater 37:398–405. https://doi.org/10.1016/j.conbuildmat.2012.07.042
10. Asli H, Karim MR (2011) Implementation of waste cooking oil as RAP rejuvenator. J East Asia Soc Transp Stud 9:1336–1350. https://doi.org/10.11175/easts.9.1336
11. Wen H, Bhusal S, Wen B (2013) Laboratory evaluation of waste cooking oil-based bioasphalt as an alternative binder for hot mix asphalt. J Mater Civ Eng 25:1432–1437. https://doi.org/10.1061/(asce)mt.1943-5533.0000713
12. Zargar M, Ahmadinia E, Asli H, Karim MR (2012) Investigation of the possibility of using waste cooking oil as a rejuvenating agent for aged bitumen. J Hazard Mater 233–234:254–258. https://doi.org/10.1016/j.jhazmat.2012.06.021
13. Azahar WNAW, Bujang M, Jaya RP, et al (2016) A review on application of waste cooking oil as rejuvenator in porous asphalt mixture. J Teknol 78:105–109. https://doi.org/10.11113/jt.v78.8005

14. Ben Zair MM, Jakarni FM, Muniandy R, Hassim S (2021) A brief review: application of recycled polyethylene terephthalate in asphalt pavement reinforcement. Sustainability 13:1303
15. Kulkarni MG, Dalai AK (2006) Waste cooking oil—an economical source for biodiesel: a review. Ind Eng Chem Res 45:2901–2913. https://doi.org/10.1021/ie0510526
16. Choe E, Min DB (2007) Chemistry of deep-fat frying oils. J Food Sci 72. https://doi.org/10.1111/j.1750-3841.2007.00352.x
17. Das AK, Babylatha R, Pavithra AS, Khatoon S (2013) Thermal degradation of groundnut oil during continuous and intermittent frying. J Food Sci Technol 50:1186–1192. https://doi.org/10.1007/s13197-011-0452-7
18. Jaarin K, Kamisah Y (2012) Repeatedly heated vegetable oils and lipid peroxidation. Lipid Peroxidation. https://doi.org/10.5772/46076
19. Majidifard H, Tabatabaee N, Buttlar W (2019) Investigating short-term and long-term binder performance of high-RAP mixtures containing waste cooking oil. J Traffic Transp Eng (English Ed) https://doi.org/10.1016/j.jtte.2018.11.002
20. Ullah Z, Bustam MA, Man Z (2014) Characterization of waste palm cooking oil for biodiesel production. Int J Chem Eng Appl 5:134–137. https://doi.org/10.7763/ijcea.2014.v5.366
21. Khalisanni K, Khalizani K, Rohani MS, Khalid PO (2008) Analysis of waste cooking oil as raw material for biofuel production. Glob J Environ Res 2:81–83
22. Joni HH, Al-Rubaee RHA, Al-zerkani MA (2019) Rejuvenation of aged asphalt binder extracted from reclaimed asphalt pavement using waste vegetable and engine oils. Case Stud Constr Mater 11:e00279. https://doi.org/10.1016/j.cscm.2019.e00279
23. Gökalp İ, Uz VE (2019) Utilizing of waste vegetable cooking oil in bitumen: zero tolerance aging approach. Constr Build Mater 227:116695. https://doi.org/10.1016/j.conbuildmat.2019.116695
24. Li H, Dong B, Wang W et al (2019) Effect of waste engine oil and waste cooking oil on performance improvement of aged asphalt. A Sect Appl Sci 1767:1–20
25. El-Shorbagy AM, El-Badawy SM, Gabr AR (2019) Investigation of waste oils as rejuvenators of aged bitumen for sustainable pavement. Constr Build Mater 220:228–237. https://doi.org/10.1016/j.conbuildmat.2019.05.180
26. Xinxin C, Xuejuan C, Boming T et al (2018) Investigation on possibility of waste vegetable oil rejuvenating aged asphalt. A Sect Appl Sci 8:765. https://doi.org/10.3390/app8050765
27. Al-Omari AA, Khedaywi TS, Khasawneh MA (2018) Laboratory characterization of asphalt binders modified with waste vegetable oil using SuperPave specifications. Int J Pavement Res Technol 11:68–76. https://doi.org/10.1016/j.ijprt.2017.09.004
28. Zhang D, Chen M, Wu S et al (2017) Low Temperature properties of waste edible vegetable oil rejuvenated asphalt binder with recycled tire rubber. J Test Eval 46:602–609
29. Eriskin E, Karahancer S, Terzi S, Saltan M (2017) Waste frying oil modified bitumen usage for sustainable hot mix asphalt pavement. Arch Civ Mech Eng 17:863–870. https://doi.org/10.1016/j.acme.2017.03.006
30. Dokandari PA, Kaya D, Sengoz B, Topal A (2017) Implementing waste oils with reclaimed asphalt pavement. In: Proceedings of 2nd World Congress on Civil, Structural, and Environmental Engineering vol 17, pp 1–12. https://doi.org/10.11159/icsenm17.142
31. Ji J, Yao H, Suo Z et al (2017) effectiveness of vegetable oils as rejuvenators for aged asphalt binders. J Mater Civ Eng 29:1–10. https://doi.org/10.1061/(ASCE)MT.1943-5533.0001769
32. Azahar WNAW, Jaya RP, Hainin MR et al (2016) Chemical modification of waste cooking oil to improve the physical and rheological properties of asphalt binder. Constr Build Mater 126:218–226. https://doi.org/10.1016/j.conbuildmat.2016.09.032
33. Azahar WNAW, Bujang M, Jaya RP et al (2016) Performance of waste cooking oil in asphalt binder modification 700:216–226. https://doi.org/10.4028/www.scientific.net/KEM.700.216
34. Rasman M, Hassan NA, Hainin MR (2018) Engineering properties of bitumen modified with bio-oil. MATEC Web Conf 250(02003):9
35. Mahrez A, Karim MR, Ibrahim MR, Katman HY (2009) Prospects of using waste cooking oil as rejuvenating agent in bituminous binder. In: Proceedings of the Eastern Asia Society for transportation studies, vol. 7 (The 8th international conference of Eastern Asia Society for transportation studies, 2009). Eastern Asia Society for transportation studies, p 289

36. Chen M, Leng B, Wu S, Sang Y (2014) Physical, chemical and rheological properties of waste edible vegetable oil rejuvenated asphalt binders. Constr Build Mater 66:286–298. https://doi.org/10.1016/j.conbuildmat.2014.05.033
37. Yang S, Meizhu C, Jin W et al (2014) Effect of waste edible animal oil on physical properties of aged asphalt. Key Eng Mater 599:130–134. 10.4028/www.scientific.net/KEM.599.130
38. Joni HH, Al-rubaee RHA, Al-zerkani MA (2019) Rejuvenation of aged asphalt binder extracted from reclaimed asphalt pavement using waste vegetable and engine oils. Case Stud Constr Mater e00279. https://doi.org/10.1016/j.cscm.2019.e00279
39. Leung DYC, Guo Y (2006) Transesterification of neat and used frying oil: Optimization for biodiesel production. Fuel Process Technol 87:883–890. https://doi.org/10.1016/j.fuproc.2006.06.003
40. Zaumanis M, Mallick RB, Frank R (2014) 100% recycled hot mix asphalt: a review and analysis. Resour Conserv Recycl 92:230–245. https://doi.org/10.1016/j.resconrec.2014.07.007
41. Azahar WNAW, Bujang M, Jaya RP et al (2016) Performance of waste cooking oil in asphalt binder modification. Key Eng Mater 700:216–226. 10.4028/www.scientific.net/KEM.700.216
42. Chen M, Leng B, Wu S, Sang Y (2014) Physical, chemical and rheological properties of waste edible vegetable oil rejuvenated asphalt binders. Constr Build Mater 66:286–298. https://doi.org/10.1016/j.conbuildmat.2014.05.033
43. Somé SC, Pavoine A, Chailleux E (2016) Evaluation of the potential use of waste sunflower and rapeseed oils-modified natural bitumen as binders for asphalt pavement design. Int J Pavement Res Technol 9:368–375. https://doi.org/10.1016/j.ijprt.2016.09.001
44. Speight JG (2016) Asphalt technology, in: asphalt materials science and technology, vol 1148. Elsevier. In: Speight JGBT-AMS and T (ed) Butterworth-Heinemann, Boston, pp 361–408
45. El-shorbagy AM, El-badawy SM, Gabr AR (2019) Investigation of waste oils as rejuvenators of aged bitumen for sustainable pavement. Constr Build Mater 220:228–237. https://doi.org/10.1016/j.conbuildmat.2019.05.180
46. Nordiana M, Aifa WAWN, Hainin MR, et al (2019) Rutting resistance of untreated and treated waste cooking oil in bitumen after aging condition. IOP Conf Ser Earth Environ Sci Inst Phys Publ 244. https://doi.org/10.1088/1755-1315/244/1/012041
47. Singh-ackbarali D, Maharaj R, Mohamed N, Ramjattan-harry V (2017) Potential of used frying oil in paving material : solution to environmental pollution problem. 12220–12226. https://doi.org/10.1007/s11356-017-8793-z`
48. Maharaj R, Ramjattan-Harry V, Mohamed N (2015) Rutting and fatigue cracking resistance of waste cooking oil modified trinidad asphaltic materials. Sci World J 2015. https://doi.org/10.1155/2015/385013
49. Zaumanis M, Mallick RB, Frank R (2014) Determining optimum rejuvenator dose for asphalt recycling based on Superpave performance grade specifications. Constr Build Mater 69:159–166. https://doi.org/10.1016/j.conbuildmat.2014.07.035
50. Durga Prashanth L, Palankar N, Ravi Shankar AU (2019) A study on the effect of rejuvenators in reclaimed asphalt pavement based stone mastic asphalt mixes. Int J Pavement Res Technol 12:9–16. https://doi.org/10.1007/s42947-019-0002-7
51. Maharaj R, Ramjattan-Harry V, Mohamed N (2015) Rutting and fatigue cracking resistance of waste cooking oil modified trinidad asphaltic materials. Sci World J 2015:
52. Zaumanis M, Mallick RB, Poulikakos L, Frank R (2014) Influence of six rejuvenators on the performance properties of Reclaimed Asphalt Pavement (RAP) binder and 100% recycled asphalt mixtures. Constr Build Mater 71:538–550. https://doi.org/10.1016/j.conbuildmat.2014.08.073
53. Chen M, Xiao F, Putman B, et al (2014) High temperature properties of rejuvenating recovered binder with rejuvenator, waste cooking and cotton seed oils. Constr Build Mater 59:10–16. https://doi.org/10.1016/j.conbuildmat.2014.02.032
54. Cao X, Wang H, Cao X et al (2018) Investigation of rheological and chemical properties asphalt binder rejuvenated with waste vegetable oil. Constr Build Mater 180:455–463. https://doi.org/10.1016/j.conbuildmat.2018.06.001

55. Zaumanis M, Mallick RB (2015) Review of very high-content reclaimed asphalt use in plant-produced pavements: state of the art. Int J Pavement Eng 16:39–55. https://doi.org/10.1080/10298436.2014.893331
56. Wang C, Xue L, Xie W, et al (2018) Laboratory investigation on chemical and rheological properties of bio-asphalt binders incorporating waste cooking oil. Constr Build Mater 167:348–358. https://doi.org/10.1016/j.conbuildmat.2018.02.038
57. TP101 AT (2014) Estimating damage tolerance of asphalt binders using the linear amplitude sweep. Am Assoc State Highw Transp Off444 North Capitol Str NW, Suite 249 Washington, DC 20001cials 444 North Capitol Str NW, Suite 249 Washington, DC444 North Capitol Str NW, Suite 249 Washington, DC 20001cials 444 North Capitol Str NW, S 20001
58. Yu X, Zaumanis M, dos Santos S, Poulikakos LD (2014) Rheological, microscopic, and chemical characterization of the rejuvenating effect on asphalt binders. Fuel 135:162–171. https://doi.org/10.1016/j.fuel.2014.06.038
59. Zhang D, Chen M, Wu S, et al (2017) Analysis of the relationships between waste cooking oil qualities and rejuvenated asphalt properties. Materials (Basel) 10. https://doi.org/10.3390/ma10050508
60. islam Gökalp VEU (2019) Utilizing of Waste Vegetable cooking oil in bitumen : zero tolerance aging approach. 227. https://doi.org/10.1016/j.conbuildmat.2019.116695
61. Bahadori A, Li R, Zhang K, et al (2019) Performance evaluation of hot mix Biobinder Amir. Airf Highw Pavements 68–76
62. Zhu J, Zhang K, Liu K, Shi X (2019) Performance of hot and warm mix asphalt mixtures enhanced by nano-sized graphene oxide. Constr Build Mater 217:273–282. https://doi.org/10.1016/j.conbuildmat.2019.05.054

Waste Clay Brick Binders for Rigid Pavement Subbase and Base Concretes

Janitha Migunthanna, Pathmanathan Rajeev, and Jay Sanjayan

Abstract Both base and subbase layers of rigid pavements are currently constructed with concrete based on ordinary Portland cement (OPC). The OPC manufacturing process is heavily carbon intensive and has severe negative impacts towards the environment and sustainability. Geopolymer as an alternative is promising, in terms of both performance and sustainability. This study used waste clay bricks (WCB) blended with fly ash and slag precursors to produce one-part geopolymer binders with the aim of replacing OPC in pavement concrete. One-part binders were prepared by combining activators and the precursor at a 1:9 ratio by weight. Within the precursor, fly ash content was varied at 0%, 10%, 20% and 30% by weight. In the remaining precursor, WCB was replaced with slag at 20%, 40%, 60% and 80% replacements. All blends of WCB with fly ash and slag showed significantly higher strengths than their unblended counterparts. The highest 28-day compressive strength of 92 MPa was recorded for the sample with 30% fly ash and with 80% replacement of WCB with slag on the remaining precursor. Additionally, the WCB based geopolymer binders showed a high early strength development, achieving more than 50% of its 28-day compressive strength within 72 h. Aggregates were then introduced to the selected binders to prepare geopolymer concrete (GPC). The compressive strength of GPC samples was recorded after 28 days of aging. Initial GPC trial mixes were able to achieve compressive strengths up to 41 MPa, demonstrating the feasibility of using WCB based binders for developing pavement concretes.

Keywords Waste clay bricks · Fly ash · Slag · One-part geopolymer · Geopolymer concrete · Pavement construction

J. Migunthanna (✉) · P. Rajeev · J. Sanjayan
Centre for Sustainable Infrastructure and Digital Construction, Swinburne University of Technology, Hawthorn, VIC 3122, Australia
e-mail: jmigunthannakariyaka@swin.edu.au

P. Rajeev
e-mail: prajeev@swin.edu.au

J. Sanjayan
e-mail: jsanjayan@swin.edu.au

© The Author(s), under exclusive license to Springer Nature Switzerland AG 2022

H. R. Pasindu et al. (eds.), *Road and Airfield Pavement Technology*, Lecture Notes in Civil Engineering 193, https://doi.org/10.1007/978-3-030-87379-0_67

1 Introduction

Typical rigid pavement structure consists of base and subbase layers, usually constructed with ordinary Portland cement (OPC) concrete [1, 2]. In the Australian context, OPC concrete with 28-day compressive strength of not less than 32 MPa is required as the design strength for base applications, while lean-mixed concrete with 28-day compressive strength of not less than 5 MPa is used for subbase applications [1, 3]. However, the manufacturing process of OPC binder used in these concretes is highly carbon intensive and has severe negative environmental impacts [4–7]. Therefore, it is vitally important to search for suitable alternate binder materials to fully or partially replace OPC in concrete.

Geopolymer binders are widely regarded as a promising alternative for OPC in terms of performance and durability [5, 8, 9]. These binders are synthesized by combining aluminosilicate source materials (i.e., precursors) with strong alkaline solutions (i.e., activators) [8, 10]. A wide range of aluminosilicate materials are available as industrial by-products and waste, agricultural by-products and as natural minerals [11]. Fly ash, slag, metakaolin and rice husk ash are some common examples for such materials [11, 12]. The mechanical and durability properties of these low carbon binders require to be comprehensively investigated and optimized based on the intended field application. The present work therefore investigates the suitability of using geopolymer binders to fully replace OPC in concrete used in rigid pavement construction.

Recent studies show the feasibility of using waste clay bricks (WCB) as an aluminosilicate precursor for geopolymer production [11, 13–16]. The most reactive sources of geopolymer precursors are calcined aluminosilicate materials [17]. For example fly ash is a by-product generated through combustion of coal and slag is a by-product in pig iron manufacturing produced at 1500°C [18, 19]. When using natural clay minerals as precursors, they are required to calcine at high temperatures prior to activation [20, 21]. Clay bricks are calcined at 850–950 °C during its manufacturing process. Hence WCB also consists of forms of calcined clays which have a high possibility of undergoing geopolymerisation [16].

A large quantity of WCB are generated in worldwide annually from construction and demotion waste [15, 22] and majority of these are going to landfills without any recycling or reuse [23]. Therefore, productive use of this unutilized and available waste in construction industry will support better waste management while reducing carbon footprint in construction industry. So far in pavement industry, application of WCB are discussed as only as aggregates. However, a comprehensive analysis of WCB based binders and development of WCB binder based geopolymer concrete (GPC) is still lacking. Therefore, this study investigates WCB recovered from construction and demolition waste as an aluminosilicate source material to first produce geopolymer binders and GPC thereafter.

Considering the practical situation when constructing rigid pavements in the field, the current work deals with producing geopolymer binders at ambient curing conditions. Previous studies show the incapability of achieving satisfactory strength at ambient curing using WCB alone as the precursor [11, 13, 15]. However, the strengths can be significantly improved by blending WCB with other aluminosilicate materials such as slag, fly ash and metakaolin. For example, the study by Zawrah et al. 2016 [13] showed a 7-day compressive strength improvement from 5 MPa to 64 MPa under ambient curing conditions when 60% of WCB in the precursor was replaced by slag. Similarly, an improvement of 7-day compressive strength from around 8 MPa to 20 MPa was observed by Rovnanik et al. 2016 after replacing 50% of WCB with fly ash.

Therefore, this study investigates the development of geopolymer binders through blends of WCB with fly ash and slag. Further, with the aim of developing more OPC like alternatives and improving occupational safety, the current work focuses on synthesising one-part geopolymer binders, where solid activators are used instead of liquid activators.

2 Material and Methods

2.1 Raw Materials

WCB powder of size 150 μm sieve passing, fly ash and slag were used as the aluminosilicate precursors for producing the geopolymer binders. WCB was extracted from construction and demolition waste and was supplied by EcoGroup Pty. Ltd, Melbourne, Australia. Anhydrous sodium metasilicate (Na_2SiO_3) powder ($SiO_2/Na_2O = 1.0$) was used as the sole activator. Figure 1 shows the physical nature

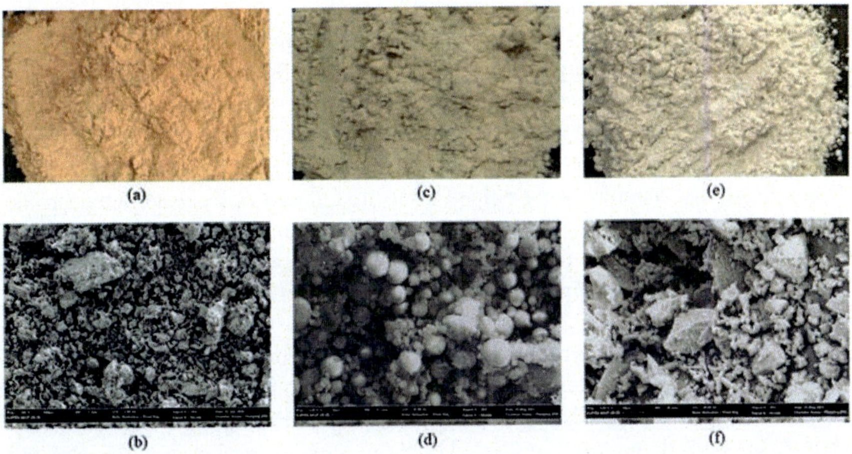

Fig. 1 Physical view and microstructure of raw precursors. **a, b** WCB powder. **c, d** Fly ash. **e, f** Slag

Table 1 Chemical composition of raw materials

Oxide	WCB (Wt.%)	Slag (Wt.%)	Fly ash (Wt.%)	Oxide	WCB (Wt.%)	Slag (Wt.%)	Fly ash (Wt.%)
SiO_2	68.97	27.54	57.13	Cr_2O_3	0.06	–	0.02
CaO	3.35	61.13	8.67	ZnO	0.05	–	–
Fe_2O_3	12.07	0.76	12.79	V_2O_5	0.03	–	0.04
Al_2O_3	8.71	2.84	16.67	Rb_2O	0.05	–	0.01
K_2O	4.48	0.60	1.13	PbO_2	0.01	0.02	0.01
TiO_2	1.68	1.05	2.74	Y_2O_3	0.01	0.01	0.03
SO_4	0.23	5.13	0.29	NiO	–	–	0.03
MnO	0.10	0.24	0.18	Ag_2O	0.01	0.01	0.01
P_2O_5	–	0.44	–	Ta_2O_5	–	–	0.00
ZrO_2	0.16	0.09	0.15	Ga_2O_3	–	–	0.02
SrO	0.02	0.14	0.10				

and microstructure of raw materials. Table 1 shows the chemical composition of precursors obtained through x-ray fluorescence (XRF) analysis.

2.2 *Preparation of Geopolymer Binders*

From initial trials, an activator content of 10% by weight of the total binder was identified as suitable for WCB based binders. In the remaining 90% of the binder, WCB blended with fly ash and slag at different replacement ratios were used. First, 0%, 10%, 20% and 30% of the total weight of the precursor was filled with fly ash. The remainder of each group was filled with WCB and slag at 20–80%, 40–60%, 60–40% and 80–20% ratios by weight. For the comparison, binder samples consisting only WCB in the precursor were also prepared. Figure 2 summarizes all these mix proportions of the geopolymer binders with different WCB, slag and fly ash compositions. Samples were named after the amount of fly ash content in the precursor and amount of slag in the remaining. For example, "FA10-S40" refers to the sample consist of 10% of fly ash in the total precursor and, 40% replacement of WCB with slag in the remaining content. Here, FA refers to fly ash and S refers to slag.

To obtain a homogeneous mixture, all the dry ingredients were dry mixed for three minutes. Then water was added to the mixture according to a pre-determined water-to-binder (w/b) ratio of 0.32 and mixing was continued for another eight minutes. Fresh pastes were then placed in 25 mm cubic moulds in two layers and each layer was subjected to a mechanical vibration for 30 s. Cast samples were covered with a thin polythene film and stored in an environmental chamber at 25 ± 2 °C. After

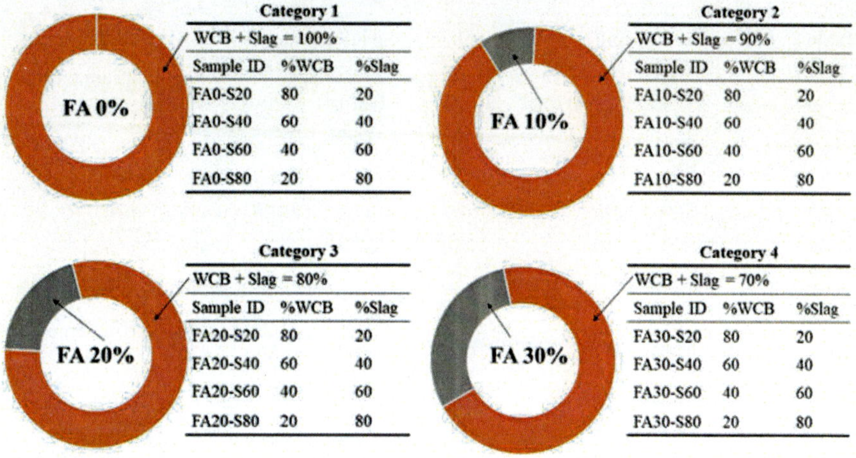

Fig. 2 Mix proportions and categories of different geopolymer binders

24 h all the samples were demoulded and stored in the same environmental chamber until the test date.

To evaluate the timeline of strength development for the different geopolymer formulations, their compressive strengths were recorded at 72 h, 7 days and 28 days of aging. X-ray diffraction analysis (XRD), Fourier transform infrared (FTIR) analysis and scanning electron microscopic (SEM) analysis were conducted for selected samples with high compressive strength development.

2.3 *Preparation of GPC*

Based on compressive strength, one binder composition from each category (Fig. 2) was selected to develop initial GPC trial mixes. A binder content of 400 kg/m^3 was used for all mixes. First, geopolymer binders were prepared by mixing all dry ingredients corresponding to the selected binder type. Subsequently, coarse and fine aggregates were added to the dry binders and all materials were mixed further for about five minutes. Finally, water was added, and the mixing was continued for another fifteen minutes.

Three different aggregate combinations were used with each selected binder, namely, 30%, 35% and 40% of fine aggregates from the total aggregates. Coarse aggregates of nominal size 14 mm and 7 mm was used in weight ratio of 3:2 for all mixes. GPC samples were named using the corresponding binder category which is shown in Fig. 2 and the percentage of fine aggregates in the total aggregate content. For example, GPC mix "C3-fine30" refers to the concrete mix prepared by using Category 3 binder (i.e., FA20-S60) and with 30% fine aggregates in total aggregates.

Table 2 Mix proportions and recorded slump of initial GPC trial mixes

Mix ID	Binder type	%Fine agg. From total agg	Fine agg. (kg/m^3)	Coarse agg. (kg/m^3)		w/b ratio	Slump (mm)
				14 mm	7 mm		
C1-fine30	FA0-S60	30	567.0	793.8	529.2	0.49	65
C1-fine35	FA0-S60	35	661.5	737.1	491.4	0.49	70
C1-fine40	FA0-S60	40	756.0	680.4	453.6	0.49	55
C2-fine30	FA10-S60	30	567.0	793.8	529.2	0.47	65
C2-fine35	FA10-S60	35	661.5	737.1	491.4	0.47	60
C2-fine40	FA10-S60	40	756.0	680.4	453.6	0.47	55
C3-fine30	FA20-S60	30	567.0	793.8	529.2	0.45	60
C3-fine35	FA20-S60	35	661.5	737.1	491.4	0.45	65
C3-fine40	FA20-S60	40	756.0	680.4	453.6	0.46	55
C4-fine30	FA30-S80	30	567.0	793.8	529.2	0.45	60
C4-fine35	FA30-S80	35	661.5	737.1	491.4	0.45	70
C4-fine40	FA30-S80	40	756.0	680.4	453.6	0.46	50

In order to evaluate the workability of different GPCs, slump tests were conducted in accordance with AS 1012.3.1 standards [24]. To satisfy the Australian rigid pavement specifications [1, 3, 25], different w/b ratios were attempted until the mixture displayed a slump around 50-70 mm. Test cylinders with diameter 100 mm and height 200 mm was casted with each successful trial (Table 2). Similar to the binders, GPC samples were also subjected to ambient curing. Compressive strengths were recorded after 28 days of aging.

3 Results and Discussion

3.1 Geopolymer Binders

Compressive strength. Figure 3 shows the compressive strength development of different WCB based geopolymer binders. All WCB blends with fly ash and slag showed significantly higher compressive strengths than the sample with WCB alone in the precursor. Since WCB is a low reactive and low amorphous material, it is difficult to gain notable strengths at ambient temperatures [11, 14]. Previous studies have experimentally shown the inability of WCB alone to achieve satisfactory compressive strength at ambient curing conditions. For example, Tuyan et al. 2018 [11] recorded compressive strengths of 0 MPa, 0 MPa and 5.3 MPa at 3, 7 and 28 days respective for WCB geopolymer binders cured at ambient conditions.

Compared to fly ash, varying the slag content in the precursor showed a greater impact on compressive strengths. This is due to the abundant presence of calcium

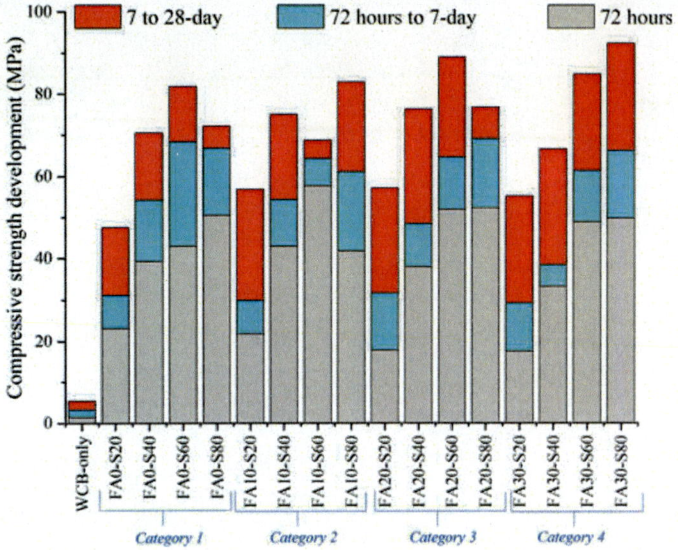

Fig. 3 Compressive strength development of different geopolymer binders

oxide in slag which forms calcium aluminosilicate hydrate (CSH) gels along with the geopolymer gels [13, 26]. In general, WCB-to-slag weight ratio of 2:3 gave the best compressive strengths. For mixes with higher fly ash content (i.e., 30% of the total precursor), the highest compressive strength of 92.3 MPa was recorded for the mixture with the highest slag content.

For pavement concretes, achieving high early strength is a crucial consideration, since pavements usually have to be opened to traffic within a few days. According to the AASHTO rigid pavement guidelines, all pavement concrete should achieve a compressive strength of at least 15 MPa before opening to the traffic [27, 28]. Fortuitously, the geopolymer binders developed in this study showed higher early strength gain, achieving more than 50% of its 28-day strength before 72 h. This early strength gain can be attributed to the formation of CSH gel due to the presence of slag in the precursor [16, 29, 30].

Sample FA0-S60 showed the highest 28-day compressive strength (81.8 MPa) in Category 1 while achieving around 52% of that strength within 72 h. Sample FA30-S80 gave the highest compressive strength among the all developed mixes and showed 49.8 MPa at 72 h, which was almost 53% of its 28-day strength. Samples FA10-S60 and FA20-S60 achieved 84% and 58% of its 28-day strength within 72 h. Therefore, considering both early strength gain and 28-day compressive strength samples FA0-S60, FA10-S60, FA20-S60 and FA30-S80 were selected for further investigations.

XRD analysis. Figure 4 shows the XRD analysis results for raw precursor and selected geopolymer binders, namely, FA0-S60, FA10-S60, FA20-S60 and FA30-S80. Quartz was found to be the major mineralogical phase in raw WCB and the

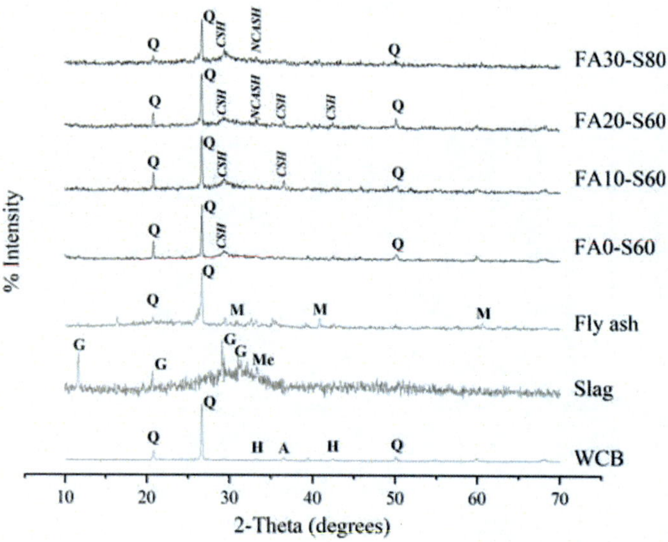

Fig. 4 XRD analysis for raw precursors and selected geopolymer binder samples. (Q-Quartz, H-Hematite, G-Gypsum, Me-Merwinite, M-mullite, CSH-Calcium silicate hydrate, NCASH-Sodium calcium aluminosilicate hydrate)

intense peaks show its crystalline nature [31]. In addition, hematite and albite was identified as other mineralogical phases present in WCB. The computed amorphousness for WCB was only 33.5%. In contrast, fly ash and slag showed higher amorphous contents, around 54.5% and 67.5% respectively.

The binder samples developed with partial replacement of WCB with fly ash and slag showed formation of new hydration products such as CSH and sodium calcium aluminosilicate hydrate (NCASH). CSH peak around 2-theta 29.2° [13] was clearly visible in all binders. These results correlated with the highest recorded compressive strengths. This was because, the coexistence of CSH gel in the matrix reinforced the geopolymer gels and increased its strength. However, quartz was identified as the dominant mineralogical phase in developed binder samples, indicating partial geopolymerisation and presence of unreacted particles [13, 32].

Fourier transform infrared (FTIR) analysis. Figure 5 shows the FTIR spectra obtained for selected geopolymer samples. The spectra of all samples exhibited peaks around 3450 and 1650 cm^{-1} corresponding to the bending and stretching vibrations of O–H bonds [11, 13, 33]. Compared to WCB-only samples, the intensity of these O–H bonds was enhanced in blended mixes, indicating presence of hydrated products such as CSH and NCASH [13]. The vibration modes corresponding to O-C-O bonds were visible in all samples at wavelength 1450 to 1473 cm^{-1} [11, 34, 35]. This can be attributed to the presence of carbonates from atmospheric carbonation [11, 33]. The strong peaks present at around 1010 cm^{-1} and 964 cm^{-1} are caused by the asymmetric stretching vibrations of Si–O and Al-O [11, 13, 33, 34]. Further, clear peaks around 451 cm^{-1} indicate the vibrations corresponding to Si–O-Si and O-Si–O

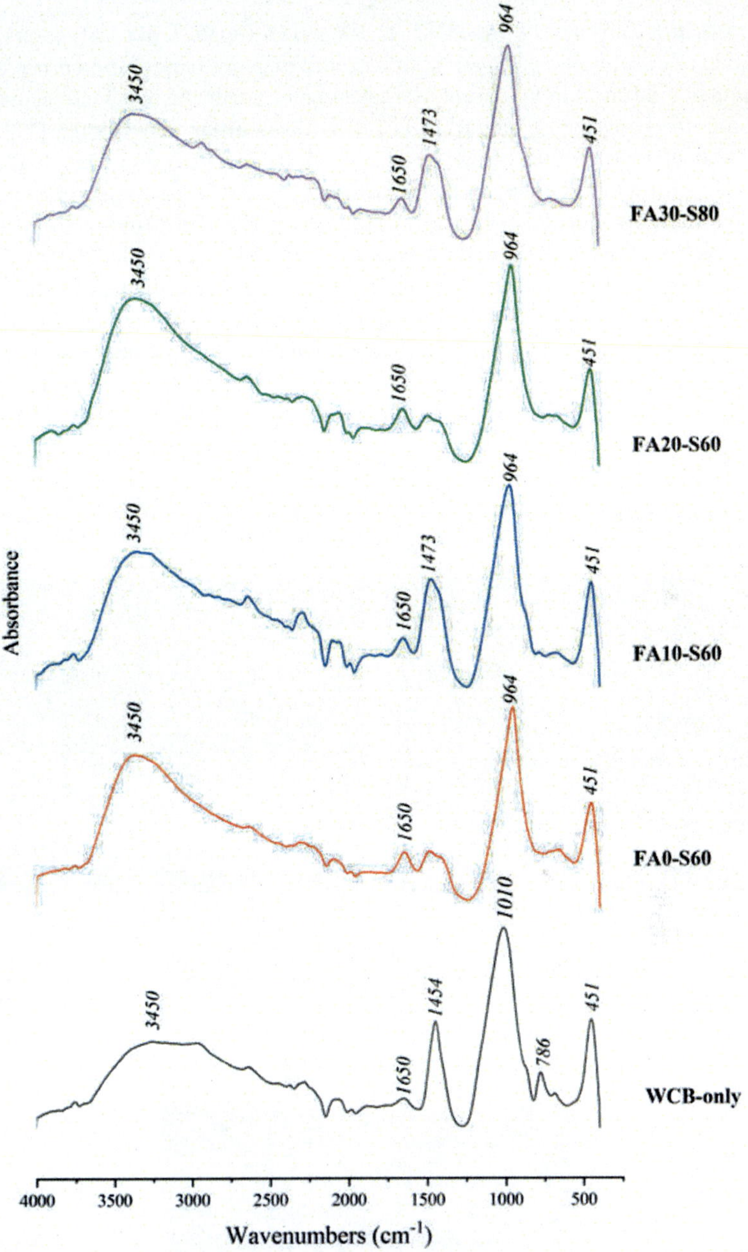

Fig. 5 FTIR spectra of the selected geopolymer samples WCB-only, FA0-S60, FA10-S60, FA20-S60 and FA30-S80

bonds [35, 36]. Existence of these O-T, T-O-T and O-T-O (T = Si or Al) points to the formation of geopolymer gels and the occurrence of geopolymerisation process. The peak around 786 cm^{-1} in WCB-only samples can be attributed to the unreacted SiO_2 indicating the presence of unreacted WCB particles in the end product [13]. This peak is attenuated in other blended mixes, indicating formation of better reaction products causing their higher strengths values.

SEM analysis. The microstructural features of selected geopolymer samples, namely, WCB-only, FA0-S60 and FA30-S80 are given in Fig. 6. All the images were obtained after 28 days of curing. Figure 6a shows the microstructure of geopolymer sample which consists WCB as the only precursor. Abundant presence of unreacted particles and the disturbed microstructure well explains the poor compressive strength of WCB-only geopolymer. In contrast, a dense and compacted microstructure could be observed when WCB was partially replaced with fly ash and slag (Fig. 6b and 6c). However, it was difficult to identify a significant difference in the microstructure of the different categories of geopolymer blends. Microcracks were clearly visible in all samples as result of high water absorption by both WCB and slag.

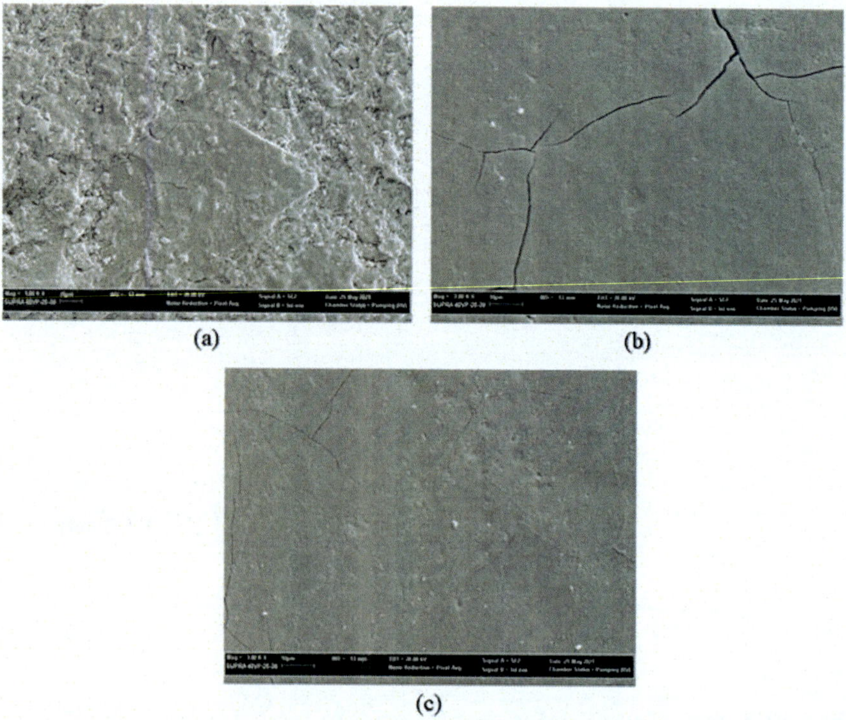

Fig. 6 Microstructure of selected geopolymer samples **a** WCB-only, **b** FA0-S60 and **c** FA30-S60

(a) (b)

Fig. 7 **a** Slump test for fresh GPC. **b** Hardened GPC with light reddish-brown colour

3.2 *Geopolymer Concrete*

WCB based one-part geopolymer binders from each category, namely, FA0-S60, FA10-S60, FA20-S60 and FA30-S80 were selected to develop initial GPC trials. Use of binders containing WCB imparted a unique reddish-brown texture to the fresh and hardened concrete (Fig. 7). All four binder samples gave dense and healthy fresh concrete mixtures without any segregation. The fresh GPC mixtures recorded a slump range in between 50 and 70 mm (Table 2).

Figure 8 indicates 28-day compressive strength recorded for different WCB based GPC samples. The strength of GPC was significantly less than the strength of the corresponding paste samples. This can be attributed to the higher w/b ratio in GPC compared to paste samples. Highest compressive strength of 41 MPa was recorded for the GPC developed using FA30-S80 binder with 35% of fine aggregates in the total aggregate content. Irrespective of the binder type and aggregate combinations, all GPC trials gave a compressive strength significantly higher than 5 MPa at 28 days, which is the minimum acceptable compressive strength for concrete subbase construction. Except for samples C1-fine40, C3-fine35 and C3-fine40, all other GPC samples showed compressive strength higher than 32 MPa at 28 days, indicating their feasibility to use in rigid pavement base applications.

The initial GPC trial mixes developed in this study demonstrates the potential of using WCB based geopolymer binders to produce pavement concretes with the minimum strength requirements (i.e., 5 MPa for subbase concrete, 32 MPa for base concrete). However, for the successful field applications of these low carbon concretes in pavement industry more research is necessary evaluating other parameters such

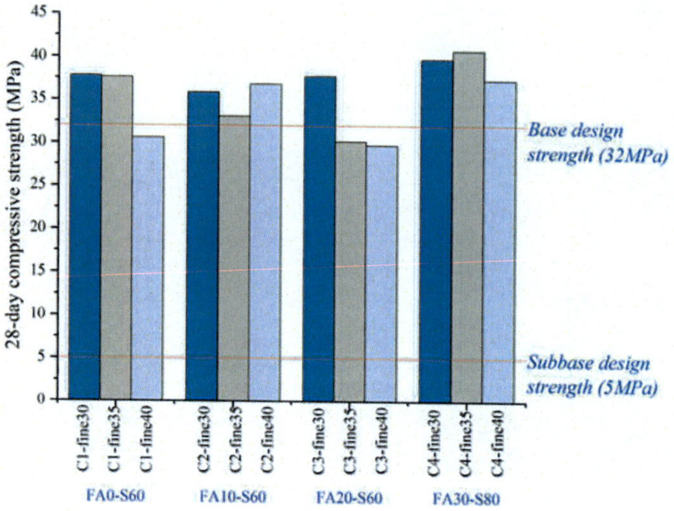

Fig. 8 28-day compressive strength of WCB based GPC

as flexural strength, creep, shrinkage, stress strain behavior, abrasion and cracking tendency.

Further investigations are recommended to optimize fresh and hardened concrete properties. In addition, comprehensive studies on the durability of WCB based GPC are essential before using as pavement construction material. For example, compared to OPC based concrete, GPC consists of slag show a higher carbonation rate [37]. This can be attributed to the rapid absorption of carbon dioxide by highly soluble alkalis present in GPC [37]. However, compared to the natural level of carbonation, accelerated carbonation test overestimates the carbonation of slag blended GPC cured at ambient conditions [38]. The laboratory tests methods to evaluate durability of conventional OPC based concrete cannot be directly applied to GPC [39]. Therefore, to evaluate the durability of GPC, newer test methods are needed by modifying existing tests methods for conventional concrete [40].

4 Conclusion

Different WCB based one-part geopolymer binders were developed by partially replacing WCB with fly ash and slag, and using anhydrous sodium metasilicate as the only activator. Based on the experimental results, following conclusion were made.

- WCB alone in the precursor cannot develop geopolymer binders with notable compressive strengths under ambient curing conditions.

- Blending WCB with slag and fly ash lead to a significant improvement in strength, recording compressive strengths up to 92 MPa.
- All blended binders showed a higher early strength gain, recording more than 50% of its 28-day compressive strength within first 72 h.
- Formation of new phases such as CSH and NCASH were visible in geopolymer samples with WCB, fly ash and slag in the precursor.
- Blended samples showed dense and compacted microstructure compared to the sample with WCB alone in the precursor.
- Preliminary GPC mixes developed using these newly synthesized one-part geopolymer binders showed acceptable fresh and hardened concrete properties, indicating the possibility of engineering new pavement construction materials.
- All the concrete mixes were able to gain 28-day strength more than 5 MPa which is the minimum strength requirement for rigid pavement subbase concrete.
- The GPC mixes, except C1-fine30, C3-fine35 and C3-fine40, were able to achieve strengths exceeding 32 MPa at 28 days, which is the minimum design strength for pavement base concrete.

WCB based GPC is a new and innovative concept introduced and investigated in this study with the aim of developing low carbon and sustainable rigid pavement construction materials. The preliminary investigations of GPC trials showed the feasibility of using WCB based binders to produce concrete with acceptable strength for rigid pavement applications. Therefore, further studies are recommended to optimize fresh and hardened properties and durability of WCB based GPC.

Acknowledgements This research work is part of a research project (Project No IH18.4.2) sponsored by the SPARC Hub (https://sparchub.org.au) at Department of Civil Engineering, Monash University funded by the Australian Research Council (ARC) Industrial Transformation Research Hub (ITRH) Scheme (Project ID: IH180100010). The financial and in-kind support from Austroads, Swinburne University of Technology and Monash University are gratefully acknowledged. Authors would like to acknowledge Ehsan Negahban and Purna Chava for the fruitful and inspired support in experimental work.

References

1. Austroads (2017) Guide to pavement technology part 2, in Pavement structural design. Austroads, Sydney, Australia
2. Mallick RB, El-Korchi T (2013) Pavement engineering: principles and practice. CRC Press
3. Transport for NSW (TfNSW) (2020) QA Specification R83, Concrete pavement base. TfNSW, NSW, Australia
4. Mehta A, Siddique R (2018) Sustainable geopolymer concrete using ground granulated blast furnace slag and rice husk ash: strength and permeability properties. J Cleaner Prod 49–57
5. Pasupathy K, Berndt M, Castel A, Sanjayan J, Pathmanathan R (2016) Carbonation of a blended slag-fly ash geopolymer concrete in field conditions after 8 years. Constr Build Mater 661–669
6. Chen C, Habert G, Bouzidi Y, Jullien A (2010) Environmental impact of cement production: detail of the different processes and cement plant variability evaluation. J Clean Prod 18(5):478–485

7. Muthukrishnan S, Gupta S, Kua HW (2019) Application of rice husk biochar and thermally treated low silica rice husk ash to improve physical properties of cement mortar. Theor Appl Fract Mech 102376
8. Nazari A, Sanjayan J (2016) Handbook of low carbon concrete. Butterworth-Heinemann
9. Sata V, Sathonsaowaphak A, Chindaprasirt P (2012) Resistance of lignite bottom ash geopolymer mortar to sulfate and sulfuric acid attack. Cem Concr Compos 34(5):700–708
10. Dhasindrakrishna K, Pasupathy K, Ramakrishnan S, Sanjayan J (2020) Effect of yield stress development on the foam-stability of aerated geopolymer concrete. Cem Concr Res 138
11. Tuyan M, Andiç-Çakir Ö, Ramyar K (2018) Effect of alkali activator concentration and curing condition on strength and microstructure of waste clay brick powder-based geopolymer. Compos Part B Eng 242–252
12. Pacheco-Torgal F, Labrincha J, Leonelli C, Palomo A, Chindaprasit P (2014) Handbook of alkali-activated cements, mortars and concretes. Elsevier (2014)
13. Zawrah MF, Gado RA, Feltin N, Ducourtieux S, Devoille L (2016) Recycling and utilization assessment of waste fired clay bricks (Grog) with granulated blast-furnace slag for geopolymer production. Process Saf Environ Prot 237–251
14. Rovnaník P, Řezník B, Rovnaníková P (2016) Blended Alkali-activated fly ash brick powder materials. Procedia Eng 108–113
15. Robayo-Salazar RA, Mejía-Arcila JM, Mejía de Gutiérrez R (2017) Eco-efficient alkali-activated cement based on red clay brick wastes suitable for the manufacturing of building materials. J Cleaner Prod 242–252
16. Migunthanna J, Rajeev P, Sanjayan J (2021) Investigation of waste clay brick as partial replacement of geopolymer binders for rigid pavement application. Constr Build Mater 305:124787. https://doi.org/10.1016/j.conbuildmat.2021.124787
17. Austroads (2016) Specification and use of geopolymer concrete in the manufacture of structural and non-structural components: review of literature. Austroads: Sydney, Australia
18. Marinković S, Dragaš J (2018) Fly ash. In: Waste and supplementary cementitious materials in concrete, pp 325–360
19. Yuksel I (2018) Blast-furnace slag, in waste and supplementary cementitious materials in concrete 361–415
20. Molino B, De Vincenzo A, Ferone C, Messina F, Colangelo F, Cioffi R (2014) Recycling of clay sediments for geopolymer binder production. A new perspective for reservoir management in the framework of Italian legislation: The Occhito reservoir case study. Mater (Basel) 5603–5616
21. Ferone C, Colangelo F, Cioffi R, Montagnaro F, Santoro L (2013) Use of reservoir clay sediments as raw materials for geopolymer binders. Adv Appl Ceram 184–189
22. Reig L, Tashima MM, Borrachero MV, Monzó J, Cheeseman CR, Payá J (2013) Properties and microstructure of alkali-activated red clay brick waste. Constr Build Mater 98–106
23. Pickin J, Randell P, Trinh J, Grant B (2018) National waste report 2018. Department of the Environment and Energy
24. AS 1012.3.1 (2014) Methods of testing concrete—determination of properties related to the consistency of concrete—slump test. Standards Australia International, NSW, Australia
25. Austroads (2017) Guide to pavement technology part 4C, in materials for concrete road pavements. Sydney, Australia, Austroads
26. Nath P, Sarker PK (2014) Effect of GGBFS on setting, workability and early strength properties of fly ash geopolymer concrete cured in ambient condition. Constr Build Mater 163–171
27. AASHTO guide for design of pavement structures. American Association of State Highway and Transportation Officials, Washington (1993)
28. FDOT (2019) Rigid pavement design manual. Florida Department of Transportation (FDOT), Office of Design, Pavement Management Section, Florida
29. Shah SFA, Chen B, Oderji SY, Haque MA, Ahmad MR (2020) Improvement of early strength of fly ash-slag based one-part alkali activated mortar. Constr Building Mater 118533
30. Ukritnukun S, Koshy P, Rawal A, Castel A, Sorrell CC (2020) Predictive model of setting times and compressive strengths for low-alkali, ambient-cured, fly ash/slag-based geopolymers. Minerals 10(10):920

31. Fořt J, Vejmelková E, Koňáková D, Alblová N, Čáchová M, Keppert M, Rovnaníková P, Černý R (2018) Application of waste brick powder in alkali activated aluminosilicates: functional and environmental aspects. J Cleaner Prod 714–725
32. Fořt J, Novotný R, Vejmelková E, Trník A, Rovnaníková P, Keppert M, Pommer V, Černý R (2019) Characterization of geopolymers prepared using powdered brick. J Market Res 8(6):6253–6261
33. Rovnaník P, Rovnaníková P, Vyšvařil M, Grzeszczyk S, Janowska-Renkas E (2018) Rheological properties and microstructure of binary waste red brick powder/metakaolin geopolymer. Constr Build Mater 924–933
34. Ouda AS, Gharieb M (2020) Development the properties of brick geopolymer pastes using concrete waste incorporating dolomite aggregate. J Build Eng 100919
35. Komnitsas K, Zaharaki D, Vlachou A, Bartzas G, Galetakis M (2015) Effect of synthesis parameters on the quality of construction and demolition wastes (CDW) geopolymers. Adv Powder Technol 368–376
36. Austroads (2017) Specification and use of geopolymer concrete in manufacturing of structural and non-structural components: experimental work. Sydney, NSW, Australia, Austroads
37. Guo X, Shi H, Wei X (2017) Pore properties, inner chemical environment, and microstructure of nano-modified CFA-WBP (class C fly ash-waste brick powder) based geopolymers. Cem Concr Compos 53–61
38. Bernal SA, San Nicolas R, Provis JL, Mejía de Gutiérrez R, van Devente JSJ (2014) Natural carbonation of aged alkali-activated slag concretes. Mater Struct 693–707 (2014)
39. Austroads (2016) Specification of geopolymer concrete: general guide., Austroads: Sydney, NSW, Australia
40. van Provis JL, Deventer JSJ (2014) Alkali Activated materials state-of-the-art report, RILEM TC 224-AAM, 1st edn. Springer, Netherlands